INTEGRATED SOLID WASTE MANAGEMENT
Engineering Principles and Management Issues

McGraw-Hill Series in Water Resources and Environmental Engineering

Consulting Editors

Paul H. King
Rolf Eliassen, Emeritus

Bailey and Ollis: *Biochemical Engineering Fundamentals*
Bouwer: *Groundwater Hydrology*
Canter: *Environmental Impact Assessment*
Chanlett: *Environmental Protection*
Chow, Maidment, and Mays: *Applied Hydrology*
Davis and Cornwell: *Introduction to Environmental Engineering*
Eckenfelder: *Industrial Water Pollution Control*
Linsley, Franzini, Freyberg, and Tchobanoglous: *Water Resources and Engineering*
McGhee: *Water Supply and Sewerage*
Mays and Tung: *Hydrosystems Engineering and Management*
Metcalf & Eddy, Inc.: *Wastewater Engineering: Collection and Pumping of Wastewater*
Metcalf & Eddy, Inc.: *Wastewater Engineering: Treatment, Disposal, Reuse*
Peavy, Rowe, and Tchobanoglous: *Environmental Engineering*
Sawyer and McCarty: *Chemistry for Environmental Engineering*
Tchobanoglous, Theisen, and Vigil: *Integrated Solid Waste Management: Engineering Principles and Management Issues*

INTEGRATED SOLID WASTE MANAGEMENT
Engineering Principles and Management Issues

George Tchobanoglous

Professor of Civil and Environmental Engineering
University of California, Davis

Hilary Theisen

Vice President
Brown & Caldwell, Consulting Engineers

Samuel Vigil

Professor of Civil and Environmental Engineering
California Polytechnic State University
San Luis Obispo, California

McGraw-Hill, Inc.
New York St. Louis San Francisco Auckland Bogotá
Caracas Lisbon London Madrid Mexico City Milan
Montreal New Delhi San Juan Singapore
Sydney Tokyo Toronto

This book was set in Times Roman by Publication Services.
The editors were B. J. Clark and John M. Morriss;
the production supervisor was Friederich W. Schulte.
The cover was designed by Nicholas Krenitsky.
R. R. Donnelley & Sons Company was printer and binder.

This book is printed on acid-free paper.

INTEGRATED SOLID WASTE MANAGEMENT
Engineering Principles and Management Issues

6 7 8 9 0 DOC DOC 9 0 9 8 7 6 5

ISBN 0-07-063237-5

Library of Congress Cataloging-in-Publication Data

Tchobanoglous, George.
Integrated solid waste management : engineering principles and management issues / George Tchobanoglous, Hilary Theisen, Samuel Vigil.
p. cm. — (McGraw-Hill series in water resources and environmental engineering)
Includes bibliographical references and index.
ISBN 0-07-063237-5
1. Refuse and refuse disposal. 2. Hazardous wastes. I. Theisen, Hilary. II. Vigil, S. A. III. Title. IV. Series.
TD791.T37 1993
363.72'85—dc20 92-35582
CIP

ABOUT THE AUTHORS

George Tchobanoglous is a professor of civil and environmental engineering at the University of California at Davis. He received a B.S. in civil engineering from the University of the Pacific, an M.S. in sanitary engineering from the University of California at Berkeley, and a Ph.D. in environmental engineering from Stanford University. His principal research interests are in the areas of solid waste management, wastewater treatment, wastewater filtration, aquatic systems for wastewater treatment, and individual on-site treatment systems. He has authored or coauthored over 200 technical publications and 10 textbooks. He is the principal author of the predecessor of this textbook. Unless otherwise noted, all of the photographs in this textbook were taken, developed, and printed by him. Professor Tchobanoglous serves nationally and internationally as a consultant to both governmental agencies and private concerns. An active member of numerous professional societies, he is past president of the Association of Environmental Engineering Professors. He has served as a member of the California Waste Management Board. He is a registered civil engineer in California and a Diplomate of the American Academy of Environmental Engineers.

Hilary Theisen is Vice President and Director of the solid waste and resource recovery program at Brown and Caldwell Consultants. He received a B.S. in civil engineering from the University of Minnesota and an MBA from the University of Santa Clara. His broad solid waste management experience started as a consultant in 1966. In the mid 1970s he directed solid waste operations in Sacramento County, California, which provided collection, transfer, recycling, and disposal services in a community of 380,000 people. At Brown and Caldwell Consultants, he oversees the production of studies, designs, and reports for public agencies and private industry. He has provided consulting services throughout the United States, Argentina, Taiwain, Puerto Rico, and Venezuela. He has authored numerous papers and articles on solid waste management and is the coauthor of the predecessor of this textbook, *Solid Waste: Engineering Principles and Management Issues*. He is a registered professional engineer in California, Hawaii, Oregon, and Washington.

Samuel Vigil is a professor of civil and environmental engineering at California Polytechnic State University, San Luis Obispo. He received a B.S. in civil engineering from the University of California at Berkeley, an M.S. in environmental engineering from Texas A&M University, and a Ph.D. in environmental engineering from the University of California at Davis. His principal research interests are in the areas of gasification of solid wastes, recycling technologies, computer modeling of integrated waste management systems, and computer-aided engineering. He has authored or coauthored 26 publications and holds a U.S. Patent in energy conversion. Professor Vigil is active as a consultant to state and local governments and has also consulted internationally in Europe, Latin America, and Southeast Asia. He is active in the Solid Waste Processing Division of the American Society of Mechanical Engineers, the Air and Waste Management Association, and the American Public Works Association. A Navy veteran, Professor Vigil is a Commander in the Naval Reserve Civil Engineer Corps. He is a registered civil engineer in California and a Diplomate of the American Academy of Environmental Engineers.

To Rosemary
Annette
Eva

CONTENTS

Part VI Solid Waste Management and Planning Issues

PREFACE

Solid wastes are all the wastes arising from human and animal activities that are normally solid and are discarded as useless or unwanted. Because of their intrinsic properties, discarded waste materials are often reusable and may be considered a resource in another setting. *Integrated Solid Waste Management* is the term applied to all of the activities associated with the management of society's waste. The basic goal of Integrated Solid Waste Management is to manage society's waste in a manner that meets public health and environmental concerns and the public's desire to reuse and recycle waste materials.

The need for a text that puts the engineering and scientific details of Integrated Solid Waste Management into the framework of resource management has grown significantly in recent years. This textbook is a response to that need. Both the student and the practitioner will find in this book the engineering principles, the data, the engineering and scientific formulas, and examples of the day-to-day issues associated with the management of municipal solid waste. The book integrates and expands the principles of solid waste management that were introduced in a predecessor text entitled, *Solid Wastes: Engineering Principles and Management Issues*.

ORGANIZATION

This book is organized into six parts. To understand the many facets of solid waste management, it is important to know how the field has evolved from the technology of horse-drawn carts to legislation-driven technology. The historical development of this field and its current perspectives are presented in Part I. To answer the question of whether solid waste is an untapped resource or a disposal problem, information must be available on the sources, composition, and properties of solid waste. These subjects are considered in Part II.

Because solid waste management has the dual functions of resource recovery and waste disposal, there is no one best place to apply the appropriate technology. In each situation, engineering principles must be applied to evaluate equipment and facility options, to make operational choices, and to develop management systems. The basic engineering principles that are an integral part of solid waste management are presented in Part III.

Advanced engineering principles related to the separation, processing, and transformation of solid waste are presented in Part IV. Separate chapters are included on materials separation and processing technologies, thermal conversion technologies, and biological and chemical conversion technologies. Because the reuse and/or sale of recovered materials is of considerable importance, a separate chapter is devoted to this subject.

The need for continuing care of the land remains after landfills are closed. Closure, restoration, and rehabilitation of landfills are presented in Part V. Both active and abandoned landfill sites are considered as there are thousands of sites that existed before current regulatory standards for closure were developed.

Important management issues that must be evaluated in the development and operation of Integrated Solid Waste Management systems are discussed in Part VI. For many communities, the critical issues arise from state mandates for waste diversion from landfills. Two chapters are devoted to this important topic. The methodology for completing solid waste management plans and documents, mandated by federal and state laws, is considered in the final chapter.

IMPORTANT FEATURES OF THIS BOOK

To illustrate the principles and facilities involved in the field of Integrated Solid Waste Management, over 530 illustrations, graphs, and diagrams are included. To help the reader understand the material presented in this textbook, detailed solved examples and case studies are presented in Chapters 3 through 20. Whenever possible, spreadsheet solutions are presented. To help the readers of this textbook hone their analytical skills, a series of discussion topics and problems are included at the end of each chapter. Selected references are also included at the end of each chapter.

To further increase the utility of this textbook, a series of appendixes have been included. A glossary of terms is presented in Appendix A. Physical characteristics of water and landfill gases are presented in Appendixes B and C, respectively. The statistical analysis of solid waste management data is delineated in Appendix D. Cost data for solid waste equipment and facilities are presented in Appendix E. The remaining appendixes deal with information related to the analysis and design of landfills.

USE OF THIS BOOK

Enough material is presented in this textbook to support up to three quarters or two semester-length courses at either the undergraduate or graduate level. Drafts of this book have been used at both levels at the University of California, Davis, and at the California Polytechnic State University. The first eleven chapters comprise a basic introduction to the field of Integrated Solid Waste Management. In the publisher's outside review process, it was suggested that the material presented in Chapters 12 through 15, which deals with materials recovery and waste transformation, be combined with the material presented in Chapter 9. To have combined these chapters would have altered the basic objective of this textbook. The material presented in Chapters 12 through 20 has been included to allow the book to be used for an advanced course in materials recovery and transformation and for a course dealing with policy issues in integrated solid waste management. A suggested outline for an introductory course in integrated solid waste management is presented below.

Topic	Reading
Introduction and evolution of solid waste management	Chapters 1 and 2
Sources, composition, and properties	Chapters 3, 4, and 5
Solid waste generation rates	Chapter 6
Waste handling, separation, storage, and processing at the source	Chapter 7
Collection of solid waste and source-separated materials	Chapter 8
Separation processing, and transformation of waste materials	Chapter 9
Waste/transfer and transport	Chapter 10
Disposal of solid wastes and residual matter	Chapter 11
Closure of landfills (added for semester course)	Chapter 16
Remedial actions for abandoned waste disposal sites (added for semester course)	Chapter 17
Recycling of waste materials (added for semester course)	Chapter 15

A suggested outline for a graduate course dealing with materials recovery, processing, and waste transformation is presented below.

Topic	Reading
Introduction and evolution of solid waste management	Chapters 1 and 2
Sources, composition, and properties	Chapters 3, 4, and 5
Solid waste generation rates	Chapter 6
Introduction to materials processing	Sections 9-1–9-6
Materials processing and recovery	Chapter 12
Introduction to thermal conversion technologies	Section 9-7
Thermal conversion technologies	Chapter 13
Introduction to biological and chemical conversion technologies	Section 9-8
Biological and chemical conversion technologies	Chapter 14
Recycling of waste materials	Chapter 15
Strategies for selecting the proper mix of technologies	Chapter 18

The following outline is appropriate for a course dealing with integrated solid waste management policy issues.

Topic	Reading
Evolution of waste management and legislation	Chapters 1 and 2
Sources, composition, and properties of solid waste	Chapters 3 and 4
Solid waste generation and collection rates	Chapter 6
Management issues: meeting mandated diversion goals/planning	Chapters 18, 19, and 20
Issues in waste handling, separation, storage, and processing at the source	Readings from Chapter 7
Issues in collection/transfer and transport	Readings from Chapters 8 and 10
Issues in materials recovery	Readings from Chapters 9, 12, and 15
Issues in the disposal of solid wastes and residuals	Readings from Chapter 11
Issues in the closure, restoration, and rehabilitation of landfills	Readings from Chapters 16 and 17
Strategies for selecting the proper mix of technologies	Selected readings

In an undertaking of the magnitude of this textbook, it is impossible to avoid errors. Any corrections, criticisms, or suggestions for improvements will be appreciated by the authors. Additional information and data are also welcomed.

ACKNOWLEDGMENTS

This textbook could not have been written without the help of a number of people. The help and support of the following individuals are acknowledged gratefully: the solid waste management classes of the senior author that worked with and corrected earlier draft versions of this textbook; Professor Michael Stallard, who reviewed several drafts of Chapter 11 and offered valuable suggestions for organizing the material; Ms. Doreen Brown Salizar for her help with the example problems in Chapter 11; Dr. Masoud Kayhanian, who reviewed and revised Chapter 14; Mr. Bill Freeman, who researched and prepared the first draft of Chapter 15; Mrs. Eva Vigil, who researched and prepared the section on landfill revegetation in Chapter 16; Professors Audrey Levine, Don Modesitt, Alan Molof, Roberto Narbaitz, Jerry Ongerth, Fred Pohland, Debra Reinhart, Kanti Shah, and Albert Yeung, who taught with draft versions of this textbook and offered many valuable suggestions; and Rosemary Tchobanoglous, who proofread much of the text. The following outside reviewers made helpful suggestions on both the content and organization of the text: Charles Cole, Pennsylvania State University–Harrisburg; Robert E. Deyle, Florida State University; and Kanti L. Shah, Ohio Northern University.

In addition, the help of the following organizations is acknowledged gratefully: Paul Geisler and Paul Hart of Davis Waste Removal for allowing us to photograph all aspects of their collection and recycling operation; Joe Garbarino of Marin Recycling for allowing us to photograph all aspects of his multifaceted operation; NORCAL Waste Systems, for allowing us to photograph their transfer station operation at San Francisco, CA; The County Sanitation Districts of Los Angeles County, CA; the County of Orange, CA; Escambia County, FL; Oakland Scavengers and Waste Management, Inc.; Yolo County, CA, who allowed us to take photographs at landfills operated under their jurisdiction; and the equipment manufacturers who supplied photographs of equipment and facilities.

We should also mention Dr. Marguerite Torrey and Mr. Jerome Colburn, our copy editors, whose concern for logic and correctness deserves special thanks. Finally, we are pleased to acknowledge the key role played by Kristina Williamson, senior production editor at Publication Services. Her attention to detail and her tireless efforts in coordinating and formatting this book went well beyond the call of duty.

George Tchobanoglous
Hilary Theisen
Samuel Vigil

INTEGRATED SOLID WASTE MANAGEMENT
Engineering Principles and Management Issues

PART I

PERSPECTIVES

What are solid wastes? What are the impacts of solid waste generation? What is the magnitude of the waste management problem? What does the future hold with respect to solid waste generation? What are the future challenges and opportunities for change? How did the field of solid waste management evolve? Why are the various activities associated with waste generation, on-site storage, collection, transfer and transport, processing and recovery, and disposal identified as functional elements? What does the term *integrated waste management* mean as applied to solid waste management? What are the day-to-day responsibilities of an operating agency? Which legislation at the federal and state level has affected the field of solid waste management? Which government agencies are responsible for administering the applicable legislation? And what are the impacts of legislation at the local level?

The answers to these questions are discussed in Part I, which also serves as an introduction to the field of solid waste management. Part I is the story of progress in this field from the use of horse-drawn carts to specially designed vehicles for the collection of wastes placed at curbside as well as the progress from open dumps (which became hazards to public health and sites on which burning and horrendous air pollution took place) to the development of environmentally safe methods of landfilling for the control of disease vectors. It is also the story of integrated solid waste management, encompassing the hierarchical elements of source reduction, recycling, waste transformation, and landfilling.

CHAPTER 1

EVOLUTION OF SOLID WASTE MANAGEMENT

Solid wastes comprise all the wastes arising from human and animal activities that are normally solid and that are discarded as useless or unwanted. The term *solid waste* as used in this text is all-inclusive, encompassing the heterogeneous mass of throwaways from the urban community as well as the more homogeneous accumulation of agricultural, industrial, and mineral wastes. This book is focused on the urban setting, where the accumulation of solid wastes is a direct consequence of life.

The purpose of this chapter is to introduce the reader to the field of solid waste management and to identify the demands that must be met by those practicing in the field. The material is presented in five sections: (1) solid wastes—a consequence of life; (2) waste generation in a technological society; (3) the evolution of solid waste management; (4) integrated solid waste management; and (5) solid waste management systems.

1-1 SOLID WASTE—A CONSEQUENCE OF LIFE

From the days of primitive society, humans and animals have used the resources of the earth to support life and to dispose of wastes. In early times, the disposal of human and other wastes did not pose a significant problem, for the population was

FIGURE 1-1
Solid waste problems are not new. (By permission of Johnny Hart and Creators Syndicate, Inc.)

small and the amount of land available for the assimilation of wastes was large. Although emphasis is currently being placed on recycling the energy and fertilizer values of solid wastes, the farmer in ancient times probably made a bolder attempt at this. Indications of recycling may still be seen in the primitive, yet sensible, agricultural practices in many of the developing nations where farmers recycle solid wastes for fuel or fertilizer values.

Problems with the disposal of wastes can be traced from the time when humans first began to congregate in tribes, villages, and communities and the accumulation of wastes became a consequence of life (see Fig. 1-1). Littering of food and other solid wastes in medieval towns—the practice of throwing wastes into the unpaved streets, roadways, and vacant land—led to the breeding of rats, with their attendant fleas carrying bubonic plague. The lack of any plan for the management of solid wastes thus led to the epidemic of plague, the Black Death, that killed half of the fourteenth-century Europeans and caused many subsequent epidemics with high death tolls. It was not until the nineteenth century that public health control measures became a vital consideration to public officials, who began to realize that food wastes had to be collected and disposed of in a sanitary manner to control rodents and flies, the vectors of disease.

The relationship between public health and the improper storage, collection, and disposal of solid wastes is quite clear. Public health authorities have shown that rats, flies, and other disease vectors breed in open dumps, as well as in poorly constructed or poorly maintained housing, in food storage facilities, and in many other places where food and harborage are available for rats and the insects associated with them. The U.S. Public Health Service (USPHS) has published the results of a study [2] tracing the relationship of 22 human diseases to improper solid waste management.

Ecological phenomena such as water and air pollution have also been attributed to improper management of solid wastes. For instance, liquid from dumps and poorly engineered landfills has contaminated surface waters and groundwaters. In mining areas the liquid leached from waste dumps may contain toxic elements, such as copper, arsenic, and uranium, or it may contaminate water supplies with unwanted salts of calcium and magnesium. Although nature has the capacity to dilute, disperse, degrade, absorb, or otherwise reduce the impact of unwanted residues in the atmosphere, in the waterways, and on the land, ecological imbalances have occurred where the natural assimilative capacity has been exceeded.

1-2 WASTE GENERATION IN A TECHNOLOGICAL SOCIETY

The development of a technological society in the United States can be traced to the beginnings of the Industrial Revolution in Europe; unfortunately, so can a major increase in solid waste disposal problems. In fact, in the latter part of the nineteenth century, conditions were so bad in England that an urban sanitary act was passed in 1888 prohibiting the throwing of solid wastes into ditches, rivers, and waters. This preceded by about 11 years the enactment of the Rivers and Harbors Act of 1899 in the United States, which was intended to regulate the dumping of debris in navigable waters and adjacent lands.

Thus, along with the benefits of technology have also come the problems associated with the disposal of the resultant wastes. To understand the nature of these problems, it will be helpful to examine the flow of materials and the associated generation of wastes in a technological society and to consider the direct impact of technological advances on the design of solid waste facilities.

Materials Flow and Waste Generation

An indication of how and where solid wastes are generated in our technological society is shown in the simplified materials flow diagram presented in Fig. 1-2. Solid wastes (debris) are generated at the start of the process, beginning with the mining of raw materials. The debris left from strip-mining operations, for example, is well known to everyone. Thereafter, solid wastes are generated at every step in the process as raw materials are converted to goods for consumption.

It is apparent from Fig. 1-2 that one of the best ways to reduce the amount of solid wastes that must be disposed of is to limit the consumption of raw materials and to increase the rate of recovery and reuse of waste materials. Although the concept is simple, effecting this change in a modern technological society has proved extremely difficult. Therefore, society has undertaken improved waste management and searched for new permanent locations in which to place solid waste. Unlike water-borne and air-dispersed wastes, solid waste will not go away. Where it is thrown is where it will be found in the future.

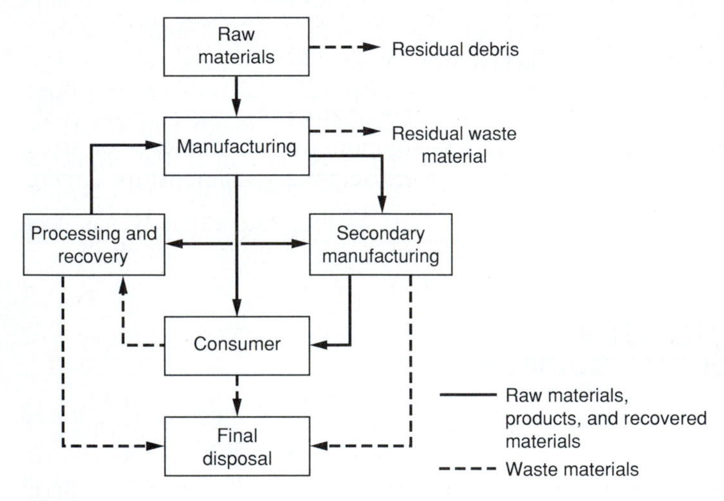

FIGURE 1-2
Materials flow and the generation of solid wastes in a technological society.

The Effects of Technological Advances

Modern technological advances in the packaging of goods create a constantly changing set of parameters for the designer of solid waste facilities [7]. Of particular significance are the increasing use of plastics and the use of frozen foods, which reduce the quantities of food wastes in the home but increase the quantities at agricultural processing plants. The use of packaged meals, for example, results in almost no wastes in the home except for the packaging materials. These continuing changes present problems to the facilities designer because engineering structures for the processing of solid wastes involves such large capital expenditures that they must be designed to be functional for approximately 25 years. Thus, the engineers responsible for the design of solid waste facilities must be aware of trends, even though they cannot, of course, predict all the changes in technology that will affect the characteristics of solid wastes.

On the other hand, every possible prediction technique must be used in this ever-changing technological society so that flexibility and utility can be designed into the facilities. Ideally, a facility should be functional and efficient over its useful life, which should coincide with the maturity of the bonds that were floated to pay for it. But important questions arise: Which elements of society generate the greatest quantities of solid waste and what is the nature of these wastes? How can the quantities be minimized? What is the role of resource recovery? Can disposal and recovery technology keep up with consumer product technology?

1-3 THE DEVELOPMENT OF SOLID WASTE MANAGEMENT

Solid waste management may be defined as the discipline associated with the control of generation, storage, collection, transfer and transport, processing, and disposal of solid wastes in a manner that is in accord with the best principles of public health, economics, engineering, conservation, aesthetics, and other environmental considerations, and that is also responsive to public attitudes. In its scope, solid waste management includes all administrative, financial, legal, planning, and engineering functions involved in solutions to all problems of solid wastes. The solutions may involve complex interdisciplinary relationships among such fields as political science, city and regional planning, geography, economics, public health, sociology, demography, communications, and conservation, as well as engineering and materials science.

Historical Development

> To describe the characteristics of the different classes of refuse, and to draw attention to the fact that, if a uniform method of nomenclature and record of quantities handled could be kept by the various cities, then the data obtained and the information so gained would be a material advance toward the sanitary disposal of refuse. Such uniformity would not put any expense upon cities, and direct comparisons and correct conclusions could be made for the benefit of others. [6]

This statement of objectives was written in 1906 by H. de B. Parsons in *The Disposal of Municipal Refuse* [6], which may have been the first book to deal solely with the subject of solid wastes from a rigorous engineering standpoint. It is interesting to note that many of the basic principles and methods underlying what is known today as the field of solid waste management were well-known even then. For example, although the motor truck has replaced the horse-drawn cart (see Fig. 1-3), the basic methods of solid waste collection remain the same; they continue to be labor intensive. (The development of uniform data for purposes of comparison is still an important need.)

The most commonly recognized methods for the final disposal of solid wastes at the turn of the century were (1) dumping on land, (2) dumping in water, (3) plowing into the soil, (4) feeding to hogs, (5) reduction, and (6) incineration [3, 6]. Not all these methods were applicable to all types of wastes. Plowing into the soil was used for food wastes and street sweepings, whereas feeding to hogs and reduction were used specifically for food wastes [3].

Enlightened solid waste management, with emphasis on controlled tipping (now known as "sanitary landfilling"), began in the early 1940s in the United States and a decade earlier in the United Kingdom [4]. New York City, under the leadership of Mayor La Guardia, and Fresno, California, with its health-minded Director of Public Works, Jean Vincenz, were the pioneers in the sanitary landfill method for large cities. During World War II, the U.S. Army Corps of Engineers, under the direction of Jean Vincenz, who then headed its Repairs and Utilities Division in Washington, DC, modernized its solid waste disposal programs to serve as model landfills for communities of all sizes. The medical Department of the Army, through Col. W. A. Hardenbergh of the Sanitary Corps' engineering group, took an active part in vector control and the prevention of disease by helping to sponsor the sanitary landfill program.

But municipalities did not follow these programs with consistency. The California Department of Health Services, along with several other progressive state health departments, established standards for municipal sanitary landfills and carried out aggressive campaigns for the elimination of conventional dumps. Still, in 1965, after a thorough review of solid waste management practices in the United States, Congress concluded that

> . . . inefficient and improper methods of disposal of solid waste result in scenic blights, create serious hazards to public health, including pollution of air and water resources, accident hazards, and increase in rodent and insect vectors of disease, have an adverse effect on land values, create public nuisances, otherwise interfere with community life and development; . . . the failure or inability to salvage and reuse such materials economically results in the unnecessary waste and depletion of natural resources.[1]

Congress also found that the trend of population concentration in metropolitan and urban areas had presented these communities with serious financial and administrative problems in the collection, transportation, and disposal of solid wastes.

(*a*)

(*b*)

(*c*)

FIGURE 1-3
Evolution of vehicles used for the collection of solid waste: (*a*) horse-drawn cart, circa 1900; (*b*) solid tire motor truck, circa 1925; and (*c*) modern collection vehicle equipped with container-unloading mechanism.

Functional Elements of a Waste Management System

The problems associated with the management of solid wastes in today's society are complex because of the quantity and diverse nature of the wastes, the development of sprawling urban areas, the funding limitations for public services in many large cities, the impacts of technology, and the emerging limitations in both energy and raw materials. As a consequence, if solid waste management is to be accomplished in an efficient and orderly manner, the fundamental aspects and relationships involved must be identified, adjusted for uniformity of data, and understood clearly.

In this text, the activities associated with the management of solid wastes from the point of generation to final disposal have been grouped into the six functional elements: (1) waste generation; (2) waste handling and separation, storage, and processing at the source; (3) collection; (4) separation and processing and transformation of solid wastes; (5) transfer and transport; and (6) disposal. The functional elements are illustrated photographically in Fig. 1-4, and the interrelationship between the elements is identified in Fig. 1-5. By considering each functional element separately, it is possible (1) to identify the fundamental aspects and relationships involved in each element and (2) to develop, where possible, quantifiable relationships for the purposes of making engineering comparisons, analyses, and evaluations. This separation of functional elements is important because it allows the development of a framework within which to evaluate the impact of proposed changes and future technological advancements. For example, the means of transport in the collection of solid wastes has changed from the horse-drawn cart to the motor vehicle (see Fig. 1-3), but the fundamental method of collection—that is, the manual physical handling required—remains the same (see Chapter 8).

The individual functional elements are described in the following discussion. Each one is considered in detail in Part III. The purpose of the following discussion is to introduce the reader to the physical aspects of solid waste management and to establish a useful framework within which to view the activities associated with management of solid wastes.

Waste Generation. *Waste generation* encompasses activities in which materials are identified as no longer being of value and are either thrown away or gathered together for disposal. For example, the wrapping of a candy bar is usually considered to be of little further value to the owner once the candy is consumed, and more often that not it is just thrown away, especially outdoors. It is important in waste generation to note that there is an identification step and that this step varies with each individual waste.

Waste generation is, at present, an activity that is not very controllable. In the future, however, more control will be exercised over the generation of wastes. In states where waste diversion goals are set by law, and must be met under threat of economic penalty, it is necessary to put in place a manifest system to

FIGURE 1-4
Views of the functional elements that constitute a solid waste management system: (*a*) waste generation; (*b*) waste handling and separation, storage, and processing at the source; (*c*) collection; (*d*) separation and processing and transformation of solid wastes; (*e*) transfer and transport; and (*f*) disposal.

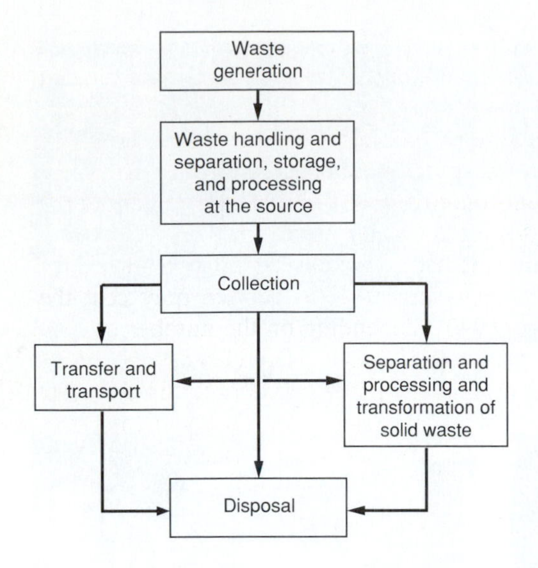

FIGURE 1-5
Simplified diagram showing the interrelationships between the functional elements in a solid waste management system.

monitor waste diversion. Source reduction, although not controlled by solid waste managers, is now included in system evaluations as a method of limiting the quantity of waste generated.

Waste Handling and Separation, Storage, and Processing at the Source. The second of the six functional elements in the solid waste management system is *waste handling and separation, storage, and processing at the source*. Waste handling and separation involves the activities associated with management of wastes until they are placed in storage containers for collection. Handling also encompasses the movement of loaded containers to the point of collection. Separation of waste components is an important step in the handling and storage of solid waste at the source. For example, from the standpoint of materials specifications and revenues from the sale of recovered materials, the best place to separate waste materials for reuse and recycling is at the source of generation. Homeowners are becoming more aware of the importance of separating newspaper and cardboard, bottles, yard wastes, aluminum cans, and ferrous materials. Currently, source separation of hazardous wastes by homeowners is being discussed widely and implemented to varying degrees.

On-site storage is of primary importance because of public health concerns and aesthetic considerations. Unsightly makeshift containers and even open ground storage, both of which are undesirable, are often seen at many residential and commercial sites. The cost of providing storage for solid wastes at the source is normally borne by the homeowner or apartment owner in the case of individuals, or by the management of commercial and industrial properties. Processing at the source involves activities such as compaction and yard waste composting.

Collection. The functional element of *collection,* as used in this book, includes not only the gathering of solid wastes and recyclable materials, but also the transport of these materials, after collection, to the location where the collection vehicle is emptied. This location may be a materials processing facility, a transfer station, or a landfill disposal site. In small cities, where final disposal sites are nearby, the hauling of wastes is not a serious problem. In large cities, however, where the haul distance to the point of disposal is often greater than 15 miles, the haul may have significant economic implications. Where long distances are involved, transfer and transport facilities are normally used.

As shown in Table 1-1, collection accounts for almost 50 percent of the total annual cost of urban solid waste management. This service may cost the individual homeowner $200/yr or more (1990), depending on the number of containers and frequency of collection. Typically, collection is provided under various management arrangements, ranging from municipal services to franchised private services conducted under various forms of contracts. In several parts of the country, large solid waste disposal companies, with contracts in many cities, own and

TABLE 1-1
Estimated cost of solid waste collection and disposal in the U.S. in 1990 and the year 2000[a]

	1990			2000 (estimated)[b]		
	Low	Medium	High	Low	Medium	High
Solid waste,[c] tons $\times 10^6$/yr						
Collected	149	190	238	149	196	257
Recycled[d]	15	24	42	50	65	86
Total	164	214	280	199	261	343
Unit costs, $/ton						
Collection	20	50	200	Future costs will depend on legislation and technological developments		
Recycling[e]	60	100	400			
Disposal	15	40	200			
Total national costs,[f] 10^6 dollars						
Collection		9,500		Future costs will depend on legislation and technological developments		
Recycling		2,400				
Disposal		7,600				
Total		19,500				

[a] Adapted in part from Ref. 8.

[b] Derived by assuming a population growth of 1 percent per year, a 1 percent per year increase in the quantity of waste generated per person, and that the entire country will achieve a 25 percent diversion goal.

[c] Municipal solid waste exclusive of wastes from municipal services, treatment plant sludges, and industrial and agricultural wastes.

[d] Based on recycle rates of 9, 11, and 15 percent in 1990.

[e] Gross cost excluding credits for revenue from the sale of recovered materials.

[f] Medium value only.

Note: tons $\times$ 907.2 = kg
$/ton $\times$ 0.001221 = $/kg

operate collection vehicles and landfill disposal sites. Collection services for industries vary widely. Some industrial wastes are handled like residential wastes; some companies have disposal sites on their own properties that use conveyor belts or water slurry transport. The latter is used for mineral wastes and agricultural wastes in many cases. Each industry requires an individual solution to its waste problems.

Separation, Processing, and Transformation of Solid Waste. The *separation, processing, and transformation of solid waste materials* is the fourth of the functional elements. The recovery of separated materials, the separation and processing of solid waste components, and transformation of solid waste that occurs primarily in locations away from the source of waste generation are encompassed by this functional element. The types of means and facilities that are now used for the recovery of waste materials that have been separated at the source include curbside collection, drop off, and buy back centers. The separation and processing of wastes that have been separated at the source and the separation of commingled wastes usually occur at a materials recovery facilities, transfer stations, combustion facilities, and disposal sites. Processing often includes the separation of bulky items, separation of waste components by size using screens, manual separation of waste components, size reduction by shredding, separation of ferrous metals using magnets, volume reduction by compaction, and combustion.

Transformation processes are used to reduce the volume and weight of waste requiring disposal and to recover conversion products and energy. The organic fraction of municipal solid waste (MSW) can be transformed by a variety of chemical and biological processes. The most commonly used chemical transformation process is combustion, which is used in conjunction with the recovery of energy in the form of heat. The most commonly used biological transformation process is aerobic composting. The selection of a given set of processes will depend on the waste management objectives to be achieved.

Transfer and Transport. The functional element of *transfer and transport* involves two steps: (1) the transfer of wastes from the smaller collection vehicle to the larger transport equipment and (2) the subsequent transport of the wastes, usually over long distances, to a processing or disposal site. The transfer usually takes place at a transfer station. Although motor vehicle transport is most common, rail cars and barges are also used to transport wastes.

For example, in the city of San Francisco the collection vehicles, which are relatively small because of the need to maneuver in city streets, haul their loads to a transfer station at the southern boundary of the city. At the transfer station, the wastes unloaded from the collection vehicles are reloaded into large tractor-trailer trucks. The loaded trucks are then driven to a disposal site located about 60 miles away.

Disposal. The final functional element in the solid waste management system is *disposal*. Today the disposal of wastes by landfilling or landspreading is the ultimate fate of all solid wastes, whether they are residential wastes collected and transported directly to a landfill site, residual materials from materials recovery facilities (MRFs), residue from the combustion of solid waste, compost, or other substances from various solid waste–processing facilities. A modern sanitary landfill is not a dump; it is an engineered facility used for disposing of solid wastes on land or within the earth's mantle without creating nuisances or hazards to public health or safety, such as the breeding of rats and insects and the contamination of groundwater.

In most cities, planning for waste disposal involves dealing with city, county, or other regional planning commissions and agencies. Thus, land-use planning becomes a primary determinant in the selection, design, and operation of processing facilities and landfills. Environmental impact statements (see Chapter 2) are required for all new landfill sites to ensure compliance with public health, aesthetics, and future use of land.

1-4 INTEGRATED SOLID WASTE MANAGEMENT

When all of the functional elements have been evaluated for use, and all of the interfaces and connections between elements have been matched for effectiveness and economy, the community has developed an integrated waste management system. In this context *integrated solid waste management* (ISWM) can be defined as the selection and application of suitable techniques, technologies, and management programs to achieve specific waste management objectives and goals. Because numerous state and federal laws have been adopted, ISWM is also evolving in response to the regulations developed to implement the various laws [9]. The waste diversion goals adopted in California (25 percent by 1995 and 50 percent by the year 2000) are an example. A hierarchy of waste management activities has also been established by recent regulations.

Hierarchy of Integrated Solid Waste Management

A hierarchy (arrangement in order of rank) in waste management can be used to rank actions to implement programs within the community. The ISWM hierarchy adopted by the U.S. Environmental Protection Agency (EPA) is composed of the following elements: source reduction, recycling, waste combustion, and landfilling [9]. The ISWM hierarchy used in this book is source reduction, recycling, waste transformation, and landfilling. The term *waste transformation* is substituted for the U.S. EPA's term *combustion,* which is too limiting. In the broadest interpretation of the ISWM hierarchy, ISWM programs and systems should be developed in which the elements of the hierarchy are interrelated and are selected to complement each other. For example, the separate collection of yard wastes can be used to effect positively the operation of a waste-to-energy combustion facility.

It is important to note that the U.S. EPA does not make a distinction between waste transformation (combustion) and landfilling; both are viewed as viable components of an integrated waste management program. Nevertheless, some states and organizations have adopted a more restrictive interpretation of the ISWM hierarchy. In the more restrictive interpretation, recycling can only be considered after all that can be done to reduce the quantity of waste at the source has been done. Similarly, waste transformation is considered only after the maximum amount of recycling has been achieved. A distinction is made between transformation and disposal in California and other states. Interpretation of the ISWM hierarchy will, most likely, continue to vary by state.

Source Reduction. The highest rank of the ISWM hierarchy, source reduction, involves reducing the amount and/or toxicity of the wastes that are now generated. Source reduction is first in the hierarchy because it is the most effective way to reduce the quantity of waste, the cost associated with its handling, and its environmental impacts. Waste reduction may occur through the design, manufacture, and packaging of products with minimum toxic content, minimum volume of material, or a longer useful life. Waste reduction may also occur at the household, commercial, or industrial facility through selective buying patterns and the reuse of products and materials [9].

Recycling. The second highest rank in the hierarchy is *recycling,* which involves (1) the separation and collection of waste materials; (2) the preparation of these materials for reuse, reprocessing, and remanufacture; and (3) the reuse, reprocessing, and remanufacture of these materials. Recycling is an important factor in helping to reduce the demand on resources and the amount of waste requiring disposal by landfilling.

Waste Transformation. The third rank in the ISWM hierarchy, *waste transformation,* involves the physical, chemical, or biological alteration of wastes. Typically, the physical, chemical, and biological transformations that can be applied to MSW are used (1) to improve the efficiency of solid waste management operations and systems, (2) to recover reusable and recyclable materials, and (3) to recover conversion products (e.g., compost) and energy in the form of heat and combustible biogas. The transformation of waste materials usually results in the reduced use of landfill capacity. The reduction in waste volume through combustion is a well-known example.

Landfilling. Ultimately, something must be done with (1) the solid wastes that cannot be recycled and are of no further use; (2) the residual matter remaining after solid wastes have been separated at a materials recovery facility; and (3) the residual matter remaining after the recovery of conversion products or energy. There are only two alternatives available for the long-term handling of solid wastes and residual matter: disposal on or in the earth's mantle, and disposal at the bottom of the ocean. Landfilling, the fourth rank of the ISWM hierarchy, involves the controlled disposal of wastes on or in the earth's mantle, and it is by far the most common method of ultimate disposal for waste residuals. Landfilling is the lowest rank in the ISWM hierarchy because it represents the least desirable means of dealing with society's wastes.

Planning for Integrated Waste Management

Developing and implementing an ISWM plan is essentially a local activity that involves the selection of the proper mix of alternatives and technologies to meet changing local waste management needs while meeting legislative mandates [5, 8, 9]. The proper mix of technologies, flexibility in meeting future changes, and the need for monitoring and evaluation are considered briefly in the following discussion and are considered in more detail throughout this text.

Proper Mix of Alternatives and Technologies. A wide variety of alternative programs and technologies are now available for the management of solid wastes. Several questions arise from this variety: What is the proper mix between (1) the amount of waste separated for reuse and recycling, (2) the amount of waste that is composted, (3) the amount of waste that is combusted, and (4) the amount of waste to be disposed of in landfills? What technology should be used for collecting wastes separated at the source, for separating waste components at MRFs, for composting the organic fraction of MSW, and for compacting wastes at a landfill? What is the proper timing for the application of various technologies in an ISWM system and how should decisions be made?

Because of the wide range of participants in the decision-making process for the implementation of solid waste management systems, the selection of the proper mix of alternatives and technologies for the effective management of wastes has become a difficult, if not impossible, task. The development of effective ISWM systems will depend on the availability of reliable data on the characteristics of the waste stream, performance specifications for alternative technologies, and adequate cost information.

Flexibility in Meeting Future Changes. The ability to adapt waste management practices to changing conditions is of critical importance in the development of an ISWM system. Some important factors to consider include (1) changes in the quantities and composition of the waste stream, (2) changes in the specifications and markets for recyclable materials, and (3) rapid developments in technology. If the ISWM system is planned and designed on the basis of a detailed analysis of the range of possible outcomes related to these factors, the local community will be protected from unexpended changes in local, regional, and larger-scale conditions [9].

Monitoring and Evaluation. Integrated solid waste management is an ongoing activity that requires continual monitoring and evaluation to determine if program objectives and goals (e.g., waste diversion goals) are being met (see also Section 2-5). Only by developing and implementing ongoing monitoring and evaluation programs can timely changes be made to the ISWM system that reflect changes in waste characteristics, changing specifications and markets for recovered materials, and new and improved waste management technologies.

1-5 OPERATION OF SOLID WASTE MANAGEMENT SYSTEMS

The facilities that compose a solid waste management system are often identified as *solid waste management system units*. The planning and engineering of solid waste management units include social, political, and technical factors. The combination of all of these factors forms a series of issues that must be addressed by community decision makers. Some contemporary solid waste management issues and future challenges and opportunities are introduced in the following discussion; these subjects will also be addressed in greater detail later in the text.

Management Issues

In addition to meeting the requirements associated with ISWM, a number of other management issues must be addressed in the operation of ISWM systems. The solid waste practitioner must acknowledge these management issues or face a high risk of failure in the implementation of solid waste management programs.

Setting Workable but Protective Regulatory Standards. Solid waste management units are subject to an increasing number of regulations. The attention is justified and timely, but strict adherence to very protective regulatory standards often causes failure of the processes by which waste management units are put in place.

Municipal solid waste management is caught in the backlash of understandable public concern over hazardous waste management. Regulatory agencies, in setting standards for construction, operation, and monitoring of units, are beset by lawyers and environmental groups recently armed with scientific data derived from experiments with massive doses of toxic compounds. Municipal waste does not contain massive quantities of toxics, but it does contain the small amounts found in the wastes from normal household activities. An unworkable regulation is one that ignores reality and deals only with certain technical data.

Nobody wants wastes. Solid waste cannot be wished away or hidden by the paper of regulations. In 1987 the attention of the U.S. public, news media, regulatory agencies, and Congress was focused on a garbage barge from New York. The episode started when a regulation forced wastes from a landfill when there was no other place to put 3000 tons of waste. The waste was loaded on a barge for shipment by sea to a place where it would be accepted. The barge was rejected by numerous states and foreign countries. After three months afloat the barge was unloaded where it started from—in New York.

Improving Scientific Methods for Interpretation of Data. The need to know about hazards in the environment has generated large amounts of data on toxics.

Billions of dollars have been invested in analytical equipment, laboratories, and data accumulation since the passage of the *Comprehensive Environmental Response Compensation and Liability Act* (CERCLA) (see Chapter 2). Even with all the data, however, there is a lack of a uniform basis for data interpretation. Analytical equipment and laboratory techniques produce data with accuracy in parts per billion or parts per trillion. What does such a detection accuracy mean to a solid waste management unit? If the component detected at a solid waste management system unit is on a regulatory list of cancer-causing agents, the unit may be shut down.

The goal is to understand the effects of very small quantities of toxic components on the environment. In the meantime, how much data should be presented to the public? When should data be delivered to the public? How does the public participate in data gathering and interpretation?

Identification of Hazardous and Toxic Consumer Products Requiring Special Waste Management Units. Municipal solid waste is a heterogeneous mass made up of every discard from homes, businesses, and institutions. Although small in quantity, some discards are hazardous, as identified on the product container. Examples are bleach, cleaning fluids, insecticides, and gasoline.

The issue is whether household hazardous waste contaminates the municipal waste management unit and whether, because of the large land areas in landfills, certain household wastes should be removed from the garbage can for disposal in smaller, highly controlled waste management units. Which products are most hazardous? How will the consumer store hazardous discards until they are picked up or delivered to the special management unit? Who will set up and operate special waste management units as such units will be defined by regulators as hazardous waste units?

Paying for Improved Waste Management Units. Solid waste management has a tradition of low cost. The improvements demanded by a concerned public are more costly than past practices (see Fig. 1-6). The increased costs must be paid by waste generators. This issue involves changing the manner in which a consumer thinks about paying for waste disposal. How is the cost of waste disposal presented to the consumer? When is the consumer asked to pay—at the time of product purchase or when the product is discarded? Since solid waste decays very slowly, who pays for long-term maintenance of land disposal waste management units—the generator at the time of discard or future users as maintenance costs are incurred?

Designating Land Disposal Units at or near Large Urban Centers. Waste management units are difficult to place in an urban environment. A suspicious public views these units as open dumps and littered transfer stations served by odorous, dripping garbage collection trucks. Yet it is within urban centers that the greatest quantity of solid waste is generated. Urban land use planning is facing a severe challenge to provide designated waste management units, especially land disposal units.

FIGURE 1-6
Landfill liner system under construction. A geomembrane liner has been placed over a compacted clay layer in the upper portion of the landfill, while the rest of the site is being prepared (background). Note: modern landfills are equipped with geomembrane liner systems for the protection of groundwater and to control the migration of landfill gases.

The issues are identifying environmentally acceptable land areas for land disposal units and then preserving lands for the intended use. Who will set a standard for "environmentally acceptable"? Will different standards apply for urban and rural areas? Can a scientific basis be identified that will satisfy a suspicious public regarding the safety of land disposal units?

Establishing and Maintaining More Qualified Managers to Develop and Operate Waste Management Units. Solid waste management units are increasing in quantity and complexity. In response, a set of managers must be trained and put in the appropriate positions to develop and operate expanded and improved management units.

The goal is to develop the human resources needed to develop and operate waste management units. Who will train the managers? How will the cost of training be paid? What standards will apply during the interim period while managers receive training?

Future Challenges and Opportunities

The multibillion-dollar industry of solid waste management can be supported only by the public, which is responsible for the generation of the vast tonnage of wastes (see Table 1-1). Public attitudes must be modified to reduce the environmental and economic burden placed on society for the disposal of solid waste. National concern must transcend the question of cost in an attempt to implement whatever individual or societal action is necessary.

Unfortunately, the standard of living in the United States is inevitably tied to the generation of solid wastes—the squandering of natural resources from this country and abroad, the one-time use of materials of so many types, and the philosophy of wastefulness and rapid obsolescence of products. It is reasonable to assume that a departure from this philosophy of wastefulness will reduce the tonnage of wastes to be managed. This concept inevitably leads to the need for source reduction and the reuse and recycling of recovered materials.

Changing Consumption Habits in Society. Product consumption is a natural activity. Society changes its standard of living by changing the quantity and quality of products it consumes. Solid wastes, the discards of product consumption, vary in quantity and quality as changes occur in the standard of living. Consumption habits must be changed if the quantity of solid wastes from consuming activities is to be reduced. The challenge is to change consumption habits that have been established over many years, as a result of advertising pressure that glamorizes increased consumption.

Reducing the Volume of Waste at the Source. Efforts must be made to reduce the quantity of materials used in both packaging and obsolescent goods and to begin the process of recycling at the source—the home, office, or factory—so that fewer materials will become part of the disposable solid wastes of a community. Source reduction is an alternative that will conserve resources and also has economic viability.

Making Landfills Safer. Landfills will always be the final disposal place for wastes that cannot be recovered. For this reason, every effort must be made to reduce the toxicity of the wastes that will ultimately be placed in landfills. The design of landfills must also improve to provide the safest possible location for the long-term storage of waste materials. The data base for existing landfills is expanding as improved construction and operations are implemented at new facilities. With an expanded data base comes the opportunity to understand how landfills function and how to manage the wastes placed in landfills more effectively.

Development of New Technologies. There are numerous opportunities to introduce new technologies into the solid waste management system. The challenge is to encourage the development of technologies that are most conservative of natural resources and that are cost-effective. Because many unproven technologies have been sold to unsuspecting cities, it may be necessary to write laws to regulate the use of technology. The testing and implementing of new technologies will be an important part of ISWM in the future.

1-6 DISCUSSION TOPICS AND PROBLEMS

1-1. Waste composition is the basis of all subsequent waste management programs. Do you feel that the changes in the composition of solid wastes will be significant in the next 10, 25, or 50 years? Explain.

1-2. Describe your present concept of resource recovery. In what ways can it affect the costs of solid waste disposal?

1-3. What is being done in your community for the recycling of bottles, cans, and paper? In your opinion, is the program successful? How can the program be improved?

1-4. From historical records, develop a brief chronology of the disposal methods used in your community during the past 50 years. Identify, where possible, the major events that led to the abandonment of one given method in favor of another.

1-5. In your opinion, what effect do the ownership and operation of landfill disposal sites by private contractors (as compared to public agencies) have on the economics, efficiency, and environmental aspects of operation?

1-6. Has your state set a hierarchy of integrated waste management? If so, identify the components of the hierarchy and explain why the hierarchy is important to a solid waste system manager.

1-7. Call the solid waste system manager in your community and determine the amount of education and formal training of the person. Do you think it is adequate for the management of an integrated waste management system?

1-8. List the functional elements of waste management. In your own terms, how does integrated waste management incorporate the functional elements?

1-9. Why have solid waste management practices been so slow in developing? Will changes come more quickly in the future? Explain.

1-10. Identify and discuss briefly the issues that you feel will be important in the field of solid waste management in the late 1990s.

1-7 REFERENCES

1. Eliassen, R.: *Solid Waste Management: A Comprehensive Assessment of Solid Waste Problems, Practices, and Needs,* Office of Science and Technology, Executive Office of the President, Washington, DC, 1969.
2. Hanks, T. G.: *Solid Waste/Disease Relationships,* U.S. Department of Health, Education, and Welfare, Solid Wastes Program, Publication SW-1c, Cincinnati, OH, 1967.
3. Hering, R. and S. A. Greeley: *Collection and Disposal of Municipal Refuse,* 1st ed., McGraw-Hill, New York, 1921.
4. Jones, B. B. and F. Owen: *Some Notes on the Scientific Aspects of Controlled Tipping,* Henry Blacklock & Co., Manchester, England, 1934.
5. Kreith, F. (ed.): *Integrated Solid Waste Management: Options For Legislative Action,* Genium Publishing Corporation, Schenectady, NY, 1990.
6. Parsons, H. de B.: *The Disposal of Municipal Refuse,* 1st ed., John Wiley & Sons, New York, 1906.
7. Selke, S. E.: *Packaging and the Environment: Alternatives, Trends and Solutions,* Technomic Publishing Company, Lanchester, PA, 1990.
8. U.S. Environmental Protection Agency: *The Solid Waste Dilemma: An Agenda for Action—Background Document,* EPA/530-SW-88-054A, U.S. EPA Office of Solid Waste, Washington, DC, 1988.
9. U.S. Environmental Protection Agency: *Decision-Makers Guide to Solid Waste Management,* EPA/530-SW89-072, Washington, DC, November 1989.

CHAPTER 2

LEGISLATIVE TRENDS AND IMPACTS

Because so much of the current activity in the field of solid waste management, especially with respect to resources recovery and groundwater quality, is a direct consequence of recent legislation, the purposes of this chapter are (1) to review the principal pieces of legislation that have affected the entire field of solid waste management and (2) to present the impacts of legislation on engineering and scientific activity for solid waste management. This information provides some perspective on the political environment in which solid waste management is now conducted and introduces the reader to some of the more important requirements associated with the preparation of planning reports, permits, and collection of scientific data. State governments adopt federal regulations as minimum standards for solid waste management. The evolution of landfill engineering in California and New Jersey is discussed in Section 2-2, which deals with the impacts of federal legislation.

2-1 MAJOR LEGISLATION

Environmental legislation has become increasingly restrictive as public health agencies, conservationists, and concerned citizens have pressured Congress and state legislatures to take action. Federal agencies have taken the lead, since solid waste has no political boundaries. The earliest legislation was passed in the nineteenth century. In 1899 the Rivers and Harbors Act directed the Army Corps of Engineers to regulate the dumping of debris in navigable waters and adjacent lands. Many USPHS regulations were enacted to permit the federal government to regulate the interstate transport of solid wastes, particularly food waste that was fed to hogs, in an attempt to control trichinosis. From the strong public health push during the latter twentieth century came legislation that encouraged the construction and operation of facilities based on the engineering properties of solid waste.

Solid Waste Disposal Act, 1965

Modern solid waste legislation dates from 1965, when the Solid Waste Disposal Act, Title II of Public Law 89-272, was enacted by Congress. The intent of this act was to

1. Promote the demonstration, construction, and application of solid waste management and resource recovery systems that preserve and enhance the quality of air, water, and land resources.
2. Provide technical and financial assistance to state and local governments and interstate agencies in the planning and development of resource recovery and solid waste disposal programs.
3. Promote a national research and development program for improved management techniques; more effective organizational arrangements; new and improved methods of collection, separation, recovery, and recycling of solid wastes; and the environmentally safe disposal of nonrecoverable residues.
4. Provide for the promulgation of guidelines for solid waste collection, transport, separation, recovery, and disposal systems.
5. Provide for training grants in occupations involving the design, operation, and maintenance of solid waste disposal systems.

Enforcement of this act became the responsibility of the USPHS, an agency of the Department of Health, Education, and Welfare, and the Bureau of Mines, an agency of the Department of the Interior. The USPHS had responsibility for the regulation of most of the municipal wastes generated in the United States. The Bureau of Mines was charged with supervision of solid wastes from mining activities and the fossil-fuel solid wastes from power plants and industrial steam plants.

National Environmental Policy Act, 1969

The National Environmental Policy Act (NEPA) of 1969 is an all-encompassing congressional law. It affects all projects that have some federal funding or that come under the regulation of federal agencies. Although the act has some shortcomings and has caused delay in the completion of some projects, it has served a useful purpose in giving the public an opportunity to participate in the decision-making process.

The act specified the creation of the Council on Environmental Quality in the Office of the President. This body has the authority to force every federal agency to submit to the council an *Environmental Impact Statement* (EIS) on every activity or project that it sponsors or over which it has jurisdiction.

Practically, the preparation of an EIS is an interdisciplinary undertaking. Every conceivable effect on the environment must be taken into account because the EIS becomes a legal document that may have to be defended in court. Consulting engineers and planners have large and diversified staffs to serve municipal, county, regional, and state agencies that must prepare environmental impact reports for solid waste management facilities. These staffs include experts in fields such as ecology, land-use planning, aquatic and terrestrial biology, soil science, economics and sociology, to name just a few. It is important that each new solid waste management project have a well-conceived, competently prepared, and thoroughly justified environmental impact report to inform the public, to invite its participation, and to seek its support. This will ensure that the project will be approved under the terms of the National Environmental Policy Act and any state and regional laws.

Resources Recovery Act, 1970

The Solid Waste Disposal Act of 1965 was amended by Public Law 95-512, the Resources Recovery Act of 1970. This act directed that the emphasis of the national solid waste management program should be shifted from disposal as its primary objective to recycling and reuse of recoverable materials in solid wastes, or to the conversion of wastes to energy. The USPHS, through its National Office of Solid Waste Management, was directed to prepare a report on the *Recovery and Utilization of Municipal Solid Waste* [1], which was completed in 1971. By that time the U.S. EPA had been formed by presidential order under Reorganizational Plan No. 3 of 1970, and all solid waste management activities were transferred from the USPHS to the EPA. Many other reports on various phases of solid waste management have been published since, including yearly reports to Congress on resource recovery and the basic reference report, *Decision-Makers Guide to Solid Waste Management* [6].

Resource Conservation and Recovery Act, 1976

Progress under the Resources Recovery Act of 1970 caused Congress to pass Public Law 94-580, the *Resource Conservation and Recovery Act* (RCRA), in 1976. The legislation had a profound effect on solid waste management. At this time EPA presented a set of guidelines for solid waste management to the public. The RCRA legislation gave the legal basis for implementation of guidelines and standards for solid waste storage, treatment, and disposal [3, 5, 7]. Because Congress had included both hazardous and solid waste in the legislation and assigned hazardous waste the earliest completion time for mandated controls, the EPA separated hazardous waste from municipal solid waste.

RCRA was amended and reauthorized by various Public Laws in 1978, 1980, 1982, 1983, 1984, 1986, and 1988. In 1980 and 1984 Congress was preoccupied with hazardous waste and the establishment of a strong regulatory framework for its storage, treatment, and disposal. Municipal solid waste management units were regulated under general environmental guidelines promulgated by EPA. Although the guidelines were intended to bring municipal solid waste landfills under stringent environmental controls, community officials and the public debated the interpretation of the legislation and community-specific needs for the environment.

The EPA published draft guidelines for municipal solid waste management in 1978 and expanded those guidelines in 1984 and 1988. Congress is holding hearings for the reauthorization of RCRA in 1992. Because hazardous waste regulations contain lists of hazardous compounds and municipal solid waste contains small quantities of these listed compounds, the hearings will probably focus on setting municipal solid waste disposal regulations that are compatible with EPA's hazardous waste regulations. The resulting legislation and EPA guidelines will greatly influence the activities of engineers and scientists dealing with municipal solid waste management.

Comprehensive Environmental Response, Compensation and Liability Act, 1980 (Superfund)

Public Law 96-510, 42 U.S.C. Article 9601, the Comprehensive Environmental Response, Compensation and Liability Act (CERCLA) of 1980 was enacted to provide a means of directly responding, and funding the activities of response, to problems at uncontrolled hazardous waste disposal sites [6]. Uncontrolled MSW landfills are facilities that have not operated or are not operating under a RCRA permit. Uncontrolled MSW landfills, both active and closed, are subject to CERCLA activities if they are demonstrated to contain hazardous waste or if to be the source of hazardous wastes. CERCLA became commonly known as the "Superfund Law."

Public Utility Regulation and Policy Act, 1981

The Public Utility Regulation and Policy Act (PURPA) of 1981 is a congressional law that, among its statutes, directs public and private utilities to purchase power from waste-to-energy facilities. The manner in which utilities set prices for the energy they purchase is also addressed. The intent is to establish a consistent method of reporting utility costs by all public utilities across the nation. This would make it possible to estimate more accurately potential revenues from the sale of electricity generated from solid waste–fueled energy plants. Even though a consistent methodology is helpful in planning solid–waste fueled energy systems, the contract for sale of energy must be negotiated on a case-by-case basis with the actual price varying from utility to utility. The legislation has been very effective in advancing the use of solid waste as a fuel in generating electricity.

Miscellaneous Laws and Executive Orders

Many other laws apply to the control of solid waste management problems. These include the Noise Pollution and Abatement Act of 1970, a federal law that limits environmental noise among workers employed in all industries, including many of the facilities of solid waste management systems (as well as the public), because many noisy operations may be involved—from collection to disposal. The Clean Air Act of 1970 (Public Law 91-604) (reauthorized in 1990) pertains where dust, smoke, and gases discharged from solid waste operations are involved. Many old incinerators have been shut down because stack emissions were exceeding newly established limits. Composting plants have been shut down because of odor emissions beyond the control of the operators. (Air pollution control is discussed in subsequent chapters of this book.) A summary of selected current legislation and legislative options for the future is presented in Ref. 2.

States have adopted their own laws and restrictive covenants and have established new agencies for the control of solid waste management systems. Thus, in the planning and design of these facilities, the consulting engineers or staff planners for the solid waste management agency must seek legal advice from attorneys who are qualified in federal and state environmental and consumer product laws and their peculiarities. This field is in a state of continuous flux.

Similarly, in the financing of new solid waste management systems, solid waste managers and planners should seek the advice of financial consultants. Many types of governmental grants, from both the federal government and state agencies, might apply to any particular project. These grants and financial aids are also subject to change and are followed closely by financial consultants.

Incentives that encourage recycling or resource recovery are often included in laws. Incentives may appear in various parts of the cycle that is made up of the manufacturer, consumer, and solid waste manager. Examples are recycled material preference allowances in product purchasing specifications and tax credits for installation of pollution control equipment at resource recovery plants. Many states have passed legislation that mandates the diversion of solid wastes from landfills. New solid waste management plans are being developed to achieve the mandated waste diversion goals.

2-2 IMPACT OF FEDERAL LEGISLATION

Public laws concerning solid waste are passed to improve solid waste management. Improvements are generally needed when solid waste disposal activity causes problems in public health, the environment, and economics. Such legislation has been structured to have a significant impact on solid waste management, starting first with hazardous solid wastes and more recently moving to nonhazardous municipal solid waste.

Legislation impacts many professions involved in solid waste management. Here the emphasis is on the impacts on the engineering and scientific professions. Summaries of engineering and scientific actions in response to RCRA and CERCLA legislation are presented in Refs. 3 and 7.

Landfills are the most commonly known solid waste disposal units. Before the enactment of RCRA in 1976, landfills were identified by two names: dump and sanitary landfill. To the professionals, the sanitary landfill was a land disposal unit built and operated with recognition and use of the engineering properties of solid waste. Dumps were not defined except as places where communities buried solid wastes. The public did not understand the engineering argument and had one name for all landfills—dumps! In response to public demands, elected officials passed legislation to control landfills and to eliminate dumps.

Impacts on States

Federal legislation and associated regulations have encouraged solid waste management programs to be implemented at the state level of government. Funding for state programs directly from the RCRA legislation is limited to planning grants and technical assistance. The impact on states was inconsistent. With no strong federal mandates or funding to direct programs, each state developed a solid waste management plan in accordance with its priorities for the environment and the ability of state residents to pay the costs of management. The result is 50 different programs.

TABLE 2-1
Evolution of landfill engineering in California

Time period	Legislation	Regulatory activity	Facilities
Pre-1970		1. Discharge permits from RWQCB	Landfills, dumps
1970 to 1976	SB 5 creates SSWMB	1. Discharge permits from RWQCB 2. County plans from SSWMB 3. Writing of standards for SWM facilities	Landfills, dumps
1977 to 1984	Federal RCRA passed in 1976	1. Discharge permits from RWQCB 2. Triennial updates of county plans 3. County LEA requires permits and inspects SW facilities for compliance with standards 4. RCRA requires survey of open dumps and their closure 5. Some landfills are placed on CERCLA Superfund list 6. Issuance of Federal guidelines for sanitary landfills	Landfills, not so many dumps
1984 to 1986	1. Reauthorization of RCRA in 1984 2. Subchapter 15 Guidelines for RWQCB 3. Various hazardous substance lists	1. RWD for RWQCB 2. RDSI for SSWMB 3. Permits for RWQCB 4. Triennial updates of COSWMP 5. LEA permits 6. County LEA inspections 7. Continued open dump closures 8. Federal guidelines under RCRA for hazardous wastes ST & D 9. Monitoring plans, well installation and sampling	Hazardous waste landfills, Municipal waste landfills, Designated waste management units

(continued)

California and New Jersey, among other states, have management programs that provide strong direction to the selection and implementation of solid waste management facilities. A general description of programs in each state is presented to illustrate the impacts of legislation.

California. California was one of the first states to go beyond questions of public health and aesthetic acceptability to develop engineering data on landfill disposal units. That first engineering concern was for the surface waters of the state, as manifested in Tentative Waste Discharge Orders written by engineering and scientific professionals for solid waste land-disposal units.

A brief representation of legislative milestones since 1970 and the resultant actions by California agencies to implement the various laws is presented in Table 2-1. Four things are noteworthy from the table:

1. Seven legislative milestones occurred in the period from 1977 to 1990.
2. The engineering and scientific design, permitting and reporting requirements have grown more than 10-fold between 1970 and 1990. As reported in Table 2-1 there was one regulatory activity before 1970, whereas 14 activities were in place after 1986.

TABLE 2-1 (*continued*)

Time period	Legislation	Regulatory activity	Facilities
1986 to Present	1. Calderon Data Gathering Bill passed in 1987 2. Tanner HWMP Bill passed in 1987 3. Solid Waste Act of 1989; RCRA reauthorization expected in 1992	1. RWD for RWQCB 2. RDSI for RWQCB 3. Calderon water reports (SWAT) for RWQCB 4. Permits from RWQCB 5. Calderon air reports for ARB 6. Triennial updates of COSWMP 7. LEA permits and operating plan 8. County LEA inspections 9. RCRA standards for inactive landfill/dump sites 10. State DOHS required COHWMP 11. Certification of real estate transactions 12. Monitoring plans and sampling 13. Designate landfill capacity for minimum 8 years 14. Mandatory waste diversion by cities and counties (25% by 1995, 50% by 2000)	Hazardous waste landfills, Municipal waste landfills, Designated waste management units

List of acronyms:

ARB	Air Resource Board
CERCLA	Comprehensive Environmental Response, Compensation, and Liability Act
COHWMP	County Hazardous Waste Management Plan
COSWMP	County Solid Waste Management Plan
DOHS	Department of Health Services
HWMP	Hazardous Waste Management Plan
LEA	Local Enforcement Agency
RCRA	Resource Conservation and Recovery Act
RDSI	Report of Disposal Site Information
RWD	Report of Waste Discharge
RWQCB	Regional Water Quality Control Board
SB 5	Senate Bill 5
SSWMB	State Solid Waste Management Board
ST & D	Storage Treatment and Disposal
SW	Solid Waste
SWAT	Solid Waste Assessment Test
SWM	Solid Waste Management

3. The land disposal units have changed from dumps into waste management units that are defined by engineering properties and scientific data mandated in construction and operating permits.
4. Waste diversion is mandated to reduce the quantity of wastes going to landfills.

New Jersey. New Jersey chose solid waste control, separation, and recycling as the first priority in solid waste management programs. Control was applied at the state level through legislation passed in 1970. This first statewide solid waste management law lacked the clear direction needed to overcome the inertia of decades of letting management facilities evolve without an implementation strategy.

In 1975 the law was amended to state clearly that county governments had control and responsibility for the siting of solid waste management facilities. Since that time, the state and counties have worked as a coordinated unit to identify, permit, build, and operate solid waste management facilities.

An outline of legislative milestones since 1970 and the resultant actions to implement the various laws in New Jersey is presented in Table 2-2. Several things are noteworthy:

TABLE 2-2
Evolution of landfill engineering in New Jersey

Time period	Legislation	Regulatory activity	Facilities
Pre-1970		1. Landfills declared a nuisance. Enforcement actions to be carried out by local boards of health pursuant to DOH's state sanitary code	Landfills, open dumps
1970 to 1976	1. NJDEP created (1970) 2. NJ Solid Waste Management Act passed 3. NJ Solid Waste Utility Control Act passed (1970) 4. NJ Solid Waste Management Act amended (1975)	1. State Sanitary Code revised (1970) to require registering of landfills 2. NJDEP assumes solid waste powers of NJDOH 3. Facilities must be registered by BPU 4. NJDEP issues new solid waste regulations (1974) 5. State divided into 22 solid waste districts. Maximum use of resource recovery	Landfills, open dumps
1976 to 1984	1. Federal RCRA passed (1976) 2. Federal CERCLA passed (1980) 3. Amended NJ Solid Waste Management Act effective (1977)	1. RCRA requires open dump survey and closure 2. Many NJ landfills placed on CERCLA Superfund list 3. Issuance of federal guidelines for sanitary landfills 4. All counties adopt and NJDEP approves solid waste management plans	Landfills, open dump closings. State issues guidelines to reduce dependency on landfills

(*continued*)

1. All solid waste powers were vested in one agency, the New Jersey Department of Environmental Protection.
2. Legislative amendments were passed to support implementation of facilities.
3. In 1982 the statewide plan advocated recycling and resource recovery; in 1987 the legislature adopted mandatory recycling.
4. The state is collecting taxes to facilitate development of resource recovery facilities.
5. The state used out-of-state disposal sites as a short-term strategy in implementing a long term strategy of disposal self-sufficiency by 1992.

Impacts on Manufacturers

Solid waste management legislation has not been written to affect manufacturers. In our consumption-oriented society the manufacturer produces salable goods, leaving the disposal of unwanted, unused, or partly consumed goods and associated packaging to the ingenuity of the solid waste practitioner. However, current legislative initiatives may have significant impacts on the quantities of solid waste

TABLE 2-2 (*continued*)

Time period	Legislation	Regulatory activity	Facilities
1976 to 1984 (*cont.*)		5. NJDEP issues statewide solid waste management plan that advocates recycling, resource recovery and 10-year county plans (1982)	
1984 to 1986	1. Reauthorization of RCRA (1984) 2. NJ passes Act-1778 (1985)	1. NJDEP issues update to statewide solid waste management plan advocating resource recovery, planning and landfilling (1986) 2. State collects taxes to facilitate development of resource recovery facilities 3. First state-of-the-art landfill opens (1984)	First controlled landfill. Open dump closings. Landfill closings cause use of out-of-state landfills
1986 to Present	1. NJ adopts Mandatory Recycling Act (1987) 2. RCRA reauthorization under way in 1992	1. NJDEP issues comprehensive solid waste regulation (1987) 2. All counties required to adopt mandatory recycling plans to achieve 25% recycling within two years 3. First major resource recovery facility opens (1988)	More controlled landfills. First major resource recovery facility. Continued use of out-of-state landfills

List of acronyms:

BPU	Bureau of Public Utilities	NJ	New Jersey
CERCLA	Comprehensive Environmental Response, Compensation, and Liability Act	NJDEP	New Jersey Department of Environmental Protection
DOH	Department of Health	NJDOH	New Jersey Department of Health
		RCRA	Resource Conservation and Recovery Act

generated from consumer products. Product manufacturers may be faced with legislation that includes (1) packaging taxes, (2) packaging standards, (3) minimum content of recycled material in consumer products, and (4) product bans for styrofoam containers, disposable diapers, and juice packs.

Some federal laws act as a disincentive to efficient solid waste management. Examples are the tax laws favoring use of virgin raw materials—through depletion allowances—and the interstate commerce regulations that penalize the hauling of recycled (secondary) materials—through the setting of higher transportation rates compared with virgin materials. However, a number of paper manufacturers have increased the percentage of recycled fiber in newsprint and higher-grade papers. Similarly, a number of plastics manufacturers have increased the capacity of their plants to use recycled plastic containers.

The greatest legislative impact is achieved through laws affecting the consumer of manufactured products. Bottle bills, typically passed at the state level, are meant to provide an economic incentive to the consumer to return liquid containers to a redemption center. The manufacturers and product distributors are affected because the legislation mandates that they receive, store, and process returned containers. In California, the legislature provided negative incentives for manufacturers by requiring that the tax for a given container increase if the degree of recycling of that container does not meet the mandated goals.

Impacts on Consumers

No federal solid waste legislation affects consumers. However, both state and local government laws that mandate source separation of solid waste by consumers are in place. Many laws—federal, state and local—cause a higher cost to the consumer in the forms of new product taxes, higher disposal fees, and special tax assessments.

2-3 GOVERNMENTAL AGENCIES

The various laws, regulations, and executive orders have created a divided responsibility among many federal departments and agencies for the regulation and financing of solid waste management.

Federal Agencies

Federal agencies interpret congressional laws and prescribe the minimum standards of waste management to be followed by all states. Some of the significant agencies and their impacts are presented in Table 2-3. Note the agencies listed under the heading "Indirect solid waste management impacts." Transportation, commerce, and government purchasing are often overlooked when one is searching for the government support necessary to install waste management facilities. Yet these

TABLE 2-3
Federal agencies with significant impacts on solid waste management

Agency	Selected significant impacts
Direct solid waste management impacts	
Environmental Protection Agency (EPA)	Sets performance criteria for sanitary landfills and discharge requirements for combustion facilities
Health, Education, and Welfare (HEW)	Sets health standards for waste storage (direct for commercial establishments, indirect for residential housing)
Department of Defense (DOD)	Army Corps of Engineers protects navigable waterways
Indirect solid waste management impacts	
Department of Commerce (DOC)	Decisions regarding restrictions on interstate commerce; rulings on tariffs for virgin versus recycled materials
Department of Transportation (DOT)	Load restrictions
General Services Administration (GSA)	Materials specifications for federal purchasing
Housing and Urban Development (HUD)	Loans/grants for waste-fuel–fired district heating systems
Department of Energy (DOE)	Development of alternative fuels
Department of Interior	Siting
Food and Drug Administration (FDA)	Testing and approval of packaging materials for food products

agencies have guidelines for waste handling and provide funding for solid waste management facilities.

State Agencies

State legislators pass laws to adopt standards equal to or more restrictive than mandated by Congress. Local community officials then work with state officials to implement solid waste management regulations, usually through the issuance of permits. All solid waste management units are required to operate within the regulations specified in a facility's permit.

Each state has a unique structure of agencies to implement and control solid waste management functions. No attempt is made here to list all agencies for all states. However, the states have substantial similarities in the functions and impacts of solid waste agencies. Typical agencies and a selected list of significant agency impacts are presented in Table 2-4. Agency names and functions for each state can be obtained from each state government.

TABLE 2-4
Typical state agencies with significant impacts on solid waste management

Typical state agency	Selected significant impacts
Environmental boards or departments	
Multiple boards Water resources Air Solid waste Health Department with multiple divisions Water Air Solid waste	Interpret laws and develop regulations to implement the law regarding discharges to surface and groundwaters, air discharges, and protection of public health; implement legal mandates for recycling and planning; provide for the monitoring of waste management units and the interpretation of collected data; assist local waste management agencies; set standards and guidelines for construction and operation of waste management units; issue permits for facilities.
Board of Equalization or Public Utility Commission	Set rates; review power purchase agreements for sale of energy from energy recovery facilities.
Highway Patrol	Issue citations for overweight collection trucks or transfer trailers.

2-4 ENFORCING THE HIERARCHY OF INTEGRATED SOLID WASTE MANAGEMENT

In establishing a hierarchy of integrated solid waste management, the federal government has provided guidelines and numerous states have mandated waste management activities for local governments. What are the consequences if the hierarchy is not adopted by local government? Source reduction, the highest rank in the hierarchy, is the most difficult activity to implement at the local level. Local government tax moneys spent on source reduction programs often do not provide visible benefits to community residents. Therefore, local elected officials are often reluctant to implement and pay for programs with the voter's money. Recycling, the second rank in the hierarchy, has the same difficulties for implementation by local governments because of a lack of local control and the inability to pay for new facilities.

In *The Solid Waste Dilemma: An Agenda for Action—Background Document* [5], the U.S. EPA has outlined some options for actions to stimulate source reduction of solid wastes. The actions requiring legislation or new regulations at the federal level include:

1. Product constituent regulation. This source reduction option would target constituents that are known to be in the waste stream and are known to be toxic. A range of regulatory actions could be used, including product bans, cautionary labeling, and approval labeling.
2. New product approval process. Manufacturers who introduce a new product or package would be required by the government to test or otherwise demon-

strate the impact of the products on the waste stream. The manufacturer would determine how the product would perform in a landfill environment, in an incinerator, and in a mixed-waste recycling/processing system.

3. Existing product reviews. The government would establish criteria to identify products and materials of concern because of waste toxicity. The actions could range from labeling to bans.
4. Procurement restrictions. Governments control the procurement of large quantities of goods. Control over acquisition of large quantities makes these institutions reasonable candidates for regulations fostering procurement of goods that result in a waste having lower toxicity and less volume. The action would be to change purchasing regulations that deal only with the performance of the goods in service.

The hierarchy of integrated solid waste management must be enforced in a manner that is flexible enough to allow local governments to implement waste management facilities that match the communities' ability to pay for them. Federal and state government can help local government by supporting markets and uses for the materials diverted from landfill disposal [4, 6].

2-5 FUTURE TRENDS

Legislation will be an important part of future solid waste management. An informed people will resist new waste management units in their communities and local elected officials will not annoy their electorate by selecting sites for new units. An acceptable alternative is a federally mandated solid waste program created by legislation, including standards and mandates for facilities. The legislated programs for clean water and air are successful examples of this approach.

The costs of solid waste management will continue to increase. More restrictive regulations will be applied to waste management units. The U.S. EPA is gathering data on the economic and environmental impacts of improved waste management practices and studying ways to improve future practices.

Today, landfills are the predominant waste management unit. Future integrated solid waste management will be balanced between source reduction, recycling, energy recovery, and land disposal. To achieve more balanced solid waste management programs, more states must adopt diversion and recycling goals and provide the financing to meet those goals. A stable set of environmental standards is needed for ash residues and air discharges from energy recovery plants if the balance is to be achieved. Congress is developing legislation to set such standards.

Finally, education and training will be expanded in the future to provide the people to manage the new facilities. Organizations such as the Solid Waste Association of North America (SWANA) and the National Solid Waste Management Association (NSWMA) are expanding certification programs at landfills, hoping to get state recognition and mandatory use of certified managers of solid waste management units.

2-6 DISCUSSION TOPICS AND PROBLEMS

2-1. Many federal agencies have responsibilities for parts of the solid waste management program in the United States. Would you favor combining the responsibilities into a single agency? Why? If not, why not?

2-2. Knowing that the U.S. EPA has established a hierarchy of integrated solid waste management with source reduction as the first rank, what do you believe to be the role of legislation in causing the implementation of source reduction? Explain your position.

2-3. If you were to choose a career in solid waste management in a governmental agency, what level of government—city, county, regional, state, federal—would you select to give you breadth of experience and a fair degree of responsibility for taking constructive action? Why?

2-4. Identify the principal state and local laws and agencies important in solid waste management in your area.

2-5. Which agency of your state government has jurisdiction over the promulgation of codes and guidelines for solid waste management? Does this agency have an adequate staff to assist municipalities and enforce regulations?

2-6. Referring to Problem 2-5, what is your opinion of the literature the agency distributes to the cities? Are the regulations responsive to the U.S. EPA Agenda for Action as outlined in Ref. 5?

2-7. Develop a chronology of the growth of solid waste management legislation in your state.

2-8. Where is the regional office of the U.S. EPA in your area? What is it doing to assist your state government in its attempts to improve solid waste management in the cities of your state?

2-9. Has your departmental library received an adequate number and variety of U.S. EPA publications on solid waste management? If not, can a team project be organized to obtain these publications from the EPA regional office? What is your opinion of these publications?

2-10. Obtain the organization chart for the agency responsible for solid waste management in your community. Does the agency have the administrative structure to implement integrated solid waste management? Explain the difference between the agency regulations and those regulations used in California and New Jersey.

2-7 REFERENCES

1. Drobny, N. L., H. E. Hull, and R. F. Testin: *Recovery and Utilization of Municipal Solid Waste*, U.S. EPA, Publication SW-10C, Washington, DC, 1971.
2. Kreith, F. (ed.): *Integrated Solid Waste Management: Options for Legislative Action,* Genium Publishing Corporation, Schenectady, NY, 1990.
3. O'Brien & Gere Engineers, Inc.: *Hazardous Waste Site Remediation: The Engineer's Perspective,* Van Nostrand Reinhold, New York, 1988.
4. Office of Technology Assessment: *Facing America's Trash*, Report to Congress, Office of Technology Assessment, Washington, DC, 1989.
5. U.S. Environmental Protection Agency: The Solid Waste Dilemma: An Agenda For Action—Background Document, EPA/530-SW-88-054A, U.S. EPA Office of Solid Waste, Washington, DC, 1988.
6. U.S. Environmental Protection Agency: *Decision-Makers Guide to Solid Waste Management*, EPA/530-SW89-072, Washington, DC, November 1989.
7. Wagner, T. P.: *Hazardous Waste Regulations*, 2nd ed., Van Nostrand Reinhold, New York, 1991.

PART II

SOURCES, COMPOSITION, AND PROPERTIES OF SOLID WASTE

The purpose of Part II is to present information and data on the sources, composition, and properties of municipal solid waste and to consider the transformations that can be used to alter the form of the materials constituting the waste. This knowledge is critical to the planning and implementation of effective source reduction programs; source separation and recycling programs; and the design of collection systems for commingled and source-separated wastes, processing and transformation facilities, transfer facilities and transport equipment, and ultimate disposal facilities.

The components of municipal solid waste and their percentage distribution are introduced in Chapter 3. The physical, chemical, and biological properties of solid waste are considered in Chapter 4, as are waste transformations. To develop an integrated waste management program, information must also be available on the small amounts of hazardous wastes that may be commingled with municipal wastes. The types and amounts of compounds that are present, their chemical properties, and the transformation these compounds can undergo must be known to assess the impact of these materials on the functional elements that make up an integrated solid waste management system. The hazardous wastes found in municipal waste are discussed in Chapter 5.

CHAPTER 3

SOURCES, TYPES, AND COMPOSITION OF MUNICIPAL SOLID WASTES

Solid wastes include all solid or semisolid materials that the possessor no longer considers of sufficient value to retain. The management of these waste materials is the fundamental concern of all the activities encompassed in solid waste management—whether the planning level is local, regional or subregional, or state and federal. For this reason, it is important to know as much about municipal solid waste (MSW) as possible. Important questions that must be answered include the following [5]:

1. What types and quantities of MSW will be received?
2. At what rates will these types arrive?
3. What types and quantities of materials have already been removed for reuse and recycling?
4. What properties does MSW have as it is received?
5. How do the properties of MSW vary: hourly, daily, weekly, and seasonally?
6. How do the properties of MSW change during processing?
7. How can the properties of MSW be changed during processing?
8. What are the properties of MSW that are of economic value?
9. What unwieldy or hazardous objects must be removed?
10. What contaminants should be removed?

11. What tests and measurements can be performed to obtain answers to the above questions?
12. What range of variations should be expected in the measured quantities and with what level of confidence?

The purpose of this chapter is to identify the sources, types, and composition of solid wastes. The physical, chemical, and biological properties and transformations of waste materials are considered in Chapter 4. The sources and properties of the small amounts of hazardous waste found in MSW are considered in Chapter 5. Data on the quantities of waste generated and their variation and information on the types and quantities of waste materials now recovered from MSW are presented in Chapter 6. Information presented in this and the following two chapters will have application throughout the remainder of this text.

3-1 SOURCES OF SOLID WASTES

Knowledge of the sources and types of solid wastes, along with data on the composition and rates of generation, is basic to the design and operation of the functional elements associated with the management of solid wastes. To avoid confusion, the term *refuse*, often used interchangeably with the term *solid wastes*, is not used in this text.

Sources of solid wastes in a community are, in general, related to land use and zoning. Although any number of source classifications can be developed, the following categories are useful: (1) residential, (2) commercial, (3) institutional, (4) construction and demolition, (5) municipal services, (6) treatment plant sites, (7) industrial, and (8) agricultural. Typical waste generation facilities, activities, or locations associated with each of these sources are reported in Table 3-1, where *municipal solid waste* (MSW) is normally assumed to include all community wastes with the exception of industrial process wastes and agricultural wastes.

3-2 TYPES OF SOLID WASTES

As a basis for subsequent discussions, it will be helpful to define the various types of solid wastes that are generated (see Table 3-1). It is important to be aware that the definitions of solid waste terms and the classifications vary greatly in the literature and in the profession. Consequently, the use of published data requires considerable care, judgment, and common sense. The following definitions are intended to serve as a guide and are not meant to be precise in a scientific sense.

Residential and Commercial

Residential and commercial solid wastes, excluding special and hazardous wastes discussed below, consist of the organic (combustible) and inorganic (noncombustible) solid wastes from residential areas and commercial establishments. Typically,

TABLE 3-1
Sources of solid wastes within a community[a]

Source	Typical facilities, activities, or locations where wastes are generated	Types of solid wastes
Residential	Single family and multifamily detached dwellings, low-, medium-, and high-rise apartments, etc.	Food wastes, paper, cardboard, plastics, textiles, leather, yard wastes, wood, glass, tin cans, aluminum, other metals, ashes, street leaves, special wastes (including bulky items, consumer electronics, white goods, yard wastes collected separately, batteries, oil, and tires), household hazardous wastes
Commercial	Stores, restaurants, markets, office buildings, hotels, motels, print shops, service stations, auto repair shops, etc.	Paper, cardboard, plastics, wood, food waste, glass, metals, special wastes (see above), hazardous wastes, etc.
Institutional	Schools, hospitals, prisons, governmental centers	As above in commercial
Construction and demolition	New construction sites, road repair/renovation sites, razing of buildings, broken pavement	Wood, steel, concrete, dirt, etc.
Municipal services (excluding treatment facilities)	Street cleaning, landscaping, catch basin cleaning, parks and beaches, other recreational areas	Special wastes, rubbish, street sweepings, landscape and tree trimmings, catch basin debris, general wastes from parks, beaches, and recreational areas
Treatment plant sites; municipal incinerators	Water, wastewater, and industrial treatment processes, etc.	Treatment plant wastes, principally composed of residual sludges
Municipal solid waste[b]	All of the above	All of the above
Industrial	Construction, fabrication, light and heavy manufacturing, refineries, chemical plants, power plants, demolition, etc.	Industrial process wastes, scrap materials, etc. Non-industrial wastes including food wastes, rubbish, ashes, demolition and construction wastes, special wastes, hazardous wastes
Agricultural	Field and row crops, orchards, vineyards, dairies, feedlots, farms, etc.	Spoiled food wastes, agricultural wastes, rubbish, hazardous wastes

[a] For comparison, the sources of waste and waste classifications used in the early 1900s are given in Table 3-12.

[b] The term *municipal solid waste* (MSW) normally is assumed to include all of the wastes generated in a community with the exception of industrial process wastes and agricultural solid wastes.

the organic fraction of residential and commercial solid waste consists of materials such as food waste (also called garbage), paper of all types, corrugated cardboard (also known as paperboard and corrugated paper), plastics of all types, textiles, rubber, leather, wood, and yard wastes. The inorganic fraction consists of items such as glass, crockery, tin cans, aluminum, ferrous metals, and dirt. If the waste components are not separated when discarded, then the mixture of these wastes is also known as *commingled residential and commercial MSW*.

Wastes that will decompose rapidly, especially in warm weather, are also known as putrescible waste. The principal source of putrescible wastes is the handling, preparation, cooking, and eating of foods. Often, decomposition will lead to the development of offensive odors (see Section 4-3) and the breeding of flies. In many locations, the putrescible nature of these wastes will influence the design and operation of the solid waste collection system (see Chapter 8).

Although there are more than 50 classifications for paper, the waste paper found in MSW is typically composed of newspaper, books and magazines, commercial printing, office paper, other paperboard, paper packaging, other non-packaging paper, tissue paper and towels, and corrugated cardboard.

The plastic materials found in MSW fall into the following seven categories:

- Polyethylene terephthalate (PETE/1)
- High-density polyethylene (HDPE/2)
- Polyvinyl chloride (PVC/3)
- Low-density polyethylene (LDPE/4)
- Polypropylene (PP/5)
- Polystyrene (PS/6)
- Other multilayered plastic materials (7)

The type of plastic container can be identified by number code (1 through 7) molded into the bottom of the container (see Fig. 3-1). *Mixed plastic* is the term used for the mixture of the individual types of plastic found in MSW.

Special Wastes. Special wastes from residential and commercial sources include bulky items, consumer electronics, white goods, yard wastes that are collected

FIGURE 3-1
Code designation used for various types of plastics.

separately, batteries, oil, and tires. These wastes are usually handled separately from other residential and commercial wastes.

Bulky items are large worn-out or broken household, commercial, and industrial items such as furniture, lamps, bookcases, filing cabinets, and other similar items. *Consumer electronics* includes worn-out, broken, and other no-longer-wanted items such as radios, stereos, and television sets. *White goods* are large worn-out or broken household, commercial, and industrial appliances such as stoves, refrigerators, dishwashers, and clothes washers and dryers. Collected separately, white goods are usually dismantled for the recovery of specific materials (e.g., copper, aluminum, etc.).

The principal sources of batteries are from households and automobile and other vehicle servicing facilities. Household batteries come in a variety of types, including alkaline, mercury, silver, zinc, nickel, and cadmium. The metals found in household batteries can cause groundwater contamination by their presence in leachate; they can also contaminate air emissions and ash from waste combustion facilities. Many states now prohibit the landfilling of household batteries. Automobiles use lead-acid batteries, each of which contains approximately 18 pounds of lead and a gallon of sulfuric acid, both hazardous materials.

The principal source of used oil is from the servicing of automobiles and other moving vehicles by their owners. Waste oil, not collected for recycling, is often poured onto the ground; down sanitary, combined, and storm water sewers; or into trash containers. Waste oil discharged onto the ground or into municipal sewers often contaminates surface water and groundwater as well as the soil. Waste oil placed in the same container as other solid waste components tends to contaminate the waste components and thus reduces their value as recycled materials.

Somewhere between 230 and 240 million rubber tires are disposed of annually in landfills or in tire stockpiles. Because tires do not compact well, their disposal in landfills is expensive and wasteful of space. Stockpiling of tires also poses serious aesthetic as well as environmental problems. Large, difficult-to-extinguish fires have occurred in a number of stockpiles. In addition, stockpiled tires form an ideal breeding place for mosquitos.

Hazardous Wastes. Wastes or combinations of wastes that pose a substantial present or potential hazard to human health or living organisms have been defined as hazardous wastes. The U.S. EPA has defined RCRA hazardous wastes in three general categories: (1) listed wastes, (2) characteristic hazardous wastes, and (3) other hazardous wastes. Hazardous wastes found in MSW are considered further in Chapter 5.

Institutional

Institutional sources of solid waste include government centers, schools, prisons, and hospitals. Excluding manufacturing wastes from prisons and medical wastes from hospitals, the solid wastes generated at these facilities are quite similar to commingled MSW. In most hospitals medical wastes are handled and processed separately from other solid wastes.

Construction and Demolition

Wastes from the construction, remodeling, and repairing of individual residences, commercial buildings, and other structures are classified as *construction wastes*. The quantities produced are difficult to estimate. The composition is variable but may include dirt; stones; concrete; bricks; plaster; lumber; shingles; and plumbing, heating, and electrical parts. Wastes from razed buildings, broken-out streets, sidewalks, bridges, and other structures are classified as *demolition wastes*. The composition of demolition wastes is similar to construction wastes, but may include broken glass, plastics, and reinforcing steel.

Municipal Services

Other community wastes, resulting from the operation and maintenance of municipal facilities and the provision of other municipal services, include street sweepings, road side litter, wastes from municipal litter containers, landscape and tree trimmings, catch-basin debris, dead animals, and abandoned vehicles. Because it is impossible to predict where dead animals and abandoned automobiles will be found, these wastes are often identified as originating from nonspecific diffuse sources. Wastes from nonspecific diffuse sources can be contrasted to that of the residential sources, which are also diffuse but specific in that the generation of the wastes is a recurring event.

Treatment Plant Wastes and Other Residues

The solid and semisolid wastes from water, wastewater, and industrial waste treatment facilities are termed *treatment plant wastes*. The specific characteristics of these materials vary, depending on the nature of the treatment process. At present, their collection is not the charge of most municipal agencies responsible for solid waste management. However, wastewater treatment plant sludges are commonly co-disposed with MSW in municipal landfills. In the future, the disposal of treatment plant sludges will likely become a major factor in any solid waste management plan.

Materials remaining from the combustion of wood, coal, coke, and other combustible wastes are categorized as *ashes and residues*. (Residues from power plants normally are not included in this category because they are handled and processed separately.) These residues are normally composed of fine, powdery materials, cinders, clinkers, and small amounts of burned and partially burned materials. Glass, crockery, and various metals are also found in the residues from municipal incinerators.

Industrial Solid Waste Excluding Process Wastes

Sources and types of solid waste generated at industrial sites, grouped according to their Standard Industrial Classification (SIC), are reported in Table 3-2 on pages 46 and 47. This list excludes industrial process wastes and any hazardous wastes that may be generated.

Agricultural Wastes

Wastes and residues resulting from diverse agricultural activities—such as the planting and harvesting of row, field, tree and vine crops; the production of milk; the production of animals for slaughter; and the operation of feedlots—are collectively called *agricultural wastes*. At present, the disposal of these wastes is not the responsibility of most municipal and county solid waste management agencies. However, in many areas the disposal of animal manure has become a critical problem, especially from feedlots and dairies.

3-3 COMPOSITION OF SOLID WASTES

Composition is the term used to describe the individual components that make up a solid waste stream and their relative distribution, usually based on percent by weight. Information on the composition of solid wastes is important in evaluating equipment needs, systems, and management programs and plans. For example, if the solid wastes generated at a commercial facility consist of only paper products, the use of special processing equipment, such as shredders and balers, may be appropriate. Separate collection may also be considered if the city or collection agency is involved in a paper-products recycling program. The potential for significant changes in composition in the future is considered at the end of this chapter.

Composition of MSW

The total solid wastes from a community are composed of the waste materials identified in Table 3-1. Typical data on the distribution of MSW are presented in Table 3-3. As noted in Table 3-3, the residential and commercial portion makes up about 50 to 75 percent of the total MSW generated in a community. The actual percentage distribution will depend on (1) the extent of the construction and demolition activities, (2) the extent of the municipal services provided, and (3) the types of water and wastewater treatment processes that are used. The wide variation in the special wastes category (3 to 12 percent) is due to the fact that in many communities yard wastes are collected separately. The percentage of construction and demolition wastes varies widely depending on the part of the country and the general health of the local, state, and national economy. The

TABLE 3-2
Sources and types of industrial wastes[a]

Code	SIC group classification[b]	Waste-generating processes	Expected specific wastes
19	Ordnance and accessories	Manufacturing and assembling	Metals, plastic, rubber, paper, wood, cloth, chemical residues
20	Food and kindred products	Processing, packaging, shipping	Meats, fats, oils, bones, offal, vegetables, fruits, nuts and shells, cereals
22	Textile mill products	Weaving, processing, dyeing, and shipping	Cloth and fiber residues
23	Apparel and other finished products	Cutting, sewing, sizing, pressing	Cloth, fibers, metals, plastics, rubber
24	Lumber and wood products	Sawmills, millwork plants, wooden container, miscellaneous wood products, manufacturing	Scrap wood, shaving, sawdust; in some instances metals, plastics, fibers, glues, sealers, paints, solvents
25a	Furniture, wood	Manufacture of household and office furniture, partitions, office and store fixtures, mattresses	Those listed under code 24; in addition, cloth and padding residues
25b	Furniture, metal	Manufacture of household and office furniture, lockers, bed-springs, frames	Metals, plastics, resins, glass, wood, rubber, adhesives, cloth, paper
26	Paper and allied products	Paper manufacture, conversion of paper and paperboard, manufacture of paperboard boxes and containers	Paper and fiber residues, chemicals, paper coatings and fillers, inks, glues, fasteners
27	Printing and publishing	Newspaper publishing, printing, lithography, engraving, and bookbinding	Paper, newsprint, cardboard, metals, chemicals, cloth, inks, glues
28	Chemical and related products	Manufacture and preparation of inorganic chemicals (ranges from drugs and soaps to paints and varnishes, and explosives)	Organic and inorganic chemicals, metals, plastics, rubber, glass, oils, paints, solvents, pigments
29	Petroleum refining and related industries	Manufacture of paving and roofing materials	Asphalt and tars, felts, asbestos, paper, cloth, fiber
30	Rubber and miscellaneous plastic products	Manufacture of fabricated rubber and plastic products	Scrap rubber and plastics, lampblack, curing compounds, dyes
31	Leather and leather products	Leather tanning and finishing; manufacture of leather belting and packing	Scrap leather, thread, dyes, oils, processing and curing compounds

32	Stone, clay, and glass products	Manufacture of flat glass, fabrication and forming of glass; manufacture of concrete, gypsum, and plaster products; forming and processing of stone and stone products, abrasives, asbestos, and miscellaneous non-mineral products	Glass, cement, clay, ceramics, gypsum, asbestos, stone, paper, abrasives
33	Primary metal industries	Melting, casting, forging, drawing, rolling, forming, extruding operations	Ferrous and nonferrous metals scrap, slag, sand, cores, patterns, bonding agents
34	Fabricated metal products	Manufacture of metal cans, hand tools, general hardware, nonelectric heating apparatus, plumbing fixtures, fabricated structural products, wire, farm machinery and equipment, coating and engraving of metal	Metals, ceramics, sand, slag, scale, coatings, solvents, lubricants, pickling liquor
35	Machinery (except electrical)	Manufacture of equipment for construction, mining, elevators, moving stairways, conveyors, industrial trucks, trailers, stackers, machine tools, etc.	Slag, sand, cores, metal scrap, wood, plastics, resins, rubber, cloth, paints, solvents, petroleum products
36	Electronic and other electrical equipment and components	Manufacture of electric equipment, appliances, and communication apparatus; machining, drawing, forming, welding, stamping, winding, painting, planting, baking, firing operations	Metal scrap, carbon black, glass, exotic metals, rubber, plastics, resins, fibers, cloth residues
37	Transportation equipment	Manufacture of motor vehicles, truck and bus bodies, motor vehicle parts and accessories, aircraft and parts, ship and boat building and repairing, motorcycles and bicycles and parts, etc.	Metal scrap, glass, fiber, wood, rubber, plastics, cloth, paints, solvents, petroleum products
38	Measuring, analyzing and controlling instruments	Manufacture of engineering, laboratory and research instruments and associated equipment	Metals, plastics, resins, glass, wood, rubber, fibers, abrasives
39	Miscellaneous manufacturing	Manufacture of jewelry, silverware, plated ware, toys, amusement, sporting and athletic goods, costume novelties, buttons, brooms, brushes, signs, advertising displays	Metals, glass, plastics, resins, leather, rubber, bone, cloth, straw, adhesives, paints, solvents, composite materials

[a] Adapted in part from Ref. 8.

[b] *Source:* Standard Industrial Classification Manual (SIC) 1987, Executive Office of the President, Office of Management and Budget, U.S. Government Printing Office, Washington, DC.

TABLE 3-3
Estimated distribution of all components of MSW generated in a typical community excluding industrial and agricultural wastes[a]

Waste category[b]	Percent by weight	
	Range	Typical
Residential and commercial, excluding special and hazardous wastes	50–75	62.0
Special (bulky items, consumer electronics, white goods, yard wastes collected separately, batteries, oil, and tires)	3–12	5.0
Hazardous[c]	0.01–1.0	0.1
Institutional	3–5	3.4
Construction and demolition	8–20	14.0
Municipal services		
Street and alley cleanings	2–5	3.8
Tree and landscaping	2–5	3.0
Parks and recreational areas	1.5–3	2.0
Catch basin	0.5–1.2	0.7
Treatment plant sludges	3–8	6.0
Total		100.0

[a] Adapted in part from Refs. 9, 14–16.

[b] See Table 6-3 for estimated quantities of waste generated.

[c] Range of reported values varies widely depending on method used to identify and classify hazardous wastes found in MSW.

percentage of treatment plant sludges will also vary widely depending on the extent and type of water and wastewater treatment provided.

Distribution of Individual Waste Components

Information and data on the physical composition of solid wastes are important in the selection and operation of equipment and facilities (see Chapters 6 through 10), in assessing the feasibility of resource and energy recovery (see Chapter 9), and in the analysis and design of landfill disposal facilities (see Chapter 11). Published distribution data should be used cautiously because the effects of recycling activities and the use of kitchen food waste grinders are often not reflected in earlier data.

Residential Portion of MSW in the United States. Components that typically make up the residential portion of MSW, excluding special and hazardous wastes, and their relative distribution are reported in Table 3-4. Although any number of components could be selected, those in Table 3-4 have been selected because they are readily identifiable and consistent with component categories reported in the literature and because they have proven adequate for the characterization

TABLE 3-4
Typical physical composition of residential MSW excluding recycled materials and food wastes discharged with wastewater (1990)

	Percent by weight			
	United States[a]			
Component	Range	Typical[b]	Packaging materials[c]	Davis, California[d]
Organic				
Food wastes	6–18	9.0	—	6.0
Paper	25–40	34.0	50–60	33.1
Cardboard	3–10	6.0		7.9
Plastics	4–10	7.0	12–16	10.7
Textiles	0–4	2.0	—	2.4
Rubber	0–2	0.5	—	2.5
Leather	0–2	0.5	—	0.1
Yard wastes	5–20	18.5	—	17.7
Wood	1–4	2.0	4–8	5.0
Misc. organics	—	—	—	0.4
Inorganic				
Glass	4–12	8.0	20–30	5.8
Tin cans	2–8	6.0	6–8	3.9
Aluminum	0–1	0.5	2–4	0.4
Other metal	1–4	3.0	—	3.6
Dirt, ash, etc.	0–6	3.0	—	0.5
Total		100.0		100.0

[a] Adapted in part from Refs. 2, 3, 9, and 14–16. Reported percentage distributions are exclusive of special and hazardous wastes.

[b] Twenty percent of the households in the United States are assumed to have food waste grinders. Additionally, it is assumed that the percentage of food waste ground up and discharged with wastewater is 25 percent. Current (1990) recycling rate for the United States assumed to be 11 percent.

[c] Adapted in part from Ref. 10.

[d] Based on measurements made over a five-year period (1985 to 1990) during the first two weeks of October (see Table 3-9).

of solid wastes for most applications. The data in Table 3-4 are derived from both the literature and the authors' experience. For the purpose of comparison, the percentage distribution of the materials used for packaging is reported in Column 3 of Table 3-4. It is estimated that packaging wastes now account for approximately one-third of the residential and commercial MSW [10].

The values given in Table 3-4 for food waste, plastics, and yard wastes are considerably different from the values given in the corresponding table (Table 4-2) in the predecessor of this text, published in 1977 [12]. The differences are due largely to (1) improved food processing techniques and the increased use of kitchen food waste grinders, (2) the increased use of plastics for food packaging and other packaging, and (3) the fact that burning of yard wastes is no longer allowed in most communities.

TABLE 3-5
Typical distribution of components in residential MSW for low-, middle-, and upper-income countries excluding recycled materials[a,b]

Component	Low-income countries	Middle-income countries	Upper-income countries[c]
Organic			
Food wastes	40–85[d]	20–65	6–30
Paper	1–10	8–30	20–45
Cardboard			5–15
Plastics	1–5	2–6	2–8
Textiles	1–5	2–10	2–6
Rubber	1–5	1–4	0–2
Leather			0–2
Yard wastes	1–5	1–10	10–20
Wood			1–4
Misc. organics	—	—	—
Inorganic			
Glass	1–10	1–10	4–12
Tin cans	1–5	1–5	2–8
Aluminum			0–1
Other metal			1–4
Dirt, ash, etc.	1–40	1–30	0–10

[a] Adapted in part from Refs. 1 and 17.

[b] Low-income countries: per capita income of less than U.S. $750 in 1990.
Middle-income countries: per capita income of more than U.S. $750 and less than U.S. $5000 in 1990.
Upper-income countries: per capita income of more than U.S. $5000 in 1990.

[c] Upper-income countries are more highly industrialized.

[d] Food wastes composed predominantly of waste from the preparation of food (corn husks, melon rinds, banana leaves, etc.).

Residential Portion of MSW in Other Countries. For the purposes of comparison, typical data on the distribution of the components in residential MSW from other countries are presented in Table 3-5. In comparing the data presented in Tables 3-5 and 3-6, note the high percentage of food waste in less-developed countries. The percentage of food wastes is high because most vegetables and fruits are not pre-trimmed, there are essentially no kitchen food waste grinders, and the amounts of the other components are quite small.

Industrial Solid Wastes Excluding Process Wastes. Data on the percentage distribution of the wastes generated from the industrial activities given in Table 3-2 are presented in Table 3-6. A range of values is given for each waste component category, because industrial operations tend to be quite variable.

Effect of Waste Diversions on Distribution of Components in Residential MSW

To assess the impact of waste diversions (resulting from the use of food waste grinders and waste recycling programs) on the distribution of waste components,

TABLE 3-6

Typical data on the distribution of solid wastes generated by major industries excluding recycled materials and industrial process wastes[a]

SIC Code		Percent by weight									
		Food wastes[b]	Paper	Wood	Leather	Rubber	Plastics	Metals	Glass	Textiles	Misc.
20	Food and kindred products	15–20	50–60	5–10	0–2	0–2	0–5	5–10	4–10	0–2	5–15
22	Textile mill products	0–2	40–50	0–2	0–2	0–2	3–10	0–2	0–2	20–40	0–5
23	Apparel and other finished products	0–2	40–60	0–2	0–2	0–2	0–2	0–2	0–2	30–50	0–5
24	Lumber and wood products	0–2	10–20	60–80	0–2	0–2	0–2	0–2	0–2	0–2	5–10
25a	Furniture, wood	0–2	20–30	30–50	0–2	0–2	0–2	0–2	0–2	0–5	0–5
25b	Furniture, metal	0–2	20–40	10–20	0–2	0–2	0–2	20–40	0–2	0–5	0–10
26	Paper and allied products	0–2	40–60	10–15	0–2	0–2	0–2	5–15	0–2	0–2	10–20
27	Printing and publishing	0–2	60–90	5–10	0–2	0–2	0–2	0–2	0–2	0–2	0–5
28	Chemicals and related products	0–2	40–60	2–10	0–2	0–2	5–15	5–10	0–5	0–2	15–25
29	Petroleum refining and related industries	0–2	60–80	5–15	0–2	0–2	10–20	2–10	0–2	0–2	2–10
30	Rubber and miscellaneous plastic products	0–2	40–60	2–10	0–2	5–20	10–20	0–2	0–2	0–2	0–5
31	Leather and leather products	0–2	5–10	5–10	40–60	0–2	0–2	10–20	0–2	0–2	0–5
32	Stone, clay, and glass products	0–2	20–40	2–10	0–2	0–2	0–2	5–10	10–20	0–2	30–50
33	Primary metal industries	0–2	30–50	5–15	0–2	0–2	2–10	2–10	0–5	0–2	20–40
34	Fabricated metal products	0–2	30–50	5–15	0–2	0–2	0–2	15–30	0–2	0–2	5–15
35	Machinery (except electrical)	0–2	30–50	5–15	0–2	0–2	1–5	15–30	0–2	0–2	0–5
36	Electrical	0–2	60–80	5–15	0–2	0–2	2–5	2–5	0–2	0–2	0–5
37	Transportation equipment	0–2	40–60	5–15	0–2	0–2	2–5	0–2	0–2	0–2	15–30
38	Professional scientific controlling instruments	0–2	30–50	2–10	0–2	0–2	5–10	5–15	0–2	0–2	0–5
39	Miscellaneous manufacturers	0–2	40–60	10–20	0–2	0–2	5–15	2–10	0–2	0–2	5–15

[a] From Ref. 13.

[b] With the exception of food and kindred products, food wastes are from company cafeterias, canteens, etc.

the distribution given in Table 3-4 for *as collected* residential MSW must be adjusted. The methodology used to adjust the waste component distribution is illustrated in Example 3-1. The adjusted component distribution data are reported in Table 3-7. As shown in Table 3-7, the distribution data for the United States do not change significantly. On the other hand, the distribution data for the city of Davis, California, would change quite a bit more because of the higher percentage of recycling.

TABLE 3-7
Typical physical composition of residential MSW in the United States in 1990[a]

	Percent by weight			
Component	Solid waste *as collected* excluding waste components now recycled and food waste that is ground up[b]	Solid waste *as collected* plus ground up food waste, but excluding waste components now recycled[c]	Solid waste *as collected* plus waste components now recycled excluding food waste that is ground up[d]	Solid waste *as collected* plus waste components now recycled and food waste that is ground up[e]
Organic				
Food wastes	9.0	9.4	8.0	8.4
Paper	34.0	33.8	35.8	35.6
Cardboard	6.0	6.0	6.4	6.4
Plastics	7.0	7.0	6.9	6.9
Textiles	2.0	2.0	1.8	1.8
Rubber	0.5	0.5	0.4	0.4
Leather	0.5	0.5	0.4	0.4
Yard wastes	18.5	18.4	17.3	17.2
Wood	2.0	2.0	1.8	1.8
Misc. organics	—	—	—	—
Inorganic				
Glass	8.0	7.9	9.1	9.0
Tin cans	6.0	6.0	5.8	5.8
Aluminum	0.5	0.5	0.6	0.6
Other metal	3.0	3.0	3.0	3.0
Dirt, ash, etc.	3.0	3.0	2.7	2.7
Total	100.0	100.0	100.0	100.0

[a] Procedure used to compute the values in this table is delineated in Example 3-1.

[b] From Table 3-4.

[c] Twenty percent of the households in the United States are assumed to have food waste grinders. Additionally, it is assumed that the percentage of food waste ground up and discharged with wastewater is 25 percent.

[d] Current (1990) recycling rate for the United States assumed to be 11 percent.

[e] Column 5 represents the percentage distribution of the total amount of residential MSW now generated, including the waste components that are now recycled and the food wastes that are ground up and discharged to the sewer.

Example 3-1 Impact of food waste grinders and waste recycling on the distribution of waste components in residential MSW. Assess the impact of the use of food waste grinders and waste recycling on the percentage distribution of the components found in residential MSW. Assume the following data apply:

1. Use of waste food grinders
   (*a*) Households in the United States that have food waste grinders = 20%
   (*b*) Percentage of the total amount of food waste that is ground up and discharged to the local sewer = 25%
2. Waste recycling
   (*a*) Percentage of the total amount of residential MSW that is now recycled, excluding food waste that is ground up = 11%
   (*b*) Percentage distribution by weight of waste components now recycled and not included in *as collected* waste distribution.

   Paper = 50%
   Cardboard = 10%
   Plastic = 6%
   Yard wastes = 8%
   Tin cans = 4%
   Glass = 18%
   Aluminum = 1%
   Nonferrous metal = 3%

Solution

1. Estimate the amount of food waste that is now ground up.
   (*a*) To account for the food wastes that are now ground up and discharged to the local sewer, the original weight of food wastes must be adjusted as follows:

$$\mathrm{FW}_a,\ \mathrm{lb} = \frac{\mathrm{FW}}{[H_{w/o} + H_w(1 - \mathrm{fw}_g)]}$$

   where FW_a = adjusted food waste (accounts for food waste that is now ground up and discharged to sewer)
   FW = food waste in "as collected MSW," lb (based on 100-lb sample of waste—see column 3 in Table 3-4)
   $H_{w/o}$ = fraction of homes without food waste grinders
   H_w = fraction of homes with food waste grinders
   fw_g = fraction of food waste that is ground up

   (*b*) Compute the adjusted food waste value using the above expression and the given data.

$$\mathrm{FW}_a,\ \mathrm{lb} = \frac{9.0}{0.80 + 0.20(1 - 0.25)} = 9.5$$

2. Set up a computation table to determine the percentage distribution of the waste components given in Table 3-4 taking into account the amount of food waste that is now

ground up (0.5 lb = 9.5 lb − 9.0 lb). The new distribution given in the following table (column 3) is obtained by adding 0.5 lb to the food waste and dividing all of the component weights by the new total weight 100.5 lb (100 lb + 0.5 lb).

	Percent by weight	
Component	**Solid waste as collected**	**Solid waste as collected plus ground up food waste**
Organic		
Food wastes	9.0 (9.5)	9.4
Paper	34.0	33.8
Cardboard	6.0	6.0
Plastics	7.0	7.0
Textiles	2.0	2.0
Rubber	0.5	0.5
Leather	0.5	0.5
Yard wastes	18.5	18.4
Wood	2.0	2.0
Misc. organics	—	—
Inorganic		
Glass	8.0	7.9
Tin cans	6.0	6.0
Aluminum	0.5	0.5
Other metal	3.0	3.0
Dirt, ash, etc	3.0	3.0
Total	100.0 (100.5)	100.0

[a] From Table 3-4.

3. Set up a computation table to determine the percentage distribution of the waste components given in Table 3-4 taking into account the amount of waste that is recycled and the food wastes that are now ground up and discharged to the sewer. The required computation table follows on page 55.

   (*a*) The percentage distribution of solid waste as collected is given in column 2.

   (*b*) The percentage distribution of the solid waste materials that are now recycled separately from the *as collected* waste is given in column 3.

   (*c*) The actual weight of material that is now recycled separately from the *as collected* waste, based on 11 percent of the total weight of solid waste (i.e., 11 lb out of 100 lb) is given in column 4.

   (*d*) The actual weight of material that is now collected, based on 89 percent of the total weight of solid waste (i.e., 89 lb) is given in column 5. The actual values given in column 5 are obtained by multiplying the distribution percentage values given in column 2 by 89 lb.

   (*e*) The percentage distribution of the solid waste as collected plus the recycled materials is given in column 6. The values reported in column 6 are obtained by summing the values in columns 4 and 5.

Component (1)	Percent by weight		Weight of solid waste components now recycled (11 lb based on a total of 100 lb excluding ground up food waste), lb[b] (4)	Solid waste *as collected* excluding waste now recycled (89 lb based on a total 100 lb excluding ground up food waste), lb[c] (5)	Percent by weight	
	Solid waste *as collected* excluding recycled waste components and ground up food waste[a] (2)	Solid waste components now recycled (not reflected in *as collected* distribution) (3)			Solid waste *as collected* plus recycled wastes[b] (6) = (4) + (5)	Solid waste *as collected* plus recycled and ground up food waste[d] (7)
Organic						
Food wastes	9.0	0.0	0.00	8.01	8.0	8.4
Paper	34.0	50.0	5.50	30.26	35.8	35.6
Cardboard	6.0	10.0	1.10	5.34	6.4	6.4
Plastics	7.0	6.0	0.66	6.23	6.9	6.9
Textiles	2.0	0.0	0.00	1.78	1.8	1.8
Rubber	0.5	0.0	0.00	0.45	0.4	0.4
Leather	0.5	0.0	0.00	0.45	0.4	0.4
Yard wastes	18.5	8.0	0.88	16.46	17.3	17.2
Wood	2.0	0.0	0.00	1.78	1.8	1.8
Misc. organics	—	—	—	—	—	—
Inorganic						
Glass	8.0	18.0	1.98	7.12	9.1	9.0
Tin cans	6.0	4.0	0.44	5.34	5.8	5.8
Aluminum	0.5	1.0	0.11	0.44	0.6	0.6
Other metal	3.0	3.0	0.33	2.67	3.0	3.0
Dirt, ash, etc.	3.0	0.0	0.00	2.67	2.7	2.7
Total	100.0	100.0	11.00[b]	89.00	100.0	100.0

[a] From Table 3-4.

[b] Amount now recycled = 11 percent or 11 lb based on 100 lb.

[c] 89.0 lb = 100 lb − 11 lb (amount now recycled).

[d] Column 7 represents the percentage distribution of the total amount of waste generated including the waste components that are now recycled and the food wastes that are ground up.

(*f*) The percentage distribution of the solid waste as collected plus recycled materials and ground up food waste is given in column 7. The adjusted value for food wastes, computed using the expression given in Step 1 is:

$$FW_a, \text{lb} = \frac{8.0}{0.80 + 0.20(1 - 0.25)} = 8.4$$

The distribution values given in column 7 are obtained by adding 0.4 lb to the food waste in column 6 and dividing all of the component percentages in column 6, expressed as lb, by the new total weight of 100.4 lb (100 lb + 0.4 lb).

Comment. The computational approach used in this example can be used to determine the percentage distribution values for any level of recycling. This approach will be especially important in determining the percentage of wastes diverted from landfills.

Variation in the Percentage Distribution of Waste Components

The percentage distribution values for the components in MSW vary with location, season, economic conditions, and many other factors. Typical seasonal variations in waste quantities are presented in Table 3-8. Because variations are known to occur, if the distribution of components is a critical factor in a particular management decision process, a special study should be undertaken if possible to assess the actual distribution. Even then, it may still be impossible to obtain an accurate assessment unless a prohibitively large number of samples are analyzed. In general, the coefficient of variation (CV) (see Appendix D) for the individual waste constituents is quite large. Typical CV values for paper in residential MSW range from about 20 to 40 percent. For the remaining components in the waste stream, CV values can vary from 40 to 100 percent. Data collected during October for the city of Davis, CA, over a 20-year period are reported in Table 3-9.

TABLE 3-8
Typical seasonal variation observed in the *as collected* composition of residential MSW[a]

	Percent by weight		Percent variation[b]	
Waste	**Winter season**	**Summer season**	**Decrease**	**Increase**
Food waste	11.1	13.5		21.6
Paper	45.2	40.0	11.5	
Plastics	9.1	8.2	9.9	
Other organics	4.0	4.6		15.0
Yard wastes	18.7	24.0		28.3
Glass	3.5	2.5	28.6	
Metals	4.1	3.1	24.4	
Inert and other waste	4.3	4.1	4.7	
Total	100.0	100.0		

[a] Adapted from Ref. 11.

[b] Based on winter season.

TABLE 3-9
Typical composition of residential MSW from Davis, CA, excluding recycled materials and food wastes discharged to sewers. All data collected during the first two weeks of October. The community of Davis also has separate collection of yard wastes.

	Percent by weight															
Component	**1971**	**1972**	**1973**	**1974**	**1975**	**1977**	**1978**	**1979**	**1981**	**1982**	**1984**	**1986**	**1987**	**1988**	**1989**	**1990**
Organic																
Food wastes	13.5	5.5	19.6	7.5	7.7	7.9	9.8	11.9	5.2	5.0	16.0	6.8	6.5	5.8	3.4	7.6
Paper	33.4	31.8	29.5	43.6	32.8	39.1	28.9	41.3	36.9	42.9	22.7	35.9	32.2	29.1	33.9	34.1
Cardboard	14.2	10.4	23.0	7.8	15.0	8.0	12.3	3.4	15.2	11.0	10.5	8.9	6.8	8.2	6.6	8.9
Plastics	3.1	4.0	3.9	1.8	5.2	5.6	6.7	5.2	7.6	8.1	11.4	8.2	10.0	10.6	12.8	11.8
Textiles	3.9	1.9	0.2	0.1	2.3	1.1	1.5	1.8	6.2	9.2	7.8	1.9	3.6	1.9	2.5	1.9
Rubber / Leather (combined)	1.3	1.6	Trace	1.4		0.9										
Rubber					0.8		1.2	5.2	3.2	1.7	1.0	1.9	2.8	0.7	4.7	2.4
Leather					0.3		0.1	0.7	—	—	—	—	—	0.3	—	0.1
Yard wastes	1.0	24.3	10.8	14.1	14.4	14.6	17.7	9.9	4.9	4.8	8.1	17.6	20.0	21.4	13.1	16.5
Wood	2.3	2.6	0.9	4.2	1.6	0.8	0.8	0.7	3.6	1.6	4.3	4.6	4.2	5.9	8.0	2.2
Misc. organics	2.0	—	—	—	—	—	3.1	1.6	—	1.5	0.5	—	—	2.2	—	—
Inorganic																
Glass	13.0	8.2	6.4	7.7	6.4	11.7	10.4	5.7	6.0	5.0	9.2	7.0	6.6	6.3	3.2	6.1
Tin cans	6.1	6.6	1.9	1.8	6.7	4.8	5.2	7.8	4.1	3.5	3.1	3.7	4.3	3.6	4.4	3.6
Aluminum cans	0.2	—	0.5	1.4	1.2	1.0	1.5	2.3	1.2	1.4	0.8	0.5	0.5	0.5	0.2	0.3
Other metal	5.8	—	2.6	7.6	5.0	0.5	0.7	—	5.2	4.3	1.6	2.2	2.5	3.5	6.2	3.7
Dirt, ashes, etc.	0.2	1.2	0.7	1.0	0.6	4.6	0.1	2.5	0.7	—	3.0	0.8	—	—	1.0	0.8

A common failing in many engineering studies is to spend far too much money collecting data that are of limited value or may never be used. This situation is often true with regard to the collection of statistical distribution data on solid waste components during one sampling period. For example, it is usually more important to have information on the seasonal variation of waste generation rates (e.g., Table 3-8) than to know whether the percentage of a given component is 8.1 versus 8.12 during any one sampling period. In the conduct of waste characterization studies, the distribution of components presented in Table 3-7 may be used as a guide in assessing the reasonableness of the findings.

3-4 DETERMINATION OF THE COMPOSITION OF MSW IN THE FIELD

Because of the heterogeneous nature of solid wastes, determination of the composition is not an easy task. Strict statistical procedures are difficult, if not impossible, to implement. For this reason, more generalized field procedures, based on common sense and random sampling techniques, have evolved for determining composition.

Residential MSW

The procedure for residential MSW involves unloading and analyzing a quantity of residential waste in a controlled area of a disposal site that is isolated from winds and separate from other operations. A representative residential sample might be a truckload resulting from a typical weekday collection route in a residential area. A mixed sample from an incinerator storage pit or the discharge pit of a shredder would also be representative. Common sense is important in selecting the load to be sampled. For example, a load containing the weekly accumulation of yard wastes (leaves) during autumn would not be typical. To ensure that the results obtained are representative, a large enough sample must be examined. It has been found that measurements made on a sample size of about 200 lb vary insignificantly from measurements made on samples of up to 1700 lb taken from the same waste load [5, 7]. The authors have obtained similar results in field studies performed in Hawaii and at Davis, California.

To obtain a sample for analysis, the load is first quartered. One part is then selected for additional quartering until a sample size of about 200 lb is obtained. It is important to maintain the integrity of each selected quarter, regardless of the odor or physical decay, and to make sure that all the components are measured (see Fig. 3-2). Only in this way can some degree of randomness and unbiased selection be maintained. (Additional information on sampling procedures is presented in Section 6-7 in Chapter 6.)

Commercial and Industrial MSW

The field procedure for component identification for commercial and nonprocess industrial solid wastes involves the analysis of representative waste samples taken directly from the source, not from a mixed waste load in a collection vehicle.

(a)

(b)

FIGURE 3-2
Determination of the percentage distribution of waste components in the field: (*a*) separated waste materials placed in separate containers to be weighed and (*b*) weighing of the wastes using a platform scale. (Courtesy of Brown and Caldwell, Consultants.)

Because commercial and industrial sources are so variable, statistically valid sampling is seldom possible. Estimation of the distribution of waste components and quantities for these activities remains an art form.

3-5 TYPES OF MATERIALS RECOVERED FROM MSW

The purpose of this section is to identify the types of materials that are now separated from MSW for recycling and introduces and discusses briefly the importance of materials specifications in the processing and marketing of recovered materials. Knowledge of the waste materials that are now recovered for reuse and recycling is important in the conduct of waste generation and diversion studies.

Materials Commonly Separated from MSW

Materials that are now (1992) separated for recycling are reported in Table 3-10. The most common ones from MSW are aluminum, paper, plastics, glass, ferrous metal, nonferrous metal, yard wastes, and construction and demolition wastes (see Fig. 3-3). Each is considered briefly in the following discussion. Other materials recovered from residential and commercial MSW as identified in Table 3-10 are considered in Chapter 15, where the actual recycling of these materials is considered in greater detail.

Aluminum. Aluminum recycling is made up of two sectors: aluminum cans and secondary aluminum. Secondary aluminum includes window frames, storm doors, siding, and gutters. Because secondary materials are of different grades, specifications for recycled aluminum should be checked, to recover the maximum value when selling separated materials to brokers. The demand for recycled aluminum cans is high, as it takes 95 percent less energy to produce an aluminum can from an existing can than from ore.

Paper. The principal types of waste paper that are recycled are old newspaper, cardboard, high-grade paper, and mixed paper. Each of these four grades consists of individual grades, which are defined according to the type of fiber, source, homogeneity, extent of printing, and physical or chemical characteristics. High-grade paper includes office paper, reproduction paper, computer printout, and other grades having a high percentage of long fibers. Mixed grades include paper with high ground-wood content, such as magazines; coated paper; and individual grades containing excessive percentages of "outthrows" (papers of lower grades than the grade specified). The types of paper found in residential solid waste before the removal of newspapers or other papers for recycling are reported in Table 3-11.

Plastics. Plastics can be classified into two general categories: clean commercial grade scrap and post-consumer scrap. The two types of post-consumer plastics that are now most commonly recycled are polyethylene terephthalate (PETE/1), which is used for the manufacture of soft drink bottles, and high-density polyethylene (HDPE/2), used for milk and water containers and detergent bottles. In 1987, more than 150 million pounds of plastic soft drink bottles were recycled. Even so, less than five percent of the available scrap plastic is being recycled. It is anticipated that all of the other types of plastics will be recycled in greater quantities in the future, however, as processing technologies improve.

Glass. Glass is also a commonly recycled material. Container glass (for food and beverage packing), flat glass (e.g., window glass), and pressed or amber and green glass are the three principal types of glass found in MSW. Glass to be reprocessed is often separated by color into categories of clear, green, and amber.

Ferrous Metals (Iron and Steel). The largest amount of recycled steel has traditionally come from large items such as cars and appliances. Many communities

TABLE 3-10
Materials that have been recovered for recycling from MSW[a]

Recyclable material	Types of materials or uses
Aluminum	Soft drink and beer cans
Paper	
Old newspaper (ONP)	Newsstand and home-delivered newspaper
Corrugated cardboard	Bulk packaging; largest single source of waste paper for recycling
High-grade paper	Computer paper, white ledger paper, and trim cuttings
Mixed paper	Various mixtures of clean paper, including newsprint, magazines, and white and colored long-fiber paper
Plastics	
Polyethylene terephthalate (PETE/1)	Soft drink bottles, salad dressing and vegetable oil bottles; photographic film
High-density polyethylene (HDPE/2)	Milk jugs, water containers, detergent and cooking oil bottles
Polyvinyl chloride (PVC/3)	Home landscaping irrigation piping, some food packaging, and bottles
Low-density polyethylene (LDPE/4)	Thin-film packaging and wraps; dry cleaning film bags; other film material
Polypropylene (PP/5)	Closures and labels for bottles and containers, battery casings, bread and cheese wraps, cereal box liners
Polystyrene (PS/6)	Packaging for electronic and electrical components, foam cups, fast food containers, tableware and microwave plates
Multilayer and other (7)	Multilayered packaging, ketchup and mustard bottles
Mixed plastics	Various combinations of the above products
Glass	Clear, green, and brown glass bottles and containers
Ferrous metal	Tin cans, white goods, and other metals
Nonferrous metals	Aluminum, copper, lead, etc.
Yard wastes, collected separately	Used to prepare compost; biomass fuel; intermediate landfill cover
Organic fraction of MSW	Used to prepare compost for soil applications; compost for use as intermediate landfill cover; methane; ethanol and other organic compounds; refuse-derived fuel (RDF)
Construction and demolition wastes	Soil, asphalt, concrete, wood, drywall, shingles, metals
Wood	Packing materials, pallets, scraps, and used wood from construction projects
Waste oil	Automobile and truck oil; reprocessed for reuse or fuel
Tires	Automobile and truck tires; road building material; fuel
Lead-acid batteries	Automobile and truck batteries; shredded to recover individual components such as acid, plastic, and lead
Household batteries	Potential recovery of zinc, mercury, and silver

[a] Detailed information on the recycling opportunities for the individual materials may be found in Chapter 15.

FIGURE 3-3
Commonly recycled waste materials: (*a*) paper, (*b*) cardboard, (*c*) plastics, (*d*) glass, (*e*) aluminum cans, (*f*) ferrous metals, (*g*) yard wastes, and (*h*) construction and demolition wastes.

TABLE 3-11
Percentage distribution of paper types found in residential solid waste[a]

Type of paper	Percent by weight: Range	Percent by weight: Typical
Newspaper	10–20	17.7
Books and magazines	5–10	8.7
Commercial printing	4–8	6.4
Office paper	8–12	10.1
Other paperboard	8–12	10.1
Paper packaging	6–10	7.8
Other nonpackaging paper	8–12	10.6
Tissue paper and towels	6–8	5.9
Corrugated materials	20–25	22.7
Total		100.0

[a] Adapted from Refs. 4 and 16.

have large scrap metal piles at the local landfill or transfer station. In many cases, the piles are unorganized and different metals are mixed together, making them unattractive to scrap metal buyers. Steel can recycling is also becoming more popular. Steel cans, used as juice, soft drink, and food containers, are easily separated from mixed recyclables or municipal solid waste using large magnets (which also separate other ferrous metals).

Nonferrous Metals. Recyclable nonferrous metals are recovered from common household items (outdoor furniture, kitchen cookware and appliances, ladders, tools, hardware); from construction and demolition projects (copper wire, pipe and plumbing supplies, light fixtures, aluminum siding, gutters and downspouts, doors, windows); and from large consumer, commercial, and industrial products (appliances, automobiles, boats, trucks, aircraft, machinery). Virtually all nonferrous metals can be recycled if they are sorted and free of foreign materials such as plastics, fabrics, and rubber.

Yard Wastes Collected Separately. In most communities yard wastes are collected separately. The composting of yard wastes has become of great interest as cities and towns seek to find ways in which to achieve mandated diversion goals. Leaves, grass clippings, bush clippings, and brush are the most commonly composted yard wastes. Stumps and wood are also compostable, but only after they have been chipped to produce a smaller more uniform size. Composting of the organic fraction of MSW is also becoming more popular.

Construction and Demolition Wastes. In many locations construction and demolition (C&D) wastes are now being processed to recover marketable items such as wood chips for use as a fuel in biomass combustion facilities, aggregate for concrete in construction projects, ferrous and nonferrous metals for remanufacture, and soil for use as fill material. The reprocessing of C&D wastes is gaining in popularity as disposal fees at landfills continue to increase. When disposal fees were below 5 dollars per ton (early 1970s), reprocessing was not economically feasible. Today (1992), with average landfill disposal fees approaching 60 dollars per ton in many parts of the country, the reprocessing of C&D wastes is economically feasible.

Specifications for Recovered Materials

As the amount of material recovered from MSW continues to increase as communities develop programs to meet waste diversion goals, materials specifications will become an important factor. In general, there is less contamination in source-separated material, but collection is more labor-intensive, and many communities are choosing to sort all materials at a central materials recovery facility (MRF). In many regions, markets for materials are not keeping pace with the volume collected, and it is expected that buyers will tighten specifications; as a result, vendors will no longer have assured markets, and will be competing to sell materials. As the specifications for recovered materials become more restrictive, recovery program managers must consider buyer specifications carefully when choosing collection and sorting systems, especially where large capital expenditures are involved.

3-6 FUTURE CHANGES IN WASTE COMPOSITION

In terms of solid waste management planning, knowledge of future trends in the composition of solid wastes is of great importance. For example, if a paper recycling program were instituted on the basis of current distribution data and if paper production were subsequently eliminated, such a program would more than likely become a costly white elephant. Although this case is extreme, it nevertheless illustrates the point that future trends must be assessed carefully in long-term planning. Another important question is whether the quantities are actually changing or only the reporting system has improved. Factors that affect the actual quantities of waste generated are considered in Chapter 6.

Impacts of Waste Diversion Programs

As more states adopt legislation mandating the development of source reduction and recycling programs, the composition of the wastes collected will change. The impact of waste diversion programs on the composition of the wastes collected will vary depending on the other types of waste management programs that are in place. For example, a solid waste combustion facility, developed to serve a commercial area, was planned and designed on the basis that the energy content of the waste would be 6500 Btu/lb. Six years later when the facility went on line, the actual energy content of the waste was 4500 Btu/lb, owing to the extensive recycling of cardboard that had developed in the intervening years. To meet a contractual requirement to generate a firm amount of power, the number of truck loads of waste had to be increased, necessitating a change in the facility permit.

Future Changes in Waste Components

In planning for future waste management systems, it will be important to consider the changes that may occur in the composition of solid waste with time. Four waste components that have an important influence on the composition of the wastes collected are food waste, paper and cardboard, yard waste, and plastics. For the purposes of comparison, the classification of materials used to define municipal refuse in the early 1900s is given in Table 3-12. In comparing the entries in Table 3-12 with those in Table 3-1, public refuse corresponds to waste from municipal sources, trade and market refuse corresponds to wastes from commercial sources, and house refuse corresponds to waste from residential sources. Stable refuse has disappeared as a waste category in current classification systems, while plastics were nonexistent in the early 1900s.

Food Wastes. The quantity of residential food wastes collected has changed significantly over the years as a result of technical advances and changes in public attitude. Two technological advances that have had a significant effect are the development of the food processing and packaging industry and the use of kitchen food waste grinders. The percentage of food waste, by weight, has decreased from about 14 percent in the early 1960s to about 9 percent in 1992.

Recently, because the public has become more environmentally aware and concerned, a trend has developed toward the use of more raw, rather than processed, vegetables. While it would appear that such a trend would increase the quantity of food wastes collected no firm data are available on this subject.

Paper and Cardboard. The percentage of paper and cardboard (also known as paperboard and corrugated paper) found in solid wastes has increased greatly over the past half century, rising from about 20 percent in the early 1940s to about 40 percent in 1992. It is expected that use of paper and cardboard will remain stable for the next few years. If the U.S. postal rate for bulk mail were increased to that for first class mail, a significant reduction would occur in the amount of paper collected for disposal.

Yard Wastes. The percentage of yard wastes in MSW has also increased significantly during the past quarter century, due primarily to passage of laws that prohibit burning of yard wastes. By weight, yard waste currently accounts for about 16 to 24 percent of the waste stream. Environmental conditions such as droughts have also affected the quantities of yard wastes collected in certain locations. For example, in Santa Barbara, CA, many areas that had lawns have been converted permanently to arid type landscaping with a concomitant decrease in the portion of yard wastes. Whether drought conditions in the south-western United States will continue to affect the quantity of yard wastes is unknown.

TABLE 3-12
Classification of refuse materials in the early 1900s

Municipal refuse	Public refuse		Street manure and litter
			Sweepings and dust
			Leaves
			Droppings from carts
			Large dead animals
			Snow
			Cleanings from public catch basins
	Trade refuse		Steam ashes
			Dry factory wastes
			Slaughter house waste
			Rubbish from office buildings and factories
			Cleanings from private catch basins
	Market refuse		Garbage from markets
			Rubbish and cleanings from markets
			Old boxes and barrels
	Stable refuse		Manure
			Straw
			Cleanings from stables
			Fly maggots
	House refuse	Garbage	Animal matter, including moisture
			Vegetable matter, including moisture
			Tin cans
			Small dead animals
		Ashes	Coal and cinders
			Clinker and slate
			Dust
			Glass
			Crockery
			Brick and stone
			Metal fragments
		Rubbish	Sweepings from buildings
			Boxes and barrels
			Wood
			Paper
			Rags
			Excelsior
			Straw
			Leather
			Rubber
			Metal ware
			Bedding
			Old furniture
		Night soil	Contents of privies

Source: From Ref. 6, adapted from paper entitled "Disposal of Municipal Refuse and Rubbish Incineration," by H. de B. Parsons, *Transactions ASCE,* Vol. LVII, p. 45, 1906.

Plastics. The percentage of plastics in solid wastes has also increased significantly during the past 50 years. The use of plastic has increased from almost nonmeasurable quantities in the early 1940s to between 7 and 8 percent, by weight, in 1992. It is anticipated the the use of plastics will continue to increase, but at a slower rate than during the past 25 years.

3-7 DISCUSSION TOPICS AND PROBLEMS

3-1. Gather data for your community or a nearby community on the total amount of MSW that is now generated, and compute the percentage distribution using the waste categories given in Table 3-3. How does the percentage distribution for your community compare with the distribution given in Table 3-3? If there are major differences, what explanation can you offer for the differences?

3-2. Obtain data for your community or a nearby community on the percentage distribution of the components found in the residential and commercial portion of the MSW. How do the values obtained compare with the typical values given in Table 3-4? Explain any major differences. If the individual component values are not within the ranges given in Table 3-4, explain why.

3-3. A community is now achieving a 25 percent by weight separation of wastes made up of the following items: mixed paper, 44%; cardboard, 6%; plastics, 10%; yard waste, 16%; glass, 12%; and mixed metal (tin cans and other metals), 12%. If the distribution of waste components given in column 3 of Table 3-4 is representative of the wastes that are now collected, determine the *as generated* percentage distribution of the waste components.

3-4. A community is proposing to achieve a 50 percent rate of separation by weight of wastes made up of the following items: mixed paper, 40%; cardboard, 8%; plastics, 8%; yard waste, 24%; glass, 12%; and tin cans, 8%. Determine the *as collected* percentage distribution for the residual waste components, assuming the typical distribution of waste components given in column 4 of Table 3-7 is representative of the wastes that are now generated.

3-5. Identify the materials that are now recovered for recycling in your community. How does your list of materials compare to the list given in Table 3-10? What other materials could be recycled that are not being recycled?

3-6. Describe the general trends you would expect in the future in the generation of the following types of wastes in your community: food wastes, paper, plastics, rags and leather, and yard wastes. What effect will improved reporting techniques have on your answer?

3-8 REFERENCES

1. Cointreau, S. J., G. G. Gunnerson, J. M. Huls, and N. N. Seldman: *Recycling from Municipal Refuse: A State of the Art Review and Annotated Bibliography*, Integrated Resource Recovery, World Bank Technical Paper No. 30, United Nations Development Program, Washington, DC, 1985.
2. Franklin Associates Limited: *Characteristics of Municipal Solid Waste in the United States*, 1960 to 2000, Franklin Associates Ltd., Prairie Village, KS, 1986.

3. Franklin, W. and M. Franklin: "III. Solid Waste Stream Characteristics," in F. Kreith (ed.): *Integrated Solid Waste Management: Options for Legislative Action*, Genium Publishing Corporation, Schenectady, NY, 1990.
4. Freeman, H. M.: "Source Reduction As an Option for Municipal Waste Management," in F. Kreith (ed.): *Integrated Solid Waste Management: Options For Legislative Action*, Genium Publishing Corporation, Schenectady, NY, 1990.
5. Hasselriis, F.: *Refuse Derived Fuel*, an Ann Arbor Science Book, Butterworth Publishers, Boston, 1984.
6. Hering, R. and S. A. Greeley: *Collection and Disposal of Municipal Refuse*, 1st ed., McGraw-Hill New York, 1921.
7. Klee, A. J. and D. Carruth: "Sample Weights in Solid Waste Composition Studies," ASCE, *Journal of the Sanitary Engineering Division*, Vol. 96, No. 5A, 1970.
8. Mantell, C. L. (ed.): *Solid Wastes: Origin, Collection, Processing, and Disposal*, Wiley-Interscience, New York, 1975.
9. Neissen, W. R.: "Properties of Waste Materials," in D. G. Wilson (ed.): *Handbook of Solid Waste Management*, Van Nostrand Reinhold, New York, 1977.
10. Selke, S. E.: *Packaging and the Environment: Alternatives, Trends and Solutions*, Technomic Publishing Company, Lanchester, PA, 1990.
11. Smith, K. T., Consultants: *City of Folsom Preliminary Waste Characterization Study*, Sacramento, CA, April 1991.
12. Tchobanoglous, G., H. Theisen, and R. Eliassen: *Solid Waste Engineering: Engineering Principles and Management Issues,* McGraw-Hill, New York, 1977.
13. Tchobanoglous, G.: "Management of Industrial Solid Wastes," in R. H. Perry, D. W. Green, and J. O. Maloney (eds.): *Perry's Chemical Engineers' Handbook*, 6th ed., McGraw-Hill, New York, 1984.
14. U.S. Environmental Protection Agency: *Characterization of Municipal Waste in the United States—1960–2000*, 1988 Update, Washington, DC, 1988. Available through National Technical Information Service (NTIS), Springfield, VA.
15. U.S. Environmental Protection Agency: *Decision-Makers Guide to Solid Waste Management*, EPA/530-SW89-072, Washington, DC, November 1989.
16. U.S. Environmental Protection Agency: *Characterization of Municipal Solid Waste in the United States: 1990 Update*, EPA/530-SW-90-04, Washington, DC, June 1990.
17. Wilson, D. C.: *Waste Management Evaluation Technologies*, Oxford University Press, New York, 1981.

CHAPTER 4

PHYSICAL, CHEMICAL, AND BIOLOGICAL PROPERTIES OF MUNICIPAL SOLID WASTE

The purpose of this chapter is to introduce the reader to the physical, chemical, and biological properties of MSW and to the transformations that can affect the form and composition of MSW. These properties must be known to develop and design integrated solid waste management systems. Further, the physical, chemical, and biological properties and transformations introduced in this chapter form the basis for topics discussed in the remaining portions of this book.

4-1 PHYSICAL PROPERTIES OF MSW

Important physical characteristics of MSW include specific weight, moisture content, particle size and size distribution, field capacity, and compacted waste porosity. The discussion is limited to an analysis of residential, commercial, and some industrial solid wastes. The hazardous wastes found in MSW are addressed separately in Chapter 5. Note, however, that the fundamentals of analysis presented in this and the following chapter are applicable to all types of solid wastes. Additional details on the various physical, chemical, and microbiological methods of testing for solid wastes may be found in the various publications of the American Society for Testing and Materials (ASTM).

Specific Weight

Specific weight is defined as the weight of a material per unit volume (e.g., lb/ft^3, lb/yd^3). (It should be noted that specific weight expressed as lb/yd^3 is commonly referred to in the solid waste literature incorrectly as density. In U.S. customary units density is expressed correctly as slug/ft^3.) Because the specific weight of MSW is often reported as *loose, as found in containers, uncompacted, compacted*, and the like, the basis used for the reported values should always be noted. Specific weight data are often needed to assess the total mass and volume of waste that must be managed. Unfortunately, there is little or no uniformity in the way solid waste specific weights have been reported in the literature. Frequently, no distinction has been made between uncompacted or compacted specific weights. Typical specific weights for various wastes as found in containers, compacted, or uncompacted are reported in Table 4-1.

TABLE 4-1
Typical specific weight and moisture content data for residential, commercial, industrial, and agricultural wastes

	Specific weight, lb/yd^3		Moisture content, % by weight	
Type of waste	**Range**	**Typical**	**Range**	**Typical**
Residential (uncompacted)				
Food wastes (mixed)	220–810	490	50–80	70
Paper	70–220	150	4–10	6
Cardboard	70–135	85	4–8	5
Plastics	70–220	110	1–4	2
Textiles	70–170	110	6–15	10
Rubber	170–340	220	1–4	2
Leather	170–440	270	8–12	10
Yard wastes	100–380	170	30–80	60
Wood	220–540	400	15–40	20
Glass	270–810	330	1–4	2
Tin cans	85–270	150	2–4	3
Aluminum	110–405	270	2–4	2
Other metals	220–1940	540	2–4	3
Dirt, ashes, etc.	540–1685	810	6–12	8
Ashes	1095–1400	1255	6–12	6
Rubbish	150–305	220	5–20	15
Residential yard wastes				
Leaves (loose and dry)	50–250	100	20–40	30
Green grass (loose and moist)	350–500	400	40–80	60
Green grass (wet and compacted)	1000–1400	1000	50–90	80
Yard waste (shredded)	450–600	500	20–70	50
Yard waste (composted)	450–650	550	40–60	50
Municipal				
In compactor truck	300–760	500	15–40	20
In landfill				
Normally compacted	610–840	760	15–40	25
Well compacted	995–1250	1010	15–40	25
Commercial				
Food wastes (wet)	800–1600	910	50–80	70
Appliances	250–340	305	0–2	1

(*continued*)

Because the specific weights of solid wastes vary markedly with geographic location, season of the year, and length of time in storage, great care should be used in selecting typical values. Municipal solid wastes as delivered in compaction vehicles have been found to vary from 300 to 700 lb/yd^3; a typical value is about 500 lb/yd^3.

Moisture Content

The moisture content of solid wastes usually is expressed in one of two ways. In the wet-weight method of measurement, the moisture in a sample is expressed

TABLE 4-1 (*continued*)

	Specific weight, lb/yd^3		Moisture content, % by weight	
Type of waste	**Range**	**Typical**	**Range**	**Typical**
Commercial (*cont.*)				
Wooden crates	185–270	185	10–30	20
Tree trimmings	170–305	250	20–80	5
Rubbish (combustible)	85–305	200	10–30	15
Rubbish (noncombustible)	305–610	505	5–15	10
Rubbish (mixed)	235–305	270	10–25	15
Construction and demolition				
Mixed demolition (noncombustible)	1685–2695	2395	2–10	4
Mixed demolition (combustible)	505–675	605	4–15	8
Mixed construction (combustible)	305–605	440	4–15	8
Broken concrete	2020–3035	2595	0–5	—
Industrial				
Chemical sludges (wet)	1350–1855	1685	75–99	80
Fly ash	1180–1515	1350	2–10	4
Leather scraps	170–420	270	6–15	10
Metal scrap (heavy)	2530–3370	3000	0–5	—
Metal scrap (light)	840–1515	1245	0–5	—
Metal scrap (mixed)	1180–2530	1515	0–5	—
Oils, tars, asphalts	1350–1685	1600	0–5	2
Sawdust	170–590	490	10–40	20
Textile wastes	170–370	305	6–15	10
Wood (mixed)	675–1140	840	30–60	25
Agricultural				
Agricultural (mixed)	675–1265	945	40–80	50
Dead animals	340–840	605	—	—
Fruit wastes (mixed)	420–1265	605	60–90	75
Manure (wet)	1515–1770	1685	75–96	94
Vegetable wastes (mixed)	340–1180	605	60–90	75

Adapted in part from Refs. 6 and 8.

Note: lb/yd^3 × 0.5933 = kg/m^3

as a percentage of the wet weight of the material; in the dry-weight method, it is expressed as a percentage of the dry weight of the material. The wet-weight method is used most commonly in the field of solid waste management. In equation form, the wet-weight moisture content is expressed as follows:

$$M = \left(\frac{w - d}{w}\right)100 \tag{4-1}$$

where M = moisture content, %
w = initial weight of sample as delivered, lb (kg)
d = weight of sample after drying at 105°C, lb (kg)

Typical data on the moisture content for the solid waste components given in Table 3-3 as well as other materials are given in Table 4-1. For most MSW in the United States, the moisture content will vary from 15 to 40 percent, depending on the composition of the wastes, the season of the year, and the humidity and weather conditions, particularly rain. The use of data in Table 4-1 to estimate the overall moisture content of solid wastes is illustrated in Example 4-1.

EXAMPLE 4-1 Estimation of moisture content of typical residential MSW. Estimate the overall moisture content of a sample of *as collected* residential MSW with the typical composition given in Table 3-4.

Solution

1. Set up the computation table to determine dry weights of the solid waste components using the data given in Table 4-1.

Component	Percent by weight	Moisture content, %	Dry weight,[a] lb
Organic			
Food wastes	9.0	70	2.7
Paper	34.0	6	32.0
Cardboard	6.0	5	5.7
Plastics	7.0	2	6.9
Textiles	2.0	10	1.8
Rubber	0.5	2	0.5
Leather	0.5	10	0.4
Yard wastes	18.5	60	7.4
Wood	2.0	20	1.6
Misc. organics	—	—	—
Inorganic			
Glass	8.0	2	7.8
Tin cans	6.0	3	5.8
Aluminum	0.5	2	0.5
Other metal	3.0	3	2.9
Dirt, ash, etc.	3.0	8	2.8
Total	100.0		78.8

[a] Based on an *as delivered* sample weight of 100 lb.

2. Determine the moisture content of the solid waste sample using Eq. (4-1).

$$\text{Moisture content}(\%) = \left(\frac{100 - 78.8}{100}\right)100 = 21.2\%$$

Particle Size and Size Distribution

The size and size distribution of the component materials in solid wastes are an important consideration in the recovery of materials, especially with mechanical means such as trommel screens and magnetic separators. The size of a waste component may be defined by one or more of the following measures:

$$S_c = l \tag{4-2}$$

$$S_c = \left(\frac{l + w}{2}\right) \tag{4-3}$$

$$S_c = \left(\frac{l + w + h}{3}\right) \tag{4-4}$$

$$S_c = (l \times w)^{1/2} \tag{4-5}$$

$$S_c = (l \times w \times h)^{1/3} \tag{4-6}$$

where S_c = size of component, in (mm)
l = length, in (mm)
w = width, in (mm)
h = height, in (mm)

A general indication of the particle size distribution (by longest dimension and ability to pass a sieve) may be obtained from the data presented in Figs. 4-1 and 4-2. Typical data on the size distribution of the individual components in MSW are presented in Fig. 4-3. Based on single linear measurement as defined by Eq. (4-2), the average size of the individual components found in residential MSW is between 7 and 8 in. Typical data on the size distribution of aluminum cans, tin cans, and glass, based on Eq. (4-5), are presented in Fig. 4-4. Because there are significant differences among the various measures on size, individual measurements should be made on the waste in question using a measure of size that will provide the information needed for the specific application.

Field Capacity

The field capacity of solid waste is the total amount of moisture that can be retained in a waste sample subject to the downward pull of gravity. The field capacity of waste materials is of critical importance in determining the formation of leachate in landfills. Water in excess of the field capacity will be released as leachate. The field capacity varies with the degree of applied pressure and the state of decomposition of the waste. A field capacity of 30 percent by volume

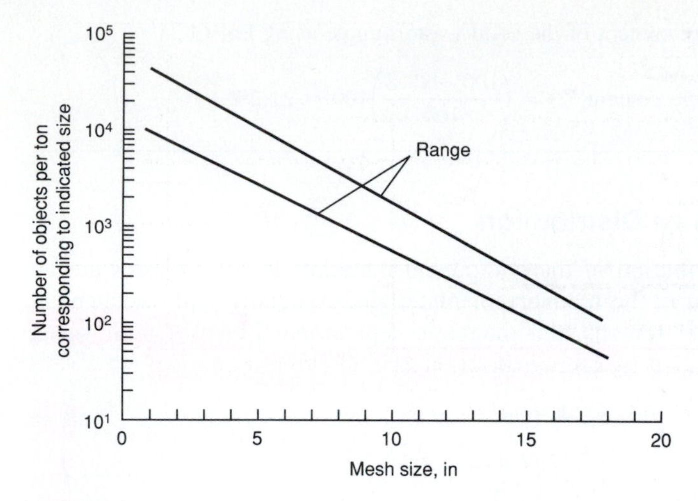

FIGURE 4-1
Typical sizes of individual components comprising residential and commercial MSW [4, 12].

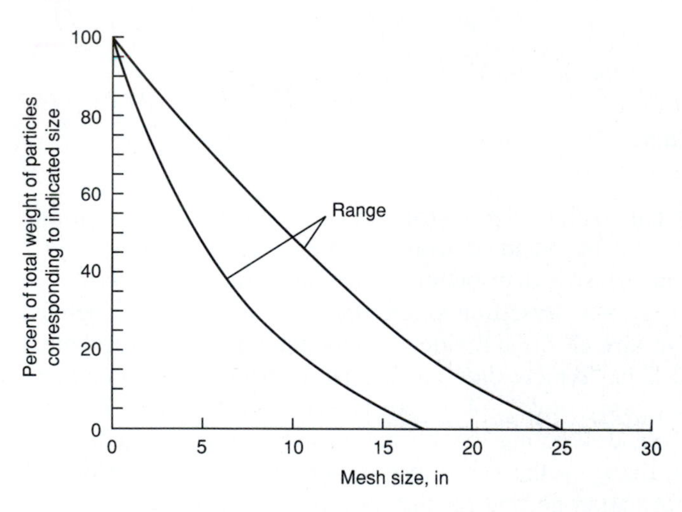

FIGURE 4-2
Percentage of total mass of residential and commercial MSW as a function of mesh size [4, 12].

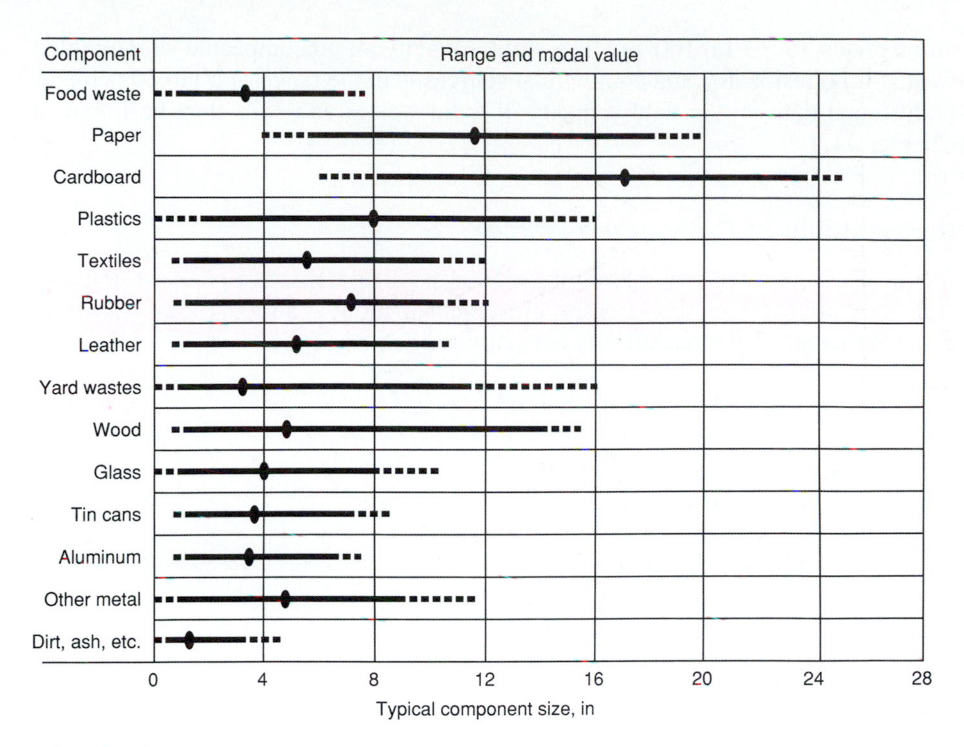

FIGURE 4-3
Typical size distribution of the components found in residential MSW (adapted, in part, from Ref. 4).

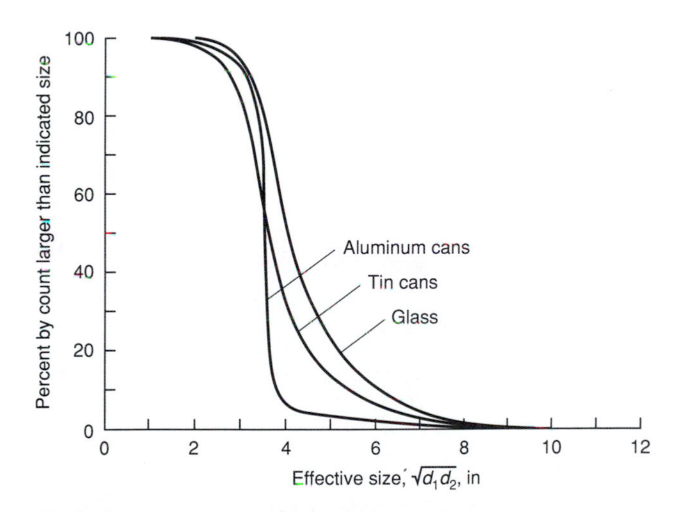

FIGURE 4-4
Typical distribution by count of effective sizes $(l \times w)^{1/2}$ of aluminum cans, tin cans, and glass containers found in residential MSW as delivered to a landfill.

corresponds to 30 in/100 in. The field capacity of uncompacted commingled wastes from residential and commercial sources is in the range of 50 to 60 percent. (Additional data on the field capacity of solid wastes and soils may be found in Chapter 11.)

Permeability of Compacted Waste

The hydraulic conductivity of compacted wastes is an important physical property that, to a large extent, governs the movement of liquids and gases in a landfill. The coefficient of permeability is normally written as [2]:

$$K = Cd^2 \frac{\gamma}{\mu} = k \frac{\gamma}{\mu} \tag{4-7}$$

where K = coefficient of permeability
C = dimensionless constant or shape factor
d = average size of pores
γ = specific weight of water
μ = dynamic viscosity of water
k = intrinsic permeability

The term Cd^2 is known as the intrinsic (or specific) permeability. The intrinsic permeability depends solely on the properties of the solid material, including pore size distribution, tortuosity, specific surface, and porosity. Typical values for the intrinsic permeability for compacted solid waste in a landfill are in the range between about 10^{-11} and 10^{-12} m^2 in the vertical direction and about 10^{-10} m^2 in the horizontal direction.

4-2 CHEMICAL PROPERTIES OF MSW

Information on the chemical composition of the components that constitute MSW is important in evaluating alternative processing and recovery options. For example, the feasibility of combustion depends on the chemical composition of the solid wastes. Typically, wastes can be thought of as a combination of semimoist combustible and noncombustible materials. If solid wastes are to be used as fuel, the four most important properties to be known are:

1. Proximate analysis
2. Fusing point of ash
3. Ultimate analysis (major elements)
4. Energy content

Where the organic fraction of MSW is to be composted or is to be used as feedstock for the production of other biological conversion products, not only will information on the major elements (ultimate analysis) that compose the waste be important, but also information will be required on the trace elements in the waste materials.

Proximate Analysis

Proximate analysis for the combustible components of MSW includes the following tests [3, 10]:

1. Moisture (loss of moisture when heated to 105°C for 1 h)
2. Volatile combustible matter (additional loss of weight on ignition at 950°C in a covered crucible)
3. Fixed carbon (combustible residue left after volatile matter is removed)
4. Ash (weight of residue after combustion in an open crucible)

Proximate analysis data for the combustible components of MSW *as discarded* are presented in Table 4-2. It is important to note that the test used to determine volatile combustible matter in a proximate analysis is different from the volatile solids test used in biological determinations (see Section 4-3).

Fusing Point of Ash

The fusing point of ash is defined as that temperature at which the ash resulting from the burning of waste will form a solid (clinker) by fusion and agglomeration. Typical fusing temperatures for the formation of clinker from solid waste range from 2000 to 2200°F (1100 to 1200°C).

Ultimate Analysis of Solid Waste Components

The ultimate analysis of a waste component typically involves the determination of the percent C (carbon), H (hydrogen), O (oxygen), N (nitrogen), S (sulfur), and ash. Because of the concern over the emission of chlorinated compounds during combustion, the determination of halogens is often included in an ultimate analysis. The results of the ultimate analysis are used to characterize the chemical composition of the organic matter in MSW. They are also used to define the proper mix of waste materials to achieve suitable C/N ratios for biological conversion processes. Data on the ultimate analysis of individual combustible materials are presented in Table 4-3. Representative data for the typical MSW components given in Table 3-4 are presented in Table 4-4. Estimation of the average chemical composition of solid waste materials using the data given in Tables 4-1 and 4-2 is illustrated in Example 4-2.

TABLE 4-2
Typical proximate analysis and energy data for materials found in residential, commercial, and industrial solid wastes[a]

	Proximate analysis, % by weight				Energy content, Btu/lb		
Type of waste	Moisture	Volatile matter	Fixed carbon	Non-combustible	*As collected*	Dry	Dry ash-free
Food and food products							
Fats	2.0	95.3	2.5	0.2	16,135	16,466	16,836
Food wastes (mixed)	70.0	21.4	3.6	5.0	1,797	5,983	7,180
Fruit wastes	78.7	16.6	4.0	0.7	1,707	8,013	8,285
Meat wastes	38.8	56.4	1.8	3.1	7,623	12,455	13,120
Paper products							
Cardboard	5.2	77.5	12.3	5.0	7,042	7,428	7,842
Magazines	4.1	66.4	7.0	22.5	5,254	5,478	7,157
Newsprint	6.0	81.1	11.5	1.4	7,975	8,484	8,612
Paper (mixed)	10.2	75.9	8.4	5.4	6,799	7,571	8,056
Waxed cartons	3.4	90.9	4.5	1.2	11,326	11,724	11,872
Plastics							
Plastics (mixed)	0.2	95.8	2.0	2.0	14,101	14,390	16,024
Polyethylene	0.2	98.5	<0.1	1.2	18,687	18,724	18,952
Polystyrene	0.2	98.7	0.7	0.5	16,419	16,451	16,430
Polyurethane	0.2	87.1	8.3	4.4	11,204	11,226	11,744
Polyvinyl chloride	0.2	86.9	10.8	2.1	9,755	9,774	9,985
Textiles, rubber, leather							
Textiles	10.0	66.0	17.5	6.5	7,960	8,844	9,827
Rubber	1.2	83.9	4.9	9.9	10,890	11,022	12,250
Leather	10.0	68.5	12.5	9.0	7,500	8,040	8,982

Wood, trees, etc.							
Yard wastes	60.0	30.0	9.5	0.5	2,601	6,503	6,585
Wood (green timber)	50.0	42.3	7.3	0.4	2,100	4,200	4,234
Hardwood	12.0	75.1	12.4	0.5	7,352	8,354	8,402
Wood (mixed)	20.0	68.1	11.3	0.6	6,640	8,316	8,383
Glass, Metals, etc.							
Glass and mineral	2.0	—	—	96–99+	84[b]	86	60
Metal, tin cans	5.0	—	—	94–99+	301[b]	319	317
Metal, ferrous	2.0	—	—	96–99+	—	—	—
Metal, nonferrous	2.0	—	—	94–99+	—	—	—
Miscellaneous							
Office sweepings	3.2	20.5	6.3	70.0	3,669	3,791	13,692
Residential MSW	21.0 (15–40)	52.0 (40–60)	7.0 (4–15)	20.0 (10–30)	5,000	6,250	8,333
Commercial MSW	15.0 (10–30)	—	—		5,500	6,470	
MSW	20.0 (10–30)	—	—		4,600	5,750	

[a] Adapted in part from Refs. 6–8.

[b] Energy content is from coatings, labels, and attached materials.

Note: Btu × 1.0551 = kJ

TABLE 4-3
Typical data on the ultimate analysis of the combustible materials found in residential, commercial, and industrial solid wastes[a]

	Percent by weight (dry basis)					
Type of waste	**Carbon**	**Hydrogen**	**Oxygen**	**Nitrogen**	**Sulfur**	**Ash**
Food and food products						
Fats	73.0	11.5	14.8	0.4	0.1	0.2
Food wastes (mixed)	48.0	6.4	37.6	2.6	0.4	5.0
Fruit wastes	48.5	6.2	39.5	1.4	0.2	4.2
Meat wastes	59.6	9.4	24.7	1.2	0.2	4.9
Paper products						
Cardboard	43.0	5.9	44.8	0.3	0.2	5.0
Magazines	32.9	5.0	38.6	0.1	0.1	23.3
Newsprint	49.1	6.1	43.0	<0.1	0.2	1.5
Paper (mixed)	43.4	5.8	44.3	0.3	0.2	6.0
Waxed cartons	59.2	9.3	30.1	0.1	0.1	1.2
Plastics						
Plastics (mixed)	60.0	7.2	22.8	—	—	10.0
Polyethylene	85.2	14.2	—	<0.1	<0.1	0.4
Polystyrene	87.1	8.4	4.0	0.2	—	0.3
Polyurethane[b]	63.3	6.3	17.6	6.0	<0.1	4.3
Polyvinyl chloride[b]	45.2	5.6	1.6	0.1	0.1	2.0
Textiles, rubber, leather						
Textiles	48.0	6.4	40.0	2.2	0.2	3.2
Rubber	69.7	8.7	—	—	1.6	20.0
Leather	60.0	8.0	11.6	10.0	0.4	10.0
Wood, trees, etc.						
Yard wastes	46.0	6.0	38.0	3.4	0.3	6.3
Wood (green timber)	50.1	6.4	42.3	0.1	0.1	1.0
Hardwood	49.6	6.1	43.2	0.1	<0.1	0.9
Wood (mixed)	49.5	6.0	42.7	0.2	<0.1	1.5
Wood chips (mixed)	48.1	5.8	45.5	0.1	<0.1	0.4
Glass, metals, etc.						
Glass and mineral[c]	0.5	0.1	0.4	<0.1	—	98.9
Metals (mixed)[c]	4.5	0.6	4.3	<0.1	—	90.5
Miscellaneous						
Office sweepings	24.3	3.0	4.0	0.5	0.2	68.0
Oils, paints	66.9	9.6	5.2	2.0	—	16.3
Refuse-derived fuel (RDF)	44.7	6.2	38.4	0.7	<0.1	9.9

[a] Adapted in part from Ref. 6.

[b] Remainder is chlorine.

[c] Organic content is from coatings, labels, and other attached materials.

TABLE 4-4
Typical data on the ultimate analysis of the combustile components in residential MSW[a]

	Percent by weight (dry basis)					
Component	**Carbon**	**Hydrogen**	**Oxygen**	**Nitrogen**	**Sulfur**	**Ash**
Organic						
Food wastes	48.0	6.4	37.6	2.6	0.4	5.0
Paper	43.5	6.0	44.0	0.3	0.2	6.0
Cardboard	44.0	5.9	44.6	0.3	0.2	5.0
Plastics	60.0	7.2	22.8	—	—	10.0
Textiles	55.0	6.6	31.2	4.6	0.15	2.5
Rubber	78.0	10.0	—	2.0	—	10.0
Leather	60.0	8.0	11.6	10.0	0.4	10.0
Yard wastes	47.8	6.0	38.0	3.4	0.3	4.5
Wood	49.5	6.0	42.7	0.2	0.1	1.5
Inorganic						
Glass[b]	0.5	0.1	0.4	<0.1	—	98.9
Metals[b]	4.5	0.6	4.3	<0.1	—	90.5
Dirt, ash, etc.	26.3	3.0	2.0	0.5	0.2	68.0

[a] Adapted in part from Ref. 6.

[b] Organic content is from coatings, labels, and other attached materials.

EXAMPLE 4-2 Estimation of the chemical composition of a solid waste sample. Determine the chemical composition of the organic fraction, without and with sulfur and without and with water, of a residential MSW with the typical composition shown in Table 3-4.

Solution

1. Set up a computation table to determine the percentage distribution of the major elements composing the waste. The necessary computations are presented below:

	Wet weight,	Dry weight,	Composition, lb					
Component	**lb**	**lb**	**C**	**H**	**O**	**N**	**S**	**Ash**
Food wastes	9.0	2.7	1.30	0.17	1.02	0.07	0.01	0.14
Paper	34.0	32.0	13.92	1.92	14.08	0.10	0.06	1.92
Cardboard	6.0	5.7	2.51	0.34	2.54	0.02	0.01	0.28
Plastics	7.0	6.9	4.14	0.50	1.57	—	—	0.69
Textiles	2.0	1.8	0.99	0.12	0.56	0.08	—	0.05
Rubber	0.5	0.5	0.39	0.05	—	0.01	—	0.05
Leather	0.5	0.4	0.24	0.03	0.05	0.04	—	0.04
Yard wastes	18.5	6.5	3.11	0.39	2.47	0.22	0.02	0.29
Wood	2.0	1.6	0.79	0.10	0.68	—	—	0.02
Total	79.5	58.1	27.39	3.62	22.97	0.54	0.10	3.48

Moisture content = 21.4 lb (79.5 lb − 58.1 lb)

2. Prepare a summary table of the percentage distribution of the elements without and with the water contained in the waste.

	Weight, lb	
Component	**Without H_2O**	**With H_2O**
Carbon	27.39	27.39
Hydrogen	3.62	6.00
Oxygen	22.97	41.99
Nitrogen	0.54	0.54
Sulfur	0.10	0.10
Ash	3.48	3.48

3. Compute the molar composition of the elements neglecting the ash.

	Atomic weight,	Moles	
Component	**lb/mole**	**Without H_2O**	**With H_2O**
Carbon	12.01	2.280	2.280
Hydrogen	1.01	3.584	5.940
Oxygen	16.00	1.436	2.624
Nitrogen	14.01	0.038	0.038
Sulfur	32.07	0.003	0.003

4. Determine an approximate chemical formula without and with sulfur and without and with water. Set up a computation table to determine normalized mole ratios.

	Mole ratio (Nitrogen = 1)		Mole ratio (Sulfur = 1)	
Component	**Without H_2O**	**With H_2O**	**Without H_2O**	**With H_2O**
Carbon	60.0	60.0	760.0	760.0
Hydrogen	94.3	156.3	1194.7	1980.0
Oxygen	37.8	69.1	478.7	874.7
Nitrogen	1.0	1.0	12.7	12.7
Sulfur	0.1	0.1	1.0	1.0

(*a*) The chemical formulas without sulfur are:

1. Without water $C_{60.0}H_{94.3}O_{37.8}N$
2. With water $C_{60.0}H_{156.3}O_{69.1}N$

(*b*) The chemical formulas with sulfur are:

1. Without water $C_{760.0}H_{1194.7}O_{478.7}N_{12.7}S$
2. With water $C_{760.0}H_{1980.0}O_{874.7}N_{12.7}S$

Comment. The fractional coefficients reported in these formulas are usually rounded off, as the original data do not warrant such precision.

Energy Content of Solid Waste Components

The energy content of the organic components in MSW can be determined (1) by using a full scale boiler as a calorimeter, (2) by using a laboratory bomb calorimeter (see Fig. 4-5), and (3) by calculation, if the elemental composition is known. Because of the difficulty in instrumenting a full-scale boiler, most of the data on the energy content of the organic components of MSW are based on the results of bomb calorimeter tests. Typical data for energy content and inert residue for the

(*a*)

(*b*)

FIGURE 4-5
Laboratory bomb calorimeter used to determine the energy content of solid waste materials: (*a*) calorimeter with access cover opened for loading oxygen bomb and (*b*) disassembled oxygen bomb with sample cup in foreground.

components of residential wastes are reported in Table 4-5. As shown, the energy content values are on an *as discarded* basis. The Btu values given in Table 4-5 may be converted to a dry basis by using Eq. (4-8).

$$\text{Btu/lb (dry basis)} = \text{Btu/lb (as discarded)}\left(\frac{100}{100 - \%\ \text{moisture}}\right) \tag{4-8}$$

The corresponding equation for the Btu per pound on a dry ash-free basis is

$$\text{Btu/lb (dry ash-free basis)} = \text{Btu/lb (as discarded)}\left(\frac{100}{100 - \%\ \text{moisture} - \%\ \text{ash}}\right) \tag{4-9}$$

Data from Table 4-5 are used to compute the energy content of an MSW in Example 4-3. Additional data on the energy content of individual waste materials on an *as collected*, dry, and dry ash-free basis are given in Table 4-2.

TABLE 4-5
Typical values for inert residue and energy content of residential MSW[a]

	Inert residue,[b] percent		Energy,[c] Btu/lb	
Component	**Range**	**Typical**	**Range**	**Typical**
Organic				
Food wastes	2–8	5.0	1,500–3,000	2,000
Paper	4–8	6.0	5,000–8,000	7,200
Cardboard	3–6	5.0	6,000–7,500	7,000
Plastics	6–20	10.0	12,000–16,000	14,000
Textiles	2–4	2.5	6,500–8,000	7,500
Rubber	8–20	10.0	9,000–12,000	10,000
Leather	8–20	10.0	6,500–8,500	7,500
Yard wastes	2–6	4.5	1,000–8,000	2,800
Wood	0.6–2	1.5	7,500–8,500	8,000
Misc. organics	—	—	—	—
Inorganic				
Glass	96–99+	98.0	50–100[d]	60
Tin cans	96–99+	98.0	100–500[d]	300
Aluminum	90–99+	96.0	—	—
Other metal	94–99+	98.0	100–500[d]	300
Dirt, ashes, etc.	60–80	70.0	1,000–5,000	3,000
Municipal solid wastes			4,000–6,000	5,000[e]

[a] Adapted in part from Refs. 6 and 8.

[b] After complete combustion.

[c] As discarded basis.

[d] Energy content is from coatings, labels, and attached materials.

[e] The typical energy value given in this table is higher than the corresponding value given in the predecessor of this text (see Table 4-10) [11]. The reason is due largely to (1) the reduced amount of raw food waste and (2) the increased percentage distribution of plastic (7 versus 4 percent) in residential MSW.

Note: Btu/lb × 2.326 = kJ/kg

EXAMPLE 4-3 Estimation of energy content of typical residential MSW. Determine the energy value of a typical residential MSW with the average composition shown in Table 3-4.

Solution

1. Assume the heating value will be computed on an *as discarded* basis.
2. Determine the total energy content using the data given in Table 4-5. The necessary computations are presented below.

Component	Solid wastes,[a] lb	Energy,[a] Btu/lb	Total energy, Btu
Organic			
Food wastes	9.0	2,000	18,000
Paper	34.0	7,200	244,800
Cardboard	6.0	7,000	42,000
Plastics	7.0	14,000	98,000
Textiles	2.0	7,500	15,000
Rubber	0.5	10,000	5,000
Leather	0.5	7,500	3,750
Yard wastes	18.5	2,800	51,800
Wood	2.0	8,000	16,000
Inorganic			
Glass	8.0	60	480
Tin cans	6.0	300	1,800
Aluminum	0.5	—	—
Other metal	3.0	300	900
Dirt, ashes, etc.	3.0	3,000	9,000
Total	100.0		506,530

[a] Data were derived from Tables 3-4 and 4-5.

Note: Btu/lb × 2.326 = kJ/kg

3. Determine the *as discarded* energy content per lb of waste

$$\text{Energy content} = \frac{506{,}530 \text{ Btu}}{100 \text{ lb}} = \frac{5065 \text{ Btu}}{\text{lb}} = \frac{11{,}782 \text{ kJ}}{\text{kg}}$$

The computed value compares well with the typical value given in Table 4-5.

Comment. Note that the value computed in this example is somewhat different from the corresponding value (4762 Btu/lb) computed in Example 4-2 in the predecessor of this text [11]. The difference is due largely to (1) the reduced amount of raw food waste and (2) the increased amount of plastic (7 versus 4 percent).

If Btu values are not available, approximate Btu values for the individual waste materials can be determined by using Eq. (4-10), known as the modified Dulong formula [8, 10], and the data in Tables 4-1 and 4-2.

$$\text{Btu/lb} = 145\text{C} + 610(\text{H}_2 - \frac{1}{8}\text{O}_2) + 40\text{S} + 10\text{N} \tag{4-10}$$

where C = carbon, percent by weight
H_2 = hydrogen, percent by weight
O_2 = oxygen, percent by weight
S = sulfur, percent by weight
N = nitrogen, percent by weight

In Eq. (4-10), the oxygen content is divided by eight and subtracted from the hydrogen to account for the amount of hydrogen that reacts with the oxygen that is present and, thus, does not contribute to the energy content of the waste. The use of Eq. (4-10) is illustrated in Example 4-4.

EXAMPLE 4-4 Estimation of energy content of typical residential MSW based on chemical composition. Determine the energy value of typical residential MSW with the average composition determined in Example 4-2 including sulfur and water.

Solution

1. The chemical composition of the waste including sulfur and water is:

$$C_{760.0}H_{1980.0}O_{874.7}N_{12.7}S$$

2. Determine the total energy content using Eq. (4-10).
   (*a*) Determine the percentage distribution by weight of the elements composing the waste, using coefficients that have been rounded off.

Component	Number of atoms per mole	Atomic weight	Weight contribution of each element	%
Carbon	760	12	9120	36.03
Hydrogen	1980	1	1980	7.82
Oxygen	875	16	14,000	55.30
Nitrogen	13	14	182	0.72
Sulfur	1	32	32	0.13
Total			25,314	100.00

   (*b*) The energy content of the waste using Eq. (4-10) is:

$$\text{Btu/lb} = 145(36.0) + 610\left(7.8 - \frac{55.3}{8}\right) + 40(0.1) + 10(0.7)$$

$$\text{Btu/lb} = 5772$$

Comment. The computed energy content of the waste is higher than the value computed in Example 4-3 because only the organic fraction of the residential MSW was considered in Example 4-2.

TABLE 4-6
Elemental analysis of the organic materials used as the feedstock for biological conversion processes

		Feed substrate (dry basis)			
Constituent	Unit	Newsprint	Office paper	Yard waste	Food waste
NH_4-N	ppm	4	61	149	205
NO_3-N	ppm	4	218	490	4278
P	ppm	44	295	3500	4900
PO_4-P	ppm	20	164	2210	3200
K	%	0.35	0.29	2.27	4.18
SO_4-S	ppm	159	324	882	855
Ca	%	0.01	0.10	0.42	0.43
Mg	%	0.02	0.04	0.21	0.16
Na	%	0.74	1.05	0.06	0.15
B	ppm	14	28	88	17
Se	ppm	—	—	<1	<1
Zn	ppm	22	177	20	21
Mn	ppm	49	15	56	20
Fe	ppm	57	396	451	48
Cu	ppm	12	14	7.7	6.9
Co	ppm	—	—	5.0	3.0
Mo	ppm	—	—	1.0	<1
Ni	ppm	—	—	9.0	4.5
W	ppm	—	—	4.0	3.3

Essential Nutrients and Other Elements

Where the organic fraction of MSW is to be used as feedstock for the production of biological conversion products such as compost, methane, and ethanol, information on the essential nutrients and elements in the waste materials is of importance with respect to the microbial nutrient balance and in assessing what final uses can be made of the materials remaining after biological conversion. The essential nutrients and elements found in the principal materials that compose the organic fraction of MSW are reported in Table 4-6.

4-3 BIOLOGICAL PROPERTIES OF MSW

Excluding plastic, rubber, and leather components, the organic fraction of most MSW can be classified as follows:

1. Water-soluble constituents, such as sugars, starches, amino acids, and various organic acids,
2. Hemicellulose, a condensation product of five- and six-carbon sugars,
3. Cellulose, a condensation product of the six-carbon sugar glucose,
4. Fats, oils, and waxes, which are esters of alcohols and long-chain fatty acids,

5. Lignin, a polymeric material containing aromatic rings with methoxyl groups ($-OCH_3$), the exact chemical nature of which is still not known (present in some paper products such as newsprint and fiberboard),
6. Lignocellulose, a combination of lignin and cellulose,
7. Proteins, which are composed of chains of amino acids.

Perhaps the most important biological characteristic of the organic fraction of MSW is that almost all of the organic components can be converted biologically to gases and relatively inert organic and inorganic solids. The production of odors and the generation of flies are also related to the putrescible nature of the organic materials found in MSW (e.g., food wastes).

Biodegradability of Organic Waste Components

Volatile solids (VS) content, determined by ignition at 550°C, is often used as a measure of the biodegradability of the organic fraction of MSW. The use of VS in describing the biodegradability of the organic fraction of MSW is misleading, as some of the organic constituents of MSW are highly volatile but low in biodegradability (e.g., newsprint and certain plant trimmings). Alternatively, the lignin content of a waste can be used to estimate the biodegradable fraction, using the following relationship [1]:

$$BF = 0.83 - 0.028\ LC \tag{4-11}$$

where BF = biodegradable fraction expressed on a volatile solids (VS) basis
0.83 = empirical constant
0.028 = empirical constant
LC = lignin content of the VS expressed as a percent of dry weight

The biodegradability of several of the organic compounds found in MSW, based on lignin content, is reported in Table 4-7. As shown in Table 4-7, wastes with

TABLE 4-7
Data on the biodegradable fraction of selected organic waste components based on lignin content

Component	Volatile solids (VS), percent of total solids (TS)	Lignin content (LC), percent of VS	Biodegradable fraction (BF)[a]
Food wastes	95–98	0.4	0.82
Paper			
Newsprint	96–99	21.9	0.22
Office paper	90–95	0.4	0.82
Cardboard	90–95	12.9	0.47
Yard wastes	85–90	4.1	0.72

[a]Computed using Eq. (4-11).

high lignin contents, such as newsprint, are significantly less biodegradable than the other organic wastes found in MSW.

The rate at which the various components can be degraded varies markedly. For practical purposes, the principal organic waste components in MSW are often classified as rapidly and slowly decomposable.

Production of Odors

Odors can develop when solid wastes are stored for long periods of time on-site between collections, in transfer stations, and in landfills. The development of odors in on-site storage facilities is more significant in warm climates. Typically, the formation of odors results from the anaerobic decomposition of the readily decomposable organic components found in MSW. For example, under anaerobic (reducing) conditions, sulfate can be reduced to sulfide (S^{2-}), which subsequently combines with hydrogen to form H_2S. The formation of H_2S can be illustrated by the following two series of reactions.

$$\underset{\text{Lactate}}{2CH_3CHOHCOOH} + \underset{\text{Sulfate}}{SO_4^{2-}} \rightarrow \underset{\text{Acetate}}{2CH_3COOH} + \underset{\text{Sulfide ion}}{S^{2-}} + 2H_2O + 2CO_2 \quad (4\text{-}12)$$

$$4H_2 + SO_4^{2-} \rightarrow S^{2-} + 4H_2O \quad (4\text{-}13)$$

$$S^{2-} + 2H^+ \rightarrow H_2S \quad (4\text{-}14)$$

The sulfide ion can also combine with metal salts that may be present, such as iron, to form metal sulfides.

$$S^{2-} + Fe^{2+} \rightarrow FeS \quad (4\text{-}15)$$

The black color of solid wastes that have undergone anaerobic decomposition in a landfill is primarily due to the formation of metal sulfides. If it were not for the formation of a variety of sulfides, odor problems at landfills could be quite significant.

The biochemical reduction of an organic compound containing a sulfur radical can lead to the formation of malodorous compounds such as methyl mercaptan and aminobutyric acid. The reduction of methionine, an amino acid, serves as an example.

$$\underset{\text{methionine}}{CH_3SCH_2CH_2CH(NH_2)COOH} \xrightarrow{+2H} \underset{\text{methyl mercaptan}}{CH_3SH} + \underset{\text{aminobutyric acid}}{CH_3CH_2CH_2(NH_2)COOH} \quad (4\text{-}16)$$

The methyl mercaptan can be hydrolyzed biochemically to methyl alcohol and hydrogen sulfide:

$$CH_3SH + H_2O \rightarrow CH_4OH + H_2S \quad (4\text{-}17)$$

Breeding of Flies

In the summertime and during all seasons in warm climates, fly breeding is an important consideration in the on-site storage of wastes. Flies can develop in less than two weeks after the eggs are laid. The life history of the common house fly from egg to adult can be described as follows [5, 9]:

Eggs develop	8–12 hours
First stage of larval period	20 hours
Second stage of larval period	24 hours
Third stage of larval period	3 days
Pupal stage	4–5 days
Total	9–11 days

The extent to which flies develop from the larval (maggot) stage in on-site storage containers depends on the following facts: If maggots develop, they are difficult to remove when the containers are emptied. Those remaining may develop into flies. Maggots can also crawl from uncovered cans and develop into flies in the surrounding environment.

4-4 PHYSICAL, CHEMICAL, AND BIOLOGICAL TRANSFORMATIONS OF SOLID WASTE

The purpose of this section is to introduce the reader to the principal transformation processes that can be used for the management of MSW. These transformations can occur either by the intervention of people or by natural phenomena. Solid waste can be transformed by physical, chemical, and biological means (Table 4-8). One must understand the transformation processes that are possible and the products that may result because they will affect directly the development of integrated solid waste management plans.

Physical Transformations

The principal physical transformations that may occur in the operation of solid waste management systems include (1) component separation, (2) mechanical volume reduction, and (3) mechanical size reduction. Physical transformations do not involve a change in phase (e.g., solid to gas), unlike chemical and biological transformation processes.

Component Separation. *Component separation* is the term used to describe the process of separating, by manual and/or mechanical means, identifiable components from commingled MSW. Component separation is used to transform a heterogeneous waste into a number of more-or-less homogeneous components. Component separation is a necessary operation in the recovery of reusable and

TABLE 4-8
Transformation processes used for the management of solid waste

Transformation process	Transformation means or method	Transformation or principal conversion product(s)
Physical		
Component separation	Manual and/or mechanical separation	Individual components found in commingled municipal waste
Volume reduction	Application of energy in the form of a force or pressure	The original waste reduced in volume
Size reduction	Application of energy in the form of shredding, grinding, or milling	The original waste components altered in form and reduced in size
Chemical		
Combustion	Thermal oxidation	Carbon dioxide (CO_2), sulfur dioxide (SO_2), other oxidation products, ash
Pyrolysis	Destructive distillation	A gas stream containing a variety of gases, tar and/or oil, and a char
Gasification	Starved air combustion	A low-Btu gas, a char containing carbon and the inerts originally in the fuel, and pyrolytic oil
Biological		
Aerobic composting	Aerobic biological conversion	Compost (humus-like material used as a soil conditioner)
Anaerobic digestion (low- or high-solids)	Anaerobic biological conversion	Methane (CH_4), carbon dioxide (CO_2), trace gases, digested humus or sludge
Anaerobic composting[a]	Anaerobic biological conversion	Methane (CH_4), carbon dioxide (CO_2), digested waste

[a] Anaerobic composting occurs in landfills (see Chapter 11).

recyclable materials from MSW, in the removal of contaminants from separated materials to improve specifications of the separated material, in the removal of hazardous wastes from MSW, and where energy and conversion products are to be recovered from processed wastes.

Mechanical Volume Reduction. *Volume reduction* (sometimes known as *densification*) is the term used to describe the process whereby the initial volume occupied by a waste is reduced, usually by the application of force or pressure. In most cities, the vehicles used for the collection of solid wastes are equipped with compaction mechanisms to increase the amount of waste collected per trip. Paper, cardboard, plastics, and aluminum and tin cans removed from MSW for recycling are baled to reduce storage and handling costs and shipping costs to

FIGURE 4-6
Baler used at materials recovery facility to bale paper, cardboard, plastic, and aluminum cans.

processing centers (see Fig. 4-6). Recently, high-pressure compaction systems have been developed to produce materials suitable for various alternative uses such as production of fireplace logs from paper and cardboard. To decrease the costs associated with the transport of waste materials to landfill disposal sites, municipalities also may use transfer stations equipped with compaction facilities. To increase the useful life of landfills, wastes are usually compacted before being covered (see Fig. 4-7).

Mechanical Size Reduction. *Size reduction* is the term applied to the transformation processes used to reduce the size of the waste materials. The objective of size reduction is to obtain a final product that is reasonably uniform and considerably reduced in size in comparison with its original form (see Fig. 4-8). Note that size reduction does not necessarily imply volume reduction. In some situations, the total volume of the material after size reduction may be greater than that of the original volume (e.g., the shredding of office paper). In practice, the terms *shredding, grinding,* and *milling* are used to describe mechanical size-reduction operations.

FIGURE 4-7
Compaction of waste at a landfill before daily cover is applied.

FIGURE 4-8
Yard trimmings before and after mechanical size reduction in a tub grinder.

Chemical Transformations

Chemical transformations of solid waste typically involve a change of phase (e.g., solid to liquid, solid to gas, etc.). To reduce the volume and/or to recover conversion products, the principal chemical processes used to transform MSW include (1) combustion (chemical oxidation), (2) pyrolysis, and (3) gasification. All three of these processes are often classified as thermal processes.

Combustion (Chemical Oxidation). *Combustion* is defined as the chemical reaction of oxygen with organic materials, to produce oxidized compounds accompanied by the emission of light and rapid generation of heat. In the presence of excess air and under ideal conditions, the combustion of the organic fraction of MSW can be represented by the following equation:

$$\text{Organic matter} + \text{excess air} \rightarrow N_2 + CO_2 + H_2O + O_2 + \text{ash} + \text{heat} \quad (4\text{-}18)$$

Excess air is used to ensure complete combustion. The end products derived from the combustion of MSW, Eq. (4-18), include hot combustion gases—composed primarily of nitrogen (N_2), carbon dioxide (CO_2), water (H_2O, flue gas), and oxygen (O_2)—and noncombustible residue. In practice, small amounts of ammonia (NH_3), sulfur dioxide (SO_2), nitrogen oxides (NO_x), and other trace gases will also be present, depending on the nature of the waste materials.

Pyrolysis. Because most organic substances are thermally unstable, they can be split, through a combination of thermal cracking and condensation reactions in an oxygen-free atmosphere, into gaseous, liquid, and solid fractions. *Pyrolysis* is the term used to describe the process. In contrast with the combustion process, which is highly exothermic, the pyrolytic process is highly endothermic. For this reason, *destructive distillation* is often used as an alternative term for pyrolysis.

The characteristics of the three major component fractions resulting from the pyrolysis of the organic portion of MSW are (1) a gas stream containing primarily hydrogen (H_2), methane (CH_4), carbon monoxide (CO), carbon dioxide (CO_2), and various other gases, depending on the organic characteristics of the waste material being pyrolyzed; (2) a tar and/or oil stream that is liquid at room temperature and contains chemicals such as acetic acid, acetone, and methanol; and (3) a char consisting of almost pure carbon plus any inert material that may have entered the process. For cellulose ($C_6H_{10}O_5$) the following expression has been suggested as being representative of the pyrolysis reaction [3]:

$$3(C_6H_{10}O_5) \rightarrow 8H_2O + C_6H_8O + 2CO + 2CO_2 + CH_4 + H_2 + 7C \quad (4\text{-}19)$$

In Eq. (4-19), the liquid tar and/or oil compounds normally obtained are represented by the expression C_6H_8O.

Gasification. The gasification process involves partial combustion of a carbonaceous fuel so as to generate a combustible fuel gas rich in carbon monoxide, hydrogen, and some saturated hydrocarbons, principally methane. The combustible fuel gas can then be combusted in an internal combustion engine or boiler. When a gasifier is operated at atmospheric pressure with air as the oxidant, the end products of the gasification process are (1) a low-Btu gas typically containing carbon dioxide (CO_2), carbon monoxide (CO), hydrogen (H_2), methane (CH_4), and nitrogen (N_2); (2) a char containing carbon and the inerts originally in the fuel, and (3) condensible liquids resembling pyrolytic oil.

Other Chemical Transformation Processes. In addition to the various combustion, pyrolysis, and gasification processes under investigation and/or construction, a variety of other public and proprietary processes are being developed and evaluated for the transformation of solid waste. The hydrolytic conversion of cellulose to glucose, followed by the fermentation of glucose to ethyl alcohol, is an example of such a process (see Chapter 14).

Biological Transformations

The biological transformations of the organic fraction of MSW may be used to reduce the volume and weight of the material; to produce compost, a humus-like material that can be used as a soil conditioner; and to produce methane. The principal organisms involved in the biological transformations of organic wastes are bacteria, fungi, yeasts, and actinomycetes. These transformations may be accomplished either *aerobically* or *anaerobically*, depending on the availability of oxygen. The principal differences between the aerobic and anaerobic conversion reactions are the nature of the end products and the fact oxygen must be provided to accomplish the aerobic conversion. Biological processes that have been used for the conversion of the organic fraction of MSW include aerobic composting, anaerobic digestion, and high-solids anaerobic digestion.

Aerobic Composting. Left unattended, the organic fraction of MSW will undergo biological decomposition. The extent and the period of time over which the decomposition occurs will depend on the nature of the waste, the moisture content, the available nutrients, and other environmental factors. Under controlled conditions, yard wastes and the organic fraction of MSW can be converted to

FIGURE 4-9
Compost produced from processed (see Fig. 4-8) yard wastes.

a stable organic residue known as *compost* (see Fig. 4-9) in a reasonably short period of time (four to six weeks).

Composting the organic fraction of MSW under aerobic conditions can be represented by the following equation:

$$\text{Organic matter} + O_2 + \text{nutrients} \rightarrow \text{new cells} + \begin{matrix}\text{resistant}\\ \text{organic}\\ \text{matter}\end{matrix} + CO_2 + H_2O + NH_3 + SO_4^{2-} + \text{heat} \quad (4\text{-}20)$$

In Eq. (4-20), the principal end products are new cells, resistant organic matter, carbon dioxide, water, ammonia, and sulfate. Compost is the resistant organic matter that remains. The resistant organic matter usually contains a high percentage of lignin, which is difficult to convert biologically in a relatively short time. Lignin, found most commonly in newsprint, is the organic polymer that holds together the cellulose fibers in trees and certain plants.

Anaerobic Digestion. The biodegradable portion of the organic fraction of MSW can be converted biologically under anaerobic conditions to a gas containing carbon dioxide and methane (CH_4). This conversion can be represented by the following equation:

$$\text{Organic matter} + H_2O + \text{nutrients} \rightarrow \text{new cells} + \begin{matrix}\text{resistant}\\ \text{organic}\\ \text{matter}\end{matrix} + CO_2 + CH_4 + NH_3 + H_2S + \text{heat} \quad (4\text{-}21)$$

Thus, the principal end products are carbon dioxide, methane, ammonia, hydrogen sulfide, and resistant organic matter. In most anaerobic conversion processes carbon dioxide and methane constitute over 99 percent of the total gas produced. The resistant organic matter (or digested sludge) must be dewatered before it can be disposed of by land spreading or landfilling. Dewatered sludge is often composted aerobically to stabilize it further before application.

Other Biological Transformation Processes. In addition to the aerobic composting and anaerobic digestion processes, a variety of other public and proprietary processes are being developed and evaluated for the biological transformation of solid waste. The high-solids anaerobic digestion process discussed in Chapter 14 is one such example.

Importance of Waste Transformations in Solid Waste Management

Typically, physical, chemical, and biological transformations are used (1) to improve the efficiency of solid waste management operations and systems, (2) to recover reusable and recyclable materials, and (3) to recover conversion products and energy. The implications of waste transformation in the design of integrated solid waste management systems can be illustrated by the following example. If composting is to be an element of a solid waste management plan, the organic fraction of the MSW must be separated from the commingled MSW. If the organic fraction must be separated, should it be done at the source of generation or at a materials recovery facility? If separation of wastes is to occur at the source, what components should be separated to produce an optimum compost?

Improving Efficiency of Solid Waste Management Systems. To improve the efficiency of solid waste management operations and to reduce storage volume requirements at medium- and high-rise apartment buildings, wastes are often baled. For example, waste paper, recovered for recycling, is baled to reduce storage volume requirements and shipping costs. In some cases, waste materials are baled to reduce haul costs to the disposal site. At disposal sites, solid wastes are compacted to use the available landfill capacity effectively. If solid wastes are to be transported hydraulically or pneumatically, some form of shredding is normally required. Mechanical size reduction (shredding) has also been used to improve the efficiency of disposal sites. Hand separation at the point of generation is now considered an efficient way to remove small quantities of hazardous waste from MSW, thereby making landfills safer. Chemical and biological processes can be used to reduce the volume and weight of waste requiring disposal and to produce useful products.

Recovery of Materials for Reuse and Recycling. As a practical matter, components that are most amenable to recovery are those for which markets exist and which are present in the wastes in sufficient quantity to justify their separation. Materials most often recovered from MSW include paper, cardboard, plastic, garden trimmings, glass, ferrous metal, aluminum, and other nonferrous metal.

Recovery of Conversion Products and Energy. The organic fraction of MSW can be converted to usable products and ultimately to energy in a number of ways, including (1) combustion to produce steam and electricity; (2) pyrolysis to produce a synthetic gas, liquid or solid fuel, and solids; (3) gasification to produce a synthetic fuel; (4) biological conversion to produce compost; and (5) biodigestion to generate methane and to produce a stabilized organic humus.

4-5 DISCUSSION TOPICS AND PROBLEMS

4-1. Using the data in Table 4-1, determine the *as discarded* specific weight of typical residential MSW. Use the typical distribution of waste components given in column 3 of Table 3-4.

4-2. Using the data in Table 4-1, estimate the specific weight of two wastes (to be selected by instructor) with the composition given in the following table:

	Percent by weight		
Component	**A**	**B**	**C**
Food wastes	15	45	70
Paper	35	25	10
Cardboard	7	4	3
Plastics	5	5	6
Textiles	3	0	0
Rubber	3	0	0
Leather	2	0	0
Yard wastes	20	18	11
Wood	10	3	0

4-3. Estimate the overall moisture content of wastes (to be selected by instructor) given in Problem 4-2.

4-4. Using the data in Problem 4-2, estimate the overall moisture content of wastes C.

4-5. Using the data in Table 4-3, derive an approximate chemical formula for waste A given in Problem 4-2.

4-6. Estimate the *as discarded* energy content for waste A in Problem 4-2. Use the typical values in Table 4-5. What is the energy content on a dry and dry ash-free basis?

4-7. Compare the *as discarded* energy content for wastes A and C in Problem 4-2. Use the typical data given in Table 4-5. What are the implications of this comparison?

4-8. Estimate the energy content of waste A in Problem 4-2 based on the chemical composition of the individual waste components. Use the data given in Table 4-3.

4-9. Hydrogen sulfide has an odor recognition threshold concentration of 0.47 ppb. Based on Eqs. (4-12) and (4-14), determine the minimum concentration of both lactate and sulfate that would lead to the recognition of H_2S. Assume that both reactions convert 100% of the starting material.

4-10. Obtain data on the breeding time for flies from your local vector control agency. How do the values you obtained compare with the values given in Section 4-3? Explain any differences.

4-11. Identify (*a*) the physical, chemical, and biological transformations that can be applied to solid waste with which you have first-hand experience and (*b*) the context of your experience—for example, the combustion of paper in a fireplace.

4-12. Identify the physical, chemical, and biological transformations that are used by the waste management agency in your community and the context (volume reduction, energy production) in which they are used.

4-13. What waste materials do you now separate where you live? What waste materials do your parents separate at their home?

4-14. What is the difference between compaction and consolidation? What effect will consolidation have in baled material that has a specific weight of 1800 lb/yd^3? *Hint:* What is the weight of one cubic yard of water?

4-15. Although compaction of waste increases the amount of solid waste that can be collected per collection trip, what are the disadvantages of compaction with respect to the separation of waste components at a materials recovery facility?

4-16. Combustion of solid waste involves a chemical transformation in which solid matter is transformed to gas. However, there will always be some undesirable products. Describe which factors and process variables affect the conversion products.

4-17. Steam and carbon dioxide are two main products of the combustion process. Is the following ratio constant for all solid wastes? Explain.

$$R = \frac{\text{moles } H_2O}{\text{moles } CO_2}$$

4-18. Explain how the nature of the waste affects the aerobic decomposition of solid waste.

4-6 REFERENCES

1. Chandler, J. A., W. J. Jewell, J. M. Gossett, P. J. Vansoset, and J. B. Robertson: "Predicting Methane Fermentation Biodegradability," *Biotechnology and Bioengineering Symposium,* No. 10, pp. 93–107, 1980.
2. Davis, S. N. and R. J. M. DeWiest: *Hydrogeology,* John Wiley & Sons, New York, 1966.
3. Drobny, N. L., H. E. Hull, and R. F. Testiu: *Recovery and Utilization of Municipal Solid Waste,* U.S. Environmental Protection Agency, Publication SW-10c, Washington, DC, 1971.
4. Hasselriis, F.: *Refuse Derived Fuel*, an Ann Arbor Science Book, Butterworth, Boston, 1984.
5. James, M. T. and R. F. Harwood: *Medical Entomology*, 6th ed., Macmillan, London, 1969.
6. Kaiser, E. R.: "Chemical Analyses of Refuse Compounds," in *Proceedings of National Incinerator Conference,* ASME, New York, 1966.
7. Mantell, C. L. (ed.): *Solid Wastes: Origin, Collection, Processing, and Disposal,* Wiley-Interscience, New York, 1975.
8. Neissen, W. R.: "Properties of Waste Materials," in D. G. Wilson (ed.): *Handbook of Solid Waste Management,* Van Nostrand Reinhold, New York, 1977.
9. Salvato, J. A.: *Environmental Engineering and Sanitation,* 4th ed., Wiley-Interscience, New York, 1992.
10. Singer, J. G. (ed.): *Combustion: Fossil Power Systems,* Combustion Engineering, Inc., Windsor, CT, 1981.
11. Tchobanoglous, G., H. Theisen, and R. Eliassen: *Solid Waste: Engineering Principles and Management Issues,* McGraw-Hill, New York, 1977.
12. Winkler, P. F. and D. C. Wilson: "Size Characteristics of Municipal Solid Wastes," *Compost Science*, Vol. 14, No. 5, April 1973.

CHAPTER 5

SOURCES, TYPES, AND PROPERTIES OF HAZARDOUS WASTES FOUND IN MUNICIPAL SOLID WASTE

As concern over the disposal of industrially derived hazardous waste has spread, concern has also grown over the disposal of MSW, which may contain small amounts of hazardous wastes from households and commercial facilities. What is unknown is the fate of hazardous materials found in MSW when products such as compost are produced from MSW, when MSW is combusted, and when MSW is placed in landfills. The long-term effects, if any, on public health and the environment from the presence of these materials also are unknown.

The purposes of this chapter are (1) to consider the properties and classification of hazardous wastes, (2) to review the sources and significance of hazardous wastes found in municipal waste, (3) to review the occurrence of hazardous wastes in various activities associated with the management of municipal solid waste, (4) to review the nature of the physical, chemical, and biological transformations that these wastes undergo, and (5) to consider the management of hazardous wastes in MSW. Because so much more needs to be known, the material in this chapter is only intended to serve as an introduction to this subject.

5-1 PROPERTIES AND CLASSIFICATION OF HAZARDOUS WASTES

Before discussing the sources, occurrences, and transformations associated with the hazardous wastes found in MSW, it will be helpful to define what constitutes a hazardous waste. Hazardous wastes have been defined as wastes or combinations of wastes that pose a substantial present or potential hazard to humans or other living organisms because (1) such wastes are nondegradable or persistent in nature, (2) they can be biologically magnified, (3) they can be lethal, or (4) they may otherwise cause or tend to cause detrimental cumulative effects [14]. Properties of waste materials that have been used to assess whether a waste is hazardous are related to questions of safety and health.

Safety-related properties

- Corrosivity
- Explosivity
- Flammability
- Ignitability
- Reactivity

Health-related properties

- Carcinogenicity
- Infectivity
- Irritant (allergic response)
- Mutagenicity
- Toxicity (poisons)
  - Acute toxicity
  - Chronic toxicity
- Radioactivity
- Teratogenicity

When dealing with the hazardous waste materials found in MSW, municipalities have most commonly used the following properties to define a hazardous waste:

- Ignitability
- Corrosivity
- Reactivity
- Toxicity
- Carcinogenicity

At present, a variety of classification systems and priority lists have been adopted by different regulatory agencies to define a hazardous waste. For this reason, all of the current operative classification systems must be considered in any assessment.

Definitions Given by the U.S. Environmental Protection Agency

In developing regulations for the implementation of legislation related to solid waste and wastewater, the U.S. Environmental Protection Agency has published

TABLE 5-1
Categories of RCRA hazardous wastes

Listed	Characteristic	Other
Nonspecific sources	Ignitable	Mixtures (hazardous and nonhazardous)
Specific sources	Corrosive	Residues derived from treatment of wastes
Commercial chemical products (acutely hazardous)	Reactive	Materials containing listed hazardous wastes
Commercial chemical products (nonacutely hazardous)	Toxic	

and refined definitions for (1) RCRA hazardous waste, and (2) priority pollutants. These definitions are reviewed briefly below.

RCRA Hazardous Wastes. The U.S. EPA has defined RCRA hazardous wastes in three general categories: (1) listed wastes, (2) characteristic hazardous wastes, and (3) other hazardous wastes (see Table 5-1). Based on criteria defined in the Code of Federal Regulations (40 CFR 261.11), the U.S. EPA has prepared a list of specific hazardous wastes. If a waste meets these criteria, it is presumed toxic regardless of the concentration. Characteristic hazardous wastes are established on the basis of their ignitability, corrosivity, reactivity, and toxicity (see Table 5-2). Toxicity is determined by an analysis of the constituents derived from an extraction test. Using "Acute hazardous waste" and "Toxic waste" classifications in addition to those reported in columns 1 and 2 of Table 5-1, more than 1200 compounds have been listed as hazardous wastes in the *Federal Register* since 1976.

Other RCRA hazardous wastes (column 3 of Table 5-1) include mixtures of hazardous and nonhazardous wastes; wastes derived from the management of other wastes, such as treatment plant residues (Derived-From Rule); and hazardous materials contained in nonhazardous wastes (Contained-In Rule) [14]. Medical wastes and low-level radioactive mixed wastes are considered special hazardous wastes. Because medical and low-level radioactive mixed wastes are regulated and managed separately, they are not considered further in this text.

Priority Pollutants. In 1979, pursuant to the Federal Water Pollution Control Act as amended by the Clean Water Act of 1977, the U.S. EPA was required to prepare a list of toxic pollutants proven to be harmful to human health. Four criteria were used to classify the pollutants.

1. Actual or potential damage that a water discharge of these materials may create by virtue of certain toxicological properties. These properties include bioaccumulation, carcinogenicity, mutagenicity, teratogenicity, or high acute toxicity.
2. Seriousness of discharge or potential discharge of the pollutant by point sources. Factors include the nature and extent of toxic effects associated

TABLE 5-2
EPA listed wastes based on hazardous characteristics

Characteristic	40 CRF subpart[a]	Considerations	Hazard code
Ignitability	261.21	1. Liquids with flashpoints of less than 140°F (60°C) 2. Nonliquids liable to cause fires through friction, spontaneous chemical change, etc. 3. Ignitable compressed gas 4. Is an oxidizer	I
Corrosivity	261.22	1. Aqueous wastes exhibiting a pH of <3 or >12.5 2. Liquid wastes capable of corroding steel at a rate greater than 0.250 in/year	C
Reactivity	261.23	1. Instability and readiness to undergo violent change 2. Violent reactions when mixed with water 3. Formation of potentially explosive mixtures when mixed with water 4. Generation of toxic fumes when mixed with water 5. Cyanide or sulfide-bearing material that generates toxic fumes when exposed to acidic conditions 6. Ease of detonation or explosive reaction when exposed to pressure or heat 7. Ease of detonation or explosive decomposition or reaction at standard temperature and pressure 8. Defined as a forbidden explosive or a Class A or B explosive by U.S. Department of Transportation	R
Toxicity characteristic (TC), as defined by the toxicity characteristics leaching procedure test (TCLP)	261.24	The following steps are required in the TCLP test: 1. If the waste is liquid (i.e., contains less than 0.5% solids), after it is filtered the waste itself is considered the extract (simulated leachate). 2. If the waste contains greater than 0.5% solid material, the solid phase is separated from the liquid phase, if any. If required, the particle size of the solid phase is reduced until it passes through a 9.5 mm sieve. 3. For analysis other than for volatiles, the solid phase is then placed in an acidic solution and rotated at 30 rev/min for 18 hours. The pH of the solution is approximately 5, unless the solid is more basic, in which case a solution with a pH of approximately 3 is used. After extraction (rotation), solids are filtered from the liquid extract and discarded. 4. For volatiles analysis a solution of pH 5 is used, and a zero headspace extraction vessel is used for liquid/solid separation, agitation, and filtration. 5. Liquid extracted from the solid/acid mixture is combined with any original liquid separated from the solid material and is analyzed for the presence of specified contaminants. 6. If *any* of the contaminants in the extract meets or exceeds any of the maximum concentration levels allowed for the specified contaminants, the waste is classified as a TC hazardous waste.	E

[a] Adapted from the Code of Federal Regulations (CFR) and Ref. 14.

with the pollutant; extent that discharges have been identified; production and distribution; and use pattern of the pollutant.

3. The setting of effluent standards for point source dischargers.
4. Overall environmental effect of the control measures available.

The initial list of priority pollutants contained 65 classes of pollutants, comprising a total of 129 specific substances.

Other Hazardous Waste Classifications

Other hazardous waste classifications have been proposed by the following agencies:

- International Agency for Research on Cancer
- National Cancer Institute
- Environmental Protection Agency—Carcinogen Assessment Group

5-2 SOURCES, TYPES, AND QUANTITY OF HAZARDOUS WASTES FOUND IN MSW

The purpose of this section is to introduce the reader to the sources and types of hazardous wastes found in municipal solid wastes. The significance of these wastes is considered in the following section. Information on the quantities of hazardous waste found in municipal solid waste is presented in Chapter 6.

Typical Hazardous Wastes from Residential Sources

Many of the products used around the home every day such as household cleaners, personal products, automotive products, paint products, and garden products are toxic and can be hazardous to health and the environment (see Fig. 5-1). Typical

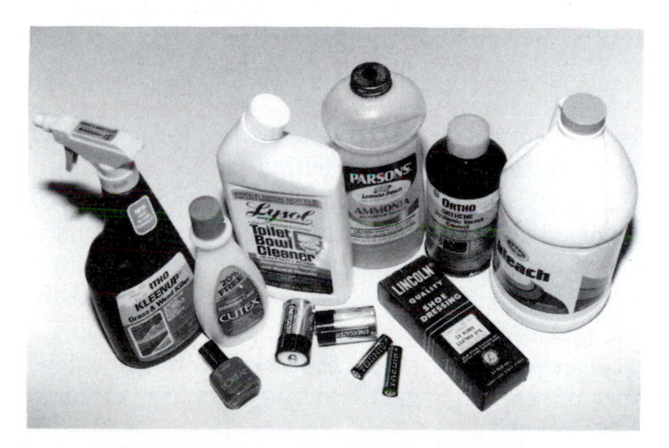

FIGURE 5-1
Typical household hazardous wastes found in small amounts in residential MSW.

TABLE 5-3
Typical hazardous household products[a]

Product	Concern	Disposal
Household cleaners		
Abrasive scouring powders	Corrosive	Hazardous waste facility
Aerosols	Flammable	Hazardous waste facility
Ammonia and ammonia-based cleaners	Corrosive[b]	Hazardous waste facility, or dilute small amounts
Chlorine bleach	Corrosive[c]	Hazardous waste facility, or dilute small amounts
Drain openers	Corrosive	Hazardous waste facility
Furniture polish	Flammable	Hazardous waste facility
Glass cleaners	Irritant	Dilute small amounts
Outdated medicines	Hazardous to others in family	Dilute small amounts and flush down toilet
Oven cleaner	Corrosive	Hazardous waste facility
Shoe polish	Flammable	Hazardous waste facility
Silver polish	Flammable	Hazardous waste facility
Spot remover	Flammable	Hazardous waste facility
Toilet bowl cleaner	Corrosive	Hazardous waste facility
Upholstery and carpet cleaner	Flammable and/or corrosive	Hazardous waste facility
Personal care products		
Hair-waving lotions	Poison	Dilute small amounts and flush down toilet
Medicated shampoos	Poison	Dilute small amounts and flush down toilet
Nail polish remover	Poison, flammable	Hazardous waste facility
Rubbing alcohol	Poison	Dilute small amounts and flush down toilet

(*continued*)

household hazardous products are identified in Table 5-3. These products are of concern owing to their corrosive, flammable, irritant, and poisonous properties.

A survey was conducted to assess the presence of household hazardous wastes in residential MSW; the distribution of hazardous wastes that was found is reported in Table 5-4. These data can be compared with data on hazardous wastes that were received in several Bay Area cities in California on days specified for hazardous waste collection (see Table 5-5). In comparing the data reported in Tables 5-4 and 5-5, it is clear that the relative distribution of hazardous wastes will be quite variable depending on time of year (e.g., greater during the spring cleanup)

TABLE 5-3 (*continued*)

Product	Concern	Disposal
Automotive products		
Antifreeze	Poison	Hazardous waste facility
Brake and transmission fluid	Flammable	Hazardous waste facility
Car batteries	Corrosive	Recycling center/repair
Diesel fuel	Flammable	Recycling center
Kerosene	Flammable	Recycling center
Gasoline	Flammable, poison	Hazardous waste facility
Waste oil	Flammable	Recycling center
Paint products		
Enamel, oil-based, latex or water-based paints	Flammable	Donate or hazardous waste facility
Paint solvents and thinners	Flammable	Reuse or hazardous waste facility
Miscellaneous products		
Batteries	Corrosive	Recycling center or hazardous waste facility
Photographic chemicals	Corrosive, poison	Hazardous waste facility, or donate to photo shop
Pool acids and chlorine	Corrosive	Hazardous waste facility
Pesticides, herbicides, and fertilizers		
Including garden insecticides, ant and roach killers, weed killers, etc.	Poison: some are flammable	Hazardous waste facility or County Department of Agriculture
Chemical fertilizers	Poison	Hazardous waste facility or County Department of Agriculture
Houseplant insecticide	Poison	Hazardous waste facility

[a] Adapted from Ref. 1.

[b] Never mix ammonia and chlorine-based products—a deadly gas can be produced.

[c] Never mix with toilet bowl cleaner.

and the number of hazardous waste collection days that have been completed in a given community.

Typical Hazardous Wastes from Commercial Sources

The hazardous wastes produced by commercial establishments (often identified as small-quantity generators) are related primarily to the services provided. Typical examples include inks from print shops, solvents from dry cleaning

TABLE 5-4
Distribution of hazardous waste materials found in residential and commercial MSW[a,b]

Item	Percent
Household and cleaning products	40.0
Personal care products	16.4
Automotive products	30.1
Paint and related products	7.5
Pesticides, insecticides, and herbicides	2.5
Other	3.5

[a] From Ref. 5.

[b] Note that the amount of hazardous waste material found in residential and commercial MSW is on the order of 0.1 percent.

TABLE 5-5
Distribution of hazardous waste materials collected during days specified for the collection of hazardous wastes[a]

Hazardous waste	Percent of total
Oil-based paint	31.9
Solvents	15.4
Latex paint	12.1
Pesticides	9.9
Empty oil cans	8.8
Cleaners	8.3
Waste oil	5.0
Acids/bases	4.0
Petroleum products	1.7
Other	2.9

[a] From Ref. 5.

establishments, cleaning solvents from auto repair shops, and paints and thinners from painting contractors. Typical compounds from commercial, industrial, and agricultural sources are presented in Table 5-6.

The U.S. EPA has defined a *small-quantity generator* as one which produces in a calendar month:

- More than 100 kg but less than 1000 kg of nonacutely hazardous waste
- Less than 100 kg of waste resulting from the cleanup of any residue or contaminated soil, water, or other debris involving the cleanup of an acutely hazardous waste
- Less than 1 kg of an acutely hazardous waste

TABLE 5-6
Hazardous waste compounds produced by commercial, industrial, and agricultural activities that are typically found in MSW[a]

Name	Formula or symbol	Use	Concern
Nonmetals			
Arsenic	As	Alloying additive for metals, especially lead and copper as shot, battery grids, cable sheaths, boiler tubes	Carcinogen and mutagen. Long-term: Sometimes can cause fatigue and loss of energy; dermatitis
Selenium	Se	Electronics, xerographic plates, TV cameras, photocells, magnetic computer cores, solar batteries, rectifiers, relays, ceramics (colorant for glass), steel and copper, vulcanizing agent, catalyst, trace element in animal feeds	Long-term: Red staining of fingers, teeth, and hair; general weakness; depression; irritation of nose and mouth
Metals			
Barium	Ba	Getter alloys in vacuum tubes, deoxidizer for copper, Frary's metal, lubricant for anode rotors in x-ray tubes, spark plug alloys	Flammable at room temperature in powder form. Long-term: Increased blood pressure and nerve block
Cadmium	Cd	Electrodeposited and dipped coatings on metals, bearing and low-melting alloys, brazing alloys, fire protection system, nickel-cadmium storage batteries, power transmission wire, TV phosphors, basis of pigments used in ceramic glazes, machinery enamels, fungicide, photography and lithography, selenium rectifiers, electrodes for cadmium-vapor lamps and photoelectric cells	Flammable in powder form. Toxic by inhalation of dust or fume. A carcinogen. Soluble compounds of cadmium are highly toxic. Long-term: Concentrates in the liver, kidneys, pancreas, and thyroid; hypertension suspected effect
Chromium	Cr	Alloying and plating element on metal and plastic substrates for corrosion resistance, chromium-containing and stainless steels, protective coating for automotive and equipment accessories, nuclear and high-temperature research, constituent of inorganic pigments	Hexavalent chromium compounds are carcinogenic and corrosive on tissue. Long-term: Skin sensitization and kidney damage
Lead	Pb	Storage batteries, gasoline additive, paint pigments, cable covering, ammunition, piping, tank linings, solder and fusible alloys, vibration damping in heavy construction, foil, babbitt, and other bearing alloys	Toxic by ingestion or inhalation of dust or fumes. Long-term: Brain, nervous system, and kidney damage; birth defects

(*continued*)

TABLE 5-6 (*continued*)

Name	Formula or symbol	Use	Concern
Metals (*cont.*)			
Mercury	Hg	Amalgams, catalyst, electrical apparatus, cathodes for production of chlorine and caustic soda, instruments, mercury-vapor lamps, mirror coating, arc lamps, boilers	Highly toxic by skin absorption and inhalation of fume or vapor. Long-term: Toxic to central nervous system; may cause birth defects
Silver	Ag	Manufacture of silver nitrate, silver bromide, photo chemicals; lining vats and other equipment for chemical reaction vessels, water distillation, etc.; mirrors, electric conductors, silver plating, electronic equipment; sterilant; water purification; surgical cements; hydration and oxidation catalyst, special batteries, solar cells, reflectors for solar towers; low-temperature brazing alloys; table cutlery; jewelry; dental, medical, and scientific equipment; electrical contacts; bearing metal; magnet windings; dental amalgams. Colloidal silver is used as a nucleating agent in photography and medicine, often combined with protein	Toxic metal. Long-term: Permanent grey discoloration of skin, eyes, and mucous membranes
Organic compounds			
Benzene[b] (Benzol)	C_6H_6	Manufacturing of ethylbenzene (for styrene monomer); dodecylbenzene (for detergents); cyclohexane (for nylon); phenol; nitrobenzene (for aniline); maleic anhydride; chlorobenzene hexachloride; benzene sulfonic acid; as a solvent	A carcinogen. Highly toxic. Flammable; dangerous fire risk
Ethylbenzene[b] (Phenylethane)	$C_6H_5C_2H_5$	Intermediate in production of styrene; solvent	Toxic by ingestion, inhalation, and skin absorption; irritant to skin and eyes. Flammable, dangerous fire risk
Toluene[b] (Methylbenzene)	$C_6H_5CH_3$	Aviation gasoline and high-octane blending stock; manufacture of benzene, phenol, and caprolactam; solvent for paints and coatings, gums, resins, most oils, rubber, vinyl organosols; diluent and thinner in nitrocellulose lacquers; adhesive solvent in plastic toys and model airplanes; chemicals (benzoic acid, benzyl and benzoyl derivatives, saccharine, medicines, dyes, perfumes); source of toluenediisocyanates (polyurethane resins); explosives (TNT); toluene sulfonates (detergents); scintillation counters	Flammable, dangerous fire risk. Toxic by ingestion, inhalation, and skin absorption

Halogenated compounds			
Chlorobenzene (Phenylchloride)	C_6H_5Cl	Phenol, chloronitrobenzene, aniline, solvent carrier for methylene diisocyanate, solvent, pesticide intermediate, heat transfer	Moderate fire risk. Toxic by inhalation and skin contact
Chloroethene[b] (Vinyl chloride)	CH_2CHCl	Polyvinyl chloride and copolymers, organic synthesis, adhesives for plastics	Extremely toxic and hazardous by all avenues of exposure. A carcinogen
Dichloromethane[b] (Methylene chloride)	CH_2Cl_2	Paint removers, solvent degreasing, plastics processing, blowing agent in foams, solvent extraction, solvent for cellulose acetate	Toxic. A carcinogen, narcotic
Tetrachloroethene (Tetrachloroethylene, perchloroethylene)	CCl_2CCl_2	Dry-cleaning solvent, vapor-degreasing solvent, drying agent for metals and certain other solids, vermifuge, heat transfer medium, manufacture of fluorocarbons	Irritant to eyes and skin
Pesticides, herbicides, insecticides[c]			
Endrin™	$C_{12}H_8OCl_6$	Insecticide and fumigant	Toxic by inhalation and skin absorption, carcinogen
Lindane™	$C_6H_6Cl_6$	Pesticide	Toxic by inhalation, ingestion, skin absorption
Methoxychlor™	$Cl_3CCH(C_6H_4OCH_3)_2$	Insecticide	Toxic material
Toxaphene™	$C_{10}H_{10}Cl_8$, approximately	Insecticide and fumigant	Toxic by ingestion, inhalation, skin absorption
Silvex	$Cl_3C_6H_2OCH(CH_3)COOH$	Herbicide, plant growth regulator	Toxic material; use has been restricted

[a] From Ref. 11.

[b] The five trace organic compounds footnoted are most common in landfill gas.

[c] Pesticides, herbicides, and insecticides are listed by trade name. The compounds listed are also halogenated organic compounds.

Generators who produce more than these quantities are considered to be *large generators*. Generators who produce less than these quantities are considered *conditionally exempt generators*. Of the three types of waste generators, conditionally exempt generators are subject to the fewest regulatory controls [9]. Under federal regulations and in a few states, conditionally exempt generators are permitted to dispose of small amounts of waste in sanitary landfills that would otherwise be regulated as hazardous waste. It should be noted that many states do not have a lower threshold for small-quantity or conditionally exempt generators and do not allow any hazardous materials to be commingled with MSW.

Quantity of Hazardous Wastes in MSW

To put the issue of hazardous wastes in MSW in perspective, it is estimated that the amount of hazardous waste found in MSW (1992) varies from 0.01 to 1 percent by weight, with a typical value of 0.1 percent. These percentages do not account for the HHW that are illegally disposed of onto land and into sewers and storm drains. The reported range is wide because the methods used to identify and classify the hazardous wastes found in MSW are quite variable and the sampling periods are not consistent between studies. For example, measured values are higher in late spring, when many residents clean their garages of materials such as small amounts of paint, unused cleaning products, and garden products including pesticides and herbicides. The authors have measured values in the range of 0.075 to 0.2 percent by weight for residential and commercial MSW. Although the exact distribution between residential and commercial sources is quite variable, approximately 75 to 85 percent of the hazardous wastes found in MSW are from residential sources. (See Section 6-6.)

5-3 SIGNIFICANCE OF HAZARDOUS WASTES IN MSW

The small amounts of hazardous wastes found in MSW are of significance because of their occurrence in all solid waste management facilities and their persistence when discharged to the environment.

Occurrence of Hazardous Wastes in Solid Waste Management Facilities

The occurrence in MSW of small amounts of solid, semisolid, and liquid hazardous wastes and gaseous compounds derived from these wastes influences the recovery of materials, conversion products (e.g., compost), combustion products, and landfills.

Hazardous Waste Constituents in Conversion Products. Trace amounts of hazardous organic constituents had been found in waste components that had been separated mechanically from commingled MSW. Trace contaminants have also been found in the compost produced from MSW. In both cases, the presence of trace amounts of hazardous constituents has rendered the materials and products unusable. Source separation is being encouraged to eliminate these constituents from solid waste–processing operations.

Hazardous Waste Constituents in Combustion Products. Hazardous waste constituents have been measured in the gaseous emissions and in the residual materials resulting from the combustion of solid wastes. Toxic heavy metals such as barium, cadmium, chromium, lead, mercury, and silver are especially troublesome.

Hazardous Waste Constituents at Landfills. Trace organic constituents have been found in the atmosphere near landfills, in extracted landfill gas, and in landfill leachate. The trace constituents that have been measured at landfills have two basic sources. They are derived from the hazardous wastes themselves and/or they may be produced by chemical and biological conversion reactions within the landfill.

Long-Term Persistence

As we noted in the introduction to this chapter, the fate of small quantities of hazardous wastes found in MSW is generally unknown. The environmental persistence of these hazardous compounds is one of the critical issues in their long-term management. Often, hazardous wastes are classified as either nonpersistent or persistent (see Table 5-7).

TABLE 5-7
Hazards associated with nonpersistent and persistent organic wastes[a]

Typical compounds	Hazards
Nonpersistent organic wastes	
Oil, low molecular-weight solvents, some biodegradable pesticides (organophosphates, carbamates, triazines, anilines, ureas), waste oils, most detergents	Toxicity problems primarily to environment and biota at the source or point of release. Toxic effects occur rapidly after exposure (acute and subacute).
Persistent organic wastes	
High molecular-weight chlorinated and aromatic hydrocarbons, some pesticides (chlorinated insecticides like hexachlorobenzene, DDT, DDE, lindane); PCBs, phthalates	Immediate toxic effects (acute and subacute) may occur at the source or point of release. Long-term chronic toxicity may result. Transport of organic waste from the source can result in widespread contamination and bioconcentration in the food chain. Environmental transport may expose biota to lower levels of the pollutant, resulting in chronic toxicity.

[a] Adapted from Ref 10.

The half-life concept can be used to characterize and compare the relative environmental persistence of various hazardous wastes [4, 13]. At the relatively low concentrations encountered in MSW, the decay (disappearance) of an individual hazardous waste constituent can be described adequately as a first-order function as follows:

$$\frac{d[C]}{dt} = -k_T C \tag{5-1}$$

where $[C]$ = concentration at time t
t = time
k_T = first order reaction rate constant

The integrated form of Eq. (5-1) is

$$\ln \frac{[C_o]}{[C]} = k_T t \tag{5-2}$$

where $[C_o]$ = concentration at time zero

When half of the initial material has decayed away, $[C_o]/[C]$ is equal to 2; the corresponding time is given by the following expression:

$$t_{1/2} = \frac{\ln 2}{k_T} = \frac{0.69}{k_T} \tag{5-3}$$

Example 5-1 illustrates the application of the half-life concept.

Example 5-1 Evaluation of contaminant persistence. Determine the time required for the concentrations of toluene and Dieldrin® spilled in a shallow leachate treatment pond to be reduced to one half their initial values. Assume the first-order removal constants for toluene and Dieldrin are 0.0665/hr and 2.665×10^{-5}/hr, respectively.

Solution. Use Eq.(5-3) to determine the time required for the concentrations in the treatment pond to reach one half their original values.

1. For toluene

$$t_{1/2} = \frac{0.69}{k_T} = \frac{0.69}{0.0665/\text{hr}}$$

$$= 10.4 \text{ hr}$$

2. For Dieldrin

$$t_{1/2} = \frac{0.69}{2.665 \times 10^{-5}/\text{hr}}$$

$$= 25{,}891 \text{ hr}$$

Comment. The time required for the concentration of Dieldrin to reach one half of the initial value can be used as an argument for the development and use of agricultural chemicals that are more readily broken down in the environment.

5-4 PHYSICAL, CHEMICAL, AND BIOLOGICAL TRANSFORMATIONS OF HAZARDOUS WASTE CONSTITUENTS FOUND IN MSW

In general, hazardous wastes in MSW are either solids, semisolids, or liquids. In addition, trace chemical compounds can exist as a solute within a liquid solvent, as a gas adsorbed onto a solid, or as a component of the gaseous emissions from MSW, particularly MSW placed in landfills. Physical, chemical, and biological transformations that are important in determining the fate and dispersal of these materials are introduced in this and the following two sections.

Physical Transformations

The principal physical transformations that alter the form of the hazardous constituents found in MSW are volatilization and phase distribution.

Volatilization. The principal mechanisms leading to the production of the gaseous substances from MSW are volatilization, biodegradation, and chemical reaction. Of these, volatilization is thought to be the most important. Hazardous wastes can occur in the gaseous state as a result of three related processes: volatilization of chemical wastes, volatilization of liquid chemical wastes in water and leachate, and volatilization of chemical wastes adsorbed on soil or other solids [7]. The first process is a function of exposed surface area, time, diffusion coefficients, vapor pressures, molecular weight, and temperature. The second process is primarily affected by the Henry's law constant for the substance in question but also by temperature, liquid turbulence, trace constituent concentration in the gas phase, and wind speed. The important factors in the third process are surface area, strength of adsorption, vapor pressure, and type of soil or solid. Other factors affecting volatilization include pH, solubility, the amount and type of organics present, the size of particles, the density of the solid wastes, reactivity, and leaching [7].

Vapor pressure. In a closed container, part of which is filled by a substance in a liquid state, a portion of the liquid will evaporate so as to fill the remaining volume with this substance in the vapor state. The pressure exerted by the vapor on the liquid when the two phases are in equilibrium is defined as the *vapor pressure* of the compound. Vapor pressure is strongly affected by temperature, increasing as the temperature increases. When the boiling temperature of a liquid has been reached the vapor pressure is equal to the atmospheric pressure. Vapor pressure is a characteristic property of the substance and is important for several reasons. The vapor pressure can be used to determine the partial pressure of each component

TABLE 5-8
Relationship between Henry's law constants and the tendency of an organic compound to volatilize[a]

Henry's law constant (K_H), $m^3 \cdot atm/mol$	Comments on transfer mechanism
$< 10^{-7}$	The substance is essentially nonvolatile.
$10^{-7} < K_H < 10^{-5}$	The gas phase resistance dominates the liquid phase resistance by a factor of 10 at least; therefore, the substance volatilizes slowly.
$10^{-5} < K_H < 10^{-3}$	Liquid phase and gas phase resistances are both important. Volatilization for compounds in this range is less rapid than for compounds in a higher range of K_H, but is still a significant transfer mechanism.
$> 10^{-3}$	The resistance of the liquid phase dominates. Thus, transfer is liquid-phase controlled and these substances are highly volatile.

[a] Adapted from Ref. 8.

in a mixture of gases. The relative proportions of the individual components in a mixture can be determined when the partial pressure is known. The vapor pressure can also be used as a measure of the volatility of the substance. Liquids with a high vapor pressure will tend to evaporate easily, while liquids with a low vapor pressure will evaporate slowly. Compounds that are considered volatile have vapor pressures greater than 0.1 mm Hg at 20°C and/or boiling points less than 100°C [12].

Henry's law. With a dilute amount of trace constituent (TC) in a mixture of compounds, the Henry's law constant (K_H) relates the partial pressure of the solute in the vapor phase to the mole fraction of the constituent in solution. Henry's law is given in Appendix F. Values of the Henry's law constant for various volatile and semivolatile compounds are presented in Appendix H. Values of Henry's law constants for most hazardous waste compounds range from 10^{-7} to $> 10^{-3}$ $m^3 \cdot atm/mol$ (see Table 5-8). When the value of the Henry's law constant is high, the resistance of the liquid phase dominates over the gas phase and these compounds are highly volatile. For compounds with Henry's law constants between 10^{-5} and 10^{-3} $m^3 \cdot atm/mol$, both the liquid- and gas-phase resistances are important. Volatilization for compounds with constants in this range is less rapid than for compounds in a higher range, but is still significant [8].

Example 5-2 Volatility of trace organic compounds found in MSW. Given the following hazardous waste compounds that may be found in MSW, arrange them in order of most volatile to least volatile, and discuss their relative volatility.

Benzene
Chloroethene
1,1,1,2-Tetrachloroethane
Tetrachloroethene
Tetrachloromethane
Toluene

Solution

1. Using the data from Appendix H arrange the compounds according to their boiling point, as in the table below.

Compounds	mw	bp, °C	vp, mm Hg	K_H, $m^3 \cdot atm/mol$
Chloroethene	62.5	−13.9	2548	1.07–6.4 × 10^{-2}
Tetrachloromethane	153.82	76.7	90	2.86 × 10^{-2}
Benzene	78.11	80.1	76	5.43–5.49 × 10^{-3}
Toluene	92.1	110.8	22	5.94–6.44 × 10^{-3}
Tetrachloroethene	165.83	121	15.6	2.85 × 10^{-2}
1,1,1,2-Tetrachloroethane	167.86	146.2	14.74	4.2–4.55 × 10^{-4}

2. Referring to the above table, the proposed arrangement according to boiling point corresponds to how the compounds would be arranged based on their vapor pressure. Based on their Henry's law constants, all of the compounds would be considered to be highly volatile.
3. Based on boiling points it can be concluded that chloroethene is significantly more volatile than any of the other compounds in this example.

Distribution of Waste between Phases. The distribution of a substance between two immiscible phases or liquids is defined by the distribution coefficient. Knowledge of the amount of a waste in each phase is important in developing waste management plans.

Distribution coefficient for two phases. When a substance that is soluble in each of two (immiscible) phases is added to a system of these two immiscible phases, the substance will be distributed in each in fixed proportions at a given temperature, independent of the quantity of the substance. This statement defines the distribution law. The ratio of the concentrations in each phase is called the distribution coefficient (or the partition coefficient). For dilute solutions, the preceding is also a working statement of Henry's law.

Distribution coefficient for two immiscible liquids. An identical relationship to that for two phases also holds for the distribution of a solute between two immiscible liquids. The coefficient is constant only when the given solute dissolves in both solvents in the same form and no association or dissociation takes place. In practice, the distribution ratio for liquids is seldom strictly a constant. Stated mathematically, the distribution coefficient for two phases is

$$\frac{C_{X/A}}{C_{X/B}} = K_D \tag{5-4}$$

where $C_{X/A}$ = concentration of solute X in solvent A, g/m^3
$C_{X/B}$ = concentration of solute X in solvent B, g/m^3
K_D = distribution or partition coefficient

If the solute remains unchanged during its distribution between the two solvents, the distribution coefficient can be used to calculate the efficiency of an extraction process in which a given solvent is used to extract a solute from another solvent.

Octanol: water distribution coefficient. For the purposes of comparison and analysis, the octanol:water solvent system is used to characterize a variety of organic substances. Initially the octanol:water distribution coefficient of a compound was used to assess the bioaccumulation potential of a compound. Values of the octanol:water distribution coefficient are available in the literature [6, 7] but vary widely. Typical values for selected compounds are presented in Appendix H.

Chemical Transformations

The organic hazardous waste constituents in MSW can be transformed by a variety of chemical reactions, some of which are considered briefly below.

Chemical Reactions in Combustion. Under ideal conditions the combustion process is an effective means of destroying the hazardous organic constituents found in municipal solid waste. On the other hand, if the combustion process is not ideal, the products of incomplete combustion (PICs) can be quite varied and, in many cases, toxic. The stoichiometric combustion of chlorobenzene (C_6H_5Cl) with oxygen can be described by the following reaction:

$$C_6H_5Cl + 7O_2 \rightarrow 6CO_2 + 2H_2O + HCl \tag{5-5}$$

If the combustion is complete, then the chlorine in C_6H_5Cl will be converted to hydrochloric acid. If, however, the combustion is incomplete, then the formation of toxic trace constituents is possible. Similar reactions can occur with most hazardous waste compounds.

Chemical Reactions in Landfills. The principal classes of chemical (abiotic) reactions that can occur to alter the composition of the hazardous waste compounds found in MSW include [13]:

1. Simple substitution
2. Dehydrogenation (hydrolysis)

3. Oxidation (Auto-oxidation)
4. Reduction

While these reactions can occur at any time, they typically occur during landfilling and in completed landfills. Most abiotic reactions involving the more persistent halogenated hazardous compounds are slow but may be significant in the time scales involved in the long-term management of landfills [13]. In an environment such as a landfill where active biological decomposition is occurring, the rates of abiotic reaction may be increased significantly by the activity of biologically produced enzymes. Typical half lives for abiotic dehydrohalogenation (hydrolysis) of halogenated compounds are indicated in Table 5-9. These abiotic half lives can vary from 20 days to 7000 years. In the absence of biotic reactions the half lives of some of the compounds reported in Table 5-9 are troublesome from the standpoint of the long-term management of these compounds in landfills.

Biological Transformations

The biological transformations of the major organic constituents, as discussed in Chapter 4, are relatively well established. Much less is known, however, about the chemical/biological transformations of the hazardous inorganic and organic wastes found in MSW. Because this subject is so complex and so many pathways are unknown, what is presented here is meant to serve as an introduction to this important subject.

TABLE 5-9
Typical half lives and products derived from the chemical hydrolysis or dehydrogenation of halogenated aliphatic compounds at 20°C

Compound	Half life, yr	Products
Methanes		
Bromomethane	0.10	
Bromodichloromethane	137	
Trichloromethane	1.3	
Tetrachloromethane	7000	
Ethanes		
Chloroethane	0.12	Ethanol
1,1,2-Trichloroethane	170	1,1-Dichloroethene
1,1,1,2-Tetrachloroethane	384	Trichloroethene
Ethenes		
Trichloroethene	0.9	
Tetrachlorethene	0.7	
Propanes		
1-Bromopropane	0.07	Bromopropene
1,2-Dibromopropane	0.88	

[a] Adapted from Ref. 13.

Transformations Involving Metals. Many of the hazardous inorganic constituents present in MSW, such as chromium, lead, and mercury, can be converted biologically into a variety of compounds, some of which are extremely toxic. For example, toxic compounds that can be produced under anaerobic conditions, such as those in landfills, include methylmercury, dimethylarsine, and dimethylselenide. It is interesting to note that as the metallic covers (shells) on household and other batteries decompose with time and mercury is released under anaerobic conditions, the biological transformation of mercury will probably occur for years to come.

Transformations Involving Biodegradable (Nonpersistent) Organic Compounds. Based on both laboratory and field studies, it is also known that a number of the hazardous organic compounds found in municipal waste are biodegradable. Typically, biodegradable chemicals undergo reactions such as:

1. Simple substitution
2. Dehydrogenation (hydrolysis)
3. Oxidation
4. Reduction

Transformations Involving Persistent Organic Compounds. Again, based on both laboratory and field studies, it is also known that a number of the so-called persistent hazardous organic compounds found in municipal waste are biodegradable, but at extremely slow rates. The slowly biodegradable chemicals may undergo one or more of the following biologically mediated reactions:

1. Amide and ester hydrolysis
2. Dealkylation
3. Deamination
4. Dehalogenation
5. Double bond reduction
6. Hydroxylation
7. Oxidation (β-oxidation)
8. Reduction
9. Ring cleavage

Many of these reactions lead to the detoxification of the original compound. Unfortunately, many of these reactions also result in the formation of new toxic compounds, some of which may be more toxic than the original.

Combined Abiotic and Biotic Transformations. In addition to the abiotic and biotic transformations considered above, a number of hazardous waste compounds are transformed by a combination of abiotic/biotic reactions. The conversion of

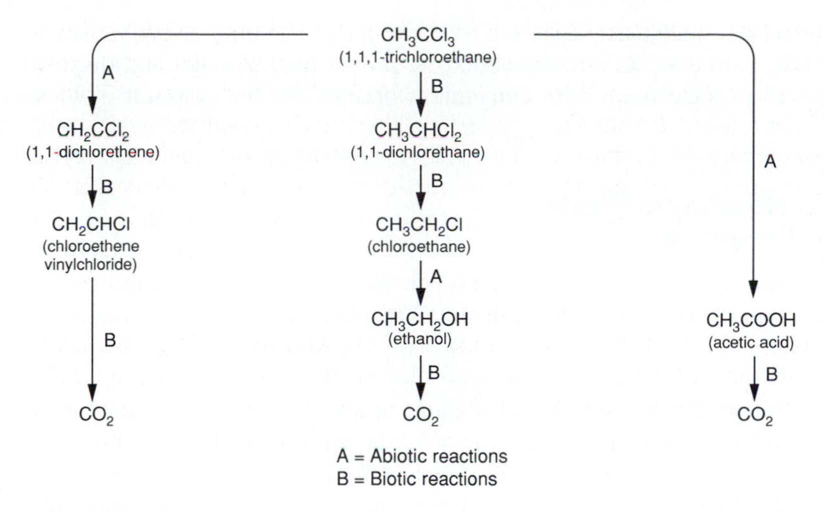

FIGURE 5-2
Pathways for the transformation of TCA (1,1,1-trichloroethane, CH_3CCl_3) under methanogenic conditions (from Refs. 3 and 13).

TCA (1,1,1-trichloroethane) to CO_2 and H_2O by a combination of abiotic and biotic reactions is illustrated in Fig. 5-2. There is some evidence that detoxification reactions, like those in Fig. 5-2, can be enhanced by successively creating an aerobic and anaerobic environment within a landfill. In the future, as these reactions become understood more clearly, it may be possible to accelerate the conversion of these compounds through more effective management of landfills.

5-5 MANAGEMENT OF HAZARDOUS WASTES IN MSW

The most effective way to eliminate the small quantities of hazardous wastes now found in municipal solid waste is to separate them at the point of generation. The number and types of hazardous components separated will depend on the hazardous waste storage, collection, treatment, and disposal facilities provided by the community.

Handling and Storage of Hazardous Wastes at Residential Dwellings

The handling and storage of household hazardous wastes (HHW) depend on the nature of the product. The principal categories of HHW were listed in Table 5-3. In reviewing the entries in Table 5-3, it is clear that HHW will be found (stored) in all parts of a residence. The disposal of HHW is, at present, unregulated in most states. As a consequence, many of the products within the various generic categories are often stored and, once used, disposed of improperly. The only effective way to deal with HHW is to educate citizens about the proper use, storage, and disposal of HHW and to provide them with convenient options for the disposal of these wastes.

Household Hazardous Waste Collection Programs

To minimize the improper disposal of HHW, product exchange programs, special collection days and permanent collection sites have been established by a number of communities.

Product Exchange Programs. Because paint products form a major portion of HHW, paint exchange programs are being used in a number of communities to reduce the cost of HHW disposal. The reuse of latex-based paints has proven to be the most successful, with up to 50 percent recovery being reported [2]. Unrecoverable paint must be either combusted in a hazardous waste combustor or disposed of in a hazardous waste landfill.

Specific Collection Days. One of the most common approaches to HHW management is to hold one or more community waste collection days. On collection days, community members are invited to bring their HHW, at little or no charge, to a specified location for recycling, treatment, or disposal by professional waste handlers. In larger communities, several locations are used on successive days. For these collection days to be successful, adequate promotion and education are critical. Even though hazardous waste collection days are well attended when properly promoted, at present (1992) it is estimated that less than 5 to 10 percent of the total available HHW is collected through such programs.

Permanent Collection Sites. To increase the convenience of the HHW collection programs and, therefore, increase participation, more and more communities are establishing permanent collection sites (e.g., fire stations, landfills, city and corporation yards). Programs involving permanent collection facilities allow citizens to drop off wastes at their own convenience. For this reason, permanent collection sites have proven to be more effective for collecting HHW than the one-day collection programs.

Elimination of Hazardous Wastes from Commercial Sources

To reduce the toxicity of commingled MSW, most communities have sought to eliminate all discharges of hazardous wastes from commercial facilities. The key to the elimination of hazardous wastes from commercial activities is the availability of community or regional facilities for handling and processing hazardous wastes.

5-6 DISCUSSION TOPICS AND PROBLEMS

5-1. List potential sources of hazardous waste generated in a university environment. Indicate what properties make such substances hazardous (i.e., flammability, irritant, etc.).

5-2. Identify two or more sources of household radioactive waste that are common in the United States.

5-3. The allowable limit for the disposal of benzene contained in a mixed sludge from a treatment process is 5 ppb. Sixty days after disposing of some sludge waste, a sample was collected and analyzed for benzene. The concentration of benzene found in the sludge was 1.37 ppb. If the first-order removal rate constant for benzene is 0.00345/hr, determine whether the sludge could have been disposed of in the landfill if a sample had been analyzed at the time the sludge was brought to the landfill.

5-4. Solve Problem 5-3 assuming that the second-order removal rate constant for benzene is 0.00029/hr.

5-5. Assuming that the half-life time ($t_{1/2}$) for hazardous substance *A* is 15 hr,

(*a*) determine $t_{1/8}$, $t_{1/4}$, $t_{3/8}$, $t_{5/8}$, $t_{3/4}$, $t_{7/8}$ and tabulate your results,

(*b*) plot $1 - (C/C_o)$ versus time (where C_o is initial concentration),

(*c*) explain what $1 - (C/C_0)$ represents. According to the first-order model, how long would it take to achieve 99% decay?

5-6. What are the principal factors that affect the rate of decay of a hazardous substance? Could the first-order model given by Eq. (5-1) be used to account for such factors?

5-7. Rank the compounds listed below in order of decreasing volatility based on the following properties: molecular weight, vapor pressure, Henry's law constant, and boiling point.

Compound
Benzene
Toluene
Chloroethene
Bromodichloromethane
Ethylbenzene

5-8. Assume an ideal gas mixture is in contact with water. Determine the equilibrium concentration of the contaminant in the liquid if the gas stream is composed of: (*a*) 200 ppm of benzene in air, (*b*) 100 ppm of trichloromethane in air, and (*c*) 700 mg/m^3 of chlorobenzene in air. Assume the air temperature is 20°C and the pressure is equal to 1 atm.

5-9. Referring to the literature on hazardous waste management, cite typical examples of (1) simple substitution, (2) dehydrogenation (hydrolysis), (3) oxidation, and (4) reduction reactions in the biological conversion of hazardous waste.

5-10. Referring to the literature on hazardous waste management, cite some typical reactions to illustrate six of the reactions given in Section 5-4 for the biological transformation of persistent organic compounds.

5-11. How are HHW now collected and disposed of in your community? What plans have been developed to reduce the toxicity of the commingled MSW from your community further?

5-12. What regulations and policies are currently in place in your community (or region) to reduce or eliminate the discharge of hazardous wastes with other solid wastes from commercial sources?

5-7 REFERENCES

1. *A Reference Guide of Alternatives to Toxic Household Products,* City of Sunnyvale, Sunnyvale, CA, 1988.
2. Anon.: "Managing Unwanted Paint," *Household Hazardous Waste Management News,* Vol. 2, No. 5, Spring 1990.
3. Bouwer, E. J. and P. L. McCarty: "Transformation of 1- and 2-Carbon Halogenated Aliphatic Organic Compounds under Methanogenic Conditions," *Applied Environmental Microbiology,* Vol. 45, No. 4, 1983.
4. Connell, D. W. and G. J. Miller: *Chemistry and Ecotoxicology of Pollution,* John Wiley & Sons, New York, 1984.
5. *Household Hazardous Waste Advisory Committee Recommendations to the California Waste Management Board,* Sacramento, CA, October 1987.
6. Howard, P. H. (ed.): *Handbook of Environmental Fate and Exposure Data for Organic Chemicals, Volume I Large Production and Priority Pollutants,* and *Volume II Solvents,* Lewis Publishers, Chelsea, MI, 1990.
7. Lang, R. J., T. A. Herrera, D. P. Y. Chang, G. Tchobanoglous, and R. G. Spicher: *Trace Organic Constituents in Landfill Gas,* prepared for the California Waste Management Board, Department of Civil Engineering, University of California–Davis, Davis, CA, November 1987.
8. Lyman, W. J., W. F. Reehl, and D. H. Rosenblatt: *Handbook of Chemical Property Estimation Methods: Environmental Behavior of Organic Compounds*, McGraw-Hill, New York, 1982.
9. Phifer, R. W. and W. R. McTigue, Jr.: *Handbook Of Hazardous Waste Management for Small Quantity Generators,* Lewis Publishers, Chelsea, MI, 1988.
10. Porteous, A.: "Hazardous Waste in the UK: An Overview," in A. Porteous (ed.) *Hazardous Waste Management Handbook,* Butterworths, London, 1985.
11. Tchobanoglous, G. and F. L. Burton: *Wastewater Engineering: Treatment, Disposal, and Reuse,* 3rd ed., McGraw-Hill, New York, 1991.
12. U.S. Environmental Protection Agency: "Control Techniques for Volatile Organic Emissions from Stationary Sources," EPA-450/2-78-022, Office of Air and Waste Management, Office of Air Quality Planning and Standards, Research Triangle Park, NC, 1978.
13. Vogel, T. M., C. S. Criddle, and P. L. McCarty: "Transformations of Halogenated Aliphatic Compounds," *Environmental Science and Technology,* Vol. 21, No. 8, 1987.
14. Wagner, T. P.: *Hazardous Waste Regulations,* 2nd ed., Van Nostrand Reinhold, New York, 1991.

PART III

ENGINEERING PRINCIPLES

The purpose of Part III is to present, discuss, and illustrate the engineering principles that must be applied in the development of integrated solid waste management systems. The chapters are organized in a logical sequence starting with solid waste generation and continuing through on-site handling, storage, and processing; collection; transfer and transport; separation, recycling, and processing; and disposal.

Although much is known about the engineering aspects of solid waste management, the field is very dynamic, especially in developing areas such as materials and energy recovery. New technologies and equipment are also being developed constantly in other areas of the field. By devoting a separate chapter to each of the functional elements that make up solid waste management systems, it is possible to identify the fundamental aspects of each one and to delineate the interrelationships involved (to the extent that they are known). Mastery of the engineering principles presented in Part III is fundamental to the understanding and assessment of existing operations and systems, to the evaluation of the impacts of new and proposed technologies, and to the proper selection and analysis of alternatives in the development of new systems. The ability to measure the impact of alternative courses of action is vital in the management of these systems.

CHAPTER 6

SOLID WASTE GENERATION AND COLLECTION RATES

Knowledge of the quantities of solid wastes generated, separated for recycling, and collected for further processing or disposal is of fundamental importance to all aspects of solid waste management. As a means to understand the material presented in subsequent chapters, the following topics are considered in this chapter:

1. Importance of waste quantities
2. Measures and methods used to quantify solid waste quantities
3. Waste generation rates
4. Factors that affect waste generation and collection rates
5. Types and quantities of materials recovered from MSW
6. Quantities of household hazardous wastes
7. Waste characterization and diversion studies

6-1 IMPORTANCE OF WASTE QUANTITIES

The quantities of solid waste generated and collected are of critical importance in determining compliance with federal and state waste diversion programs; in selecting specific equipment; and in designing of waste collection routes, materials recovery facilities (MRFs), and disposal facilities.

Compliance with Federal and State Diversion Programs

Information on the total quantity of MSW as well as the quantity of waste that is now recycled or otherwise does not become part of the waste stream will be required to establish and assess the performance of mandated recycling programs. For example, if a 25 percent level of recycling is mandated, the following question must be answered: Is the 25 percent based on the actual quantity generated, or is it based on the amount currently collected? If a high percentage of the waste now generated is already recycled, then a 25 percent reduction in the amount collected may be difficult to achieve. The impacts of recycling on the quantity of waste collected are considered further in Section 6-3.

Design of Solid Waste Management Facilities

As the diversion and recycling of waste materials increase, the quantities of waste generated, separated for recycling, collected, and ultimately requiring disposal in landfills become determinants in planning and designing solid waste management facilities. For example, the design of special vehicles for the curbside collection of source-separated wastes depends on the quantities of the individual waste components to be collected. The sizing of MRFs depends on the amount of waste to be collected as well as the variations in the quantities delivered hourly, daily, weekly, and monthly. Similarly, the sizing of landfills depends on the amount of residual waste that must be disposed of after all the recyclable materials have been removed.

6-2 MEASURES AND METHODS USED TO ASSESS SOLID WASTE QUANTITIES

The purpose of this section is to introduce the reader to the measures and methods used to quantify solid waste quantities, the materials balance approach for estimating solid waste quantities, and the statistical techniques used to analyze waste generation rates.

Measures Used to Quantify Solid Waste Quantities

The principal reason for measuring the quantities of solid waste generated, separated for recycling, and collected for further processing or disposal is to obtain data that can be used to develop and implement effective solid waste management programs. Therefore, in any solid waste management study, extreme care must be exercised in deciding what actually needs to be known and in allocating funds for data collection. The measures and units used to quantify solid waste quantities are discussed below.

Volume and Weight Measurements. Both volume and weight are used for the measurement of solid waste quantities. Unfortunately, the use of volume as a measure of quantity can be misleading. For example, a cubic yard of loose wastes is a different quantity from a cubic yard of wastes that has been compacted in a collection vehicle, and each of these is different from a cubic yard of wastes that has been compacted further in a landfill. Accordingly, if volume measurements are to be used, the measured volumes must be related to either the degree of compaction of the wastes or the specific weight of the waste under the conditions of storage.

To avoid confusion, solid waste quantities should be expressed in terms of weight. Weight is the only accurate basis for records because tonnages can be measured directly, regardless of the degree of compaction. Weight records are also necessary in the transport of solid wastes because the quantity that can be hauled usually is restricted by highway weight limits rather than by volume. On the other hand, volume and weight are equally important with respect to the capacity of landfills.

Expressions for Unit Waste Generation Rates. In addition to knowing the sources and composition of the solid wastes that must be managed, it is equally important to be able to develop meaningful ways to express the quantities generated. Suggested units of expression for different generation sources are considered in Table 6-1. Note, however, that unit generation data for commercial and

TABLE 6-1
Suggested units of expression for solid waste quantities

Type of waste	Discussion
Residential	Because of the relative stability of residential wastes in a given location, the most common unit of expression used for their generation rates is lb/capita · d. However, should the waste composition vary significantly from typical municipal wastes (Table 3-4), the use of lb/capita · d may be misleading, especially when quantities are being compared.
Commercial	In the past, commercial waste generation rates have also been expressed in lb/capita · d. Although this practice has been continued as an expedient, it adds little useful information about the nature of solid waste generation at commercial sources. A more meaningful approach would be to relate the quantities generated to the number of customers, the dollar value of sales, or some similar unit. Use of such factors would allow comparisons to be made throughout the country.
Industrial	Ideally, wastes generated from industrial activities should be expressed on the basis of some repeatable measure of production, such as pounds per automobile for an automobile assembly plant or pounds per case for a packing plant. When and if such data are developed, it will be possible to make meaningful comparisons between similar industrial activities throughout the country.
Agricultural	Where adequate records have been kept, solid wastes from agricultural activities are now most often expressed in terms of some repeatable measure of production, such as lb of manure/l400-lb cow · d and lb of waste/ton of raw product. At present, data are available on the amounts of solid waste generated from a number of agricultural activities associated with field and row crops.

industrial activities are somewhat limited. Consequently, it has been expedient in many cases to use the same units for these activities as those used for residential wastes, as opposed to the more rational units cited in Table 6-1. The most comprehensive waste records now available are those kept at MRFs, transfer stations, and landfills.

Methods Used to Estimate Waste Quantities

Waste quantities are usually estimated on the basis of data gathered by conducting a waste characterization study, using previous waste generation data, or some combination of the two approaches. Methods commonly used to assess solid waste quantities are a (1) load-count analysis, (2) weight-volume analysis, and (3) materials-balance analysis. In this discussion, it will be helpful to remember that most measurements of waste quantities do not accurately represent what they are reported or assumed to represent. For example, in predicting residential waste generation rates, the measured rate seldom reflects the true rate because there

(*a*)

(*b*)

FIGURE 6-1
Solid waste hauled by individuals (*a*) to a convenience transfer station and (*b*) to a recycling center. In the load-count method of analysis used to determine waste quantities, the type, estimated volume, and weight of waste brought in with each vehicle is recorded. If scales are not available, average specific weight values are used to estimate the weight.

are confounding factors (e.g., onsite storage and the use of alternative disposal locations) that make the true rate difficult to assess. Most solid waste generation rates reported in the literature prior to about 1990 are actually based on measurement of the amount of waste collected, not the actual amount generated.

Load-Count Analysis. In this method, the number of individual loads and the corresponding waste characteristics (types of waste, estimated volume) are noted over a specified time period (see Fig. 6-1). If scales are available, weight data are also recorded. Unit generation rates are determined by using the field data and, where necessary, published data. The application of this method is illustrated in Example 6-1.

Example 6-1 Estimation of unit solid waste generation rates for a residential area. From the following data estimate the unit waste generation rate per week for a residential area consisting of 1200 homes. The observation location is a local transfer station that receives all of the wastes collected for disposal. The observation period was one week.

1. Number of compactor truck loads = 9
2. Average size of compactor truck = 20 yd^3
3. Number of flatbed loads = 7
4. Average flatbed volume = 2 yd^3
5. Number of loads from individual residents' private cars and trucks = 20
6. Estimated volume per domestic vehicle = 8 ft^3

Solution

1. Set up the computation table to estimate the total weight. Use the specific weight data given in Table 4-1 to convert the measured waste volumes to weight.

Item	Number of loads	Average volume, yd^3	Specific weight,[a] lb/yd^3	Total weight, lb
Compactor truck	9	20	500	90,000
Flatbed truck	7	2	225	3,150
Individual private vehicle	20	0.30	150	900
Total, lb/wk				94,050

[a] Based on limited on-site weight and volume measurements.

2. Determine the unit waste collection rate based on the assumption that each household is comprised of 3.5 people.

$$\text{Unit rate} = \frac{94{,}050 \text{ lb/wk}}{(1200 \times 3.5)(7 \text{ d/wk})}$$

$$= 3.2 \text{ lb/capita} \cdot \text{d} = 1.45 \text{ kg/capita} \cdot \text{d}$$

Comment. The difficulty in using such data is knowing whether they are truly representative of what needs to be measured. For example, how many loads were hauled elsewhere? How much waste material was separated for recycling? How much material was stored on the homeowner's premises? How much of the food waste was ground up and discharged to the sewer? All such questions tend to confound the observed data in a statistical sense.

Weight-Volume Analysis. Although the use of detailed weight-volume data obtained by weighing and measuring each load (see Fig. 6-2) will certainly provide better information on the specific weight of the various forms of solid wastes at a given location, the question remains: What information is needed in terms of study objectives?

Materials Mass Balance Analysis

The only way to determine the generation and movement of solid wastes with any degree of reliability is to perform a detailed materials balance analysis for each generation source, such as an individual home or a commercial or industrial activity. In some cases, the materials balance method of analysis will be required to obtain the data needed to verify compliance with state-mandated recycling programs.

Preparation of Materials Mass Balances. The approach to be followed in the preparation of a materials mass balance analysis is as follows. First, draw a system boundary around the unit to be studied (see Fig. 6-3). The proper selection of the system boundary is important because, in many situations, it will be possible to simplify the mass balance computations. Second, identify all the activities that cross or occur within the boundary and affect the generation of wastes. Third, identify the rate of waste generation associated with each of these activities. Fourth, using appropriate mathematical relationships, determine the

FIGURE 6-2
Weighing of collection vehicle at entrance to transfer station using platform scales. In the weight/volume method of analysis used to determine waste quantities, the volume of each truck is estimated.

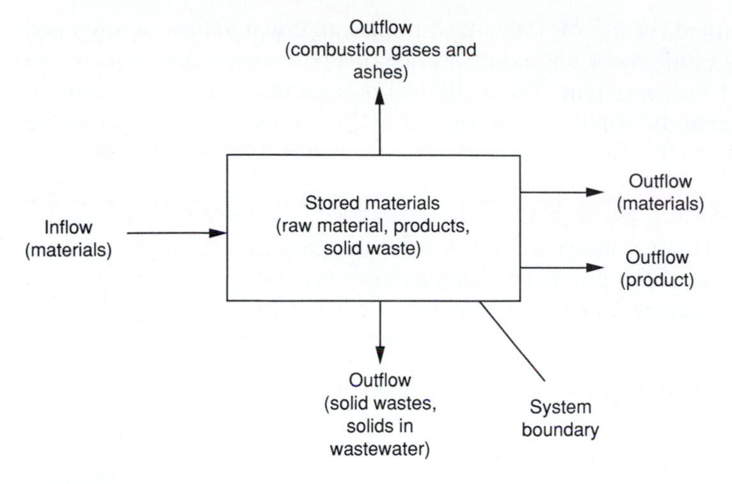

FIGURE 6-3
Definition sketch for materials balance analysis used to determine solid waste generation rates.

quantity of wastes generated, collected, and stored. The materials mass balance can be formulated as follows:

1. General word statement:

Rate of accumulation of material within the system boundary	=	rate of flow of material into the system boundary	−	rate of flow of material out of the system boundary	+	rate of generation of waste material within the system boundary	(6-1)

2. Simplified word statement:

$$\text{Accumulation} = \text{inflow} - \text{outflow} + \text{generation} \tag{6-2}$$

3. Symbolic representation (refer to Fig. 6-3):

$$\frac{dM}{dt} = \Sigma M_{\text{in}} - \Sigma M_{\text{out}} + r_w \tag{6-3}$$

where dM/dt = rate of change of the weight of material stored (accumulated) within the study unit, lb/d
ΣM_{in} = sum of all of the material flowing into study unit, lb/d
ΣM_{out} = sum of all of the material flowing out of study unit, lb/d
r_w = rate of waste generation, lb/d
t = time, d

In some biological transformation processes (e.g., composting) the weight of organic matter will be reduced and, therefore, the term r_w will be negative. In writing the mass balance equation the rate should always be written as a positive term. The correct sign for the term will be added when the appropriate rate expression is substituted for r_w. Before substituting numerical values in any mass balance expression, a unit check should always be made to ensure that units of the individual quantities are consistent. The analytical procedures used for the solution of mass balance equations usually are governed by the mathematical form of the final expression [6].

Application of Materials Mass Balance. In practice, the most difficult aspect of applying a mass balance analysis for the determination of waste quantities is defining adequately all of the inputs and outputs crossing the system boundary. A simplified materials-balance analysis is illustrated in Example 6-2.

Example 6-2 Materials-balance analysis. A cannery receives on a given day 12 tons of raw produce, 5 tons of cans, 0.5 tons of cartons, and 0.3 tons of miscellaneous materials. Of the 12 tons of raw produce, 10 tons become processed product, 1.2 tons end up as produce waste, which is fed to cattle, and the remainder is discharged with the wastewater from the plant. Four tons of the cans are stored internally for future use, and the remainder is used to package the product. About 3 percent of the cans used are damaged. Stored separately, the damaged cans are recycled. The cartons are used for packaging the canned product, except for 3 percent that are damaged and subsequently separated for recycling. Of the miscellaneous materials, 25 percent is stored internally for future use; 50 percent becomes waste paper, of which 35 percent is separated for recycling with the remainder being discharged as mixed waste; and 25 percent becomes a mixture of solid waste materials. Assume the materials separated for recycling and disposal are collected daily. Prepare a materials balance for the cannery on this day and a materials flow diagram accounting for all of the materials. Also determine the amount of waste per ton of product.

Solution

1. On the given day, the cannery receives

    12 tons of raw produce
    5 tons of cans
    0.5 tons of cartons
    0.3 tons of miscellaneous materials

2. As a result of internal activity,
    (*a*) 10 tons of product is produced, 1.2 tons of produce waste is generated, and the remainder of the produce is discharged with the wastewater
    (*b*) 4 tons of cans are stored and the remainder is used, of which 3 percent are damaged
    (*c*) 0.5 tons of cartons are used of which 3 percent are damaged
    (*d*) 25 percent of the miscellaneous materials is stored; 50 percent becomes paper waste, of which 35 percent is separated and recycled, with the remainder disposed of as mixed solid waste; the remaining 25 percent of the miscellaneous materials are disposed of as mixed waste.

3. Determine the required quantities
    (*a*) Wastes generated from raw produce
        i. Solid waste fed to cattle = 1.2 ton (1089 kg)
        ii. Waste produce discharged with wastewater = (12 − 10 − 1.2) ton = 0.8 ton (726 g)

(*b*) Cans
  i. Damaged and recycled = (0.03)(5 − 4) ton = 0.03 ton (27 kg)
  ii. Used for production of product = (1 − 0.03) ton = 0.97 ton (880 kg)

(*c*) Cartons
  i. Damaged and recycled = (0.03)(0.5 ton) = 0.015 ton (14 kg)
  ii. Cartons used in product = (0.5 − 0.015) ton = 0.485 ton (440 kg)

(*d*) Miscellaneous material
  i. Amount stored = (0.25)(0.3 ton) = 0.075 ton (68 kg)
  ii. Paper separated and recycled = (0.50)(0.35)(0.3 ton) = 0.053 ton (48 kg)
  iii. Mixed waste = (0.3 − 0.075) − 0.053) ton = 0.172 ton (156 kg)

(*e*) Total weight of product = (10 + 0.97 + 0.485) ton = 11.455 ton (10,392 kg)

(*f*) Total material stored = (4 + 0.075) ton = 4.075 ton (3696 kg)

4. Prepare a materials balance and flow diagram for the cannery for the day in question
   (*a*) The appropriate materials balance equation is

$$\text{Amount of material stored} = \text{inflow} - \text{outflow} - \text{waste generation}$$

   (*b*) The materials balance quantities are as follows:
     i. Material stored = (4.0 + 0.075) ton = 4.075 ton
     ii. Material input = (12.0 + 5.0 + 0.5 + 0.3) ton = 17.8 ton
     iii. Material output = (10.0 + 0.97 + 0.485 + 1.2 + 0.03 + 0.015 + 0.053) ton = 12.753 ton
     iv. Waste generation = (0.8 + 0.172) ton = 0.972 ton
     v. The final materials balance is

$$4.075 = 17.8 - 12.753 - 0.972 \text{ (mass balance checks)}$$

   (*c*) Materials balance flow diagram is given below

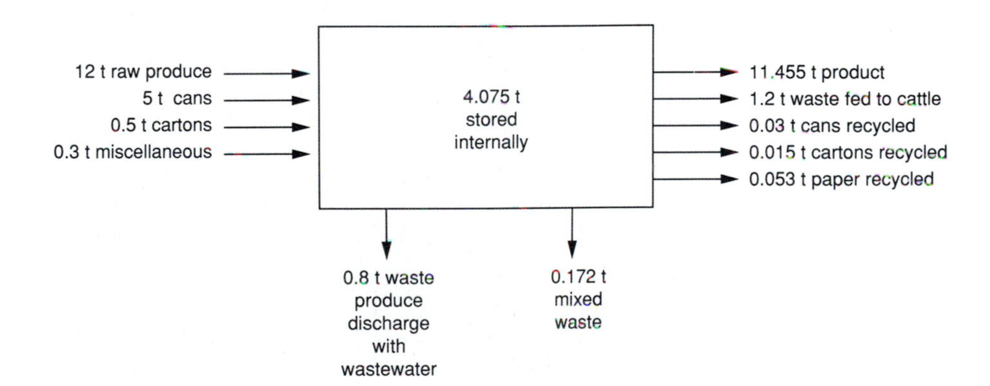

5. Determine the amount of waste per ton of product:
   (*a*) Recyclable material = (1.2 + 0.03 + 0.015 + 0.053) ton/11.455 ton = 0.11 ton/ton
   (*b*) Mixed waste = (0.8 + 0.172) ton/11.455 ton = 0.085 ton/ton

Comment. This simple example illustrates the computational approach involved in the preparation of a materials-balance analysis used to quantify solid waste generation rates. If the internal processing activities are more complex, the amount of work involved in arriving at a materials balance is significant. Because materials balance computations involve a considerable amount of bookkeeping, they are best done using a spreadsheet.

Statistical Analysis of Measured Waste Quantities

In developing solid waste management systems, it is often necessary to determine the statistical characteristics of the observed solid waste generation rates. For example, for many large industrial activities it would be impractical to provide container capacity to handle the largest conceivable quantity of solid wastes to be generated in a given day. The container capacity to be provided must be based on a statistical analysis of the generation rates and the characteristics of the collection system.

The first step in assessing the statistical characteristics of a series of observations is to determine whether the observations are distributed normally or are skewed (log normal). The nature of the distribution can be determined most readily by plotting the data on arithmetic and logarithmic probability graph paper (see Appendix D). Once the nature of the distribution is known, statistical measures that are used to describe the distribution include: the mean, median, mode, standard deviation, coefficient of variation, coefficient of skewness, and coefficient of kurtosis. The definitions of these measures are given in Appendix D. The determination of statistical measures for waste production data is illustrated in Example 6-3.

Example 6-3 Statistical analysis of solid waste collection data. Determine the statistical characteristics of the weekly waste production data obtained from an industrial account for a calendar quarter of operation.

Week no.	Waste, yd^3/wk	Week no.	Waste, yd^3/wk
1	29	8	37
2	30	9	38
3	35	10	35
4	34	11	33
5	38	12	32
6	41	13	31
7	40		

Solution

1. Determine graphically whether the waste production data are distributed normally or are skewed (log normal) using probability paper.

(*a*) Set up a data analysis table with three columns as described below.
  i. In column 1, enter the rank serial number starting with number 1
  ii. In column 2, arrange the waste production data in ascending order
  iii. In column 3, enter the probability plotting position (see Appendix D)

Rank serial no., *m*	Waste, yd^3/wk	Plotting position,[a] %
1	29	7.1
2	30	14.3
3	31	21.4
4	32	28.6
5	33	35.7
6	34	42.9
7	35	50.0
8	35	57.1
9	37	64.3
10	38	71.4
11	38	78.6
12	40	85.7
13	41	92.9

[a] Plotting position $= [m/(n+1)]100, n = 13$

(*b*) Plot the weekly quantity of waste, expressed in yd^3/wk, versus the plotting position (determined above) on both arithmetic and logarithmic probability graph paper. The resulting plots are presented below. Because the data fall on a straight line in both plots, the waste production data can be described adequately by either type of distribution. The fact that the data can be described adequately with both distributions is often the case with waste production data.

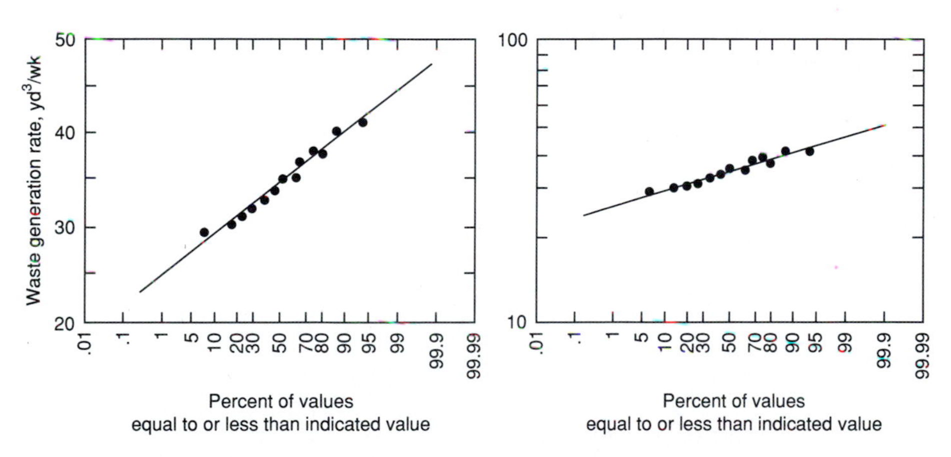

2. Determine the statistical characteristics of the waste collection data.
   (*a*) Set up a data analysis table to obtain the quantities needed to determine the statistical characteristics (refer to Appendix D for equations and definitions).

Waste, yd^3/wk	$(x_i - \bar{x})$	$(x_i - \bar{x})^2$	$(x_i - \bar{x})^4$
29	−5.8	33.6	1131.6
30	−4.8	23.0	530.8
31	−3.8	14.4	208.5
32	−2.8	7.8	61.5
33	−1.8	3.2	10.2
34	−0.8	0.6	0.4
35	0.2	0.0	0.0
35	0.2	0.0	0.0
37	2.2	4.8	23.4
38	3.2	10.2	104.9
38	3.2	10.2	104.9
40	5.2	27.0	731.2
41	6.2	38.4	1477.6
453		173.2	4385.0
$\bar{x} = 453/13 = 34.8$			

(*b*) Determine the statistical characteristics

i. Mean

$$\bar{x} = \frac{\Sigma x}{n}$$

$$\bar{x} = \frac{453}{13} = 34.8 \text{ yd}^3/\text{wk}$$

ii. Median (the middle value)

Median = 35 yd^3/wk (see data table above)

iii. Mode

Mode = 3 Med − $2\bar{x}$ = 3(35) − 2(34.8) = 35.4

iv. Standard deviation

$$s = \sqrt{\frac{\Sigma(x - \bar{x})^2}{n}}$$

$$s = \sqrt{\frac{173.2}{13}} = 3.65$$

v. Coefficient of variation

$$\text{CV} = \frac{100s}{\bar{x}}$$

$$\text{CV} = \frac{100(3.65)}{34.8} = 10.5$$

vi. Coefficient of skewness

$$\alpha_3 = \frac{2(\bar{x} - \text{Mod})}{s}$$

$$\alpha_3 = \frac{2(34.8 - 35.4)}{3.65} = -0.33$$

vii. Coefficient of kurtosis

$$\alpha_4 = \frac{\Sigma(x_i - \bar{x})^4/n}{s^4}$$

$$\alpha_4 = \frac{4385.0/13}{(3.65)^4} = 1.9$$

Comment. The term $(n + 1)$ is used to obtain the plotting positions in Step 1, as opposed to just n, to account for the fact that there may be an observation that is either larger or smaller than the largest or smallest in the data set. Reviewing the statistical characteristics it can be seen that the distribution is skewed ($\alpha_3 = -0.33$ versus 0 for a normal distribution) and is considerably flatter than a normal distribution would be ($\alpha_4 = 1.9$ versus 3.0 for a normal distribution).

6-3 SOLID WASTE GENERATION AND COLLECTION RATES

Data and information on the quantities of solid waste generated and collected are discussed in this section. Factors that affect the quantities generated are considered in Section 6-4.

Total Waste Generation in the United States

Based on an analysis of a number of data sources [1–4, 5, 8, 10] and on personal computations, the authors' estimate of the amount of municipal, industrial, and agricultural solid waste generated per capita in the United States for the year 1990 is reported in Table 6-2. A little over a ton of MSW is generated per capita per year in the United States. In using data on waste generation reported in the literature before about 1990, great care must be exercised because in most cases the data reported do not reflect the amount of waste generated, but are actually the amount of waste collected. As such, the quantities reported often do not

TABLE 6-2
Estimated total per capita solid waste quantities generated in the United States and selected states for the year 1990[a]

	Waste generation rate, lb/capita · yr					
	United States		California		Florida	
Waste	Range	Typical	Range	Typical	Range	Typical
MSW[b]	1450–3000	2225	1850–3500	2500	1350–2400	2200
Industrial waste	500–1750	750	750–1500	1000	250–750	500
Agricultural waste	250–3000	—[c]	1500–4000	3000	500–1500	1000
Total	2500–7750		3700–9000	6500	2050–4650	3500

[a] Developed in part from Refs. 1–4, 5, 8, 10.

[b] A detailed analysis of the waste categories that compose MSW are presented in Table 6-3.

[c] Must be estimated separately for each location.

Note: lb/capita · yr × 0.4536 = kg/capita · yr

reflect the amount of waste material that was: (1) recycled (directly and indirectly), (2) ground up in kitchen food waste grinders, (3) burned in fireplaces, (4) composted, and (5) stored temporarily.

Unit Solid Waste Generation Rates

Often it is necessary to estimate the quantities of solid waste that will be generated, by waste category, within a community. Estimates of MSW quantities are usually based on the amount of waste generated per person per day. For industrial and agricultural wastes, the amounts of waste generated are, as noted in Table 6-1, based on some unit of production. The general values for industrial and agricultural waste reported in Table 6-2 must be verified locally because of the significant variations that exist in these categories.

Municipal Solid Wastes. The distribution of wastes that constitute the MSW of a community are reported in Table 6-3. If actual data are not available, the unit

TABLE 6-3
Estimated quantities for the waste categories comprising MSW generated per capita in the United States for the year 1990[a]

	Distribution of MSW, percent of total		Solid waste generation rate			
			lb/capita · yr		lb/capita · d	
Waste category[b]	Range	Typical	Range	Typical	Range	Typical
Residential and commercial, excluding special and hazardous wastes	50–75	62.0	1125–1700	1395.0	3.1–4.7	3.82
Special (bulky items, consumer electronics, white goods, yard wastes collected separately, batteries, oil, and tires)	3–12	5.0	65–180	112.5	0.2–0.5	0.31
Hazardous	0.01–1.0	0.1	0.15–30	2.3	0.0004–0.082	0.0063
Institutional	3–5	3.4	65–110	76.5	0.2–0.3	0.21
Construction and demolition	8–20	14.0	180–450	315.0	0.5–1.2	0.86
Municipal services						
Street and alley cleanings	2–6	3.8	45–135	85.5	0.1–0.4	0.23
Tree and landscaping	2–5	3.0	45–110	67.5	0.1–0.3	0.19
Parks and recreational areas	1.5–3	2.0	30–65	45.0	0.08–0.2	0.12
Catch basin	0.5–1.2	0.7	10–30	15.7	0.03–0.08	0.04
Treatment plant sludges	3–8	6.0	68–180	135.0	0.2–0.5	0.37
Total		100.0		2250.0		6.16

[a] Data derived from Tables 3-3 and 6-2.

[b] See Table 6-2 for industrial and agricultural waste quantities.

Note: lb/capita · yr × 0.4536 = kg/capita · yr.

waste generation rates for MSW given in Table 6-3 can be used for estimating purposes.

Residential and commercial. As reported in Table 6-3, residential and commercial wastes, excluding special and hazardous wastes, comprise about 50 to 75 percent of the total MSW of a community. The individual components that comprise the residential and commercial portion of MSW were presented and discussed in Chapter 3 (see Table 3-7). The residential and commercial MSW generation data presented in Table 6-3 should be adjusted to reflect local conditions. For example, if kitchen food waste grinders are used extensively, the total quantity of MSW should be reduced by an appropriate amount. Information and data to use for estimating the distribution of special wastes are given in Table 6-4.

Institutional. As noted in Chapter 3, institutional sources of solid waste include schools, prisons, and hospitals. Excluding manufacturing wastes from prisons and medical wastes from hospitals, the distribution of waste components in the solid wastes generated at these facilities is quite similar to commingled residential and commercial MSW.

Construction and demolition. The quantities of construction and demolition waste are difficult to estimate and variable in composition, but typically

TABLE 6-4
Information and data that can be used to estimate the generation of residential and commercial special wastes[a]

Special waste	Information for estimating quantities
Bulky items, consumer electronics, and white goods	Best approach is to determine the number of homes, estimate the number of items per home, and use an average useful life for each item. Items recycled through charitable organizations must also be accounted for.
Household batteries	2.5 billion household batteries are purchased annually in the United States. A value of 10 household batteries/capita · yr can be used for estimating purposes.
Automotive (lead-acid) batteries	70 to 80 million automotive batteries are consumed and replaced annually in the United States. A value of 0.4 automotive batteries/capita · yr can be used for estimating purposes.
Used oil	200 million gallons of waste oil are generated annually by do-it-yourself oil changers. A value of 0.80 gal waste oil/capita · yr can be used for estimating purposes.
Automotive tires (passenger vehicles and light trucks)	190 million tires are discarded annually in the United States. A value of 0.80 tire/capita · yr can be used for estimating purposes. An alternative approach to estimating the number of waste tires generated per year is to estimate the number of cars and to use an average value for miles traveled per year (12,000 mi) and an average tire life (e.g., 35,000 mi/tire).

[a] Information abstracted from Chapter 15 of this text.

the distribution is about 40 to 50 percent rubbish (concrete, asphalt, bricks, blocks, and dirt), 20 to 30 percent wood and related products (pallets, stumps, branches, forming and framing lumber, treated lumber, and shingles), and 20 to 30 percent miscellaneous wastes (painted or contaminated lumber, metals, tar-based products, plaster, glass, white goods, asbestos and other insulation materials, and plumbing, heating and electrical parts).

Selected Industrial and Agricultural Wastes. Unit waste generation rates for selected industrial and agricultural activities are reported in Table 6-5. The rates given in this Table are related to some unit of production. By relating the unit rates of waste generation to some unit of production, comparisons can be made between facilities producing the same product in different parts of the country.

Solid Waste Collection Rates

Wastes that are collected include commingled wastes (in communities without recycling programs) and commingled wastes and source-separated wastes (in communities with recycling programs). The difference between the amount of residential and commercial MSW generated and the amount of waste collected for processing and/or disposal will typically vary from 4 to 15 percent. The differences can be accounted for by the amount of material (1) composted, (2) burned in fireplaces, (3) discharged to sewers, (4) given to charitable agencies, (5) sold at garage sales, (6) delivered to drop-off and recycling centers, and (7) recycled directly. In general, the percentage difference between the amount generated

TABLE 6-5
Unit solid waste generation rates for selected industrial and agricultural sources

Source	Unit	Range
Industrial		
Canned and frozen foods	ton/ton of raw product	0.04–0.06
Printing and publishing	ton/ton of raw paper	0.08–0.10
Automotive	ton/vehicle produced	0.7–0.9
Petroleum refining	ton/employee · d	0.04–0.05
Rubber	ton/ton of raw rubber	0.01–0.3
Agricultural		
Manures		
Chickens (fryers)	ton/1000 birds · yr	45–50
Hens (layers)	tons/1000 birds · yr	45–50
Cattle	lb/head · d	85–120
Fruit and nut crops	tons/acre · yr	1.3–2.5
Field and raw crops	tons/acre · yr	1.5–4.5

Note: tons × 907.2 = kg
lb × 0.4536 = kg

and collected will be smaller (4 to 6 percent) for apartments than for individual residences with adequate space for backyard composting (8 to 15 percent).

Variation in Generation and Collection Rates

The quantities of solid waste generated and collected vary daily, weekly, monthly, and seasonally. Information on the variations to be expected in the peak and minimum waste generation rates for solid wastes generated from individual residences, small commercial establishments, and small and large communities is presented in Table 6-6. This information can be used as a guide in the selection of equipment and in the sizing of solid waste management units. As shown in Table 6-6, the largest variation in waste generation rates occurs with individual residences and small commercial establishments. Residential waste generation rates usually peak during Christmas holiday season and during spring housecleaning days. In many communities, unlimited collection service is provided on designated clean-up days. In general, as the size of the waste source increases (e.g., from individual residences to a community) the variation in the peak day, week, and month decreases. Typical monthly weight data from a transfer station are shown in Fig. 6-4. Because of the random waste generation pattern illustrated in Fig.6-4, total weight data collected on a given month can be significantly in error. For example, if the weight for the month of February were used for estimating waste quantities, the actual average quantity would have been underestimated by 40 percent.

TABLE 6-6
Peak and minimum waste generation factors for solid wastes generated from individual residences, small commercial establishments, and small and large communities[a]

	Individual residence		Small commercial establishment		Small community		Large community	
Factor[b]	**Range**	**Typical**	**Range**	**Typical**	**Range**	**Typical**	**Range**	**Typical**
Peak waste generation factors								
Peak day	2.0–4.0	3.0	1.75–3.5	2.5	1.5–2.5	2.0	1.5–2.25	1.9
Peak weeks	1.5–3.5	2.5	1.5–2.5	2.25	1.25–2.0	1.75	1.25–2.0	1.5
Peak month	1.25–2.5	2.0	1.25–2.0	1.75	1.25–1.75	1.5	1.15–1.75	1.25
Minimum waste generation factors								
Minimum day	0.15–0.5	0.20	0.25–0.5	0.4	0.35–0.6	0.5	0.5–0.7	0.6
Minimum week	0.25–0.6	0.5	0.4–0.6	0.5	0.5–0.7	0.6	0.6–0.8	0.7
Minimum month	0.5–0.7	0.6	0.5–0.7	0.65	0.6–0.8	0.7	0.7–0.9	0.8

[a] The reported factors are exclusive of extreme waste generation events (i.e., values greater than the 99- or less than the 1-percentile value).

[b] Ratio of peak or minimum values to average value.

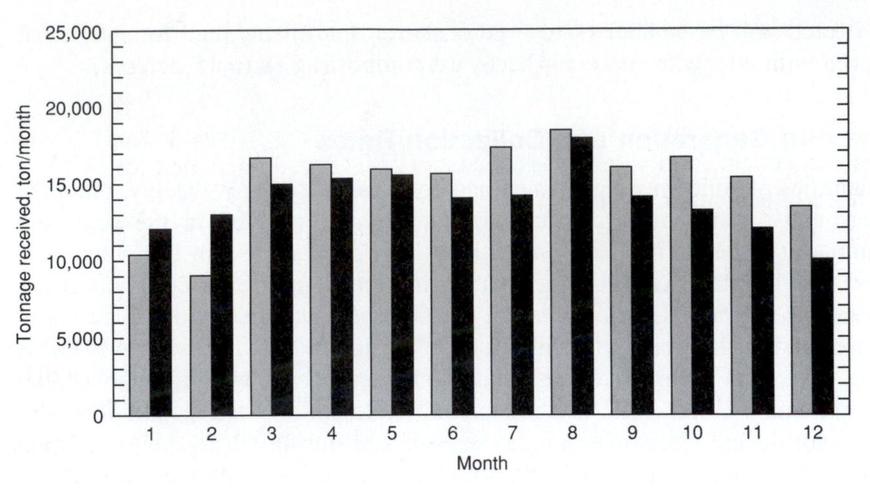

FIGURE 6-4
Variation in the monthly quantity of solid waste received at a transfer station over a two-year period.

6-4 FACTORS THAT AFFECT WASTE GENERATION RATES

The affect of (1) source reduction and recycling activities, (2) public attitudes and legislation, and (3) geographic and physical factors on the generation of solid waste are considered in the following discussion.

Effect of Source Reduction and Recycling Activities on Waste Generation

The effects of source reduction and the extent of recycling activities on waste generation are considered in the following discussion.

Source Reduction. Waste reduction may occur through the design, manufacture, and packaging of products with minimum toxic content, minimum volume of material, and/or a longer useful life. Waste reduction may also occur at the household, commercial or industrial facility through selective buying patterns and the reuse of products and materials. Because source reduction is not a major element in waste reduction at the present time, it is difficult to estimate the actual impact that source reduction programs have had (or will have) on the total quantity of waste generated. Nevertheless, source reduction will likely become an important factor in reducing the quantity of waste generated in the future. For example, if the postage rate for bulk mail were increased significantly, the quantity of bulk mail would be reduced sharply. Some of the other ways in which source reduction can be achieved follow:

- Decrease unnecessary or excessive packaging (see Fig. 6-5)
- Develop and use products with greater durability and repairability (e.g., more durable appliances and tires)
- Substitute reusable products for disposable, single-use products (e.g., reusable plates and cutlery, refillable beverage containers, cloth diapers and towels)
- Use fewer resources (e.g., two-sided copying)
- Increase the recycled materials content of products
- Develop rate structures that encourage generators to produce less waste [8]

Extent of Recycling. The existence of recycling programs within a community definitely affects the quantities of wastes collected for further processing or disposal. Whether such operations affect the quantities of waste generated is another question. Until more information is available, no definite statement can be made on this issue.

Effect of Public Attitudes and Legislation on Waste Generation

Along with source reduction and recycling programs, public attitudes and legislation also significantly affect the quantities generated.

Public Attitudes. Ultimately, significant reductions in the quantities of solid wastes generated occur when and if people are willing to change—of their own volition—their habits and lifestyles to conserve natural resources and to reduce the economic burdens associated with the management of solid wastes. A program of continuing education is essential in bringing about a change in public attitudes.

States with Beverage Container Deposit Laws. A number of states now have beverage container deposit laws. The first was enacted in Oregon in 1972. In the

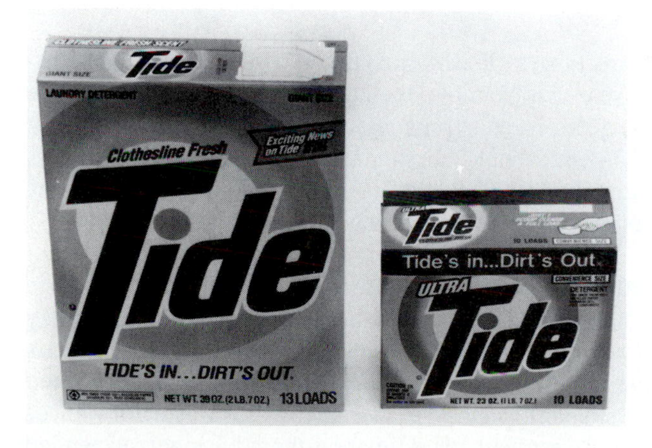

FIGURE 6-5
Reducing the size of the packaging required by reformulating product strength. Original strength on left and concentrated product strength on right.

states with container deposit laws, the return rates for bottles and cans vary from 93 to 96 percent and 90 to 96 percent, respectively [10]. It is anticipated that more states will enact container deposit laws covering a wide variety of container types not now included.

Legislation. Perhaps the most important factor affecting the generation of certain types of wastes is the existence of local, state, and federal regulations concerning the use of specific materials. Legislation dealing with packaging and beverage container materials is an example. Encouraging the purchase and use of recycled materials by allowing a price differential (typically 5 to 10 percent) for recycled materials is another method.

Effect of Geographic and Physical Factors on Waste Generation

Geographic and physical factors that affect the quantities of waste generated and collected include location, season of the year, the use of kitchen waste food grinders, waste collection frequency, and the characteristics of the service area. Because broad generalizations are of little or no value, the impact of these factors must be evaluated separately in each situation.

Geographic Location. Different climates influence both the amount of certain types of solid wastes generated and the time period over which the wastes are generated. For example, substantial variations in the amount of yard and garden wastes generated in various parts of the country are related to climates. That is, in the warmer southern areas, where the growing season is considerably longer than in the northern areas, yard wastes are collected not only in considerably greater amounts but also over a longer time. Because of the variations in the quantities of certain types of solid wastes generated under different climates, special studies should be conducted when such information will have a significant impact on the system. Often, the necessary information can be obtained from a load-count analysis.

Season of the Year. The quantities of certain types of solid wastes are also affected by the season of the year. For example, the quantities of food wastes relate to the growing season for vegetables and fruits. Because of the wide variations reported previously in Table 3-8, seasonal sampling also will be required to assess changes in the percentage distribution of the waste materials comprising MSW, especially in areas of the country with extensive vegetation.

Use of Kitchen Food Waste Grinders. While the use of kitchen food waste grinders definitely reduces the quantity of kitchen wastes collected, whether they affect quantities of wastes generated is not clear. Because the use of home grinders varies widely throughout the country, the effects of their use must be evaluated separately in each situation if such information is warranted. Unit waste allowances made in the field of wastewater treatment for estimating the additional suspended solids per capita contributed from homes with food grinders varies from 0.03 to 0.08 lb/capita · d [8]. Typically, the values used in the wastewater field only reflect the increase in solids removed at wastewater treatment facilities and do not reflect the material that has solubilized in the process of being transported. More realistic values for estimating the effect of food waste grinders are 0.08 to 0.12 lb/capita · d. Alternatively, for homes with food waste grinders one can assume that 25 to 33 percent of the total amount of food waste generated is ground up. The impact of kitchen food waste grinders on the weight and volume of waste collected is illustrated in Example 6-4.

Example 6-4 Impact of food waste grinders on weight and volume of solid wastes collected. The kitchen food waste grinder in a detached single-family dwelling has gone "on the fritz." Estimate the increase in both the volume and the weight of solid wastes to be collected over a one-week period. Also estimate what percentage of the food waste is ground up. Assume that the family has four members and that the collection frequency is once per week. Assume the amount of food waste that is normally ground up equals 0.10 lb/capita · d.

Solution

1. Estimate the total weekly solid waste production rate using the data from Table 6-3.

   Solid waste production, lb/wk

   $$= 4 \text{ persons } \times 7 \text{ d/wk} \times 3.82 \text{ lb/capita} \cdot \text{d} = 107.0 \text{ lb/wk}$$

2. Set up a computation table to determine the volume occupied by the total amount of solid waste generated in one week. Use the weight distribution data given in Table 3-7, column 3 (solid waste as collected plus ground up food waste) and the specific weight values given in Table 4-1.

Component	Solid wastes, %	Solid wastes, lb	Specific weight, lb/ft³	Volume, ft³
Organic				
Food wastes	9.4	10.1	18.1	0.56
Paper	33.8	36.3	5.6	6.48
Cardboard	6.0	6.4	3.1	2.06
Plastics	7.0	7.5	4.1	1.83
Textiles	2.0	2.1	4.1	0.51
Rubber	0.5	0.5	8.1	0.06
Leather	0.5	0.5	10.0	0.05
Yard wastes	18.4	19.7	6.7	2.94
Wood	2.0	2.1	14.8	0.14
Inorganic				
Glass	7.9	8.5	12.2	0.70
Tin cans	6.0	6.4	5.6	1.14
Aluminum	0.5	0.5	10.0	0.05
Other metal	3.0	3.2	20.0	0.16
Dirt, ashes, etc.	3.0	3.2	30.0	0.11
Total	100.0	107.0		16.79

3. Estimate the total amount of food waste that is discharged weekly to the sewer.

   Food waste discharged to sewer, lb/wk

$$= 4 \text{ persons} \times 7 \text{ d/wk} \times 0.10 \text{ lb/capita} \cdot \text{d} = 2.8 \text{ lb/wk}$$

4. Compute the percentage of food waste that is ground up.

$$\text{Percentage of food waste that is ground up} = (2.8 \text{ lb}/10.1 \text{ lb}) \times 100 = 27.7\%$$

5. Compute the weight and volume of solid waste without the food waste.

$$\text{Weight} = 107.0 \text{ lb/wk} - 2.8 \text{ lb/wk} = 104.2 \text{ lb/wk}$$

$$\text{Volume} = 16.8 \text{ ft}^3 - (2.8 \text{ lb}/18.1 \text{ lb/ft}^3) = 16.6 \text{ ft}^3$$

6. Compute the percent increase in weight and volume when the food waste that is now ground up is added to the solid waste.

$$\text{Increase in weight,\%} = (2.8 \text{ lb}/104.2 \text{ lb}) \times 100 = 2.7$$

$$\text{Increase in volume, \%} = [(16.8 - 16.6)/16.6] \times 100 = 1.2$$

Comment. Note that the removal of food waste from the residential waste stream results in a minor reduction in weight and has a minimal effect on volume. In fact, it could be argued that the food waste would take up even less space because it would tend to fill the void space that already exists in the solid waste placed in storage containers.

Frequency of Collection. In general, where unlimited collection service is provided, more wastes are collected. This observation should not be used to infer that more wastes are generated. For example, if a homeowner is limited to one or two containers per week, he or she may, because of limited container capacity, store newspapers or other materials; with unlimited service, the homeowner would tend to throw them away. In this situation the quantity of wastes generated may actually be the same, but the quantity collected is considerably different. Thus, the fundamental question of the effect of collection frequency on waste generation remains unanswered.

Characteristics of Service Area. Pecularities of the service area can influence the quantity of solid wastes generated. For example, the quantities of yard wastes generated on a per capita basis are considerably greater in many of the wealthier neighborhoods than in other parts of town. Other factors that will affect the amount of yard waste include the size of the lot, the degree of landscaping, and the frequency of yard maintenance.

6-5 QUANTITIES OF MATERIALS RECOVERED FROM MSW

The estimated percentages of total amount of waste material that is now (1992) recovered from MSW by waste category are reported in Table 6-7. For residential and commercial recycling in the United States estimates vary from about 12 to 16 percent. Information on the other sectors is so site-specific that few general-

TABLE 6-7
Materials and estimated amounts currently recovered for recycling in the United States (1992)[a]

Material	Percentage of total amount of waste material generated that is now recovered for recycling	Remarks
Aluminum	60–70	Primarily beverage containers
Paper	30–40	
Cardboard	40–50	
Plastics	4–5	Overall; 150×10^6 lb of plastic soft drink bottles were recycled in 1987 (about 20%)
Glass	6–10	
Ferrous metal	15–25	
Nonferrous metals	10–15	
Yard waste (compost)	5–10	Compost; biomass fuel
Refuse-derived fuel	<1	Produced from organic fraction of MSW
Construction and demolition wastes	15–25	
Wood	5–10	
Waste oil	20–30	
Tires	40–50	
Lead–acid batteries	75–85	
Household batteries	<1	
Overall total for United States based on weight[b]	12–16	

[a] Adapted in part from Refs. 1–4, 8–10.

[b] Reported values vary depending on the basis used for the computation.

izations are possible. At the present time, the degree of recycling depends on the type of recycling program that is in effect and on local regulations. Methods for determining existing recycling rates are considered further below.

6-6 QUANTITIES OF HOUSEHOLD HAZARDOUS WASTES

Special household wastes—including hazardous waste, used oil, old tires, and white goods—and construction and demolition wastes, are not usually collected with other municipal solid waste. As noted in Chapter 5, small amounts of hazardous waste compounds are, at present, normal constituents of solid wastes from residential (see Fig. 5-1), commercial, and light industry sources. Data on the quantities of hazardous waste are quite variable, depending on the method used to classify the hazardous waste materials. The results of a recently completed study of the hazardous waste constituents present in municipal waste are summarized in Table 6-8 [1]. The quantities reported in Table 6-8 were determined on the basis of their toxicity using the criteria given In Table 6-9. Only those compounds with

TABLE 6-8
Estimated annual tonnage of designated household hazardous waste in California solid waste[a]

Generic hazardous material	Concentration, ppm	Annual weight,[b] tons
Non-chlorinated organics	87	2,697
Chlorinated organics	0.2	6
Other organics	9	279
Pesticides	0.1	3
Latex (water base) paint	394	12,214
Oil-based paints	76	2,356
Waste oil	61	1,891
Automobile battery	1,661	51,491
Household battery	668	20,708
Total		91,645

[a] From Ref. 1.

[b] Based on a generation rate of 31 million tons per year.

a toxicity rating greater than four were considered in developing the data reported in Table 6-8. Using the data given in Table 6-8, the percentage of hazardous waste in the MSW generated in California is about 0.3 percent. If automotive batteries are excluded from the data reported in Table 6-7, the percentage of hazardous waste in the MSW is about 0.13 percent, essentially the same as the average value given in Table 6-3. If automotive and household batteries are excluded, the corresponding percentage is about 0.06 percent. Depending on the definition used for hazardous wastes, it is clear that a wide range of percentages can be reported.

Because different definitions are used in assessing the quantities of hazardous waste in municipal waste it is difficult to draw any firm conclusions concerning the actual quantities involved. It is interesting to note that if the tons of household

TABLE 6-9
Criteria used to classify hazardous wastes found in MSW with respect to toxicity[a,b]

Toxicity rating	Probable oral lethal dose
6—Super toxic	< 5 mg/kg
5—Extremely toxic	5–50 mg/kg
4—Very toxic	50–500 mg/kg
3—Moderately toxic	0.5–5 g/kg
2—Slightly toxic	5–15 g/kg
1—Probably nontoxic	> 15 g/kg

[a] From Ref. 1.

[b] Materials with a toxicity rating of 4 and above were used to develop the quantity data reported in Table 6-8.

batteries reported in Table 6-8 are projected for the United States using the data given in Table 6-2, the total amount would be about 160,000 tons per year. Given that a typical household battery weighs 50 grams, the corresponding number of batteries is 2,910,000,000. By comparison, it is estimated that more than 2,700,000,000 battery units were purchased in the United States in 1990 (see Chapter 15). If half the household batteries were alkaline, and it is assumed that each battery contains about 1200 mg of mercury, then, based on the data reported in Table 6-8, 1923 tons of mercury would enter the environment each year in California alone. It should be noted, that begining in 1993, the mercury content of alkaline and carbon-zinc batteries has been essentially eliminated as a result of manufacturing changes. Clearly, the proper disposal of household batteries is an important issue that must be addressed.

6-7 WASTE CHARACTERIZATION AND DIVERSION STUDIES

As communities across the United States strive to comply with federal- and state-mandated diversion goals (25 percent by 1995 and 50 percent by the year 2000 in California), information must be available on the types and quantities of waste generated; on the types and quantities of waste currently separated for recycling, or otherwise diverted from landfill disposal; and on the types and quantities of waste collected for further processing or disposal. This information must be developed to define the current situation and to demonstrate that the mandated recycling goals will be met in the future.

Waste Characterization

The goal of a waste characterization study is to identify the sources, characteristics, and quantities of the waste generated. Waste characterization studies are difficult to perform because of the large number of sources and the limited number of waste samples that can be analyzed. The typical steps involved in a waste characterization study are as follows.

1. Gather Existing Information
   The use of existing information can save money, time, and serve as a cross reference. Existing information sources include
   - Previous solid waste management and planning studies and documents
   - Waste collection company records (public and private)
   - Processing facility records (e.g., composting facilities, incineration facilities, etc.)
   - Landfill and transfer station records
   - Previous waste disposal studies
   - Information from comparable communities
   - Department of Public Works
   - Utilities
   - Retail trade reports
   - Community employment records (Chamber of Commerce)

2. Identify Waste Generation Sources and Waste Characteristics
   - Sources
     - Residential
     - Commercial
     - Institutional
     - Construction and demolition
     - Municipal services
     - Water and wastewater treatment plants
     - Industrial
     - Agricultural
   - Develop waste categories (see Table 6-10, see also Table 3-10). The need for a detailed analysis of the individual waste components within each waste

TABLE 6-10
Typical waste categories that have been used for MSW characterization studies

Waste category	Types of waste
Residential and commercial	
Food waste	Wastes from the handling, preparation, cooking, and eating of foods
Paper	Old newspaper, high-grade (e.g., office, computer, etc.) paper, magazines, mixed paper, and other nonusable paper (e.g., wax impregnated, carbon paper, thermal FAX paper)
Cardboard	Old corrugated cardboard/kraft (recyclable, contaminated)
Plastics	PETE (soft drink bottles), HDPE (milk and water containers and detergent bottles), mixed (comingled) plastics, other plastics (PVC, LDPE, PP, and PS), film plastic
Textiles	Clothing, rags, etc.
Rubber	All types of rubber products excluding motor vehicle tires
Leather	Shoes, coats, jackets, upholstery
Yard wastes	Grass clippings, leaves, bush and tree trimmings, other plant materials
Wood	Waste building materials, wooden pallets
Miscellaneous	Disposable diapers
Glass	Container glass (clear, amber, green), flat glass (e.g., window glass), other noncontainer glass materials
Aluminum	Beverage containers, secondary aluminum (window frames, storm doors, siding, and gutters)
Ferrous metals	Tin cans, appliances and cars, other iron and steel
Special wastes	
Bulky items	Furniture, lamps, bookcases, file cabinets, etc.
Consumer electronics	Radios, stereos, television sets, etc.
White goods	Large appliances (stoves, refrigerators, washers, dryers)

(continued)

category presented in Table 6-10 will depend on the uses to be made of the information that is to be gathered.

3. Develop Sampling Methodology
   - Sample identification and characteristics including
     Source(s)
     Size of sample (e.g., pounds of waste separated)
     Number of samples needed for statistical significance
   - Duration of sampling period
   - Time of year
4. Conduct Field Studies
5. Conduct Market Surveys for Special Wastes
6. Assess Factors Affecting Waste Generation Rates

TABLE 6-10 (*continued*)

Waste category	Types of waste
Residential and commercial (*cont.*)	
Special wastes (*cont.*)	
Yard wastes collected separately	Grass clippings, leaves, bush and tree trimmings, tree stumps
Batteries	Household (alkaline, carbon-zinc, mercury, silver, zinc, and nickel-cadmium). Motor vehicle (lead-acid batteries)
Oil	Used oil from automobiles and trucks
Tires	Worn-out tires from automobiles and trucks
Hazardous	See generic product listings in Chapter 5
Institutional	Same types of waste as cited above under the categories of residential and commercial
Construction and demolition	Dirt; stones; concrete; bricks; plaster; lumber; shingles; and plumbing, heating, and electrical parts. Wastes from razed buildings, broken-out streets, sidewalks, bridges, and other structures
Municipal services	
Street and alley cleanings	Dirt, rubbish, dead animals, abandoned automobiles
Tree and landscaping	Grass clippings, bush and tree trimmings, tree stumps, old metal and plastic pipe
Parks and recreational areas	Food wastes, newspaper, cardboard, mixed paper, soft drink bottles, milk and water containers, mixed plastics, clothing, rags, etc.
Catch basin	General debris, sand, used oil mixed with debris, etc.
Treatment plant residuals	Water and wastewater treatment plant sludges, ash from combustion facilities
Industrial	Varies with each region of the country
Agricultural	Varies with each region of the country

Assessment of Current Waste Diversions

The goal of a waste diversion study is to identify the types and quantities of waste materials that are now separated for recycling or otherwise diverted from disposal in landfills (see Table 3-10). The typical steps involved in a waste diversion study are as follows.

1. Gather Existing Information. Existing information sources include
   - Previous solid waste management studies
   - Previous waste diversion studies
   - Curbside recycling programs (public and private)
   - Materials recovery facilities (MRFs)
   - Buy-back centers
   - Drop-off centers
   - Tire and oil recycling centers
   - Private haulers (special wastes)
   - Charitable and service organizations
2. Develop Methodology for Estimating the Quantities of Waste Now Diverted
   - Residential
   - Commercial
   - Institutional
   - Construction and demolition
   - Municipal services
   - Water and wastewater treatment plants
   - Industrial
   - Agricultural
3. Identify Other Existing Activities
4. Conduct Field Studies
5. Assess Factors Affecting Waste Diversion Rates

Analysis of Total Waste Generated and Diverted

To assess the quantity of waste that is currently diverted, it will be necessary to first develop data on the total quantity of waste generated. The total waste generated will be made up of the amount of waste now placed in a landfill and the amount of waste now diverted. In determining the amount of waste diverted, a number of ambiguities will arise in the interpretation of what exactly is a waste material. Some states have ruled that the federal- and state-mandated diversion percentages (i.e., 25 and 50 percent) must be based on waste materials that are now discharged to landfills. Thus, if a material is considered a waste by a discharger, but is now totally recycled it could not be considered in determining the percentage diversion.

6-8 DISCUSSION TOPICS AND PROBLEMS

6-1. Consider a household that generates a certain amount of waste per day. Of this amount, bottles and cans represent 20 percent (by weight) and are recycled by the family. Twenty percent of the paper waste (32 percent total) is burned in the fireplace. The remaining paper along with the rest of the waste is put into containers for collection. On a given day, 20 lb of consumer goods (food, newspapers, magazines, etc.) are brought into the house. The family consumes 7 lb of food that day, and 5 lb of food is stored. The magazines received represent 5 percent of the paper wastes of the day, and they are not thrown away. Draw a materials flow diagram of this problem and calculate the amount of solid wastes disposed of during this day.

6-2. Each week Segovia's grocery receives several shipments of fresh produce from a produce wholesaler. Even in midwinter, shipments of fresh produce arrive from the Southern Hemisphere. (As a loss leader Segovia's will sell grapes from Chile during the Northern Hemisphere's winter at a price close to that of local grapes during the late summer. The low price of these grapes will induce shoppers to come to the store, and they will buy other items while shopping there. Naturally, many grapes are sold at these prices.)

The consumption of fresh produce generates very little waste compared with more highly processed foods, but there is still waste. A shipment of Chilean grapes arrives at Segovia's in wooden boxes that contain 20 lbs of grapes. The boxes arrive strapped to pallets, 72 boxes to a pallet. Each box weighs 2.5 lbs empty and each pallet weighs 20 lbs. The boxes and pallets are crushed, bundled and sent to the landfill. In addition, there are a certain number of pieces of fruit that are damaged, loose, or otherwise unsalable. These are also thrown away.

Segovia's purchases paper bags with the store logo printed on them. These are distributed one per customer. Produce is packaged in plastic bags that are purchased by weight. The produce is weighed at the cash register, and the weight is recorded on the cash register tape. The weight of the plastic bags containing the produce is included in this weight.

One midweek shipment to Segovia's from the wholesaler consists of the following:

7200 lbs of Chilean seedless grapes
5000 lbs of Honduran bananas
20 gross of paper bags (1 gross = 12 dozen)
50 lbs of plastic produce bags

Assume that these plastic bags are used only in the banana and grape section of the store and, in the interest of the environment, shoppers do not use plastic bags for their bananas. At week's end it is found that all of the plastic bags have been used. There were 522 customers who shopped; they purchased 7142 lbs of grapes and 4544 lbs of bananas. Five customers were so overwhelmed at the bargain price of grapes that they bought an entire box (taking the box). Bananas are packaged in boxes of 50 lbs net. The banana boxes are cardboard and weigh 2 lbs. Bananas arrive in loose boxes—not on pallets. Compute a materials balance for the midweek produce shipment. (Courtesy of Robert Anex.)

6-3. Consider a privately owned retail business that sells metal accessories. The business receives approximately 250 lbs of new merchandise every day. Since much of the merchandise sold must be matched to an old sample for verification, most customers bring their old merchandise into the store. A number of customers leave their old parts behind to be disposed of by the business. It is estimated that about 20 percent of the total amount of metal sold is brought in and left by customers.

Of the amount of total merchandise received, approximately 9 percent of the weight is in packaging material (paper and cardboard). Eighty-seven percent of the packaging material is cardboard, of which 60 percent is recycled after the merchandise is unpacked. About 7 percent of the paper and 15 percent of the cardboard is sold with merchandise over the counter. The remaining paper and cardboard is disposed of in the dumpster.

The remaining weight of merchandise is metal parts and chemicals. Chemicals make up 11 percent of the total and all but 10 percent are sold daily. The remaining 10 percent is used within the business for cleaning equipment and is disposed of as hazardous waste. Seventy-eight percent of the metal parts are sold per day with the remainder stored internally. Perform a mass balance and a flow diagram. (Courtesy of Paul Renter.)

6-4. A small family-owned and operated orchard sells apples and cider daily during the fall. On a particular day in October, the store processing facility receives 300 bushels of unsorted apples. Each bushel of apples is transported in a wooden crate. Crates weigh 4 lbs each, and a bushel of apples weighs 40 lbs.

Also delivered to the store are containers in which the apples and cider will be sold: for the apples, 500 one-bushel cardboard boxes with a total weight of 100 lbs, and for the cider, 200 one-gallon plastic jugs with a total weight of 20 lbs. As the containers are unloaded, it is found that 2 percent of the contents of each container are damaged and must be discarded.

During the day, all of the apples harvested are sorted into one of the following categories by weight:

1. Cider apples—20 percent (average)
2. Utility apples—20 percent (average)
3. USDA Grade A apples—remainder of the apples

. The sales for this day were:

1. 75 gallons of cider
2. All bushels of utility apples
3. 140 bushels of Grade A apples

A container is provided with each gallon of cider or bushel of apples. Discounts are given to customers with their own containers. On the average, 20 percent of the apple and cider customers have their own containers. At the end of the day, all remaining unsold cider is stored in a bulk tank. The unsold apples are stored in wooden crates, except for 10 bushels that are placed in cardboard bushel containers, displayed and ready for sale for the next day. All empty wooden crates are removed and transported back to the orchard except for damaged crates, which are stored and repaired at the facility. On average, 5 percent of the wooden crates need repair.

From the above information, draw a materials flow diagram and calculate the amount of solid waste disposed of during this day. Assume the density of apple cider is 8.5 lbs/gal. (Courtesy of Linda Allen.)

6-5. An auto salvage yard receives each day, on the average, two junked cars weighing 4000 lbs apiece. Each car includes

5 qts motor oil
12 qts transmission oil
1 qt steering oil
5 qt clutch oil
4 gal dilute antifreeze mixture

2 gal gasoline
1 oil filter
1 transmission oil filter
4 rubber tires
4 sets asbestos brake shoes

Once the cars have been stripped of these, the remaining weight distribution is

20 percent nonrecyclable materials
80 percent recyclable steel, aluminum, etc.

Each day the yard sells 2 complete engines, which represent 20 percent of the car's weight once the fluids, tires, oil filters, and brake shoes have been removed. Also sold each day are

80 lbs of misc. metal parts
15 lbs of misc. nonmetal parts
2 rubber tires

Each day after sales, 50 percent of the remaining nonrecyclable material is shipped to the landfill and 25 percent of recyclable metals and 2 rubber tires are recycled. The rest are stored for future sales.

Prepare a materials mass balance analysis assuming that the oils have a specific weight of 1 lb/qt, other toxic liquids have a specific weight of 5 lb/gal, the tires weigh 25 lbs each, the asbestos brake shoes, oil filter, and transmission filter weigh 20 lbs combined. (Courtesy of William Sutcliffe.)

6-6. The residential and commercial solid wastes of a city of 25,000 people are collected on Tuesday and Saturday mornings. The volume of wastes collected has been recorded for one year, and the data are given below. Prepare a frequency histogram for each collection day. Find the mean, median, mode, standard deviation, and coefficient of variation for each distribution (see Appendix D). Discuss briefly the nature of the distribution and its significance.

Generation rate, yd^3/collection	Frequency	
	Tuesday	Saturday
800–899	0	0
900–999	0	0
1000–1099	4	1
1100–1199	9	3
1200–1299	14	4
1300–1399	11	9
1400–1499	7	11
1500–1599	4	10
1600–1699	2	7
1700–1799	0	4
1800–1899	1	2
1900–1999	0	1

6-7. Given the following daily solid waste generation data for a period of 10 days for location A, determine the type of distribution, mean, standard deviation, and coefficient of variation.

	Generation rate, yd³/d			
	Location			
Day	A	B	C	D
1	34	10	20	38
2	48	20	28	46
3	290	110	33	52
4	61	120	39	54
5	205	70	34	40
6	170	140	24	41
7	120	60	25	48
8	75	50	30	57
9	110	100	35	62
10	90	30	30	50
11		40		60
12		130		44

6-8. Solve Problem 6-7 using the waste generation data for location B, C, or D (location to be selected by instructor).

6-9. What conclusions can be drawn from frequency plots (histograms) of solid waste generation?

6-10. The shape of a solid waste generation frequency curve reflects the nature of the generating facility. From the following frequency curves, what can be deduced about the facility's activity and operation?

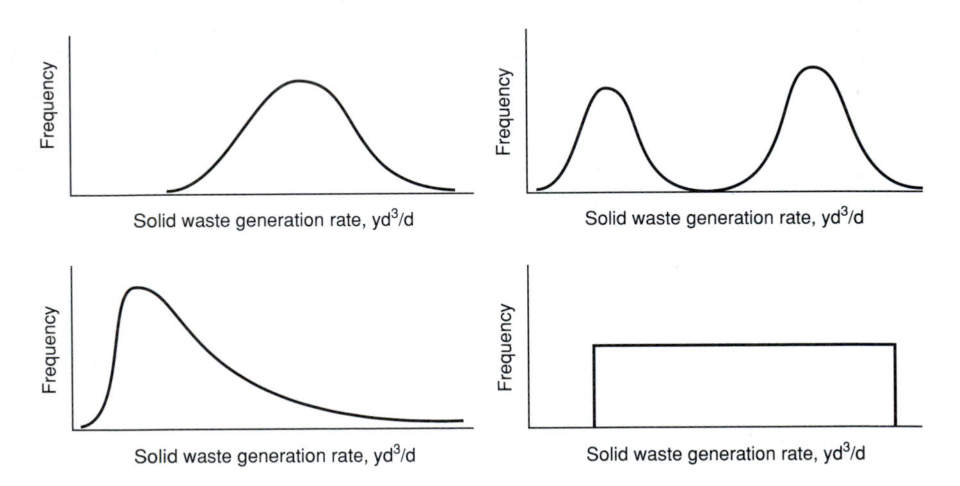

6-11. What makes up white goods at your university or work location? How are the white goods now collected and disposed of?

6-12. How are bulky items, consumer electronics, white goods, oil, tires, and batteries now handled in your community? What fraction of the individual components are now recycled?

6-13. Determine the nature of the waste distribution, the average (mean) value, and the peak-to-average value for the following monthly tonnage data recorded at a transfer station: 250, 218, 209, 276, 315, 300, 260, 293, 246, 289, 189, 334.

6-14. Obtain the monthly weight data from your local transfer station or disposal site for the past year. Plot the data on arithmetic and or logarithmic probability graph paper and determine the nature of the waste distribution, the mean value, and the peak-to-average ratio.

6-15. One of the first steps in conducting a solid waste management study is the identification of factors contributing to the generation of solid wastes now and in the future. In outline form, list the factors that affect the generation of municipal, industrial, and agricultural solid wastes in your community, and list those that may affect generation in the future.

6-16. In your first position as a junior city engineer, you are assigned by your supervisor to report on the generation rates and composition of solid wastes for various sources of your community. How would you go about it? If these data were needed in 30 days and thus you had no time to assess seasonal effects, how would you estimate this factor?

6-17. Should construction and demolition wastes be counted in determining the waste diversion percentage for a community?

6-18. How should the waste categories given in Table 6-10 be modified if you were to conduct a waste characterization for your community, university, work activity, or other? (To be selected by the instructor.)

6-9 REFERENCES

1. Bomberger, D. C., R. Lewis, and A. Valdes: *Waste Characterization Study: Assessment of Recyclable and Hazardous Components,* report prepared for California Waste Management Board, SRI International, Menlo Park, CA, 1988.
2. California Solid Waste Management Board: *Solid Waste Generation Factors in California,* Technical Information Series, Bulletin 2, State of California, Sacramento, 1989.
3. Franklin, W. (ed.), and M. Franklin: "III. Solid Waste Stream Characteristics," in F. Kreith (ed.), *Integrated Solid Waste Management: Options for Legislative Action,* Genium Publishing Corporation, Schenectady, NY, 1990.
4. Freeman, H. M.: "Source Reduction As an Option for Municipal Waste Management," in F. Kreith (ed.), *Integrated Solid Waste Management: Options for Legislative Action,* Genium Publishing Corporation, Schenectady, NY, 1990.
5. Neissen, W. R.: "Properties of Waste Materials," in D. G. Wilson (ed.), *Handbook of Solid Waste Management,* Van Nostrand Reinhold, New York, 1977.
6. Tchobanoglous, G. and E. D. Schroeder: *Water Quality: Characteristics, Modeling, Modifications,* Addison-Wesley, Reading, MA, 1985.
7. Tchobanoglous, G. and F. L. Burton: *Wastewater Engineering: Treatment Disposal and Reuse* 3rd. ed., McGraw-Hill, New York, 1991.
8. U.S. Environmental Protection Agency: *Characterization of Municipal Waste in the United States: 1960–2000,* 1988 Update, Washington, DC, 1988. Available through National Technical Information Service (NTIS), Springfield, VA.
9. U.S. Environmental Protection Agency: *Decision-Makers' Guide To Solid Waste Management,* EPA/530-SW89-072, Washington, DC, November 1989.
10. U.S. Environmental Protection Agency: *Characterization of Municipal Solid Waste in the United States: 1990 Update,* EPA/530-SW-90-04, Washington, DC, June 1990.

CHAPTER 7

WASTE HANDLING AND SEPARATION, STORAGE, AND PROCESSING AT THE SOURCE

The handling and separation, storage, and processing of solid wastes at the source before they are collected is the second of the six functional elements in the solid waste management system. Because this element can have a significant effect on the characteristics of the waste, on subsequent functional elements, on public health, and on public attitudes concerning the operation of the waste management system, it is important to understand what this element involves. This chapter includes a description and discussion of the handling and separation, storage, and processing of waste materials at the source, with particular emphasis on residential sources of waste generation. Processing at the source may take place at any time before collection (before, during, or after storage) and is, therefore, discussed as appropriate throughout the chapter. (The handling and storage of hazardous wastes at the source was discussed in Chapter 5.)

7-1 HANDLING AND SEPARATION OF SOLID WASTE AT THE SOURCE

The handling and separation of solid wastes at the source before they are collected is a critical step in the management of residential solid waste. Because waste diversion has been mandated by law in a number of states, the separation of waste components at the source has also become an important element of solid waste management programs.

Waste Handling

In general, handling refers to the activities associated with managing solid wastes until they are placed in the containers used for their storage before collection or return to drop-off and recycling centers. The specific activities associated with handling waste materials at the source of generation will vary depending on the types of waste materials that are separated for reuse and recycling and the extent to which these materials are separated from the waste stream. Depending on the type of collection service, handling may also be required to move the loaded containers to the collection point and to return the empty containers to the point where they are stored between collections. Waste handling and separation methods used at residential and commercial sources are described and discussed in the following two sections.

Separation for Recycling

The separation of solid waste components including wastepaper, cardboard, aluminum cans, glass, and plastic containers at the source of generation is one of the most positive and effective ways to achieve the recovery and reuse of materials. Once the waste component is separated, the question facing the homeowner is what to do with the wastes until they are collected or taken to a local buy-back or recycling center. Some homeowners store the separated components within the home, periodically transferring the accumulated wastes to larger containers used for the storage of these materials between collections. Other homeowners prefer to take separated waste components and place them directly in the containers used for the storage of these materials. Waste separation is considered in greater detail in the following two sections.

7-2 WASTE HANDLING AND SEPARATION AT RESIDENTIAL DWELLINGS

While residential dwellings and building types can be classified in various ways, a classification based on the number of stories is adequate for the purpose of discussing the handling and separation of wastes at residential dwellings. The three classifications most often used, and adopted for this text, are these: low-rise—under four stories; medium-rise—from four to seven stories; and high-rise—over seven stories. In the discussion of handling and separation, low-rise residential dwellings are further subdivided into the following categories: single-family detached; single-family attached—such as row or town houses; and multifamily. Because of the significant differences in the solid waste handling operations for low-rise dwellings and medium- and high-rise apartments, each is considered separately in the following discussion. Kitchen waste food grinders and home waste compactors are used in all types of residential units. The use of these processing units is considered further in Section 7-5.

At Low-Rise Detached Dwellings

The residents or tenants of low-rise detached dwellings are responsible for placing the solid wastes and recyclable materials that are generated and accumulated in and around their dwellings in storage containers (see Table 7-1). The types of storage containers used depend on whether waste separation is mandated. In many communities, the decision has been made not to require the residents to separate their wastes, but rather to reach the mandated separation goals using materials recovery facilities (MRFs). In some collection systems, mixed wastes are placed in a variety of storage containers with little or no standardization (see Fig. 7-1). In other systems, mixed wastes are placed in large 90-gal containers equipped with wheels (see Fig. 7-2). In both of these systems, the homeowner or tenants

TABLE 7-1
Persons responsible for and auxiliary equipment used in the handling and separation of solid waste at the source

Source	Persons responsible	Auxiliary equipment and facilities
Residential		
Low-rise	Residents, tenants	Household compactors, large-wheeled containers, small-wheeled handcarts
Medium-rise	Tenants, building maintenance crews, janitorial services, unit managers	Gravity chutes, service elevators, collection carts, pneumatic conveyors
High-rise	Tenants, building maintenance crews, janitorial services	Gravity chutes, service elevators, collection carts, pneumatic conveyors
Commercial	Employees, janitorial services	Wheeled or castered collection carts, container trains, burlap drop cloths, service elevators, conveyors, pneumatic conveyors
Industrial	Employees, janitorial services	Wheeled or castered collection carts, container trains, service elevators, conveyors
Open areas	Owners, park officers, municipal employees	Vandalproof containers
Treatment plant sites	Plant operators	Various conveyors and other manually operated equipment and facilities
Agricultural	Owners, workers	Varies with the individual commodity

FIGURE 7-1
Curb collection service with no standardization of container sizes.

FIGURE 7-2
Typical storage containers (90-gal capacity), equipped with wheels for ease of movement, which are emptied into the collection vehicle mechanically.

are responsible for transporting the containers filled with wastes to the street curb for collection.

In systems in which the waste components are separated, the solid wastes remaining after the recyclable materials have been separated are placed in one or more large containers. The separated wastes are placed in special containers or bags. In some residences, waste compactors are being used to reduce the volume of wastes to be collected. Compacted wastes are usually placed in waste containers or in sealed plastic bags. The homeowner or tenants are responsible for transporting the containers used for the storage of the solid wastes remaining after the recyclable materials have been separated, along with the containers used for the recyclable materials, to the street curb for collection. A number of different collection systems, with and without recycling, are delineated in Table 7-2. Two residential waste separation systems are compared in Example 7-1.

TABLE 7-2
Typical options used for the collection of residential MSW from detached dwellings without and with separation of waste components at the source

Options	Remarks
1. *Without separation of waste components at the source*	
a. One 60- to 90-gal container with curbside collection; separate collection of garden trimmings	Separation of components occurs at a materials recovery facility (MRF)
b. All types of containers; unlimited curbside collection service; separate collection of garden trimmings	Separation of waste components occurs at a MRF
2. *With separation of waste components at the source*	
a. Unlimited curbside collection service; separated newspaper placed in bundles; separate collection of garden trimmings	Regular collection vehicles equipped with bins for newspaper; newspaper unloaded separately at a MRF or paper recovery facility
b. Unlimited curbside collection service; separated waste components placed in three specially designed plastic bins; separate collection of garden trimmings	One bin is for newspaper, one is for glass and plastic, and one is for aluminum and tin cans; glass, plastic, aluminum, and tin cans separated at a MRF
c. Curbside collection with four containers for separated waste components. (This option is best suited to communities with limited amounts of garden trimmings such as San Francisco, CA.)	One container is for all types of uncontaminated paper and cardboard, one is for recyclable materials including plastic containers, glass, aluminum and tin cans, one is for garden trimmings, and one is for the leftover materials; individual components separated at a MRF
d. Curbside collection with one 90-gal container and two heavy-duty plastic bags; separate collection of garden trimmings. Plastic bags are placed in the 90-gal container for collection (see Fig. 7-2); separate collection of garden trimmings	One plastic bag, colored or clear, is for all types of uncontaminated paper, cardboard, magazines, junk mail, and all other paper; the clear plastic bag is for other recyclable materials including plastic bottles and containers, glass bottles and jars, aluminum and tin cans; other materials are placed in 90-gal container; individual components separated at a MRF. It should be noted that see-through bags may be considered a violation of privacy.
e. Curbside collection with three see-through or coated heavy-duty plastic bags and one container; separate collection of garden trimmings. Plastic bags and other waste are collected with the same collection vehicle; separate collection of garden trimmings	One plastic bag is for all types of uncontaminated paper and cardboard, one is for recyclable materials including plastic containers, glass, aluminum and tin cans, one is for garden trimmings; leftover materials are placed in the container; individual components separated at a MRF (one option that has been proposed is to use prisoners to do the sorting)
f. Any of options 2a through 2e but with garden trimmings placed in plastic bags and collected in the same collection vehicle with other wastes	Bagged garden trimmings are placed to one side of the collection vehicle hopper and then separated by hand at the unloading point. It should be noted that this option has limited application, primarily to rear-loaded collection vehicles used in conjunction with appropriate unloading points

Example 7-1 Comparison of residential waste separation programs. The effectiveness of residential waste separation programs depends on the type of system used for the collection of separated wastes. A number of communities use a collection system in which three containers are used for recycled materials in addition to one or more containers for non-recyclable materials. In the three-container system (system 1), newspaper is placed in one container. Aluminum cans, glass, and plastics are placed in the second container. The remaining wastes are placed in the third container. The separated materials, placed in special containers, are collected at the curb. In another system (system 2), four containers are used. All paper and cardboard materials are placed in one container. All plastic, glass, tin cans, aluminum, and any other metals are placed in a second container. Garden wastes are placed in the third container, and all remaining waste materials are placed in the fourth container. Compare these two systems. Assume newspaper represents 25 percent of the total amount of paper.

Solution. Determine realistically how much of the waste stream can be separated for recycling using the two systems described above. Assume 80 percent of the available material is separated and the participation rate is 100 percent. An estimate of the amount of solid waste that can be separated using the above systems is presented in the following table.

	Percent by weight			
	Recycling system 1		**Recycling system 2**	
Component	**As generated[a,b]**	**Separated for recycle[c]**	**As generated[a,b]**	**Separated for recycle[d]**
Organic				
Food wastes	9.0 (3)[e]		9.0 (4)[e]	
Paper	**34.0** (1)	6.8[f]	**34.0** (1)	27.2
Cardboard	6.0 (3)		**6.0** (1)	4.8
Plastics	**7.0** (2)	5.6	**7.0** (2)	5.6
Textiles	2.0 (3)		2.0 (4)	
Rubber	0.5 (3)		0.5 (4)	
Leather	0.5 (3)		0.5 (4)	
Yard wastes	18.5 (3)		**18.5** (3)	14.8
Wood	2.0 (3)		2.0 (4)	
Misc. organics	—	—	—	
Inorganic				
Glass	**8.0** (2)	6.4	**8.0** (2)	6.4
Tin cans	6.0 (3)		**6.0** (2)	4.8
Aluminum	**0.5** (2)	0.4	**0.5** (2)	0.4
Other metal	3.0 (3)		**3.0** (2)	2.4
Dirt, ash, etc.	3.0 (3)		3.0 (4)	
Total	100.0	19.2	100.0	66.4

[a] From Table 3-4.

[b] Waste components that are to be recycled are shown in bold.

[c] Based on 80 percent recovery with 100 percent participation. If only 50 percent of the homes participate, the recycling rate drops to about 9.6 percent.

[d] Based on 80 percent recovery with 100 percent participation. If only 50 percent of the homes participate, the recycling rate drops to about 33.2 percent.

[e] Container number.

[f] $6.8 = 34.0 \times 0.25 \times 0.8$

Comment. As shown in above computation table, the amount of material separated for recycling with system 1 is 19.2 percent versus 66.4 percent for system 2. If the participation rate were to drop to 50 percent, the corresponding amounts are 9.6 versus 33.2 percent. Using system 1, it will be difficult to achieve the 25 percent recycling goal without a high degree of homeowner participation. Additional separation, possibly at a MRF, will be required to reach the 50 percent goal by the year 2000. Using system 2, both the 25 and 50 percent diversion goals are achievable with a reasonable amount of homeowner participation.

At Low- and Medium-Rise Apartments

Handling methods in most low- and medium-rise apartment buildings resemble those used for low-rise dwellings, but methods may vary somewhat depending on the waste storage location and collection method. Typical solid waste storage locations include basement storage, outdoor storage, and, occasionally, compactor storage. (The use of compactors is considered in the discussion of waste handling in high-rise apartments.) Different types of nonrecycling and recycling waste handling and separation systems for low- and medium-rise apartment buildings are delineated in Table 7-3. The recycling of wastes from apartments is considered in detail in Ref. 1.

Basement Storage/Curbside Collection. Curbside collection service is common for most low- and medium-rise apartments. Where curbside collection is used, the building owner provides a basement storage room or area for the storage of solid waste. Typically, the containers used for recycling are located within or next to the solid waste storage area. Residents carry their waste and recyclable materials to the storage area and deposit them in the appropriate containers. The maintenance staff is responsible for transporting the containers to the street for curbside collection. Alternatively, in many apartments the maintenance staff is responsible for the collection of wastes and recyclable materials left outside of the doorway to the apartment or from a utility room located on each floor.

Outdoor Storage/Mechanized Collection. In many low- and medium-rise apartments, large waste storage containers are located outdoors in special enclosures (see Fig. 7-3). The large containers are emptied mechanically using collection vehicles equipped with unloading mechanisms. The containers used for recycling are often located within the outdoor storage area or next to the waste storage area. Residents carry their waste and recyclable materials to the storage area and deposit them in the appropriate containers. If needed, the maintenance staff is responsible for moving the containers to the collection point. As noted above, in some apartments the maintenance staff is responsible for the collection of wastes and recyclable materials left outside of the doorway to the apartment or from a utility room located on each floor.

TABLE 7-3

Typical options used for the collection of residential MSW from low-, medium-, and high-rise residential apartments without and with separation of waste components at the source

Options	Remarks
1. *Without separation of waste components at the source*	
a. Standard-size (20- to 30-gal capacity) containers stored in service area or in outdoor enclosed storage areas; curbside collection; separate collection of yard wastes (Low-rise apartments)	Apartment owners, tenants, or building maintenace crews transport wastes to street curb for collection; separation of materials occurs at a MRF
b. Large (up to 300-gal capacity) mechanically unloaded containers located in basement service areas or in outdoor enclosed storage areas; separate collection of landscape wastes by grounds maintenance contractor (Low- and medium-rise apartments)	If needed, apartment maintenance crews move containers for unloading; separation of materials occurs at a MRF
c. Wastes placed outside of individual apartments or in service areas located on each floor; waste chutes used in newer high-rise apartments; large containers and processing equipment (e.g., waste balers) stored in service areas until collection, usually in the basement of high-rise apartments; separate collection of landscape wastes by grounds maintenance contractor (Medium- and high-rise apartments)	If needed, apartment maintenance crews move containers for unloading; separation of materials occurs at a MRF
2. *With separation of waste components at the source*	
a. 1a and 1b from above; separated waste components placed in conventional or specially designed containers located in the basement or in outdoor enclosed storage areas	Residents or maintenance crews move containers to the designated locations for emptying; individual components separated at a MRF
b. 1c from above; separated materials put outside individual apartments for collection, taken to service areas located on each floor, placed in separate chutes, or taken to service area and placed in separate containers	Separate chutes are normally installed in new construction where required by local building codes; separated wastes collected by building crews stored in service areas; individual components separated at a MRF

FIGURE 7-3
Typical large (4 to 8 yd^3) container used for the storage of commingled wastes and smaller (90 gal) containers used for the storage of source-separated materials (glass and aluminum cans, paper, and plastic) at low-rise apartments.

At High-Rise Apartments

In high-rise apartment buildings the most common methods of handling solid wastes involve one or more of the following: (1) wastes are picked up by building maintenance personnel or porters from the various floors and taken to the basement or service area; (2) wastes are taken to the basement or service area by tenants; or (3) wastes, usually bagged, are placed by the tenants in specially designed vertical chutes (usually circular) with openings located on each floor (see Fig. 7-4). Wastes discharged in chutes are collected in large containers, compacted into large containers, or baled directly. Recyclable materials may be put outside in the hall or entry way for pickup, or they may be taken by the tenants to the service area located on each floor for pickup. The entrance to the waste chute is usually located in the service area. Bulky items are usually taken to the service area by

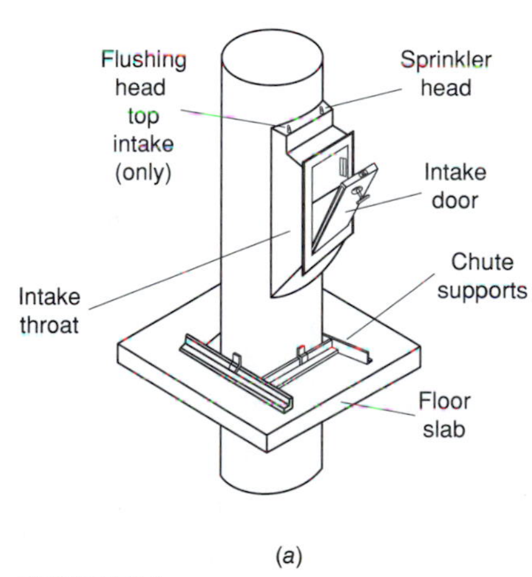

(*a*)

(*b*)

FIGURE 7-4
Typical chute openings for the discharge of waste materials in high-rise apartment buildings: (*a*) isometric view of waste chute opening on individual floor (courtesy of Cutler Manufacturing Corp.) and (*b*) outdoor type used in some older high-rise apartment buildings.

the tenants or the building maintenance crew. The latter are responsible for handling or processing the wastes accumulated in the service areas. In many high-rise apartments, solid waste chutes are used in conjunction with large compactors. The building maintenance personnel are responsible for handling the compressed wastes and any other waste or recyclable materials that tenants bring to the service areas.

Chutes for use in apartment buildings are available in diameters from 12 to 36 in. The most common chute diameter is 24 in. All the available chutes can be furnished with suitable intake doors, either side- or bottom-hinged, for installation on various floor levels (see Fig. 7-4). Draft baffles at the intake doors, door locks, sprinklers, disinfection systems, sound insulation, and roof vents are among the many accessories that are available. The use of a disinfecting and sanitizing unit is recommended because the cleanliness of the chute and the absence of odors generally depend to a large extent on their use. In designing chutes for high-rise buildings, one must consider variations in the rate at which solid wastes are discharged. Typical discharge rates in apartments with chutes are shown in Fig. 7-5. In sizing chutes it is common to assume that (1) the bulk specific weight of the solid wastes equals 175 lb/yd^3, (2) all the daily wastes will be discharged within a 4-h period, and (3) between 1 and 2 lb of wastes will be generated by each tenant each day.

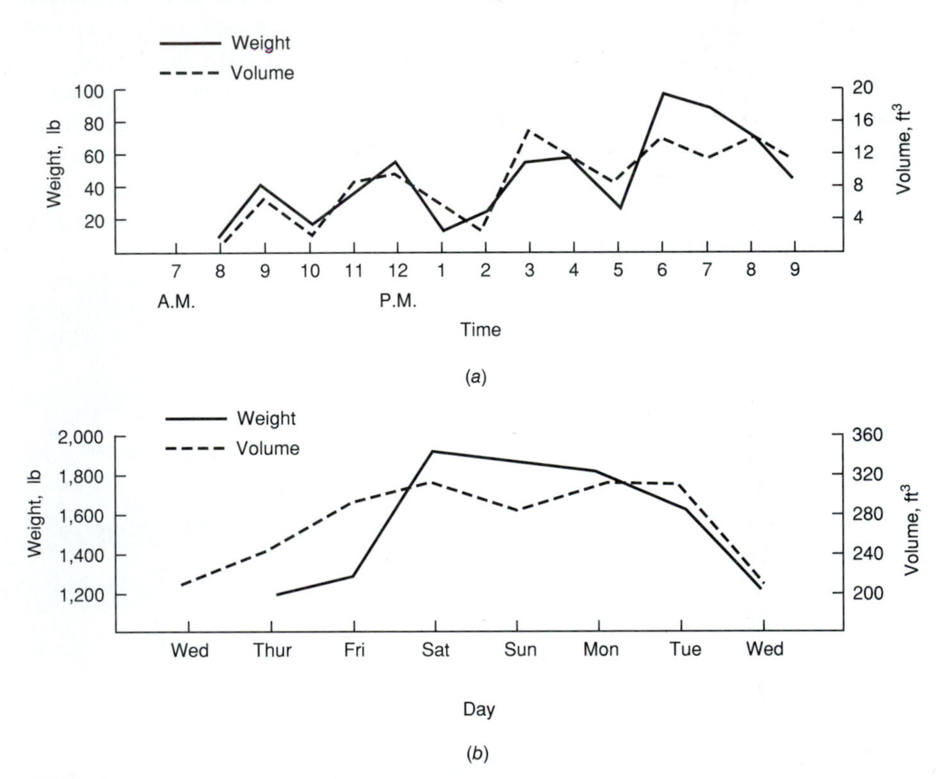

FIGURE 7-5
Typical waste discharge rates in apartments with waste chutes: (*a*) hourly and (*b*) daily.

In some of the more recently constructed apartment buildings, underground pneumatic transport systems have been used in conjunction with individual apartment chutes (see Fig. 7-6). The underground pneumatic systems transport wastes from the chute discharge points to centralized processing facilities. Both air pressure and vacuum transport systems have been used in this application.

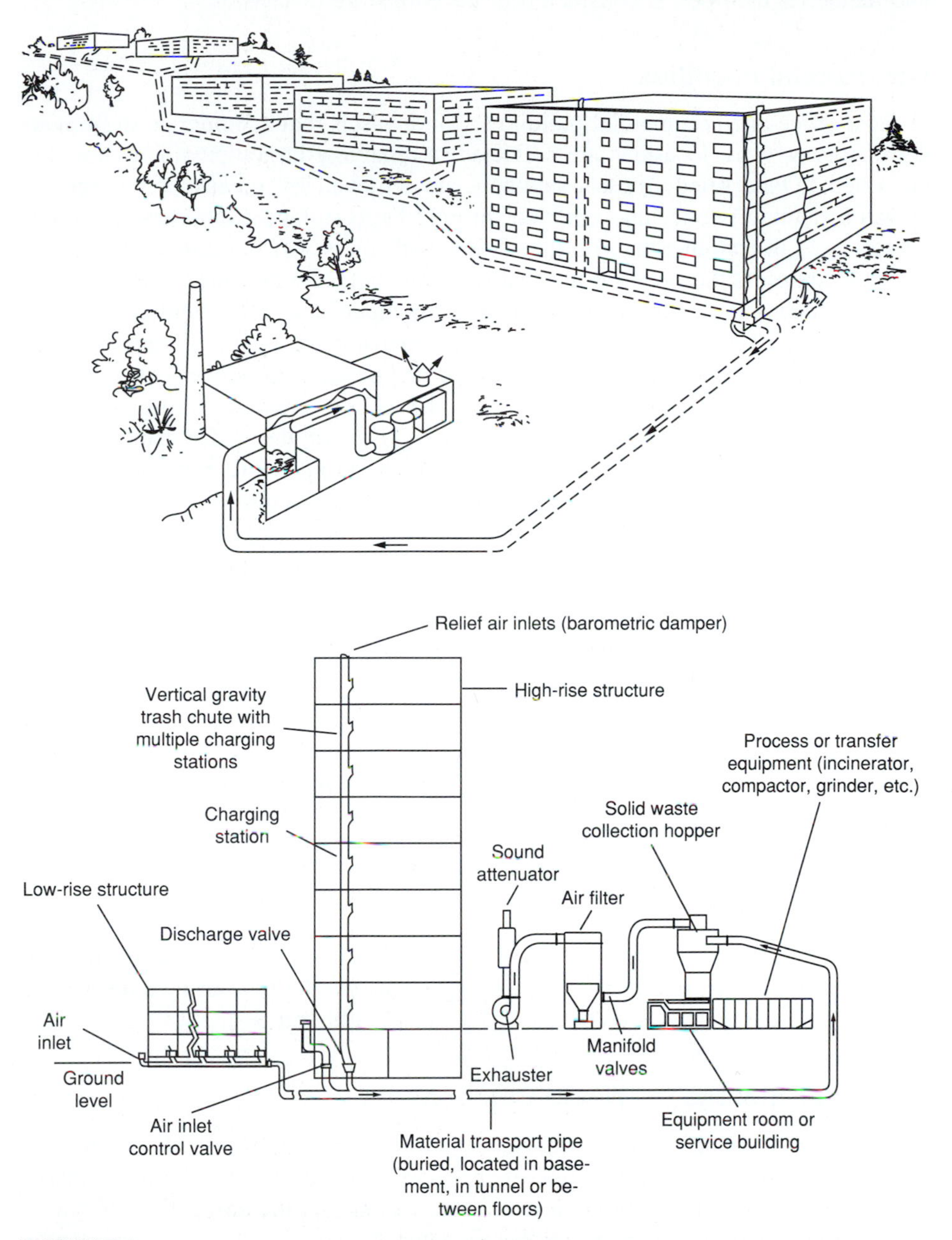

FIGURE 7-6
Typical underground pneumatic waste transport system for high-rise apartment buildings.

7-3 WASTE HANDLING AND SEPARATION AT COMMERCIAL AND INDUSTRIAL FACILITIES

The handling of solid wastes and the recycling of waste materials at commercial and industrial facilities is considered in the following discussion.

Commercial Facilities

In most office and commercial buildings, solid wastes that accumulate in individual offices or work locations are collected in relatively large containers mounted on rollers. Once filled, these containers are removed by means of the service elevator, if there is one, and emptied into (1) large storage containers, (2) compactors used in conjunction with the storage containers, (3) stationary compactors that can compress the material into bales or into specially designed containers, or (4) other processing equipment. Because many older large office and commercial buildings were designed without adequate provision for onsite storage of solid wastes and recyclable materials, the storage and processing equipment now used is often inadequate due to space limitations and tends to create handling problems.

In many office and commercial buildings, all of the office paper is now collected for recycling. The facilities used for collecting waste material for recycling are essentially the same as those used for the collection of other wastes as described above. The wastes to be recycled are stored in separate containers. In large commercial facilities baling equipment is used for the paper and can crushers are used for the aluminum cans.

Industrial Facilities

The handling and separation of nonindustrial solid wastes at industrial facilities is essentially the same as for commercial facilities. Because the handling of industrial wastes is industry- and site-specific, few generalizations are possible.

7-4 STORAGE OF SOLID WASTES AT THE SOURCE

Factors that must be considered in the onsite storage of solid wastes include (1) the effects of storage on the waste components (2) the type of container to be used, (3) the container location, and (4) public health and aesthetics.

Effects of Storage on Waste Components

An important consideration in the onsite storage of wastes are the effects of storage itself on the characteristics of the wastes being stored. These effects of storing wastes include (1) biological decomposition, (2) the absorption of fluids, and (3) the contamination of waste components.

Microbiological Decomposition. Food and other wastes placed in onsite storage containers will almost immediately start to undergo microbiological decomposition (often called *putrefaction*) as a result of the growth of bacteria and fungi. As discussed in Chapter 3, if wastes are allowed to remain in storage containers for extended periods of time, flies can start to breed and odorous compounds can develop.

Absorption of Fluids. Because the components that comprise solid wastes have differing initial moisture contents (see Table 3-7), re-equilibration takes place as wastes are stored onsite in containers. Where mixed wastes are stored together, paper will absorb moisture from food wastes and fresh garden trimmings. The degree of absorption that takes place depends on the length of time the wastes are stored until collection. If wastes are allowed to sit for more than a week in enclosed containers, the moisture will become distributed throughout the wastes. If watertight container lids are not used, wastes can also absorb water from rainfall that enters partially covered containers. Saturation of wastes to their field capacity is a common occurrence in tropical regions where it rains on most days.

Contamination of Waste Components. Perhaps the most serious effect of onsite storage of wastes is the contamination that occurs. The major waste components may be contaminated by small amounts of wastes such as motor oils, household cleaners, and paints. The effect of this contamination is to reduce the value of the individual components for recycling. While the contamination that occurs during onsite storage decreases the value of the individual waste components, one can also argue that this contamination is beneficial with respect to the disposal of these wastes in a landfill. That is, the concentrations of the individual contaminants are reduced considerably when the contaminated waste components are spread out and compacted for landfilling.

Types of Containers

To a large extent, the types and capacities of the containers used depend on the characteristics and types of solid wastes to be collected, the type of collection system in use, the collection frequency, and the space available for the placement of containers. The types and capacities of containers now commonly used for on-site storage of commingled MSW and separated waste components are summarized in Table 7-4. Typical container applications and limitations are presented in Table 7-5. Some of the more common types of containers are shown in Figs. 7-1 and 7-2.

Low-Rise Dwellings with Manual Curbside Waste Collection Service. Because solid wastes are collected manually at curbside for most low-rise detached residential dwellings, the containers should be light enough to be handled easily by one collector when they are full. Injuries to collectors have resulted from handling containers that were loaded too heavily. In general, the upper weight limit should

TABLE 7-4
Data on the types and sizes of containers used for onsite storage of solid wastes

Type	Capacity			Dimensions[a]	
	Unit	Range	Typical	Unit	Typical
Small					
Container, plastic or galvanized metal	gal	20–40	30	in	20D × 26H (30 gal)
Barrel, plastic, aluminum, or fiber	gal	20–65	30	in	20D × 26H (30 gal)
Disposable paper bags					
Standard	gal	20–55	30	in	15W × 12d × 43H (30 gal)
Leak-resistant	gal	20–55	30	in	as above
Leakproof	gal	20–55	30	in	as above
Disposable plastic bag				in	18W × 15d × 40H (30 gal)
				in	30W × 40H (30 gal)
Medium					
Container	yd^3	1–10	4	in	72W × 42d × 65H (4 yd^3)
Large					
Container					
Open top, roll off (also called debris boxes)	yd^3	12–50	—[b]	ft	8W × 6H × 20L (35 yd^3)
Used with stationary compactor	yd^3	20–40	—[b]	ft	8W × 6H × 18L (30 yd^3)
Equipped with self-contained compaction mechanism	yd^3	20–40	—[b]	ft	8W × 8H × 22L (30 yd^3)
Container, trailer-mounted					
Open top	yd^3	20–50	—[b]	ft	8W × 12H × 20L (35 yd^3)
Enclosed, equipped with self-contained compaction mechanism	yd^3	20–40	—[b]	ft	8W × 12H × 24L (35 yd^3)

[a] D = diameter, H = height, L = length, W = width, d = depth.

[b] Size varies with waste characteristics and local site conditions.

Note: gal × 0.003785 = m^3

in × 2.54 = cm

yd^3 × 0.7646 = m^3

ft × 0.3048 = m

TABLE 7-5
Typical applications and limitations of containers used for the onsite storage of solid wastes

Container type	Typical applications	Limitations
Small		
Container, plastic or galvanized metal	Very low-volume waste sources, such as individual homes, walkways in parks, and small isolated commercial establishments; low-rise residential areas with setout collection service	Containers are damaged over time and degraded in appearance and capacity; containers add extra weight that must be lifted during collection operations; containers are not large enough to hold bulky wastes.
Disposable paper bags	Individual homes with setout collection service; can be used alone or as a liner inside a household container; low- and medium-rise residential areas	Bag storage is more costly; if bags are set out on streets or curbside, dogs or other animals tear them and spread their contents; paper bags themselves add to the waste load.
Disposable plastic bags	Individual homes with setout collection service; can be used alone or as a liner inside household and storage containers; useful in holding wet food wastes inside household containers as well as in commercial containers; low-, medium-, and high-rise residential areas; commercial areas and industrial areas	Bag storage is more costly; bags tear easily, causing litter and unsightly conditions; bags become brittle in very cold weather, causing breakage; plastic lightness and durability causes later disposal problems. Bags stretch and break in warm climates.
Medium		
Container	Medium-volume waste sources that might also have bulky wastes; location should be selected for direct-collection truck access; high-density residential areas; commercial areas; industrial areas	Snow inside the containers forms ice and lowers capacity while increasing weight; containers are difficult to get to after heavy snows.
Large		
Container, open top	High-volume commercial areas; bulky wastes in industrial areas; low-density rural residential areas; location should be within a covered area but with direct-collection truck access	Initial cost is high; snow inside containers lowers capacity.
Container, used with stationery compactor	Very high-volume commercial areas; location should be outside buildings with direct-collection truck access	Initial cost is high; if container is compacted too much, it is difficult to unload at the disposal site.

be between 40 and 65 lb. The 30-gal galvanized metal or plastic container has proved to be the least expensive means of storage for low-rise dwellings.

The choice of container materials depends on the preferences of the homeowner. Galvanized metal containers tend to be noisy when being emptied and, in time, can be damaged so that a proper lid seal cannot be achieved. Although less noisy in handling, some containers constructed of plastic materials tend to crack under exposure to the ultraviolet rays of the sun and to freezing temperatures, but the more expensive plastic containers apparently do not present these problems.

Temporary and disposable containers such as paper bags, cardboard boxes, plastic containers and bags, and wooden boxes are routinely used as temporary and disposable containers of accumulated wastes (see Fig. 7-7). Under normal circumstances, temporary containers are removed along with the wastes. The principal problem in the use of temporary containers is the difficulty involved in loading them. Paper and cardboard containers tend to disintegrate because of the leakage of liquids. Where disposable plastic bags are used for lawn trimmings, plastic containers frequently stretch or break at the seams when the collector lifts the loaded bag. Such breakage is potentially hazardous and may lead to injuries to the collector because of the presence of broken glass or other dangerous items in the wastes.

With the widespread availability of paper and plastic products, the use of container liners is now common. All types of thicknesses and grades of material are available. In most areas, the homeowner decides what type of liner to use, if any. A disadvantage in the use of liners is that if the wastes are to be separated by component, or if they are to be combusted, it is necessary to break up the liner bags in a preprocessing step. Thus, although their use may be a convenience for the homeowner, liners may not be ideal from the standpoint of materials recovery and recycling.

Low-Rise Dwellings with Mechanized Curbside Waste Collection Service. Over the past 10 years, there has been a significant increase in the use of mechanical collection systems for residential service. This increase is expected

FIGURE 7-7
Temporary and disposable containers used for the storage and collection of wastes from the curb.

FIGURE 7-8
Typical containers used for the storage of commingled wastes. Containers are designed for curbside collection with mechanized collection vehicles.

to continue, especially as labor and insurance premium costs rise. Where mechanized collection systems are used, the container used for the onsite storage of wastes is an integral part of the collection system. The containers are designed specifically to work with the container-unloading mechanism attached to the collection vehicle. Typical examples are shown in Fig. 7-8. The containers used for residential service with most mechanized systems vary in size from about 75 to 120 gals, with 90 gal being the most common. Although the containers shown in Fig. 7-8 appear bulky and difficult to manage, they are designed so that they can be tilted back and moved quite easily by residents.

Low- and Medium-Rise Apartments. In low-rise apartment complexes, a number of different storage containers have been used [1]. The two most common types are (1) individual plastic or galvanized metal containers and (2) large portable or fixed containers. Where apartments are grouped in close proximity, containers assigned to the individual apartments are often located in a common area. Although individual containers are used in some low-rise apartment buildings, the most common practice is to use one or more large containers for a group of apartments. Typically, these containers are kept in enclosed areas with easy access to a nearby street (see Fig. 7-9). Often, the container enclosures are covered. In most locations the containers are equipped with casters or rollers so that they can be moved easily for emptying into collection vehicles or onsite processing equipment. The containers used for recyclable materials depend on the types of wastes that are separated and the type of waste collection systems used.

High-Rise Apartments. Where solid waste chutes are available, separate storage containers are not used. In some older medium- and high-rise apartments without chutes, wastes are stored in containers on the premises between collections. The most common means of storage for wastes accumulated from the individual apartments include: (1) enclosed storage containers or disposable bags used in conjunction with compaction equipment; (2) large open-top containers for

FIGURE 7-9
Typical containers used for the storage of commingled wastes and recyclable materials at apartment buildings.

uncompacted waste, bulky items, and white goods; and (3) large open-top containers for recycled materials.

Commercial Facilities. The types of containers used for commercial facilities will depend, to a large extent, on the methods used for collecting the wastes produced at various locations within the facility and on the available space (see Fig. 7-10).

FIGURE 7-10
Typical container-storage locations for containers used at commercial facilities.

(*a*)

(*b*)

FIGURE 7-11
Typical compaction facilities used for waste management in commercial establishments: (*a*) large container equipped with internal compaction mechanism and (*b*) compactor used in conjunction with stationary compactor.

Typically, large open-top containers are used for unseparated wastes. The use of containers equipped with compaction mechanisms or in conjunction with stationary compactors is increasing (see Fig. 7-11). Where a considerable amount of recoverable materials is generated, special onsite processing equipment may also be used. Selection of a container size for a commercial facility is illustrated in Example 7-2.

Example 7-2 Selection of container size for use at a commercial facility. A commercial facility produced the following quantities of solid waste each week for a calendar quarter of operation. Using these waste production data, determine the size of container at which it becomes more cost effective to make extra pickup trips, on call, instead of using a larger-sized container. Assume the data given below are applicable.

1. Waste production data

Week no.	Waste, yd^3/wk	Week no.	Waste, yd^3/wk
1	41	8	27
2	30	9	37
3	35	10	36
4	34	11	33
5	39	12	28
6	25	13	31
7	32		

2. Collection system data
   (*a*) The capital cost of containers and annual operation and maintenance (O & M) costs

Capacity, yd^3	Capital cost, $	Annual O & M cost, $/yr
30	3000	150
35	3500	175
40	4000	225
45	4900	300
50	6100	400

   (*b*) Cost per trip = $50.00/trip
   (*c*) Useful life of containers = 10 yr
   (*d*) Discount rate = 10%
   (*e*) Capital recovery factor (i = 10%, n = 10 yr) = 0.16275

Solution

1. Plot the waste production data on arithmetic probability graph paper. If the data are not normally distributed, it may be necessary to plot them on log-probability paper. The data

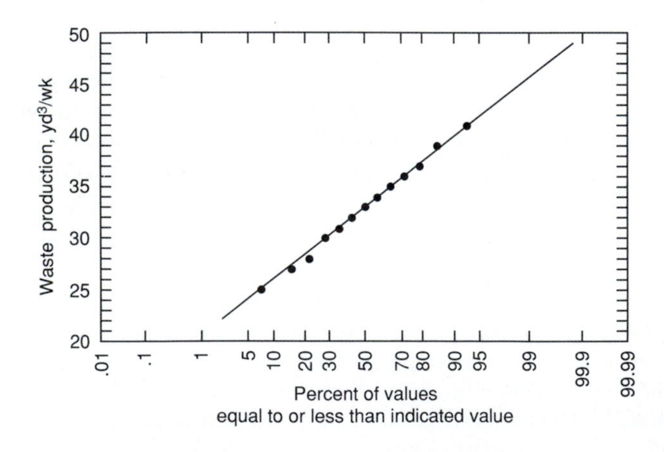

are plotted on the accompanying graph using the techniques outlined in Appendix D (see also Example 6-3). Because a straight line can be fitted to the plotted points, it can be assumed, for all practical purposes, that the waste production data are distributed normally.

2. From the probability plot prepared in Step 1, determine the number of additional trips that would be required per year for each of the container sizes.
   (*a*) The percentage of the time that production rates of 30, 35, 40, 45, and 50 yd^3/wk will be exceeded is as follows:

yd^3/wk	Percent
30	70.0
35	35.0
40	10.0
45	1.6
50	0.1

   (*b*) The number of extra trips required is

yd^3/wk	Extra trips, trip/yr[a]
30	36.4 (37)
35	18.2 (19)
40	5.2 (6)
45	0.8 (1)
50	0.05 (1)

[a] Percent from 2(*a*) × 52

3. Determine the annual cost of providing collection service to the commercial facility as a function of the container size.
   (*a*) Determine the annual capital cost. The annual capital cost for a 30 yd^3 container is computed as follows:

$$\text{Annual capital cost} = \text{capital cost} \times \text{capital recovery factor}$$

$$= \$3000 \times 0.16275 = \$488.25$$

   (*b*) The cost for the regular collection trips is calculated as follows:

$$\text{Regular collection cost} = 52 \text{ trip/yr} \times \$50.00\text{/trip} = \$2600.00$$

   (*c*) The cost for the extra collection trips if a 30 yd^3 container is used is computed as follows:

$$\text{Extra collection cost} = 37 \text{ trip/yr} \times \$50.00\text{/trip} = \$1850.00$$

   (*d*) Summarize the costs for each container size. The required cost summary is presented in the following table:

	Cost, dollars				
	Container size, yd^3				
Cost item	30	35	40	45	50
Annual capital cost	488.25	569.62	651.00	797.48	992.78
Annual maintenance cost	150.00	175.00	225.00	300.00	400.00
Regular haul trips	2600.00	2600.00	2600.00	2600.00	2600.00
Extra haul trips	1850.00	950.00	300.00	50.00	50.00
Total	5085.25	4294.62	3776.00	3747.48	4042.78

4. From the preceding analysis it can be concluded that a container size of 45 yd^3 will result in the lowest cost.

Comment. The approach delineated in this example is applicable to a variety of problems in solid waste management where choices must be made between the expenditure of capital cost versus an increase in operational costs.

Container Storage Locations

Container storage locations depend on the type of dwelling or commercial and industrial facilities, the available space, and access to collection services.

Residential Dwellings. Between collections, containers used in low-rise detached dwellings usually are placed (1) at the sides or rear of the house (see Fig. 7-12), (2) in alleys, where alley collection is used, (3) in or next to the garage/carport or, where available, some common location specifically designated for that purpose. When two or more dwellings are located in close proximity, a concrete pad may be constructed at some convenient location between them. The pad may be either open or surrounded by a wooden enclosure. Unless enclosed

FIGURE 7-12
Typical storage locations for containers between collections at residential dwellings.

pads are supervised properly, however, unsightly and unsanitary conditions may develop.

Typical solid waste container storage locations for low- and medium-rise apartment buildings include basement storage and outdoor storage. In large high-rise apartments, the waste storage and processing equipment is located in the basement of the building.

Commercial and Industrial Facilities. The siting of containers at existing commercial and industrial facilities depends on both the location of available space and service-access conditions (see Fig. 7-11). In many of the newer designs, specific service areas are included for this purpose. Often, because the containers are not owned by the commercial or industrial activity, the locations and types of containers to be used for onsite storage must be worked out jointly between the building owners and the public or private collection agency.

Public Health and Aesthetics

Although residential solid wastes account for a relatively small proportion of the total wastes generated in the United States (10 to 15 percent), they are perhaps the most important because they are generated in areas with limited storage space. As a result, they can have significant public health and aesthetic impacts.

Public health concerns are related primarily to the infestation of areas used for the storage of solid wastes with vermin and insects that often serve as potential disease vectors. By far the most effective control measure for both rats and flies is proper sanitation. Typically, proper sanitation involves the use of containers with tight lids, the periodic washing of the containers as well as of the storage areas, and the periodic removal of biodegradable materials (usually within less than 8 days), which is especially important in areas with warm climates. An excellent description of solid waste–disease relationships may be found in Ref. 2.

Aesthetic considerations are related to the production of odors and the unsightly conditions that can develop when adequate attention is not given to the maintenance of sanitary conditions. Most odors can be controlled through the use of containers with tight lids and with the maintenance of a reasonable collection frequency. If odors persist, the contents of the container can be sprayed with a masking deodorant as a temporary expedient. To maintain aesthetic conditions, the container should be scrubbed and washed periodically.

7-5 PROCESSING OF SOLID WASTES AT RESIDENTIAL DWELLINGS

Waste processing is used to (1) reduce the volume, (2) recover usable materials, or (3) alter the physical form of the solid wastes. The most common onsite processing operations used at low-rise detached residential dwellings include food waste grinding, component separation, compaction, incineration (in fireplaces), and composting. Backyard incineration, once a common processing technique used to reduce the volume of waste, is no longer allowed in most urban areas. Processing operations used at low-, medium-, and high-rise apartments include food waste grinding, component separation, and compaction. Typical onsite processing operations and facilities are listed by source in Table 7-6.

Grinding of Food Wastes

In the past 20 years, the use of kitchen food waste grinders has gained such wide acceptance that, in some areas, nearly all new homes are equipped with them and retrofitting of older homes is common. Food waste grinders are used primarily for wastes from the preparation, cooking, and serving of foods. Most grinders sold for home use cannot be used for large bones or other bulky items.

Functionally, grinders render the material that passes through them suitable for transport through the sewer system. Because the organic material added to the wastewater has resulted in overloading many treatment facilities, it has been necessary in some communities to forbid the installation of waste food grinders in new developments until additional treatment capacity becomes available.

Where food waste grinders are used extensively, the weight of waste collected per person will tend to be lower (see Example 6-4). In terms of the col-

TABLE 7-6
Typical operations and facilities used for the processing of solid waste at the source of generation

Source	Persons responsible	Operations and facilities
Residential dwellings		
Low-rise detached	Residents, tenants	Grinding, component separation, compaction, composting
Low- and medium-rise	Tenants	Grinding, component separation, compaction, combustion (fireplace)
	Building maintenance crews, contract services	Compaction, component separation, composting
High-rise	Tenants	Grinding, component separation, compaction, combustion (fireplace)
	Building maintenance crews, contract services	Compaction, component separation, combustion, shredding, hydropulping
Commercial	Janitorial services	Component separation, compaction, shredding, combustion, hydropulping
Industrial	Janitorial services	Component separation, compaction, shredding, combustion, hydropulping
Open areas	Owners, park operators	Compaction, separation of waste components
Treatment plant sites	Plant operators	Dewatering facilities
Agricultural	Owner, workers	Varies with individual commodity

lection operation, the use of home grinders does not have a significant impact on the volume of solid wastes collected. Even the weight difference is not major. In some cases where grinders are used, it has been possible to increase the time period between collection pickups because wastes that might readily decay are not stored.

Separation of Wastes

As noted earlier, the separation of solid waste components at the source of generation is one of the most effective ways to achieve the recovery and reuse of materials. Where the remaining wastes are to be combusted, the question that must be answered is, what is the energy content of residual solid waste? The effect of home separation on the energy content of the residential MSW is considered in Example 7-3.

Example 7-3 Effect of home separation of waste on energy content of *as-collected* residential MSW. Using the computation table prepared in Example 4-3 (reproduced below), estimate the energy content in Btu/lb of the remaining solid wastes if 60 percent of the paper and 90 percent of the cardboard are separated by the homeowner.

Component	Solid wastes,[a] lb	Energy,[b] Btu/lb	Total energy, Btu
Organic			
Food wastes	9.0	2,000	18,000
Paper	34.0	7,200	244,800
Cardboard	6.0	7,000	42,000
Plastics	7.0	14,000	98,000
Textiles	2.0	7,500	15,000
Rubber	0.5	10,000	5,000
Leather	0.5	7,500	3,750
Yard wastes	18.5	2,800	51,800
Wood	2.0	8,000	16,000
Inorganic			
Glass	8.0	60	480
Tin cans	6.0	300	1,800
Aluminum	0.5	—	—
Other metal	3.0	300	900
Dirt, ash, etc.	3.0	3,000	9,000
Total	100.0		506,530

[a] From Table 3-4.

[b] From Table 4-5.

Solution

1. The total energy content of 100 lb of solid waste, with the composition given in Table 3-4, is equal to 506,530 Btu.

2. Determine the energy content and weight of 60 percent of the paper in the original sample.
   (*a*) Energy content, 60% paper

$$0.60 \times 244{,}800 \text{ Btu} = 146{,}880 \text{ Btu}$$

   (*b*) Weight, 60% paper

$$0.60 \times 34 \text{ lb} = 20.4 \text{ lb}$$

3. Determine the energy content and weight of 90 percent of the cardboard in the original sample.
   (*a*) Energy content, 90% cardboard

$$0.90 \times 42{,}000 \text{ Btu} = 37{,}800 \text{ Btu}$$

   (*b*) Weight, 90% cardboard

$$0.90 \times 6 \text{ lb} = 5.4 \text{ lb}$$

4. Determine the total energy content, weight, and energy content per pound of the original sample after paper and cardboard have been separated.
   (*a*) Total energy after recovery

$$(506{,}530 - 146{,}880 - 37{,}800) = 321{,}850 \text{ Btu}$$

   (*b*) Total weight after recovery

$$(100 - 20.4 - 5.4) \text{ lb} = 74.2 \text{ lb}$$

   (*c*) Energy content of waste per lb after separation

$$\frac{321{,}850 \text{ Btu}}{74.2 \text{ lb}} = 4338 \text{ Btu/lb } (10{,}090 \text{ kJ/kg}) \text{ versus } 5065 \text{ Btu/lb } (11{,}781 \text{ kJ/kg})$$

   in original sample

Comment. In this example, the removal by weight of approximately 26 percent of the wastes reduced the per-pound energy content of the original sample by approximately 14 percent.

Compaction

The two principal types of compactors used for the processing of wastes at residential dwellings are (1) small (individual) home and apartment compaction units and (2) large compactors used to compact wastes from a large number of apartments.

Home and Apartment Compaction Units. In the past few years, a number of small compactors designed for home use have appeared on the market. Manufacturers' claims for these units in terms of the compaction ratio usually are based on the compaction of loose paper and corrugated paper. Although compactors can reduce the original volume of wastes placed in them by up to 70 percent, they can be used for only a small proportion of the solid wastes generated. The effect of the use of home compactors on the volume of wastes collected is illustrated in Example 7-4.

Example 7-4 Effect of home compactors on volume of collected solid wastes. Assume that home compaction units are to be installed in a residential area. Estimate the volume reduction that could be achieved in the solid wastes collected if the compacted specific weight is equal to 540 lb/yd^3 and the data given in Tables 3-4 and 4-1 are applicable.

Solution

1. Set up a computation table to determine the volume of wastes as discarded in containers, using the data given in Tables 3-4 and 4-1.

Component	Weight,[a] lb	Specific weight,[b] lb/yd^3	Volume, yd^3 $\times 10^{-2}$
Organic			
Food wastes	9.0	490	1.84
Paper	34.0	150	22.67
Cardboard	6.0	167[c]	3.59
Plastics	7.0	110	6.36
Textiles	2.0	110	1.82
Rubber	0.5	220	0.23
Leather	0.5	270	0.19
Yard wastes	18.5[d]	170	10.88[d]
Wood	2.0[d]	400	0.50[d]
Inorganic			
Glass	8.0	330	2.42
Tin cans	6.0	150	4.00
Aluminum	0.5	270	0.19
Other metal	3.0[d]	540	0.56[d]
Dirt, ash, etc.	3.0[d]	810	0.37[d]
Total	100.0		55.62
			43.31[e]

[a] Data from Table 3-4.

[b] Data from Table 4-1.

[c] Cardboard partially compressed by hand before being placed in waste compactor.

[d] Components not usually placed in home waste compactors.

[e] Total excluding components not usually placed in home waste compactors.

2. Determine the volume of compacted wastes, excluding yard wastes; wood; metals other than aluminum and tin cans; and dirt, ashes, etc.

$$\text{Compacted volume} = \frac{(100 - 18.5 - 2 - 3 - 3)\ \text{lb}}{540\ \text{lb/yd}^3}$$

$$= \frac{73.5\ \text{lb}}{540\ \text{lb/yd}^3} = 0.136\ \text{yd}^3$$

3. Determine the volume reduction for the compressible material.

$$\text{Volume reduction} = \left(\frac{(0.433 - 0.136)\ \text{yd}^3}{0.433\ \text{yd}^3}\right) \times 100$$

$$= 69\%$$

4. Determine the overall volume reduction achieved with a home compactor, taking into account garden trimmings; wood; metals other than aluminum and tin cans; dirt, ashes, etc.

Overall volume reduction

$$= \left(\frac{0.556\ \text{yd}^3 - (0.136 + 0.109 + 0.005 + 0.006 + 0.004)\ \text{yd}^3}{0.556\ \text{yd}^3}\right) \times 100$$

$$= \left(\frac{0.556 - 0.260}{0.556}\right) \times 100$$

$$= 53\%$$

Comment. When the overall volume reduction is assessed, the significance of a home compactor is reduced. This finding is especially true as the percentages of the components not compacted, such as garden trimmings, increase.

The use of home compactors may also be counterproductive from the standpoint of subsequent processing operations. For example, if the wastes are to be separated mechanically into components at a MRF, the compacted wastes will have to be broken up again before sorting. Also by compacting, the wastes may become so saturated with the liquids present in the food wastes that the recovery of paper or other components may not be feasible because product specifications may not be met.

Compactors for Large Apartment Buildings. To reduce the volume of solid wastes that must be handled, compaction units commonly are installed in large apartment buildings. Typically, a compactor is installed at the bottom of a solid waste chute (see Fig. 7-13). Wastes falling through the chute activate the compactor by means of photoelectric cells or limit switches. Once these switches are activated, the wastes are compressed. Depending on the design of the compactor, the compressed wastes may be formed into bales or extruded and loaded automatically into metal containers or paper bags.

When a bale has been formed or a container or bag is filled, the compactor shuts down automatically and a warning light turns on. The operator must then tie and remove the bale from the compactor, or remove the full bag and replace it with an empty one. In some applications the use of completely automatic equipment may be warranted. In sizing compaction equipment for use in conjunction with solid waste chutes in apartments, it is common to use the same assumptions used in sizing the chutes (see Section 7-2).

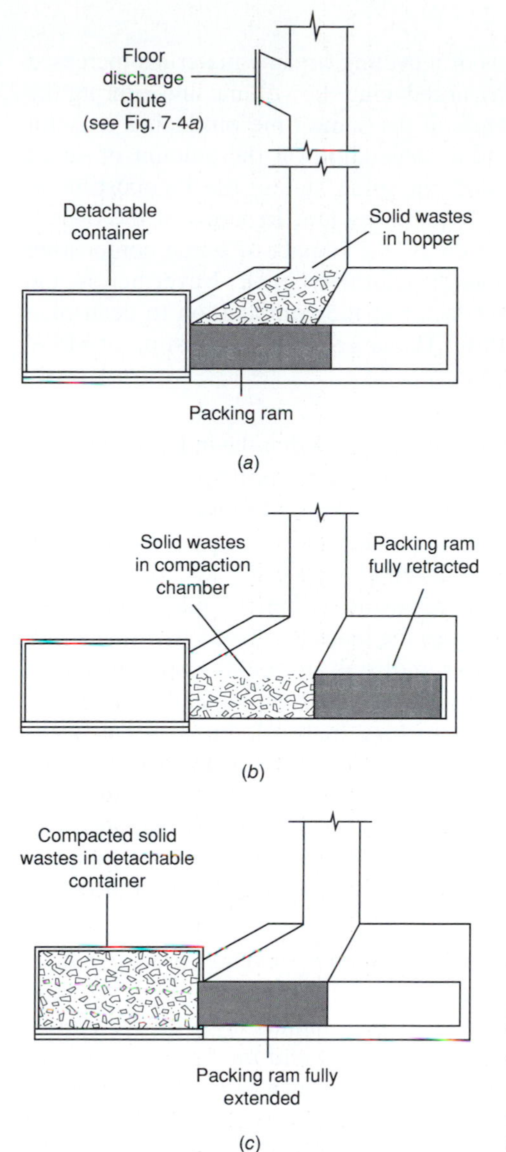

FIGURE 7-13
Typical compactor used in conjunction with waste chutes in large apartment buildings: (*a*) start of compaction cycle; (*b*) loading of compaction chamber; and (*c*) compaction into container.

While the use of compactors reduces the bulk volume of the wastes to be handled, the weight of course remains the same. Typically, the compacted volume will vary from 20 to 60 percent or less of the original volume. Unless baled solid wastes are broken up, it is impossible to recover individual components from compacted wastes. Where solid waste incinerators are used, the compacted wastes must be broken up to avoid delayed combustion in the furnace and high losses of unburned combustible materials. All these factors must be considered when the use of onsite compactors is being evaluated.

Composting

In the 1970s home composting as a means of recycling organic materials increased in popularity [4, 5]. It is an effective way of reducing the volume and altering the physical composition of solid wastes while at the same time producing a useful by-product. A variety of methods are used, depending on the amount of space available and the wastes to be composted. In some states, the composting of leaves by individual homeowners is now required by law. In terms of the overall waste management problems facing most cities, the impact of home composting on the volume of solid wastes to be handled is relatively small. Nevertheless, the composting of leaves can be a significant factor in the computation to determine the quantity of waste diverted from landfills. The large-scale composting of MSW is considered further in Chapters 9 and 14.

Backyard Composting. Backyard composting requires that the individual homeowner develop some method of composting yard wastes, principally leaves and grass clippings [3]. If they are chipped, brush, stumps, and wood are also compostable. The simplest backyard composting method involves placement of the material to be composted in a pile and occasionally watering and turning it to provide moisture and oxygen to the microorganisms within the pile. During the composting period, which can take up to year, the material placed in the pile will undergo bacterial and fungal decompostion until only a humus material known as *compost* remains. The composted material, which is biologically stabilized, can be used as a soil amendment or as a mulching material. Examples of composting units that can be built by homeowners are illustrated in Fig. 7-14. Many homeowners prefer a more organized approach to composting, using one of the many prefabricated composting units now available (see Fig. 7-15). A number of additives are also available to enhance the composting process. Other imaginative backyard compost systems are being developed even as you read this discussion.

Lawn Mulching. Another type of composting involves leaving grass clippings from a newly mowed lawn where they were cut. If the grass clippings are short enough they will fall through the upright grass to the humus layer on the ground surface. In time, the grass clippings will be composted and incorporated into the humus. Allowing the grass clippings to remain on the lawn not only reduces the amount of waste generated at the source, but also allows for the the recycling of nutrients.

Combustion

In the past, the burning of combustible materials in fireplaces and the burning of rubbish in backyard incinerators was common practice. Backyard incineration is now banned in most parts of the country. The impact of burning combustible wastes in home fireplaces on the amount of waste collected depends on the location and the length of the burning season. Elimination of backyard burning significantly

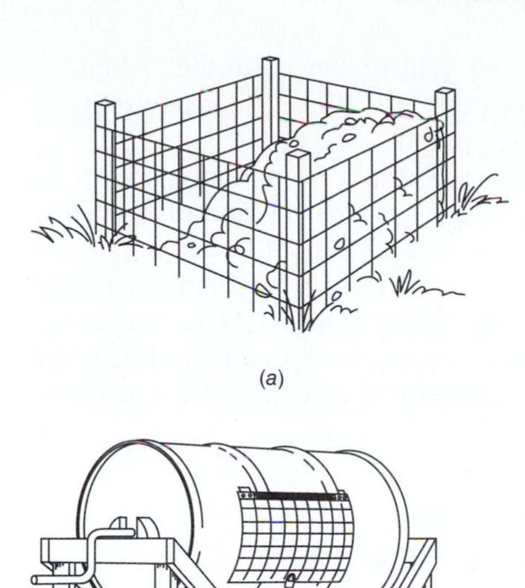

(a)

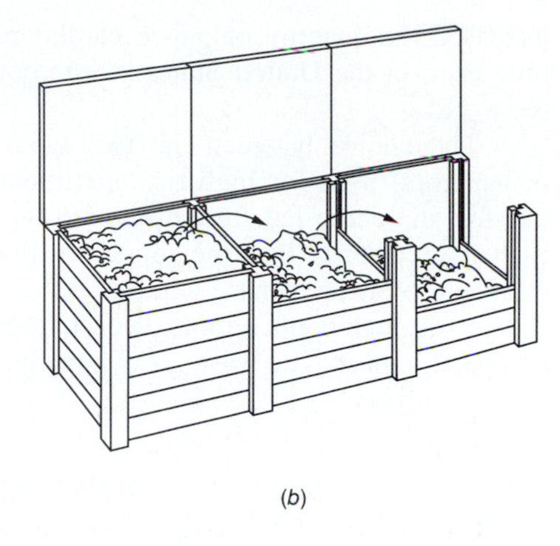

(b)

(c)

FIGURE 7-14
Composting units that can be built by homeowners: (*a*) simple fenced enclosure, (*b*) three-bin system used to accelerate the compost process, and (*c*) composter constructed from 55-gal drum. (Courtesy of Minnesota Pollution Control Agency.)

FIGURE 7-15
Commercial composting unit used for the production of compost at residences.

increased the quantity of paper, cardboard, and yard wastes collected. Although little used in the United States, waste combustion is still used in many parts of the world.

The choice between the two types of combusters (commonly known as incinerators) used for high-rise apartments depends on the method of charging: flue-fed and chute-fed. In the flue-fed type, wastes are charged through doors on each floor directly into the refractory flue, the bottom of which opens directly into the top of the furnace combustion chamber. In the chute-fed type, wastes are charged through hopper doors on each floor into a metal chute, and they collect in a basement hopper. The wastes are then either manually or mechanically transferred into the furnace.

7-6 PROCESSING OF SOLID WASTES AT COMMERCIAL AND INDUSTRIAL FACILITIES

Onsite processing operations carried out at commercial-industrial facilities are generally similar to those described for residential sources. However, compaction is very important in commercial facilities. Other differences occur primarily in industrial facilities. Because most of the processes tend to be industry-specific, no attempt has been made to document the various processes that have been used.

Compaction

The baling of waste cardboard at markets and other commercial establishments is quite common. The bales vary in size, but typically are about $36 \times 48 \times 60$ in. Baled cardboard is reprocessed for the production of packing materials or shipped overseas for remanufacture into a variety of products.

Shredding and Hydropulping

Shredding and pulping are alternative processing operations that have been used, both in conjunction with the previous methods and by themselves, for reducing the volume of wastes that must be handled. Shredding is used most commonly in commercial establishments and by governmental agencies to destroy sensitive documents that are no longer of value. In some cases, the volume of wastes has been observed to increase after shredding.

Although hydropulping systems work well, they are expensive and typically involve discharge to the local wastewater collection system. Because the discharge of pulped material increases the organic loading on local treatment facilities, the use of pulverizers may be restricted if treatment capacity is limited.

7-7 DISCUSSION TOPICS AND PROBLEMS

7-1. Tour your community and make a brief survey of the different types of containers now used for the collection of solid waste.

7-2. Estimate the maximum amount (by weight) of material that could be separated for recycling from residential MSW using a two-container system. What should be the size of the two containers? Use the waste composition given in Table 3-4, and assume that there are 3.1 residents per residence.

7-3. Obtain a component distribution of the solid wastes generated in your community, and determine the percentage reduction that could be achieved if home compactors were installed. Assume the specific weight of the compacted wastes to be 20 lb/ft^3. Compare your answer with that derived in Example 7-4.

7-4. List the advantages and disadvantages associated with the separation of solid wastes in high-rise apartment buildings.

7-5. Assuming that the quantities of solid wastes generated daily at a commercial facility are distributed normally (see Appendix D), with a mean value of 10 yd^3 and a standard deviation of 7 yd^3, what size container would you recommend for this facility? What are the important tradeoffs in the selection of container size?

7-6. Using the data presented in Fig. 7-5, estimate the size of a container to be used with a gravity chute for a 24-story apartment building with 192 individual living units if the container will be emptied daily. Assume that the average occupancy rate for each living unit is 3.1 persons.

7-7. What compactor volume displacement (e.g., capacity), expressed in terms of cubic yards per hour, would you recommend for use in the 24-story apartment building of Problem 7-6?

7-8. As a consulting engineer, you have been commissioned to develop a comprehensive solid waste system for a community interested in achieving a greater recovery and reuse of its solid wastes. Two of the possible alternatives are separation at the home or separation at a materials recovery facility. What important factors must be considered in evaluating these two alternatives?

7-9. Find out what is now being done in your community with respect to the collection of household hazardous wastes. Do you believe the program is effective? What can you suggest to improve the effectiveness of the program?

7-8 REFERENCES

1. Bergen County Utilities Authority: *Bergen County Apartment Recycling Manual,* Little Ferry, NJ, 1988.
2. Hanks, T. G.: *Solid Waste/Disease Relationships,* U.S. Department of Health, Education, and Welfare, Solid Wastes Program, Publication SW–1c, Cincinnati, OH, 1967.
3. Strom, P. and M. Finstein: *Leaf Composting Manual for New Jersey Municipalities,* New Jersey Department of Environmental Protection, Division of Solid Waste Management, Office of Recycling, Trenton, NJ, 1986.
4. *The BioCycle Guide to Composting Municipal Wastes,* BioCycle, Emmaus, PA, 1989.
5. U.S. Environmental Protection Agency: *Decision-Makers Guide to Solid Waste Management,* EPA/530-SW89-072, Washington, DC, November 1989.

CHAPTER 8

COLLECTION OF SOLID WASTE

Collection of unseparated (commingled) and separated solid waste in an urban area is difficult and complex because the generation of residential and commercial-industrial solid waste takes place in every home, every apartment building, and every commercial and industrial facility as well as in the streets, parks, and even vacant areas. The mushroom-like development of suburbs all over the country has further complicated the collection task.

As the patterns of waste generation become more diffuse and the total quantity of waste increases, the logistics of collection become more complex. Although these problems have always existed to some degree, they have now become more critical because of the high costs of fuel and labor. Of the total amount of money spent for collection, transportation, and disposal of solid wastes in 1992, approximately 50 to 70 percent was spent on the collection phase. This fact is important because a small percentage improvement in the collection operation can effect a significant savings in the overall cost.

The collection operation is considered from four aspects in this chapter: (1) the types of collection services that are provided, (2) the types of collection systems and some of the equipment now used as well as the associated labor requirements, (3) an analysis of collection systems, including the component relationships that can be used to quantify collection operations, and (4) the general methodology involved in setting up collection routes.

8-1 WASTE COLLECTION

The term *collection,* as noted in Chapter 1, includes not only the gathering or picking up of solid wastes from the various sources, but also the hauling of these wastes to the location where the contents of the collection vehicles are emptied. The unloading of the collection vehicle is also considered part of the collection operation. While the activities associated with hauling and unloading are similar for most collection systems, the gathering or picking up of solid waste will vary with the characteristics of the facilities, activities, or locations where wastes are generated (see Table 3-1) and the methods used for onsite storage of accumulated wastes between collections. The principal types of collection services now used for unseparated (commingled) and separated wastes are described below.

Collection of Unseparated (Commingled) Waste

The collection of wastes from low-rise detached dwellings, from medium-rise apartments, from high-rise apartments, and from commercial/industrial facilities is considered in the following discussion. The collection of wastes separated at the source is considered following this discussion.

From Low-Rise Detached Dwellings. The most common types of residential collection services for low-rise detached dwellings include (1) curb, (2) alley, (3) setout-setback, and (4) setout. Where curb service is used, the homeowner is responsible for placing the containers to be emptied at the curb on collection day and for returning the empty containers to their storage location until the next collection. Where alleys are part of the basic layout of a city or a given residential area, alley storage of containers used for solid waste is common. In setout-setback service, containers are set out from the homeowner's property and set back after being emptied by additional crews that work in conjunction with the collection crew responsible for loading the collection vehicle. Setout service is essentially the same as setout-setback service, except that the homeowner is responsible for returning the containers to their storage location. These services are compared in Table 8-1.

Manual methods used for the collection of residential wastes include (1) the direct lifting and carrying of loaded containers to the collection vehicle for emptying, (2) the rolling of loaded containers on their rims to the collection vehicle for emptying, and (3) the use of small lifts for rolling loaded containers to the collection vehicle. Large containers (referred to as *tote containers*) or drop cloths (often called *tarps*), into which wastes from small containers were emptied before being carried and/or rolled to the collection vehicle, are still used in some communities. For curb collection where collection vehicles with low loading heights are used, the collection crew transfers the wastes directly from the containers in which they are stored or carried to the collection vehicle (see Fig. 8-1). In other cases, collection vehicles are equipped with auxiliary containers into which the wastes are emptied. The auxiliary containers are then emptied into the collection vehicle by mechanical means. Still another variant involves the use of small satellite vehicles. Wastes are emptied into a large container carried by a satellite vehicle. When the container is loaded, the satellite vehicle is driven to

TABLE 8-1
Comparison of residential MSW collection services

Considerations	Type of service					
	Curb	Curb (mechanized)	Alley	Setout-setback	Setout	Backyard carry
Requires homeowner cooperation:						
To carry full containers	Yes	Yes	Optional	No	No	No
To carry empty containers	Yes	Yes	Optional	No	Yes	No
Requires scheduled service for homeowner cooperation	Yes	Yes	No	No	Yes	No
Poor aesthetically:						
Spillage and litter problem	High	Moderate	High	Low	High	Low
Containers visible	Yes	Yes	No	No	Yes	No
Attractive to scavengers	Yes	Yes	Highest	No	No	No
Prone to upsets	Yes	No	Yes	No	Yes	No
Number of persons in crew						
Typical	2	2	1	3	3	3
Range	1 to 3	1 to 2	1 to 3	3 to 7	1 to 5	3 to 5
Crew time	Low	Low	Low	Great	Medium	Medium
Collector injury rate due to lifting and carrying	Low	Low	Low	High	Medium	High
Trespassing complaints	Low	Low	Low	High	High	High
Special considerations		Requires standardized containers	Requires alleys and vehicles that can maneuver in them; less prone to block traffic; high vehicle and container depreciation rate			Requires wheeled caddy to roll filled barrels or the use of burlap carry cloth or hand-carry bin; works best with driveway
Cost due to size and time requirements	Low	Low	Low	High	Medium	Medium

(a)

(b)

FIGURE 8-1
Collection of wastes placed at curb by homeowner: (*a*) Davis, California and (*b*) Venice, Italy.

the collection vehicle where the container is emptied into the truck by mechanical means (see Fig. 8-2).

Where mechanically loaded collection vehicles are used, the container used for the onsite storage of waste must be brought to the curb or other suitable collection site. Typically, 90-gal (340-L) containers are used in conjunction with mechanized collection vehicles (see Fig. 8-3).

From Low- and Medium-Rise Apartments. Curbside collection service is common for most low- and medium-rise apartments. Typically, the maintenance staff is responsible for transporting the containers to the street for curbside collection by manual or mechanical means. Where large containers are used, the containers are emptied mechanically using collection vehicles equipped with unloading mechanisms.

From High-Rise Apartments. Typically, large containers are used to collect wastes from large apartment buildings. Depending on the size and type of container used, the contents of the containers may be emptied mechanically using

(a)

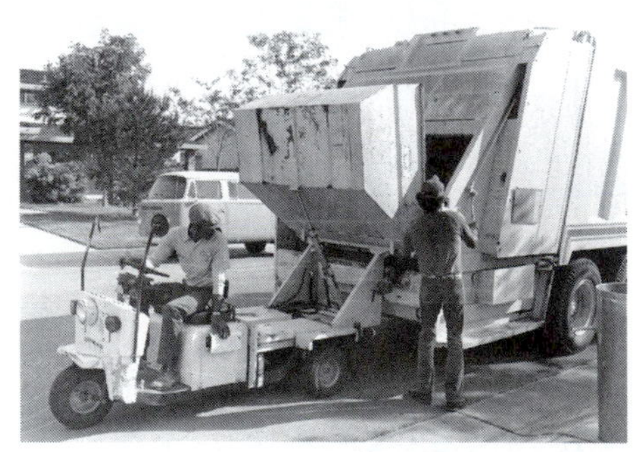

(b)

FIGURE 8-2
Satellite vehicle collection system: (*a*) loading of satellite vehicle equipped with 2-yd^3 container (note that in high-wind conditions, blowing of wastes is a problem) and (*b*) mechanical unloading of container contents of satellite vehicle.

(a) (b)

FIGURE 8-3
Typical example of mechanically loaded collection vehicle used for the collection of residential wastes: (*a*) collection of domestic wastes placed in large 90-gal (340-L) containers with mechanical articulated pickup mechanism and (*b*) closeup of pickup mechanism. Containers are brought to the curb by the homeowner.

collection vehicles equipped with unloading mechanisms (see Fig. 8-4), or the loaded containers may be hauled to an off-site location (e.g., a materials recovery facility) where the contents are unloaded (see Fig. 8-5).

From Commercial-Industrial Facilities. Both manual and mechanical means are used to collect wastes from commercial facilities. To avoid traffic congestion during the day, solid wastes from commercial establishments in many large cities are collected in the late evening and early morning hours. Where manual collection is used, wastes from commercial establishments are put into plastic bags, cardboard boxes and other disposable containers that are placed at the curb for collection. Waste collection is usually accomplished with a three- or, in some cases, four-person crew, consisting of a driver and two or three collectors who load the wastes from the curbside into the collection vehicle. In most off-hour collection operations, the driver remains with the collection vehicle for reasons of safety.

(a) (b)

FIGURE 8-4
Self-loading collection vehicle equipped with internal compactor: (*a*) approaching container to be emptied and (*b*) contents of container being emptied.

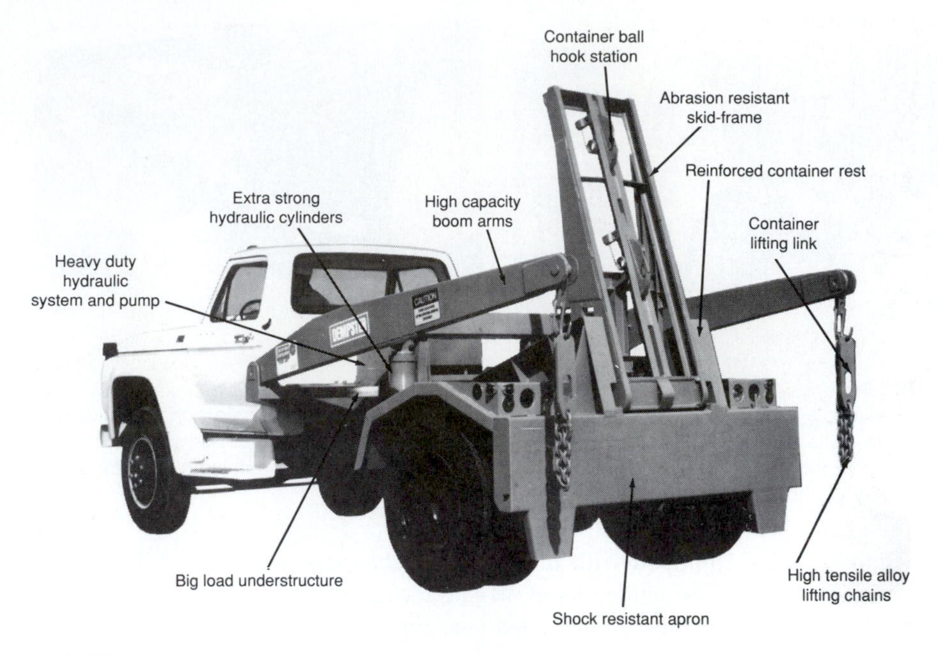

FIGURE 8-5
Collection vehicle used to haul and empty large containers (2 to 12 yd^3). Container hoist and unloading mechanism is mounted on truck frame. (Courtesy of Dempster Dumpster Systems.)

If congestion is not a major problem and space for storing containers is available, the collection service provided to commercial-industrial facilities centers on the use of movable containers (see Fig. 8-6), containers that can be coupled to large stationary compactors (see Fig. 8-7), and large-capacity open-top containers (see Fig. 8-8). Again, depending on the size and type of container used, the contents of the containers may be emptied mechanically

(*a*)

(*b*)

FIGURE 8-6
Emptying sequence for containers used at commercial complex: (*a*) loaded containers are brought and attached to collection vehicle and (*b*) contents of container are emptied mechanically.

FIGURE 8-7
Stationary compactor used in conjunction with large container for the collection of wastes from commercial establishments.

or the loaded containers hauled to an off-site location where the contents are unloaded. To minimize the difficulties due to traffic congestion, mechanized collection can also be accomplished during the evening hours with a driver and helper.

Collection of Wastes Separated at the Source

Waste materials that have been separated at the source must be collected or gathered together before they can be recycled. The principal methods now used for the collection of these materials include curbside collection using conventional and specially designed collection vehicles, incidental curbside collection by charitable organizations, and delivery by homeowners to drop-off and buy-back centers. The curbside collection of wastes separated at the source is considered in the following discussion. The recovery of materials at drop-off and buy-back centers is considered in Chapter 9.

FIGURE 8-8
Large open-top containers used for the collection of wastes from commercial establishments.

Residential Curbside Collection. In a curbside system, source-separated recyclables are collected separately from commingled waste at the curbside, alley, or commercial facility. Because residents and businesses do not have to transport the recyclables any further than the curb, participation in curbside programs is typically much higher than for drop-off programs. Curbside programs vary greatly from community to community. Some programs require residents to separate several different materials (e.g., newspaper, plastic, glass, and metals), which are then stored in their own containers and collected separately. Other programs use only one container, to store commingled recyclables, or two containers, one for paper and the other for "heavy" recyclables, such as glass, aluminum and tin cans. Clearly, the method used to collect source-separated wastes will affect directly the layout and design of separation and processing facilities. The principal types of collection vehicles used for the collection of separated wastes are (1) standard collection vehicles and (2) specialized collection vehicles, including closed-body recycling trucks, recycling trailers, modified flatbed trucks, open-bin recycling trucks, and compartmentalized trailers. Four commonly used vehicles for the collection of source-separated waste are shown in Fig. 8-9. The characteristics of these specialized vehicles are reviewed in Table 8-2. Determination of the amount of each waste to be collected is illustrated in Example 8-1.

(*a*)

(*b*)

(*c*)

(*d*)

FIGURE 8-9
Specially designed vehicles for the collection of source-separated waste: (*a*) open bin manually loaded, (*b*) open bin manually loaded mechanically emptied, (*c*) containers mechanically emptied, and (*d*) mobile container system.

TABLE 8-2
Characteristics of vehicles used for the collection of wastes separated at the source[a]

Item	Comment
Standard compactor trucks	Compactor trucks used for the collection of commingled waste can also be used for collection of recyclables. Many communities use compactor trucks in their recycling programs. Rear-loader compactors have been used for newspaper, cardboard and magazines, with trailers attached to them for cans and glass. Front-end loaders have been used to service large containers containing newspaper recovered from apartment buildings. Some cities used side- and rear-loading compactor trucks to pick up newspaper one week and glass and cans the next. When glass and cans are collected, the compacting mechanism is not used because glass is highly abrasive and would damage the packer plate. Also, by not compacting, the majority of the glass remains unbroken and is, therefore, easier to sort into different colors at the processing site.
Mobile container system	The mobile container system is essentially a steel frame with sets of hydraulic forks that can be used to transport large bins. The trailers range in size from three bins to six bins and have a low pull or gooseneck (fifth wheel) style. To load the trailer, the fork lifts are lowered to the ground and the bins are wheeled over them so that the forks slide into channels on the underside of the bins. The bins are then hydraulically raised and secured to the trailer frame. An empty set of bins can be left to replace the full ones. A pickup truck is used to pull the trailer.
Modified flatbed truck	Some curbside programs utlize a standard flatbed truck with a hydraulic dumping box mounted on the truck bed. The box is usually divided into three or four compartments and has a standard capacity of approximately 15 cubic yards.
Open-bin recycling truck	The open-bin recycling truck is a specially designed vehicle with two or three open-top, self-dumping bins. The front bins are typically 4 to 5 cubic yards and can be specified to unload right or left. The back bin, which dumps to the rear, has a capacity of 8 to 9 cubic yards. The cab can be designed for the right-hand stand-up drive to allow the loading function to be performed by the driver.
Closed-body recycling truck	This truck consists of an enclosed steel body installed on a lowered truck chassis, and a low entry walk-in cab with dual left- and right-hand driving controls (which allows for one-person operation). There are adjustable hinged dividers on the body that can be used to create from two to four compartments for different materials. One or both sides are open for manual loading. Removable aluminum side panels contain the load as the level of material rises. The overall capacity of the truck can range from 27 to 31 cubic yards, although operational capacity when manually loading is 20 to 25 cubic yards. The truck is equipped with a front-mounted telescopic hoist and rear body hinge for dumping. Each compartment is discharged separately by opening the rear door, unlocking the appropriate divider, and tipping the body.

[a] Adapted from Ref. 1.

Example 8-1 Home separation and curbside collection of recyclables. A community is purchasing specialized vehicles for the curbside collection of source-separated wastes. Three recycling containers are to be provided to each residence and residents will be asked to separate newspaper and cardboard, plastics and glass, and aluminum and tin cans. The homeowner is to place the separated materials in the appropriate containers and then move the recycling containers to curbside once per week for collection by special recycling vehicles. Estimate the relative volumetric capacity required for each material in recycling collection vehicles. Assume 80 percent of the recyclable material will be separated and that newsprint represents 20 percent of the total paper waste. The number of homes that will participate in the separation program is estimated to be 60 percent. If the separated wastes are to be collected from a subdivision of 1200 homes, determine the number of trips that will be required if the size of the collection vehicle is 15 yd^3. Assume 3.5 residents per residence.

Solution

1. Set up a computation table to calculate the relative volume of the recycled materials. Use the weight distribution data given in column 4, Table 3-7, and the specific weight values given in Table 4-1.

Component	Total solid wastes, lb	Waste materials separated, lb	Specific weight, lb/ft^3	Volume, ft^3
Organic				
Food wastes	8.0	—	18.0	—
Paper	35.8	5.7[a]	5.6	1.02
Cardboard	6.4	5.1	3.1	1.65
Plastics	6.9	5.5	4.1	1.34
Textiles	1.8	—	4.1	—
Rubber	0.4	—	8.1	—
Leather	0.4	—	10.0	—
Garden trimmings	17.3	—	6.3	—
Wood	1.8	—	14.8	—
Inorganic				
Glass	9.1	7.3	12.2	0.60
Tin cans	5.8	4.6	5.6	0.82
Aluminum	0.6	0.5	10.0	0.05
Other metal	3.0	—	20.0	—
Dirt, ashes, etc.	2.7	—	30.0	—
Total	100.0	28.7		5.48

[a] 5.7 lb = [(35.8 lb × 0.20) × 0.8]

2. Determine the relative volume of the recycled materials.
   (*a*) The volume in each category is:
      i. Newspaper + cardboard = 1.02 + 1.65 = 2.67 ft^3
      ii. Plastics + glass = 1.34 + 0.60 = 1.94 ft^3
      iii. Aluminum and tin cans = 0.82 + 0.05 = 0.87 ft^3

(*b*) The relative volumes of waste compared with aluminum plus tin cans are

i. Newspaper + cardboard = 3.1 (= 2.67 ft^3/0.87 ft^3)

ii. Plastics + glass = 2.2 (= 1.94 ft^3/0.87 ft^3)

iii. Aluminum and tin cans = 1.0

(*c*) Thus, if a 15 yd^3 collection vehicle is to be used, 7.3 yd^3 [(2.67/5.48) × 15] of the capacity would be allocated for newspaper and cardboard, 5.3 yd^3 for plastic and glass, and 2.4 yd^3 for aluminum and tin cans.

3. Determine the number of trips required to collect the separated materials.

(*a*) Estimate the total weekly solid waste production rate using the data from Table 6-3.

Solid waste production, lb/wk

$$= 3.5 \text{ persons } \times 7 \text{ d/wk } \times 3.82 \text{ lb/capita} \cdot \text{d } = 93.6 \text{ lb/wk}$$

(*b*) Estimate the total weekly quantity of separated newspaper and cardboard.

i. Separated newspaper, lb/wk

$$= 93.6 \text{ lb/wk} \times (5.7/100) = 5.3 \text{ lb/wk}$$

ii. Separated cardboard, lb/wk

$$= 93.6 \text{ lb/wk} \times (5.1/100) = 4.8 \text{ lb/wk}$$

(*c*) Estimate the total weekly volume of separated newspaper and cardboard.

i. Separated newspaper, ft^3/wk

$$= (5.3 \text{ lb/wk})/(5.6 \text{ lb/ft}^3) = 1.0 \text{ ft}^3\text{/wk}$$

ii. Separated cardboard, ft^3/wk

$$= (4.8 \text{ lb/wk})/(3.1 \text{ lb/ft}^3) = 1.5 \text{ ft}^3\text{/wk}$$

(*d*) Estimate the total number of weekly collection trips

Number of trips

$$= [(1.0 + 1.5) \text{ ft}^3\text{/wk} \cdot \text{ home}] \times 1200 \text{ homes}$$

$$\times 0.60 \text{ (percent participation rate)}/(27 \text{ ft}^3\text{/yd}^3)/(7.3 \text{ yd}^3\text{/trip})$$

$$= 9.1 \text{ trips/wk, say 9 trips/wk}$$

Comment. Although the numbers in this example will change, the approach is valid for any collection operation. In applying such an analysis, it will be important to prepare a sensitivity analysis to assess how variable the relative volumes may become as waste characteristics change and new regulations are implemented.

Commercial Facilities. Source-separated materials from commercial establishments are usually collected by private haulers. In many cases, the haulers have contracts with the establishments for the separated material. The wastes to be recycled are stored in separate containers. In some cities, cardboard is bundled and left at curbside where it is collected separately. In large commercial facilities baling equipment may be used for the paper and cardboard, and can crushers are used for the aluminum cans. Commingled MSW, generated in addition to the separated materials, is most commonly collected by private haulers or by city crews, if the city provides collection services.

8-2 TYPES OF COLLECTION SYSTEMS, EQUIPMENT, AND PERSONNEL REQUIREMENTS

Over the past 10 years a wide variety of systems and equipment have been used for the collection of solid wastes. These systems may be classified from several points of view, such as the mode of operation, the equipment used, and the types of wastes collected. In this text, collection systems have been classified according to their mode of operation into two categories: (1) hauled container systems (HCS) and (2) stationary container systems (SCS) [4]. In the former, the containers used for the storage of wastes are hauled to the disposal site, emptied, and returned to either their original location or some other location. In the latter, the containers used for the storage of wastes remain at the point of generation, except when they are moved to the curb or other location to be emptied. These two types of collection systems, and the corresponding personnel requirements for these systems are described in this section.

Hauled Container Systems

Hauled container systems are ideally suited for the removal of wastes from sources where the rate of generation is high, because relatively large containers are used (see Table 8-3). The use of large containers reduces handling time as well as the unsightly accumulations and unsanitary conditions associated with the use of numerous smaller containers. Another advantage of hauled container systems is their flexibility: containers of many different sizes and shapes are available for the collection of all types of wastes.

Because containers used in this system usually must be filled manually, the use of very large containers often leads to low-volume utilization unless loading aids, such as platforms and ramps, are provided. In this context, container utilization is defined as the fraction of the total container volume actually filled with wastes.

While hauled container systems have the advantage of requiring only one truck and driver to accomplish the collection cycle, each container picked up requires a round trip to the disposal site (or other transfer point). Therefore, container size and utilization are of great economic importance. Further, when highly compressible wastes are to be collected and hauled over considerable distances, the economic advantages of compaction are obvious.

There are three main types of hauled container systems: (1) hoist truck, (2) tilt-frame container, and (3) trash-trailer. Typical data on the collection vehicles used with these systems are reported in Table 8-4. Cost data for these vehicles are presented in Appendix E.

TABLE 8-3
Representative data on the capacities of containers available for use with various collection systems

Vehicle	Container type	Typical range of container capacities,[a] yd^3
Hauled container system		
Hoist truck	Used with stationary compactor	6–12
Tilt-frame	Open top, also called drop or debris boxes	12–50
	Used with stationary compactor	15–40
	Equipped with self-contained compaction mechanism	20–40
Truck-tractor	Open-top trash-trailers	15–40
	Enclosed trailer-mounted containers equipped with self-contained compaction mechanism	20–40
Stationary container systems		
Compactor, mechanically loaded	Enclosed top and side-loading	1–8
Compactor, mechanically loaded	Special containers used for the collection of residential wastes from individual residences	0.23–0.45 (60–120 gal)
Compactor, manually loaded	Small plastic or galvanized metal containers, disposable paper and plastic bags	0.08–0.21 (20–55 gal)

[a] See Table 7-2 for typical container dimensions.

Note: $yd^3 \times 0.7646 = m^3$

$gal \times 0.003785 = m^3$

Hoist Truck Systems. In the past, hoist trucks were widely used with containers varying in size from 2 to 12 yd^3 (see Fig. 8-5). With the advent of large capacity mechanically loaded collection vehicles, however, this system appears to be applicable in only a limited number of cases, the most important of which are as follows:

1. For the collection of wastes by a collector who has a small operation and collects from only a few pickup points at which considerable amounts of wastes are generated. Generally, for such operations the purchase of newer and more efficient collection equipment cannot be justified economically.
2. For the collection of bulky items and industrial rubbish such as scrap metal and construction debris that are not suitable for collection with compaction vehicles.

Tilt-Frame Container Systems. Systems that use tilt frame–loaded vehicles (see Fig. 8-10) and large containers, often called *drop* or *debris boxes,* are ideally suited for the collection of all types of solid waste and rubbish from locations where the generation rate warrants the use of large containers. As noted in Table 8-3, various types of large containers are available for use with tilt-frame collection vehicles. Open-top containers are used routinely at warehouses and construction

TABLE 8-4
Vehicles used for the collection of solid waste

Collection vehicle			Typical overall collection vehicle dimensions				
Type	Available container or truck body capacities,[a] yd^3	Number of axles	With indicated container or truck body capacity,[b] yd^3	Width, in	Height, in	Length,[c] in	Unloading method
Hauled container systems							
Hoist truck	6–12	2	10	94	80–100	110–150	Gravity, bottom opening
Tilt-frame	12–50	3	30	96	80–90	220–300	Gravity, inclined tipping
Truck-tractor trash-trailer	15–40	3	40	96	90–150	220–450	Gravity, inclined tipping
Stationary container system							
Compactor (mechanically loaded)							
Front loading	20–45	3	30	96	140–150	240–290	Hydraulic ejector panel
Side loading	10–36	3	30	96	132–150	220–260	Hydraulic ejector panel
Rear loading	10–30	2	20	96	125–135	210–230	Hydraulic ejector panel
Compactor (manually loaded)							
Side loading	10–37	3	37	96	132–150	240–300	Hydraulic ejector panel
Rear loading	10–30	2	20	96	125–135	210–230	Hydraulic ejector panel

[a] See Tables 8-2 and 7-2.

[b] See Table 7-2 for dimensions of typical containers.

[c] From front of the truck to the rear of container or truck body.

Note: $yd^3 \times 0.7646 = m^3$

$in \times 0.0254 = m$

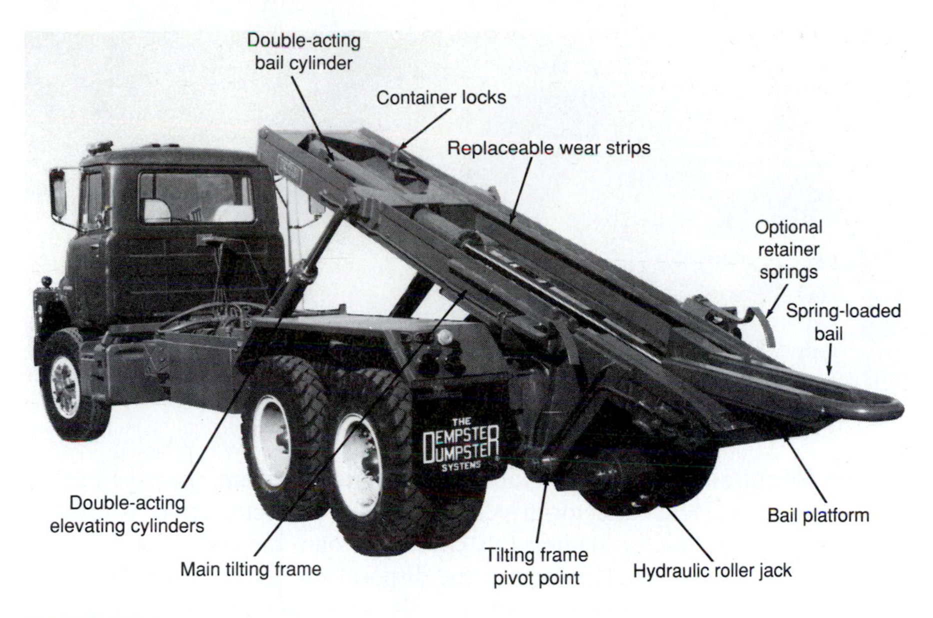

FIGURE 8-10
Truck with tilt-frame loading mechanism used to haul and unload large-capacity containers. (Courtesy of Dempster Dumpster Systems.)

sites (see Fig. 8-11). Large containers, in conjunction with stationary compactors, are common at apartment complexes, commercial services, and transfer stations. Because of the large volume that can be hauled, the use of the tilt-frame hauled container system has become widespread, especially among private collectors servicing commercial accounts.

Trash-Trailer Systems. The application of trash-trailers is similar to that for tilt-frame container systems. Trash-trailers are better for the collection of especially heavy rubbish, such as sand, timber, and metal scrap, and often are used for the collection of demolition wastes at construction sites (see Fig. 8-12).

FIGURE 8-11
Contents of large tilt frame–loaded container being emptied at landfill.

FIGURE 8-12
Contents of trash-trailer, used for demolition wastes, being unloaded at landfill.

Personnel Requirements for the Hauled Container System. In most hauled container systems a single collector-driver is used. The collector-driver is responsible for driving the vehicle, loading full containers onto the collection vehicle, emptying the contents of the containers at the disposal site (or transfer point), and redepositing (unloading) the empty containers. In some cases, for safety reasons, both a driver and helper are used. The helper usually is responsible for attaching and detaching any chains or cables used in loading and unloading containers on and off the collection vehicle; the driver is responsible for the operation of the vehicle. A driver and helper should always be used where hazardous wastes are to be handled.

Stationary Container Systems

Stationary container systems may be used for the collection of all types of wastes. The systems vary according to the type and quantity of wastes to be handled, as well as the number of generation points. There are two main types: (1) systems in which mechanically loaded collection vehicles are used (see Figs. 8-3, 8-4, and 8-6), and (2) systems in which manually loaded collection vehicles are used (see Figs. 8-1 and 8-13). Because of the economic advantages involved, almost all of the collection vehicles now used are equipped with internal compaction mechanisms. Data on the collection vehicles used in this system are reported in Table 8-4. Cost data for the collection vehicles used in the stationary container system are presented in Appendix E.

Systems with Mechanically Loaded Collection Vehicles. Container size and utilization are not so critical in stationary container systems using collection vehicles equipped with a compaction mechanism as they are in hoist-truck systems. Trips to the materials recovery facility (MRF), transfer station, or disposal site are made after the contents of a number of containers have been collected and compacted, and the collection vehicle is full. For this reason, the utilization of the driver in terms of the quantities of wastes hauled is considerably greater for these systems than for hauled container systems.

(a)

(b)

FIGURE 8-13
Types of collection vehicles: (*a*) standup right-hand–drive side-loaded collection vehicle and (*b*) collector manually emptying the contents of a container into a rear-loaded compaction-type collection vehicle (this type of vehicle is commonly used with two- and three-person crews for the collection of residential and commercial wastes in many parts of the United States).

A variety of container sizes is available for use with these systems (see Table 8-3 and Figs. 8-3, 8-6, and 8-8). Containers vary from relatively small (60 gal) to sizes comparable with those handled with a hoist truck (see Table 8-3). Smaller containers offer greater flexibility in terms of shape, ease of loading, and special features available, and also lead to considerably increased utilization. These systems can also be used for the collection of residential wastes, substituting one large container for a number of small containers.

Because truck bodies are difficult to maintain and because of the weight involved, these systems are not well suited for the collection of heavy industrial wastes and bulk rubbish, such as that produced at construction and demolition sites. Locations where high volumes of rubbish are produced are also difficult to service because of the space requirements for the large number of containers.

Systems with Manually Loaded Collection Vehicles. The major application of manual loading methods is in the collection of residential wastes and litter (see Fig. 8-1*a*). Manual loading can compete effectively with mechanical loading in residential areas because the quantity picked up at each location is small and the loading time is short. In addition, manual methods are used for residential collection because many individual pickup points are inaccessible to mechanized self-loading collection vehicles.

Special attention must be given to the design of the collection vehicle intended for use with a single collector-driver. At present, it appears that a side-loaded compactor, such as the one shown in Figs. 8-1*a* and 8-13*a* equipped with standup right-hand drive, is best suited for curb and alley collection.

Transfer Operations. Transfer operations, in which the wastes, containers, or collection vehicle bodies holding the wastes are transferred from a collection vehicle to a transfer or haul vehicle, are used primarily for economic considerations. Transfer operations may prove economical when (1) relatively small, manually loaded collection vehicles are used for the collection of residential wastes and long haul distances are involved, (2) extremely large quantities of wastes must be hauled over long distances, and (3) one transfer station can be used by a number of collection vehicles. Transfer and transport operations are considered in detail in Chapter 10.

Personnel Requirements for Stationary Container Systems. The personnel requirements for the stationary collection system will vary depending on whether the collection vehicle is loaded mechanically or manually. Labor requirements for mechanically loaded stationary container systems are essentially the same as for hauled container systems. Where a helper is used, the driver often assists the helper in bringing loaded containers mounted on rollers to the collection vehicle and returning the empty containers. Occasionally, a driver and two helpers are used where the containers to be emptied must be rolled (transferred) to the collection vehicle from inaccessible locations, such as in congested downtown commercial areas.

In stationary container systems where the collection vehicle is loaded manually, the number of collectors varies from one to three in most cases, depending on the type of service and the collection equipment. Typically, two persons, a collector and a driver, are used for curb and alley service, and a multiperson crew is used for backyard carry service (see Table 8-1). In satellite-vehicle collection systems, one collector-driver is used for the main collection vehicle and one collector-driver is used with each satellite collection vehicle. While the satellite vehicles are being loaded, the collector-driver of the main vehicle picks up wastes from curb locations along the route. Although the aforementioned crew sizes represent current practices, there are many exceptions. In many cities multiperson crews are used for curb service as well as for backyard carry service.

8-3 ANALYSIS OF COLLECTION SYSTEMS

To establish vehicle and labor requirements for the various collection systems and methods, the unit time required to perform each task must be determined. By separating the collection activities into unit operations, it is possible (1) to develop design data and relationships that can be used universally and (2) to evaluate both the variables associated with collection activities and the variables related to, or controlled by, the particular location. The discussion that follows is intended to serve as an introduction to the types of information and data that are needed to evaluate waste collection operations and systems properly.

Definition of Terms

Before the relationships for collection systems can be modeled effectively, the component tasks must be delineated. The operational tasks for the hauled container and stationary container systems are shown schematically in Figs. 8-14 and 8-15, respectively. The activities involved in the collection of solid wastes can be resolved into four unit operations: (1) pickup, (2) haul, (3) at-site, and (4) off-route [5, 7].

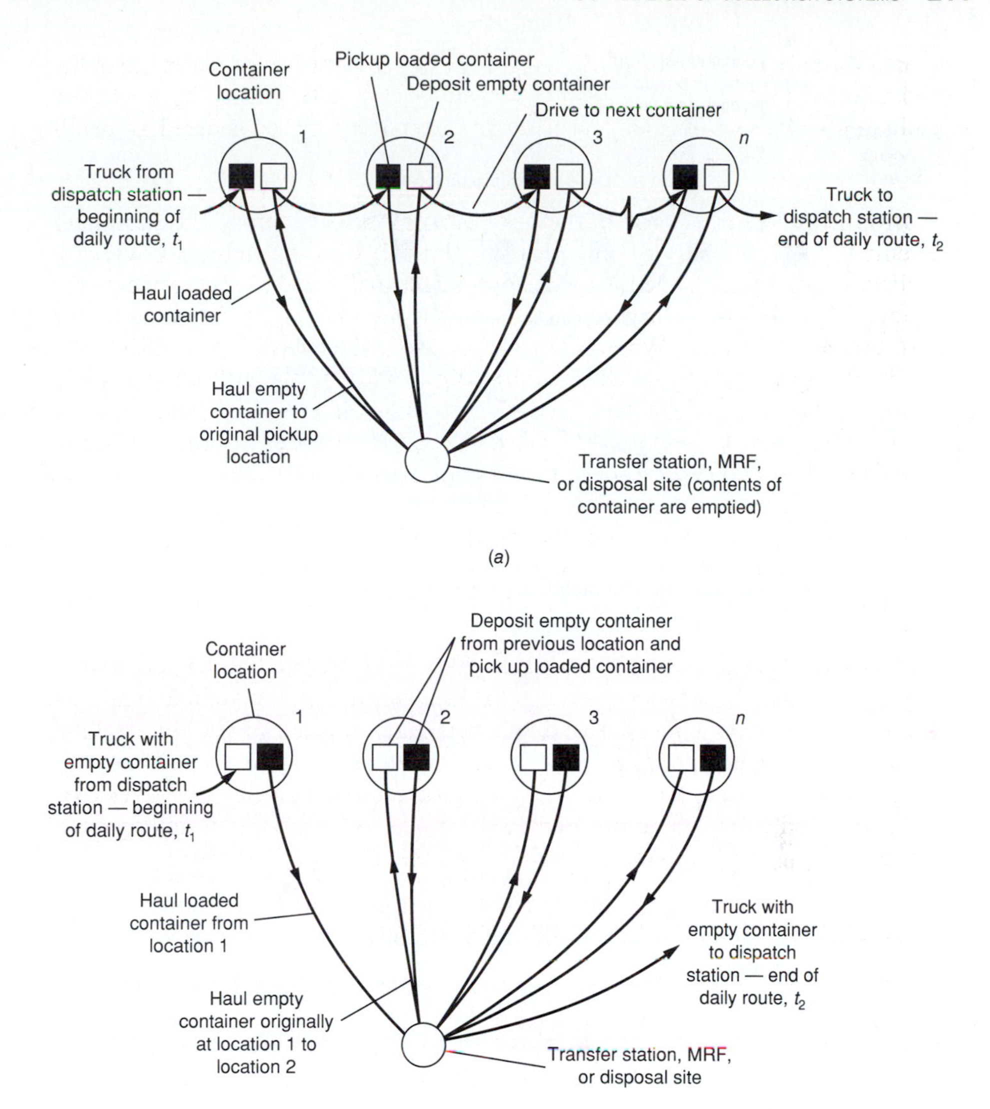

FIGURE 8-14
Schematic of operational sequence for hauled container system: (*a*) conventional mode and (*b*) exchange container mode.

Pickup. The definition of the term *pickup* depends on the type of collection system used.

1. For hauled container systems operated in the conventional mode (see Fig. 8-14*a*), pickup (P_{hcs}) refers to the time spent driving to the next container after an empty container has been deposited, the time spent picking up the loaded container, and the time required to redeposit the container after its contents

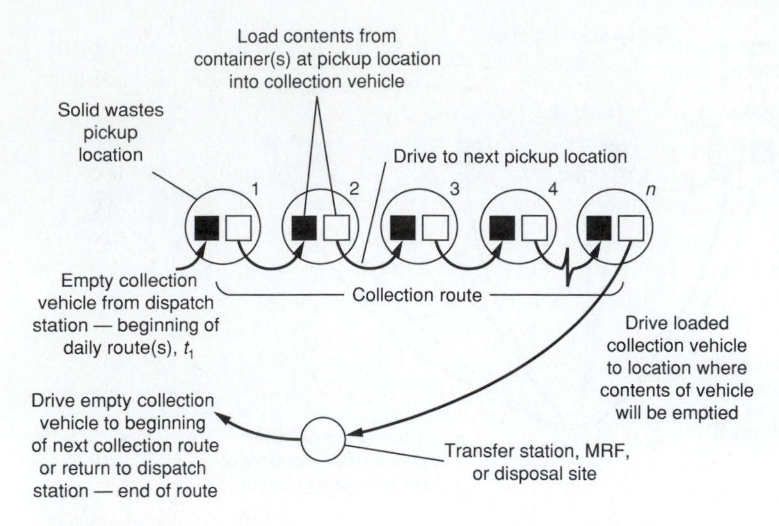

FIGURE 8-15
Schematic of operational sequence for stationary container system.

have been emptied. For hauled container systems operated in the exchange-container mode (see Fig. 8-14*b*), pickup includes the time required to pick up a loaded container and to redeposit the container at the next location after its contents have been emptied.

2. For stationary container systems (see Fig. 8-15), *pickup* (P_{scs}) refers to the time spent loading the collection vehicle, beginning with stopping the vehicle before loading the contents of the first container and ending when the contents of the last container to be emptied have been loaded. The specific tasks in the pickup operation depend on the type of collection vehicle as well as the collection methods used.

Haul. The definition of the term *haul* (h) also depends on the type of collection system used.

1. For hauled container systems, *haul* represents the time required to reach the location where the contents of the container will be emptied (e.g., transfer station, MRF, or disposal site), starting when a container whose contents are to be emptied has been loaded on the truck and continuing through the time after leaving the unloading location until the truck arrives at the location where the empty container is to be redeposited. Haul time does not include any time spent at the location where the contents of the container are unloaded.
2. For stationary container systems, haul refers to the time required to reach the location where the contents of the collection vehicle will be emptied (e.g., transfer station, MRF, or disposal site), starting when the last container on the route has been emptied or the collection vehicle is filled and continuing through the time after leaving the unloading location until the truck arrives at

the location of the first container to be emptied on the next collection route. Haul time does not include the time spent at the location where the contents of the collection vehicle are unloaded.

At-Site. The unit operation *at-site* (s) refers to the time spent at the location where the contents of the container (hauled container system) or collection vehicle (stationery container system) are unloaded (e.g., transfer station, MRF, or disposal site) and includes the time spent waiting to unload as well as the time spent unloading the wastes from the container or collection vehicle.

Off-Route. The unit operation *off-route* (W) includes all time spent on activities that are nonproductive from the point of view of the overall collection operation. Many of the activities associated with off-route times are sometimes necessary or inherent in the operation. Therefore, the time spent on off-route activities may be subdivided into two categories: necessary and unnecessary. In practice, however, both necessary and unnecessary off-route times are considered together because they must be distributed equally over the entire operation.

Necessary off-route time includes (1) time spent checking in and out in the morning and at the end of the day, (2) time lost due to unavoidable congestion, and (3) time spent on equipment repairs, maintenance, and so on. Unnecessary off-route time includes time spent for lunch in excess of the stated lunch period and time spent on taking unauthorized coffee breaks, talking to friends, and the like.

Hauled Container Systems

The time required per trip, which also corresponds to the time required per container, is equal to the sum of the pickup, at-site, and haul time and is given by the following equation:

$$T_{hcs} = (P_{hcs} + s + h) \tag{8-1}$$

where T_{hcs} = time per trip for hauled container system, h/trip
P_{hcs} = pickup time per trip for hauled container system, h/trip
s = at-site time per trip, h/trip
h = haul time per trip, h/trip

For hauled container systems the pickup and at-site times are relatively constant, but the haul time depends on both haul speed and distance. From an analysis of a considerable amount of haul data for various types of collection vehicles (see Fig. 8-16), it has been found [5, 7] that the haul time h may be approximated by the following expression:

$$h = a + bx \tag{8-2}$$

where h = total haul time, h/trip
a = empirical haul-time constant, h/trip
b = empirical haul-time constant, h/mi
x = average round-trip haul distance, mi/trip

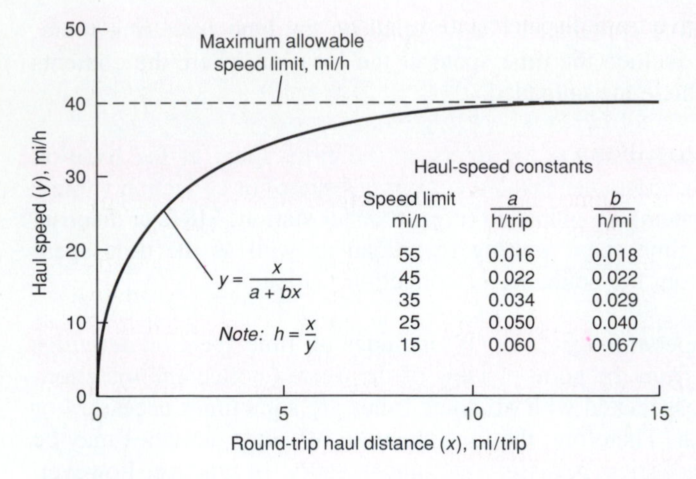

FIGURE 8-16
Correlation between average haul speed and round-trip haul distance for waste collection vehicles [6].

For places where a number of pickup locations are located in a given service area, the average round-trip haul distance from the center of gravity of the service area to the disposal site can be used in Eq. (8-2). Determination of the haul-time constants is illustrated in Example 8-2, presented at the end of this discussion.

Substituting in Eq. (8-1) the expression for h given in Eq. (8-2), the time per trip can be expressed as follows:

$$T_{\text{hcs}} = (P_{\text{hcs}} + s + a + bx) \tag{8-3}$$

The pickup time per trip, P_{hcs}, for the hauled container system is equal to

$$P_{\text{hcs}} = \text{pc} + \text{uc} + \text{dbc} \tag{8-4}$$

where P_{hcs} = pickup time per trip, h/trip
pc = time required to pick up loaded container, h/trip
uc = time required to unload empty container, h/trip
dbc = time required to drive between container locations, h/trip

If the average time required to drive between containers is unknown, the time can be estimated by using Eq. (8-2). The distance between containers is substituted for the round-trip haul distance and the haul-time constants for 15 mi/h (see Fig. 8-16) should be used.

The number of trips that can be made per vehicle per day with a hauled container system, taking into account the off-route factor W, can be determined by using Eq. (8-5):

$$N_d = [H(1 - W) - (t_1 + t_2)]/T_{\text{hcs}} \tag{8-5}$$

where N_d = number of trips per day, trips/d
H = length of work day, h/d
W = off-route factor, expressed as a fraction

t_1 = time to drive from dispatch station (garage) to first container location to be serviced for the day, h

t_2 = time to drive from the last container location to be serviced for the day to the dispatch station (garage), h

T_{hcs} = time per trip, h/trip

In deriving Eq. (8-5), it is assumed that off-route activities can occur at any time during the day. Data that can be used in the solution of Eq. (8-5) for various types of hauled container systems are given in Fig. 8-16 and Table 8-5. The off-route factor in Eq. (8-5) varies from 0.10 to 0.40; a factor of 0.15 is representative for most operations. Application of Eqs. (8-3) through (8-5) is illustrated in Example 8-2.

The number of trips that can be made per day, computed from Eq. (8-5), can be compared with the number of trips required per day (or week), which can be computed using the following expression:

$$N_d = V_d/(cf) \qquad (8\text{-}6)$$

where N_d = number of trips per day, trips/d

V_d = average daily quantity of waste collected, yd^3/d

c = average container size, yd^3/trip

f = weighted average container utilization factor

As noted previously, the container utilization factor may be defined as the fraction of the container volume occupied by solid wastes. Because this factor will vary with the size of the container, a weighted container utilization factor is used in Eq. (8-6). The weighted factor is found by dividing the sum of the values obtained by multiplying the number of containers in each size by their corresponding utilization factor by the total number of containers.

TABLE 8-5
Representative data to use for computing equipment and labor requirements for various collection systems[a]

Collection data					
Vehicle	**Loading method**	**Compaction ratio, *r***	**Time required to pick up loaded container and to deposit empty container, h/trip**	**Time required to empty contents of loaded container, h/container**	**At-site time, h/trip**
Hauled container system					
Hoist truck	Mechanical	—	0.067		0.053
Tilt-frame	Mechanical	—	0.40		0.127
Tilt-frame	Mechanical	2.0–4.0[a]	0.40		0.133
Stationary container system					
Compactor	Mechanical	2.0–2.5		0.008–0.05[b]	0.10
Compactor	Manual	2.0–2.5		—	0.10

[a]Containers used in conjunction with stationary compactor.

[b]Time required varies depending on the size of the container.

Example 8-2 Determination of haul-speed constants. The following average speeds were obtained for various round-trip distances to a disposal site (see Comment statement at end of problem). Find the haul-speed constants a and b and the round-trip haul time for a site that is located 11.0 mi away.

Round-trip distance (x), mi/trip	Average haul speed (y), mi/h	Total time ($h = x/y$), h
2	17	0.12
5	28	0.18
8	32	0.25
12	36	0.33
16	40	0.40
20	42	0.48
25	45	0.56

Solution

1. Linearize the haul-speed equation given in Fig. 8-16. The basis haul-speed equation (a rectangular hyperbola) is

$$y = \frac{x}{a + bx}$$

The linearized form of this equation is

$$\frac{x}{y} = h = a + bx$$

2. Plot x/y, which is the total haul travel time versus the round-trip distance as shown below.

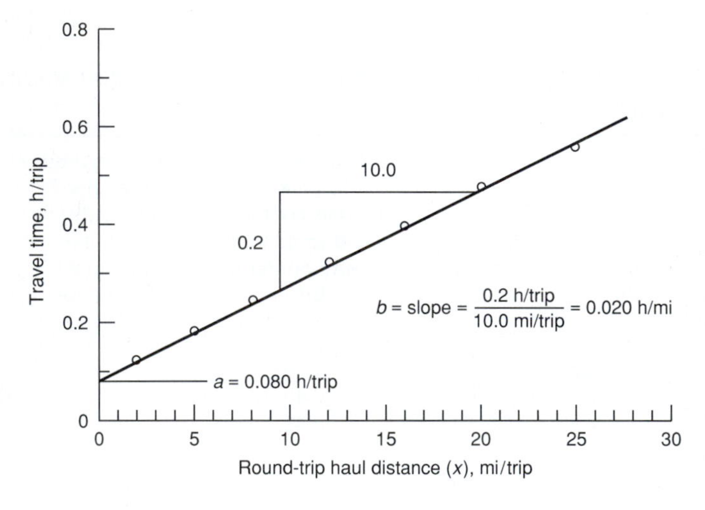

3. Determine the haul-time constants a and b. When $x = 0$, a = intercept value = 0.080 h/trip, b = slope of line = (0.2 h/trip)/(10 mi/trip) = 0.020 h/mi (0.012 h/km).

4. Find the round-trip haul time for a site that is located 11.0 mi away.

$$\text{Round-trip distance} = 2(11.0 \text{ mi/trip}) = 22 \text{ mi/trip}$$

$$\begin{aligned}\text{Round-trip haul time } h &= a + bx \\ &= 0.080 \text{ h/trip} + (0.020 \text{ h/mi})(22 \text{ mi/trip}) \\ &= 0.52 \text{ h/trip}\end{aligned}$$

Comment. When determining the time required to travel to the disposal site in the field the times should be determined at approximately the same times the collection vehicles will be traveling to and from the unloading location. Haul time data collected during working hours will include the effects of traffic congestion, weather conditions, and so on.

Example 8-3 Analysis of a hauled container system. Solid waste from a new industrial park is to be collected in large containers (drop boxes), some of which will be used in conjunction with stationary compactors. Based on traffic studies at similar parks, it is estimated that the average time to drive from the garage to the first container location (t_1) and from the last container location (t_2) to the garage each day will be 15 and 20 min, respectively. If the average time required to drive between containers is 6 min and the one-way distance to the disposal site is 15.5 mi (speed limit: 55 mi/h), determine the number of containers that can be emptied per day, based on an 8-h workday. Assume the off-route factor, W, is equal to 0.15.

Solution

1. Determine the pickup time per trip using Eq. (8-4).

$$P_{\text{hcs}} = \text{pc} + \text{uc} + \text{dbc}$$

Use pc + uc = 0.4 h/trip (see Table 8-5)

$$\begin{aligned}\text{dbc} &= 0.1 \text{ h/trip (given)} \\ P_{\text{hcs}} &= (0.4 + 0.1) \text{ h/trip} \\ &= (0.4 + 0.1) \text{ h/trip} \\ &= 0.5 \text{ h/trip}\end{aligned}$$

2. Determine the time per trip using Eq. (8-3).

$$\begin{aligned}T_{\text{hcs}} &= (P_{\text{hcs}} + s + a + bx) \\ P_{\text{hcs}} &= 0.5 \text{ h/trip (from Step 1)} \\ s &= 0.133 \text{ h/trip (see Table 8-5)} \\ a &= 0.016 \text{ h/trip (see Fig. 8-16)} \\ b &= 0.018 \text{ h/trip (see Fig. 8-16)} \\ T_{\text{hcs}} &= [0.5 + 0.133 + 0.016 + 0.018(31)] \text{ h/trip} \\ &= 1.21 \text{ h/trip}\end{aligned}$$

3. Determine the number of trips that can be made per day using Eq. (8-5).

$$N_d = [H(1 - W) - (t_1 + t_2)]/T_{hcs}$$

$$\text{Use } H = 8 \text{ h (given)}$$

$$W = 0.15 \text{ (assumed)}$$

$$t_1 = 0.25 \text{ h (given)}$$

$$t_2 = 0.33 \text{ h (given)}$$

$$T_{hcs} = 1.21 \text{ h/trip}$$

$$N_d = [8(1 - 0.15) - (0.25 + 0.33)]/(1.21 \text{ h/trip})$$

$$= (6.8 - 0.58)/(1.21 \text{ h/trip})$$

$$= 5.14 \text{ trips/d}$$

$$\text{Use } N_d = 5.0 \text{ trips/d}$$

4. Determine the actual length of the workday.

$$5 \text{ trips/d} = [H(1 - 0.15) - 0.58]/(1.21 \text{ h/trip})$$

$$H = 7.80 \text{ h (essentially 8 h)}$$

Comment. Where fractional equipment and labor requirements are obtained, the use of large containers and reduced collection frequency should be investigated. If it is assumed that no off-route activities occur during times t_1 and t_2, then theoretically 5.21 trips/d could be made. Again, only 5 trips/d would be made in an actual operation. If, however, the number of trips per day that could be made were 5.8, for example, it may be cost-effective to pay the driver for the overtime and make 6 trips/d.

Stationary Container Systems

Because of differences in the loading processes, mechanically and manually loaded stationary container systems are considered separately in the following discussion.

Mechanically Loaded Collection Vehicles. For systems using self-loading collection vehicles, the time per trip is expressed as

$$T_{scs} = (P_{scs} + s + a + bx) \tag{8-7}$$

where T_{scs} = time per trip for stationary container system, h/trip
P_{scs} = pickup time per trip for stationary container system, h/trip
s = at-site time per trip, h/trip
a = empirical constant, h/trip
b = empirical constant, h/mi
x = average round-trip haul distance, mi/trip

As with the hauled container system, if no other information is available the average round-trip distance from the center of gravity of the service area to the disposal site can be used in Eq. (8-7).

The only difference between Eq. (8-7) and Eq. (8-3) for hauled container systems is the pickup term. For the stationary container system, the pickup time is given by

$$P_{scs} = C_t(\mathrm{uc}) + (n_p - 1)(\mathrm{dbc}) \tag{8-8}$$

where P_{scs} = pickup time per trip for stationary container system, h/trip
C_t = number of containers emptied per trip, containers/trip
uc = average unloading time per stationary container for stationary container systems, h/container
n_p = number of container pickup locations per trip, locations/trip
dbc = average time spent driving between container locations, h/location

The term $(n_p - 1)$ accounts for the fact that the number of times the collection vehicle will have to be driven between container locations is equal to the number of container locations less 1. As in the case of the hauled container system, if the time spent driving between container locations is unknown, it can be estimated by using Eq. (8-2) where the distance between containers is substituted for the round-trip haul distance and the haul-time constants for 15 mi/h (see Fig. 8-16) are used.

The number of containers that can be emptied per collection trip is related directly to the volume of the collection vehicle and the compaction ratio that can be achieved. This number is given by

$$C_t = vr/cf \tag{8-9}$$

where C_t = number of containers emptied per trip, containers/trip
v = volume of collection vehicle, yd^3/trip
r = compaction ratio
c = container volume, yd^3/container
f = weighted container utilization factor

The number of trips required per day can be estimated by using the following equation:

$$N_d = V_d/vr \tag{8-10}$$

where N_d = number of collection trips required per day, trips/d
V_d = average daily quantity of waste collected, yd^3/d

The time required per day, taking into account the off-route factor W, can be expressed as follows:

$$H = [(t_1 + t_2) + N_d(T_{scs})]/(1 - W) \tag{8-11}$$

where t_1 = time to drive from dispatch station (garage) to the location of the first container to be picked up on the first route of the day, h
t_2 = time to drive from the approximate location of the last container pickup on last route of the day to the dispatch station (garage), h
other terms = as defined previously

In defining t_2, the term *approximate location* is used because in the stationary container system, the collection vehicle is normally driven directly back to the dispatch station after the wastes collected on the last route have been emptied. If the travel time from the disposal site (or transfer point) to the dispatch station is less than one half the average round-trip haul time, t_2 is assumed to be equal to zero. If the travel time from the disposal site (or transfer point) to the dispatch station is greater than the travel time from the last pickup location to the disposal site, the time t_2 is assumed to be equal to the difference between the time to drive to the dispatch station from the disposal site and one half the average round-trip haul time.

Where an integer number of trips are to be made each day, the proper combination of trips per day and the size of the vehicle can be determined by using Eq. (8-11) in conjunction with an economic analysis. To determine the required truck volume, substitute two or three different values for N_d in Eq. (8-11) and determine the available pickup times per trip. Then, by successive trials, using Eqs. (8-8) and (8-9), determine the truck volume required for each value of N_d. From the available truck sizes, select the ones that most nearly correspond to the computed values. If available truck sizes are smaller than the required values, compute the actual time per day that will be required using these sizes. The most cost-effective combination can then be selected. The application of the above equations is illustrated in Example 8-3.

When the truck size is fixed and an integer number of trips must be made each day, the length of the required workday can be estimated using Eqs. (8-8), (8-9), and (8-11). A hauled and a stationary container system are analyzed and compared in Example 8-4.

Once the labor requirements for each combination of truck size and number of trips per day have been determined, the most cost-effective combination can be selected. For example, where long haul distances are involved, it may be more economical to use a large collection vehicle and make two trips/day (even though some time at the end of the day may not be used) than to use a smaller vehicle and make three trips/day by using all the available time.

Example 8-4 Comparison between the hauled container and stationary container systems. A private solid waste collector wishes to locate a MRF near a commercial area. The collector would like to use a hauled container system but fears that the haul costs might be prohibitive. What is the maximum distance away from the commercial area that the MRF can be located so that the weekly costs of the hauled container system do not exceed those of a stationary container system? Assume that one collector-driver will be used with each system and that the following data are applicable. For the purpose of this example assume the travel times t_1 and t_2 are included in the off-route factor.

1. Hauled container system
   (*a*) Quantity of solid wastes = 300 yd^3/wk
   (*b*) Container size = 8 yd^3/trip
   (*c*) Container utilization factor = 0.67

(*d*) Container pickup time = 0.033 h/trip
(*e*) Container unloading time = 0.033 h/trip
(*f*) Haul-time constants: a = 0.022 h/trip and b = 0.022 h/mi
(*g*) At-site time = 0.053 h/trip
(*h*) Overhead costs = $400/wk
(*i*) Operational costs = $15/h of operation

2. Stationary container system
(*a*) Quantity of solid wastes = 300 yd^3/wk
(*b*) Container size = 8 yd^3/location
(*c*) Container utilization factor = 0.67
(*d*) Collection vehicle capacity = 30 yd^3/trip
(*e*) Collection vehicle compaction ratio = 2
(*f*) Container unloading time = 0.05 h/container
(*g*) Haul-time constants: a = 0.022 h/trip and b = 0.022 h/mi
(*h*) At-site time = 0.10 h/trip
(*i*) Overhead costs = $750/wk
(*j*) Operational costs = $20/h of operation

3. Location characteristics
(*a*) Average distance between container locations = 0.1 mi
(*b*) Constants for estimating driving time between container locations for both the hauled container and stationary container systems are a' = 0.060 h/trip and b' = 0.067 h/mi

Solution

1. Hauled container system
(*a*) Determine the number of trips per week, using Eq. (8-6).

$$N_w = V_w/cf = (300 \text{ yd}^3/\text{wk})/(8\text{yd}^3/\text{trip})(0.67)$$
$$= 56.0 \text{ trips/wk}$$

(*b*) Estimate the average pickup time for the hauled container system, using Eq. (8-4).

$$P_{\text{hcs}} = \text{pc} + \text{uc} + \text{dbc} = \text{pc} + \text{uc} + a' + b'x'$$
$$= 0.033 \text{ h/trip} + 0.033 \text{ h/trip} + 0.060 \text{ h/trip}$$
$$+ (0.067 \text{ h/mi})(0.1 \text{ mi/trip})$$
$$= 0.133 \text{ h/trip}$$

(*c*) Estimate the time required per week, T_w, as a function of the round-trip haul distance, using the following expression.

$$T_w = N_w(P_{\text{hcs}} + s + a + bx)/[H(1 - W)]$$
$$T_w = (56 \text{ trips/wk})[0.133 \text{ h/trip} + 0.053 \text{ h/trip} + 0.022 \text{ h/trip}$$
$$+ (0.022 \text{ h/mi})(x)]/[(8 \text{ h/d})(1 - 0.15)]$$
$$= [1.71 + (0.181/\text{mi})(x)] \text{ d/wk}$$

(*d*) Determine the weekly operational cost as a function of the round-trip haul distance.

$$\text{Operational cost} = (\$15/\text{h})(8\ \text{h/d})[1.71 + (0.181/\text{mi})(x)]\ \text{d/wk}$$
$$= [205.20 + (21.7/\text{mi})(x)]\ \$/\text{wk}$$

2. Stationary container system

(*a*) Determine the number of containers emptied per trip, using Eq. (8-9).

$$C_t = vr/cf = (30\ \text{yd}^3/\text{trip})(2)/(8\ \text{yd}^3/\text{container})(0.67)$$
$$= 11.19\ \text{containers/trip} = 11\ \text{containers/trip}$$

(*b*) Estimate the pickup time per container by using Eq. (8-8).

$$P_{\text{scs}} = C_t(\text{uc}) + (n_p - 1)(\text{dbc})$$
$$= C_t(\text{uc}) + (n_p - 1)(a' + b'x')$$
$$= (11\ \text{containers/trip})(0.050\ \text{h/container})$$
$$+ (11 - 1\ \text{locations/trip})[(0.06\ \text{h/locations})$$
$$+ (0.067\ \text{h/mi})(0.1\ \text{mi/location})]$$
$$= 1.22\ \text{h/trip}$$

(*c*) Determine the number of trips required per week by using Eq. (8-10).

$$N_w = V_w/vr = (300\ \text{yd}^3/\text{wk})/(30\ \text{yd}^3/\text{trip})(2)$$
$$= 5\ \text{trips/wk}$$

(*d*) Determine the time required per week, T_w, as a function of the round-trip haul distance using the following expression. The term T_w represents the integer number of trips made to the location where the contents of the collection vehicle will be unloaded. The numerical value of T_w is obtained by rounding up the value of N_w to an integer value.

$$T_{\text{w(scs)}} = [(N_w)P_{\text{scs}} + t_w(s + a + bx)]/[H(1 - W)]$$
$$= \{(5\ \text{trips/wk})(1.22\ \text{h/trip}) + (5\ \text{trips/wk})$$
$$\times [0.10\ \text{h/trip} + 0.022\ \text{h/trip} + (0.022\ \text{h/mi})(x)]\}$$
$$/[(8\ \text{h/d})(1 - 0.15)]$$
$$= [0.99 + (0.016/\text{mi})(x)]\ \text{d/wk}$$

(*e*) Determine the weekly operational costs as a function of the round-trip haul distance.

$$\text{Operational cost} = (\$20/\text{h})(8\ \text{h/d}) \times [0.99 + (0.016/\text{mi})(x)]\ \text{d/wk}$$
$$= [158.40 + (2.56/\text{mi})(x)]\ \$/\text{wk}$$

3. Comparison of systems

(*a*) Determine the maximum round-trip haul distance at which the cost for hauled container systems equals the cost for the stationary container systems by equating the total costs for the two systems and solving for x.

$$\$400/\text{wk} + [205.20 + (21.7/\text{mi})(x)]\ \$/\text{wk} = \$750/\text{wk}$$
$$+[158.40 + (2.56/\text{mi})(x)]\ \$/\text{wk}$$
$$(19.1/\text{mi})(x) = 303.20$$
$$x = 15.9 \text{ mi (one-way distance} \sim 8.0 \text{ mi)}$$

(*b*) Plot the weekly cost versus round-trip haul distance for each system. The required plot is presented below.

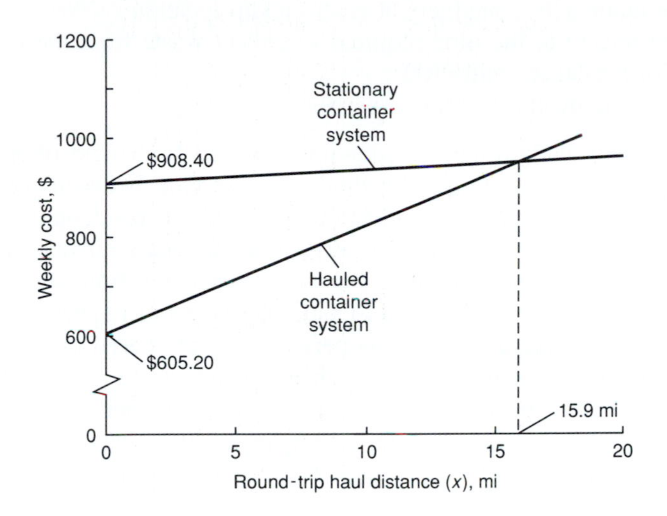

Comment. The curves shown in the figure given above are characteristic of those obtained when hauled container systems are compared with stationary container systems. In most cases the round-trip haul distance at which hauled container systems are no longer competitive is much shorter than in this example.

Manually Loaded Vehicles. The analysis and design of residential collection systems using manually loaded vehicles may be outlined as follows. If H hours are worked per day and the number of trips to be made per day is known or fixed, the time available for the pickup operation can be computed by using Eq. (8-11) because either all the factors are known or they can be assumed. Once the pickup time per trip is known, the number of pickup locations from which wastes can be collected per trip can be estimated as follows:

$$N_p = 60 P_{\text{scs}} n / t_p \tag{8-12}$$

where N_p = number of pickup locations per trip, locations/trip
60 = conversion factor from hours to minutes, 60 min/h
P_{scs} = pickup time per trip, h/trip
n = number of collectors, collectors
t_p = pickup time per pickup location, collector-min/location

The pickup time t_p per location depends on the time required to drive between container locations, the number of containers per pickup location, and the percent of rear-of-house pickup locations. The corresponding relationship is

$$t_p = \text{dbc} + k_1 C_n + k_2(\text{PRH}) \tag{8-13}$$

where t_p = average pickup time per pickup location, collector-min/location
dbc = average time spent driving between container locations, h/location
k_1 = constant related to the pickup time per container, min/container
C_n = average number of containers at each pickup location
k_2 = constant related to the time required to collect waste from the backyard of a residence, min/PRH
PRH = rear-of-house pickup locations, percent

Equation (8-13) is typical of the types of equations derived from field observations for the pickup time per location. The time spent driving between pickup locations will, of course, depend on the characteristics of the residential area. Typical pickup-time data derived from field observations for a two-person collection crew are reported in Fig. 8-17. If once-per-week curb collection service is provided, the data in Table 8-5 may be used to estimate the labor requirements.

The data reported in Table 8-6 for a one-person crew are based on the use of a side-loaded collection vehicle equipped with a standup drive [6] (see Fig. 8-1*a*). If conventional trucks are used for curb collection, the pickup time per service reported in Table 8-6 should be increased by 5 to 10 percent. Although Eq. (8-13) and the data in Table 8-6 can be used to estimate the time per pickup

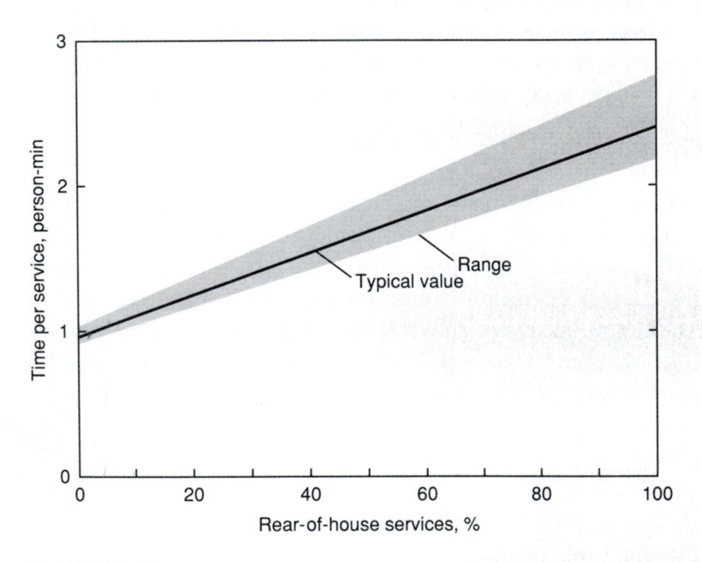

FIGURE 8-17
Relationship between time requirements for pickup and percent of rear-of-house services for a two-person crew [6].

TABLE 8-6
Labor requirements for manual curbside collection using a one-person crew

Average number of containers and/or boxes per pickup location	Pickup time, collector · min/location
1 or 2	0.50–0.60
3 or more, or unlimited service	0.92

location, it is recommended that field measurements be made whenever possible because residential collection operations are so variable. The use of Eq. (8-13) is illustrated in Example 8-5.

Once the number of pickup locations per trip is known, the proper size of collection vehicle can then be estimated as follows:

$$v = V_p N_p / r \tag{8-14}$$

where v = volume of collection vehicle, yd^3/trip
V_p = volume of solid wastes collected per pickup location, yd^3/location
N_p = number of pickup locations per trip, locations/trip
r = compaction ratio

In many housing areas, the collection frequency is twice per week. The effect of twice-per-week collection on the amount of wastes collected was discussed in Chapter 6. In terms of labor requirements, it has been found that the requirements for the second weekly collection are about 0.9 and 0.95 times those of the first weekly collection. In general, the labor requirements are not significantly different, because container handling time is about the same for both full and partly full containers. Often this difference is neglected in computing the labor requirements.

Example 8-5 Design of residential collection system. Design a solid waste curb collection system to service a residential area with 1000 single-family dwellings. Two manually loaded collection systems are to be evaluated. The first involves the use of a side-loaded collection vehicle with a one-person crew; the second involves the use of a rear-loaded collection vehicle with a two-person crew. Determine the size of collection vehicle required and compare the labor requirements for each collection system. Assume the following data are applicable:

1. Average number of residents per service = 3.5
2. Solid waste generation rate per capita = 2.5 lb/capita · d
3. Density of solid wastes (at containers) = 200 lb/yd^3
4. Containers per service = two 32-gal containers plus 1.5 cardboard containers (20 gal on average)
5. Collection frequency = once per week

6. Collection vehicle compaction ratio, $r = 2.5$
7. Round-trip haul distance, $x = 35$ mi
8. Nominal length of workday, $H = 8$ h
9. Trips per day, $N_d = 2$
10. Travel time to first pickup location, $t_1 = 0.3$ h
11. Travel time from last pickup location, $t_2 = 0.4$ h
12. Off-route factor, $W = 0.15$
13. Haul-time constants: $a = 0.016$ h/trip and $b = 0.018$ mi/h
14. At-site time per trip, $s = 0.10$ h/trip

Solution

1. Determine the time available for the pickup operation using Eq. (8-11). Substituting Eq. (8-7) for the term T_{scs} in Eq. (8-11) yields

$$
\begin{aligned}
H &= [(t_1 + t_2) + N_d(P_{scs} + s + a + bx)]/(1 - W) \\
P_{scs} &= [H(1 - W) - (t_1 + t_2)]/N_d - (s + a + bx) \\
&= [(8 \text{ h/day})(1 - 0.15) - (0.3 \text{ h/day} + 0.4 \text{ h/day})]/(2 \text{ trips/day}) \\
&\quad - [0.10 \text{ h/trip} + 0.016 \text{ h/trip} + (0.018 \text{ h/mi})(35 \text{ mi/trip})] \\
&= (3.05 - 0.75) \text{ h/trip} \\
&= 2.30 \text{ h/trip}
\end{aligned}
$$

2. Determine the pickup time required per pickup location using Eq. (8-13).

(*a*) One-person crew

$$t_p = 0.92 \text{ min/location (see Table 8-6)}$$

(*b*) Two-person crew (based on field observations)

$$
\begin{aligned}
t_p &= 0.72 + 0.18(C_n) \\
&= 0.72 + 0.18(3.5) \\
&= 1.35 \text{ collector-min/location}
\end{aligned}
$$

3. Determine the number of pickup locations from which wastes can be collected, using Eq. (8-12).

(*a*) One-person crew

$$
\begin{aligned}
N_p &= 60P_{scs}n/t_p \\
&= (60 \text{ min/h})(2.30 \text{ h/trip})(1 \text{ collector})/(0.92 \text{ collector-min/location}) \\
&= 150 \text{ locations/trip}
\end{aligned}
$$

(*b*) Two-person crew

$$
\begin{aligned}
N_p &= 60P_{scs}n/t_p \\
&= (60 \text{ min/h})(2.30 \text{ h/trip})(2 \text{ collector})/(1.35 \text{ collector-min/location}) \\
&= 204 \text{ locations/trip}
\end{aligned}
$$

4. Determine the volume of wastes generated per pickup location per week.

$$\text{Volume per week per location} = (2.5\ \text{lb/person/day})(3.5\ \text{persons/pickup location})(7\ \text{days/wk})/(200\ \text{lb/yd}^3)(1/\text{wk})$$
$$= 0.306\ \text{yd}^3/\text{location}$$

5. Determine the required truck volume using Eq. (8-14).

(*a*) One-person crew

$$v = V_p N_p / r$$
$$= (0.306\ \text{yd}^3/\text{location})(150\ \text{locations/trip})/2.5$$
$$= 18.4\ \text{yd}^3/\text{trip (use an 18 yd}^3\text{ collection vehicle)}$$

(*b*) Two-person crew

$$v = V_p N_p / r$$
$$= (0.306\ \text{yd}^3/\text{location})(204\ \text{locations/trip})/2.5$$
$$= 25.0\ \text{yd}^3/\text{trip (use a 25 yd}^3\text{ collection vehicle or nearest standard size, if available)}$$

6. Determine the number of trips required per week.

(*a*) One-person crew

$$N_w = (1000\ \text{locations})(1/\text{wk})/(150\ \text{locations/trip})$$
$$= 6.67\ \text{trips/wk}$$

(*b*) Two-person crew

$$N_w = (1000\ \text{locations})(1/\text{wk})/(204\ \text{locations/trip})$$
$$= 4.90\ \text{trips/wk}$$

7. Determine the labor requirements. Note that even though a partial trip is computed, a full trip will have to be made to the location where the contents of the collection vehicle will be unloaded.

(*a*) One-person crew

$$1.0\ \text{collector}\{(6.67\ \text{trips/wk})(2.3\ \text{h/trip}) + (7\ \text{trips/wk})[0.10\ \text{h/trip} + 0.016\ \text{h/trip} + (0.018\ \text{h/mi})(35\ \text{mi/trip})]\}/(1 - 0.15)(8\ \text{h/day})$$
$$= 3.02\ \text{collector-day/wk}$$

(*b*) Two-person crew

$$2.0\ \text{collectors}\{(4.9\ \text{trips/wk})(2.3\ \text{h/trip}) + (5\ \text{trips/wk})[0.10\ \text{h/trip} + 0.016\ \text{h/trip} + (0.018\ \text{h/mi})(35\ \text{mi/trip})]\}/(1 - 0.15)(8\ \text{h/day})$$
$$= 4.41\ \text{collector-day/wk}$$

Comment. As determined in this problem, the labor requirements for the one-person crew are approximately 25 percent less than corresponding requirements for the two-person collection crew. The results of this example illustrate why the trend in collection is towards the use of curb collection with one collector-driver and collection vehicles that are either manually or mechanically loaded.

8-4 COLLECTION ROUTES

Once equipment and labor requirements have been determined, collection routes must be laid out so that both the collectors and equipment are used effectively. In general, the layout of collection routes involves a series of trials. There is no universal set of rules that can be applied to all situations. Thus, collection vehicle routing remains today a heuristic (common sense) process [4].

Some heuristic guidelines that should be taken into consideration when laying out routes are as follows:

1. Existing policies and regulations related to such items as the point of collection and frequency of collection must be identified.
2. Existing system characteristics such as crew size and vehicle types must be coordinated.
3. Wherever possible, routes should be laid out so that they begin and end near arterial streets, using topographical and physical barriers as route boundaries.
4. In hilly area, routes should start at the top of the grade and proceed downhill as the vehicle becomes loaded.
5. Routes should be laid out so that the last container to be collected on the route is located nearest to the disposal site.
6. Wastes generated at traffic-congested locations should be collected as early in the day as possible.
7. Sources at which extremely large quantities of wastes are generated should be serviced during the first part of the day.
8. Scattered pickup points (where small quantities of solid waste are generated) that receive the same collection frequency should, if possible, be serviced during one trip or on the same day.

Layout of Collection Routes

The general steps involved in establishing collection routes include (1) preparation of location maps showing pertinent data and information concerning the waste generation sources, (2) data analysis and, as required, preparation of information summary tables, (3) preliminary layout of routes, and (4) evaluation of the preliminary routes and the development of balanced routes by successive trials. Step 1, as discussed below, is essentially the same for all types of collection systems. Because the application of Steps 2, 3, and 4 is different for the hauled and stationary container systems, each is discussed separately. The layout of collection routes is illustrated in Examples 8-5 and 8-6.

Note that the balanced routes prepared in the office (Step 4) are given to the collector-drivers, who implement them in the field. Based on the field experience of the collector-driver, each route is modified to account for specific local conditions. In large municipalities, route supervisors are responsible for the preparation of collection routes. In most cases, the routes are based on the operating experience of the route supervisor, gained over a period of years working in the same section of the city. The following discussion is designed to quantify on paper what most route supervisors do in their heads.

Collection Route Layout—Step 1. On a relatively large-scale map of the commercial, industrial, or residential housing area to be served, the following data should be plotted for each solid waste pickup point: location, collection frequency, number of containers. If a mechanically loaded stationary container system is used for commercial and industrial services, the estimated quantity of wastes to be collected at each pickup location should also be entered on the map. For residential sources it is generally assumed that approximately the same average quantity of waste will be collected from each source. Often, for residential sources only the number of homes per block will be shown.

Because the layout of collection routes involves a series of successive trials, tracing paper should be used once the basic data have been entered on the work map. Depending on the size of the area and the number of pickup points, the area should be subdivided into areas corresponding roughly to similar land-use areas (e.g., residential, commercial, industrial). For locations with less than 20 to 30 pickup points, this step is usually not necessary. For larger areas it may be necessary to subdivide further each of the similar land-use areas into smaller areas, taking into account factors such as waste generation rates and collection frequency.

Collection Route Layout—Steps 2, 3, and 4 for Hauled Container Systems. Steps 2, 3, and 4 for the hauled container system can be outlined as follows.

Step 2. On a spreadsheet program first enter the following headings: collection frequency, times/wk; number of pickup locations; total number of containers; number of trips, trips/wk; and a separate column for each day of the week during which wastes will be collected. Second, determine the number of pickup locations requiring multiple pickups during the week (e.g., Monday through Friday or Monday, Wednesday, Friday) and enter the information on the spreadsheet. Start the listing with the locations requiring the highest number of pickups per week (e.g., 5 times/wk). Third, distribute the number of containers requiring once per week service so that the number of containers emptied per day is balanced for each collection day. Preliminary collection routes can be laid out once this information is known.

Step 3. Using the information from Step 2, the layout of collection routes can be outlined as follows. Starting from the dispatch station (or where the collection vehicles are parked), a route should be laid out that connects all the pickup points (containers) to be serviced during each collection day. The next step is to modify the basic route to include the additional containers that will be serviced on each collection day. Each daily route should be laid out so it begins and ends near the dispatch station. The collection operation should proceed in a logical manner, taking into account the guidelines cited previously and specific local constraints.

Step 4. When preliminary routes have been laid out, the average distance to be traveled between containers should be computed. If the routes are unbalanced with respect to the distance traveled (> 15 percent), they should be redesigned so that each route covers approximately the same distance. In general, a number of collection routes must be tried before the final ones are selected. When more than one collection vehicle is required, collection routes for each functional-use or service area must be laid out, and work loads for each driver must be balanced.

Collection Route Layout—Steps 2, 3, and 4 for Stationary Container System (with Mechanically Loaded Collection Vehicles). Steps 2, 3, and 4 for stationary container systems that are loaded mechanically can be outlined as follows.

Step 2. On a spreadsheet program first enter the following heads: collection frequency, times/wk; number of pickup locations; total waste, yd^3/wk; and a separate column for each day of the week during which wastes will be collected. Second, determine the amount of waste to be collected from pickup locations requiring multiple pickups during the week (e.g., Monday through Friday or Monday, Wednesday, Friday) and enter the information on the spreadsheet. Start the listing with the locations requiring the highest number of pickups per week (e.g., 5 times/wk). Third, using the effective volume of the collection vehicle (nominal collection vehicle volume × compaction ratio), determine the amount of additional waste that can be collected each day from locations receiving once per week service. Distribute the amount of waste collected so that the amount of waste collected (and the number of containers emptied) per trip is balanced for each collection route. Preliminary collection routes can be laid out once this information is known.

Step 3. Once the foregoing information is known, the layout of collection routes can proceed as follows. Starting from the dispatch station (or where the collection vehicles are parked), a route should be laid out that connects all the pickup points to be serviced during each collection day. Depending on the quantity of waste to be collected, several basic routes may have to be laid out.

The next step is to modify the basic route (routes) to include the additional pickup locations that will have to be serviced to complete the load. These modifications should be made so that the same general area is serviced with each collection route. For large areas that have been subdivided and that are serviced daily, it will be necessary to establish basic routes in each subdivided area; in some cases, between them, depending upon the number of trips to be made per day.

Step 4. When the collection routes have been laid out, the quantity of waste to be collected and the haul distance for each route should be determined. In some cases it may be necessary to readjust the collection routes to balance the work load. After the routes have been established, they should be drawn on the master map.

Example 8-6 Layout of collection routes for an industrial park. Lay out collection routes for both a hauled container and a stationary container collection system for the industrial service area shown in the accompanying map. There are, as shown on the map, a total of 28 pickup locations and 32 containers. The total quantity of waste to be collected each week is 277 yd^3. The map and the information it contains would be prepared as the first step in the layout of collection routes. Assume the following conditions apply:

1. Containers with a collection frequency of twice per week must be picked up on Tuesday and Friday.
2. Containers with a collection frequency of three times per week must be picked up on Monday, Wednesday, and Friday.
3. Containers may be picked up from any side of the intersection where they are stationed.
4. Start and finish each day at the dispatch station.
5. For the hauled container system, collection will be provided Monday through Friday.

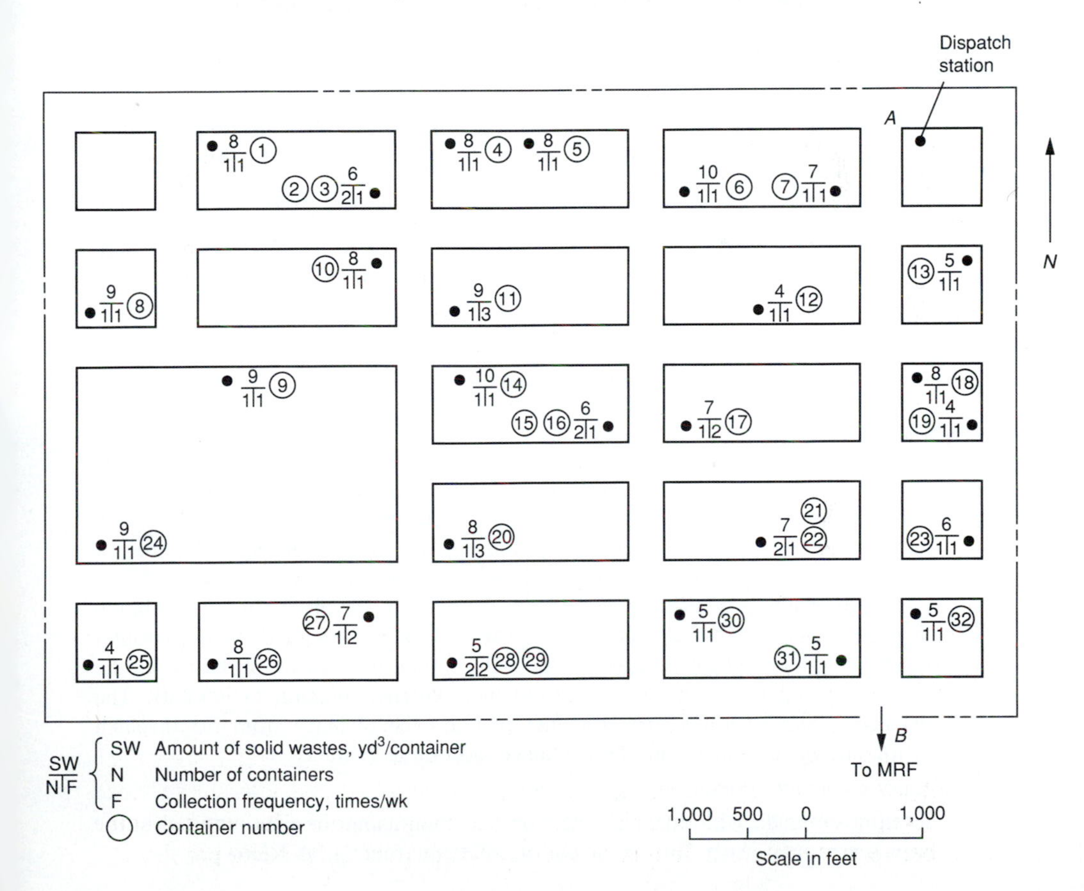

6. Hauled containers are exchanged rather than being returned to the location from which they were picked up (see Fig. 8-14*b*).
7. For the stationary container system, collection will be provided only 4 days/wk (Monday, Tuesday, Wednesday, and Friday) with only 1 trip/day.
8. For the stationary container system, the collection vehicle will be a self-loading compactor with a capacity of 35 yd^3 and a compaction ratio of 2.

Solution

1. Hauled container system
   (*a*) Set up a summary table for the collection operation using the data reported in the service area map (Step 2 in the layout of collection routes). The summary table and a brief description of the entries in the table are presented below.
      i. The number of pickup locations and containers requiring three collections per week are entered in Row 1. As noted in the problem statement, these containers must be emptied on Monday, Wednesday, and Friday.
      ii. The number of pickup locations requiring two collections per week are entered in Row 2. These containers must be emptied on Tuesday and Friday.
      iii. The additional number of containers receiving once per week service that must be emptied each collection day are entered in Row 3. The containers to be emptied are distributed so that an equal number of containers are emptied each work day.

Collection frequency, times/wk	Number of pickup locations	Total no. of containers	Number of trips/wk[a]	Number of containers (receiving the same collection frequency) emptied per day				
				Mon.	Tues.	Wed.	Thurs.	Fri.
3	2	2	6	2	—	2	—	2
2	4	4	8	—	4	—	—	4
1	22	26	26	6	4	6	8	2
Total	28	32	40	8	8	8	8	8

[a] In the hauled container system each container to be emptied corresponds to a trip.

   (*b*) Lay out balanced collection routes for each day of the week by successive trials (Steps 3 and 4 in the layout of collection routes). The routes will vary from one solution to another, but containers 11 and 20 must be picked up on Monday, Wednesday, and Friday, and containers 17, 27, 28, and 29 must be picked up on Tuesday and Friday. The optimum solution will be to have an equal number of containers picked up on each day as well as equal distances driven on each day.

   The resulting weekly routes and travel distances are shown in the tabulation on page 233. With the exception of the first container emptied on each route, the distance reported for each container includes the distance from Point B to the container location and the distance from the container location to Point B. The distance reported for the first container includes the distance from the dispatch station and the distance from the container location to Point B.

2. Stationary container system
   (*a*) Set up a summary table for the collection operation using the data reported in the service area map (Step 2 in the layout of collection routes), as follows:

Order of container pickup	Mon.		Tues.		Wed.		Thurs.		Fri.	
	Cont. no.	Dist.[a]	Cont. no.	Dist.[a]	Cont. no.	Dist.[a]	Cont. no.	Dist.[a]	Cont. no.	Dist.[a]
	A→1	6.2	A→7	1.1	A→3	5.9	A→2	5.9	A→13	1.6
1	1→B	11.2	7→B	4.5	3→B	8.8	2→B	8.8	13→B	4.9
2	B→8→B	20.7	B→10→B	17.6	B→9→B	15.3	B→6→B	12.7	B→5→B	16.3
3	B→11→B	14.1	B→14→B	14.0	B→4→B	17.6	B→18→B	6.0	B→11→B	14.1
4	B→20→B	10.0	B→17→B	9.3	B→11→B	14.1	B→15→B	9.6	B→17→B	9.3
5	B→22→B	4.4	B→26→B	12.1	B→12→B	8.8	B→16→B	9.6	B→20→B	10.0
6	B→30→B	5.6	B→27→B	10.9	B→20→B	10.0	B→24→B	16.0	B→27→B	10.9
7	B→19→B	6.9	B→28→B	8.0	B→21→B	4.4	B→25→B	14.0	B→28→B	8.0
8	B→23→B	4.7	B→29→B	8.0	B→31→B	1.1	B→32→B	1.7	B→29→B	8.0
	B→A	5.0	B→A	5.0	B→A	5.0	B→A	5.0	B→A	5.0
Total distance[b]		88.8		90.5		91.0		89.3		88.1

[a] Distance, in thousand feet.

[b] Total distance driven between Points A and B, in thousand feet, during each collection day.

i. The quantity of waste to be collected from the locations requiring three collections per week is entered in Row 1. As noted in the problem statement, the waste from these locations must be collected on Monday, Wednesday, and Friday.
ii. The quantity of waste to be collected from the locations requiring two collections per week is entered in Row 2. These containers must be emptied on Tuesday and Friday.
iii. The additional quantity of waste that can be collected on each collection route is determined and entered in Row 3. Note that the maximum quantity of wastes that can be collected per day is 70 yd^3 [35 yd^3 × 2 (compaction ratio)].

Collection frequency, times/wk	Number of pickup locations	Total waste, yd^3/wk	Quality of wastes collected per day, yd^3			
			Mon.	Tues.	Wed.	Fri.
3	2	51	17	—	17	17
2	4	48	—	24	—	24
1	22	178	53	44	52	29
Total	28	277	70	68	69	70

(*b*) Lay out balanced collection routes by successive trials in terms of quantity of wastes collected (Steps 3 and 4 in the layout of collection routes). Collection routes for the stationary container system will vary, but containers 11 and 20 must be picked up on Monday, Wednesday, and Friday, and containers 17, 27, 28, and 29 must be picked up on Tuesday and Friday. Again, the optimum solution will be to have an equal amount of waste collected on each collection route as well as equal distances driven on each route.

The resulting routes and travel distance are shown in the following tabulation. The travel distance between Points A (dispatch station) and B includes the distance

Order of pickup	Mon.		Tues.		Wed.		Fri.	
	Cont. no.	yd^3	Cont. no.	yd^3	Cont. no.	yd^3	Cont. no.	yd^3
1	5	8	2	6	7	7	13	5
2	4	8	3	6	6	10	11	9
3	1	8	10	8	11	9	17	7
4	8	9	24	9	15	6	18	8
5	9	9	25	4	16	6	19	4
6	11	9	26	8	20	8	23	6
7	14	10	28	5	30	5	20	8
8	20	8	29	5	31	7	27	7
9	—	—	27	7	222	7	28	5
10	—	—	17	7	31	5	29	5
11	—	—	12	4	—	—	32	5
Total		69		69		69		70
Dist.[a]	19,000		22,000		17,000		21,000	

[a] Total distance driven between Points A and B in feet on each collection route.

from Point A to the first container pickup location, the distance traveled on the collection route, and the distance from the last container pickup location to Point B.

Comment. The economic advantage of the stationary container system is apparent in this example. However, if container sizes greater than about 12 yd^3 are needed, the stationary container system can no longer be used.

Collection Route Layout—Steps 2, 3, and 4 for Stationary Container System (with Manually Loaded Collection Vehicles). Steps 2, 3, and 4 for a stationary container system that is manually loaded can be outlined as follows.

Step 2. Estimate the total quantity of wastes to be collected from pickup locations serviced each day that the collection operation is conducted. Using the effective volume of the collection vehicle (nominal collection vehicle volume × compaction ratio), determine the average number of residences from which wastes are to be collected during each collection trip.

Step 3. Once the foregoing data are known, the layout of collection routes can proceed as follows. Starting from the dispatch station (or garage) lay out collection routes that include all of the pickup locations to be serviced during each collection route. These routes should be laid out so that the last of these locations is nearest the disposal site.

Step 4. When the collection routes have been laid out, the actual container density and haul distance for each route should be determined. Using these data, the labor requirements per day should be checked against the available work time per day. In some cases it may be necessary to readjust the collection routes to balance the work load. After the routes have been established, they should be drawn on the master map.

Example 8-7 Layout of residential collection routes. Lay out collection routes for the residential area shown in the accompanying figure on page 236. The service area map would be prepared as the first step in the layout of collection routes. Assume that the following conditions apply:

1. General
   (*a*) Occupants per residence = 3.5
   (*b*) Solid waste collection rate = 3.5 lb/person · d
   (*c*) Collection frequency = once/wk
   (*d*) Type of collection service = curb
   (*e*) Collection crew size = one person
   (*f*) Collection vehicle capacity = 28 yd^3
   (*g*) Compacted specific weight of solid waste in collection vehicle = 540 lb/yd^3

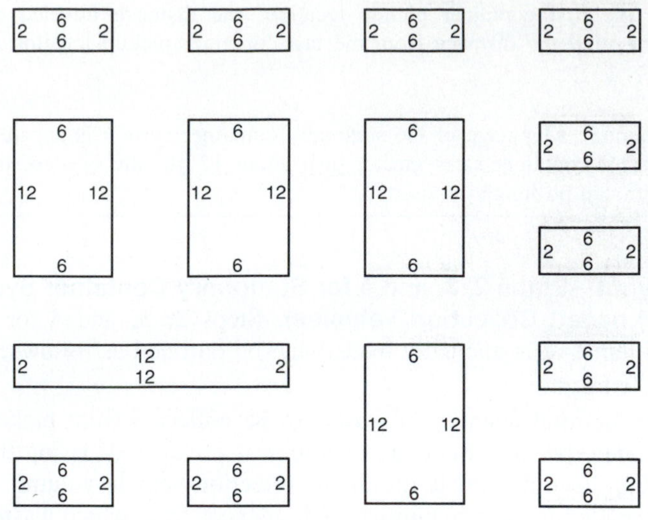

2. Collection route constraints
   (*a*) No U-turns in streets
   (*b*) Collection from each side of the street with stand-up right-hand-drive collection

Solution

1. Develop data needed to establish collection routes (Step 2 in the layout of collection routes).
   (*a*) Determine total number of residences from which wastes are to be collected.

$$\text{Residences} = 10(16) + 4(36) + 1(28) = 332$$

   (*b*) Determine the compacted volume of solid waste to be collected per week.

$$\begin{aligned}\text{Vol/wk} &= [(332 \text{ residences} \times 3.5 \text{ persons/residence} \\ &\quad \times 3.5 \text{ lb/person} \cdot \text{d} \times 7 \text{ d/wk})]/(540 \text{ lb/yd}^3) \\ &= 52.7 \text{ yd}^3\end{aligned}$$

   (*c*) Determine the number of trips required per week.

$$\text{Trip/wk} = \frac{52.7 \text{ yd}^3/\text{wk}}{28 \text{ yd}^3/\text{trip}} = 1.88; \text{ Use } 2$$

   (*d*) Determine the average number of residences from which wastes are to be collected on each collection trip.

$$\text{Residences/trip} = 332/2 = 166$$

2. Lay out collection routes by successive trials using the route constraints cited above as a guide (Step 3 in the layout of collection routes). Two typical routes are shown in the following figure.

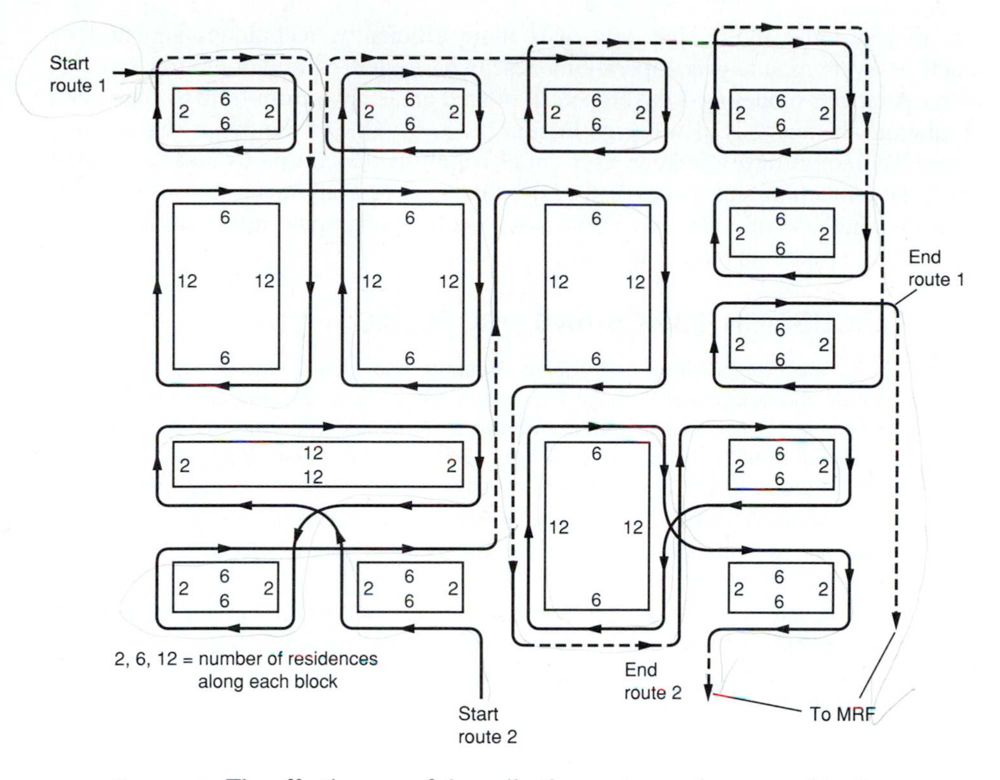

Comment. The effectiveness of the collection routes can be assessed by the amount of route overlap shown by the dotted lines. An interesting problem for the reader is to determine if two collection routes can be laid out without any overlap.

Schedules

A master schedule for each collection route should be prepared for use by the engineering department and the transportation dispatcher. A schedule for each route, which includes the location and order of each pickup point to be serviced, should be prepared for the driver. In addition, a route book should be maintained by each truck driver. The driver uses the route book to check the location and status of accounts. It is also a convenient place in which to record any problems with the accounts. The information contained in the route book is useful in modifying the collection routes.

8-5 ALTERNATIVE TECHNIQUES FOR ANALYSIS OF COLLECTION SYSTEMS

Interest in the analysis of solid waste collection systems arises from the desire to improve (optimize) the operation of existing systems and to develop data and techniques that can be used to design or evaluate new or future systems. In the past and at present the design and operation of most solid waste collection systems are based largely on experience and intuition. In an effort to operate existing systems and design new ones more efficiently, techniques and tools—such as systems analysis, operations research, system simulation, and systems and operations modeling—developed in related areas have from time to time been applied to the analysis of waste collection [2, 3, 6]. One of the major reasons that these techniques have not been used more widely is the enormous cost associated with the collection and processing of field data. A new approach to routing combines a data base of collection information with a geographic information system containing street mapping data.

8-6 DISCUSSION TOPICS AND PROBLEMS

8-1. Drive, walk, or pedal around your community and identify the principal types of systems and equipment used for the collection of residential and commercial solid wastes. Select two of the more common systems and time the various activities associated with the collection of wastes. How do your values compare with those given in this chapter? If your values are significantly different, explain why.

8-2. Drive, walk, or pedal around your community and identify the principal types of service and equipment used for the collection of source-separated wastes from residential and commercial sources.

8-3. How are yard wastes (e.g., grass clippings, brush, and tree trimmings) collected in your community? If your community does not collect yard wastes separately, is the separate collection of yard wastes feasible given the waste management system that is now used? What modifications would be required if separate collection of yard wastes were to be instituted? If your community collects yard wastes separately, describe the operation in terms of its basic operations (e.g., pickup, haul, at-site time, etc.).

8-4. Determine the haul equation constants a and b for the following data (A, B, or C to be selected by your instructor.)

Average haul speed (y), mi/h	Round-trip distance(x), mi/trip		
	A	B	C
10.0	0.9	1.0	1.1
20.0	2.9	3.1	4.6
25.0	4.0	4.8	8.0
29.0	5.5	6.5	16.0
30.0	6.0	7.0	21.0
35.0	8.0	10.5	
39.0	10.0	16.0	
40.0	11.0	20.0	
45.0	15.5		
50.0	28.0		

8-5. You are the city engineer in a small rural town. During a council meeting you are asked to compare the satellite method of collection with the more traditional curb-and-alley collection service that the city is currently providing. Startled, because you had fallen asleep during the preceding 4-h debate concerning the merits of the city slogan, you try to collect your thoughts. What are some of the important considerations that must be brought out in your discussion?

8-6. Develop an equation similar to those presented in this chapter that can be used to determine the labor requirements for a stationary container system employing satellite collection vehicles (see Fig. 8-2).

8-7. Develop an expression similar to Eq. (8-13) that can be used to estimate the time required for the curbside collection of source-separated wastes. Assume the following conditions apply:

(*a*) Three separate containers will be used for (i) mixed paper, (ii) commingled plastics and glass, and (iii) commingled aluminum and tin cans.

(*b*) Some homeowners will also periodically put out cardboard for separate collection.

8-8. Because of a difference of opinion among city staff members, you have been retained as an outside consultant to evaluate the collection operation of the city of Davisville. The basic question centers around the amount of time spent on off-route activities by the collectors. The collectors say that they spend less than 15 percent of each 8-h workday on off-route activities; management claims that the amount of time spent is more than 15 percent. You are given the following information that has been verified by both the collectors and management:

(*a*) A hauled container system, without container exchange, is used.

(*b*) The average time spent driving from yard to the first container is 20 min, and no off-route activities occur.

(*c*) The average pickup time per container is 6 min.

(*d*) The average time to drive between containers is 6 min.

(*e*) The average time required to empty the container at the disposal site is 6 min.

(*f*) The average round-trip distance to the disposal site is 10 mi/trip, and the haul equation $(a + bx)$ constants are $a = 0.004$ h/trip and $b = 0.02$ h/mi.

(*g*) The time required to redeposit a container after it has been emptied is 6 min.

(*h*) The average time spent driving from the last container to the corporation yard is 15 min, and no off-route activities occur.

(*i*) The number of containers emptied per day is 10.

From this information, determine whether the truth is on the side of the collectors or the management.

8-9. The amount of solid wastes generated per week in a large residential complex is about 600 yd^3. There are two containers, each with a capacity of 40 gal, at the rear of every house. The solid wastes are collected by a two-person crew using a 35-yd^3 manually loaded compactor truck once a week. Determine the time per trip and the weekly labor requirements in person-days. The disposal site is located 15 mi away; haul-time constants a and b are 0.022 h/trip and 0.022 h/mi, respectively; the container utilization factor is 0.7; and the compaction ratio is 2. Assume that collection is based on an 8-h day.

8-10. A city desires to determine the impact of a new subdivision on solid waste collection services. The subdivision will add 150 new houses. A two-person crew will collect the wastes twice a week, using 24-m^3 manually based loaded compactor trucks. The allowable container size is 0.14 m^3. It is estimated that there will be 3.2 persons per household and that each person will dispose of 2.5 kg of waste daily. Determine the number of containers that will be needed per household, the average container utilization factor, and the weekly labor requirement in person-days. The compaction ratio for the collection vehicle is 2.5, the average density of the solid wastes in the containers is 120 kg/m^3, the disposal site is located 25 km away, and the haul-time constants a and b are 0.08 h/trip and 0.015 h/km, respectively. Collection is during an 8-h day. Collection is at curbside except for elderly persons (about 5 percent) who receive backyard service.

8-11. A new residential area composed of 800 low-rise detached dwellings is about to be occupied. Assuming that either two or three trips per day will be made to the disposal site, design the collection system and compare the two alternatives. The following data are applicable:

(*a*) Solid waste generation rate = 0.032 yd^3/home · d

(*b*) Containers per service = 2

(*c*) Type of service = 75 percent curbside and 25 percent rear of house

(*d*) Collection frequency = once per week

(*e*) Collection vehicle is a rear-loaded compactor with a compaction ratio of 2.5.

(*f*) Length of workday = 8 h

(*g*) Collection crew = 2 persons

(*h*) Round-trip haul distance = 20 mi

(*i*) Haul constants: a = 0.08 h/trip and b = 0.025 h/mi

(*j*) At-site time per trip = 0.083 h/trip

8-12. TT&E Corporation has four business locations that are each conveniently located 5 mi apart and 5 mi from the disposal site. TT&E presently uses a conventional hauled container system with large open-top containers. It has been suggested to TT&E that money could be saved by renting a fifth container from the waste collection company at a cost of $120/month and switching the operation to the container-exchange mode (see Fig. 8-14*b*). Each location will be serviced 8 times per month. The extra container will be stored at the collection company's dispatch station. Assuming that the operating costs are $20/h, compute the costs for both systems. Is it a wise decision for TT&E to rent the fifth container? Assume that a = 0.034 h/trip and b = 0.029 h/mi for all cases. State clearly any additional assumptions.

8-13. You and your friend are looking for some part-time work. You live in a small rural community that does not receive regular waste collection service. Your friend thinks it would be a good idea to provide waste collection service using your new $\frac{3}{4}$-ton four-wheel-drive pickup truck. There are 30 houses, and each one uses two 32-gal containers. All the houses would receive backyard carry service once per week. The haul constants are 0.08 h/trip and 0.025 h/mi. Assume that the at-site time equals 0.5 h. The round-trip haul distance to the disposal site is 32 mi. The size of the pickup truck bed is 6 × 8 × 3 ft. Assuming that, working together, you and your friend can devote 10 h/wk to this project, is it operationally feasible?

8-14. You have been called to submit a proposal to evaluate your university's solid waste collection operation. Prepare a proposal, in outline form, to be submitted to the university. Note clearly the major divisions or tasks into which the work effort would be divided. Based on your knowledge to date, estimate the person-months of effort that would be required to do the work outlined in your proposal. Use an outline format in answering this question.

8-15. Lay out collection routes for both a hauled container and stationary container collection system for the industrial service area shown in the accompanying map. There are a total of 28 pickup locations and 35 containers. The total quantity of waste to be collected each week is 289 yd^3. Using an arbitrary scale, determine the distance

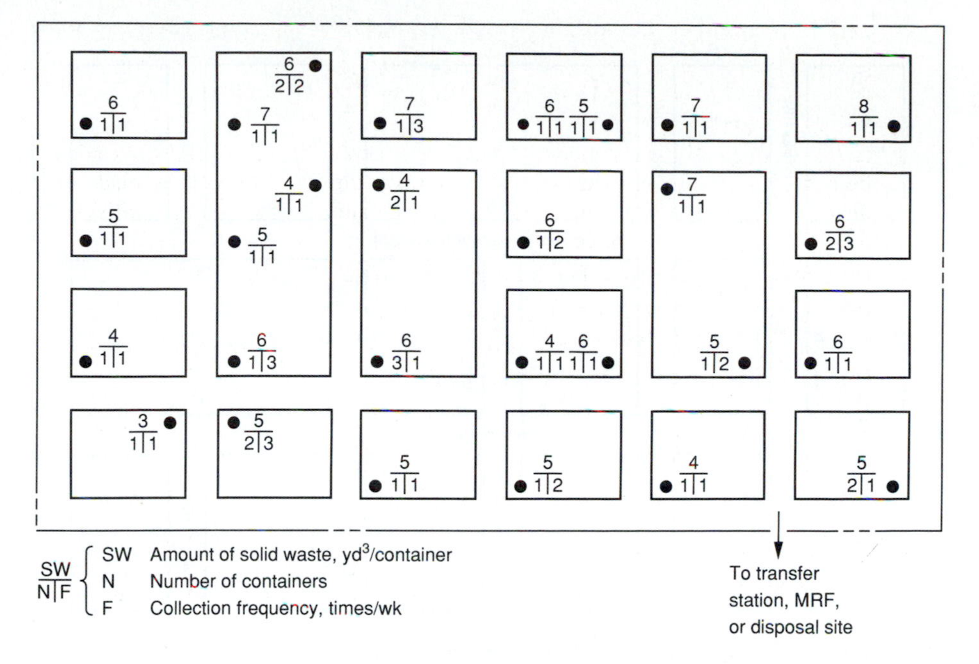

traveled during the collection operation for each route for each collection system. The map and the information it contains would be prepared as the first step in the layout of collection routes. Assume the following conditions apply:

(*a*) Containers with a collection frequency of twice per week must be picked up on Tuesday and Friday.

(*b*) Containers with a collection frequency of three times per week must be picked up on Monday, Wednesday, and Friday.

(*c*) Containers may be picked up from any side of the intersection where they are located.

(*d*) The hauled container system is of the type where the empty containers are returned to the same location where they were picked up full (see Fig. 8-14*a*).

(*e*) For both collection systems, collection will be provided Monday through Friday, as required.

(*f*) For the stationary container system, the collection vehicle will be a self-loading compactor with a capacity of 35 yd^3 and a compaction ratio of 2.8.

8-16. Lay out collection routes for the residential area shown in the figure on page 242. Assume that the following data are applicable:

(*a*) Occupants per residence = 3.1

(*b*) Solid waste generation rate = 3.75 lb /capita · d

(*c*) Collection frequency = once per week

(*d*) Type of collection service = curb

(*e*) Collection crew size = 1 person

(*f*) Collection vehicle capacity = 26 yd^3

(*g*) Compacted specific weight of solid wastes in collection vehicle is equal to 590 lb/yd^3.

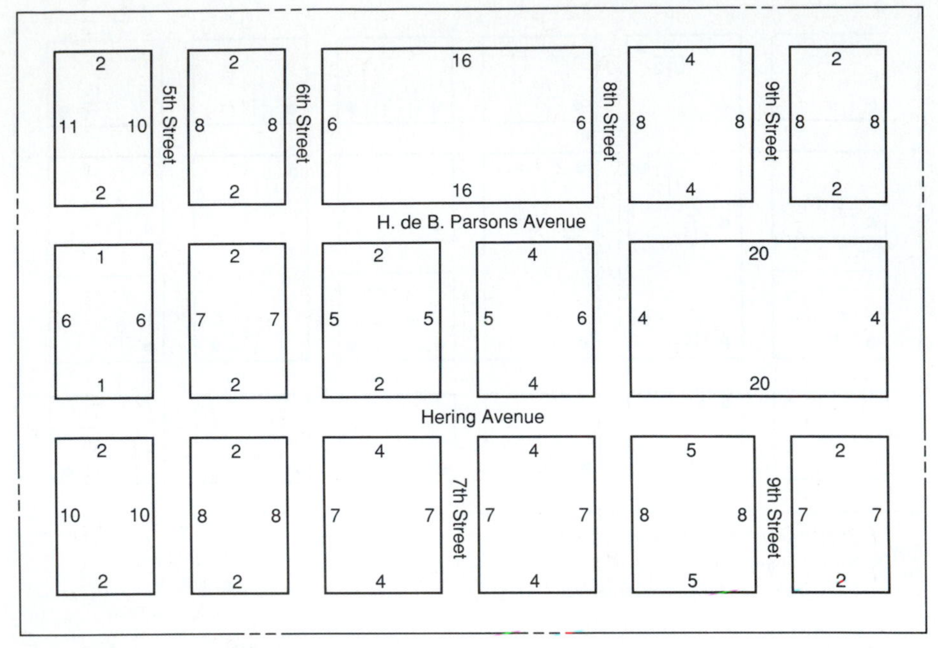

3, 6, 10, 15 = number of residences along each block

8-17. Lay out collection routes for the area shown in the figure given in Problem 8-16 using data given in Problem 8-16, assuming that 5th and 8th streets are one-way running from the south to north and that 6th street is one-way running from north to south. All of the other streets and avenues are two-way.

8-18. Lay out collection routes for the area shown in the figure given in Problem 8-16 using the data given in Problem 8-16, assuming that H. de. B Parsons Street is one-way running form the west to east and that Hering Street is one-way running from east to west. All of the other streets are two-way.

8-19. Lay out collection routes for the residential area shown in the figure on page 243. Assume that the following data are applicable:

(*a*) Occupants per residence = 2.8

(*b*) Solid waste generation rate = 3.5 lb/capita · d

(*c*) Collection frequency = once per week

(*d*) Type of collection service = curb

(*e*) Collection crew size = 1 person

(*f*) Collection vehicle capacity = 24 yd^3

(*g*) Compacted specific weight of solid wastes in collection vehicle is equal to 650 lb/yd^3.

(*h*) Collection is from each side of the street using a standup right-hand-drive collection vehicle.

(*i*) No U-turns in streets

8-20. Solve Problem 8-19 using a collection vehicle with a capacity of 20 yd^3, assuming the compacted specific weight is equal to 575 lb/yd^3.

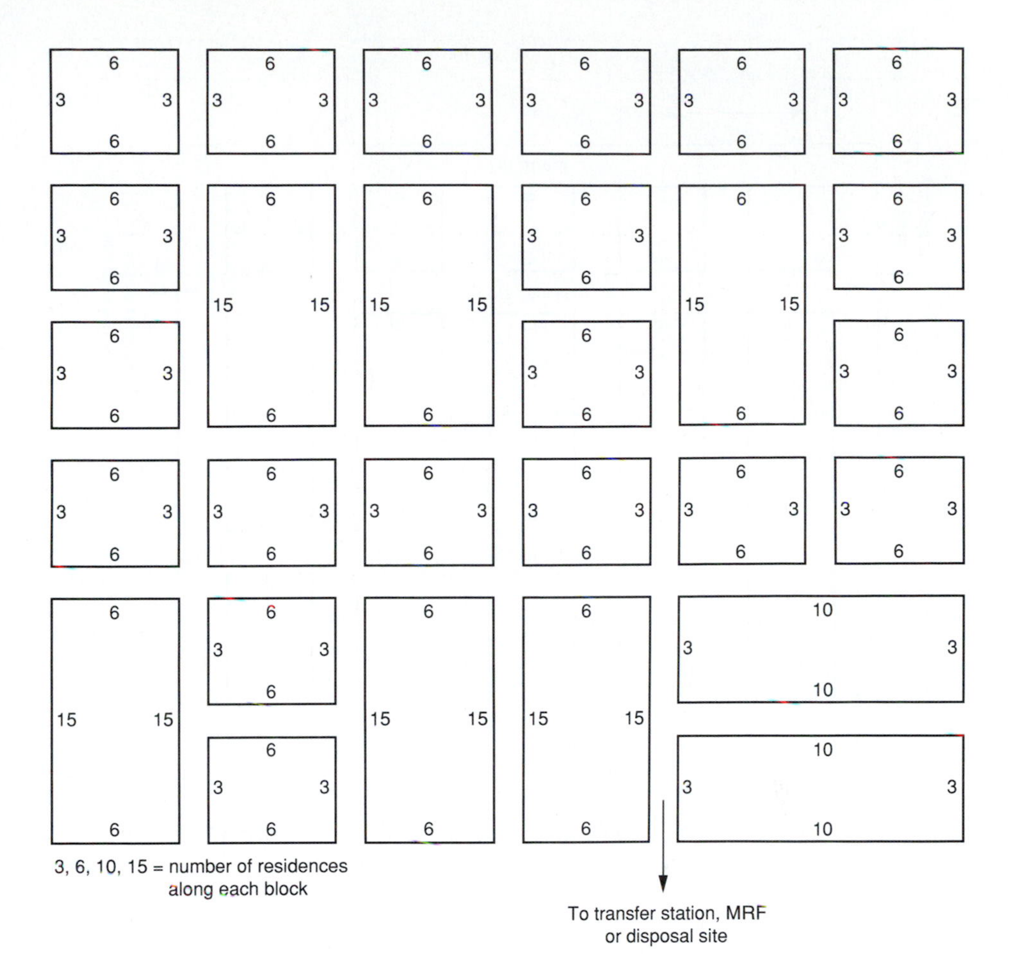

8-21. Lay out collection routes for the residential area shown in the figure on page 244. Assume that the following data are applicable:

(*a*) Occupants per residence = 3.1

(*b*) Solid waste generation rate = 3.8 lb/capita · d

(*c*) Collection frequency = once per week

(*d*) Type of collection service = curb

(*e*) Collection crew size = 1 person

(*f*) Collection vehicle capacity = 30 yd^3

(*g*) Compacted specific weight of solid wastes in collection vehicle is equal to 625 lb/yd^3.

(*h*) Collection is from each side of the street using a standup right-hand-drive collection vehicle.

(*i*) No U-turns in streets

8-22. Solve Problem 8-21 using a collection vehicle with a capacity of 24 yd^3 assuming the compacted specific weight is equal to 585 lb/yd^3.

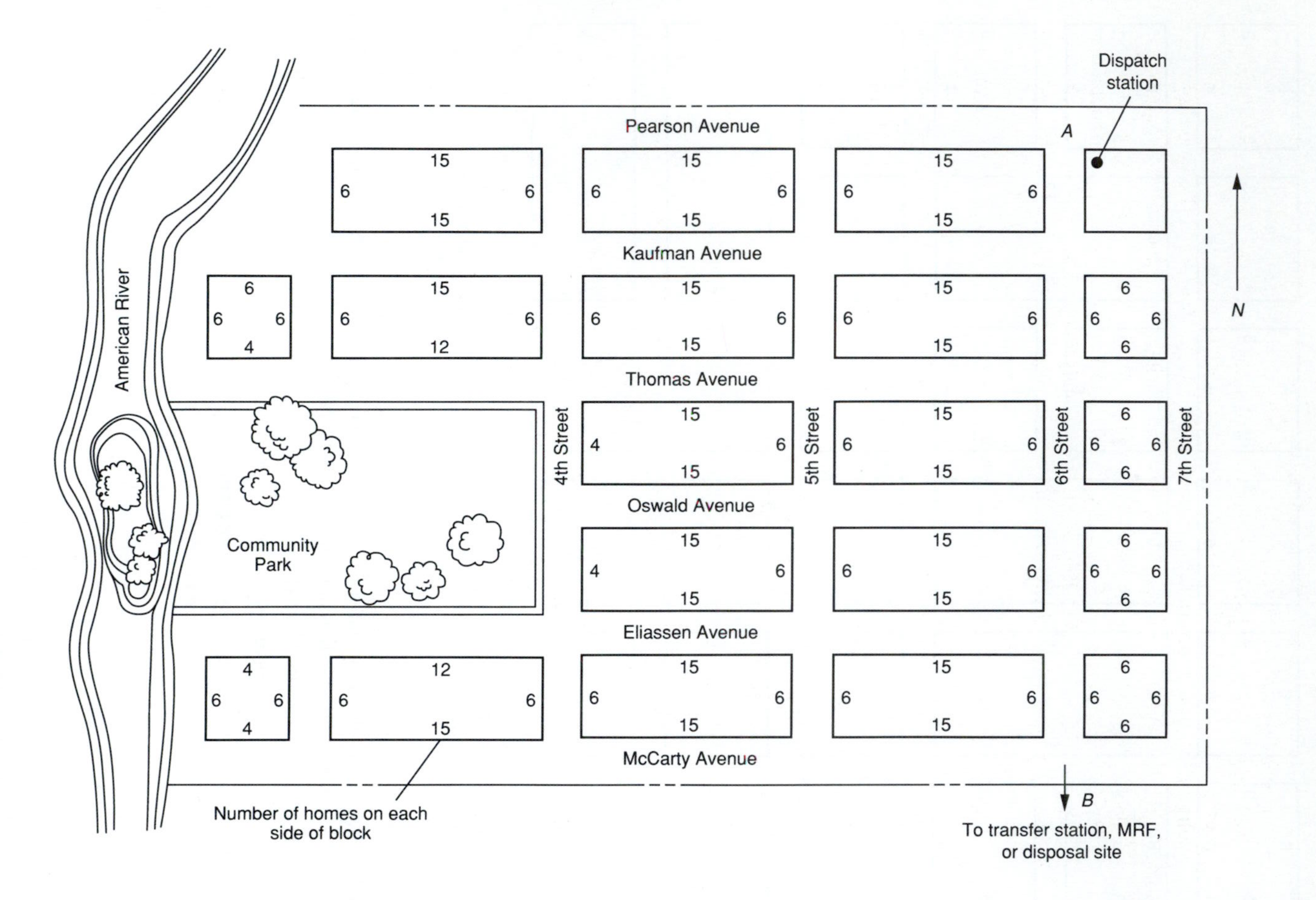
Dispatch station
A
N
Pearson Avenue
Kaufman Avenue
Thomas Avenue
Oswald Avenue
Eliassen Avenue
McCarty Avenue
4th Street
5th Street
6th Street
7th Street
American River
Community Park
Number of homes on each side of block
B
To transfer station, MRF, or disposal site

8-23. The household waste generation rates given below were observed over a period of time on a typical collection route. Assuming that curbside density of the waste is 120 kg/m^3, estimate the percentage of the time a 24-m^3 collection truck with a compaction ratio of 2.5 will need more than one trip to service 82 households. The observed waste generation rates are 42, 60, 35, 27, 50, 94, 72 kg/household · wk.

8-24. In the late 1960s and early 1970s much was written about the use of simulation and other techniques from the field of operations research for the routing of collection vehicles (e.g., Refs. 3, 4, and 7). Based on a review of one or two articles from that period (i.e., 1965 to 1975), prepare an analysis of why the techniques proposed have not been adopted to any major extent. Was the research of the period ahead of its time?

8-25. During the late 1980s the processing speed and capacity of relatively inexpensive desktop computers made the use of geographic information systems feasible for the routing of collection vehicles. Prepare a review of such systems using recent issues of trade publications such as *Waste Age* and *Public Works*. Why is this approach more feasible than the mainframe computer-based techniques discussed in Problem 8-24?

8-7 REFERENCES

1. Bergen County Utilities Authority: *Bergen County Apartment Recycling Manual,* Little Ferry, NJ, 1988.
2. Liebman, J. C.: *Routing of Solid Waste Collection Vehicles,* Final Report on Project 801289, Office of Research and Monitoring, U.S. Environmental Protection Agency, Washington, DC, 1973.
3. Quon, J. E., A. Charnes, and S. J. Wenson: Simulation and Analysis of a Refuse Collection System, *Proceedings ASCE, Journal of the Sanitary Engineering Division,* Vol. 91, No. SA5, 1965.
4. Shuster, K. A. and D. A. Schur: *Heuristic Routing for Solid Waste Collection Vehicles,* Publication SW-113, U.S. Environmental Protection Agency, Washington, DC, 1974.
5. Tchobanoglous, G. and G. Klein: *An Engineering Evaluation of Refuse Collection Systems Applicable to the Shore Establishment of the U.S. Navy,* Sanitary Engineering Research Laboratory, University of California, Berkeley, 1962.
6. Truitt, M. M., J. C. Liebman, and C. W. Kruse: *Mathematical Modeling of Solid Waste Collection Policies,* Vols. 1 and 2, Public Health Service Publication 2030, U.S. Department of Health, Education, and Welfare, Washington, DC, 1970.
7. University of California: *An Analysis of Refuse Collection and Sanitary Landfill Disposal,* Technical Bulletin 8, Series 37, University of California Press, Berkeley, 1952.

CHAPTER 9

SEPARATION AND PROCESSING AND TRANSFORMATION OF SOLID WASTE

The purpose of this chapter is to introduce the reader to the topics of the recovery of separated materials, the separation and processing of solid waste components, and the transformation processes used to alter the form of the waste and to recover useful products. The separation and processing and transformation of waste materials make up the fourth of the functional elements. Because the many details associated with the design and implementation of the unit operations and facilities used for the separation, processing, and transformation of waste materials would impede the introduction to these important subjects, design details have been grouped together and are presented in Part IV in Chapters 12 through 15.

The methods now used to recover source-separated waste materials include curbside collection and homeowner delivery of separated materials to drop-off and buy-back centers. The further separation and processing of wastes that have been source-separated, as well as the separation of commingled wastes, usually occur at *materials recovery facilities* (MRFs) or at large integrated *materials recovery/transfer facilities* (MR/TFs). Integrated MR/TFs may include the functions of a drop-off center for separated wastes, a materials separation facility, a facility for the composting and bioconversion of wastes, a facility for the production of refuse-derived fuel, and a transfer and transport facility.

Chemical and biological transformation processes are used to reduce the volume and weight of waste requiring disposal and to recover conversion products and energy. The most commonly used chemical transformation process is combustion, which is used in conjunction with the recovery of energy in the form of heat. The most commonly used biological transformation process is aerobic composting. Technical details on these processes and descriptions of other chemical and biological processes that have been applied to the transformation of solid waste are considered in Chapters 13 and 14, respectively. Issues that must be addressed in the implementation of MRFs and MR/TFs, combustion facilities, and composting facilities are also discussed in this chapter and in Chapters 18 and 19.

9-1 REUSE AND RECYCLING OPPORTUNITIES FOR WASTE MATERIALS

Materials that have been recovered from MSW for reuse and recycling have been discussed earlier (see Tables 3-10 and 6-7). Materials separated from MSW can be used directly, as a raw material for remanufacturing (see Fig. 9-1) and reprocessing, as feedstock for the production of compost and other chemical and biological conversion products, as a fuel source for the production of energy, and for land reclamation. Reuse opportunities for the materials separated from MSW are reported in Table 9-1.

In assessing the opportunities for recycling, the available options for the separation and processing of waste materials, the economics of materials recovery, and materials specifications are critical. For example, even though it may be possible to separate the various components, finding buyers for the materials may be difficult if they do not meet the buyers' specifications. Some typical materials specification items that can affect the components separated from MSW are presented in Table 9-2.

FIGURE 9-1
Representative products manufactured from recycled plastics. (Courtesy of National Association for Plastic Container Recovery.)

TABLE 9-1
Uses for materials that have been recovered from MSW

Use/application	Remarks
Direct reuse	Many of the materials separated from MSW can be reused directly. Examples of such materials include lumber, wooden pallets, 55-gal drums, furniture, etc. Whenever possible, direct reuse should be encouraged.
Raw materials for remanufacturing and reprocessing	Typical specifications for eight different materials derived from municipal wastes are presented in Table 9-2. Specific details, such as product purity, density, and shipping conditions, must be worked out with each potential buyer. Whenever possible, it is beneficial to develop a range of product specifications and product prices. In this way, processing costs to achieve a higher-quality product can be evaluated with respect to the higher market price obtainable for the higher-quality product.
Feedstock for production of biological and chemical conversion products.	Many communities have elected to meet their diversion goals by producing compost that can be marketed directly, given to the residents of the community, used on city property (e.g., greenbelts, highway dividers, etc.), or used as intermediate landfill cover. Each of these uses requires a different quality of compost, especially with respect to the type and amount of contaminant materials that may be present (e.g., plastic, pieces of metal, etc.). The production of methane in controlled reactors, ethanol, and other organic compounds will require that the materials that make up the organic fraction of the MSW be separated from the commingled MSW.
Fuel source	Energy can be derived from municipal wastes in two forms: (1) by combusting (burning) the organic fraction of MSW and/or yard wastes and recovering the heat that is given off and (2) by converting the wastes to some type of fuel (oil, gas, pellets, etc.) that can be stored and used locally or transported to distant energy markets. Specifications for direct use of wastes for the production of steam are usually not so restrictive as those for the production of a fuel. However, as firing and storage techniques improve, specifications for direct use may become more stringent. Note that in many states the use of waste materials as a fuel source is not considered an appropriate means of waste diversion or recycling.
Land reclamation	Applying wastes to land is one of the oldest and most used techniques in solid waste management. Land disposal technology has developed to the point that communities can now plan land reclamation projects without fear of the development of health problems. Typically, land reclamation will be accomplished with clean or processed demolition wastes (see Fig. 9-30). Land reclamation using wastes should not be started until a final land use has been designated.

TABLE 9-2
Typical materials specifications that affect the selection and design of processing operations for MSW[a]

Reuse category and materials components	Typical specification items
Direct reuse	Must be usable for original or related function. Degree of cleanliness (e.g., bicycles, processesd construction and demolition wastes)
Raw material for remanufacturing and reprocessing	
Aluminum	Particle size; degree of cleanliness; moisture content; density; quantity, shipment means, and delivery point
Paper and cardboard	Source; grade; no magazines; no adhesives; moisture content; quantity; storage; and delivery point
Plastics	Type (e.g., PETE/1, HDPE/2, PVC/3, LDPE/4, PP/5, PS/6, and multilayer/7); degree of cleanliness, moisture content
Glass	Amount of cullet material; color, no labels or metal; degree of cleanliness; freedom from metallic contamination; no noncontainer glass; no broken crockery; quantity, storage and delivery point
Ferrous metals	Source (domestic, industrial, etc.); specific weight; degree of cleanliness; degree of contamination with tin, aluminum, and lead; quantity; shipment means; and delivery point
Nonferrous metals	Vary with local needs and markets
Rubber (e.g., waste tires)	Recapping standards; specifications for other uses not well defined
Textiles	Type of material; degree of cleanliness
Feedstock for bioconversion products	
Yard wastes	Composition of material, particle sizes, particle size distribution, degree of contamination
Organic fraction of MSW	Composition of material, degree of contamination
Fuel source	
Yard wastes	Composition, particle size, moisture content
Organic fraction of MSW	Composition, Btu content; moisture content; storage limits; firm quantities; sale and distribution of energy and/or by-products
Plastics	Depends on application and design of combustion equipment
Wastepaper	Use as fuel will vary with local needs and markets
Wood	Composition, degree of contamination
Tires	Tire-to-energy plants; or pulp and paper mills and cement manufacturing facilities that use tire fuel
Waste oil	Depends on application and design of combustion equipment
Land reclamation	
Construction and demolition waste	Composition; degree of contamination. Local and state regulations; final land-use designation

[a] Detailed specifications on the individual materials may be found in Chapter 15.

9-2 MATERIALS RECOVERED AT DROP-OFF AND BUY-BACK CENTERS

Waste materials that have been source-separated must be collected or gathered together before they can be recycled. The principal methods now used for the collection of these materials include curbside collection using specially designed collection vehicles (see Section 8-2) and delivery by homeowners to drop-off and buy-back centers.

Drop-Off Centers

A drop-off program requires residents or businesses to separate recyclable materials at the source and bring them to a specified drop-off or collection center. Drop-off centers range from single material collection points (e.g., easy-access "igloo" containers such as shown in Fig. 9-2) to staffed, multimaterial collection centers. Because residents and businesses are responsible for not only separating their recyclable materials but also taking them to a drop-off center, low participation can be a problem in achieving the diversion rates desired from these programs. Drop-off centers also require residents and businesses to store the materials until sufficient material is collected to warrant a trip to the drop-off center. The storage of multiple material types is a problem in densely populated areas, where residences typically do not have much storage space available.

To encourage participation, most successful programs have made drop-off centers as convenient to use as possible. For example, drop-off points at shopping centers and supermarkets (see Fig. 9-3) or other convenient locations are common. In many communities, combination drop-off and buy-back centers are located at the MRF (see Fig. 9-4). Mobile collection centers, which can be moved to new locations periodically, also increase convenience. Other incentives, such as donating portions of proceeds to a local charity, can also foster greater participation.

(a)

(b)

FIGURE 9-2
Typical drop-off containers: (*a*) igloo type, Davis, CA and (*b*) pedestal type, Stockholm, Sweden.

FIGURE 9-3
Typical drop-off centers located at shopping centers with grocery stores. As shown in photo on left, drop-off centers are often operated to benefit charities.

FIGURE 9-4
Typical drop-off and buy-back centers for recyclable materials located at supermarkets and shopping centers. Trailer is used for the storage of recyclable materials to minimize transportation costs. In many communities, drop-off and buyback centers are also located at materials recovery facilities, transfer stations, and at disposal sites.

Example 9-1 Home separation and delivery to drop-off centers. A community of 1200 homes cannot pay for the initial and operating costs of the recycling collection vehicles that were to be used. Instead, residents are to haul recycling containers to a drop-off center operated by the community. Calculate the number of vehicles from which recyclable materials must be unloaded per hour at the recycling drop-off center. Assume the center is open for eight hours per day, two days per week, and that 40 percent of the residents will deliver recycling containers. Also assume that 75 percent of the participants will take their separated materials to the drop-off center once per week and that the remaining 25 percent of the participants will bring their separated materials to the drop-off center once every two weeks.

Solution

1. Determine the average number of trips per week.

$$\begin{aligned}\text{Trips/wk} &= [1200 \text{ homes } \times 0.40 \text{ (participation rate)} \times 0.75 \times 1 \text{ trip/home} \cdot \text{wk} \\ &\quad + 1200 \text{ homes } \times 0.40 \text{ (participation rate)} \times 0.25 \times 0.5 \text{ trip/home} \cdot \text{wk} \\ &= 420 \text{ trips/wk}\end{aligned}$$

2. Determine the average number of cars per hour.

$$\text{Cars/hr} = [420 \text{ trips (cars)/wk}]/[(2 \text{ d/wk}) \times (8 \text{ hr/d})] = 27 \text{ cars/hr}$$

Comment. Clearly, a small drop-off center cannot accommodate 27 cars/hr (equivalent to one car unloading every 2.2 minutes). Also, it is unlikely that the cars would arrive at a uniform rate. The most viable solution is to increase the number of hours per week that the drop-off center will be open.

Buy-Back Centers

Buy-back refers to a drop-off program that provides a monetary incentive to participate (see Fig. 9-4). In this type of program, the residents are paid for their recyclables either directly (e.g., price per pound) or indirectly through a reduction in monthly collection and disposal fees. Other incentive systems include contests or lotteries.

9-3 OPTIONS FOR THE SEPARATION OF WASTE MATERIALS

Separation is a necessary operation in the recovery of reusable and recyclable materials from MSW. Separation can be accomplished either at the source of generation or at MRFs. Depending on the separation objectives, a variety of MRFs or MR/TFs can be developed. The reuse and recycling opportunities and the options available for the separation of materials will affect the type of waste management program implemented by a community. Various waste management options for meeting diversion goals are considered in Chapter 18.

Waste Separation at the Source of Generation

Waste separation at the source is usually accomplished by manual means. The number and types of components separated will depend on the waste diversion goals established for the program. Even though waste materials have been separated at the source, additional separation and processing will usually be required before these materials can be reused or recycled.

Waste Separation at MRFs and MR/TFs

MRFs and MR/TFs are used for (1) the further processing of source-separated wastes obtained from curbside collection programs and drop-off and buy-back centers without processing facilities, (2) the separation and recovery of reusable and recyclable materials from commingled MSW, and (3) improvements in the quality (specifications) of the recovered waste materials. In the simplest terms, a MRF can function as a centralized facility for the separation, cleaning, packaging, and shipping of large volumes of materials recovered from MSW.

Manual versus Mechanical Separation. The separation of waste materials from MSW can be accomplished manually or mechanically. Manual separation is used almost exclusively for the separation of wastes at the source of generation. Many of the early MRFs built in the 1970s were designed to separate the waste components mechanically. Unfortunately, none of these early facilities is currently in operation, primarily because of mechanical problems. The current trend is to design MRFs based on the integration of both manual and mechanical separation functions.

MRFs for Source-Separated Wastes. The types of source-separated materials that are separated further at MRFs may include paper and cardboard from mixed paper and cardboard; aluminum from commingled aluminum and tin cans; plastics by class from commingled plastics; aluminum cans, tin cans, plastics, and glass from a mixture of these materials; glass by color (clear, amber, and green). The processing of wastes that have been separated at the source is considered in Section 9-6.

MRFs for Commingled MSW. All types of waste components can be separated from commingled MSW. Wastes are typically separated both manually and mechanically. The sophistication of the MRF will depend on (1) the number and types of components to be separated, (2) the waste diversion goals established for the waste recovery program, and (3) the specifications to which the separated product must conform (see Chapter 15).

9-4 INTRODUCTION TO THE UNIT OPERATIONS USED FOR THE SEPARATION AND PROCESSING OF WASTE MATERIALS

The unit operations and facilities used for the separation and processing of waste materials at MRFs are introduced in this section. Unit operations used for the separation and processing of separated and commingled wastes are designed (1) to modify the physical characteristics of the waste so that waste components can be removed more easily, (2) to remove specific components and contaminants from the waste stream, and (3) to process and prepare the separated materials for subsequent uses. Commonly used unit operations for the processing of MSW are summarized in Table 9-3. Facilities used for the handling, moving, and storage of waste materials are considered further in Section 9-5. Application of the unit operations introduced in this section is illustrated in Section 9-6, which deals with MRFs for processing MSW. Technical details on the unit operations described in this section are presented in Chapter 12. Additional details may be found in Refs. 2, 5, 6, 9, 12, and 14. Cost information for the equipment and facilities described in this section is outlined in Appendix E.

Size Reduction

Size reduction is the unit operation in which *as collected* waste materials are mechanically reduced in size. In practice, the terms shredding, grinding, and milling are used interchangeably to describe mechanical size-reduction operations. The objective of size reduction is to obtain a final product that is reasonably uniform and considerably reduced in size in comparison with its original form (see Fig. 9-5). Note that size reduction does not necessarily imply volume reduction. In some situations, the total volume of the material after size reduction may be greater than that of the original volume. Size reduction equipment used for the processing of wastes includes shredders, glass crushers, and wood grinders.

Shredders. The three most common types of shredding devices used to reduce the size of MSW are the hammer mill, the flail mill or shredder, and the shear shredder (see Fig. 9-6). Other examples of shredding devices include cutters, cage disintegrators, drum pulverizers, and wet pulpers. In operation, the hammers in the hammer mill (see Fig. 9-6*a*), attached to a rotating element, strike the waste material as it enters and eventually force the shredded material through the discharge of the unit, which may or may not be equipped with bottom grates of varying sizes. The flail mill (see Fig. 9-6*b*) is similar to the hammer mill, but provides only coarse shredding, as the hammers are spaced further apart. Operationally, flail mills are single-pass devices, whereas material remains in a hammer mill until it will pass through the openings in the bottom grate. Flail mills are often used as bag breakers. The shear shredder (see Fig. 9-6*c*) is composed of two parallel counterrotating shafts with a series of discs mounted perpendicularly that serve as cutters. The waste material to be shredded is directed to the center

TABLE 9-3

Commonly used unit operations and facilities for the separation and processing of separated and commingled MSW

Item	Function/material processed	Preprocessing
	Unit operations	
Shredding		
Hammer mills	Size reduction/all types of wastes	Removal of large bulky items, removal of contaminants
Flail mills	Size reduction, also used as bag breaker/all types of wastes	Removal of large bulky items, removal of contaminants
Shear shredder	Size reduction, also used as bag breaker/all types of wastes	Removal of large bulky items, removal of contaminants
Glass crushers	Size reduction/all types of glass	Removal of all nonglass materials
Wood grinders	Size reduction/yard trimmings/all types of wood wastes	Removal of large bulky items, removal of contaminants
Screening	Separation of over- and under-sized material; trommel also used as bag breaker/all types of waste	Removal of large bulky items, large pieces of cardboard
Cyclone separator	Separation of light combustible materials from air stream/prepared waste	Material is removed from air stream containing light combustible materials
Density separation (air classification)	Separation of light combustible materials from air stream	Removal of large bulky items, large pieces of cardboard, shredding of waste
Magnetic separation	Separation of ferrous metal from commingled wastes	Removal of large bulky items, large pieces of cardboard, shredding of waste
Densification		
Balers	Compaction into bales/paper, cardboard, plastics, textiles, aluminum	Balers are used to bale separated components
Can crushers	Compaction and flattening/aluminum and tin cans	Removal of large bulky items
Wet separation	Separation of glass and aluminum	Removal of large bulky items
Weighing facilities		
Platform scales	Operational records	
Small scales	Operational records	
	Handling, moving, and storage facilities	
Conveyor belts	Materials transport/all types of materials	Removal of large bulky items
Picking belts	Manual separation of waste materials/source-separated and commingled MSW	Removal of large bulky items
Screw (auger) conveyors (Use not well established)	Materials transport; also used as bag breaker/all types of waste	Removal of large bulky items
Movable equipment	Materials handling and moving/all types of waste	
Storage facilities	Materials storage/all types of recovered materials	Densification, glass crushing, etc.

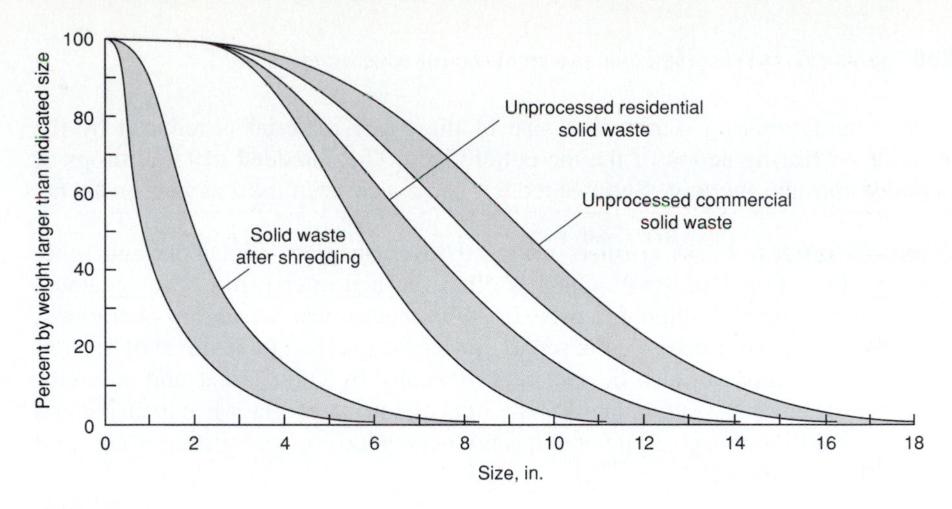

FIGURE 9-5
Typical distribution of particle sizes for the organic fraction of MSW excluding yard waste before and after shredding.

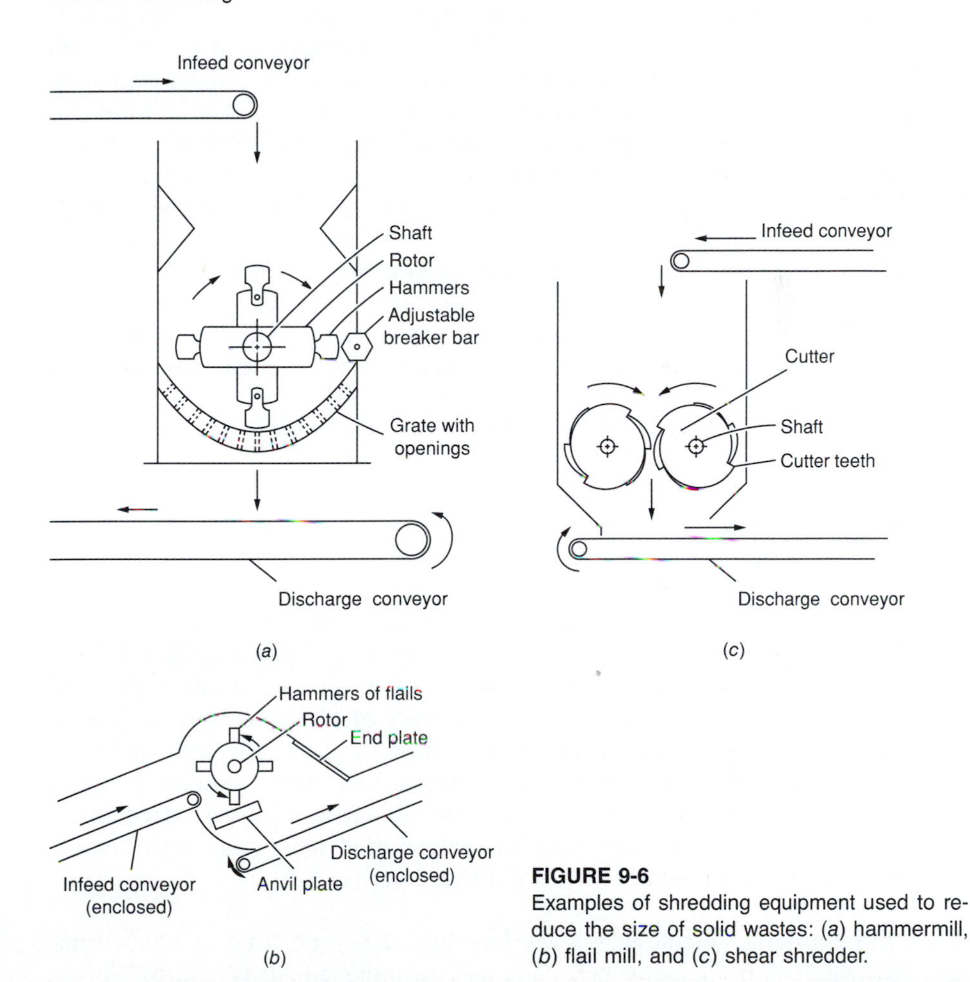

FIGURE 9-6
Examples of shredding equipment used to reduce the size of solid wastes: (*a*) hammermill, (*b*) flail mill, and (*c*) shear shredder.

of the counterrotating shafts. The size of the waste material is reduced by the shearing or tearing action of the the cutter discs. The shredded material drops or is pulled through the unit. Shear shredders have also been used as bag breakers.

Glass Crushers. Glass crushers are used to crush glass containers and other glass products found in MSW. Glass is often crushed after it has been separated to reduce storage and shipping costs. In some mechanical separation operations, glass is crushed, after one or more separation steps, to effect its removal by screening. Crushed glass can also be separated optically by color (clear and colored). However, because the equipment for the optical sorting of glass is expensive and on-line reliability of such equipment has not been good, optical sorting is not used commonly at present.

Wood Grinders. Typically, most wood grinders are wood chippers, used to shred large pieces of wood (e.g., large branches, broken pallets) into chips (see Fig. 9-7), which can be used as a fuel, and finer material, which can be composted. Tub grinders (see Fig. 9-27 in Section 9-6) are used to process yard wastes. A tub grinder consists of a large tub having a revolving upper section, and a stationary lower section containing a hammer mill. The tub grinder is fed by a front-end loader (or by a stationary grapple, if the wastes are piled close enough to the tub), and the revolving action of the tub ensures that material flows continuously to the hammer mill. The continuous stream of shredded material is carried away from the grinder by a conveyor. The wood that has been ground in a tub grinder is usually sorted by size using trommel, disc, or vibrating screens. Optical sorters have also been developed but are not used commonly because of their maintenance requirements and high initial cost. The larger-size material is used as a biomass fuel or as a bulking agent in composting operations. The fine material is usually composted.

Screening

Screening is a unit operation used to separate mixtures of materials of different sizes into two or more size fractions by means of one or more screening surfaces. Screening may be accomplished either dry or wet, with the former being more common in solid waste processing systems. The principal applications of screening devices in the processing of MSW include (1) removal of oversized materials, (2) removal of undersized materials, (3) separation of the waste into light combustibles and heavy noncombustibles, (4) recovery of paper, plastics, and other light materials from glass and metal, (5) separation of glass, grit, and sand from combustible materials, (6) separation of rocks and other oversized debris from soil excavated at construction sites, and (7) removal of oversized materials from combustion ash. The types of screens used most commonly for the separation of solid waste materials are illustrated in Fig. 9-8.

Vibrating Screens. Vibrating screens (see Fig. 9-8*a*) are used to remove undersized materials from source-separated and commingled MSW and to process

FIGURE 9-7
Large commercial wood grinder. (Courtesy of SSI Shredding Systems, Inc.)

construction and demolition wastes. Vibrating screens can be designed to vibrate from side to side, vertically, or lengthwise. Vibrating screens used for the separation of MSW are inclined and use a vertical motion. The vertical motion allows the material that is to be separated to contact the screen at different locations each time.

Rotary Screens. The most common type of rotary screen used in the processing of wastes is a trommel screen. Trommels (see Figs. 9-8*b* and 9-9*a*), also known as rotary drum screens, were first developed in England in the 1920s. Trommels are used to separate waste materials into several size fractions. Operationally, the material to be separated is introduced at the front end of the inclined rotating trommel. As the screen rotates, the material to be separated tumbles and contacts the screen numerous times as it travels down the length of the screen. Small particles will fall through the holes in the screen, while the oversized material will pass through the screen. The material falling through the screen is collectively known as *unders*, *undersize*, and *underflow*; material retained by the screen is known as *overs*, *oversize*, and *overflow*. Trommels equipped with metal blades or teeth that protrude into the drum (see Fig. 9-8*b*) are also used as bag breakers. Typically, the blades are located in the first third of the trommel. Having passed through the trommel, the oversize wastes are then sorted manually. In some systems, magnetic separation of ferrous metals will occur before manual separation. Ferrous metals will also be removed from the undersize waste materials passing through the trommel.

Disc Screens. Disc screens consist of sets of parallel horizontal shafts equipped with interlocking lobed (or star-shaped) discs (see Fig. 9-8*c* and 9-9*b*). The undersized materials to be separated fall between the spaces in the discs, and oversized materials ride over the top of the discs as in a conveyor belt. Different-sized materials can be separated using the same screen by adjusting the spacing between the rotating discs. Disc screens have several advantages over other types of screens, including self-cleaning and adjustability with respect to the spacing of the discs on the drive shafts. Disc screens are used in the same applications as trommels.

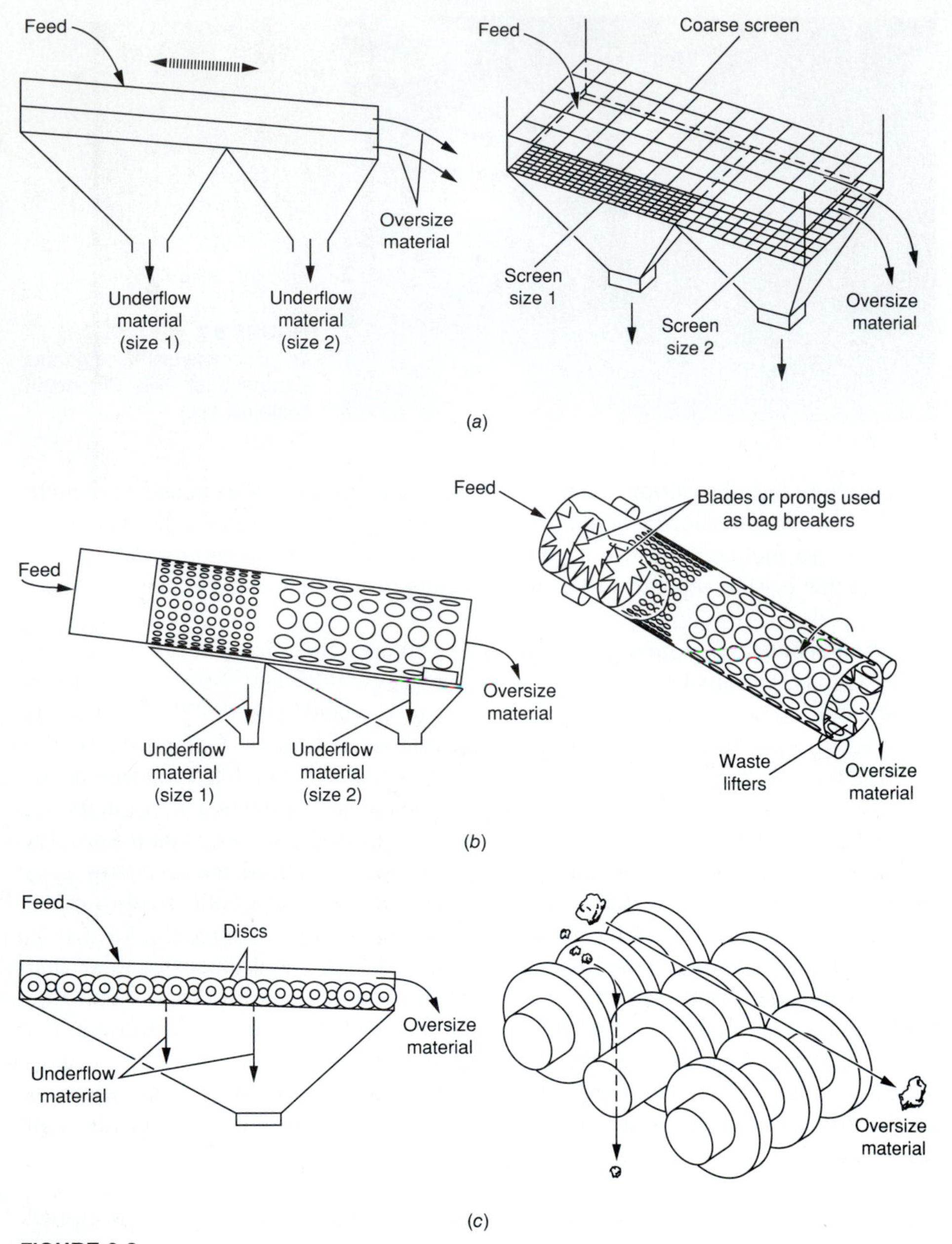

FIGURE 9-8
Typical screens used for the separation of solid wastes: (*a*) vibrating screen, (*b*) rotary drum (trommel) screen, and (*c*) disc screen.

(a)

(b)

FIGURE 9-9
Views of screens in operation with MSW: (*a*) large trommel screen and (*b*) disc screen. (Courtesy of Triple/S Dynamics Systems, Inc.)

Density Separation (Air Classification)

Air classification is the unit operation used to separate light materials such as paper and plastic from heavier materials such as ferrous metal, based on the weight difference of the material in an air stream. If materials of different weights are introduced into an air stream moving with sufficient velocity, the light materials will be carried away with the air while the heavier materials will fall in the counter-current direction. Air classification has been used for a number of years in industrial operations for the separation of various components from dry mixtures.

In MRFs, air classification is used to separate the organic material—or, as it is often called, the *light fraction*—from the heavier, inorganic material, which is called the *heavy fraction*. Air classification has also been used for the separation of commingled glass and plastic. A complete air classification system is comprised of the air classifier and cyclone separator, which is used to separate the solid materials from the air stream (see Fig. 9-10). Because there is movement away from the shredding of commingled MSW, air classification systems of the type shown in Fig. 9-10 are not commonly used today. In installations where one or more trommels are used, a device known as a *stoner*, which also involves the use of air to fluidize the wastes to be separated (see Chapter 12), is used to separate the heavy grit from the organic material in the trommel underflow (undersize material) waste stream.

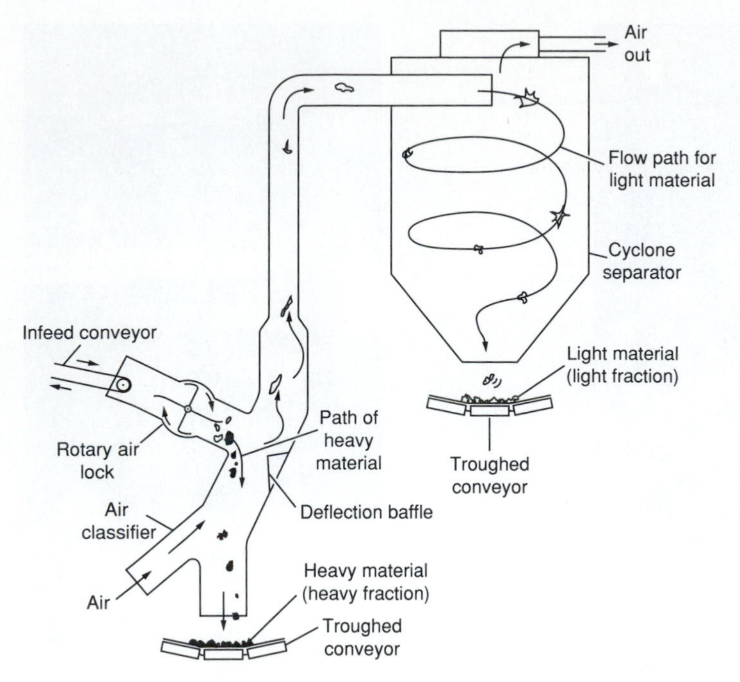

FIGURE 9-10
Typical air classification system used to separate solid waste into light and heavy fractions.

Magnetic Separation

Magnetic separation is a unit operation whereby ferrous metals are separated from other waste materials by utilizing their magnetic properties. Magnetic separation is used to recover ferrous materials from source-separated, commingled, and shredded MSW (see Fig. 9-11). Magnetic separation is used commonly to separate aluminum cans from tin cans in source-separated waste where the two types of metals are mixed. Ferrous materials are usually recovered either after shredding and before air classification or after shredding and air classification. In some large installations, overhead magnetic systems have been used to recover ferrous materials before shredding (this operation is known as *scalping*).

When commingled MSW is burned in combustors, magnetic separation is used to remove the ferrous materials from combustion residue. Magnetic recovery systems have also been used at landfill disposal sites. The specific location(s) where ferrous materials are recovered will depend on the objectives to be achieved, such as the reduction of wear and tear on processing and separation equipment, the degree of product purity to be achieved, and the required recovery efficiency.

Densification

Densification (also known as *compaction*) is a unit operation that increases the density of waste materials so that they can be stored and transported more efficiently

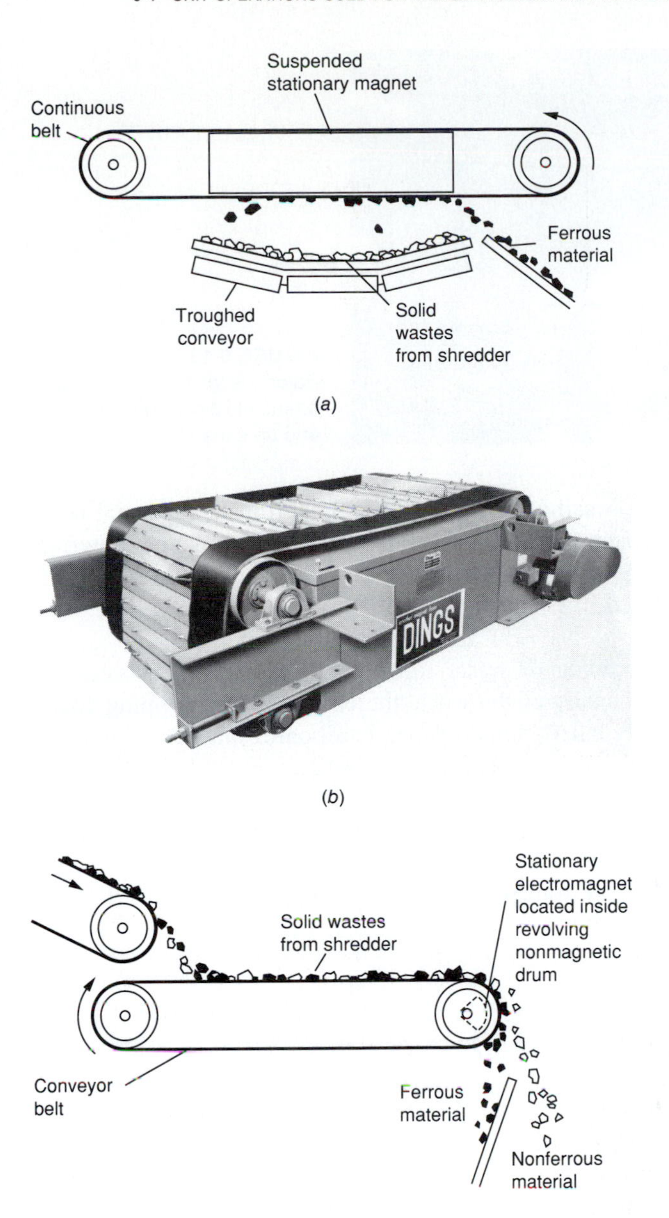

FIGURE 9-11
Typical magnet separators: (*a*) schematic overhead magnet and (*b*) view of commercial overhead magnet. The unit shown is equipped with an armored stainless steel self-cleaning belt for severe duty applications such as solid waste. (Courtesy of Dings Co., Magnetic Group), and (*c*) pulley magnet.

FIGURE 9-12
Baler used for paper, cardboard, plastics, aluminum cans, and tin cans.

and as a means of preparing densified refuse-derived fuels (dRDF). Several technologies are available for the densification of solid wastes and recovered materials including baling, cubing, and pelleting. Equipment for the densification of landfilled solid wastes is discussed in Chapter 11.

Balers. Balers reduce the volume of waste for storage, prepare the wastes for marketing, and increase the density of the waste thereby reducing shipping costs. The materials most commonly baled include paper, cardboard, plastics, aluminum and tin cans, and large metal components (see Fig. 9-12).

Can Crushers. These are used to crush aluminum and tin cans, thus increasing their density and reducing handling and shipping costs. Typically, aluminum cans are crushed and blown into large transport trailers for shipping (see Fig. 9-13).

FIGURE 9-13
Can crusher for aluminum cans used in conjunction with pneumatic discharge system. Crushed cans are blown into trailer.

9-5 FACILITIES FOR HANDLING, MOVING, AND STORING WASTE MATERIALS

To handle, move, and store waste materials at MRFs, the following are used: conveyors, conveyor facilities (picking belts) in conjunction with the manual separation of wastes, pneumatic conveyors, movable and fixed waste-handling equipment, scales, and storage facilities.

Conveyors

Conveyors transfer wastes from one location to another. The principal types of conveyors used for the management of solid waste may be classified as hinge, apron, bucket, belt drag, screw, vibrating, and pneumatic. Horizontal and inclined belt conveyors, where the material is carried above the belt, and drag conveyors, equipped with flights or crossbars to drag the material, are the most commonly used for handling solid wastes (see Fig. 9-14*a*, *b*, *c*).

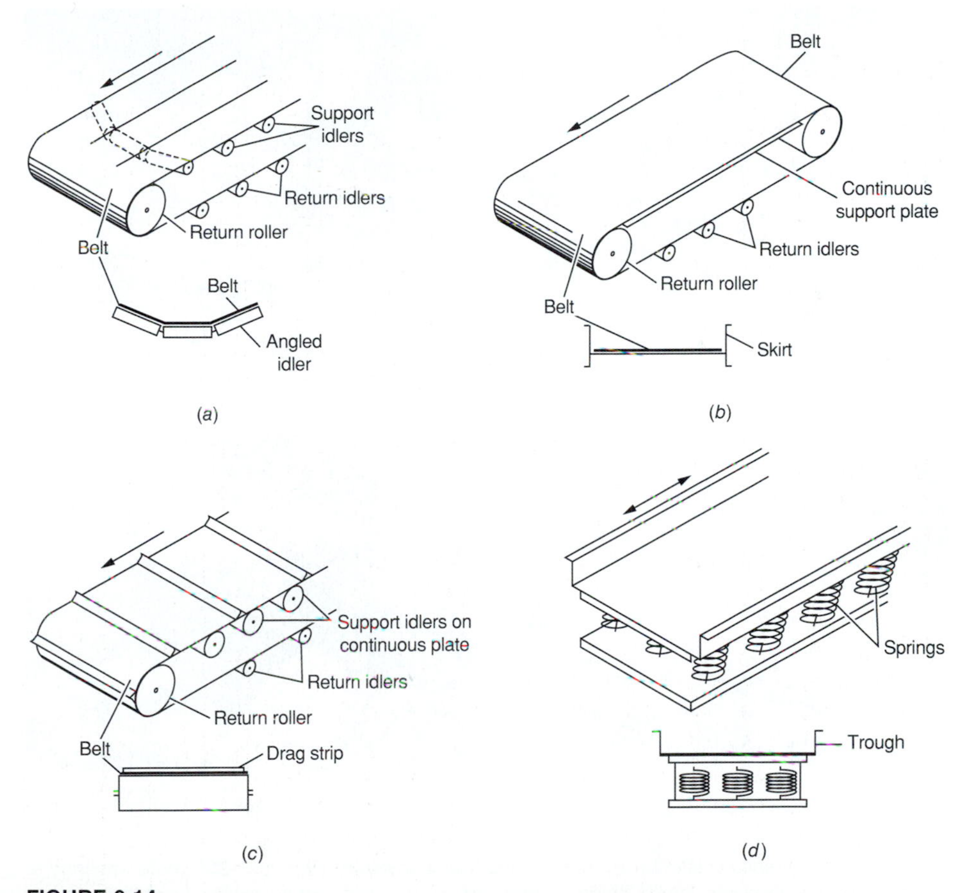

FIGURE 9-14
Conveyor belts used to transport solid waste: (*a*) troughed belt on angled idlers, (*b*) flat belt on continuous plate, (*c*) drag conveyor on idlers, and (*d*) mechanical vibrating conveyor.

The conveyance of unprocessed commingled solid wastes with conveyors has not been trouble-free. Conveyors have been damaged by solid wastes dropped onto them, especially those containing some of the heavier components often found in municipal wastes. Problems have also developed at transfer points (e.g., where the wastes are discharged from one conveyor to another or to other processing facilities). Wire and cords in the wastes become snagged on the equipment, and waste spillage and overflows are common. Binding and wedging of conveyor systems have also been a problem.

Conveyor Facilities Used in Conjunction with the Manual Sorting of Wastes

The manual separation of wastes at a MRF is usually accomplished by removing (picking) individual waste components as the waste stream, which is transported on an endless conveyor (picking) belt, moves by (see Fig. 9-15). Most facilities

(*a*) (*b*)

(*c*) (*d*)

FIGURE 9-15
Manual sorting of solid waste: (*a*, *b*) men sorting from conveyor belt, circa 1905 (note absence of plastic materials in waste that is being sorted); (*c*) modern belt-sorting facility, located in an air-conditioned room, for commingled waste; and (*d*) modern belt-sorting facility for source-separated waste materials.

used for the separation of waste components are elevated so that the separated components can be dropped into chutes that direct the material to receiving containers located below the chute (see Fig. 9-16). To improve the separation of waste components from commingled MSW, plastic bags used for the on-site storage of wastes must be broken open and the contents spread out on the belt. In Fig. 9-15, note that the sorting facilities used at the turn of the century are essentially the same as those used today. However, one important difference is in the characteristics of the wastes to be sorted. The absence of plastic materials and the ubiquitous plastic bags found in MSW today are readily apparent. When one is considering the use of sorting facilities, one can find much useful information in the early literature [7, 8]. Another major difference is that modern sorting lines are usually located in a well-lighted and air-conditioned facility designed to meet Occupational Safety and Health Administration (OSHA) requirements.

The design of facilities for sorting waste components depends to a large extent on the characteristics of the waste, on the number of commingled recyclable items that are to be separated, and on the throughput capacity of the facility. Critical factors in the design of a picking facility are the width of the belt, the speed of the belt, and the average thickness of the waste material on the belt (often referred to as the average burden depth). The maximum belt width where separation is carried out from either side of the belt is about 4 ft. Belt speeds vary from about 15 to 90 ft/min, depending on the material to be sorted and the degree of presorting to which the material has been subjected. It is interesting to note that a belt speed of 60 ft/min was used in sorting facilities at the turn of the century [7]. The average thickness of the waste material on the belt for effective picking is about 6 in. Data on the amount of waste that can be sorted per worker are presented in Table 9-4.

Pneumatic Conveyors

Pneumatic conveying can be defined as materials transport using air as the transport medium. Two types of pneumatic transport systems (positive pressure and

FIGURE 9-16
Storage containers for separated material, located below elevated sorting lines. (Courtesy of RYCO Manufacturing, Inc.)

TABLE 9-4
Picking rates for commingled materials from moving belts

Type of material	Waste sorted, ton/person · hr Range	Typical	Remarks
Commingled MSW			
Residential and commercial	0.3–4	2.5	Relatively low efficiency of recovery per ton of feedstock at the higher sorting rates
Commercial	0.4–6	3.0	
Source-separated commingled materials			
Mixed paper	0.5–4	2.5	
Paper and cardboard	0.5–3	1.5	Two products
Mixed plastics	0.1–0.4	0.2	PETE and HDPE
Mixed glass and plastic	0.2–0.6	0.5	Two products: mixed glass and mixed plastic
Glass	0.2–0.8	0.4	Clear, green, amber
Plastics, glass, aluminum and tin cans	0.1–0.5	0.3	Four products

vacuum) are illustrated in Fig. 9-17. Pneumatic conveyors offer considerable design flexibility because the piping can be routed as required. As noted in the discussion of air classification, if light materials are introduced into an air stream moving with sufficient velocity, they will be carried away with the air. Shredded materials such as newsprint, plastic, or refuse-derived fuel as well as other light materials such as crushed aluminum cans have been conveyed pneumatically. Air

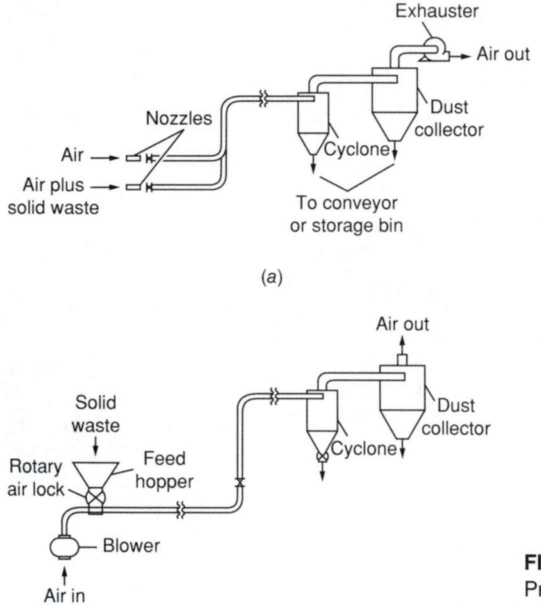

FIGURE 9-17
Pneumatic conveyor systems: (*a*) vacuum and (*b*) positive pressure. (Adapted from Ref. 15.)

velocities needed for the pneumatic transport of unprocessed solid waste are in the range of 4800 to 6000 ft/min.

Movable Waste-Handling Equipment

The use of front-end loaders and forklifts to move materials is universal in the operation of MRFs. For example, in a typical application, commingled MSW dumped on the receiving floor of the MRF by the collection vehicles is then pushed or loaded with a front-end loader onto a belt conveyor for further processing (see Fig. 9-18*a*). Front-end loaders are also used to load materials after processing, such as the loading of shredded wood waste into trucks for shipment to off-site customers. Forklifts are used almost exclusively to move baled materials from baling machines to storage areas, and then onto trucks for transport to market (see Fig. 9-18*b*).

Facilities for Weighing

Weighing facilities are an important and necessary part of any MRF. Scales of various types are used to weigh the amount of waste materials delivered, recovered, sold, and disposed of. The types of weighing facilities used at MRFs vary from the small scales used to weigh the amounts of wastes brought in by individuals (see Fig. 9-19*a*) to the platform scales used for weighing collection vehicles (see Fig. 9-19*b*).

Storage Facilities

Materials that have been separated and processed must be stored until a buyer picks them up. In some facilities, space is provided for materials to be displayed for viewing by purchasers, usually on a weekly or monthly schedule. The amount of storage space to be provided at the MRF is established by the MRF system operator

(*a*)

(*b*)

FIGURE 9-18
Movable equipment used for waste handling at MRFs: (*a*) front-end loader equipped with solid rubber tires and (*b*) forklift. Although the solid rubber tires are considerably more expensive than conventional tires, they have proven cost-effective and have reduced downtime considerably.

(a)

(b)

FIGURE 9-19
Weighing facilities at MRFs: (*a*) small platform scale for weighing material brought in by homeowners and (*b*) platform scale for collection and other large vehicles.

in coordination with the material buyers. Key considerations are these: Will the buyer provide storage containers for recovered materials? With what frequency will the buyer pick up and remove prepared materials from the MRF? Is it possible to rent temporary storage facilities for the processed materials away from the MRF? What backup facilities exist in the community that can be used to store recovered materials when there is no available storage space at the MRF? Although each MRF must be sized for materials storage in accordance with individual site-specific criteria, it is prudent to provide sufficient storage capacity to hold processed materials for one to three months.

9-6 DEVELOPMENT AND IMPLEMENTATION OF MRFS

As we noted in the introduction to this chapter, the further separation and processing of wastes that have been source-separated, as well as the separation of commingled wastes usually occurs at materials recovery facilities (MRFs) or at large integrated materials recovery/transfer facilities (MR/TFs). The successful development and implementation of a MRF or MR/TF require that proper attention be paid to both engineering considerations and nonengineering implementation issues. An introduction to the engineering considerations involved in the implementation of MRFs, some typical examples of MRFs, planning and design considerations, and the nonengineering implementation issues are considered in this section. Further engineering details are in Chapter 12.

Engineering Considerations

Engineering considerations involved in the implementation of MRFs include (1) definition of the functions of the MRF, (2) selection of the materials to be separated (now and in the future), (3) identification of the material specifications that must be met now and in the future, (4) development of separation process flow diagrams, (5) determination of process loading rates, (6) layout and design of the physical facilities, (7) selection of the equipment and facilities that will be used, (8) environmental controls, and (9) aesthetics considerations (see also Ref. 10). The adaptability of the facility to potential changes in the characteristics or quantities of the waste must also be assessed carefully. Consideration of the functions of a MRF and the materials to be separated, the development of process flow diagrams, the development of process loading rates, the layout and design of the physical facilities, planning and design considerations, and the selection of the equipment and facilities that will be used are considered in the following discussion. These topics are also examined in greater detail in Chapter 12. Selection of the materials to be separated and the specifications for the separated materials are considered in Chapter 15.

Functions of a MRF, and Materials to Be Recovered. The functions of a MRF depend directly on (1) the role the MRF is to serve in the waste management system, (2) the types of material to be recovered, (3) the form in which the materials to be recovered will be delivered to the MRF, and (4) the containerization and storage of processed materials for the buyer. For example, the function of equipment and facilities for the separation of aluminum cans from tin cans will differ significantly from that for the separation of aluminum and tin cans from commingled MSW.

MRFs for source-separated materials. The types of materials that are commonly processed at MRFs for source-separated wastes are summarized in column 1 of Table 9-5. The functions that must be carried out at MRFs to process the source-separated materials are identified in column 2 of Table 9-5. The particular combination of materials to be separated will depend on the nature of the source separation program the community has adopted. For example, a typical source separation program might involve the use of three separate containers for recyclable materials in conjunction with one or more additional containers for other wastes; yard wastes will be collected separately. The materials separated would be as follows:

Recycle container 1	Mixed paper (cardboard is often stacked alongside)
Recycle container 2	Mixed plastics and glass
Recycle container 3	Mixed aluminum and tins cans
Yard wastes	Collected separately

With this mix of source-separated materials, four separate process lines will be required to separate and/or to process the individual components. The processing of the yard wastes could be done at a separate facility or at a large integrated MRF.

TABLE 9-5
Typical examples of the materials, functions, and equipment and facility requirements of MRFs used for the processing of source separated materials

Materials	Function/operation	Equipment and facility requirements
Mixed paper and cardboard/1	Manual separation of high-value paper and cardboard or of contaminants from commingled paper types. Baling of separated materials for shipping. Storage of separated materials	Front-end loader, conveyors, baler, forklift (see Figs. 9-20*a* and 9-21)
Mixed paper and cardboard/2	Manual separation of cardboard and mixed paper. Baling of separated materials for shipping. Storage of baled materials	Front-end loader, conveyors, open picking station, baler, forklift
Mixed paper and cardboard/3	Manual separation of old newspaper, old corrugated cardboard, and mixed paper from commingled mixture. Baling of separated materials for shipping. Storage of baled materials	Front-end loader, conveyors, enclosed picking station, baler, forklift
PETE and HDPE plastics	Manual separation of PETE and HDPE from commingled plastics. Baling of separated materials for shipping. Storage of baled materials	Receiving hopper, picking conveyor, storage bins, baler, forklift
Mixed plastics	Manual separation of PETE, HDPE and other plastics from commingled mixed plastics. Baling of plastics for shipping. Storage of separate materials	Receiving hopper, picking conveyor, storage bins, baler, forklift
Mixed plastics and glass	Manual separation of PETE, HDPE, and glass by color from commingled mixture. Baling of plastics for shipping. Storage of separated materials	Receiving hopper, picking conveyor, glass crusher, storage bins, baler, forklift
Mixed glass	Manual separation of clear, green, and amber glass. Storage of separated materials	Receiving hopper, picking conveyor, glass crusher, storage bins, forklift

(*continued*)

MRFs for commingled MSW. For commingled MSW, the materials to be separated and the function and equipment requirements for the MRF will depend directly on the role the MRF is to serve in the waste management system (see Table 9-6). A MRF may be used to recover materials from commingled MSW to meet mandated diversion goals. A MRF can be used to separate and process source-separated materials along with the separation of materials from commingled MSW to meet mandated diversion goals. Another common use of a MRF for commingled MSW is to remove contaminants from the waste and to prepare the waste for subsequent uses such as a fuel for combustion facilities or a feedstock for composting. The removal of contaminants from waste materials is also known as a *negative sort*. Another MRF might be used to recover only high-value items and to process the residual waste for the production of compost to be used as intermediate landfill cover. Clearly, an endless number of variations of a MRF are possible. The types of materials and/or contaminants removed and the associated

TABLE 9-5 (*continued*)

Materials	Function/operation	Equipment and facility requirements
Aluminum and tin cans	Magnetic separation of tin cans from commingled mixture of aluminum and tin cans. Baling of separated materials for shipping. Storage of baled materials	Receiving hopper, conveyor, overhead suspended magnet, magnet pulley, storage containers, baler or can crusher and pneumatic transport system, forklift
Plastic, glass, aluminum cans, and tin cans	Manual or pneumatic separation of PETE, HDPE, and other plastics. Manual separation of glass by color. Magnetic separation of tin cans from commingled mixture of aluminum and tin cans. Baling of plastic, aluminum cans, and tin cans, and crushing of glass and shipping. Storage of baled and crushed materials	Receiving hopper, conveyor, picking conveyor, overhead suspended magnet, magnet pulley, glass crusher, storage containers, baler or can crusher and pneumatic transport system, forklift
Yard wastes/1	Manual separation of plastic bags and other contaminants from commingled yard wastes, grinding of clean yard waste, size separation of waste that has been ground up, storage of oversized waste for shipment to biomass facility, and composting of the undersized material	Front-end loader, tub grinder, conveyors, trommel or disc screen, storage containers, compost-turning machine
Yard wastes/2	Manual separation of plastic bags and other contaminants from commingled yard wastes followed by grinding and size separation to produce landscape mulch. Storage of mulch and composting of undersized materials	Front-end loader, tub grinder, conveyors, trommel or disc screen, storage containers, compost-turning machine
Yard wastes/3	Grinding of yard waste to produce a biomass fuel. Storage of ground material	Front-end loader, tub grinder, conveyors, storage containers or transport trailers

activities carried out at the different types of MRFs identified above are also summarized in Table 9-6.

Development of Separation Process Flow Diagrams. Once a decision has been made on how recyclable materials are to be recovered from MSW (e.g., source separation or separation from commingled MSW), flow diagrams must be developed for the separation of the desired materials and for processing the materials, subject to predetermined specifications. A *process flow diagram* for a MRF is defined as the assemblage of unit operations, facilities, and manual operations to achieve a specified waste separation goal or goals. The following factors must be considered in the development of process flow diagrams: (1) identification of the characteristics of the waste materials to be processed, (2) consideration of the specifications for recovered materials now and in the future, and (3) the available

TABLE 9-6
Examples of the functions, materials recovered or contaminants removed, and activities associated with MRFs used for the processing of commingled MSW

Function of MRF	Materials recovered or contaminants removed	Activities
Recovery of recyclable materials to meet mandated first-stage diversion goals (25%)	Bulky items, cardboard, paper, plastics (PETE, HDPE, and other mixed plastic), glass (clear and mixed), aluminum cans, tin cans, other ferrous materials	Manual separation of bulky items, cardboard, plastics, glass by color, aluminum cans, and large ferrous items. Magnetic separation of tin cans and other ferrous materials not removed manually. Baling of separated materials for shipping. Storage of baled materials
Recovery of recyclable materials and the further processing of source-separated materials to meet second-stage diversion goals (50%)	Bulky items, cardboard, paper, plastics (PETE, HDPE, and other mixed plastic), glass (clear and mixed), aluminum cans, tin cans, other ferrous materials. Additional separation of source-separated materials including paper, cardboard, plastic (PETE, HDPE, other), glass (clear and mixed), aluminum cans, tin cans	Manual separation of bulky items, cardboard, plastics, glass by color, aluminum cans, and large ferrous items. Magnetic separation of tin cans and other ferrous materials not removed manually. Baling of separated materials for shipping. Storage of baled materials
Preparation of MSW for use as a fuel	Bulky items, cardboard (depending on market value), glass (clear and mixed), aluminum cans, tin cans, other ferrous materials	Manual separation of bulky items, cardboard, and large ferrous items. Mechanical separation of glass, aluminum cans. Magnetic separation of tin cans and other ferrous materials not removed manually. Fuel preparation. Storage of fuel feedstock. Baling of cardboard for shipping. Storage of baled materials
Preparation of MSW for use as a feedstock for composting	Bulky items, cardboard (depending on market value), plastics (PETE, HDPE, and other mixed plastic), glass (clear and mixed), aluminum cans, tin cans, other ferrous materials	Manual separation of bulky items, cardboard, plastics, glass by color, aluminum cans, and large ferrous items. Magnetic separation of tin cans and other ferrous materials not removed manually. Baling of separated materials for shipping. Storage of baled materials. Storage of compost feedstock
Selective recovery of recyclable materials	Bulky items, office paper, old telephone books, aluminum cans, PETE and HDPE, and ferrous materials. Other materials depending on local markets	Manual separation of bulky items, cardboard. Manual separation of selected materials depending on market demands. Baling facilities, can crushers, and other equipment depending on the materials to be separated

types of equipment and facilities. For example, specific waste materials cannot be separated effectively from commingled MSW unless bulky items such as lumber and white goods and large pieces of cardboard are first removed and the plastic and paper bags in which waste materials are placed are broken open and the contents exposed. The specifications for the recovered material will affect the degree of separation to which the waste material is subjected. A typical process flow diagram for the separation of source-separated paper and cardboard is shown in Fig. 9-20*a* and discussed below. Other process flow diagrams are presented and discussed later in this section.

In Fig. 9-20*a*, mixed paper and limited amounts of cardboard from residential sources as well as cardboard from commercial sources are unloaded from the collection vehicles in separate areas of the tipping floor. Cardboard, bulky items, and nonrecyclable paper items such as spiral-bound notebooks, books, telephone books and other contaminants are removed from the mixed paper. Brown paper bags, often used to hold newspapers, are also removed and processed with the cardboard, because they command a higher market value. Bulky items and other contaminants are also removed from the cardboard. Once the mixed paper and cardboard have been sorted, a front-end loader is used to load the mixed paper onto a floor conveyor, which discharges to an inclined conveyor which, in turn, discharges into a baler. Paper bales are typically $30 \times 40 \times 60$ in and weigh about 1400 lb. Once the paper has been baled, the cardboard is then baled. Cardboard bales are the same size as the paper bales and weigh about 1100 lb. The paper bales are stored indoors to avoid deterioration from exposure to sunlight (paper becomes brown and brittle when exposed to ultraviolet light) and water damage due to rain. The cardboard bales are stored outdoors. The same baler is also used to bale the aluminum and tin cans and the separated plastic materials processed in other parts of the facility.

Materials Balances and Loading Rates. Once the process flow diagram has been developed, the next step in the design of the MRF is to estimate the quantities of materials that can be recovered and the appropriate design loading rates. The expected process loading rates must be known in order to select and size equipment properly. Loading rates for a given process are based on a mass balance (see Chapter 6) for the preceding process. For example, if in Fig. 9-20*a* the baler is to be used for three hours per day for baling paper the loading rate would be 5.83 ton/h [(17.5 ton/d)/(3 h/d)]. The value of 17.5 ton/d is based on a materials balance analysis of the sorting operation (20 ton/d − 0.15 ton/d bulky items − 0.35 ton/d contaminants − 2 ton/d cardboard). If more operations were involved, loading rates on subsequent processes would be determined similarly, taking into account the amount of material diverted at each processing step. The development of loading rates for a variety of process flow diagrams is illustrated in detail in Chapter 12.

Loading rates for most processes are expressed in tons per hour. In determining the design loading rates, one should make a careful analysis to determine the actual number of hours per day and year the equipment will be operated. Based

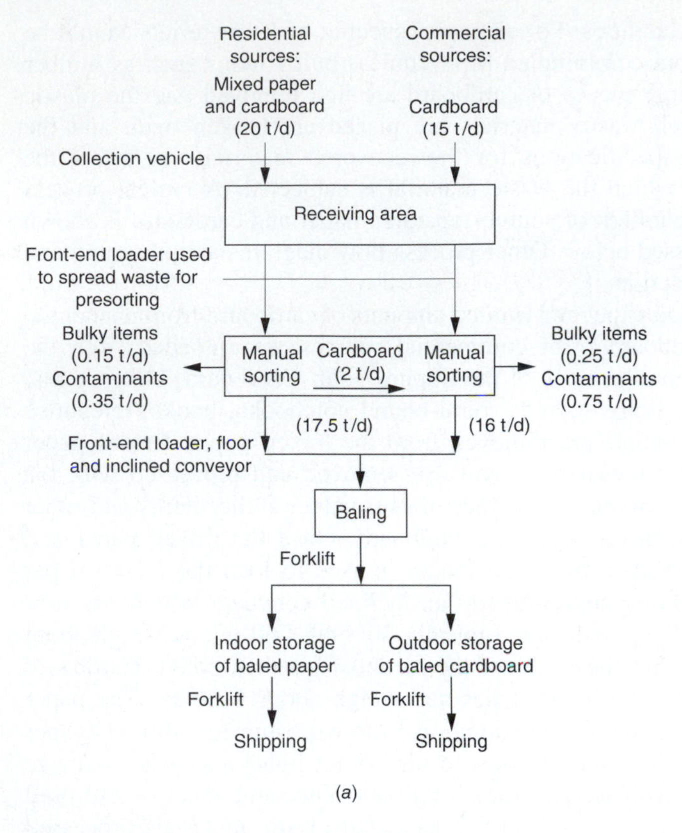

(*a*)

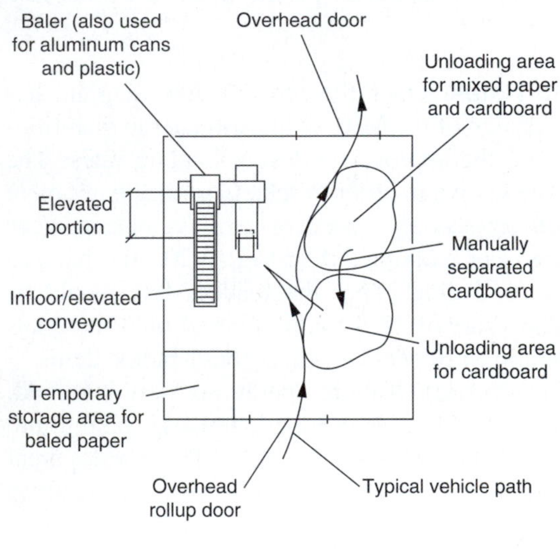

(*b*)

FIGURE 9-20
MRF for the processing of source-separated paper and cardboard: (*a*) process flow diagram and (*b*) layout of MRF.

on 1820 operating hours per year, the base hourly loading (or processing) rate is given by the following expression:

$$\text{Loading rate, ton/h} = \frac{\text{Number of ton/yr (or ton/d)}}{\text{1820 processing h/yr (or h/d)}} \tag{9-1}$$

Usually, it is assumed that the separation process at the MRF will be operational for seven hours per day, where one nominal eight-hour shift will be used per day.

System Layout and Design. The layout and design of the physical facilities that make up the processing facilities will depend on the types and amounts of materials to be processed. Important factors in the layout and design of such systems include (1) the methods and means by which the wastes will be delivered to the facility, (2) estimates of materials delivery rates, (3) definition of the materials loading rates, (4) development of materials flow and handling patterns within the MRF facility, and (5) development of performance criteria for the selection of equipment and facilities. Because there are so many combinations in which the separation processes can be grouped, it is extremely important to inspect as many operating facilities as possible before settling on a final design. The layout of a facility for the processing of source-separated mixed paper and cardboard from residential sources and cardboard from commercial sources is shown in Fig. 9-20*b*. The process flow diagram for this facility appeared in Fig. 9-20*a*. The facility shown in Fig. 9-20*b* is also used for the processing of mixed plastics and glass.

Typical Materials Recovery Facilities for Source-Separated Wastes

To illustrate the many different types of MRFs that have been developed to process source-separated materials, two different types of MRFs are considered in the following discussion: (1) a MRF designed to process source-separated wastes and (2) a MRF designed to process garden trimmings and wood wastes. These two examples illustrate the general features of MRFs used in conjunction with source-separated wastes. A MRF used for the separation of mixed paper into several grades is considered in detail in Chapter 12.

MRF for Source-Separated Wastes. The process flow diagrams and the layout of the facility that is considered in the following discussion are shown in Figs. 9-21 and 9-22, respectively. The materials to be processed are mixed newspaper and cardboard, mixed plastic and glass, and aluminum and tin cans. In addition, the facility also serves as a buy-back center. The vehicle used for the collection of the separated wastes is shown in Fig. 9-23. The processing of the separated waste materials is as follows.

Paper and cardboard. Mixed paper and cardboard are unloaded onto the tipping floor. There, cardboard and nonrecyclable paper items are removed. The mixed paper is then loaded onto a floor conveyor with a front-end loader. The

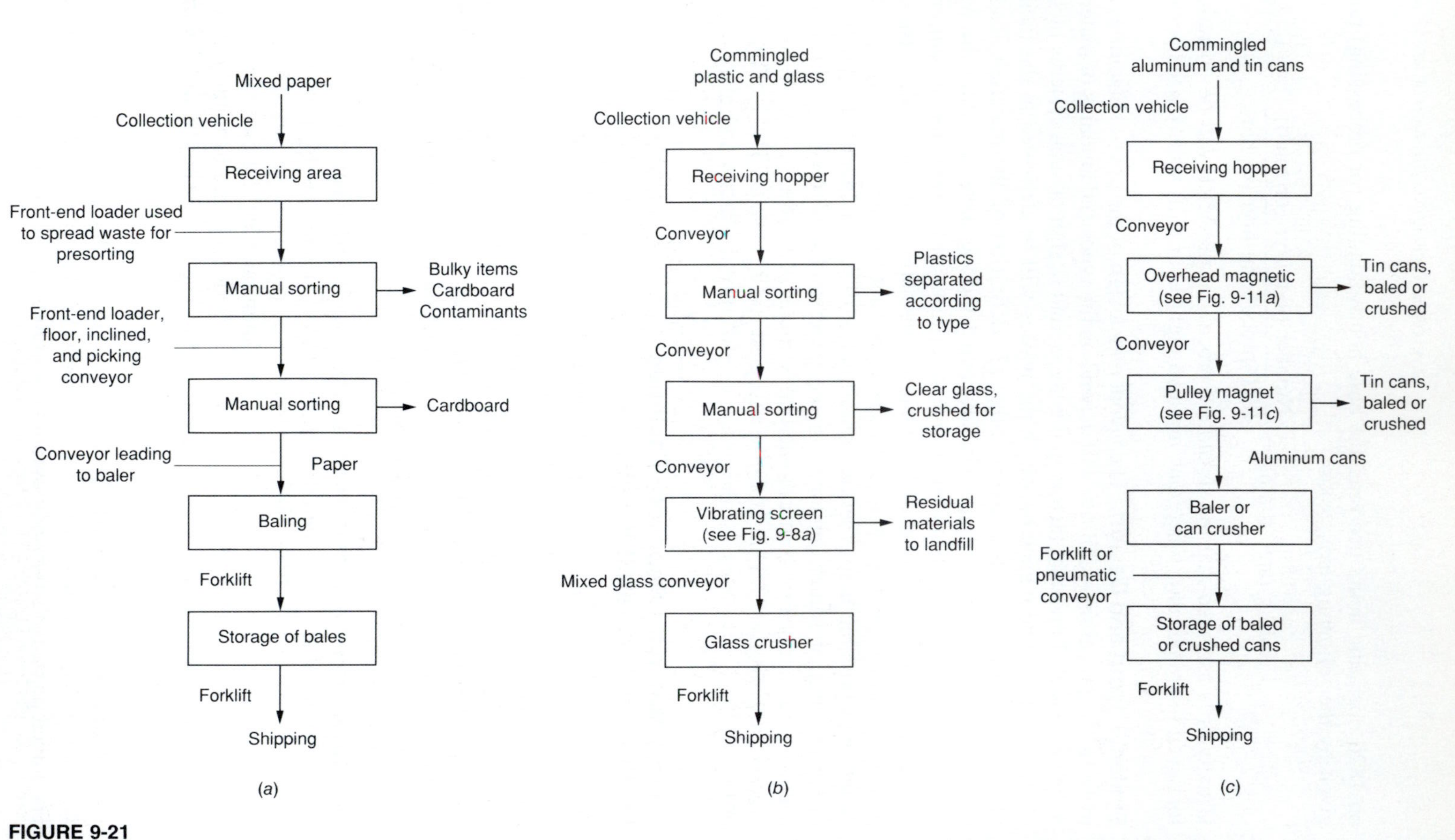

FIGURE 9-21
Flow diagrams for the separation of source-separated waste: (*a*) mixed paper, (*b*) commingled plastics and glass, and (*c*) aluminum and tin cans.

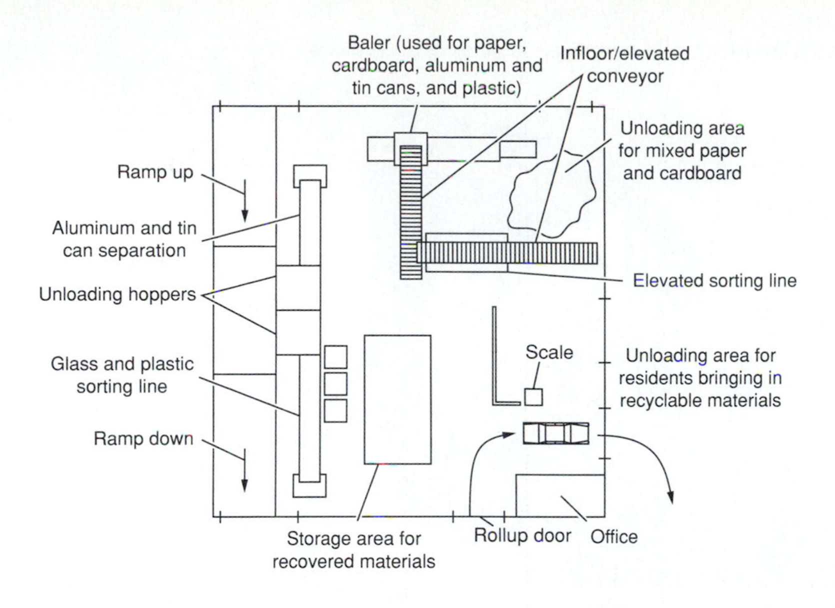

(*a*)

(*b*)

(*c*)

(*d*)

FIGURE 9-22
Layout for a MRF used to process source-separated materials: (*a*) mixed paper and cardboard deposited on tipping floor to be sorted; (*b*) elevated sorting line for paper and cardboard; (*c*) unloading aluminum and tin cans into receiving hopper; and (*d*) weighing facilities at buyback center.

FIGURE 9-23
Specially designed collection vehicle used for the collection of source-separated wastes used in conjunction with MRF shown in Fig. 9-22.

floor conveyor discharges to an inclined conveyor, that, in turn, discharges to a horizontal conveyor, which transports the mixed paper past workers who remove any remaining cardboard from the mixed paper. The paper remaining on the belt is discharged to a conveyor, located below the picking platform, that is used to feed the baler. Once the paper has been baled, the cardboard is baled. The baler is also used to bale the aluminum and tin cans and the separated plastic materials.

Aluminum and tin cans. The commingled aluminum and tin cans are discharged into a hoppered bin, which discharges to a conveyor belt. The conveyor transports the commingled cans past an overhead magnet separator (see Fig. 9-11*a*) where tin cans are removed. The endless belt continues past a pulley magnet separator (see Fig. 9-11*c*), where any tin cans not removed with the overhead magnet are taken out. The aluminum and tin cans, collected separately, are baled for shipment to markets.

Plastic and glass. The commingled plastic and glass are also discharged into a hoppered bin, which discharges to a conveyor belt. The material is transported to a sorting area, where the plastic and clear glass are separated manually from the other materials. The remaining glass is then sent to a glass crusher. The wastes are then discharged to vibrating screens where the broken glass falls through the openings in the screen. The crushed glass is loaded onto large trailers to be transported to the purchaser. Any residual materials are collected at the end of the vibrating screen. The residual materials are disposed of in a landfill. The commingled plastic is then separated further by visual inspection or according to the type (PETE and HDPE) using the imprinted codes adopted by the plastics industry (see Chapter 3).

Buy-back center. The MRF shown in Fig. 9-22*d* also serves as a buy-back center for aluminum cans, plastic, glass, and newsprint. Operationally, homeowners drive up to the electronic scale located within the facility. Materials brought in are unloaded and weighed, and the homeowner is given a printout listing the weights of the materials he or she has brought in. The homeowner is paid immediately on the basis of the weight printout.

MRF for Source-Separated Yard Wastes. The facility to be considered in this discussion is used for processing yard wastes that are collected separately. The flow diagram for this MRF is given in Fig. 9-24. Yard wastes set out in the street by homeowners are collected using a device known as a *claw*, which clamps around these piles of wastes (see Fig. 9-25). The collected wastes are emptied into a specially equipped compactor-type collection vehicle (see Fig. 9-26). The collected wastes, along with other green wastes collected by city crews and private

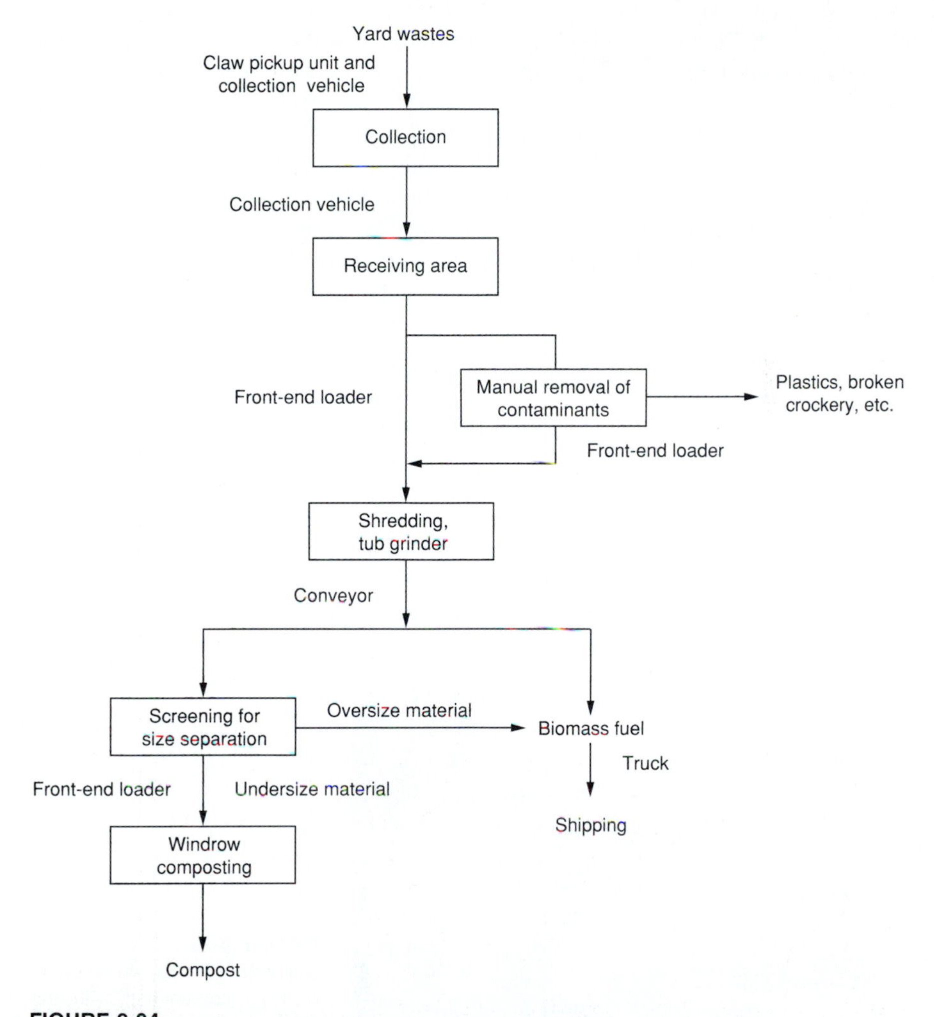

FIGURE 9-24
Flow diagram for MRF for processing yard and other green wastes.

FIGURE 9-25
Claw device mounted on a wheeled tractor used to pick up yard wastes left in the street by homeowner.

haulers, are taken to a large paved area (formerly a drive-in theater). There the wastes are ground up using a tub grinder (see Fig. 9-27). The material from the tub grinder is passed through a trommel screen to separate pieces of wood larger than one-half inch. Wood chips larger than one-half inch are sold to a local biomass waste-to-energy facility. Green wastes and wood chips smaller than one-half inch are composted using the windrow method (see Section 9-8). The resulting compost is given to the local residents free for the taking. Any remaining compost is used by the city in landscaping. Depending on the market for processed biomass fuel, all of the ground-up yard wastes have, at times, been sold for use as a biomass fuel without any further processing.

To make higher-quality compost would require that the yard wastes first be spread out on the paved area and contaminants such as plastic bags, broken concrete, and metals be removed manually. The yard wastes would then be ground up for the production of compost. Unfortunately, the price offered for high quality compost in 1992 does not usually warrant the extra processing and handling costs involved in the removal of contaminants normally found in source-separated yard wastes. In the future, as material specifications become more stringent (see Chapter 15), the removal of contaminants may become necessary if the compost produced from yard wastes is to be sold commercially.

FIGURE 9-26
Collection vehicle modified to work in conjunction with the claw (see Fig. 9-25) for the collection of yard wastes.

FIGURE 9-27
Tub grinder used to process yard wastes collected separately.

Typical Materials Recovery Facilities for Commingled Wastes

The separation of waste components from commingled wastes and their processing are necessary operations in the recovery of materials for direct reuse and recycling and for the production of a feedstock that can be used for the recovery of energy and the production of compost. The purpose of this section is to illustrate how the unit operations and facilities discussed previously are grouped together to achieve the separation of materials from commingled MSW. A MRF designed to process commingled construction and demolition wastes is also described.

MRF for Recovery of Materials from Commingled MSW and for the Processing of Source-Separated Materials. Recognizing that meeting mandated waste diversion goals with source separation programs alone will be difficult, many communities have developed plans for MRFs that can be used both to separate materials from commingled MSW and to process materials from source separation programs. A typical process flow diagram for a MRF employing manual and mechanical separation of materials from commingled MSW and manual separation of source-separated wastes is illustrated in Fig. 9-28. Commingled MSW from residential and other sources are discharged in the receiving area. Recyclable, reusable, and oversized materials such as cardboard, lumber, white goods, and broken furniture are removed in the first-stage presorting operation before the commingled waste is loaded on to an inclined conveyor. Source-separated materials in see-through plastic bags also are removed from the commingled MSW. Additional cardboard and large items are handpicked from the conveyor at the second-stage presorting station as the waste material is transported to the bag-breaking station. The next step involves breaking open the plastic bags, which can be accomplished either manually or mechanically. In some facilities, a short enclosed trommel equipped with protruding blades is used as a bag breaker (see Fig. 9-9). As noted in Table 9-3, flail mills, shear shredders, and screw augers have also been used as bag breakers.

The next step in the process involves the first stage of manual separation of specific waste materials. Materials typically removed include paper, cardboard, all

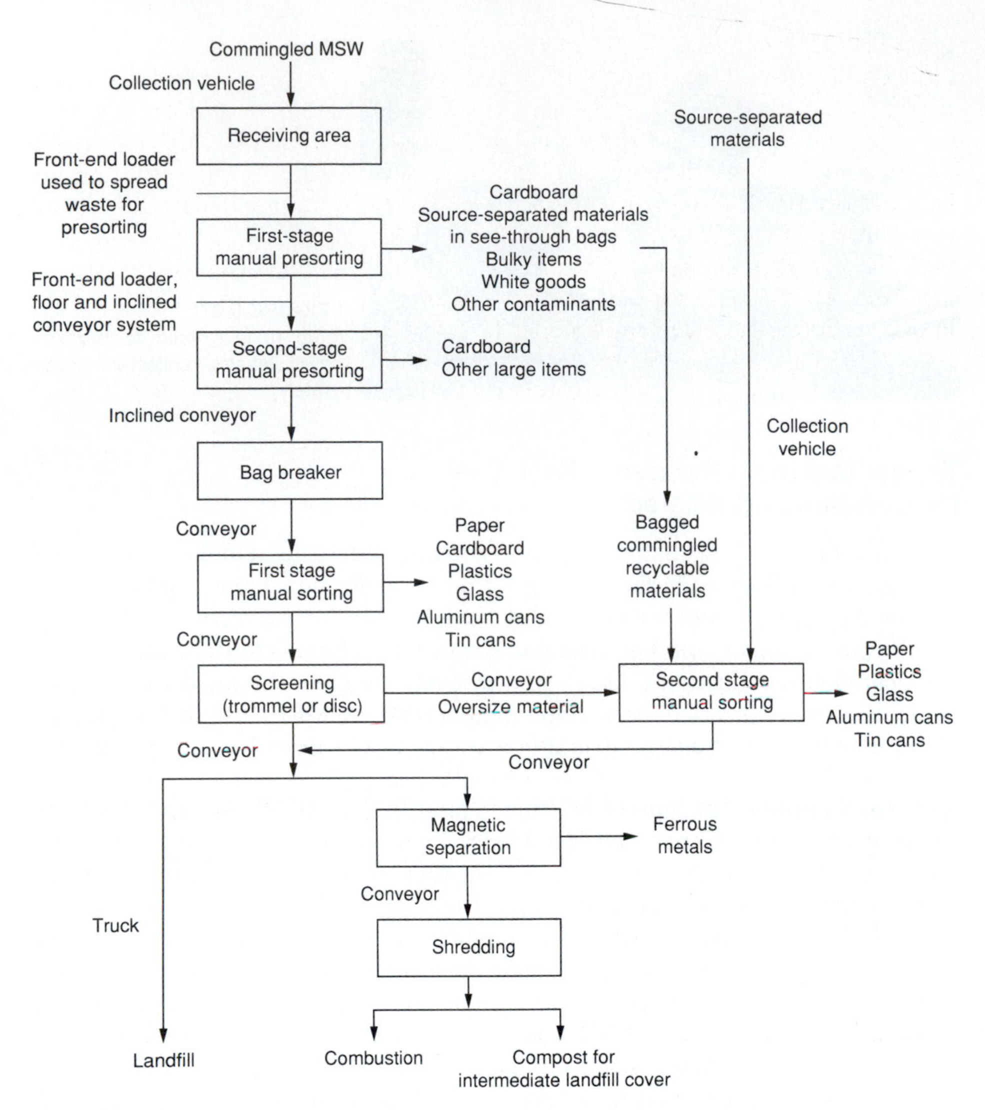

FIGURE 9-28
Flow diagram for the recovery of waste materials from commingled MSW.

types of plastic, glass, and metals. In some operations, different types of plastic are separated simultaneously. Mixed plastics are usually separated by type in a secondary separation process. Material remaining on the conveyor is discharged into a trommel (or disc screen) for size separation. The oversized material is sorted manually a second time (second-stage sorting). Commingled source-separated materials, collected separately from residential and commercial sources, and the source-separated materials contained in see-through bags removed from the commingled MSW in the first-stage presorting operation are further sorted using the second-stage sorting line. Source-separated mixed paper and cardboard would be processed separately using a flow diagram such as the one given in Fig. 9-24. It should be noted that both the first- and second-stage sorting activities would normally be carried out in an air-conditioned facility. Depending on the extent of the first- and second-stage sorting operations, the undersized material from the trommel and the material remaining after the second-stage sorting operation are hauled away for disposal in a landfill, processed further and combusted, or used to produce compost to be used as intermediate landfill cover. As shown in Fig. 9-28, further processing of the residual materials usually involves shredding and magnetic separation. A detailed materials balance analysis of the MRF described in this section is presented in Example 12-6 in Chapter 12.

The following excerpt from a text published in 1921 provides an historical perspective on current materials separation activities at MRFs:

> The most developed case of sorting refuse in Europe is at Puchheim, a suburb of Munich, where the refuse from a population of more than 600,000 is picked over and finally disposed of. First, the finer materials and dust are sifted out on a moving and vibrating belt, and the bulky salable articles are picked out. In the adjoining room, about 40 women stand on each side of the belt, each one picking out a designated material and throwing it into a designated wire basket. The substances thus removed are chiefly: Paper, white and green glass, rags, leather, bones, tinned cans, iron, brass, copper, tin, etc. The bones are treated with benzine, and, on the premises, are converted into grease, glue, bone meal, or charcoal. Garbage is cleaned, sterilized, and fed to hogs in an adjoining building. Paper is freed from dust, pressed into bales, and utilized for the manufacture of pasteboard. Wood is burned under the boilers. Bottles are cleaned, disinfected, and sold. Tinned cans are sold as iron. No one enters the works until after donning working clothes, nor leaves them until after a good wash or bath. The working rooms are washed twice a day with dilute carbolic acid. It is reported by De Fodor that this very effective sorting contains the germ of faulty economics, in the fact that the total revenue hardly covers three-quarters of the necessary expenditure.[7]

Except for many modern sorting facilities being located in air-conditioned facilities, the similarities are striking. The economic issue remains the same today, but environmental costs were not generally considered in the 1920s.

MRFs for Preparation of Feedstock from Commingled MSW. The separation of commingled MSW in highly mechanized systems is illustrated in Fig. 9-29*a* and *b*. As shown in both these process flow diagrams, the commingled MSW is first discharged in the receiving area where lumber, white goods, and oversized items are usually removed manually before the material is loaded onto the first conveyor. In Fig. 9-29*a*, the commingled MSW is shredded as the first step in the process. Air classification is then used to recover the mainly organic fraction of the MSW. In Fig. 9-29*b*, a trommel is used to achieve a better separation of the organic fraction of the MSW and to remove small contaminants more effectively. The flow diagrams in Fig. 9-29 represent two of many different approaches that have been, and continue to be, used for the mechanical separation of waste components from commingled MSW for the production of a feedstock for the production of energy.

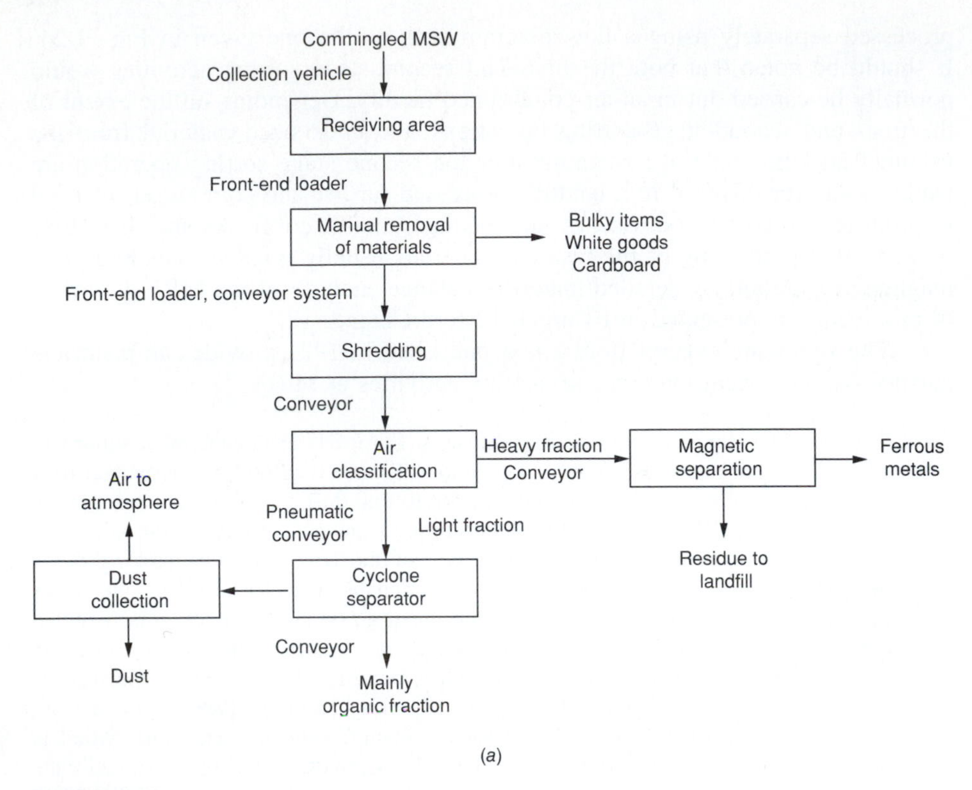

FIGURE 9-29
Flow diagrams for the recovery of waste components from solid waste: (*a*) conventional (with shredder) and (*b*) shear shredder and trommel used to replace shredder in flow diagram.

The mainly organic fraction of MSW remaining after processing is known as fluff refuse-derived fuel, commonly known as fluff-RDF. In some operations, the mainly organic fraction is used to produce a densified refuse-derived fuel known as d-RDF.

Flow diagrams similar to those shown in Fig. 9-29 have also been used to describe the preprocessing of MSW for the production of compost. Unfortunately, shredding commingled MSW before metal objects and other contaminants have been removed results in the production of compost contaminated with heavy metals and trace organic compounds. Acceptable contaminant levels for the various constituents that may be found in composted MSW are given in Table 15-8. Because of the serious problems associated with the production of contaminated compost, many communities have developed MRF process flow diagrams similar to the one given in Fig. 9-28 for the production of feedstock for composting. The picking stations are used to remove plastics, glass, aluminum and tin cans, and other contaminants before the waste is shredded to reduce the particle size for composting.

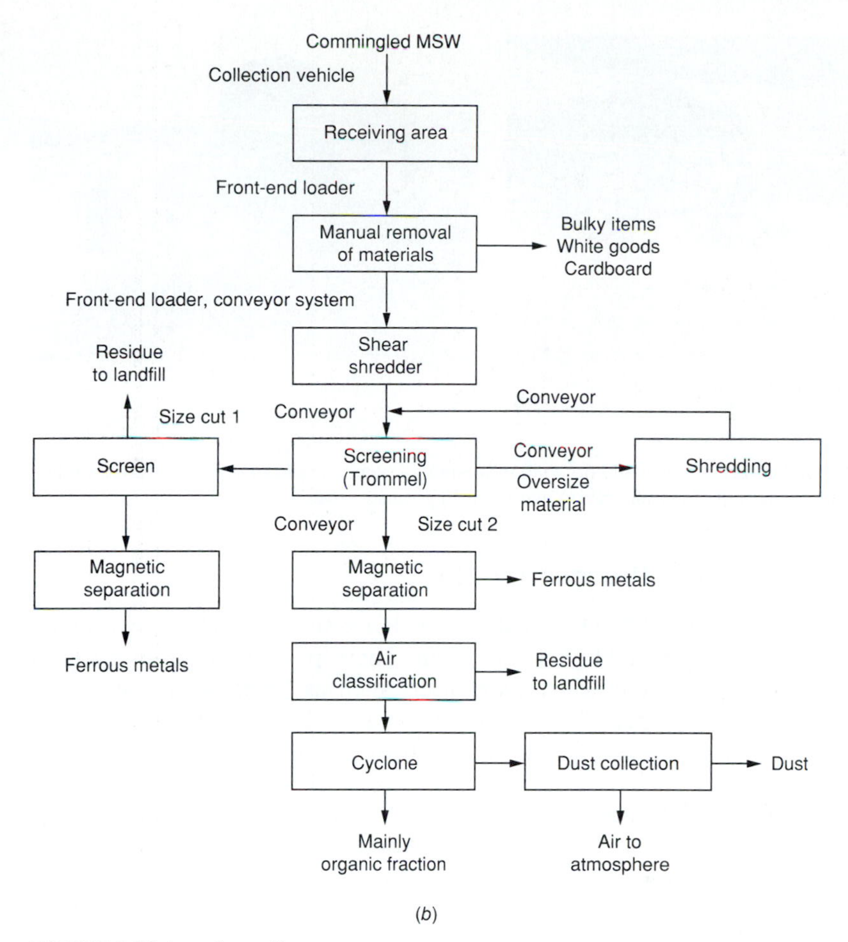

(*b*)

FIGURE 9-29 (*continued*)

MRF for Commingled Construction and Demolition Wastes. An overview of a MRF designed to process commingled construction and demolition wastes is shown in Fig. 9-30. Commingled construction and demolition wastes are brought to the site and dumped in an open area. The wastes are then spread out and all of the wood is removed manually (see Fig. 9-30*a*). The wood is taken to a large wood grinder, where it is converted to wood chips. After the wood has been removed, the waste is picked up with a front-end loader and discharged onto a two-stage vibrating screen (see Fig. 9-30*b*). The first screen is used to eliminate large pieces of concrete, roots and similar materials. The second screen, located immediately below the first screen, is used to remove finer pieces of broken concrete and other smaller-sized contaminants. The fine material passing the two screens is then conveyed to a second vibrating screen, where additional fine contaminants are removed. The final product, relatively clean dirt, is stockpiled for sale. The material removed by the screens is stockpiled and eventually hauled to the landfill for disposal.

(a)

(b)

FIGURE 9-30
Views of an MRF for construction and demolition wastes: (*a*) wastes spread on ground where wood is removed manually and (*b*) waste being screened to produce usable product.

Planning and Design Process for MRFs

The planning and design of MRFs involve three basic steps: (1) feasibility analysis, (2) preliminary design, and (3) final design. These planning and design steps are common to all major public works projects such as landfills or wastewater treatment plants. In some cases, the feasibility analysis has already been accomplished as part of the integrated waste management planning process.

Feasibility Analysis. The purpose of the feasibility analysis is to decide whether the MRF should be built. The feasibility study should provide the decision makers with clear recommendations on the technical and economic merits of the planned MRF. A typical feasibility analysis may contain sections or chapters dealing with the following topics.

The integrated waste management plan. The coordination of the MRF with the integrated waste management plan for the community is delineated in this section. A clear explanation of the role of the MRF in achieving landfill waste diversion and recycling goals is a key element of this section.

Conceptual design. What type of MRF should be built, which materials will be processed now and in the future, and what the design capacity of the MRF should be are discussed. Plan views and renderings of what the final MRF might look like are often included in this section.

Economics. Capital and operating costs (see Appendix E) are presented and discussed. Estimates of revenues available to finance the MRF (sales of recyclables, tipping fees, subsidies) are presented. A sensitivity analysis of the effects of fluctuating prices for recyclables and the impacts of changes in the composition of the waste should be included.

Ownership and operation. An analysis of how the MRF should be owned and operated is presented. Typical options to be considered include public ownership, private ownership, or public ownership with contract operation.

Procurement. The approach to be used in the design and construction of the MRF is discussed. Several options exist including (1) the traditional architect-engineer design and contractor construct process; (2) the turnkey contracting process in which design and construction are performed by a single firm; and (3) a full-service contract in which a single contractor designs, constructs, and operates the MRF.

Preliminary Design. The preliminary design includes development of the materials flow diagram, development of materials mass balances and loading rates for the unit operations (conveyors, screens, shredders, etc.) that make up the MRF, and the layout of the physical facilities. The cost estimate developed in the feasibility study is refined in the preliminary design report using actual price quotations from vendors.

Final Design. Final design includes preparation of final plans and specifications that will be used for construction. A detailed engineers' cost estimate is made based on materials take offs and vendor quotes. The cost estimate will be used for the evaluation of contractor bids if the traditional procurement process is used.

Issues in the Implementation and Operation of MRFs

The principal nonengineering issues associated with the implementation of MRFs are related to (1) siting, (2) environmental emissions, (3) public health and safety, and (4) economics. The importance of these issues cannot be over-emphasized and special attention must be devoted to their resolution before proceeding with final plans for any proposed facility.

Siting. Although it has been possible to build and operate MRFs in close proximity to both residential and industrial developments, operators must take extreme care if MRFs are to be environmentally and aesthetically acceptable. Ideally, to minimize the impact of the operation of MRFs, they should be sited in more remote locations where adequate buffer zones surrounding the facility can be maintained. In many communities, MRFs are located at the landfill site.

Environmental Emissions. Regardless of where a MRF is located, extreme care must be taken in its operation if it is to be environmentally acceptable with respect to traffic, noise, odor, dust, airborne debris, liquid discharges, visual unsightliness, and vector control. The best approach to these design issues is to visit many operating MRFs before settling on a final design. Proper housecleaning practices and proper storage of recyclable materials will reduce public complaints.

An attractive, well-maintained and -operated MRF can be a community asset and an incentive to citizen participation in recycling programs.

Public Health and Safety. Materials recovery facilities are a relatively new type of industrial facility and do not have a long history of experience in terms of public health and safety issues. Nevertheless, one must devote special attention to these issues during the design of the process. Two principal types of public health and safety issues are involved in the design of MRFs. The first is related to the public health and safety of the employees of the MRF. The second issue is related to the health and safety of the general public, especially for MRFs that will also be used as drop-off and buy-back centers.

Worker issues. Materials recovery facilities are potentially dangerous work environments unless proper precautions are taken during design and operation. Some of the most important safety and health issues are summarized in Table 9-7. Because of the moving equipment and conveyors used in most MRFs, special attention must be devoted to materials flow and worker involvement at each stage of the process. Where the manual separation of waste materials from commingled MSW is used, careful attention must be given to the types of protective clothing, air-filtering head gear, and puncture-proof gloves supplied to the workers. In addition, worker fatigue is another important issue that must be addressed. Where

TABLE 9-7
Health and safety issues in design and operation of MRFs

Component	Safety issue
Mechanical	High-speed rotating and reciprocating parts Exposed drive shafts and belts High-intensity noise Broken glass, sharp metal objects Explosive hazards
Electrical	Exposed wiring, switches, and controls Ground faults
Architectural	Ladders, stairways, and railings Vehicle routing and visibility Ergonomics of handpick conveyor belts Lighting Ventilation and air conditioning Drainage
Operational	Housekeeping practices Safety training Safety and first aid equipment
Hazardous materials	Hazardous wastes from households and small-quantity generators Biohazards such as human blood products and pathogenic organisms
Personal safety equipment	Punctureproof, impermeable gloves; safety shoes, uniforms, eye protection, noise protection

sorting from moving belts is used, the height of the worker relative to the moving belt must be adjustable. The federal government through OSHA and state OSHA-type programs now require the development of comprehensive health and safety programs for workers at MRFs.

Public access issues. Because the activities involved with the operations of a MRF are potentially dangerous, the public should be excluded from access except under careful control, as during conducted tours. Convenience stations for the deposit of recyclables should be provided for public access away from the main traffic pattern.

Economics. Because many MRFs are often underfunded owing to local economic constraints, the facilities needed for the control of environmental emissions and the management of public health and safety issues are often not incorporated in the design of these facilities. Because a MRF can be shut down for unacceptable environmental and health and safety issues, careful attention must be devoted to the design and implementation of environmental control facilities. For example, the purchase of expensive mechanical separation equipment may be postponed or delayed in favor of manual separation and the purchase of dust control facilities. Further, because of the uncertainties concerning the future quantities and characteristics of the waste, a detailed sensitivity analysis must be performed to assess the economic viability of a MRF or MR/TF subject to a wide range of projected future changes in the characteristics or quantity of the waste.

9-7 WASTE TRANSFORMATION THROUGH COMBUSTION

The separation and processing of waste materials has been considered in Section 9-6. In this and the following sections the focus is on the transformation of waste materials. Transformation processes are used to reduce the volume and weight of waste requiring disposal and to recover conversion products and energy. The organic fraction of MSW can be transformed by a variety of chemical and biological processes. The most commonly used chemical transformation process is combustion, which can be used to reduce the original volume of the combustible fraction of MSW by 85 to 95 percent. In addition, the recovery of energy in the form of heat is another attractive feature of the combustion process. Although combustion technology has advanced in the past two decades, air pollution control remains a major concern in implementation. Even if stricter air pollution control requirements can be met through the use of existing and developing technology (see Chapter 13), the problem of siting such facilities remains monumental (see Chapter 18).

Description of Combustion Process

The basic operations involved in the combustion of commingled MSW are identified in Fig. 9-31. The operation begins with the unloading of solid wastes from

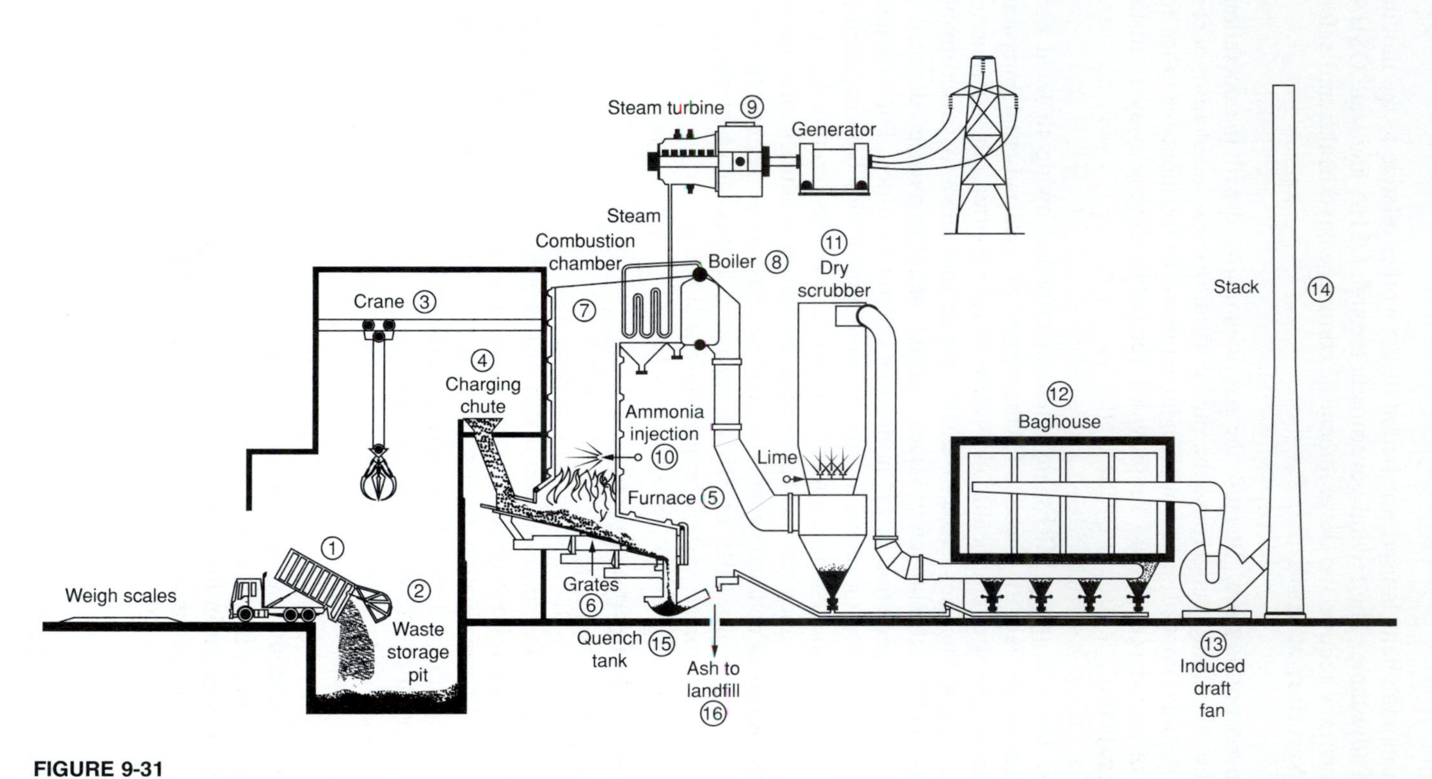

FIGURE 9-31
Section through a typical continuous-feed mass-fired municipal combustor used for the production of energy from MSW. (Courtesy of County Sanitation Districts of Los Angeles County.)

collection trucks (1) into a storage pit (2). The width of the unloading platform and storage bin is a function of the size of the facility and the number of trucks that must unload simultaneously. The depth and width of the storage bin are determined by both the rate at which waste loads are received and the rate of burning. The capacity of the storage pit is usually equal to the volume of waste for two days. The overhead crane (3) is used to batch load wastes into the feed (charging) chute (4), which directs the wastes to the furnace (5). The crane operator can select the mix of wastes to achieve a fairly even moisture content in the charge. Large or noncombustible items are also removed from the wastes with the overhead crane. Solid wastes from the feed (charging) chute fall onto the grates (6) where they are mass-fired. Several different types of mechanical stokers are commonly used.

Air may be introduced from the bottom of the grates (under-fire air) by means of a forced-draft fan or above the grates (over-fire air) to control burning rates and furnace temperature. Because most organic wastes are thermally unstable, various gases are driven off as the combustion process takes place in the furnace. These gases and small organic particles rise into the *combustion chamber* (7), and burn at temperatures in excess of 1600°F. Heat is recovered from the hot gases using water-filled tubes in the walls of the combustion chamber and with a boiler (8) that produces steam, which is converted to electricity by a turbine-generator (9).

Air pollution control equipment may include ammonia injection for NO_x (nitrogen oxides) control (10), a dry scrubber for SO_2 and acid gas control (11), and a baghouse (fabric filter) for particulate removal (12). To secure adequate air flows to provide for head losses through air pollution control equipment, as well as to supply air to the combuster itself, an induced-draft fan (13) may be needed. The end products of combustion are hot combustion gases and ash. The cleaned gases are discharged to the stack (14) for atmospheric dispersion. Ashes and unburned materials from the grates fall into a residue hopper (15) located below the grates, where they are quenched with water. Fly ash from the dry scrubber and the baghouse is mixed with the furnace ash and conveyed to ash treatment facilities (16). Details on combustion design, air pollution control equipment, and ash treatment and disposal are discussed in Chapter 13.

Combustion Products

The principal elements of solid wastes are carbon, hydrogen, oxygen, nitrogen, and sulfur (see Chapter 4). Smaller amounts of other elements will also be found in the ash. Under ideal conditions, the gaseous products derived from the combustion of municipal solid wastes with stoichiometric amounts of air, would include carbon dioxide (CO_2), water (H_2O, flue gas), nitrogen (N_2), and small amounts of sulfur dioxide (SO_2). In actuality, many different reaction sequences are possible, depending on the exact nature of the wastes and the operating characteristics of the combustion reactor. The basic reactions for the oxidation (combustion) of the carbon, hydrogen, and sulfur (and their atomic masses) contained in the organic fraction of MSW are as follows:

For carbon

$$\underset{12}{C} + \underset{32}{O_2} \rightarrow CO_2 \tag{9-2}$$

For hydrogen

$$\underset{4}{2H_2} + \underset{32}{O_2} \rightarrow 2H_2O \tag{9-3}$$

For sulfur

$$\underset{32.1}{S} + \underset{32}{O_2} \rightarrow SO_2 \tag{9-4}$$

If it is assumed that dry air contains 23.15 percent oxygen by weight, then the amount of air required for the oxidation of 1 lb of carbon would be equal to 11.52 lb [(32/12)(1/0.2315)]. The corresponding amounts for hydrogen and sulfur are 34.56 and 4.31 lb, respectively. Computation of the amount of air required for the complete combustion of an organic waste is illustrated in Example 9-2.

Example 9-2 Determination of the stoichiometric amount of air required for the combustion of an organic solid waste. Determine the amount (lbs and ft^3) of air required for the complete combustion of one ton of an organic solid waste. Assume that the composition of the organic waste to be combusted is given by C_5H_{12}. Assume the specific weight of air is 0.075 lb/ft^3.

Solution

1. Write a balanced stoichiometric equation for the oxidation of the organic compound based on oxygen:

$$\underset{72}{C_5H_{12}} + \underset{256}{8O_2} \rightarrow 5CO_2 + 6H_2O$$

2. Write a balanced equation for the oxidation of the organic compound with air. In combustion calculations, dry air is assumed to be comprised of 21 percent oxygen and 79 percent nitrogen. Thus, the corresponding reaction to that given in Step 1 for air is

$$C_5H_{12} + 8O_2 + 30.1N_2 \rightarrow 5CO_2 + 6H_2O + 30.1N_2$$

3. Determine the amount of air required for combustion, assuming air contains 23.15 percent oxygen by weight.

$$O_2 \text{ required} = \frac{256}{72} \times (2000 \text{ lb/ton}) = 7111 \text{ lb/ton}$$

$$\text{Air required} = \frac{7111 \text{ lb/ton}}{0.2315} = 30{,}717 \text{ lb/ton}$$

4. The amount of air required for combustion can also be computed using the factors, given previously.

$$\text{Air required for carbon, C} = \frac{60}{72} \times (2000 \text{ lb/ton}) \times 11.52 = 19{,}200 \text{ lb/ton}$$

$$\text{Air required for hydrogen, H} = \frac{12}{72} \times (2000 \text{ lb/ton}) \times 34.56 = 11{,}520 \text{ lb/ton}$$

$$\text{Total air required} = 19{,}200 + 11{,}520 = 30{,}720 \text{ lb/ton}$$

5. Determine the volume of air required for combustion.

$$\text{Volume of air} = (30{,}717 \text{ lb/ton})/(0.075 \text{ lb/ft}^3) = 409{,}560 \text{ ft}^3\text{/ton}$$

Comment. In Step 2, nitrogen is retained on both sides of the equation because it does not enter into the reaction. Although complete combustion was assumed in this example for the purposes of illustrating the computations involved in stoichiometric calculations, complete combustion is seldom achieved in practice. Typically, from 3 to 5 percent of the organic matter in the input feed will be found in the ash from a combustion facility.

Types of Combustors

Solid waste combustors can be designed to operate with two types of solid waste fuels: unseparated commingled MSW (mass-fired) and processed MSW known as refuse-derived fuel (RDF). Mass-fired (also known as mass-burn) combustors are the predominant type. In 1987, 68 percent of the operational combustor capacity in the United States was provided by mass-fired units and 23 percent by RDF-fired units. The remaining 9 percent of the capacity was provided by mass-fired modular combustion units.

Mass-Fired Combustors. In a mass-fired combustor (see Fig. 9-32), minimal processing is given to solid waste before it is placed in the hopper used to feed the combustor. The crane operator in charge of loading the charging hopper can reject obviously unsuitable items. However, one must assume that anything in the MSW stream may ultimately enter the combustor including bulky oversize noncombustible objects (e.g., broken tricycles, etc.) and even potentially hazardous wastes deliberately or inadvertently delivered to the system. For these reasons, the combustor must be designed to handle these objectionable wastes without damage to equipment or injury to operating personnel. The energy content of mass-fired waste can be extremely variable, depending on the climate, season, and source of waste. In spite of these potential disadvantages, mass-fired combustors have become the technology of choice for most existing and planned combustion facilities (see Fig. 9-33).

RDF-Fired Combustors. Compared with the uncontrolled nature of unprocessed commingled MSW, RDF can be produced from the organic fraction of MSW (see Chapter 12) with fair consistency to meet specifications for energy content, moisture, and ash content. The RDF can be produced in shredded or fluff form, or as densified pellets or cubes. Densified RDF (d-RDF) is more costly to produce but is easier to transport and store. Either form can be burned by itself, or mixed with coal.

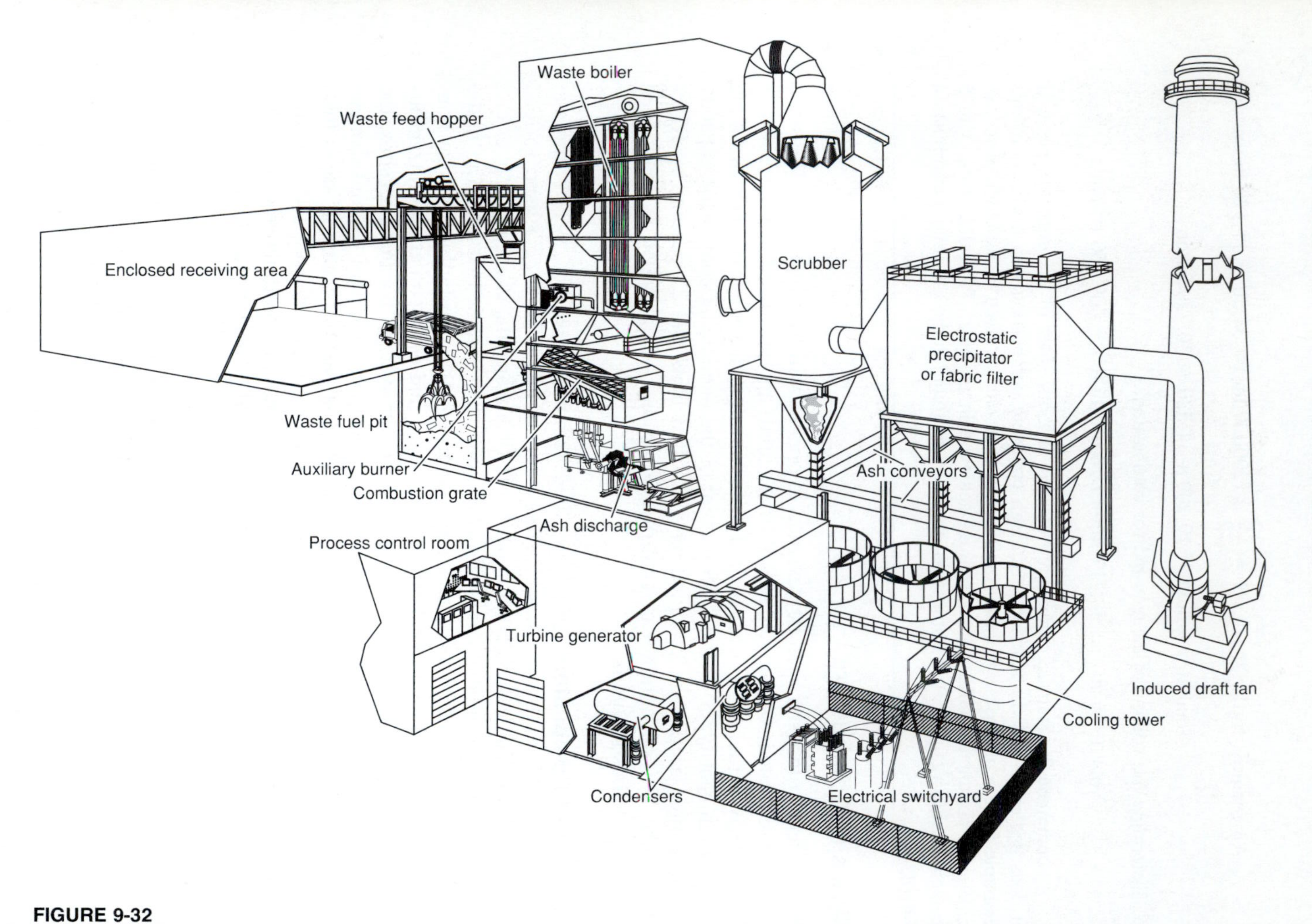

FIGURE 9-32
Section through water-wall mass-fired combustor used for the production of energy from MSW. (Courtesy of Wheelabrator Environmental Systems, Inc.)

(a)

(b)

FIGURE 9-33
Views of modern continuous-feed mass-fired municipal combustor: (*a*) collection vehicles waiting to unload and (*b*) unloading the contents of collection vehicles onto the unloading platform. The unloading platform is used to provide temporary waste storage (surge) capacity for weekend operation.

Because of the higher energy content of RDF compared with unprocessed MSW, RDF combustion systems (see Fig. 9-34) can be physically smaller than comparably rated mass-fired systems. However, more space will be required if the front-end processing system needed to prepare the RDF is to be located adjacent to the combustor. A RDF-fired system can also be controlled more effectively than a mass-fired system because of the more homogeneous nature of RDF, allowing for better combustion control and better performance of air pollution control devices. Additionally, a properly designed system for the preprocessing of MSW can effect the removal of significant portions of metals, plastics, and other materials that may contribute to harmful air emissions.

Energy Recovery

Virtually all new combustors currently under construction in the United States and Europe employ some form of energy recovery to help offset operating costs and to reduce the capital costs of air pollution control equipment. Energy can be recovered from the hot flue gases generated by combusting processed MSW, from solid fuel pellets (e.g., RDF), or from unprocessed MSW by two methods: (1) the use of a water-wall combustion chamber, and (2) the use of waste heat boilers, or both. Either hot water or steam can be generated. Hot water can be used for low-temperature industrial or space heating applications. Steam is more versatile, as it can be used for both heating and generating electricity. Perhaps the most common flow diagram for the production of electric energy using steam involves the use of a steam turbine–generator combination as shown in Fig. 9-35.

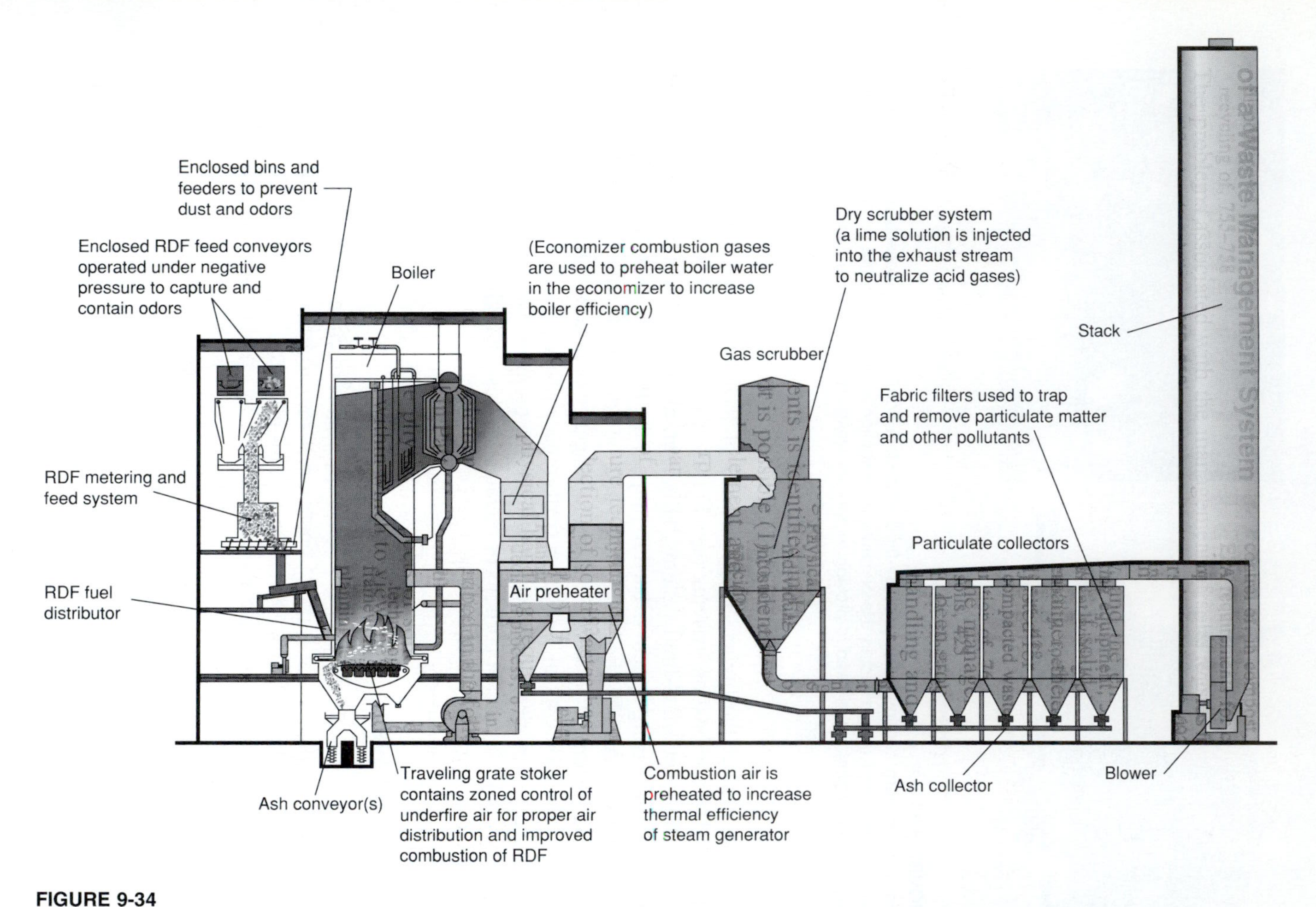

FIGURE 9-34
View of industrial water-wall boiler combustion system used for the production of energy from processed solid wastes, natural gas, oil, and coal. (Courtesy of ABB Resource Recovery Systems.)

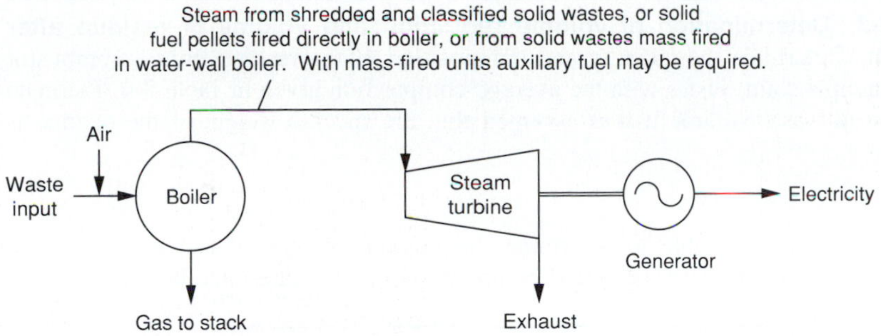

Steam from shredded and classified solid wastes, or solid fuel pellets fired directly in boiler, or from solid wastes mass fired in water-wall boiler. With mass-fired units auxiliary fuel may be required.

Air

Waste input

Boiler

Steam turbine

Electricity

Generator

Gas to stack

Exhaust

FIGURE 9-35
Schematic of energy recovery system using a steam turbine-generator combination. (See also Fig. 13-25*a*.)

Volume Reduction

Among the factors that must be considered in assessing the combustion process for MSW are the amount of residue remaining after combustion and whether auxiliary fuel will be required when heat recovery is not of primary concern. (The need for auxiliary fuel is considered in Chapter 13.) The amount of residue depends on the nature of the wastes to be combusted. Representative data on the residue from various solid waste components are reported in Table 9-8. The computations required to assess the quantity and composition of the residue after combustion are illustrated in Example 9-3.

TABLE 9-8
Composition of residue from the combustion of commingled MSW

	Percent by weight	
Component	**Range**	**Typical**
Partially burned or unburned organic matter	3–10	5
Tin cans	10–25	18
Other iron and steel	6–15	10
Other metals	1–4	2
Glass	30–50	35
Ceramics, stones, bricks	2–8	5
Ash	10–35	25
Total		100

Example 9-3 Determination of volume reduction and volume of residue after combustion. Determine the quantity and composition of the residue from a combustor used for municipal solid wastes with the average composition given in Table 3-4. Estimate the reduction in waste volume if it is assumed that the specific weight of the residue is 1000 lb/yd^3.

Solution

1. Set up a computation table to determine the amount of residue and its percentage distribution by weight. The completed computation table is presented below:

Component	Solid waste,[a] lb	Inert residue,[b] %	Residue lb	Residue %
Organic				
Food wastes	90	5	4.5	1.9
Paper	340	6	20.4	8.6
Cardboard	60	5	3.0	1.3
Plastics	70	10	7.0	2.9
Textiles	20	6.5	1.3	0.5
Rubber	5	9.9	0.5	0.2
Leather	5	9.0	0.5	0.2
Yard wastes	185	4.5	8.3	3.5
Wood	20	1.5	0.3	0.1
Misc. organics	—	—	—	—
Inorganic				
Glass	80	98	78.4	33.0
Tin cans	60	98	58.8	24.7
Aluminum	5	96	4.8	2.0
Other metal	30	98	29.4	12.4
Dirt, ash, etc.	30	68	20.4	8.6
Total	1000		237.6	100.0

[a] Based on 1000 lb of solid waste (see Table 3-4).

[b] Data from Tables 4-3 and 4-4.

Note: lb × 0.4536 = kg

2. Estimate the original and final volumes before and after combustion. To estimate the approximate initial volume, assume that the average specific weight of the solid wastes in the combustor storage pit is about 375 lb/yd^3.

$$\text{Original volume} = \frac{1000 \text{ lb}}{375 \text{ lb/yd}^3} = 2.67 \text{ yd}^3$$

$$\text{Residue volume} = \frac{237.6 \text{ lb}}{1000 \text{ lb/yd}^3} = 0.24 \text{ yd}^3$$

3. Estimate the volume reduction by using Eq. (4-1).

$$\text{Volume reduction} = \left(\frac{2.67 - 0.24}{2.67}\right) 100 = 91\%$$

Issues in the Implementation of Combustion Facilities

The principal issues associated with the use of combustion facilities for the transformation of MSW are related to (1) siting, (2) air emissions, (3) disposal of residues, (4) liquid emissions, and (5) economics. Unless the questions related to these issues are resolved, the use of combustion may face an uncertain future. These subjects are introduced below and examined in detail in Chapter 13. A decision-maker's perspective is presented in Ref. 11.

Siting. As with the siting of MRFs, it has been possible to build and operate combustion facilities in close proximity to both residential and industrial developments; however, extreme care must be taken in their operation if they are to be environmentally and aesthetically acceptable. Ideally, to minimize the impact of the operation of combustion facilities, they should be sited in more remote locations where adequate buffer zones surrounding the facility can be maintained. In many communities, combustion facilities are located in remote locations within the city limits or at the landfill site.

Air Emissions. The operation of combustion facilities results in the production of a variety of gaseous and particulate emissions, many of which are thought to have serious health impacts. The demonstrated ability of combustion facilities and equipment to effectively control gaseous and particulate emissions is of fundamental importance in the siting of these facilities. The proper design of control systems for these emissions is a critical part of the design of combustion systems. In some cases, the cost and complexity of the environmental control system(s) are equal to or even greater than the cost of the combustion facilities.

Disposal of Residues. Several solid residuals are produced by combustion facilities, including (1) bottom ash, (2) fly ash, and (3) scrubber product. Management of these solid residuals is an integral part of the design and operation of a combustion facility. Typically, bottom ash is disposed of by landfilling. The primary concern with landfilling of the ash is that it may, under certain conditions, leach contaminants into the groundwater. Consequently, ash from combustion facilities is now disposed of in lined MSW landfills or in double-lined monofills devoted solely to the disposal of ash.

Liquid Emissions. Liquid emissions from combustion facilities can arise from one or more of the following sources: (1) wastewater from the ash removal facilities, (2) effluent from wet scrubbers, (3) wastewater from pump seals, cleaning, flushing, and general housekeeping activities, (4) wastewater from treatment systems used to produce high-quality boiler water, and (5) cooling tower blowdown. The proper handling and disposal of these liquid emissions is also an important part of the design of combustion facilities.

Economics. The economics of a proposed combustion system must be evaluated carefully to permit a choice between competing systems. The best way to compare alternatives is by the use of life cycle costing, which accounts for operating and maintenance costs over the lifetime of the system. The solid waste industry has developed a standardized approach to life cycle costing through the use of the pro forma income statement.

9-8 WASTE TRANSFORMATION THROUGH AEROBIC COMPOSTING

With the exception of plastic, rubber, and leather components, the organic fraction of most MSW can be considered to be composed of proteins, amino acids, lipids, carbohydrates, cellulose, lignin, and ash [3]. If these organic materials are subjected to aerobic microbacterial decomposition, the end product remaining after microbiological activity has essentially ceased is a humus material commonly known as *compost*. In equation form the process can be described as follows:

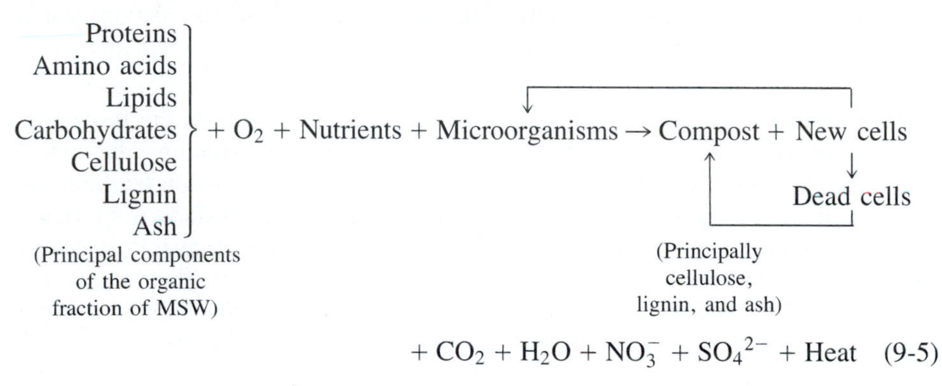

As shown in Eq. (9-5), the new cells that are produced become part of the active biomass involved in the conversion of the organic matter and on death ultimately become part of the compost. The general objectives of composting are (1) to transform the biodegradable organic materials into a biologically stable material, and in the process reduce the original volume of waste; (2) to destroy pathogens, insect eggs, and other unwanted organisms and weed seeds that may be present in MSW; (3) to retain the maximum nutrient (nitrogen, phosphorous, and potassium) content, and (4) to produce a product that can be used to support plant growth and as a soil amendment [3, 4, 13].

In general, the chemical and physical characteristics of compost vary according to the nature of the starting material, the condition under which the composting operation was carried out, and the extent of the decomposition. Some of the properties of compost that distinguish it from other organic materials are these:

1. A brown to very dark brown color
2. A low carbon-nitrogen ratio

3. A continually changing nature due to the activities of microorganisms
4. A high capacity for cation exchange and water absorption

When added to soil, compost has been found to lighten heavy soils, to improve the texture of light sandy soils, and to increase the water retention capacity of most soils. Details on the theory and practice of composting are presented in Chapter 14.

Process Description

The composting process has always occurred in nature. One of the first organized composting operations to be reported upon in the literature was carried out in India in the early 1930s under the direction of Howard and associates [3]. The process they developed, known as the Indore process, was named for the location in India where it was developed. In its simplest form, the process involves excavating a trench in the ground 2 to 3 ft deep in which successive layers of putrescible materials such as solid waste, night soil, animal manure, earth, and straw are placed. The earliest procedure was to turn the material only twice during the composting process, which lasted six months or longer [3]. The liquid released from the decomposing waste was recirculated or added to other drier composting wastes. Because of the limited turning, one can assume that the composting mass was anaerobic for most of the composting process. The Indore process has been modified extensively, with the most important innovation being the more frequent turning of the composting material to maintain aerobic or facultative conditions and to accelerate the composting period.

Most modern composting operations consist of three basic steps: (1) preprocessing of the MSW, (2) decomposition of the organic fraction of the MSW, and (3) preparation and marketing of the final compost product. A generalized process flow diagram for the composting process is shown in Fig. 9-36. Receiving, removal of recoverable materials, size reduction, and the adjustment of the waste properties (e.g., carbon-nitrogen ratio, addition of moisture and nutrients) are essential steps in the preprocessing of MSW for composting. The degree of preprocessing depends on the specific composting process employed and the specifications for the final compost product.

To accomplish the decomposition step, several techniques have been developed including windrow, static pile, and in-vessel composting. In windrow composting, for example, prepared MSW is placed in windrows in an open field (see Fig. 9-37). The windrows are turned once or twice per week for a composting period of 4 to 5 weeks. During this time, the biodegradable portion of the organic fraction of MSW is decomposed by a variety of microorganisms, which utilize the organic matter as a carbon (food) source (see Eq. 9-5). The metabolic activity of the microorganisms alters the chemical composition of the original organic matter, reduces the volume and weight of the waste, and increases the heat of the material being composted. Turning the compost pile serves to provide oxygen for the decomposition process and to control the temperature of the

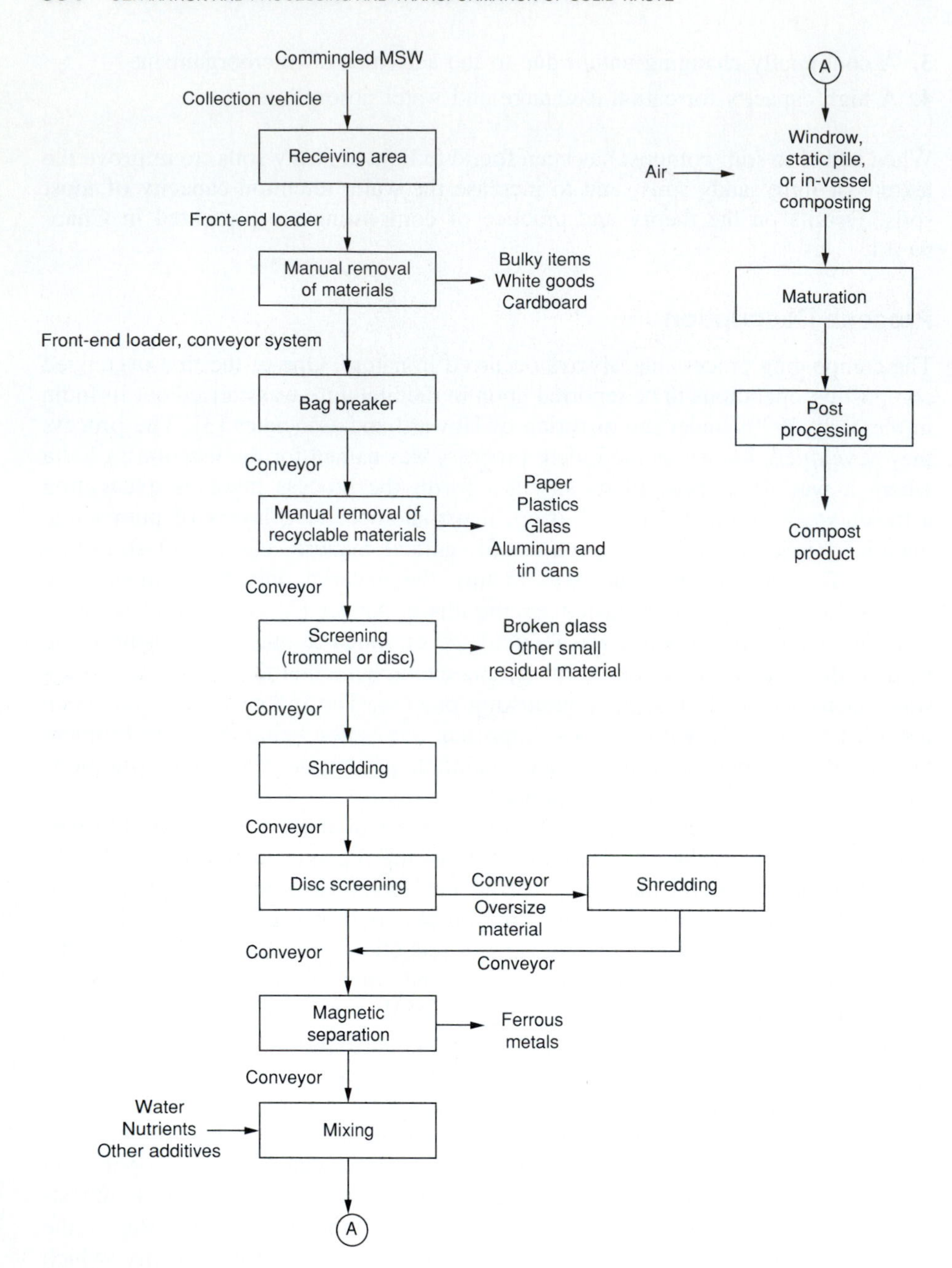

FIGURE 9-36
Generalized flow diagram for the composting process.

FIGURE 9-37
Compost produced from yard waste to be used in plant soil mixes, as a soil amendment, or as intermediate landfill cover material.

composting waste. When the readily biodegradable organic material is depleted, bacterial activity is reduced, the temperature of the composting material begins to drop, and the first stage of the composting process is complete (see Fig. 9-38). The composted material is usually cured for an additional 2 to 8 weeks in open windrows to ensure complete stabilization.

Preparation and marketing of the compost, the third step in the composting process, occurs once the compost has been cured and stabilized. At the present time, there is no universally accepted definition of what constitutes fully stabilized compost. Product preparation and marketing may include fine grinding, screening,

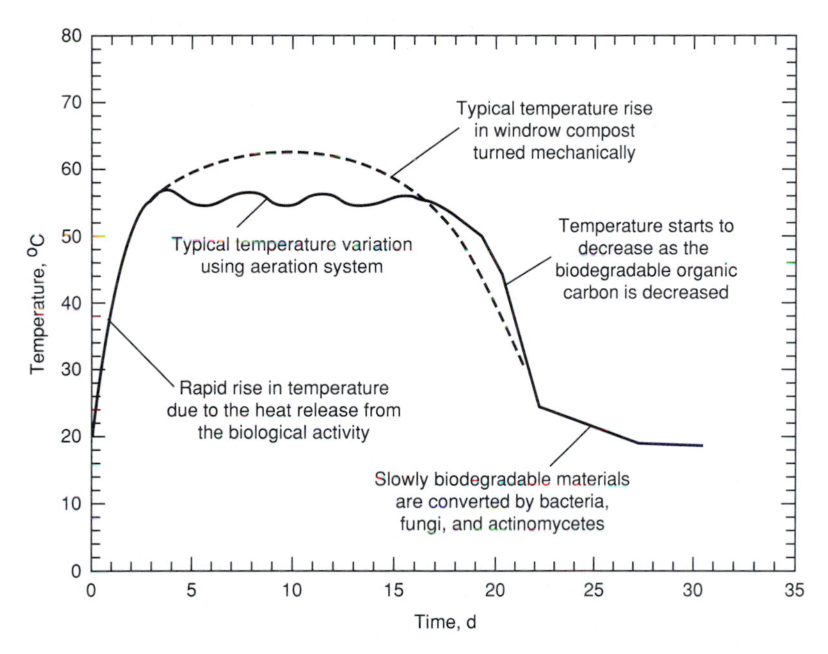

FIGURE 9-38
Variation of temperature during the composting process.

air classification, blending with various additives, granulation, bagging, storage, shipping, and in some cases, direct marketing. Typical specifications for compost produced from yard wastes, collected separately, and from the organic fraction of MSW are given in Chapter 15.

Process Design and Control

Although the composting process is easy to grasp conceptually, the actual design and control of the process are quite complex. Important process variables that must be considered in the design and operation of composting facilities include particle size and particle size distribution of the material to be composted, seeding and mixing requirements, the required mixing/turning schedule, total oxygen requirements, moisture content, temperature and temperature control, carbon-nitrogen ratio of the waste to be composted, pH, degree of decomposition, respiratory quotient (RQ), and control of pathogens.

Composting Techniques

The two principal methods of composting now in use in the United States may be classified as *agitated* and *static*. In the agitated method, the material to be composted is agitated periodically to introduce oxygen, to control the temperature, and to mix the material to obtain a more uniform product. In the static method, the material to be composted remains static and air is blown through the composting material. The most common agitated and static methods of composting are known as the windrow and static pile methods, respectively. Proprietary composting systems in which the composting operation is carried out in a reactor of some type are known as in-vessel composting systems.

Windrow Composting. Windrow composting is one of the oldest methods of composting. In its simplest form, a windrow compost system can be constructed by forming the organic material to be composted into windrows 8 to 10 ft high by 20 to 25 feet wide at the base. A minimal system could use a front-end loader to turn the windrow once per year. While such a minimal system would work, it could take up to three to five years for complete degradation. Also such a system would probably emit objectionable odors, as parts of the windrow will likely be anaerobic.

A high-rate windrow composting system employs windrows with a smaller cross section, typically 6 to 7 ft high by 14 to 16 ft wide. The actual dimensions of the windrows depend on the type of equipment that will be used to turn the composting wastes (see Fig. 9-39). Before the windrows are formed organic material is processed by shredding and screening it to approximately 1 to 3 in and the moisture content is adjusted to 50 to 60 percent. High-rate systems are turned up to twice per week while the temperature is maintained at or slightly above 131°F (55°C). Turning of the windrows is often accompanied by the release of offensive odors. Complete composting can be accomplished in three to four weeks. After

FIGURE 9-39
Specially designed machine used to turn composting material placed in windrows.

the turning period, the compost is allowed to cure for an additional three to four weeks without turning. During the curing period, residual decomposable organic materials are further reduced by fungi and actinomycetes.

Aerated Static Pile Composting. The aerated static pile composting process was developed by the U.S. Department of Agriculture Agricultural Research Service Experimental Station at Beltsville, Maryland; thus, the process is sometimes referred to as the Beltsville or ARS process. Originally developed for the aerobic composting of wastewater sludge, the process can be used to compost a wide variety of organic wastes including yard waste or separated MSW. The aerated static pile system, as shown in Fig. 9-40, consists of a grid of aeration or exhaust piping over which the processed organic fraction of MSW is placed. Typical pile

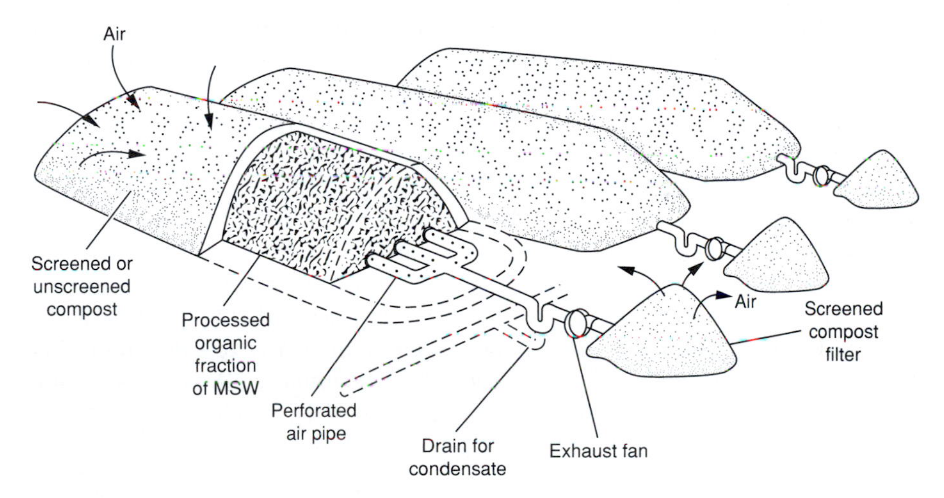

FIGURE 9-40
Schematic of aerated static pile composting system.

heights are about 7 to 8 ft (2 to 2.5 m). A layer of screened compost is often placed on top of the newly formed pile for insulation and odor control.

Each pile is usually provided with an individual blower for more effective aeration control. Disposable corrugated plastic drainage pipe is used commonly for air supply. Air is introduced to provide the oxygen needed for biological conversion and to control the temperature within the pile. Blower operation is typically controlled by a timer, or in some systems by a microcomputer to match a specific temperature profile. The material is composted for a period of three to four weeks. The material is then cured for a period of four weeks or longer. Shredding and screening of the cured compost usually is done to improve the quality of the final product. For improved process and odor control, all or significant portions of the system in newer facilities are covered or enclosed.

Where dewatered wastewater treatment plant sludge is to be composted, some type of bulking agent is required to maintain the porosity of the composting material. Wood chips are used most commonly as the bulking agent. They can also be used to absorb excess moisture. The mixture of sludge and wood chips is placed in piles over the aeration piping and covered with previously composted material. The blower can be operated either to push or to pull air through the pile. As in the high-rate windrow process, composting time is about three to four weeks. After composting, the pile is taken apart, and the bulking agent is recovered by screening. Note that a bulking agent is not usually required for composting dry materials like MSW or yard waste or mixtures of MSW and wastewater treatment plant sludge.

In-Vessel Composting Systems. In-vessel composting is accomplished inside an enclosed container or vessel. Every imaginable type of vessel has been used as a reactor in these systems, including vertical towers, horizontal rectangular and circular tanks, and circular rotating tanks (see Fig. 9-41). In-vessel composting systems can be divided into two major categories: plug flow and dynamic (agitated bed). In plug flow systems, the relationship between particles in the composting mass stays the same throughout the process, and the system operates on a first-in, first-out principle. In a dynamic system, the composting material is mixed mechanically during the processing. Examples of plug flow reactors are shown in Figs. 9-41*a*, *b* and examples of dynamic systems are illustrated in Figs. 9-41*c*, *d*.

Mechanical systems are designed to minimize odors and process time by controlling environmental conditions such as air flow, temperature, and oxygen concentration. The popularity of in-vessel composting systems has increased in recent years. Reasons for this increased use are process and odor control, faster throughput, lower labor costs, and smaller area requirements. The detention time for in-vessel systems varies from 1 to 2 weeks, but virtually all systems employ a 4- to 12-week curing period after the active composting period.

Process Applications

Composting is an increasingly popular waste management option as communities look for ways to divert portions of the local waste stream from landfills. The prin-

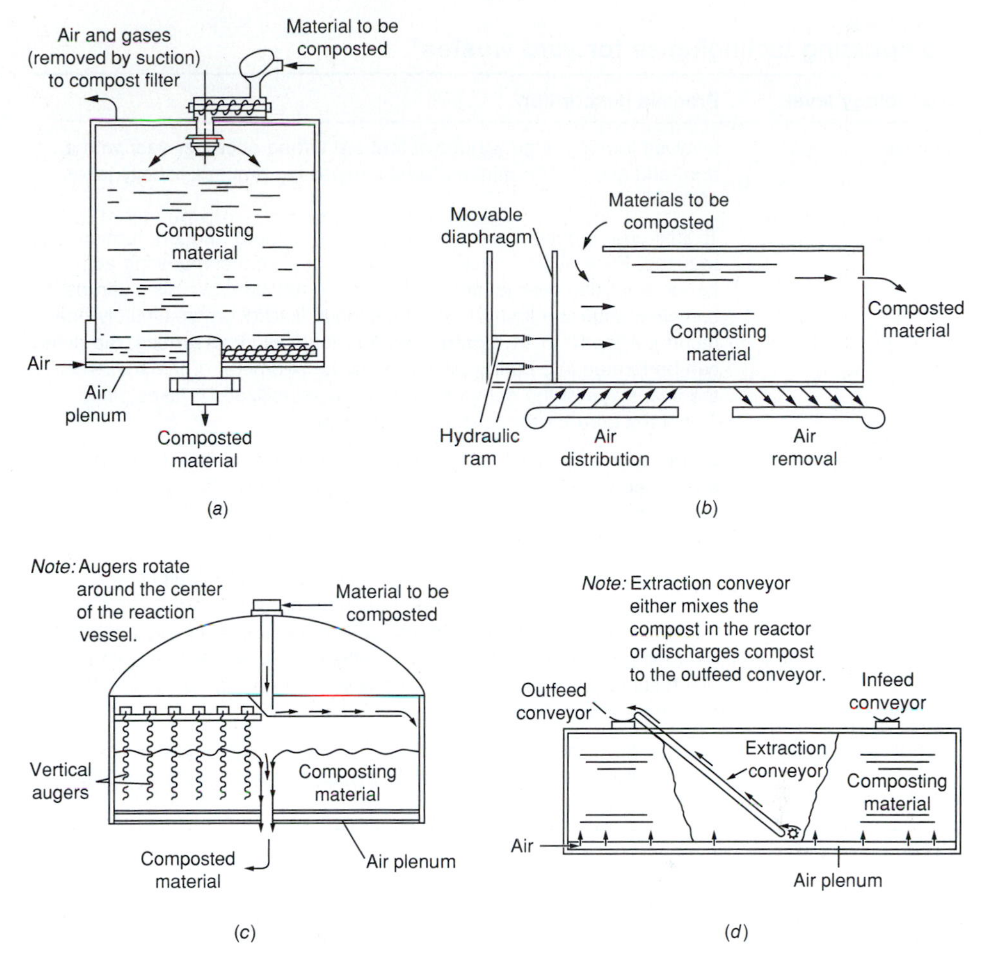

FIGURE 9-41
In-vessel composting units: (*a*) unmixed vertical plug flow reactor, (*b*) unmixed horizontal plug flow reactor, (*c*) mixed (dynamic) vertical reactor, and (*d*) mixed (dynamic) horizontal reactor.

cipal applications of composting are for (1) yard wastes, (2) the organic fraction of MSW, (3) partially processed commingled MSW, and (4) co-composting of the organic fraction of MSW with wastewater sludge. Because composting can be important in meeting mandated waste diversion goals, each of these applications is considered below. Backyard composting was considered in Chapter 7.

Composting of Yard Wastes Collected Separately. Leaves, grass clippings, bush clippings, and brush are the most commonly composted yard wastes. Stumps and wood are also compostable, but only after they have been chipped to produce a smaller, more uniform size. Five levels of technology that can be used for composting yard wastes are described in Table 9-9. Operating parameters for these five levels of composting technology are reported in Table 9-10.

TABLE 9-9
Composting technologies for yard wastes[a]

Technology level	Process description
Minimal	Involves forming large windrows that are turned once per year with a front-end loader. The minimal-level composting process usually takes 18 to 36 months.
Low-level	To limit odor problems, smaller windrows and more frequent turning are required. Piles of moderate size allow for sufficient composting activity, while limiting overheating and odors. In addition, two piles can be combined after the first "burst" of microbial activity (approximately one month). After 10 to 11 months and additional windrow turning, the piles can be formed into curing piles around the perimeter of the site, where the final stage of the composting process (stabilization) takes place. This frees area for the formation of new piles.
Intermediate-level	Similar to the low-level technology approach, except that the windrows are turned weekly with a windrow-turning machine. Use of windrow-turning machines will usually limit the size of the piles, thus increasing the total land area required.
High-level	In the high-level approach, forced aeration is used to optimize the compost process. The most common forced air approach is the static pile method. The blower in the forced aeration method is usually controlled by a temperature feedback system. When the temperature within the pile reaches a predetermined value, the blower turns on, cooling the pile and removing water vapor.
High-level in-vessel	Mechanical systems are designed to minimize odors and process time by controlling environmental conditions such as air flow, temperature, and oxygen concentration.

[a] Adapted from Ref. 11.

The collection of yard and other green wastes in specially designed containers is another recent innovation. The containers, provided with air holes, are equipped with a screened bottom (see Fig. 9-42*a*) that allows air to circulate. The yard and green wastes are collected once every two weeks. Because of the special design of the container, the material is usually dried by the time it is collected with specially equipped collection vehicles (see Fig. 9-42*b*).

Composting of Organic Fraction of MSW. Recognizing that product quality is the key to public acceptance of compost produced from municipal wastes, most operators of municipal composting systems base their efforts on separated wastes. Where mechanical means are used to separate the noncompostable materials from the compostable materials, the resulting compost often is still unacceptable because of metal contamination and the presence of trace amounts of household hazardous waste. Increasingly, professionals are recognizing that, to produce the highest-quality compost, source-separated materials should be used as the feedstock.

Composting of Partially Processed Commingled MSW. The composting of partially processed commingled municipal wastes has been suggested as a means

TABLE 9-10
Operating parameters for various technology levels for the composting of yard wastes[a]

Technology level	Windrow dimensions, ft		Turning frequency	Time to obtain finished product, months
	Height	Width		
Minimal	10–12	20–24	1 time/yr	24–36
Low	5–7	12–14	3–5 times/yr	14–18
Intermediate	5–8	12–18	Weekly	4–6
High	8–10	16–20	Aerated static pile[b]	3–4
High-level in-vessel				2.0–2.5[c]

[a] Adapted in part from Ref. 11.

[b] Forced aeration is used for a period of 2 to 10 weeks, at which time the blowers are turned off and the piles are turned periodically.

[c] In-vessel composting times vary from 8 hours to 20 days, depending on the process. The composted material is then cured in open windrows for an additional 6 to 8 weeks.

(*a*)

(*b*)

FIGURE 9-42
Specially designed system for the collection of yard and other green waste: (*a*) container equipped with air holes and a false bottom for air circulation and (*b*) specially equipped collection vehicle for emptying the containers.

of reducing the volume of wastes placed in landfills. Use of the composted material as intermediate landfill cover material has also been suggested. The composting of shredded residential and commercial MSW for these purposes is currently being done at the Escambia County landfill in Cantonment, FL. This composting operation is described in Section 11-6.

Co-composting Wastewater Treatment Plant Sludge with Organic Fraction of MSW. Composting of wastewater treatment plant sludge has been practiced since the early 1970s, and the number of composting facilities more than doubled in the 1980s [4]. Co-composting of wastewater treatment plant sludge and MSW is a relatively recent development. Mixing the sludge with the organic fraction of MSW is beneficial, as sludge dewatering may not be required and the overall metals content of the composted material will be considerably less than the composted sludge alone. Treatment plant sludges typically have a solids content ranging from 3 to 8 percent. A 2:1 mixture of compostable MSW to sludge is recommended as a minimum starting point. Both static and agitated compost systems have been

TABLE 9-11
Representative composting technologies for MSW and yard wastes[a]

Technology level	Process description
Bangalore (Indore)	Trench in ground; 2 to 3 ft deep. Material placed in alternate layers of refuse, night soil, earth, straw, etc. No grinding. Turned by hand as often as possible. Detention time of 120 to 180 days.
Casperi (briquetting)	Ground material (waste) is compressed into blocks and stacked for 30 to 40 days. Aeration by natural diffusion and air flow through stacks. Curing follows initial composting. Blocks are later ground.
DANO Biostabilizer	Rotating drum, slightly inclined from the horizontal, 9 ft to 12 ft in diameter; up to 150 ft long. One to 5 days digestion followed by windrowing. No grinding. Forced aeration into drum.
Earp-Thomas	Silo type with 8 decks stacked vertically. Ground waste is moved downward from deck to deck by ploughs. Air passes downward through the silo. Uses a patented inoculum. Digestion 2 to 3 days, followed by windrowing.
Fairfield-Hardy	Circular tank. Vertical screws, mounted on two rotating radial arms, keep ground material agitated. Forced aeration through tank bottom and holes in screws. Detention time of 5 days.
Fermascreen	Hexagonal drum, three sides of which are screens. Waste is ground. Batch loaded. Screens are sealed for initial composting. Aeration occurs during rotation with screens open. Detention time of 5 days.
Frazer-Eweson	Ground waste placed in vertical bin having 4 or 5 perforated decks and special arms to force composting material through perforations. Air is forced through bin. Detention time of 4 to 5 days.
Jersey (also known as the John Thompson system)	Structure with 6 floors, each equipped to dump ground waste onto the next lower floor. Aeration effected by dropping from floor to floor. Detention time of 6 days.

(continued)

tried. At this time, there is relatively little experience with co-composting in the United States owing to unmarketability of the final product.

Commercial Composting Systems

Over the past 50 years, more than 50 major different types of proprietary commercial composting systems have been developed and applied worldwide [3, 4, 13]. The general characteristics of the most common of these are summarized in Table 9-11. The various composting systems arranged by function or reactor type are reported in Table 9-12. A typical flow diagram for a commercial composting process is shown in Fig. 9-43. In the process shown in Fig. 9-43, MSW is first processed to remove recyclables and to reduce the size of the materials to be composted. Water and any other needed additives are added to the waste in the mixing unit. The prepared waste is then placed in plug flow compost reactors (also known as tunnel reactors). As more waste is added each day, material that has been composting is discharged from the reactor. The composted waste is aged

TABLE 9-11 ***(continued)***

Technology level	Process description
Metrowaste	Open tanks, 20 ft wide by 10 ft deep by 200 ft to 400 ft long. Processed MSW shredded in pretreatment processing. Equipped to give one or two turnings during digestion period (7 days). Air is forced through perforations in bottom of tank.
Naturizer or International	Five 9 ft wide steel conveyer belts arranged to pass material from belt to belt. Each belt is an insulated cell. Air passes through digester. Detention time of 5 days.
Riker	Four-story bins with clamshell floors. Ground waste is dropped from floor to floor. Forced air aeration. Detention time of 20 to 28 days.
Ashbrook-Simon-Hartley	Tunnel reactor typically 18 ft wide by 12 ft high by 65 ft long. Plug flow with wastes pushed into and out of reactor. Pressure and vacuum blowers used to supply and exhaust air through air diffusers located in floor of reactor. Detention time of 18 to 20 days.
T. A. Crane	Two cells consisting of 3 horizontal decks. Horizontal ribbon screws extending the length of each deck recirculate ground waste from deck to deck. Air is introduced in bottom of cells. Composting followed by curing in a bin.
Tollemache	Similar to the Metrowaste digesters.
Triga	Towers or silos called "hygienisators." In sets of 4 towers. Waste is ground. Forced air aeration. Detention time of 4 days.
Windrowing (normal, aerobic process)	Open windrows, with a "haystack" cross-section. Waste is ground. Aeration by turning windrows. Detention time depends upon number of turnings and other factors.
Van Maanen process	Underground waste in open piles, 120 to 180 days.

[a] Adapted from Refs. 3, 4, 13.

TABLE 9-12
Municipal composting systems grouped by function or reactor configuration[a]

Function or configuration	Commercial process
Heaps and windrows, natural aeration, batch operation	Indore/Bangalore
	Artsiely
	Baden-Baden (hazemag)
	Buhler
	Disposals Associates
	Dorr-Oliver
	Spohn
	Tollemache
	V.A.M.
Cells with natural or forced aeration, batch operation	Beccari
	Biotank (Degremont)
	Boggiano-Pico
	Kirkconnel (Dumfriesshire)
	Metrowaste
	Prat (Sofranie)
	Spohn
	Verdier
	Westinghouse/Naturizer
Horizontal rotating and inclined drums, continuous operation	Dano Biostabilizer
	Dun Fix
	Fermascreen (batch)
	Head Wrightson
	Vickers Seerdrum
Vertical flow reactors, continuous operation, agitated bed, natural or forced aeration	Earp-Thomas
	Fairfield Hardy
	Frazer-Eweson
	Jersey (John Thompson)
	Multibacto
	Nusoil
	Snell
	Triga
Agitated vertical bed	Fairfield-Hardy

[a] Adapted from Refs. 3, 4, 13.

in aerated static piles. After maturation, the compost is screened and shredded to produce a uniform product.

Issues in the Implementation of Composting Facilities

The principal issues associated with the use of the compost process are (1) the production of odors, (2) the presence of pathogens, (3) the presence of heavy metals, and (4) the definition of what constitutes an acceptable compost. Blowing

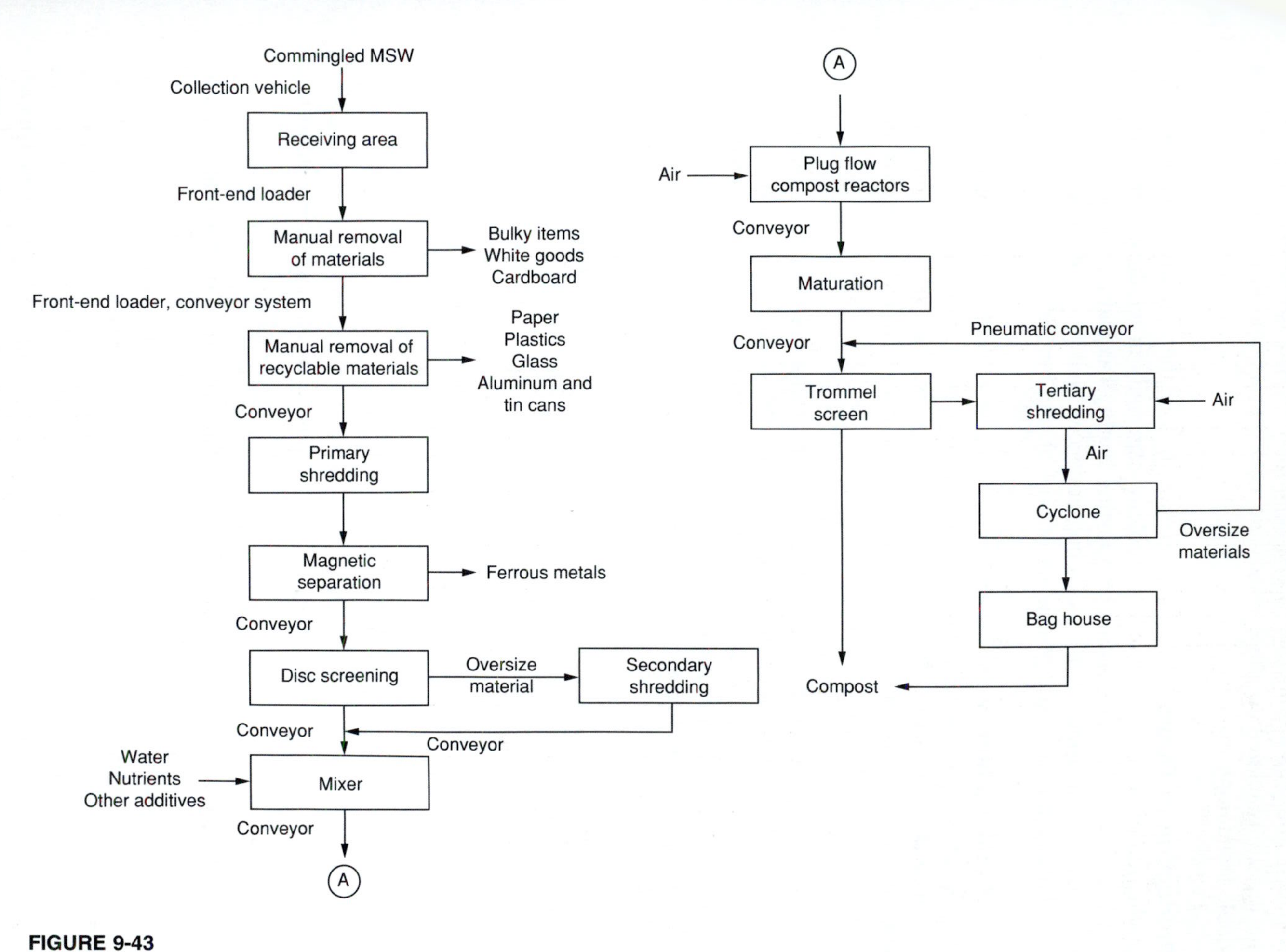

FIGURE 9-43
Flow diagrams for the Ashbrook Simon-Hartley composting process.

of papers and plastic materials is also a problem in windrow composting. Unless the questions related to these issues are resolved, composting may never be a viable technology.

Production of Odors. Without proper control of the composting process, the production of odors can become a problem, especially in windrow composting. It is fair to say that every existing composting facility has had an odor event and in some cases numerous events. As a consequence, facility siting, process design, and biological odor management are of critical importance.

Facility siting. Important issues in siting as related to the production and movement of odors include proper attention to local microclimates as they affect the dissipation of odors, distance to odor receptors, the use of adequate buffer zones, and the use of split facilities (use of different locations for composting and maturation operations).

Proper process design and operation. Proper process design and operation are critical in minimizing the potential for the production of odors. If composting operations are to be successful, special attention must be devoted to the following items: preprocessing, aeration requirements, temperature control, and turning (mixing) requirements. The facilities used to prepare the waste materials for the composting process must be capable of mixing completely and effectively any required additives, such as nutrients, inoculum (if used), and moisture with the waste material to be composted. The aeration equipment must be sized to meet peak oxygen demand requirements with an adequate margin of safety. In the static pile method of composting, the aeration equipment must also be sized properly to provide the volume of air required for cooling the composting material. The composting facilities must be instrumented adequately to provide for positive and effective temperature control. The equipment used to turn and mix the compost to provide oxygen and to control the temperature must be effective in mixing all portions of the composting mass. Unmixed compost will undergo anaerobic decomposition, leading to the production of odors. Because all of the operations cited above are critical to the operation of an odor-free composting facility, standby equipment should be available.

Biological odor management. Because occasional odor events are impossible to eliminate, special attention must be devoted to the factors that may affect biological production of odors. Causes of odors in composting operations include low carbon to nitrogen (C/N) ratios, poor temperature control, excessive moisture, and poor mixing. For example, in composting operations where the compost is not turned and the temperature is not controlled (see Fig. 9-38), the compost in the center of the composting pile can become pyrolyzed (see Chapter 13). When the composting pile is subsequently moved, the odors released from the pyrolyzed compost have been *extremely severe*. In enclosed facilities, odor control facilities such as packed towers, spray towers, activated carbon contactors, biological filters, and compost filters have been used for odor management.

In some cases, odor-masking agents and enzymes, which can split some odorous organic compound, have been used for the temporary control of odors.

Public Health Issues. If the composting operation is not conducted properly, the potential exists for pathogenic organisms to survive the composting process. The absence of pathogenic organisms is critical if the product is to be marketed for use in applications where the public may be exposed to the compost. Although pathogen control can be achieved easily with proper operation of the composting process, not all composting operations are instrumented sufficiently to allow for the reliable production of pathogen-free compost. In general, most pathogenic organisms found in MSW and other organic material to be composted will be destroyed at the temperatures and exposure times used in controlled composting operations (typically 55°C for 15 to 20 days). Temperatures required for the control of various pathogens are given in Table 14-8.

Heavy Metal Toxicity. A concern that may affect all composting operations, but especially those where mechanical shredders are used, involves the possibility of heavy metal toxicity. When metals in solid wastes are shredded, metal dust particles are generated by the action of the shredder. In turn, these metal particles may become attached to the materials in the light fraction. Ultimately, after composting, these metals would be applied to the soil. Although many of them would have no adverse effects, metals such as cadmium (because of its toxicity) are of concern. In general, the heavy metal content of compost produced from the organic fraction of MSW is significantly lower than the concentrations found in wastewater treatment plant sludges. The metal content of source-separated wastes is especially low. The co-composting of wastewater treatment plant sludges and the organic fraction of MSW is one way to reduce the metal concentrations in the sludge.

Product Quality. Product quality for compost material can be defined in terms of the nutrient content, organic content, pH, texture, particle size distribution, moisture content, moisture-holding capacity, the presence of foreign matter, the concentration of salts, residual odor, the degree of stabilization or maturity, the presence of pathogenic organisms, and the concentration of heavy metals. Unfortunately, at this time there is no universal agreement on suitable values for these parameters. This lack of agreement has been and continues to be a major impediment to the development of a uniform compost product from location to location. Some specifications that have been developed for compost are presented in Table 15-8. If compost materials are to have wide acceptance, public health issues must be resolved in a satisfactory manner.

9-9 IMPACT OF SOURCE REDUCTION AND WASTE RECYCLING ON WASTE TRANSFORMATION PROCESSES

As more states adopt legislation mandating the development of waste diversion and recycling programs, the quantities and composition of the wastes collected will change. The impact of change in composition will vary depending on the other types of waste management programs that are in place. For example, a recycling program would, as illustrated in Example 9-4, result in a reduction in the energy content of the waste. Such a reduction in the energy content could affect a waste-to-energy facility.

Example 9-4 Estimate the change in the energy content of municipal solid waste for various levels of recycling. Determine the energy content of the typical municipal solid waste given in Table 3-4 for the following levels of recycling. Also determine the overall recycle percentage, by weight, represented by each level of recycling.

	Level of recycling,[a] %		
Component	**One**	**Two**	**Three**
Organic			
Food wastes	0	0	0
Paper	20	35	50
Cardboard	20	30	40
Plastics	20	30	40
Textiles	10	20	30
Rubber	10	20	30
Leather	10	20	30
Yard wastes	0	15	30
Wood	10	20	30
Inorganic			
Glass	20	30	40
Tin cans	10	20	30
Aluminum	50	70	90
Other metal	10	20	30
Dirt, ash, etc.	0	0	0

[a] The levels of recycling are based on the total amount of material in each category.

Solution

1. Set up a computation table to determine the weight and percentage distribution of the waste remaining after various levels of recycling have been achieved. (See the computation table at the top of page 319.)
2. Set up a computation table to determine the energy content of 100 lb of the waste remaining after various levels of recycling have been achieved. The Btu values are from Table 4-5. (See the computation table at the bottom of page 319.)

Comment. As shown in the above computation, the level of recycling will effect the energy content of the waste. For example, if a contract had been signed to deliver a firm amount of power from a waste-to-energy facility, additional amounts of waste would be needed to make up for the loss of Btu content. Without additional sources of waste the facility could easily go into default on its energy contract. Although the recycling percentages may change, the general approach developed in this example can be used to assess the impacts of alternative recycling strategies.

Component	Weight, lb (percent by weight)							
	Level of recycling							
	None		One		Two		Three	
Organic								
Food wastes	9.0	(9.0)	9.0	(10.3)	9.0	(11.9)	9.0	(13.9)
Paper	34.0	(34.0)	27.2	(31.1)	22.1	(29.1)	17.0	(26.3)
Cardboard	6.0	(6.0)	4.8	(5.5)	4.2	(5.5)	3.6	(5.6)
Plastics	7.0	(7.0)	5.6	(6.4)	4.9	(6.4)	4.2	(6.5)
Textiles	2.0	(2.0)	1.8	(2.1)	1.6	(2.1)	1.4	(2.2)
Rubber	0.5	(0.5)	0.5	(0.6)	0.4	(0.5)	0.4	(0.6)
Leather	0.5	(0.5)	0.5	(0.6)	0.4	(0.5)	0.4	(0.6)
Yard wastes	18.5	(18.5)	18.5	(21.1)	15.7	(20.7)	13.0	(20.1)
Wood	2.0	(2.0)	1.8	(2.0)	1.6	(2.1)	1.4	(2.2)
Inorganic								
Glass	8.0	(8.0)	6.4	(7.3)	5.6	(7.4)	4.8	(7.4)
Tin cans	6.0	(6.0)	5.4	(6.2)	4.8	(6.3)	4.2	(6.5)
Aluminum	0.5	(0.5)	0.3	(0.3)	0.2	(0.3)	0.1	(0.2)
Other metal	3.0	(3.0)	2.7	(3.1)	2.4	(3.2)	2.1	(3.3)
Dirt, ash, etc.	3.0	(3.0)	3.0	(3.4)	3.0	(4.0)	3.0	(4.6)
Total	100.0	(100.0)	87.5	(100.0)	75.9	(100.0)	64.6	(100.0)
Amount of waste recycled, % by weight			12.5		24.1		35.4	

Component	Total energy content, Btu			
	Level of recycling, %			
	None	One	Two	Three
Organic				
Food wastes	18,000	18,000	18,000	18,000
Paper	244,800	195,840	159,120	122,400
Cardboard	42,000	33,600	29,400	25,200
Plastics	98,000	78,400	68,600	58,800
Textiles	15,000	13,500	12,000	10,500
Rubber	5,000	5,000	4,000	4,000
Leather	3,750	3,750	3,000	3,000
Yard wastes	51,800	51,800	43,960	36,400
Wood	16,000	14,400	12,800	11,200
Inorganic				
Glass	480	380	340	290
Tin cans	1,800	1,620	1,440	1,260
Aluminum	—	—	—	—
Other metal	900	810	720	630
Dirt, ash, etc.	9,000	9,000	9,000	9,000
Total	506,530	426,100	362,380	300,680
Energy content, Btu/lb	5,065	4,870	4,774	4,654

9-10 SELECTION OF PROPER MIX OF TECHNOLOGIES

As the types and quantities of wastes that are to be diverted change in the future, it is imperative that the proper type and mix of technologies be adopted in a waste management system. Until the commodity markets for recyclable materials become stabilized, it will be prudent not to commit to capital-intensive equipment and processes with long-term contracts. Selection of the proper mix of technologies is considered in greater detail in Chapter 19.

9-11 DISCUSSION TOPICS AND PROBLEMS

9-1. Drive, walk, or pedal around your community and identify the different available types of facilities to which homeowners and others can deliver source-separated materials (e.g., igloo containers, drop-off centers, buy-back centers, redemption centers, MRFs, etc.). Are the facilities located conveniently?

9-2. Collect some aluminum cans and deliver them to a buy-back center. What is your assessment of the operation of the buy-back center? Can you suggest some improvements to enhance the operation of the facility?

9-3. What fraction of the total amount of material that is now recovered for recycling from MSW in your community is recovered at drop-off and buy-back centers versus curbside programs and materials recovery facilities?

9-4. Based on your own experience, discuss the advantages and disadvantages of using see-through plastic bags for the collection of source-separated wastes from individual residences. Do you think the use of see-through bags is feasible in your community?

9-5. Discuss the advantages and disadvantages of collecting yard wastes separately. If your community does not collect yard wastes separately, is the separate collection of yard wastes feasible given the waste management system that is now used? If your community collects yard wastes separately, develop the process flow diagram for the handling and processing of the yard wastes.

9-6. Develop a process flow diagram to separate cardboard, newsprint, high-grade office and computer paper, magazines, and contaminants such as carbon paper and FAX paper and wire-bound notebooks from mixed paper that is to be collected separately from residential and commercial sources. Prepare a listing of the equipment that will be required. Assume 50 ton/d of mixed paper will be processed 5 d/wk.

9-7. Develop a process flow diagram to process source-separated mixed paper and mixed recyclables composed of plastics, glass, tin cans, and aluminum cans.

9-8. Develop a process flow diagram to separate cardboard, paper, plastics, glass by color, and ferrous metals from commingled waste that has the composition given in Table 3-4. Your flow diagram should incorporate the use of a trommel screen and one or more magnetic separators.

9-9. Using the process flow diagram developed in Problem 9-7, prepare a layout of a materials recovery facility.

9-10. Using the process flow diagram developed in Problem 9-8, prepare a layout of a materials recovery facility.

9-11. Prepare a layout of a materials recovery facility to be used for the processing of waste materials that have been recovered from MSW at the source of generation. Assume that mixed paper and cardboard, glass and plastics, and aluminum and tin cans are to be collected separately. Assume the separated paper, cardboard, aluminum cans, tin cans, and plastics are to be baled for shipment to markets.

9-12. A small salvage firm disposes of car batteries by transporting the batteries to a smelter. Being quite safety conscious, the owners of the salvage company remove the electrolyte from the batteries to reduce the possibility of employee injury and of accidental spills due to improper handling. The electrolyte is diluted with large amounts of water and discharged to the sewer. Are there any problems associated with the removal and discharge of the electrolyte?

9-13. The maximum amount of solid wastes collected per day for one week is presented below. All the solid wastes are to be burned at a municipal waste-to-energy combustion facility at a constant rate of 100 tons/d. What is the required capacity of the storage pit that should be designed to accommodate 1.15 times the required capacity?

Day	Solid wastes collected, tons/d
Monday	150
Tuesday	130
Wednesday	120
Thursday	120
Friday	100
Saturday	80
Sunday	0

9-14. Arrange in order of importance the following waste characteristics for a mass-burn and an RDF combustor.

(*a*) Paper content

(*b*) Fixed carbon

(*c*) As-delivered energy content

(*d*) Moisture content

(*e*) As-delivered specific weight

(*f*) As-delivered metal content

9-15. Because it will be practically impossible to prevent all hazardous materials from entering a MSW mass-fired combustion facility, what precautions should be taken to minimize the danger?

9-16. Estimate the composition of the residue if packaging material wastes with the component distribution reported in Table 3-4 are to be combusted. What would be the corresponding volume reduction?

9-17. The local waste management agency has proposed to set up a waste combustion facility next to the existing landfill to maximize the life span of the landfill. Given the following information, determine how much the life span of the landfill is increased by the combustion.

(*a*) Composition of MSW as given in column 1, Table 3-4

(*b*) Estimated landfill capacity remaining = 300,000 yd^3

(*c*) Capacity of combustion facility = 50 ton waste/h

(*d*) Effective on-line combustion time per day = 22 h

(*e*) Initial specific weight of waste = 287 lb/yd^3

(*f*) Final specific weight of waste (MSW and ash) in landfill = 1200 lb/yd^3

9-18. How much land area would be required per ton to compost 100 ton/d of processed (sorted and shredded) residential and commercial MSW subject to the following conditions? How does your value for the area required per ton compare with values given in the literature? Cite at least two literature references.

(*a*) Windrow composting with mechanical turning is to be used.

(*b*) Maximum width = 18 ft

(*c*) Maximum length of windrows = 500 ft

(*d*) Average distance between windrows = 8 ft

(*e*) Angle of repose of waste in windrows = 45° (1 to 1 slope)

(*f*) Specific weight of the processed MSW as placed in windrows before water is added = 550 lb/yd^3

(*g*) Active composting period = 1 mo

(*h*) Curing period = 4 mo

(*i*) A staging/storage area equal to 15 percent of the area used for composting will be required.

9-19. Solve Problem 9-18, but assume

(*a*) Active composting period = 21 d

(*b*) Curing period = 3 mo

9-20. Using the distribution of waste given in column 1 of Table 3-4, estimate the tons of compost that can be produced from 1000 tons of MSW. Assume that all of the inorganic materials and that 50 percent of the yard wastes and wood will be removed. Assume also that the initial moisture content of the MSW is 20 percent and that the final moisture content of the compost is about 35 percent. Cite all of the assumptions used in solving this problem.

9-21. When residential and commercial MSW from a community of 250,000 persons arrives at a composting facility, the moisture content is usually below the desired range of 45 to 60 percent for optimum composting. Rather than adding city water to obtain the necessary moisture content, it has been suggested that wastewater treatment plant sludge be added to achieve the same result. Determine the required moisture content of the sludge to achieve the desired moisture content of 55 percent for the combined mixture and determine whether the sludge will have adequate moisture. Assume the following conditions apply:

Wastewater treatment plant sludge = 0.25 lb capita · d (dry sludge)

Residential and commercial MSW = 5.5 lb capita · d (at 20% moisture)

9-22. Using the data from Problem 9-21, determine the amount of sludge with a solids content of 6 percent that must be added to the MSW to achieve a final moisture content of 58 percent.

9-23. Estimate the total theoretical amount of air that would be required under aerobic conditions to oxidize completely an organic waste with a chemical formula of $C_{120}H_{180}O_{80}N_2$.

9-24. Using the data and information from Example 9-4, what would you suggest as a compromise that would allow a recycling program to be functional and at the same time allow the waste-to-energy facility to remain viable? Assume the mimimum energy content of the waste has to be 4000 Btu/lb for the waste-to-energy facility to remain viable.

9-12 REFERENCES

1. Dallavale, J. M.: *The Industrial Environment and Its Control*, Pitman, NY, 1958.
2. Endahl, R. B.: *Solid Waste Processing: A State-of-the-Art Report on Unit Operations*, U.S. Department of Health, Education, and Welfare, Bureau of Solid Waste Management, Publication SW-4c, Washington, DC, 1969.
3. Gotaas, H. B.: *Composting—Sanitary Disposal and Reclamation of Solid Wastes*, World Health Organization, Geneva, 1956.
4. Haug, R. T.: *Compost Engineering: Principles and Practices*, Ann Arbor Science Publishers Inc., Ann Arbor, MI, 1980.
5. Hasselriis, F.: *Refuse Derived Fuel*, an Ann Arbor Science book, Butterworth Publishers, Boston, 1984.
6. Henderson, S. M. and R. L. Perry: *Agricultural Process Engineering*, 3rd ed., AVI, Westport, CT, 1976.
7. Hering, R. and S. A. Greeley: *Collection and Disposal of Municipal Refuse*, 1st ed., McGraw-Hill, New York, 1921.
8. Parsons, H. de B.: *The Disposal of Municipal Refuse*, 1st ed., John Wiley & Sons, New York, 1906.
9. Robinson, W. D. (ed.): *The Solid Waste Handbook*, A Wiley-Interscience Publication, John Wiley & Sons, New York, 1986.
10. Savage, G. M. and L. F. Diaz: "Key Issues Concerning Waste Processing Design," *Proceedings of the 1986 National Waste Processing Conference*, American Society of Mechanical Engineers, New York, 1986.
11. U.S. Environmental Protection Agency: *Decision-Makers Guide to Solid Waste Management*, EPA/530-SW-89-072, Washington, DC, November 1989.
12. Vesilind, P. A. and A. E. Rimer: *Unit Operations in Resource Recovery Engineering*, Prentice-Hall, Englewood Cliffs, NJ, 1981.
13. Wiles, C. C.: "Composting of Refuse," in *Composting of Municipal Residues and Sludges*, Information Transfer, Inc., Rockville, MD, 1978.
14. Wilson, D. G. (ed.): *Handbook of Solid Waste Management*, Van Nostrand Reinhold, New York, 1977.
15. Zandi, I: "Pipeline Transport of Solid Wastes," in D. G. Wilson (ed.): *Handbook of Solid Waste Management*, Van Nostrand Reinhold, New York, 1977.

CHAPTER 10

TRANSFER AND TRANSPORT

In the field of solid waste management, the functional element of transfer and transport refers to the means, facilities, and appurtenances used to effect the transfer of wastes from one location to another, usually more distant, location. Typically, the contents of relatively small collection vehicles are transferred to larger vehicles that are used to transport the waste over extended distances either to MRFs or to disposal sites. Transfer and transport operations are also used in conjunction with MRFs to transport recovered materials to markets or waste-to-energy facilities and to transport residual materials to landfills.

10-1 THE NEED FOR TRANSFER OPERATIONS

Transfer and transport operations become a necessity when haul distances to available processing centers or disposal sites increase so that direct hauling is no longer economically feasible [5]. They also become a necessity when processing centers or disposal sites are sited in remote locations and cannot be reached directly by highway. Transfer operations are an integral part of all types of MRFs. Transfer stations are also an integral part of large integrated MR/TFs. For reasons of public safety, the use of a small transfer station, for individuals hauling wastes in automobiles and pickups and other noncommercial haulers, at landfills is gaining in popularity.

Transfer operations can be used successfully with all types of collection vehicles and conveyor systems. Additional factors that tend to make the use of transfer operations attractive include (1) the occurrence of illegal dumping due to excessive haul distances, (2) the location of disposal sites relatively far from collection routes (typically more than 10 mi), (3) the use of small-capacity collection vehicles (generally under 20 yd^3), (4) the existence of low-density residential service areas, (5) the use of a hauled container system with relatively small containers for the collection of wastes from commercial sources, and (6) the use of hydraulic or pneumatic collection systems.

Excessive Haul Distances

In the early days when horse-drawn carts were used for the collection of solid wastes, it was common practice to empty the contents of the loaded carts into some auxiliary vehicle for transport to some intermediate point for processing or to the disposal site. However, with the advent of the modern motor truck and the availability of low-cost fuel, transfer operations in most cities were abandoned and direct hauling was adopted. Today, with rising labor, operating, and fuel costs and the absence of nearby solid waste disposal sites the trend is reversing, and transfer stations are again becoming common. For example, wastes from the city of Portland, OR, are hauled to a disposal site 150 mi away.

Usually, the decision to use a transfer operation is based on economics. For example, in Examples 8-2 and 8-5 the time and economic advantages of the stationary container system over the hauled container system were demonstrated clearly. Simply stated, it is cheaper to haul a large volume of wastes in large increments over a long distance than it is to haul a large volume of wastes in small increments over a long distance. The economic advantage of a transfer operation is illustrated in Example 10-1.

Example 10-1 Economic comparison of transport alternatives. Determine, based on operating costs, the break-even points for a hauled and a stationary container system as compared with a system using transfer and transport operations for transporting wastes collected from a metropolitan area to a landfill disposal site. Assume that the following cost data are applicable:

1. Operating costs
   (*a*) Haul container system using a hoist truck with an 8-yd^3 container = \$25/h
   (*b*) Stationary container system using a 20-yd^3 compactor = \$40/h
   (*c*) Tractor-semitrailer transport unit with a capacity of 105 yd^3 = \$40/h
   (*d*) Transfer station operation cost = \$2.75/yd^3

Solution

1. Convert the haul cost data to units of dollars per cubic yard per minute (see comment at end of this example).
   (*a*) Hoist truck = \$0.052/$yd^3 \cdot$ min
   (*b*) Compactor = \$0.033/$yd^3 \cdot$ min
   (*c*) Transfer station transport equipment = \$0.0063/$yd^3 \cdot$ min

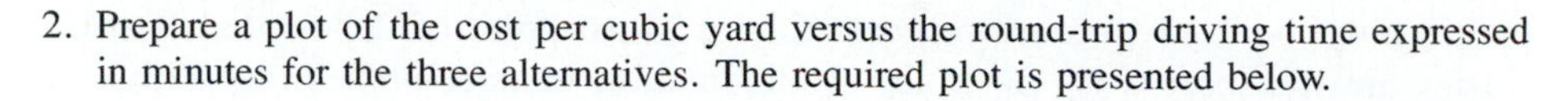

2. Prepare a plot of the cost per cubic yard versus the round-trip driving time expressed in minutes for the three alternatives. The required plot is presented below.

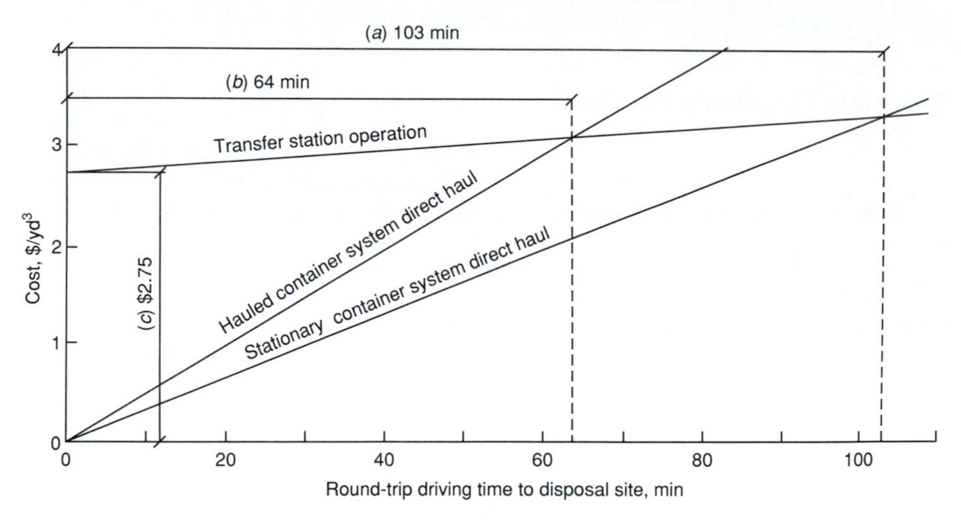

(*a*) Break-even time for stationary container system
(*b*) Break-even time for hauled container system
(*c*) Transfer station operating cost

3. Determine the break-even times for the hauled and stationary container systems using the plot prepared in Step 2.
   (*a*) Hauled container system = 64 min
   (*b*) Stationary container system = 103 min

   Thus, for example, if a stationary container system is used and the round-trip driving time to the disposal site is more than 103 min, the use of a transfer station should be investigated.

Comment. In most cases, articles, and reference books dealing with the long-distance hauling of solid wastes, cost data are expressed in terms of dollars per ton per minute or dollars per ton per mile. This practice is widely accepted for transfer station analysis because weight is the most critical measure for efficient highway or rail movement. Such cost data can be misleading, however, when the densities of solid wastes vary significantly from location to location or container to container. For example, if the density of the wastes in two hoist-truck containers varies by a factor of three, then comparing the costs of hauling two containers of the same size on a per-ton basis would tend to be misleading because the actual cost is the same for both. On the other hand, a comparison based on dollars per cubic yard per minute or dollars per minute would be valuable in comparing the two operations.

Remote Processing Facilities or Disposal Sites

Transfer operations must be used when the processing facilities or disposal sites are in such a remote location that conventional highway transportation alone is not feasible. For example, transfer stations are required when rail cars or ocean-going barges must be used to transport wastes to the final point of deposition. If solid wastes are transported by pipeline, a combination transfer-processing station is usually necessary. These subjects are considered further in Section 10-2.

Materials Recovery Facilities

In the materials recovery flow diagrams in Chapter 9, it is clear that the transfer of waste components is an integral part of the operation of a MRF. Because much of the material has been removed from the waste stream, the transfer facilities tend to be smaller.

Materials Recovery/Transfer Facilities

A recent trend in the waste management field is the development of large integrated MR/TFs. Integrated MR/TFs are multipurpose facilities that may include the functions of drop-off center, separation, composting, bioconversion processes, production of refuse-derived fuel, and transport. The use of large integrated MR/TFs is attractive because of the cost savings that are possible by combining several waste management activities in a single facility.

Convenience Transfer Station at Landfill

Because of safety concerns and the many new restrictions governing the operation of landfills, many landfill operators have constructed convenience transfer stations at the landfill site for the unloading of wastes brought to the site by individuals and small-quantity haulers. By diverting private individuals and small-quantity haulers to a separate transfer facility, the potential for accidents at the working face of the landfill is reduced significantly.

10-2 TYPES OF TRANSFER STATIONS

Transfer stations are used to accomplish transfer of solid wastes from collection and other small vehicles to larger transport equipment. Depending on the method used to load the transport vehicles, transfer stations may be classified into three general types: (1) direct-load, (2) storage-load, and (3) combined direct-load and discharge-load (see Fig. 10-1). Transfer stations may be classified with respect to throughput capacity (the amount of material that can be transferred and hauled) as follows: small, less than 100 ton/d; medium, between 100 and 500 ton/d; and large, more than 500 ton/d.

Direct-Load Transfer Stations

At direct-load transfer stations, the wastes in the collection vehicles are emptied directly into the vehicle to be used to transport them to a place of final disposition or into facilities to compact the wastes into transport vehicles or into waste bales

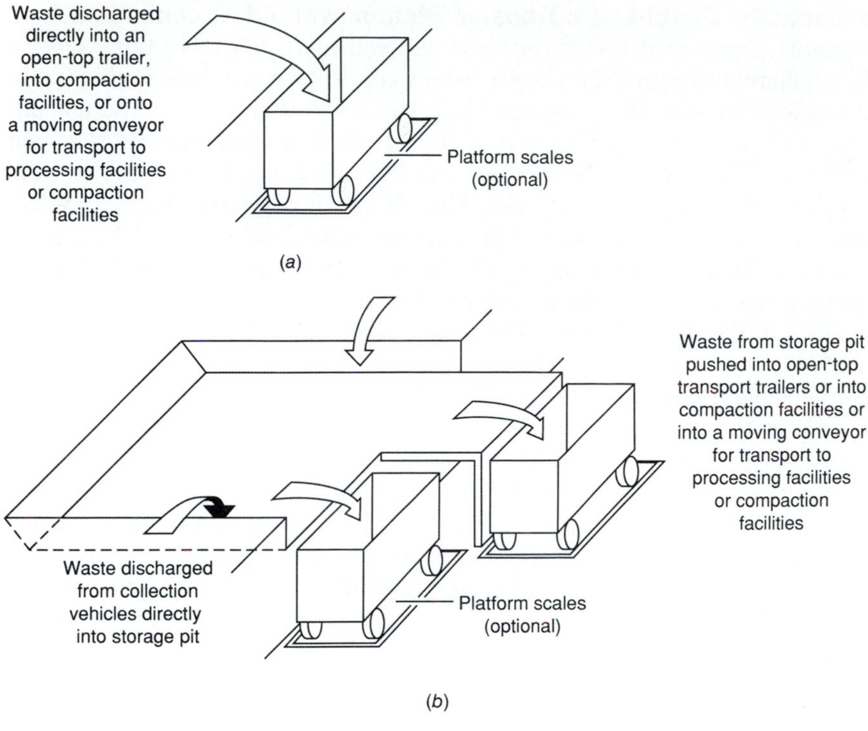

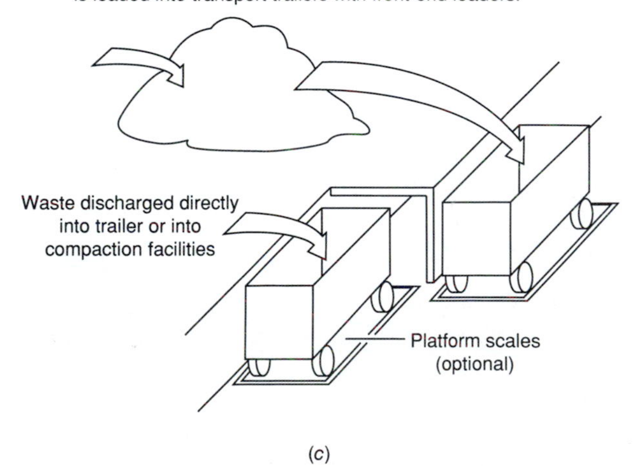

FIGURE 10-1
Definition sketch for the types of transfer stations: (*a*) direct-load, (*b*) storage-load, and (*c*) combined direct-load and discharge-load.

that are transported to the disposal site (see Fig. 10-1*a*). In some cases, the wastes may be emptied onto an unloading platform and then pushed into the transfer vehicles, after recyclable materials have been removed. The volume of waste that can be stored temporarily on the unloading platform is often defined as the *surge capacity* or the *emergency storage capacity* of the station.

Large Capacity Direct-Load Transfer Station without Compaction. In a large-capacity direct-load transfer station, the wastes in the collection vehicles usually are emptied directly into the transport vehicle. To accomplish this, these transfer stations usually are constructed in a two-level arrangement. The unloading dock or platform from which wastes from collection vehicles are discharged into the transport trailers can be elevated (see Fig. 10-2), or the transport trailers can be located in a depressed ramp (see Fig. 10-3). Photographs of a facility like that shown in Fig. 10-3 and some of the equipment used are shown in Fig. 10-4. In some direct-load transfer stations, the contents of the collection vehicles can be emptied temporarily onto the unloading platform if the trailers are filled or are being hauled to the disposal sites. The wastes are then pushed into the transport trailers.

The operation of the direct-load transfer station shown in Fig. 10-3 may be summarized as follows. Upon arrival at the transfer station, all vehicles hauling wastes are weighed by the weighmaster, who then indicates where the wastes should be unloaded by giving the driver an appropriate stall number. After the collection vehicles have been unloaded, they are reweighed and the disposal fee is determined. Commercial vehicles that regularly use the transfer station are issued credit cards showing the firm name and the truck tare weight, thereby eliminating the second weighing for these vehicles. As the trailers become loaded, the wastes in the trailer are shifted and compacted with a clamshell mounted on a rubber-tired tractor (see Fig. 10-4*b*). When the trailers are full or the maximum allowable tonnage has been placed in them, as indicated by the weighmaster, they are removed and prepared for the haul operation. Trailer volume and weight are the variables that must be checked by the operator before sending out loaded trailers.

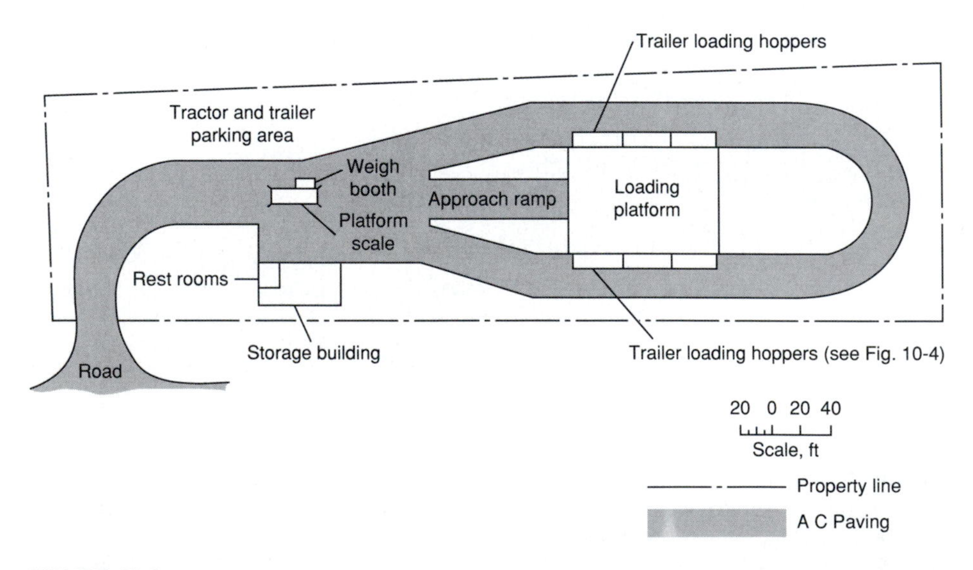

FIGURE 10-2
Typical direct-load transfer station with elevated unloading platform.

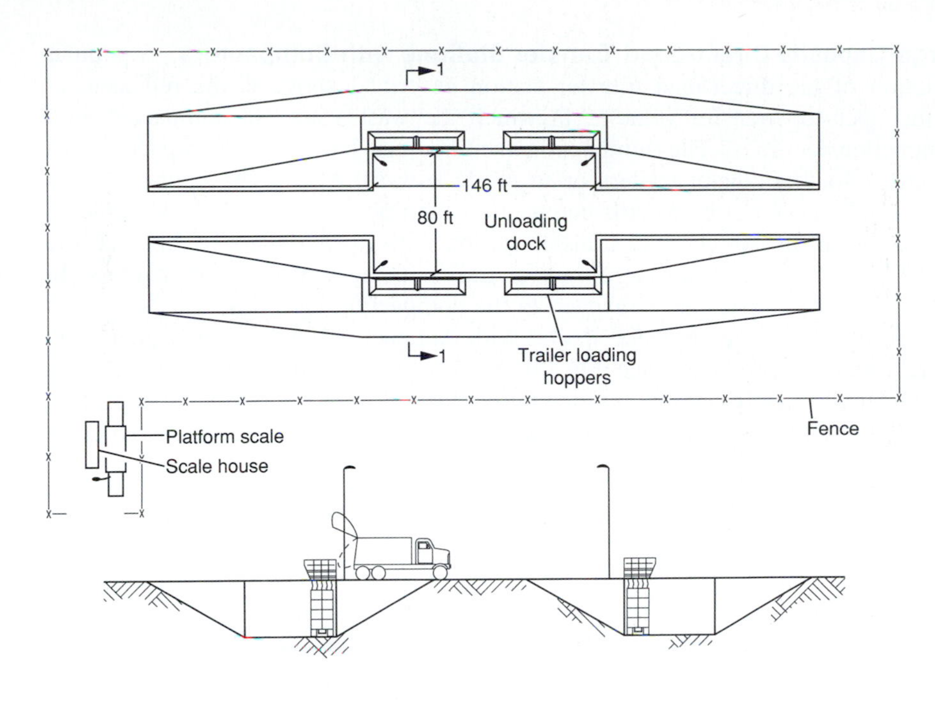

FIGURE 10-3
Typical direct-load transfer station with transport trailers located in depressed ramps. Note the heavy chain sections hung from bottom of hopper are used to direct wastes into transfer trailers.

(*a*) (*b*)

FIGURE 10-4
Facilities and equipment used at transfer station shown in Fig. 10-3: (*a*) end view showing trailers positioned under loading hopper in depressed ramp and (*b*) clamshell mounted on rubber-tired tractor on ground-level unloading platform is used to distribute and compact wastes in trailers and to pick up wastes spilled on unloading platform.

Large-Capacity Direct-Load Transfer Stations with Compactors. A popular variation of the direct-load transfer station described above is the replacement of the open-top transfer vehicles in which the wastes are not compacted with compaction facilities. The compaction facilities can be used to compact wastes directly into the transfer trailers or to produce waste bales. The operation of a direct-load transfer station with compaction facilities is essentially the same as the operation of a direct-load transfer station with open trailers except that the wastes are compacted into large transfer trailers using stationary compactors. In some cases, the wastes are conveyed to the compaction facilities.

In the direct-load transfer station with compaction facilities in which large waste bales are produced (see Fig. 10-5), wastes from the collection vehicles are unloaded directly onto the unloading platform or directly into the compaction pit hopper. After recyclable materials have been removed, a rubber-tired vehicle is

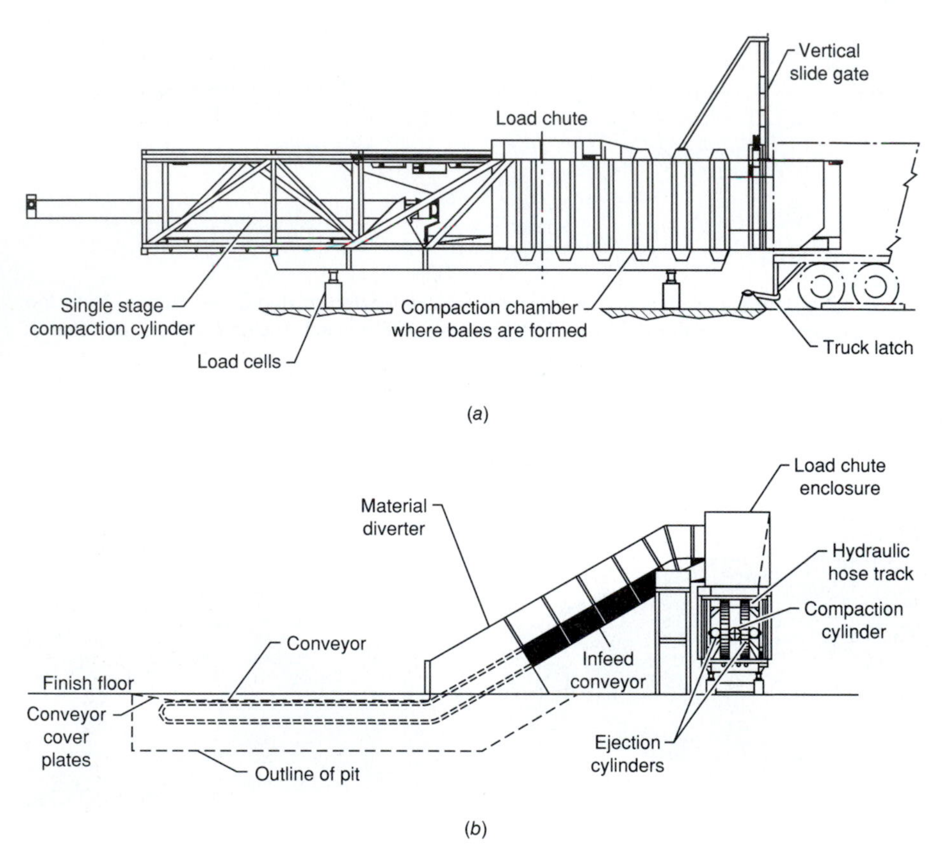

FIGURE 10-5
Compaction facilities used to produce waste bales at transfer stations: (*a*) side view of compactor with open-top direct-load chute (once a bale has been formed by compacting waste into the compaction chamber, the vertical slide gate is lifted and the bale is pushed into the transport trailer or semitrailer) and (*b*) end view of large-capacity baler equipped with enclosed load chute. Wastes are loaded into the baler with a continuous feed conveyor. (Courtesy of SSI Shredding Systems, Inc.)

used to push the wastes discharged on the unloading platform into the compactor. The compressed waste bale is loaded onto the semitrailer for transport to the disposal site. By producing a bale which, after partial expansion, is smaller than the inside dimensions of the leak-proof semitrailer transport vehicle, the cost of the transfer can be minimized.

Medium- and Small-Capacity Direct-Load Transfer Stations with Compactors. A typical medium-capacity direct-load transfer station with compaction facilities is shown in Fig. 10-6. Operationally, after the trucks are weighed, they enter the transfer station where they are directed to an unloading location. The

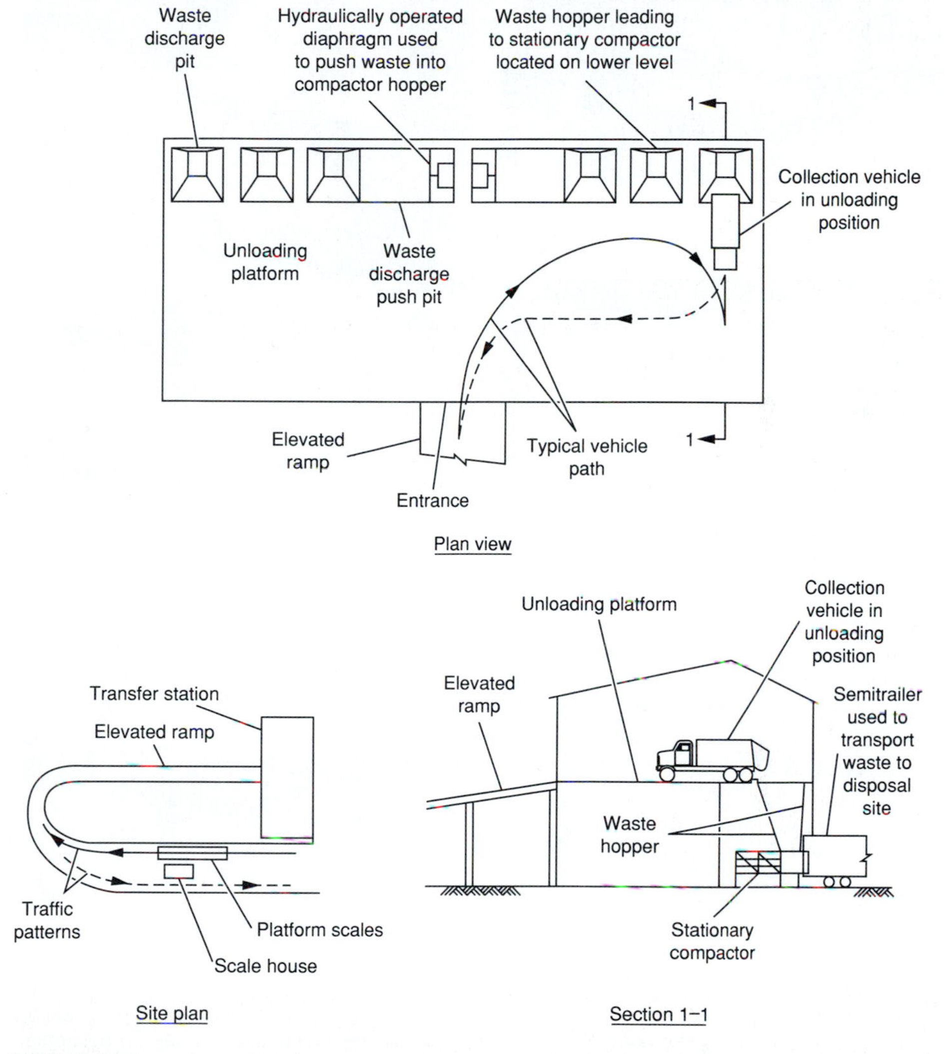

FIGURE 10-6
Enclosed medium-capacity direct-load transfer station equipped with stationary compactors.

unloading location may be one of the individual hoppers leading to a compactor or one of the rectangular waste receiving pits. Each pit is equipped with a hydraulically powered diaphram that is used to push the accumulated waste to the compactor hopper located at the opposite end of the pit. If there are no semitrailers to load, wastes are discharged temporarily on the unloading platform, from where they are loaded into the compactor hoppers with a rubber-tired front-end loader. Views of the transfer station of Fig. 10-6 are shown in Fig. 10-7.

FIGURE 10-7
View of transfer station shown in Fig. 10-6: (*a*) collection vehicle being weighed at entrance to transfer station, (*b*) unloading contents of collection vehicle into compactor hopper, (*c*) horizontal compactor used to compress wastes into transport trailers, and (*d*) transfer trailer being backed up to horizontal compactor, Bogota, Colombia.

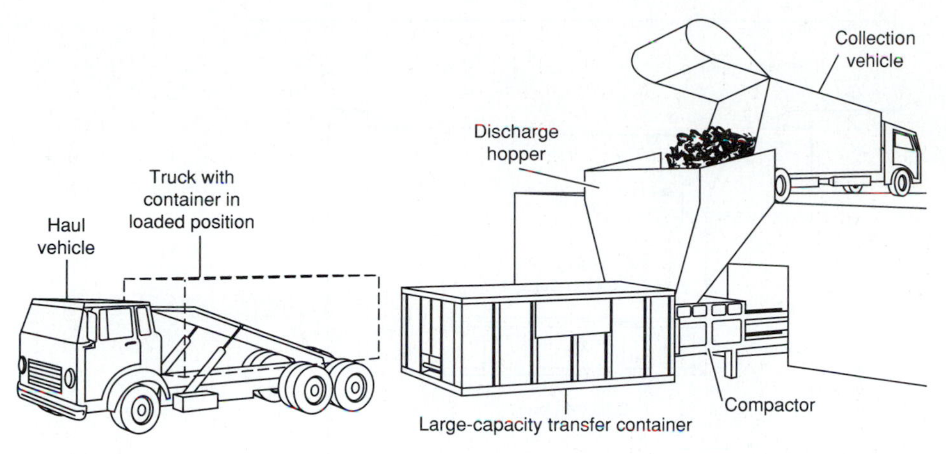

FIGURE 10-8
Small-capacity direct-load a transfer station equipped with a stationary compactor. (Adapted from Schindler Waggon AG, Prattein.)

A small-capacity direct-load transfer station with compaction facilities is shown in Fig. 10-8. As shown, a large container is used with this type of transfer station as opposed to a transfer trailer. The container is hauled to the disposal site using a tilt-frame vehicle (see Fig. 8-10). Depending on the length of time required to haul the loaded container to the disposal site and to return, an empty container may be attached to the compactor before the full container is hauled to the disposal site.

Small-Capacity Direct-Load Transfer Stations Used in Rural Areas. Used in rural and recreational areas, small-capacity direct-load transfer stations like those shown in Figs. 10-9 and 10-10 are designed so that the loaded containers are emptied into a collection vehicle for transport to the disposal site. In the design and layout of such stations, which are usually unattended, the key consideration should be simplicity. Complex mechanical systems are not suitable in such locations. The number of containers used depends on the area served and the collection frequency that can be provided. To facilitate unloading, the tops of the containers may be set about 3 ft above the top of the unloading-area platform (see Fig. 10-9). Alternatively, the tops of the containers may be set level with the unloading area (see Fig. 10-10), and the area behind the containers can be excavated to provide space for maneuvering the collection vehicles when the contents of the containers are emptied.

Small-Capacity Direct-Load Transfer Station Used at Landfill Disposal Site. The transfer station shown in Fig. 10-11 is of the type used at landfill disposal sites for individuals and small-quantity haulers. The transfer facilities are also used for the recovery of recyclable materials. After any recyclable items are dropped off, waste materials are emptied into two large transfer trailers each of which is hauled to the disposal site, emptied, and returned to the transfer station.

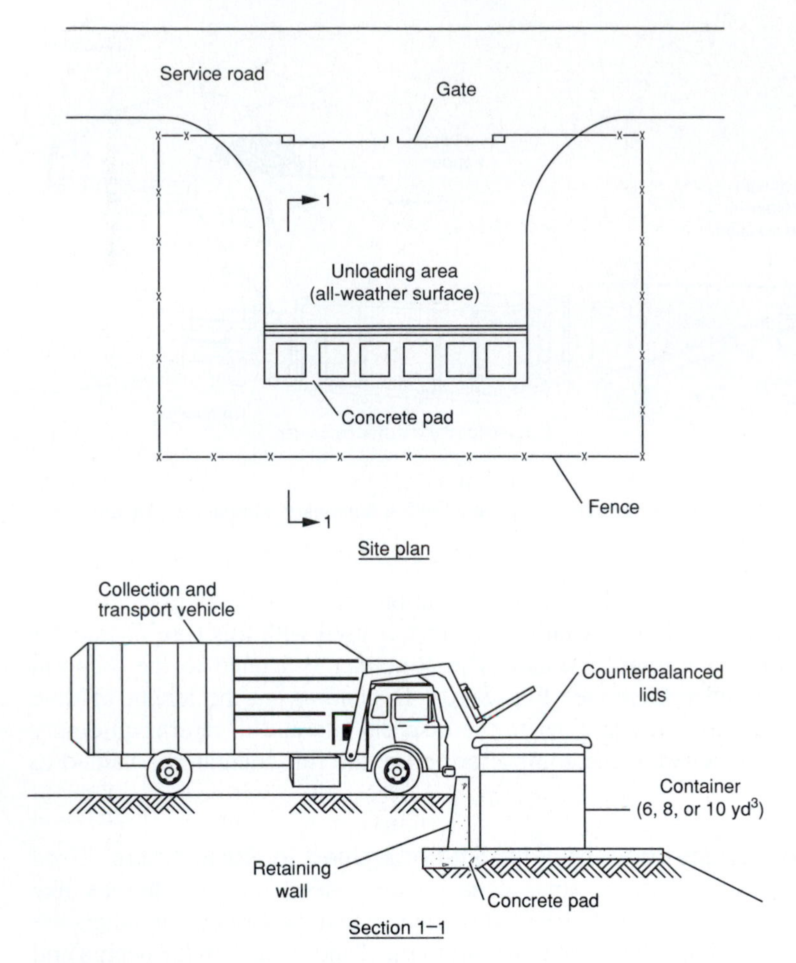

FIGURE 10-9
Small-capacity direct-load transfer station for rural or recreational areas.

Storage-Load Transfer Station

In the storage-load transfer station, wastes are emptied directly into a storage pit from which they are loaded into transport vehicles by various types of auxiliary equipment (see Fig. 10-1*b*). The difference between a direct-load and a storage-load transfer station is that the latter is designed with a capacity to store waste (typically 1–3 days).

Large-Capacity Storage-Load Transfer Station without Compaction. Perhaps the best known example of the storage-load type of transfer station is the San Francisco facility, shown schematically in Fig. 10-12 and pictorially in Fig. 10-13. In this station, all incoming collection trucks are routed to a computerized weigh station for weighing. In addition, the weighmaster records the name of

FIGURE 10-10
Small direct-load rural public convenience transfer stations. In transfer station shown on left, the open top containers are placed against retaining wall at same level as unloading platform for ease of loading.

(*a*)

(*b*)

FIGURE 10-11
Convenience transfer station located at landfill disposal site. Transfer station is used by individuals hauling wastes in automobiles and pickup trucks and other noncommercial haulers: (*a*) white goods and other recyclable materials are unloaded first and (*b*) other wastes are placed in large open-top containers or trailers for transport to the landfill.

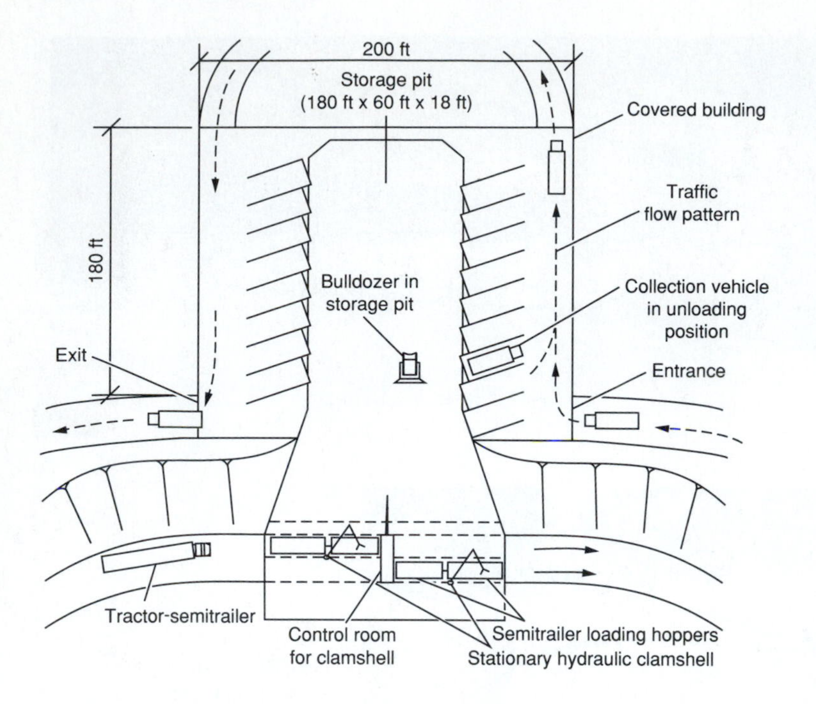

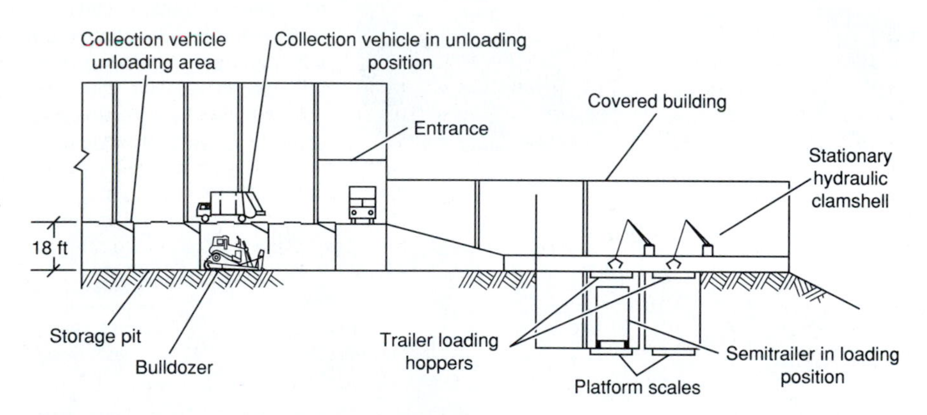

FIGURE 10-12
Enclosed large-capacity (2000 ton/d) storage-load transfer station, San Francisco, CA.

the unloading company, the identification of the particular truck, and the time it entered. Then the weighmaster directs the driver to either the east or the west side of the main entrance of the enclosed transfer station. Once inside, the driver backs up the collection vehicle at a 50° angle to the edge of a depressed central waste storage pit. The contents of the vehicle are emptied into the pit (see Fig. 10-13*a*), and the empty vehicle is driven out of the transfer station.

Within the pit, two bulldozers are used to break up the wastes and to push them into loading hoppers that are located at one end of the pit (see Fig. 10-13*b*).

(a) (b) (c) (d)

FIGURE 10-13
Operation details for storage-load transfer station shown in Fig. 10-12: (*a*) inside the transfer station, contents of collection vehicles are emptied into storage pit (two crawler tractors are used to break up wastes and push them to the hoppers used to load the semi-trailers), (*b*) solid wastes being pushed into loading hoppers where they fall by gravity into truck-trailer rig parked on platform scales located on lower level (a stationary articulated hydraulic clamshell is used to assist in the loading operation), (*c*) exterior view showing tractor-semitrailer positioned under loading hopper and tractor semitrailer waiting to be filled, and (*d*) screen being placed over loaded drop-bottom semitrailer to prevent the blowing of paper and other materials during the haul operation.

Two articulated bucket-type hoists, located on the other side of the hoppers, are used to remove any wastes that could damage the transport trailers. The wastes fall through the hoppers into trailers located on scales on a lower level (see Fig. 10-13*c*). When the allowable weight limit has been reached, the hoist operator signals the truck driver. The loaded trailers are then driven out of the loading area, and wire screens are placed over the open trailer tops to prevent any papers or other solid wastes from blowing away during transport.

Medium-Capacity Storage-Load Transfer Station with Processing and Compaction Facilities. In the transfer station shown in Fig. 10-14, the wastes are first discharged into a storage pit (also identified as a surge pit). From the storage pit, the wastes are pushed onto a conveyor system to be transported to the shredder. After shredding, ferous metal is removed and the wastes are compacted into transfer trailers for transport to the disposal site.

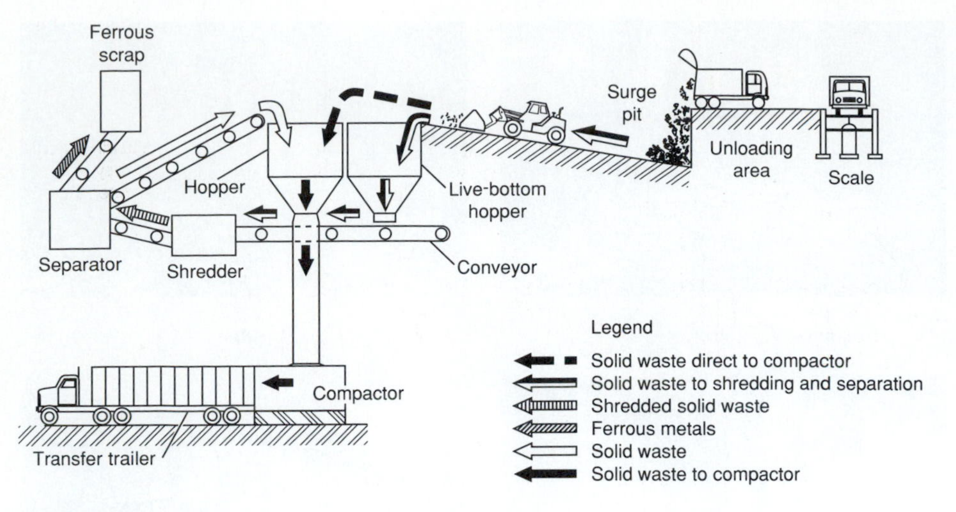

FIGURE 10-14
Storage-load transfer station with processing and compaction facilities. (Courtesy of Municipality of Metropolitan Toronto, Department of Public Works.)

Combined Direct-Load and Discharge-Load Transfer Station

In some transfer stations, both direct-load and discharge-load methods are used (see Fig. 10-1*c*). Usually these are multipurpose facilities that service a broader range of users than a single-purpose facility. A multipurpose transfer station can also house a materials recovery operation. The layout of a multipurpose transfer station, designed for use by the general public and by various waste collection agencies, is shown in Fig. 10-15.

The operation may be described as follows. All waste haulers (general public as well as commercial haulers) wishing to use the transfer station must check in at the scale house. Large commercial collection vehicles are weighed, and a commercial customer ticket is stamped and given to the vehicle driver. The driver then proceeds to the unloading platform and empties the contents of the collection vehicle directly into the transport trailer. After unloading the collection vehicle, the driver returns the vehicle to the scale house for reweighing and turns in her or his customer ticket. The weight of the empty vehicle is recorded while a discharge fee is calculated.

Individual residents as well as small independent noncommercial haulers haul significant quantities of yard wastes, tree trimmings, and bulky wastes (stoves, lawn mowers, refrigerators, etc.) to the transfer station. All automobiles pulling trailers and pickup trucks containing wastes must be checked in at the scale house. These vehicles are not weighed, but users do pay a discharge fee that is collected at the scale house by the attendant, who gives the user a cash receipt. The scale attendant visually checks the waste load to determine if it contains any recyclable materials. If it does, the attendant instructs the driver to deposit the

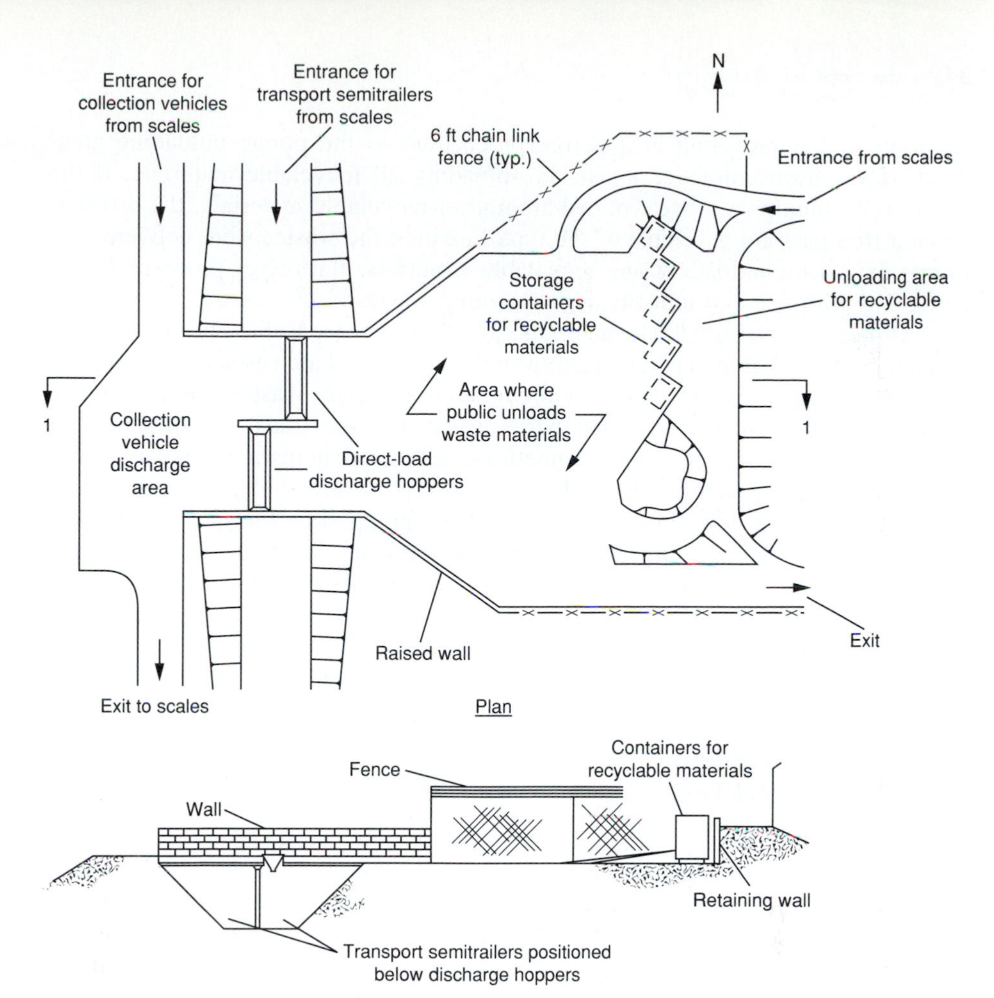

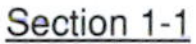

(*a*) (*b*)

FIGURE 10-15
Combination direct-load and discharge-load transfer station with materials recovery activities: (*a*) view of depressed ramp where transfer trailers are located and (*b*) view of individuals unloading wastes in the public unloading area.

materials at the recycling area before proceeding to the public unloading area. A transfer station employee assists in unloading all recyclable materials. If the waste load contains a predetermined amount of recyclable materials, the driver is given a free pass for a vehicle of the type in which the wastes were delivered for future use. After unloading any recyclable materials, the driver proceeds to the unloading platform and unloads any remaining wastes.

If there are no recyclable materials, the driver proceeds directly to the public unloading area. This area is separated from the direct-load area used by commercial vehicles, by the two 40-ft trailer-loading hopper openings. Wastes that accumulate in the unloading area are pushed periodically into the transfer trailer loading hoppers by a rubber-tired loader. Sometimes additional items are recovered from the unloading area used by the public.

Caution must be used in selecting and designing such transfer stations, for the cost of adding multipurpose facilities is often not justified in terms of the benefits achieved. Station users should be separated to prevent interferences and accidents between the large collection trucks and the smaller private vehicles. The physical separation of the discharge areas usually is the only positive way to maintain system efficiency.

Transfer and Transport Operations at MRFs and MR/TFs

In general, the transfer operations at MRFs involve the loading of trailers with materials that have been separated, materials that have been separated and processed (e.g., baled paper, cardboard, and plastics), and residual materials for landfill disposal. Where open top trailers are used, the transfer operation would be classed as a direct-load. Where processed wastes such as bales are loaded onto transfer trailers, the transfer operation would be classified as storage-load.

FIGURE 10-16
Conveyor discharging residual waste materials remaining after sorting into storage-load transfer station at a MRF.

The transfer operations at an integrated MR/TF will involve both direct-load and storage-load transfer operations. Direct loading of trailers will occur as described above from the materials recovery portion of the facility. In addition, conveyors will be used to transport the residual material remaining after sorting to the transfer portion of the facility. Where a waste storage pit is used, residual wastes from the materials recovery portion of the facility will be discharged from conveyors directly into the storage pit (see Fig. 10-16). Collection vehicles are also able to discharge directly into the storage pit. Because of processing equipment failures, the capacity of the transfer facility in an integrated MR/TF should be adequate to handle the peak daily waste quantities. Two operating shifts may be required to handle the peak daily waste quantity.

10-3 TRANSPORT MEANS AND METHODS

Motor vehicles, railroads, and ocean-going vessels are the principal means now used to transport solid wastes. Pneumatic and hydraulic systems have also been used.

Motor Vehicle Transport

Where the point of final disposition can be reached by motor vehicles, the most common means used to transport solid wastes from transfer stations are trailers, semitrailers, and compactors. All types of vehicles can be used in conjunction with either type of transfer station. In general, vehicles used for hauling on highways should satisfy the following requirements: (1) wastes must be transported at minimum cost, (2) wastes must be covered during the haul operation, (3) vehicles must be designed for highway traffic, (4) vehicle capacity must be such that the allowable weight limits are not exceeded, and (5) methods used for unloading must be simple and dependable. The principal types of vehicles used in conjunction with transfer stations for the transport of wastes are identified in Fig. 10-17.

Transport Vehicles for Uncompacted Wastes

In recent years, because of their simplicity and dependability, open-top semitrailers have found wide acceptance for the hauling of uncompacted wastes from direct-load transfer stations (see Fig. 10-18*a* and *c*). The semitrailers are of monoque construction, in which the bed of the trailer also serves as the frame of the trailer. Using monoque construction allows greater waste volumes and weights to be hauled. A further development in transport trailers is the drop-bottom trailer, in which the bottom of the center portion of the trailer is lowered (see Figs. 10-17*b* and 10-18*c*) to obtain additional capacity without increasing the length of the trailer.

The maximum volume that can be hauled in highway transport vehicles depends on the regulations in force in the state in which they are operated. These regulations usually limit the outside dimensions of the vehicles or combinations

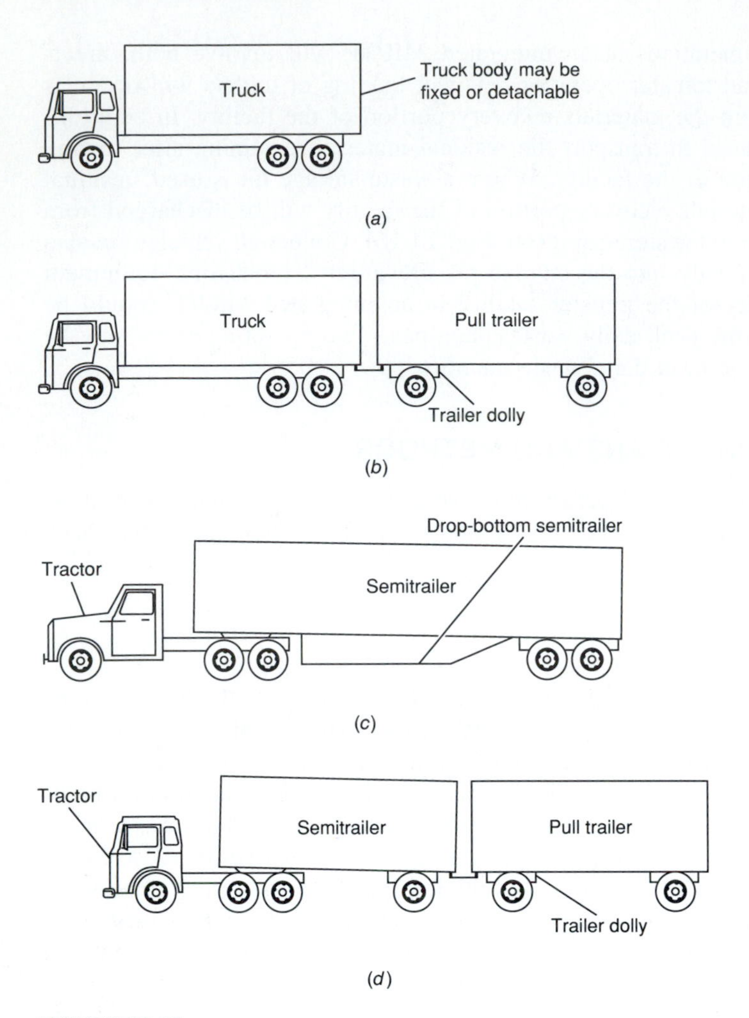

FIGURE 10-17
Definition sketch for the types of vehicles used in conjunction with transfer stations for the transport of wastes: (*a*) truck (also truck chassis with detachable body), (*b*) truck-trailer combination, (*c*) tractor-semitrailer combination (see Fig. 10-18*c*), and (*d*) tractor-semitrailer-pull trailer combination (often identified as a *set of doubles*).

of vehicles, as well as the weight per axle and the total weight. To maximize the payload, transport trailers are often designed with additional axles. Typical data on physical characteristics of the transport trailers are summarized in Table 10-1. Typical values for the amount of solid waste transported by various tractor-trailer, tractor-semitrailer, and tractor-semitrailer-pull trailer combinations are reported in Table 10-2.

Methods used to unload the transport trucks, trailers, semitrailers, and pull trailers may be classified as (1) self-emptying and (2) requiring the aid of auxiliary equipment. Self-emptying transport trucks and semitrailers have mechanisms such

(*a*)

(*b*)

(*c*)

FIGURE 10-18
Typical transport vehicles used in conjunction with transfer facilities: (*a*) 105-yd^3 open-top semitrailer with moving-floor unloading mechanism (see Figs. 10-19*b* and 10-20), (*b*) 85-yd^3 enclosed semitrailer used with stationary compactor (see Fig. 10-7*c*). Semitrailer is unloaded with movable internal diaphragm (see Fig. 10-19*a*), and (*c*) 100-yd^3 drop-bottom open-top semitrailer unloaded with hydraulic tipping ramp (see Fig. 10-21).

as hydraulic dump beds, powered internal diaphragms, and moving floors that are part of the vehicle (see Fig. 10-19). The use of powered internal diaphragms (using a hydraulic piston or moving cables) is the most common method of unloading trucks, semitrailers, and trailers. Moving floors are an adaptation of equipment used in the construction industry for unloading trailers that carry gravel and asphalt. The moving floor usually has two or more sections extending across the

TABLE 10-1
Data on haul vehicles used at large- and medium-capacity transfer stations

Station			Capacity per trailer		Dimensions for single trailer					
Location	Capacity tons/day	Type of trailer	yd^3	tons	Width, ft	Length, ft	Approx. height, empty, ft	Approx. length of tractor and trailer units, ft	Method used for covering wastes	Method used for unloading trailers
Dade Co., FL	4200	Semitrailer	85	20–25[a]	8	41	13	55–57[b]	Nylon-mesh hinged cover and enclosed	Internal diaphragm
Marin, CA	960	Semitrailer	105	25	8	45	13.5	60	Nylon-mesh hinged cover	Moving floor
Portland, OR	3500	Semitrailer	96	29	8	48	12.75	68	Completely enclosed	Internal diaphragm
San Francisco, CA	2000	Semitrailer	100	25	8	47.5	13.5	61	Wire-screen hinged cover	Tilting ramp at disposal site
Seattle, WA	2000	Semitrailer	96	19	8	40	13.5	60	Completely enclosed	Tilting ramp at disposal site

Note: $yd^3/day \times 0.7646 = m^3/day$
$yd^3 \times 0.7646 = m^3$
$ft \times 0.3048 = m$

[a] Light loads are for general debris, including yard wastes, tree trimmings, etc.
[b] Length varies depending on the type of tractor used.

TABLE 10-2
Solid waste payload hauled by various truck tractor-semitrailer combinations

	Weight, lb		
	Type of trailer		
	No compaction top load (conventional construction)	No compaction top load (monoque construction)[a]	Waste bale
Truck tractor	16,000	16,000	16,000
Semitrailer	26,000	18,000	13,000
Solid waste payload	36,000	46,000	50,000[b]
Total	78,000	80,000	79,000

[a] In monoque construction the bed of the trailer also serves as the frame of the trailer (see Fig. 10-18).
[b] Increased payloads can be hauled by increasing the number of axles.

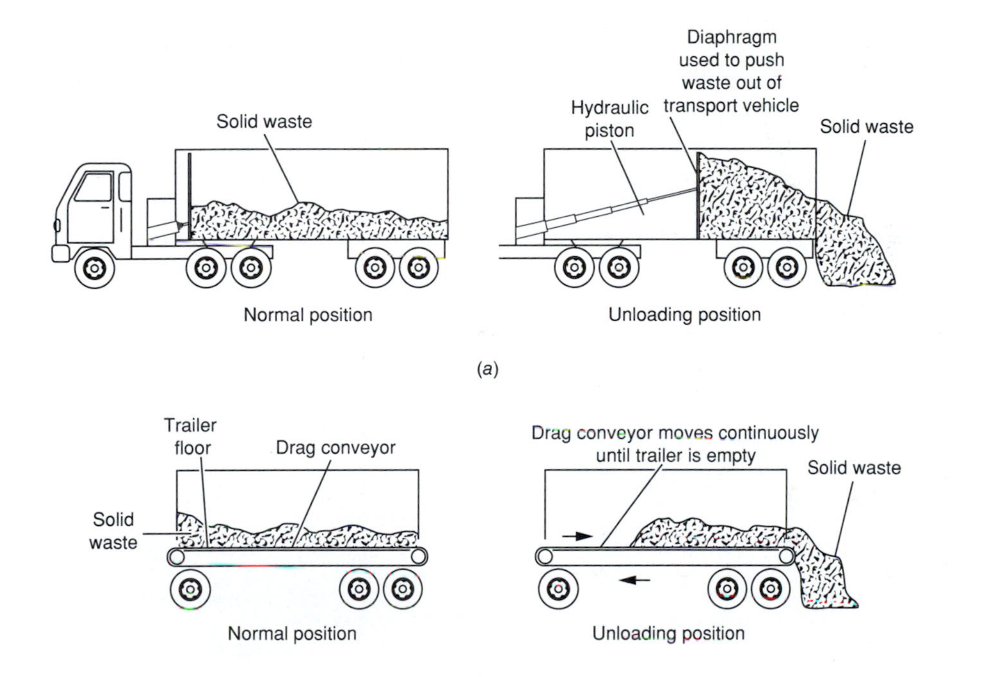

FIGURE 10-19
Methods used to unload semitrailers: (*a*) semitrailer equipped with movable internal diaphragm and (*b*) semitrailer equipped with continuous moving drag conveyor.

width of the trailer (see Fig. 10-20). Thus, if one section becomes inoperable, the moving floor does not prevent unloading because the system will function with the remaining operable section(s). Another type of moving floor uses reciprocating panels, 3 to 6 in wide, that alternate in the backward and forward directions. The use of multiple floor sections is an important feature in terms of system reliability. Another advantage of the moving-floor trailer is the rapid turnaround time (typically 6 to 10 min) achieved at the disposal site without the need for auxiliary equipment. In some designs the rear of the trailer is made larger to facilitate the unloading operation. Trailers such as those shown in Fig. 10-19 are sometimes equipped with sumps to collect any liquids that accumulate from the solid wastes. The sumps are equipped with drains so that they can be emptied at the disposal site.

Unloading systems that require auxiliary equipment are usually of the *pull-off* type, in which the wastes are pulled out of the truck by either a movable bulkhead or wire-cable slings placed forward of the load. The disadvantage of requiring auxiliary equipment and work force to help in the unloading operation at the disposal site is relatively minor in view of the simplicity and reliability of the method. An additional disadvantage, however, is the unavoidable waiting time during which the haul vehicle remains idle at the disposal site until the auxiliary equipment can be placed in the required position.

Another auxiliary unloading system that has proved very effective and efficient involves the use of movable, hydraulically operated tipping ramps (see Fig. 10-21). Operationally, where a truck-trailer combination is used the trailer is backed up onto a tipping ramp and uncoupled from the truck. Once uncoupled, the truck is backed up onto a second tipping ramp. The backs of the trailer and truck are opened, and the units are then tilted up until the wastes fall by gravity onto the disposal area. After being emptied, the truck and trailer are returned to their original positions. The truck is driven off the ramp and is backed up to the ramp used

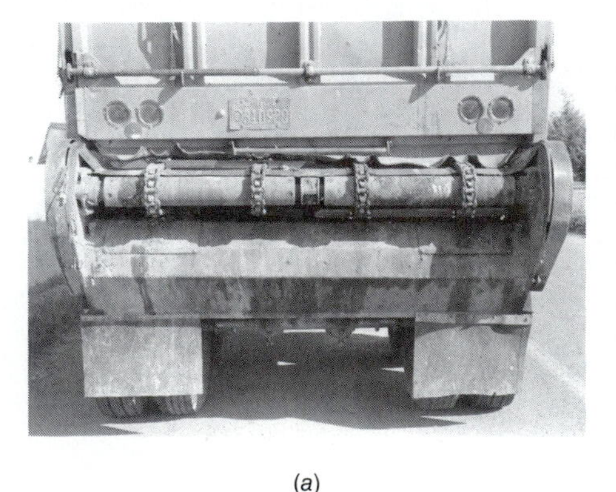

(*a*)

(*b*)

FIGURE 10-20
Views of continuous moving chain-type drag conveyor used in self-unloading waste transport trailers: (*a*) drive sprockets for chain conveyors located at rear of trailer and (*b*) chain-type drag conveyor in bottom of trailer.

(a)

(b)

FIGURE 10-21
Unloading operations using hydraulically operated tipping ramps: (*a*) drop-bottom semitrailer is backed onto tipping ramp and disconnected from the tractor for unloading and (*b*) tipping ramp is elevated and solid wastes fall out by gravity.

for the trailer. The trailer is reattached, and the transfer rig is returned to the transfer station. The time required for the entire unloading operation typically is about 6 min/trip.

Transport Vehicles and Containers Used in Conjunction with Waste Compaction Facilities. The semitrailers used in conjunction with compaction facilities are designed to function together. Typically, the stationary compactor (see Fig. 10-6*c*) will compact the wastes against the internal diaphram of the trailer. When the pressure of the diaphram reaches a predetermined value, the diaphram moves inward allowing more waste to be compacted into the trailer. The diaphram is also used to unload the semitrailer at the disposal site. An example of the type of semitrailer used in conjunction with a transfer station in which wastes are compacted into a bale is shown in Fig. 10-22.

In smaller transfer stations, large-capacity containers are often used in conjunction with stationary compactors (see Fig. 10-8). In some cases, the compaction mechanism is an integral part of the container. Representative data for such units are reported in Table 10-3. When containers are equipped with a self-contained

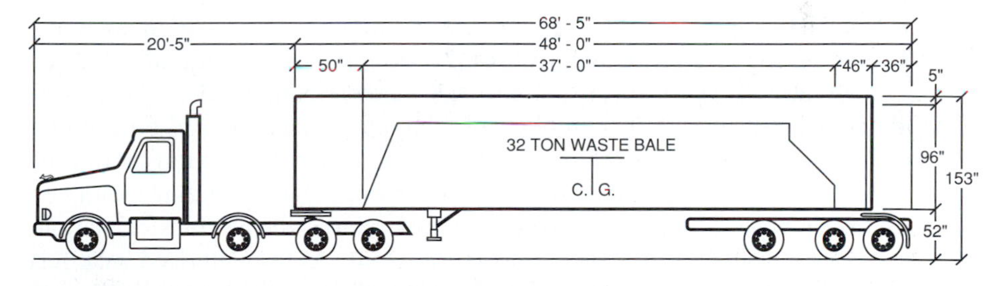

FIGURE 10-22
Typical semitrailer used to transport waste bales. (Courtesy of Jack Gray Transport, Inc.)

TABLE 10-3
Typical data on containers used with stationary compactors and container-compactor units for medium and small transfer stations

Type	Rated capacity, yd^3	Dimensions, ft			Approx. tare weight, lb	Remarks
		Width	Length	Height		
Container						
Small	20	8	14	6	8,000	Door openings where containers are attached to stationary compactor are usually reinforced.
Medium	30	8	18	6	10,000	
Large	45	8	22	9	10,000	
Container-compactor						
Small	3.7	6.5	6.5	4.5	1,500	Available with water-tight sumps and leak-proof doors. Other features on request.
Medium	15	7.5	15	6	6,000	
Large	30	8	22	8	10,000	

Note: $yd^3 \times 0.7646 = m^3$
$ft \times 0.3048 = m$
$lb \times 0.4536 = kg$

compaction mechanism, the movable bulkhead used to compress the waste is also used to discharge the compacted wastes. The contents of containers used with stationary compactors usually are unloaded by tilting the container and allowing the contents to fall out by gravity. If the wastes are compressed too tightly, unloading can be a problem. Various ejection devices also are available to empty the contents of the containers. The most common device is the movable hydraulically operated diaphragm.

Railroad Transport

Although railroads were commonly used for the transport of solid wastes in the past, they are now used by only a few communities. However, renewed interest is again developing in the use of railroads for hauling solid wastes, especially to remote landfill areas where highway travel is difficult and railroad lines now exist. One of the largest rail haul operations currently in use is used to transport wastes from the city of Seattle, WA to the Columbia Ridge Landfill located approximately 300 miles away in Gilliam, OR. Operationally, 25 to 28 tons of waste are compacted into a 40 ft, sealed shipping container mounted on a trailer chassis. The loaded containers are transported to the Union Pacific railyard in the city where they are loaded onto railcars. After dropping off the full container, the truck driver picks up an empty container and returns to the transfer station. Each container is weighed as it goes into the railyard. A computerized manifest and container tracking system allow the city to track the location and status of every container in the system.

The train is made up of approximately 50 railcars, carrying 100 "piggy-backed" containers. It leaves Seattle three evenings a week. The train travels south to Portland and then continues eastward along the Columbia River Gorge to Arlington, OR where it proceeds southward approximately 10 miles to an intermodal siding at the Columbia Ridge Landfill, arriving early in the morning. The containers are unloaded and placed on a truck chassis for the short trip up to the landfill operating face. Hydraulic tippers (see Fig. 10-21) tilt the chassis and container to discharge the waste. The waste is immediately spread and compacted, and is covered each day with six inches of compacted dirt [3].

Water Transport

Barges, scows, and special boats have been used in the past to transport solid wastes to processing locations and to seaside and ocean disposal sites. It should be noted that ocean disposal is no longer practiced by the United States. Although some self-propelled vessels (such as United States Navy garbage scows and other special boats) have been used, the most common practice is to use vessels towed by tugs or other special boats. In England, river barges are used to transport wastes [2].

One of the major problems encountered when ocean vessels are used for the transport of solid wastes is that it is often impossible to move the barges and boats during times of heavy seas. In such cases, the wastes must be stored, entailing the construction of costly storage facilities.

Pneumatic, Hydraulic, and Other Systems of Transport

Both low-pressure air and vacuum conduit transport systems have been used to transport solid wastes (see Fig. 9-17). The most common application is the transport of wastes from high-density apartments or commercial activities to a central location for processing or for loading into transport vehicles. The largest pneumatic system in the United States was installed at the Walt Disney World amusement park in Orlando, FL. The layout of this system is shown schematically in Fig. 10-23. A pneumatic system used for the collection of wastes from an apartment complex is shown in Fig. 7-6.

From a design and operational standpoint, pneumatic systems are more complex than hydraulic systems because of the complex control valves and ancillary mechanisms that are required. The need to use blowers or high-speed turbines further complicates the installation from a maintenance standpoint. Because installation costs for such systems are quite high, they are most cost-effective when used in new facilities.

The concept of using water to transport wastes is not new. Hydraulic transport is now commonly used for the transport of a portion of food wastes (where home grinders are used). One of the major problems with this method is that ultimately the water or wastewater used for transporting the wastes must be treated. As a result of solubilization, the concentration of organics in this wastewater is

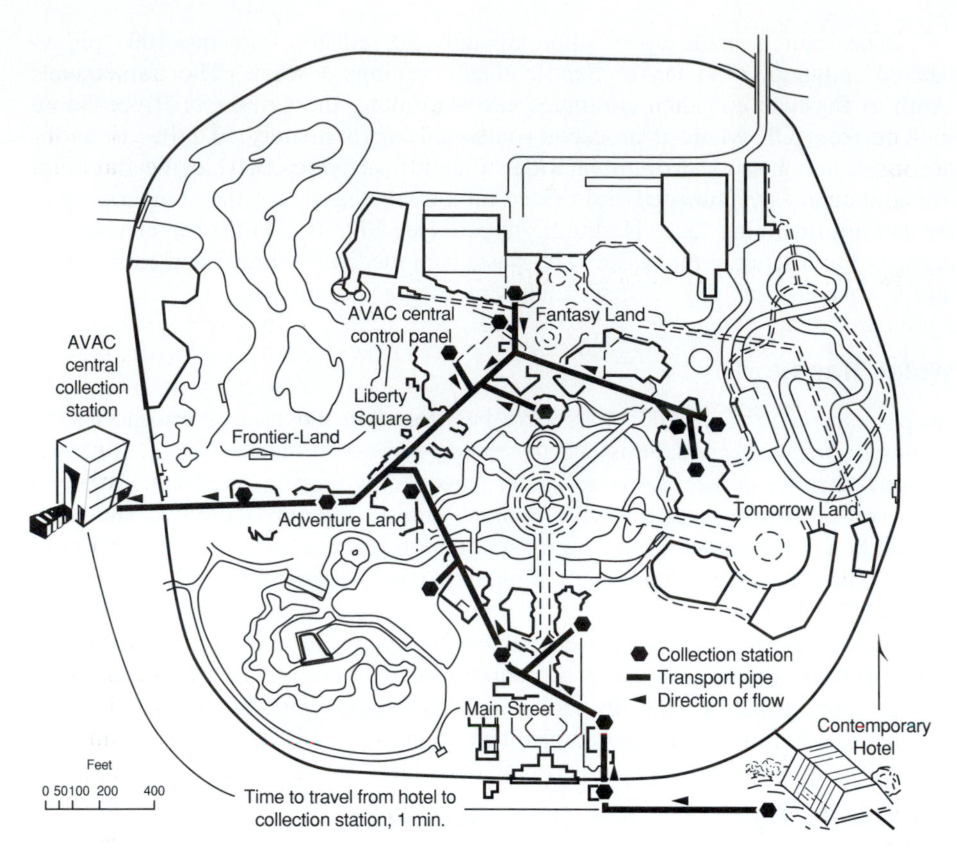

FIGURE 10-23
Pneumatic solid waste collection system for Walt Disney World, Orlando, FL.

considerably greater than in other domestic waste water. Hydraulic systems may be practical in areas where proper preprocessing and postprocessing facilities are incorporated into the treatment system. Usually, such applications are limited to areas with high population densities.

Other systems that have been suggested for the transport of solid wastes include various types of conveyors, air-cushion and rubber-tired trolleys, and underground conduits with magnetically transported gondolas, but these systems have never been put into operation.

10-4 TRANSFER STATION DESIGN REQUIREMENTS

Although specific details vary with size, important factors that must be considered in the design of transfer stations include (1) the type of transfer operation to be used, (2) storage and throughput capacity requirements, (3) equipment and accessory requirements, and (4) sanitation requirements.

Type of Transfer Station

The basic types of transfer station have been described in the previous sections. From a design standpoint the key issue is whether waste recovery operations will be incorporated into the transfer station facility. If waste recovery is to be accomplished at the transfer station, then an adequate area must be available for the collection vehicles to unload.

Transfer Station Capacity Requirements

Both the throughput and storage capacity requirements must be evaluated carefully in planning and designing transfer facilities. The throughput capacity of a transfer station must be such that the collection vehicles do not have to wait too long to unload. In most cases, it will not be cost-effective to design the station to handle the ultimate peak number of hourly loads. Ideally, an economic trade-off analysis should be made. For example, for both types of transfer stations, the annual cost of the time spent by the collection vehicles waiting to unload must be traded off against the incremental annual cost of a larger transfer station and/or the use of more transport equipment.

Because of the increased cost of transport equipment, a trade-off analysis must also be made between the capacity of the transfer station and the cost of the transport operation, including both equipment and labor components. For instance, in a given situation it may be more cost-effective to increase the capacity of a transfer station and to operate with fewer transport vehicles by increasing the working hours than to use a smaller transfer station and purchase more transport vehicles. In a storage-load transfer station, the equivalent storage capacity varies from about one-half to one day's volume of wastes. The capacity also varies with the type of auxiliary equipment used to load the transport vehicles. Seldom will the nominal storage capacity exceed three days' volume of waste.

Equipment and Accessory Requirements

The equipment and accessories used in conjunction with a transfer station depend on the function of the transfer station in the waste management system. In a direct-load transfer station, some sort of rig, usually rubber-tired, is required to push the wastes into the transfer vehicles. Another rig is required to push the wastes and to equalize the load in the transfer vehicles. The types and amounts of equipment required vary with the capacity of the station. In a pit type storage-load transfer station, one or more tractors are required to break up the wastes and to push them into the loading hopper. Additional equipment is required to distribute the wastes and to equalize the loads. In some installations an overhead clamshell crane has been used successfully for both purposes.

Scales (see Fig. 10-7*a*) should be provided at all medium- and large-sized transfer stations, both to monitor the operation and to develop meaningful management and engineering data. Scales are also necessary when the transfer station is to be used by the public and the charges are to be based on weight. If scales are to be used, it will usually be necessary to provide an enclosure for them. The scale house, as it is commonly called, should also have an office equipped with a telephone and a two-way speaker system so that the weighmaster can talk with the drivers.

If the transfer station is to be used as a dispatch center or district headquarters for a solid waste collection operation, a more complete facility should be constructed. For a headquarters facility, a lunch room, meeting rooms, offices, locker rooms, showers, and toilets should be provided. Facilities for providing equipment maintenance may also be incorporated.

Environmental Requirements

By proper construction and operation, the objectionable features of transfer stations can be minimized. Most of the modern, large transfer stations are enclosed and are constructed of materials that can be maintained and cleaned easily. To eliminate inadvertent emissions, enclosed facilities should have air-handling equipment that creates a negative pressure within the facility. In most cases, fireproof construction is used for direct-load transfer stations with open loading areas. Special attention must be given to the problem of blowing papers. Wind screens or other barriers are commonly used. Regardless of the type of station, the design and construction should be such that all areas where rubbish or paper can accumulate are eliminated [2]. The best way to maintain overall sanitation of a transfer station is to monitor the operation continually. Spilled solid wastes should be picked up immediately or in any case should not be allowed to accumulate for more than 1 or 2 h. The area should also be washed down. In some large facilities, wastewater pretreatment facilities may be required to treat plant wastewater before it is discharged to the local sewer. In remote areas, complete wastewater treatment facilities may be required.

Health and Safety

Health and safety issues at transfer stations are related to dust inhalation and other OSHA requirements. Overhead water sprays are used to keep the dust down in the storage area of a storage-load transfer station. To prevent dust inhalation, workers should wear dust masks. In storage-load transfer stations, tractors in the pit area should have enclosed cabs equipped with air-conditioning and dust-filtering units (see Fig. 10-13*b*). For reasons of safety, the public should not be allowed to discharge wastes directly into the pit at large storage-load transfer stations.

10-5 LOCATION OF TRANSFER STATIONS

Whenever possible, transfer stations should be located (1) as near as possible to the weighted center of the individual solid waste production areas to be served, (2) within easy access of major arterial highway routes as well as near secondary or supplemental means of transportation, (3) where there will be a minimum of public and environmental objection to the transfer operations, and (4) where construction and operation will be most economical [2]. Additionally, if the transfer station site is to be used for processing operations involving materials recovery and/or energy production, the requirements for those operations must also be assessed. In some cases, these latter requirements may be controlling.

Because all the above considerations can seldom if ever be satisfied simultaneously, it is usually necessary to perform a trade-off analysis among these factors. The analysis of different locations based on haul cost is described in this section. This method is applicable in those cases where a selection must be made from among several potential transfer station locations. A more complex situation in which two or more transfer stations and disposal sites are to be used is also considered.

Site Selection Based on Transportation Costs

Under ideal conditions, the transfer stations should be located so as to minimize transportation costs. However, given the difficulty most waste management agencies have had in recent times in locating transfer stations, transportation costs have become somewhat less important in the selection of an appropriate location of a transfer station. The siting of transfer stations is considered further in Chapter 18.

Site Selection Based on Operational Constraints

In situations where two or more transfer stations and disposal sites are to be used the basic question that must be answered is this: What is the optimum allocation of wastes from each transfer station to each disposal site? In the discussion that follows, this allocation problem is described, and methods of solution are suggested.

The waste allocation problem can be analyzed as follows. Assume that a determination must be made of the amount of solid wastes that should be hauled to each of three disposal sites from each of three transfer stations, so that the total haul cost will be the minimum possible value. A definition sketch for this situation is shown in Fig. 10-24. Also assume (1) that the total amount of wastes hauled to all the disposal sites must be equal to the amount delivered to the transfer station (materials-balance requirements), (2) that only specified amounts of wastes can be accepted at each disposal site (this constraint could arise as a result of limited highway access to a given disposal site), and (3) that the amount of wastes hauled from each transfer station is equal to or greater than zero. In the symbolic form, the allocation problem is set up as follows:

1. Let the transfer station sites be designated by i.
2. Let the disposal sites be designated by j.
3. Then let X_{ij} = the amount of wastes hauled from transfer station i to disposal site j.
4. Let C_{ij} = the cost of hauling wastes from transfer station i to disposal site j.

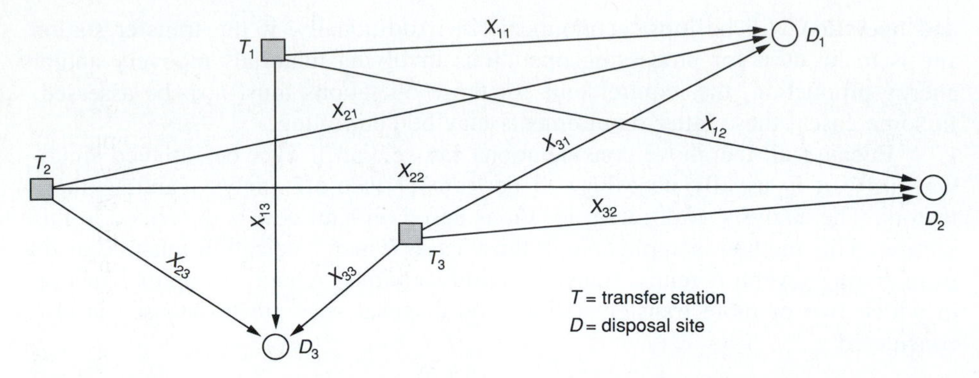

FIGURE 10-24
Definition sketch for allocation of solid wastes from three transfer stations to three disposal sites.

5. Let R_i = the total amount of wastes delivered to transfer station i.
6. Let D_j = the total amount of wastes that can be accepted at disposal site j.
7. If the total haul costs are to be minimized, then an objective function, which is defined as the sum of the following terms, must be minimized subject to the problem constraints:

$$X_{11}C_{11} + X_{12}C_{12} + X_{21}C_{21} + X_{22}C_{22} + X_{23}C_{23} + X_{31}C_{31} + X_{32}C_{32} + X_{33}C_{33} = \text{objective function}$$

8. Expressed in mathematical summation form, the problem is to minimize the function

$$\text{Objective function} = \sum_{j=1}^{3}\sum_{i=1}^{3} X_{ij}C_{ij} \qquad (10\text{-}1)$$

subject to the following constraints:

$$\sum_{i=1}^{3} X_{ij} = R_i \qquad i = 1 \text{ to } 3 \qquad (10\text{-}2)$$

$$\sum_{j=1}^{3} X_{ij} \leq D_j \qquad j = 1 \text{ to } 3 \qquad (10\text{-}3)$$

$$X_{ij} \geq 0 \qquad (10\text{-}4)$$

The fact that the amount of waste hauled to the disposal sites must be equal to the amount brought to the transfer station is given by the first constraint. The condition that the total amount of waste hauled from the transfer station must be equal to or less than the capacity of the disposal sites is given by the second constraint. The third constraint is that the amount of waste hauled from the transfer station must be equal to or greater than zero.

Solutions to Waste Allocation Problem

The problem as set up in Step 8 is commonly known as the *transportation problem* in the field of operations research. At present, a number of solution methods are available. However, most of the methods require the aid of microcomputers. As an alternative, several approximate solution techniques have been developed [1, 4]. Because the solution obtained with the approximate methods will be close to the optimum solution (within 10 percent), they are sufficiently accurate for most practical applications in the field of solid waste management. The optimum solution may be obtained by any number of methods outlined in standard texts on linear programming.

10-6 DISCUSSION TOPICS AND PROBLEMS

10-1. Given the following data on transportation costs, determine the break-even times for the two stationary container systems versus the use of a transfer and transport system. Base your computations on dollars per ton per minute.

Operating costs
  Stationary container systems
    4-ton capacity = \$25.00/h
    10-ton capacity = \$36.00/h
  Truck-semitrailer combination (25-ton capacity) = \$55.00/h
  Transfer station costs = \$3.00/ton

10-2. Determine the round-trip break-even time for solid waste collection systems in which the 30-yd^3 self-loading compactors used for collection are driven to the disposal site and compare that with using a transfer and transport system. Assume that the following data are applicable.

(*a*) Specific weight of wastes in self-loading compactor = 600 lb/yd^3
(*b*) Specific weight of wastes in transport trailers = 325 lb/yd^3
(*c*) Volume of tractor-semitrailer transport unit = 105 yd^3
(*d*) Operational cost for self-loading compactor = \$40/h
(*e*) Operational cost for tractor-semitrailer transport unit = \$60/h
(*f*) Transfer station operational costs including amortization = \$3.25/ton
(*g*) Extra unloading time cost for transport units, compared with compactors = \$0.40/ton

10-3. What would the graph in Example 10-1 and Problems 10-1 and 10-2 look like based on total cost versus round-trip haul distance?

10-4. Using the following data and information and the cost information given in Appendix E, estimate how far away a disposal site can be located from the city before the use of a transfer station is economical.

(*a*) Population = 50,000 persons
(*b*) Waste generation = 4 lb/capita · d
(*c*) Capacity of rear (manually) loaded waste collection vehicles = 20 yd^3
(*d*) Specific weight of wastes in collection vehicles = 525 lb/yd^3
(*e*) Time required to load collection vehicles = 2.4 h/trip

(*f*) Current number of collection trips made per day = 2
(*g*) Number of collection vehicles currently owned by city = 15
(*h*) Operational cost for collection vehicles = \$35/h
(*i*) Transfer station type = direct-load with compactor that produces bales with a total weight of 26 tons
(*j*) Transfer station transport vehicle = tractor-semitrailer
(*k*) Operational cost for tractor-semitrailer unit = \$50/h
(*l*) Discount rate = 7.5%
(*m*) Return period = 10 yr

10-5. Estimate the peak hourly capacity of the transfer station shown in Fig. 10-2. Express your answer in vehicles and in tons per hour. Assume the average volume per vehicle and the specific weight of the waste is 15.0 yd^3 and 475 lb/yd^3, respectively. State all of the assumptions made in solving this problem.

10-6. Estimate the peak hourly capacity of the transfer station shown in Fig. 10-3. Assume the average volume per vehicle and the specific weight of the waste is 16.5 yd^3 and 510 lb/yd^3, respectively. State all of the assumptions made in solving this problem.

10-7. How would you increase the peak hourly capacity of the transfer stations shown in Fig. 10-2 and 10-6 by 25 percent?

Do either Problem 10-8 or 10-9, depending on whether your community has a transfer station.

10-8. If your community does not have a transfer station, estimate the break-even time at which a transfer station operation would become feasible. How does this time compare to the actual time now spent by the collection vehicles in the haul operation? State all your assumptions.

10-9. If your community has a transfer station, determine what the break-even time would be for a direct-haul operation. How does this time compare to the actual time now spent by the transport units in transport operation? State clearly all your assumptions.

10-10. A 1000 ton/d transfer station is to be constructed. Consideration is being given to both a direct-load transfer station employing stationary compactors such as shown in Fig. 10-4 and a storage-load type such as shown in Fig. 10-8. Identify and discuss the important factors that must be considered in selecting one of these two choices.

10-11. Discuss the advantages and disadvantages of developing a single large MR/TF for a community compared with developing source separation programs and the use of smaller MRFs in conjunction with a transfer station.

10-12. Given the following information, determine—by the long-hand method of evaluating every possibility—the most economical allocation of wastes from each of two disposal sites on the basis of transportation cost only. Check your answer with a computer or spreadsheet program.

Transfer station	Waste, units/d	Disposal site	Capacity, units/d
1	4	1	4
2	2	2	4

The round-trip haul distance from transfer station 1 to disposal sites 1 and 2 is 10 and 20 mi, respectively. The distances from transfer station 2 to disposal sites 1 and 2 are 30 and 40 mi, respectively. Assume that the transport time in hours per trip is given by the expression [0.08 h/trip + 0.025 h/mi (x)], where x is the round-trip-haul distance in miles per trip, and that the transportation cost is $35/h.

10-13. The city shown in the accompanying figure has four disposal sites D1, D2, D3, and D4 and needs four transfer stations to handle the solid waste. The location of transfer sites T1, T2, and T3 have already been selected. The fourth site has been narrowed to two possibilities, T4 and T5 as shown. The following disposal site and transfer station data were collected for the city.

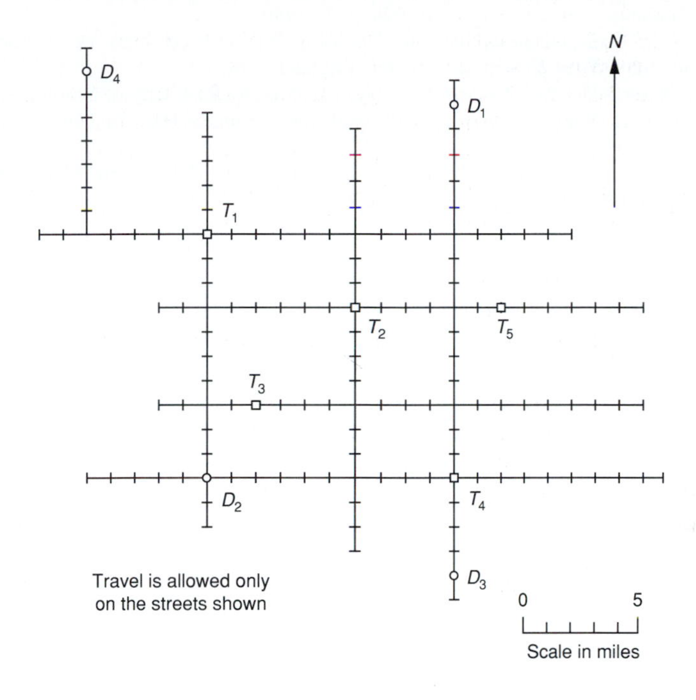

Disposal site	Capacity, units/d	Transfer station	Waste, units/d
D_1	4	T_1	3
D_2	10	T_2	3
D_3	3	T_3	5
D_4	8	T_4 or T_5	2

On the basis of transport cost alone, determine the more economical location for transfer station 4 (T4 or T5). Assume that the transport time in hours per trip is given by the expression [0.08 h/trip + 0.025 h/mi (x)], where x is the round-trip-haul distance in miles per trip, and that the transport cost is $35/h.

10-14. Once collected, how are household and other hazardous wastes from your community transported to treatment or disposal facilities. What is your assessment of the arrangements that are now used? Can you suggest any improvements?

10-15. In the late 1960s and early 1970s much was written about the use of simulation and other techniques from the field of operations research for optimizing the location of transfer stations. Based on a review of one or two articles from that period (i.e., 1965 to 1975) and any other pertinent information, prepare an analysis of why the techniques proposed during that period have not been adopted to any major extent in the siting of transfer stations.

10-7 REFERENCES

1. Golden, B., L. Brown, T. Doyle, and W. Stewart, Jr.: "Approximate Traveling Salesman Algorithms," *Applied Research,* Vol. 23, No. 3, pp. 694–711, 1980.
2. Holmes, J. R.: "Waste Management Options and Decisions," In J. R. Holmes (ed.), *Practical Waste Management,* John Wiley & Sons, Chichester, England, 1983.
3. Parker, L. and A. Ostrom: *Seattle's Road to Recovery: Garbage by Rail,* City of Seattle, 1990.
4. Reinfeld, N. V. and W. R. Vogel: *Mathematical Programming,* Prentice-Hall, Englewood Cliffs, NJ, 1958.
5. U.S. Environmental Protection Agency, *Decision-Makers Guide to Solid Waste Management,* EPA/530-SW89-072, Washington, DC, November, 1989.

CHAPTER 11

DISPOSAL OF SOLID WASTES AND RESIDUAL MATTER

The safe and reliable long-term disposal of solid waste residues is an important component of integrated waste management. Solid waste residues are waste components that are not recycled, that remain after processing at a materials recovery facility, or that remain after the recovery of conversion products and/or energy. Historically, solid waste has been placed in the soil in the earth's surface or deposited in the oceans. Although ocean dumping of municipal solid waste was officially abandoned in the United States in 1933, it is now argued that many of the wastes now placed in landfills or on land could be used as fertilizers to increase productivity of the ocean or the land. It is also argued that the placement of wastes in ocean trenches where tectonic folding is occurring is an effective method of waste disposal. Nevertheless, landfilling or land disposal is today the most commonly used method for waste disposal by far. Disposal of solid waste residues in landfills is the primary subject of this chapter.

The planning, design, and operation of modern landfills involve the application of a variety of scientific, engineering, and economic principles. The major topics covered in this chapter include: (1) a description of the landfill method of solid waste disposal, including environmental concerns and regulatory requirements; (2) a description of types of landfills and landfilling methods; (3) landfill siting considerations; (4) landfill gas management; (5) landfill leachate control; (6) surface water control; (7) landfill structural characteristics and settlement;

(8) environmental quality monitoring; (9) the layout and preliminary design of landfills; (10) development of landfill operation plan; (11) landfill closure and post-closure care; and (12) landfill design computations. Example design computations for landfills have been grouped together in the final section of this chapter. Reference is made to specific example problems as appropriate. Management policies and regulations for landfill closure and postclosure maintenance are discussed in Chapter 16.

11-1 THE LANDFILL METHOD OF SOLID WASTE DISPOSAL

Historically, landfills have been the most economical and environmentally acceptable method for the disposal of solid wastes, both in the United States and throughout the world. Even with implementation of waste reduction, recycling, and transformation technologies, disposal of residual solid waste in landfills still remains an important component of an integrated solid waste management strategy. Landfill management incorporates the planning, design, operation, closure, and postclosure control of landfills. The purposes of this section are (1) to introduce the reader to the landfilling process, (2) to review the principal reactions occurring in landfills, (3) to identify environmental concerns associated with landfills, and (4) to review briefly some federal and state regulations governing the disposal of solid waste in landfills. Many of the subjects introduced in this section are examined in greater detail later in this chapter.

The Landfilling Process

The purpose of the following discussion is to introduce the reader to the landfilling of solid waste by (1) defining some terms that are commonly used when discussing the landfilling of solid waste, (2) reviewing landfill operations and processes, (3) describing the life of a landfill, and (4) reviewing some of the reactions occurring in landfills.

Definition of Terms. *Landfills* are the physical facilities used for the disposal of residual solid wastes in the surface soils of the earth. In the past, the term sanitary landfill was used to denote a landfill in which the waste placed in the landfill was covered at the end of each day's operation. Today, *sanitary landfill* refers to an engineered facility for the disposal of MSW designed and operated to minimize public health and environmental impacts (see Fig. 11-1). Landfills for the disposal of hazardous wastes are called *secure landfills*. A sanitary landfill is also sometimes identified as a *solid waste management unit*. *Landfilling* is the process by which residual solid waste is placed in a landfill. Landfilling includes monitoring of the incoming waste stream, placement and compaction of the waste, and installation of landfill environmental monitoring and control facilities.

The term *cell* is used to describe the volume of material placed in a landfill during one operating period, usually one day (see Fig. 11-2). A cell includes

FIGURE 11-1
Views of operating landfills.

the solid waste deposited and the daily cover material surrounding it. *Daily cover* usually consists of 6 to 12 in of native soil or alternative materials such as compost that are applied to the working faces of the landfill at the end of each operating period. The purposes of daily cover are to control the blowing of waste materials; to prevent rats, flies, and other disease vectors from entering or exiting the landfill; and to control the entry of water into the landfill during operation.

A *lift* is a complete layer of cells over the active area of the landfill (see Fig. 11-2). Typically, landfills are comprised of a series of lifts. A *bench* (or *terrace*) is commonly used where the height of the landfill will exceed 50 to 75 ft. Benches are used to maintain the slope stability of the landfill, for the placement of surface water drainage channels, and for the location of landfill gas

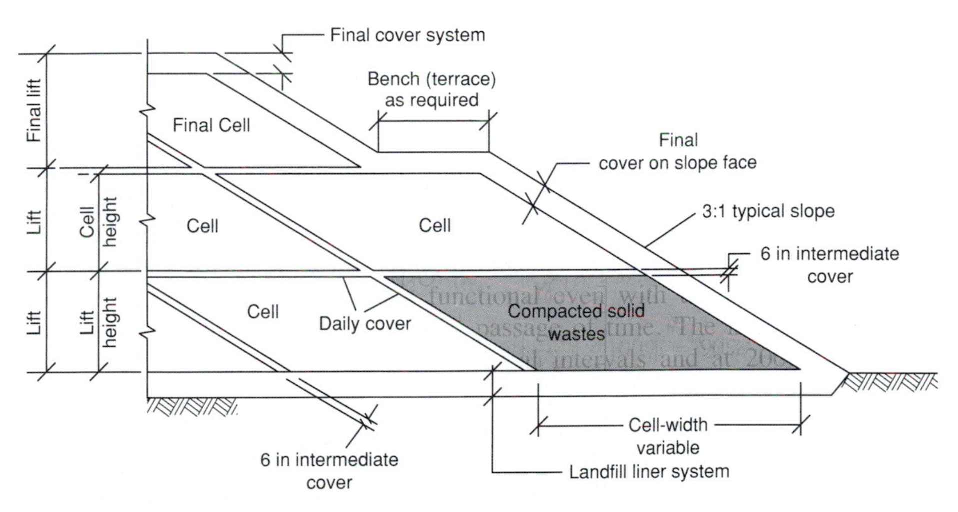

FIGURE 11-2
Sectional view through a sanitary landfill.

recovery piping. The *final lift* includes the cover layer. The *final cover layer* is applied to the entire landfill surface after all landfilling operations are complete. The final cover usually consists of multiple layers of soil and/or geomembrane materials designed to enhance surface drainage, intercept percolating water, and support surface vegetation.

The liquid that collects at the bottom of a landfill is known as *leachate*. In deep landfills, leachate is often collected at intermediate points. In general, leachate is a result of the percolation of precipitation, uncontrolled runoff, and irrigation water into the landfill. Leachate can also include water initially contained in the waste as well as infiltrating groundwater. Leachate contains a variety of chemical constituents derived from the solubilization of the materials deposited in the landfill and from the products of the chemical and biochemical reactions occurring within the landfill.

Landfill gas is the mixture of gases found within a landfill. The bulk of landfill gas consists of methane (CH_4) and carbon dioxide (CO_2), the principal products of the anaerobic decomposition of the biodegradable organic fraction of the MSW in the landfill. Other components of landfill gas include atmospheric nitrogen and oxygen, ammonia, and trace organic compounds.

Landfill liners are materials (both natural and manufactured) that are used to line the bottom area and below-grade sides of a landfill. Liners usually consist of layers of compacted clay and/or geomembrane material designed to prevent migration of landfill leachate and landfill gas. *Landfill control facilities* include liners, landfill leachate collection and extraction systems, landfill gas collection and extraction systems, and daily and final cover layers.

Environmental monitoring involves the activities, associated with collection and analysis of water and air samples, that are used to monitor the movement of landfill gases and leachate at the landfill site. *Landfill closure* is the term used to describe the steps that must be taken to close and secure a landfill site once the filling operation has been completed. *Postclosure* care refers to the activities associated with the long-term monitoring and maintenance of the completed landfill (typically 30 to 50 years).

Overview of Landfill Planning, Design, and Operation. The principal elements that must be considered in the planning, design, and operation of landfills are identified in Fig. 11-3, and include (1) landfill layout and design; (2) landfill operations and management; (3) the reactions occurring in landfills; (4) the management of landfill gases; (5) the management of leachate; (6) environmental monitoring; and (7) landfill closure and postclosure care. Each of the elements is considered in greater detail in this chapter.

The Life of a Modern Landfill. The following description of the life of a modern landfill is generic. Specific details of the operation will vary with the type of material being landfilled and the configuration of the landfill. Landfill types and configurations are described in Section 11-2, where significant departures from the generic operation scheme are noted. The development of a modern landfill is illustrated in Fig. 11-4.

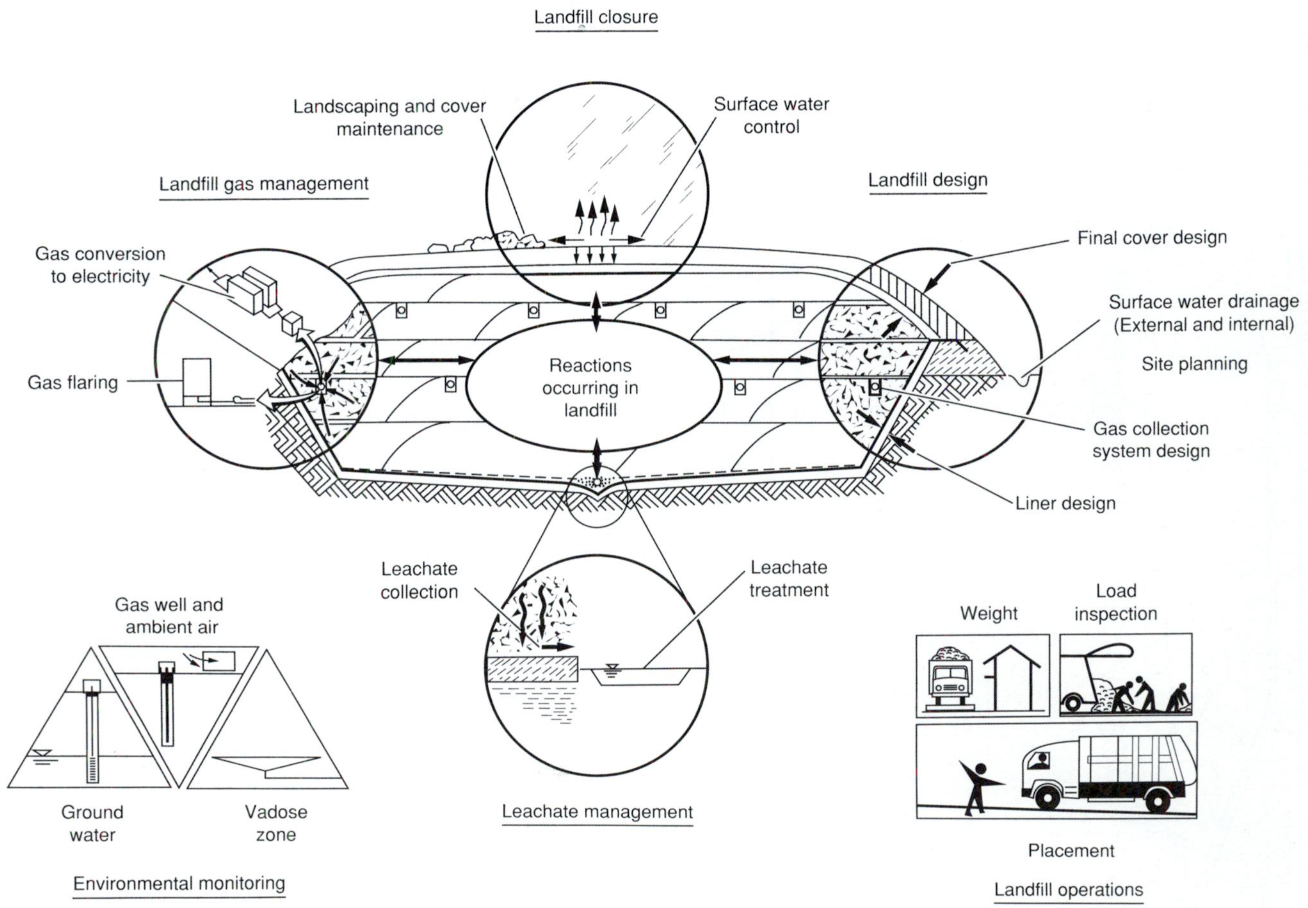

FIGURE 11-3
Definition sketch for landfill operations and processes.

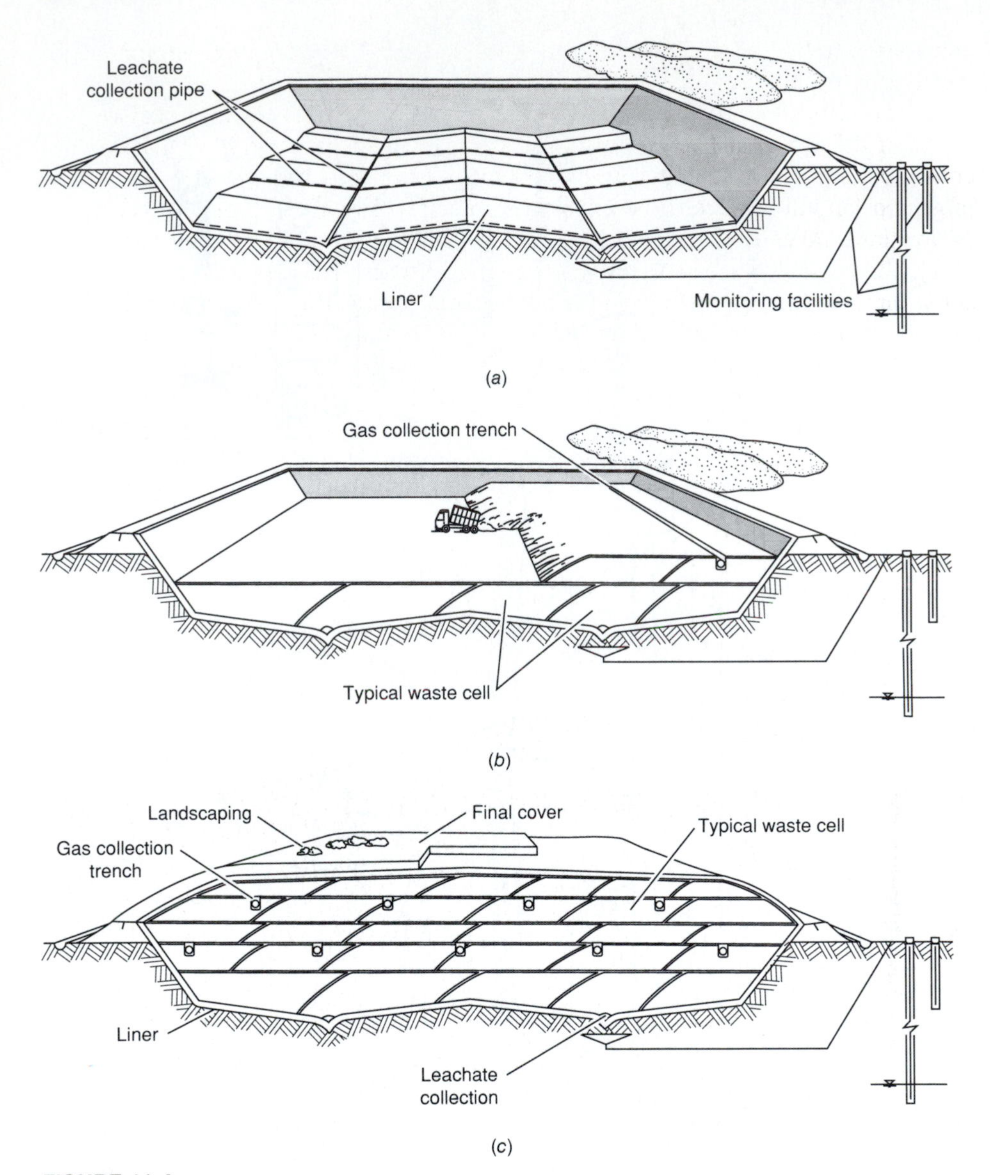

FIGURE 11-4
Development and completion of a solid waste landfill: (*a*) excavation and installation of landfill liner, (*b*) placement of solid waste in landfill, and (*c*) cutaway through completed landfill.

Preparation of the site for landfilling. The first step in the process involves the preparation of the site for landfill construction. Existing site drainage must be modified to route any runoff away from the intended landfill area. Rerouting of drainage is particularly important for ravine landfills where a significant watershed may drain through the site. In addition, drainage of the landfill area itself must be modified to route water away from the initial fill area. Other site preparation tasks include construction of access roads and weighing facilities, and installation of fences.

The next step in the development of a landfill is the excavation and preparation of the landfill bottom and subsurface sides. Modern landfills typically are constructed in sections. Working by sections allows only a small part of the unprotected landfill surface to be exposed to precipitation at any time. In addition, excavations are carried out over time, rather than preparing the entire landfill bottom at once. Excavated material can be stockpiled on unexcavated soil near the active area and the problem of precipitation collecting in the excavation is minimized. Where the entire bottom of the landfill is lined at once, provision must be made to remove stormwater runoff from the portion of the landfill that is not being used.

To minimize costs, it is desirable to obtain cover materials from the landfill site whenever possible. The initial working area of the landfill is excavated to the design depth, and the excavated material stockpiled for later use. Vadose zone (zone between ground surface and permanent groundwater) and groundwater monitoring equipment is installed before the landfill liner is laid down. The landfill bottom is shaped to provide drainage of leachate, and a low-permeability liner is installed (see Fig. 11-5). Leachate collection and extraction facilities are placed within or on top of the liner. Typically, the liner extends up the excavated walls of the landfill.

FIGURE 11-5
Aerial view of area type landfill. Geomembrane liner is being placed in front part of the landfill site (foreground). (Courtesy of Brown and Caldwell Consultants.)

Horizontal gas recovery trenches may be installed at the bottom of the landfill, particularly if emissions of volatile organic compounds (VOCs) from the newly placed waste is expected to be a problem. To minimize the release of VOCs, a vacuum is applied and air is drawn through the completed portions of the landfill. The gas that is removed must be burned under controlled conditions to destroy the VOCs. Before the fill operation begins, a soil berm is constructed at the downwind side of the planned fill area. The berm serves as a windbreak to control blowing materials and as a face against which the waste can be compacted. For excavated landfills, the wall of the excavation usually serves as the initial compaction face.

The placement of wastes. Once the landfill site has been prepared, the next step in the process involves the actual placement of waste material. As shown in Fig. 11-4*b*, the waste is placed in cells beginning along the compaction face, continuing outward and upward from the face. The waste deposited in each operating period, usually one day, forms an individual cell. Wastes deposited by the collection and transfer vehicles are spread out in 18- to 24-in layers and compacted. Typical cell heights vary from 8 to 12 ft. The length of the working face varies with the site conditions and the size of the operation (see Fig. 11-1). The working *face* is the area of a landfill where solid waste is being unloaded, placed and compacted during a given operating period. The width of a cell varies from 10 to 30 ft, again depending on the design and capacity of the landfill. All exposed faces of the cell are covered with a thin layer of soil (6 to 12 in) or other suitable material at the end of each operating period.

After one or more lifts have been placed, horizontal gas recovery trenches can be excavated in the completed surface (see Fig. 11-3). The excavated trenches are filled with gravel, and perforated plastic pipes are installed in the trenches. Landfill gas is extracted through the pipes as the gas is produced. Successive lifts are placed on top of one another until the final design grade is reached. Depending on the depth of the landfill, additional leachate collection facilities may be placed in successive lifts. A cover layer is applied to the completed landfill section. The final cover is designed to minimize infiltration of precipitation and to route drainage away from the active section of the landfill. The cover is landscaped to control erosion. Vertical gas extraction wells may be installed at this time through the completed landfill surface. The gas extraction system is tied together and the extracted gas may be flared or routed to energy recovery facilities as appropriate.

Additional sections of the landfill are constructed outward from the completed sections, repeating the construction steps outlined above. As organic materials deposited within the landfill decompose, completed sections may settle. Landfill construction activities must include refilling and repairing of settled landfill surfaces to maintain the desired final grade and drainage. The gas and leachate control systems also must be extended and maintained. Upon completion of all fill activities, the landfill surface is repaired and upgraded with the installation of a final cover. The site is landscaped appropriately and prepared for other uses.

Postclosure management. Monitoring and maintenance of the completed landfill must continue by law for some time after closure (30 to 50 years). It is particularly important that the landfill surface be maintained and repaired to enhance drainage, that gas and leachate control systems be maintained and operated, and that the pollution detection system be monitored (see Chapter 16).

Reactions Occurring in Landfills. Solid wastes placed in a sanitary landfill undergo a number of simultaneous and interrelated biological, chemical, and physical changes, which are introduced in this section. The various reactions are considered in greater detail in subsequent sections of this chapter.

Biological reactions. The most important biological reactions occurring in landfills are those involving the organic material in MSW that lead to the evolution of landfill gases and, eventually, liquids. The biological decomposition process usually proceeds aerobically for some short period immediately after deposition of the waste until the oxygen initially present is depleted. During aerobic decomposition CO_2 is the principal gas produced. Once the available oxygen has been consumed, the decomposition becomes anaerobic and the organic matter is converted to CO_2, CH_4, and trace amounts of ammonia and hydrogen sulfide. Many other chemical reactions are biologically mediated as well. Because of the number of interrelated influences, it is difficult to define the conditions that will exist in any landfill or portion of a landfill at any stated time.

Chemical reactions. Important chemical reactions that occur within the landfill include dissolution and suspension of landfill materials and biological conversion products in the liquid percolating through the waste, evaporation and vaporization of chemical compounds and water into the evolving landfill gas, sorption of volatile and semivolatile organic compounds into the landfilled material, dehalogenation and decomposition of organic compounds, and oxidation-reduction reactions affecting metals and the solubility of metal salts. The dissolution of biological conversion products and other compounds, particularly of organic compounds, into the leachate is of special importance because these materials can be transported out of the landfill with the leachate. These organic compounds can subsequently be released into the atmosphere either through the soil (where leachate has move away from an unlined landfill) or from uncovered leachate treatment facilities. Other important chemical reactions include those between certain organic compounds and clay liners, which may alter the structure and permeability of the liner material. The interrelationships of these chemical reactions within a landfill are not well understood.

Physical reactions. Among the more important physical changes in landfills are the lateral diffusion of gases in the landfill and emission of landfill gases to the surrounding environment, movement of leachate within the landfill and into underlying soils, and settlement caused by consolidation and decomposition of landfilled material. Landfill gas movement and emissions are particularly important considerations in landfill management. As gas is evolved within a landfill, internal pressure may build, causing the landfill cover to crack and leak. Water entering the landfill through the leaking cover may enhance the gas production rate, causing still more cracking. Escaping landfill gas may carry trace carcinogenic and teratogenic compounds into the surrounding environment. Because landfill gas usually has a high methane content, there may be a combustion and/or explosion hazard. Leachate migration is another concern. As leachate migrates downward in the landfill, it may transfer compounds and materials to new locations where they may react more readily. Leachate occupies pore spaces in the landfill and in doing so may interfere with the migration of landfill gas.

Concerns with the Landfilling of Solid Wastes

Concerns with the landfilling of solid waste are related to (1) the uncontrolled release of landfill gases that might migrate off-site and cause odor and other potentially dangerous conditions, (2) the impact of the uncontrolled discharge of landfill gases on the greenhouse effect in the atmosphere, (3) the uncontrolled release of leachate that might migrate down to underlying groundwater or to surface waters, (4) the breeding and harboring of disease vectors in improperly managed landfills, and (5) the health and environmental impacts associated with the release of the trace gases arising from the hazardous materials that were often placed in landfills in the past. The goal for the design and operation of a modern landfill is to eliminate or minimize the impacts associated with these concerns (see Fig. 11-6).

(*a*)

(*b*)

FIGURE 11-6

Views taken from completed landfills: (*a*) city of Sacramento, CA, skyline in background, about 30 blocks away and (*b*) area-type landfill next to housing area.

Federal and State Regulations for Landfills

In planning for the implementation of a new landfill, attention must be paid to the many federal and state regulations that have been enacted to improve the performance of sanitary landfills. The principal federal requirements for municipal solid waste landfills are contained in Subtitle D of the Resource Conservation and Recovery Act (RCRA) and in EPA Regulations on Criteria for Classification of Solid Waste Disposal Facilities and Practices (40 CFR 258). The final version of Part 258—Criteria for Municipal Solid Waste Landfills (MSWLFs) was signed on September 11, 1991. The subparts of Part 258 deal with the following issues:

Subpart A	General
Subpart B	Location restrictions
Subpart C	Operating criteria
Subpart D	Design criteria
Subpart E	Groundwater monitoring and corrective action
Subpart F	Closure and postclosure care
Subpart G	Financial assurance criteria

The Clean Air Act also contains provisions dealing with gas emissions from landfills. In addition to the federal regulations, many states have adopted regulations governing the design, operation, closure and long-term maintenance of landfills. In many cases, the regulations that have been adopted by the individual states have been more restrictive than the federal requirements. Permitting of landfills is considered in Chapter 20.

11-2 LANDFILL CLASSIFICATION, TYPES, AND METHODS

The purpose of this section is to introduce the reader to (1) a commonly used landfill classification system, (2) the different types of landfills that are now used, and (3) the different landfilling methods that are used in various parts of the country.

Classification of Landfills

Although a number of landfill classification systems have been proposed over the years, the classification system adopted by the state of California in 1984 is perhaps the most widely accepted classification system for landfills. In the California system, reported below, three classifications are used:

Classification	Type of waste
I	Hazardous waste
II	Designated waste
III	Municipal solid waste (MSW)

Designated wastes are nonhazardous wastes that may release constituents in concentrations that exceed applicable water quality objectives or those wastes which have been granted a variance by the State Department of Health Services (DOHS). Note that this classification system focuses primarily on the protection of surface and groundwater rather than landfill gas migration or air quality.

Types of Landfills

The principal types of landfills can be classified as (1) conventional landfills for commingled MSW, (2) landfills for milled solid wastes, and (3) monofills for designated or specialized wastes. Other types of landfills and landfill operations, including the recycle of leachate, are also discussed.

Landfills for Commingled MSW. The majority of the landfills throughout the United States are designed for commingled MSW. In many of these Class III landfills, limited amounts of nonhazardous industrial wastes and sludge from water and wastewater treatment plants are also accepted. In many states, treatment plant sludges are accepted if they are dewatered to a solids content of 51 percent or greater. For example, in California the deposition of sludge in MSW landfills is restricted to a ratio of five parts solid waste to one part sludge by weight. Many municipalities have adopted even more restrictive limitations on the amount of sludge that can be accepted.

In most cases, native soil is used as the intermediate and final cover material. However, in locations such as Florida and New Jersey where the amount of native soil available for use as intermediate cover material is limited, alternative materials such as compost produced from yard wastes and MSW, foam, old rugs and carpeting, dredging spoils, and demolition wastes have been used for the purpose. To obtain additional landfill capacity, abandoned and or closed landfills in some locations are being reused by excavating the decomposed material to recover the metals and using the decomposed residue as daily cover for the new wastes. In some cases, the decomposed wastes are excavated and stockpiled, and a liner is installed before the landfill is reactivated.

Landfills for Shredded Solid Wastes. An alternative method of landfilling that is being tried in several U.S. locations involves shredding of the solid wastes before placement in a landfill. Shredded (or milled) waste can be placed at up to 35 percent greater density than unshredded waste, and without daily cover in some state regulations. Blowing litter, odors, flies, and rats have not been significant problems. Because shredded waste can be compacted to a tighter and more uniform surface, a reduced amount of soil cover or some other cover material may be sufficient to control infiltration of water during the fill operation.

Disadvantages of the method include the need for a shredding facility and the need to operate a conventional landfill section for wastes that cannot be easily shredded. The shredded waste method has potential application in areas where landfill capacity is very expensive (because of the greater compaction obtainable), where suitable cover material is not readily available, and where precipitation is very low or highly seasonal. Shredded waste can also be used to produce compost that can be used as intermediate cover material.

Landfills for Individual Waste Constituents. Landfills for individual waste constituents are known as *monofills*. Combustion ash, asbestos, and other similar wastes, often identified as designated wastes, are typically placed in monofills to isolate them from materials placed in MSW landfills. Because combustion ash contains small amounts of unburned organic material, the production of odors from the reduction of sulfate (see Eq. 4-12) has been a problem in monofills used for combustion ash. In monofills used for combustion ash, the installation of a gas recovery system is recommended to control odor problems.

Other Types of Landfills. In addition to the conventional methods of landfilling already described, other specialized methods of landfilling designed to enhance different goals of landfill management are being developed. Alternative operational methods that are being used include (1) landfills designed to maximize the rate of landfill gas generation and (2) landfills operated as integrated solid waste treatment units. The practice of landfilling in wetland areas, now prohibited, is also described.

Landfills designed to maximize gas production. If the quantity of landfill gas that is produced and recovered from the anaerobic decomposition of solid wastes is to be maximized, specialized landfill designs will be required. For example, the use of deep, individually lined cells, in which the wastes are placed without intermediate layers of cover material and leachate is recycled to enhance the biological decomposition process, is a viable option. A possible disadvantage of such a landfill is that excess leachate must ultimately be disposed of.

Landfills as integrated treatment units. In this method of operation, the organic constituents would be separated out and placed in a separate landfill where the biodegradation rates would be enhanced by increasing the moisture content of the waste, either by recycling leachate or by seeding with digested wastewater treatment plant sludge or animal manure. The degraded material would be excavated and used as cover material for new fill areas, and the excavated cell would be filled with new waste.

Landfills in wetland areas. In the past, landfilling in wetland areas, such as swamps, marshes, and tidal areas, was considered acceptable if adequate drainage was provided and if nuisance conditions did not develop. Under current federal regulations, such destruction of wetland areas is prohibited, although the expansion of an existing landfill may be allowed under special conditions. Because many landfills already exist in these areas, a brief description of the methods typically used in these fills is presented.

The usual practice in filling wetlands was to divide the area into cells or lagoons and schedule the filling operations so that one individual cell or lagoon would be filled each year. Often, solid wastes were placed directly in the water.

Alternatively, clean fill material was added up to, or slightly above, the water level before waste filling operations were started. To withstand mud waves and to increase structural stability, dikes dividing the cells or lagoons were constructed with riprap, trees, tree limbs, lumber, demolition wastes, and similar materials in addition to clean fill material. In some cases, to prevent the movement of leachate and gases from completed cells or lagoons, clay and lightweight interlocking steel or wood-sheet piling has been used.

Landfilling Methods

The principal methods used for the landfilling of MSW are (1) excavated cell/trench, (2) area, and (3) canyon. The principal features of these types of landfills, illustrated in Figs. 11-7 and 11-8, are described below. Landfill design details are presented later in the chapter.

Excavated Cell/Trench Method. The excavated cell/trench method of landfilling (see Fig. 11-7*a*) is ideally suited to areas where an adequate depth of cover material is available at the site and where the water table is not near the surface. Typically, solid wastes are placed in cells or trenches excavated in the soil (see Fig. 11-8*a*). The soil excavated from the site is used for daily and final cover. The excavated cells or trenches are usually lined with synthetic membrane liners or low-permeability clay or a combination of the two to limit the movement of both landfill gases and leachate (see Fig. 11-8). Excavated cells are typically square, up to 1000 ft in width and length, with side slopes of 1.5:1 to 2:1. Trenches vary from 200 to 1000 ft in length, 3 to 10 ft in depth, and 15 to 50 ft in width.

In some states, landfills constructed below the high-groundwater level are allowed if special provisions are made to prevent groundwater from entering the landfill and to contain or eliminate the movement of leachate and gases from completed cells. Usually the site is dewatered, excavated, and then lined in compliance with local regulations. The dewatering facilities are operated until the site is filled to avoid the creation of uplift pressures that could cause the liner to heave and rupture. The use of clay and membrane liners is considered further in Section 11-5.

Area Method. The area method is used when the terrain is unsuitable for the excavation of cells or trenches in which to place the solid wastes (see Figs. 11-7*b* and 11-8*b*). High-groundwater conditions, which occur in many parts of Florida and elsewhere too, necessitate the use of area-type landfills. Site preparation includes the installation of a liner and leachate control system. Cover material must be hauled in by truck or earthmoving equipment from adjacent land or from borrow-pit areas. As noted above, in locations with limited availability of material that can be used as cover, compost produced from yard wastes and MSW has been used successfully as intermediate cover material. Other techniques that have been used include the use of movable temporary cover materials such as soil and geomembranes. Soil and geomembranes, placed temporarily over a completed cell, can be removed before the next lift is begun.

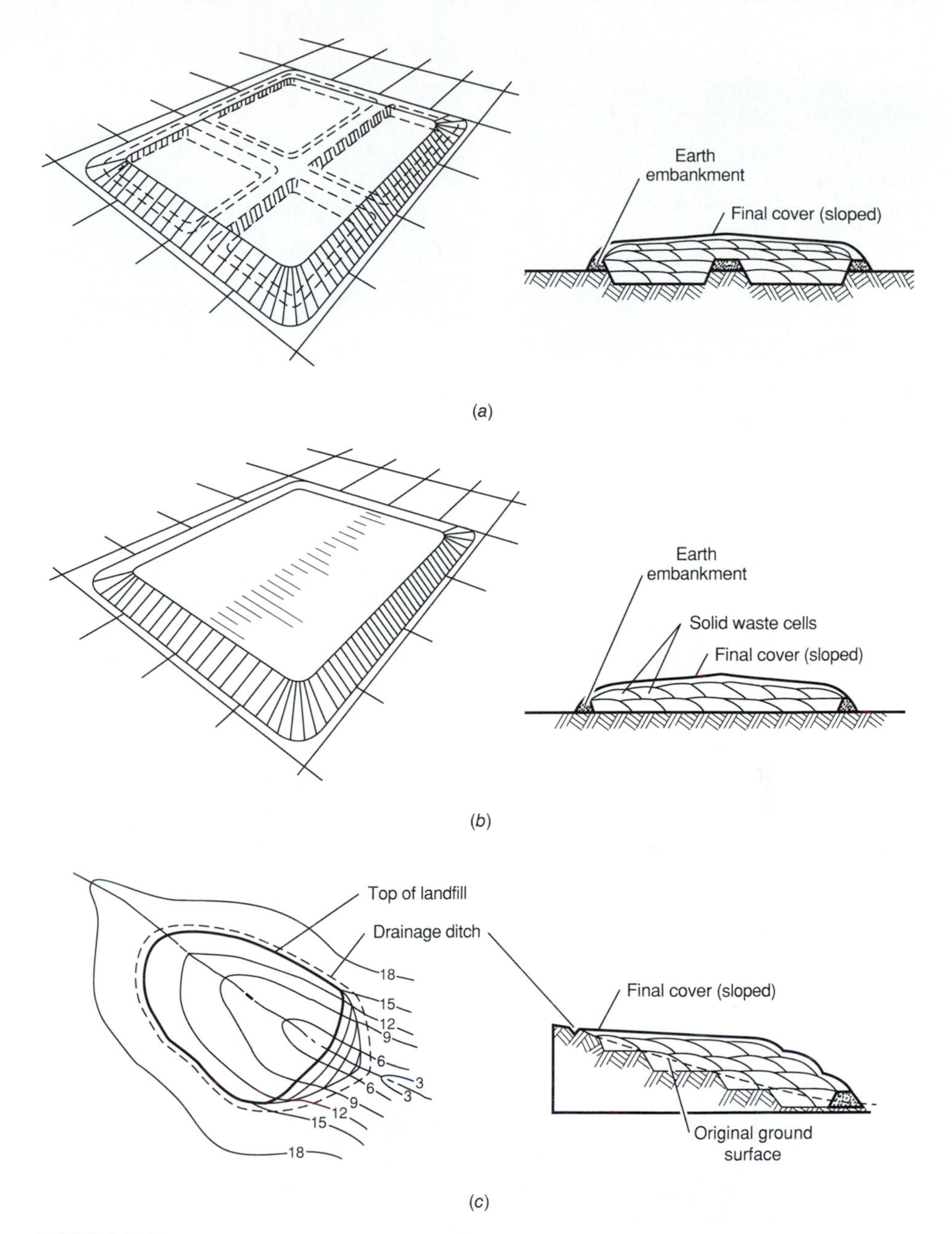

FIGURE 11-7
Commonly used landfilling methods: (*a*) excavated cell/trench, (*b*) area, and (*c*) canyon/depression.

(a) (b)

FIGURE 11-8
Pictorial views of the construction of different types of landfills: (*a*) excavated cell landfill and (*b*) area landfill.

Canyon/Depression Method. Canyons, ravines, dry borrow pits, and quarries have been used for landfills (see Figs. 11-7*c* and 11-9). The techniques to place and compact solid wastes in canyon/depression landfills vary with the geometry of the site, the characteristics of the available cover material, the hydrology and geology of the site, the type of leachate and gas control facilities to be used, and the access to the site.

Control of surface drainage often is a critical factor in the development of canyon/depression sites. Typically, filling for each lift starts at the head end of the canyon (see Fig. 11-7*c*) and ends at the mouth, so as to prevent the accumulation of water behind the landfill. Canyon/depression sites are filled in multiple lifts, and the method of operation is essentially the same as the area method described above. If a canyon floor is reasonably flat, the initial landfilling may be carried out using the excavated cell/trench method discussed previously.

FIGURE 11-9
Landfilling in a canyon site. Site is being prepared for placement of geomembrane landfill liner.

A key to the successful use of the canyon/depression method is the availability of adequate material to cover the individual lifts as they are completed and to provide a final cover over the entire landfill when the final height is reached. Cover material is excavated from the canyon walls or floor before the liner system is installed. Borrow pits and abandoned quarries may not contain sufficient soil for intermediate cover, so that cover material may have to be imported. Compost produced from yard waste and MSW can be used for the intermediate cover layers.

11-3 LANDFILL SITING CONSIDERATIONS

One of the most difficult tasks faced by most communities in implementing an integrated solid waste management program is the siting of new landfills. This section introduces the factors that must be considered in siting a new landfill. Greater detail in Chapter 20 is provided. Factors that must be considered in evaluating potential sites for the long-term disposal of solid waste include (1) haul distance, (2) location restrictions, (3) available land area, (4) site access, (5) soil conditions and topography, (6) climatological conditions, (7) surface water hydrology, (8) geologic and hydrogeologic conditions, (9) local environmental conditions, and (10) potential ultimate uses for the completed site. Final selection of a disposal site usually is based on the results of a detailed site survey, engineering design and cost studies, and an environmental impact assessment. It is interesting that the up-front development costs for new landfills in California now vary from $10 million to $20 million (1992) before the first load of waste is placed in the landfill.

Haul Distance

The haul distance is one of the important variables in the selection of a disposal site. From computations presented in Chapters 8 and 10, it is clear that the length of the haul can significantly affect the overall design and operation of the waste management system. Although minimum haul distances are desirable, other factors must also be considered. Because the siting of landfills is usually determined by environmental and political concerns, long-distance hauling, discussed in Chapter 10, is now becoming more routine.

Location Restrictions

Location restrictions refer to where landfills can be located. Restrictions now apply with respect to siting landfills near airports, in floodplains, in wetlands, in areas with known faults, in seismic impact zones, and in unstable areas (see Table 11-1). The specific federal requirements are contained in Subpart B—Location Restrictions of Part 258 of Subtitle D of the Resource Conservation and Recovery Act (RCRA). In addition, many states have adopted additional location restrictions. All current restrictions must be reviewed carefully during the preliminary siting process to avoid expending time and money evaluating a site that will not conform with the regulatory requirements.

TABLE 11-1
Siting limitations contained in Subtitle D of the Resources Conservation and Recovery Act as adopted by the EPA

Location	Siting limitation
Airports	10,000 ft from an airport used by turbojet aircraft; 5000 ft from an airport used by piston-type aircraft. Any landfills closer will have to demonstrate that they do not pose a bird hazard to aircraft.
Flood plains	100-year flood plain. Landfill located within the 100-year floodplain will have to be designed so as not to restrict flood flow, reduce the temporary water storage capacity of the floodplain, or result in washout of solid waste, which would pose a hazard to human health and the environment.
Wetlands	New landfills will not be able to locate in wetlands unless the following conditions have been demonstrated: (1) No practical alternative with less environmental risk exists. (2) Violations of other state and local laws will not occur. (3) The unit would not cause or contribute to significant degradation of the wetland. (4) Appropriate and practicable steps have been taken to minimize potential adverse impacts. (5) Sufficient information to make determination is available.
Fault areas	New landfill units cannot be sited within 200 ft of a fault line that has had a displacement in Holocene time (past 10,000 years).
Seismic impact zone	New landfill unit located within a seismic impact zone will have to demonstrate that all contaminant structures (liners, leachate collection systems, and surface water control structures) are designed to resist the maximum horizontal acceleration in lithified materials (liquid or loose materials consolidated into solid rock) for the site.
Unstable areas	Landfill units located in unstable areas must demonstrate that the design ensures stability of structural components. The unstable areas include areas that are landslide prone, that are in karst geology susceptible to sinkhole formation, and that are undermined by subsurface mines. Existing facilities that cannot demonstrate the stability of the structural components will be required to close within five years of the regulation's effective date.

Available Land Area

In selecting potential land disposal sites, it is important to ensure that sufficient land area is available. Although there are no fixed rules concerning the area required, it is desirable to have sufficient area, including an adequate buffer zone, to operate for at least five years at a given site. For shorter periods, the disposal operation becomes considerably more expensive, especially with respect to site preparation, provision of auxiliary facilities such as platform scales and storage facilities, and completion of the final cover. In the initial assessment of potential disposal sites, it is important to project the extent of the waste diversion that is likely to occur in the future and determine the impact of that diversion on the quantity and condition of the residual materials to be disposed of. For preliminary planning purposes, the amount of land area required can be estimated as illustrated in Example 11-1.

Example 11-1 Estimation of required landfill area. Estimate the required landfill area for a community with a population of 31,000. Assume that the following conditions apply:

1. Solid waste generation = 6.4 lb/capita · d
2. Compacted specific weight of solid wastes in landfill = 800 lb/yd^3
3. Average depth of compacted solid wastes = 20 ft

Solution

1. Determine the daily solid wastes generation rate in tons per day.

$$\text{Generation rate} = \frac{(31{,}000 \text{ people})(6.4 \text{ lb/capita} \cdot \text{d})}{2000 \text{ lb/ton}}$$

$$= 99.2 \text{ ton/d } (89{,}994 \text{ kg/d})$$

2. Computationally, the required area is determined as follows:

$$\text{Volume required/d} = \frac{99.2 \text{ ton/d} \times 2000 \text{ lb/ton}}{800 \text{ lb/yd}^3}$$

$$= 248 \text{ yd}^3\text{/d } (190 \text{ m}^3\text{/d})$$

$$\text{Area required/yr} = \frac{(248 \text{ yd}^3\text{/d})(365 \text{ d/yr})(27 \text{ ft}^3\text{/yd}^3)}{(20 \text{ ft})(43{,}560 \text{ ft}^2\text{/acre})}$$

$$= 2.81 \text{ acre/yr } (1.14 \text{ hectare/yr})$$

Comment. The actual site requirements will be greater than the value computed because additional land is required for a buffer zone, office and service buildings, access roads, utility access, and so on. Typically, this allowance varies from 20 to 40 percent. A more rigorous approach to the determination of the required landfill area involves consideration of the contours of the completed landfill (see Example 11-7 in Section 11-12) and the effects of gas production and overburden compaction (see Example 11-13 in Section 11-12).

Site Access

As the number of operating landfills continues to decrease, new landfills that are being sited are increasing in size. Because land areas of suitable size are often not near existing developed roadways and cities, construction of access roadways and the use of long haul equipment has become a fact of life and an important part of landfill siting. Rail lines often pass nearby remote sites that are suitable for use as landfills; thus, there is renewed interest in the use of rail haul for transporting wastes to these remote sites.

Soil Conditions and Topography

Because it is necessary to cover the solid wastes placed in the landfill each day and to provide a final cover layer after the landfilling operation is completed, data must be obtained on the amounts and characteristics of the soils in the area. If the soil under the proposed landfill area is to be used for cover material, data must be developed on its geologic and hydrogeologic characteristics. If cover material is to be obtained from a borrow pit, test borings will be needed to characterize the material. The local topography must be considered because it will affect the type of landfill operation to be used, the equipment requirements, and the extent of work necessary to make the site usable. If suitable cover material is limited or an effort is being made to extend the useful life of the landfill, it may be necessary to consider the use of compost or other materials for intermediate cover.

Climatologic Conditions

Local weather conditions must also be considered in the evaluation of potential sites. In many locations, winter conditions will affect access to the site. Wet weather may necessitate the use of separate landfill areas. Where freezing is severe, landfill cover material must be available in stockpiles when excavation is impracticable. Wind strength and wind patterns must also be considered carefully. To avoid blowing or flying debris, windbreaks must be established. The specific form of windbreak depends on local conditions.

Surface Water Hydrology

The local surface water hydrology of the area is important in establishing the existing natural drainage and runoff characteristics that must be considered. Other conditions of flooding (e.g., the limits of the 100-year flood) must also be identified. Because mitigation measures must be developed to divert surface runoff from the landfill site, planners must take great care in defining existing and intermittent flow channels and the area and characteristics of the contributing watershed.

Geologic and Hydrogeologic Conditions

Geologic and hydrogeologic conditions are perhaps the most important factors in establishing the environmental suitability of the area for a landfill site. Data on these factors are required to assess the pollution potential of the proposed site and to establish what must be done to the site to ensure that the movement of leachate or gases from the landfill will not impair the quality of local groundwater or contaminate other subsurface or bedrock aquifers. In the preliminary assessment of alternative sites, it may be possible to use U.S. Geological Survey maps and state or local geologic information. Geologic drilling logs of nearby wells can also be used for a preliminary assessment.

Local Environmental Conditions

Although it has been possible to build and operate landfill sites in close proximity to both residential and industrial developments, they must be operated very

(a)

(b)

FIGURE 11-10
Views from well-managed completed landfills: (*a*) next to an expensive residential area and (*b*) adjacent to an industrial park.

carefully if they are to be environmentally acceptable with respect to traffic, noise, odor, dust, airborne debris, visual impact, vector control, hazards to health, and property values (see Fig. 11-10). To minimize the impact of landfilling operations, landfills are now sited in more remote locations where adequate buffer zones surrounding the landfill can be maintained.

Ultimate Use for Completed Landfills

One of the advantages of a landfill is that, once it is completed, a sizable area of land becomes available for other purposes. Because the ultimate use affects the design and operation of the landfill, this issue must be resolved before the layout and design of the landfill is begun. Choices for the ultimate use of completed landfills are becoming more limited by state and federal regulations dealing with landfill closure and postclosure maintenance. If the completed landfill is to be used for some municipal function, a staged planting program should be initiated and continued as portions of the landfill are completed. The ultimate use and long-term management of landfill sites are considered in Chapters 16 and 17.

11-4 COMPOSITION AND CHARACTERISTICS, GENERATION, MOVEMENT, AND CONTROL OF LANDFILL GASES

A solid waste landfill can be conceptualized as a biochemical reactor, with solid waste and water as the major inputs, and with *landfill gas* and *leachate* as the principal outputs. Material stored in the landfill includes partially biodegraded organic material and the other inorganic waste materials originally placed in the landfill. Landfill gas control systems are employed to prevent unwanted movement of landfill gas into the atmosphere or the lateral and vertical movement through the surrounding soil. Recovered landfill gas can be used to produce energy or can be flared under controlled conditions to eliminate the discharge of harmful constituents to the atmosphere.

Composition and Characteristics of Landfill Gas

Landfill gas is composed of a number of gases that are present in large amounts (the principal gases) and a number of gases that are present in very small amounts (the trace gases). The principal gases are produced from the decomposition of the organic fraction of MSW. Some of the trace gases, although present in small quantities, can be toxic and could present risks to public health.

Principal Landfill Gas Constituents. Gases found in landfills include ammonia (NH_3), carbon dioxide (CO_2), carbon monoxide (CO), hydrogen (H_2), hydrogen sulfide (H_2S), methane (CH_4), nitrogen (N_2), and oxygen (O_2). The typical percentage distribution of gases found in a MSW landfill is reported in Table 11-2. Data on molecular weight and density are presented in Table 11-3. Data that can be used to determine the solubility of these gases in water (leachate) are presented in Appendix F. Methane and carbon dioxide are the principal gases produced from the anaerobic decomposition of the biodegradable organic waste components in MSW. When methane is present in the air in concentrations between 5 and 15

TABLE 11-2
Typical constituents found in MSW landfill gas[a]

Component	Percent (dry volume basis)[b]
Methane	45–60
Carbon dioxide	40–60
Nitrogen	2–5
Oxygen	0.1–1.0
Sulfides, disulfides, mercaptans, etc.	0–1.0
Ammonia	0.1–1.0
Hydrogen	0–0.2
Carbon monoxide	0–0.2
Trace constituents	0.01–0.6
Characteristic	**Value**
Temperature, °F	100–120
Specific gravity	1.02–1.06
Moisture content	Saturated
High heating value, Btu/sft^3	400–550

[a] Adapted from Refs. 16, 24, 34.

[b] Exact percentage distribution will vary with the age of the landfill.

TABLE 11-3
Molecular weight, density, and specific weight of gases found in sanitary landfill at standard conditions (0°C, 1 atm)

Gas	Formula	Molecular weight	Density, g/L	Specific weight, lb/ft^3
Air		28.97	1.2928	0.0808
Ammonia	NH_3	17.03	0.7708	0.0482
Carbon dioxide	CO_2	44.00	1.9768	0.1235
Carbon monoxide	CO	28.00	1.2501	0.0781
Hydrogen	H_2	2.016	0.0898	0.0056
Hydrogen sulfide	H_2S	34.08	1.5392	0.0961
Methane	CH_4	16.03	0.7167	0.0448
Nitrogen	N_2	28.02	1.2507	0.0782
Oxygen	O_2	32.00	1.4289	0.0892

[a] Adapted from Ref. 35.

Note: For ideal gas behavior, the density is equal to mp/RT where m is the molecular weight of the gas, p is the pressure, R is the *universal* gas constant, and T is the temperature using a consistent set of units.

percent, it is explosive. Because only limited amounts of oxygen are present in a landfill when methane concentrations reach this critical level, there is little danger that the landfill will explode. However, methane mixtures in the explosive range can form if landfill gas migrates off-site and mixes with air. The concentration of these gases that may be expected in the leachate will depend on their concentration in the gas phase in contact with the leachate, as estimated using Henry's law, given in Appendix F. Because carbon dioxide will affect the pH of the leachate, carbonate equilibrium data that can be used to estimate the pH of the leachate are given in Appendix G.

Trace Landfill Gas Constituents. The California Integrated Waste Management Board has performed an extensive landfill gas sampling program as part of its landfill gas characterization study. Summary data on the concentrations of trace compounds found in landfill gas samples from 66 landfills are reported in Table 11-4. In another study conducted in England, gas samples were collected from three different landfills and analyzed for 154 compounds. A total of 116 organic compounds were found in landfill gas [54]. Many of the compounds found would be classified as volatile organic compounds (VOCs). The data presented in Table 11-4 are representative of the trace compounds found at most MSW landfills. The presence of these gases in the leachate that is removed from the landfill will depend on their concentrations in the landfill gas in contact with the leachate. Expected concentrations of these constituents in the leachate can be estimated using Henry's law as outlined in Appendix F. Note that the occurrence of significant concentrations of VOCs in landfill gas is associated with older landfills that accepted industrial and commercial wastes containing VOCs. In newer landfills in which the disposal of hazardous waste has been banned, the concentrations of VOCs in the landfill gas have been extremely low.

TABLE 11-4
Typical concentrations of trace compounds found in landfill gas at 66 California MSW landfills[a]

	Concentration, ppbV[b]		
Compound	**Median**	**Mean**	**Maximum**
Acetone	0	6,838	240,000
Benzene	932	2,057	39,000
Chlorobenzene	0	82	1,640
Chloroform	0	245	12,000
1,1-Dichloroethane	0	2,801	36,000
Dichloromethane	1,150	25,694	620,000
1,1-Dichloroethene	0	130	4,000
Diethylene chloride	0	2,835	20,000
trans-1,2-Dichloroethane	0	36	850
2,3-Dichloropropane	0	0	0
1,2-Dichloropropane	0	0	0
Ethylene bromide	0	0	0
Ethylene dichloride	0	59	2,100
Ethylene oxide	0	0	0
Ethyl benzene	0	7,334	87,500
Methyl ethyl ketone	0	3,092	130,000
1,1,2-Trichloroethane	0	0	0
1,1,1-Trichloroethane	0	615	14,500
Trichloroethylene	0	2,079	32,000
Toluene	8,125	34,907	280,000
1,1,2,2-Tetrachloroethane	0	246	16,000
Tetrachloroethylene	260	5,244	180,000
Vinyl chloride	1,150	3,508	32,000
Styrenes	0	1,517	87,000
Vinyl acetate	0	5,663	240,000
Xylenes	0	2,651	38,000

[a] Adapted from Ref. 5.
[b] ppbV = parts per billion by volume.

Generation of Landfill Gases

The generation of the principal landfill gases, the variation in their rate of generation with time, and the sources of trace gases in landfills is considered in the following discussion.

Generation of the Principal Landfill Gases. The generation of the principal landfill gases is thought to occur in five more or less sequential phases, as illustrated in Fig. 11-11. Each of these phases is described below; additional details may be found in Refs. 6, 12, 13, 34, 37, and 38. A more detailed description of the anaerobic digestion process, including the organisms and the principal reactions involved in the formation of methane is presented in Chapter 14.

Phase I—initial adjustment. Phase I is the *initial adjustment phase*, in which the organic biodegradable components in MSW undergo microbial decom-

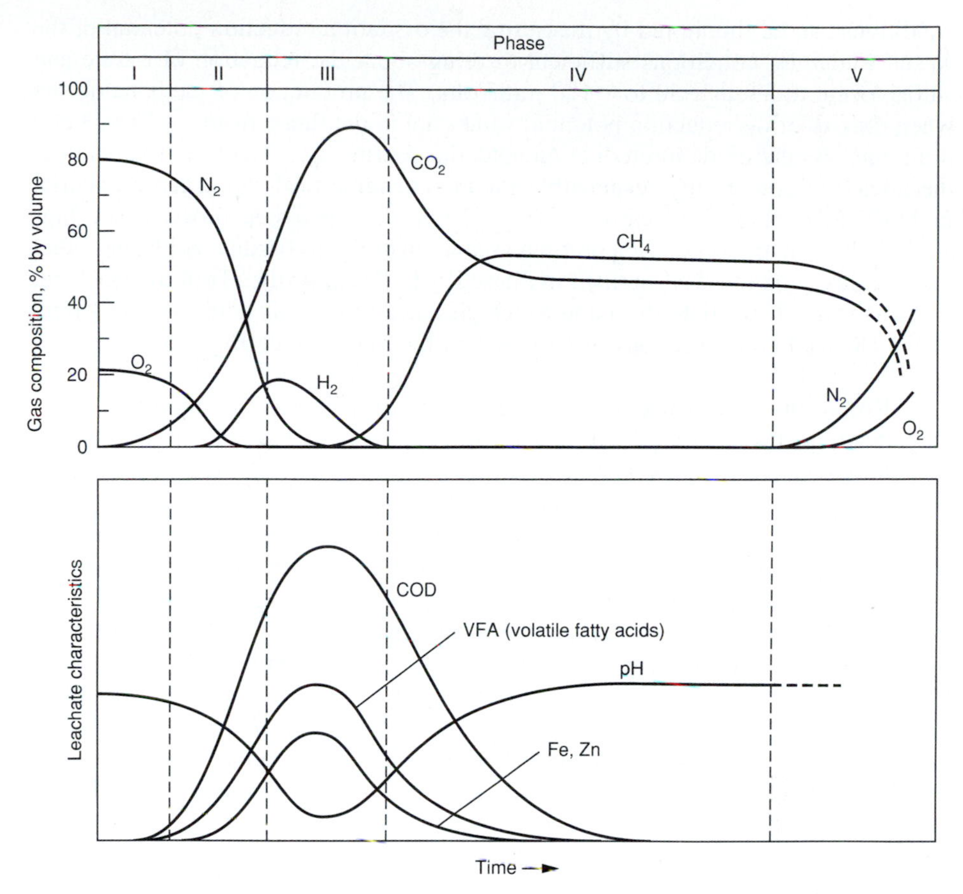

FIGURE 11-11
Generalized phases in the generation of landfill gases (I = initial adjustment, II = transition phase, III = acid phase, IV = methane fermentation, and V = maturation phase). (Adapted from Refs. 13, 34, 37, and 38.)

position as they are placed in a landfill and soon after. In Phase I, biological decomposition occurs under aerobic conditions, because a certain amount of air is trapped within the landfill. The principal source of both the aerobic and the anaerobic organisms responsible for waste decomposition is the soil material that is used as a daily and final cover. Digested wastewater treatment plant sludge, disposed of in many MSW landfills, and recycled leachate are other sources of organisms.

Phase II—transition phase. In Phase II, identified as the *transition phase*, oxygen is depleted and anaerobic conditions begin to develop. As the landfill becomes anaerobic, nitrate and sulfate, which can serve as electron acceptors (see Table 14-2) in biological conversion reactions, are often reduced to nitrogen gas and hydrogen sulfide (see Eqs. 4-12, 4-13, and 4-14). The onset of anaerobic conditions can be monitored by measuring the oxidation/reduction potential of the waste. Reducing conditions sufficient to bring about the reduction of nitrate and sulfate occur at about −50 to −100 millivolts. The production of methane occurs when the oxidation/reduction potential values are in the range from −150 to −300 millivolts. As the oxidation/reduction potential continues to decrease, members of the microbial community responsible for the conversion of the organic material in MSW to methane and carbon dioxide begin the three-step process (see Fig. 14-1), with conversion of the complex organic material to organic acids and other intermediate products as described in Phase III. In Phase II, the pH of the leachate, if any is formed, starts to drop due to the presence of organic acids and the effect of the elevated concentrations of CO_2 within the landfill (see Fig. 11-11).

Phase III—acid phase. In Phase III, the *acid phase*, the microbial activity initiated in Phase II accelerates with the production of significant amounts of organic acids and lesser amounts of hydrogen gas. The first step in the three-step process involves the enzyme-mediated transformation (hydrolysis) of higher-molecular mass compounds (e.g., lipids, polysaccharides, proteins, and nucleic acids) into compounds suitable for use by microorganisms as a source of energy and cell carbon. The second step in the process (acidogenesis) involves the microbial conversion of the compounds resulting from the first step into lower-molecular mass intermediate compounds as typified by acetic acid (CH_3COOH) and small concentrations of fulvic and other more complex organic acids. Carbon dioxide (CO_2) is the principal gas generated during Phase III. Smaller amounts of hydrogen gas (H_2) will also be produced. The microorganisms involved in this conversion, described collectively as nonmethanogenic, consist of facultative and obligate anaerobic bacteria. These microorganisms are often identified in the engineering literature as *acidogens* or *acid formers*.

The pH of the leachate, if formed, will often drop to a value of 5 or lower because of the presence of the organic acids and the elevated concentrations of CO_2 within the landfill. The biochemical oxygen demand (BOD_5), the chemical oxygen demand (COD), and the conductivity of the leachate will increase significantly during Phase III due to the dissolution of the organic acids in the leachate. Also, because of the low pH values in the leachate, a number of inorganic constituents, principally heavy metals, will be solubilized during Phase III. Many essential nutrients are also removed in the leachate in Phase III. If leachate is not recycled, the essential nutrients will be lost from the system. It is important to note that if leachate is not formed, the conversion products produced during Phase III will remain within the landfill as sorbed constituents and in the water held by the waste as defined by the field capacity (see Section 11-5).

Phase IV—methane fermentation phase. In Phase IV, the *methane fermentation phase*, a second group of microorganisms, which convert the acetic acid and hydrogen gas formed by the acid formers in the acid phase to CH_4 and CO_2, becomes more predominant. In some cases, these organisms will begin to develop toward the end of Phase III. The microorganisms responsible for this conversion are strict anaerobes and are called methanogenic. Collectively, they are identified in the literature as *methanogens* or *methane formers*. In Phase IV, both methane and acid formation proceed simultaneously, although the rate of acid formation is considerably reduced.

Because the acids and the hydrogen gas produced by the acid formers have been converted to CH_4 and CO_2 in Phase IV, the pH within the landfill will rise to more neutral values in the range of 6.8 to 8. In turn, the pH of the leachate, if formed, will rise, and the concentration of BOD_5 and COD and the conductivity value of the leachate will be reduced. With higher pH values, fewer inorganic constituents can remain in solution; as a result, the concentration of heavy metals present in the leachate will also be reduced.

Phase V—maturation phase. Phase V, the *maturation phase*, occurs after the readily available biodegradable organic material has been converted to CH_4 and CO_2 in Phase IV. As moisture continues to migrate through the waste, portions of the biodegradable material that were previously unavailable, will be converted. The rate of landfill gas generation diminishes significantly in Phase V, because most of the available nutrients have been removed with the leachate during the previous phases and the substrates that remain in the landfill are slowly biodegradable. The principal landfill gases evolved in Phase V are CH_4 and CO_2. Depending on the landfill closure measures, small amounts of nitrogen and oxygen may also be found in the landfill gas. During maturation phase, the leachate will often contain humic and fulvic acids, which are difficult to process further biologically.

Duration of phases. The duration of the individual phases in the production of landfill gas will vary depending on the distribution of the organic components in landfill, the availability of nutrients, the moisture content of waste, moisture routing through the fill, and the degree of initial compaction. For example, if several loads of brush are compacted together the carbon/nitrogen ratio and the nutrient balance may not be favorable for the production of landfill gas (see Chapter 14). Likewise, the generation of landfill gas will be retarded if sufficient moisture is not available. Increasing the density of the material placed in the landfill will decrease the possibility of moisture reaching all parts of the waste and, thus, reduce the rate of bioconversion and gas production. Typical data on the percentage distribution of principal gases found in a newly completed landfill as a function of time are reported in Table 11-5.

Volume of Gas Produced. The generalized chemical reaction for the anaerobic decomposition of solid waste can be written as

$$\underset{\text{(solid waste)}}{\underset{\text{matter}}{\text{Organic}}} + H_2O \xrightarrow{\text{bacteria}} \underset{\text{matter}}{\underset{\text{organic}}{\text{biodegraded}}} + CH_4 + CO_2 + \underset{\text{gases}}{\text{other}} \qquad (11\text{-}1)$$

TABLE 11-5
Percentage distribution of landfill gases observed during the first 48 months after the closure of a landfill cell[a]

Time interval since cell completion, months	Average, percent by volume		
	Nitrogen, N_2	Carbon dioxide, CO_2	Methane, CH_4
0–3	5.2	88	5
3–6	3.8	76	21
6–12	0.4	65	29
12–18	1.1	52	40
18–24	0.4	53	47
24–30	0.2	52	48
30–36	1.3	46	51
36–42	0.9	50	47
42–48	0.4	51	48

[a] From Ref. 32.

Note that the reaction requires the presence of water. Landfills lacking sufficient moisture content have been found in a "mummified" condition, with decades-old newsprint still in readable condition. Hence, although the total amount of gas that will be produced from solid waste derives straightforwardly from the reaction stoichiometry, local hydrologic conditions affect significantly the rate and the period of time over which that gas production takes place.

The volume of the gases released during anaerobic decomposition can be estimated in a number of ways. For example, if the individual organic constituents found in MSW (with the exception of plastics) are represented with a generalized formula of the form $C_aH_bO_cN_d$, then the total volume of gas can be estimated using Eq. (11-2), assuming the complete conversion of the biodegradable organic waste to CO_2 and CH_4.

$$C_aH_bO_cN_d + \left(\frac{4a - b - 2c + 3d}{4}\right)H_2O \rightarrow$$

$$\left(\frac{4a + b - 2c - 3d}{8}\right)CH_4 + \left(\frac{4a - b + 2c + 3d}{8}\right)CO_2 + dNH_3 \qquad (11\text{-}2)$$

In general, the organic materials present in solid wastes can be divided into two classifications: (1) those materials that will decompose rapidly (three months to five years) and (2) those materials that will decompose slowly (up to 50 years or more). The rapidly and slowly decomposable components of the organic fraction of MSW are identified in Table 11-6. A procedure that can be used to estimate the amount of gas that can be generated from the biodegradable portion of the organic waste in MSW is illustrated in Example 11-2. Assuming the formula

TABLE 11-6
Rapidly and slowly biodegradable organic constituents in MSW

Organic waste component	Rapidly biodegradable	Slowly biodegradable
Food wastes	✓	
Newspaper	✓	
Office paper	✓	
Cardboard	✓	
Plastics[a]		
Textiles		✓
Rubber		✓
Leather		✓
Yard wastes	✓[b]	✓[c]
Wood		✓
Misc. organics	–	✓

[a] Plastics are generally considered nonbiodegradable.

[b] Leaves and grass trimmings. Typically, 60 percent of the yard wastes are considered rapidly biodegradable.

[c] Woody portions of yard wastes.

$C_{68}H_{111}O_{50}N$, as developed in Example 11-2 can be used to describe the rapidly biodegradable organic fraction of the MSW, then, as computed in Example 11-2, the maximum amount of gas that would be expected under optimum conditions is 14.0 ft^3/lb of biodegradable organic solids destroyed. The biodegradable fraction of the organic waste depends to a large extent on the lignin content of the waste (see Chapter 3). The biodegradability of various organic constituents, based on lignin content, is reported in Table 11-7. As shown, newspaper is only 22 percent biodegradable.

TABLE 11-7
Biodegradability of the organic constituents in MSW

Organic waste component	Lignin content, % of VS	Biodegradable fraction,[a] % of VS
Food wastes	0.4	0.82
Newspaper	21.9	0.22
Office paper	0.4	0.82
Cardboard	12.9	0.47
Yard wastes	4.1	0.72

[a] Biodegradable fraction = 0.83 − (0.028) × LC, where LC = % of VS (volatile solids).

Example 11-2 Estimate the chemical composition and the amount of gas that can be derived from the organic constituents in MSW. Determine the chemical composition and the amount of gas that can be derived from the rapidly and slowly decomposable organic constituents in MSW as given in Table 3-4. Assume 60 percent of the yard wastes will decompose rapidly.

Solution

1. Set up a computation table to determine the percentage distribution of the major elements composing the waste. The necessary computations for the rapidly and slowly decomposable organic constituents are presented below. The moisture content of the waste constituents is taken from Table 4-1.

Component	Wet weight,[a] lb	Dry weight,[b] lb	Composition,[c] lb					
			C	H	O	N	S	Ash
Rapidly decomposable organic constituents								
Food wastes	9.0	2.7	1.30	0.17	1.02	0.07	0.01	0.14
Paper	34.0	32.0	13.92	1.92	14.08	0.10	0.06	1.92
Cardboard	6.0	5.7	2.51	0.34	2.54	0.02	0.01	0.29
Yard wastes	11.1[d]	4.4	2.10	0.26	1.67	0.15	0.01	0.20
Total	60.1	44.8	19.83	2.69	19.31	0.34	0.09	2.55
Slowly decomposable organic constituents								
Textiles	2.0	1.8	0.99	0.12	0.56	0.08	—	0.05
Rubber	0.5	0.5	0.39	0.05	—	0.01	—	0.05
Leather	0.5	0.4	0.24	0.03	0.05	0.04	—	0.04
Yard wastes	7.4[e]	3.0	1.43	0.18	1.14	0.10	0.01	0.13
Wood	2.0	1.6	0.79	0.10	0.69	—	—	0.02
Total	12.4	7.3	3.84	0.48	2.44	0.23	0.01	0.29

[a] See Table 3-4.

[b] See Table 4-1.

[c] See Table 4-3.

[d] 11.1 = 18.5 × 0.60.

[e] 7.4 = 18.5 − 11.1.

2. Compute the molar composition of the elements neglecting the ash.

	C	H	O	N	S
lb/mole	12.01	1.01	16.00	14.01	32.06
Total moles					
Rapidly decomp.	1.6511	2.6634	1.2069	0.0241	0.0028
Slowly decomp.	0.3197	0.4752	0.1525	0.0164	0.0003

3. Determine an approximate chemical formula without sulfur. Set up a computation table to determine normalized mole ratios.

	Mol. ratio (nitrogen = 1)	
Component	Rapidly decomposable	Slowly decomposable
Carbon	68.5	19.5
Hydrogen	110.5	29.0
Oxygen	50.1	9.2
Nitrogen	1.0	1.0

The chemical formulas without sulfur are

$$\text{Rapidly decomposable} = C_{68.5}H_{110.5}O_{50.1}N \qquad (\text{use } C_{68}H_{111}O_{50}N)$$

$$\text{Slowly decomposable} = C_{19.5}H_{29}O_{9.2}N \qquad (\text{use } C_{20}H_{29}O_{9}N)$$

4. Estimate the amount of gas that can be derived from the rapidly and slowly decomposable organic constituents in MSW.
   (*a*) Using Eq. (11-2), the resulting equations are
   i. Rapidly decomposable

$$\underset{1741.0}{C_{68}H_{111}O_{50}N} + \underset{288.0}{16H_2O} \rightarrow \underset{560.0}{35CH_4} + \underset{1452.0}{33CO_2} + \underset{17}{NH_3}$$

   ii. Slowly decomposable

$$\underset{427}{C_{20}H_{29}O_{9}N} + \underset{162}{9H_2O} \rightarrow \underset{176}{11CH_4} + \underset{396}{9CO_2} + \underset{17}{NH_3}$$

   (*b*) Determine the volume of methane and carbon dioxide produced. The specific weights of methane and carbon dioxide are 0.0448 and 0.1235 lb/ft^3, respectively (see Table 11-3).
   i. Rapidly decomposable

$$\text{Methane} = \frac{(560.0)(44.8)}{(1741.0)(0.0448\ \text{lb/ft}^3)} = 321.7\ \text{ft}^3 \text{ at STP}$$

$$\text{Carbon dioxide} = \frac{(1452.0)(44.8\ \text{lb})}{(1741.0)(0.1235\ \text{lb/ft}^3)} = 302\ \text{ft}^3 \text{ at STP}$$

   ii. Slowly decomposable

$$\text{Methane} = \frac{(176)(7.3\ \text{lb})}{(427)(0.0448\ \text{lb/ft}^3)} = 67.2\ \text{ft}^3 \text{ at STP}$$

$$\text{Carbon dioxide} = \frac{(396)(7.3\ \text{lb})}{(427)(0.1235\ \text{lb/ft}^3)} = 54.8\ \text{ft}^3 \text{ at STP}$$

   (*c*) Determine the total theoretical amount of gas generated per unit dry weight of organic matter destroyed.
   i. Rapidly decomposable

$$\text{Vol/lb} = \frac{321.7\ \text{ft}^3 + 302.5\ \text{ft}^3}{44.8\ \text{lb}} = 13.9\ \text{ft}^3\text{/lb}$$

ii. Slowly decomposable

$$\text{Vol/lb} = \frac{67.2\ \text{ft}^3 + 54.8\ \text{ft}^3}{7.3\ \text{lb}} = 16.7\ \text{ft}^3/\text{lb}$$

Comment. The landfill gas generation values computed in this example represent the maximum amount of gas that could be produced under optimum conditions from the destruction of the biodegradable volatile solids (BVS) in the organic fraction of MSW. The range for the individual organic constituents varies from about 10 to 17 ft^3/lb BVS destroyed. Gas generation values of 12 ft^3/lb BVS destroyed have been reported in the literature for mixed organic waste. The actual quantities of gas generated will be lower because not all of the biodegradable organic matter is available for decomposition. For example, paper contained in plastic bags, while biodegradable, is typically not available for biological conversion. Biodegradable organic wastes that are not exposed to sufficient moisture to sustain biological activity will not be converted.

Variation in Gas Production with Time. Under normal conditions, the rate of decomposition, as measured by gas production, reaches a peak within the first two years and then slowly tapers off, continuing in many cases for periods up to 25 years or more. If moisture is not added to the wastes in a well-compacted landfill, it is not uncommon to find materials in their original form years after they were buried.

The variation in the rate of gas production from the anaerobic decomposition of the rapidly (five years or less—some highly biodegradable wastes are decomposed within days of being placed in a landfill) and slowly (5 to 50 years) biodegradable organic materials in MSW can be modeled as shown in Fig. 11-12. As shown in Fig. 11-12, the yearly rates of decomposition for rapidly and slowly decomposable material are based on a triangular gas production model in which

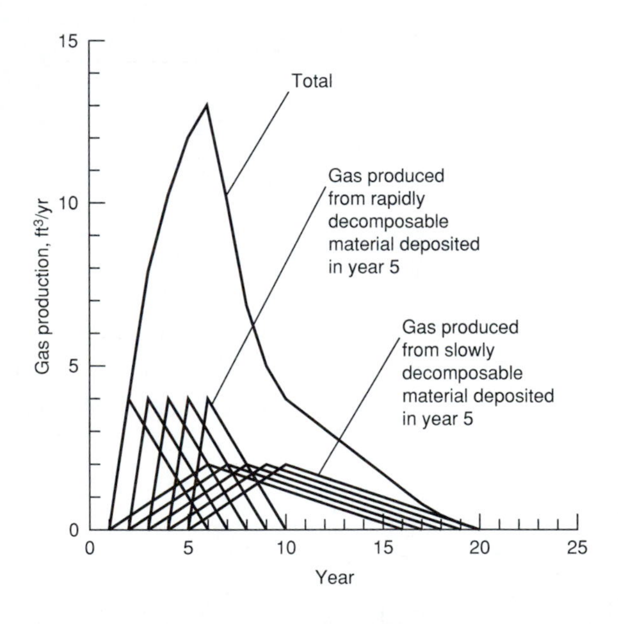

FIGURE 11-12
Graphical representation of gas production over a five-year period from the rapidly and slowly decomposable organic materials placed in a landfill.

the peak rate of gas production occurs one and five years, respectively, after gas production starts. Gas production is assumed to start at the end of the first full year of landfill operation. The area under the triangle is equal to one half the base times the altitude, therefore, the total amount of gas produced from the waste placed the first year of operation is equal to

Total gas produced, ft^3/lb

$$= 1/2 \text{ (base, yr)} \times \text{(altitude, peak rate of gas production, } ft^3/lb \cdot yr) \quad (11\text{-}3)$$

Using a triangular gas production model, the total rate of gas production from a landfill in which wastes were placed for a period of five years is obtained graphically by summing the gas produced from the rapidly and slowly biodegradable portions of the MSW deposited each year (see Fig. 11-12). The total amount of gas produced corresponds to the area under the rate curve. Determination of the total amount of gas produced in a landfill is illustrated in Example 11-8 in Section 11-12.

As noted previously, in many landfills the available moisture is insufficient to allow for the complete conversion of the biodegradable organic constituents in the MSW. The optimum moisture content for the conversion of the biodegradable organic matter in MSW is on the order of 50 to 60 percent. Also in many landfills, the moisture that is present is not uniformly distributed. When the moisture content of the landfill is limited, the gas production curve is more flattened out and is extended over a greater period of time. An example of the effect of reduced moisture content on the production of landfill gas is presented in Fig. 11-13. The production of landfill gas over extended periods of time is of great significance with respect to the management strategy to be adopted for postclosure maintenance.

Sources of Trace Gases. Trace constituents in landfill gases have two basic sources. They may be brought to the landfill with the incoming waste or they may

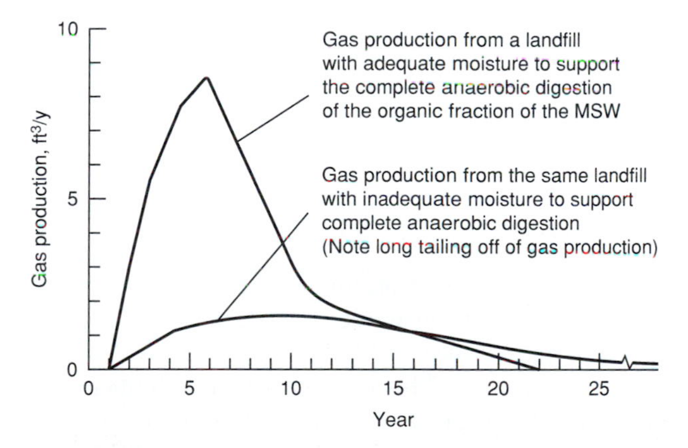

FIGURE 11-13
Effect of reduced moisture content on the production of landfill gas.

TABLE 11-8
Estimated times for the complete volatilization of selected volatile liquids found in landfills[a]

Compound	Evaporation time, d[b]
Chloroethene	0.0
Dichloromethane	1.2
Trichloromethane	4.4
Benzene	6.4
Tetrachloromethane	9.6
Trichloroethene	13.6
Toluene	23.4
Tetrachloroethene	62.6
Chlorobenzene	76.0
1,2-Dibromoethane	128.2
o-Dichlorobenzene	497.6

[a] Excerpted from Ref. 26.

[b] Based on a 10 mm sphere of volatile liquid at 25°C, in a landfill with a porosity of 0.5.

be produced by biotic and abiotic reactions occurring within the landfill [25]. Of the trace compounds found in landfill gases, many are mixed into the incoming waste in liquid form, but tend to volatilize. The tendency to volatilize can be shown to be approximately proportional to the vapor pressure of the liquid, and inversely proportional to the surface area of a sphere of the volatile liquid within the landfill [26]. The wide variation in volatilization times that are expected from some selected volatile liquids that may be found in landfills is illustrated in Table 11-8. In newer landfills where the disposal of hazardous waste has been banned, the concentrations of VOCs in the landfill gas have been reduced significantly.

Complex biochemical pathways can exist for the production or consumption of any of the trace constituents. For example, vinyl chloride is a byproduct of the degradation of di- and trichloroethene. Because of the organic nature of these gases they can be sorbed by waste constituents in the landfill. At present, very little can be stated definitively about the rates of biochemical transformation of the trace compounds. Half-lives varying from a fraction of a year to over a thousand years have been reported for various compounds.

Movement of Landfill Gas

Under normal conditions, gases produced in soils are released to the atmosphere by means of molecular diffusion. In the case of an active landfill, the internal pressure is usually greater than atmospheric pressure and landfill gas will be released by both convective (pressure-driven) flow and diffusion. Other factors influencing the movement of landfill gases include the sorption of the gases into liquid or solid components [47] and the generation or consumption of a gas component through chemical reactions or biological activity. The following general equation relates

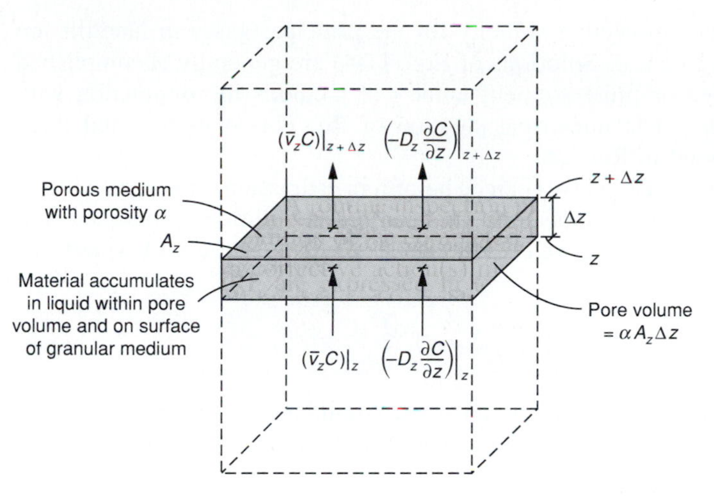

FIGURE 11-14
Control volume for the vertical movement of landfill gas.

these factors in a one-dimensional (vertical) control volume (see Fig. 11-14) [26]. Note that the following discussion of the movement of landfill gases is given in metric units with U.S. customary units in parentheses, as most of the available constants and coefficients for landfill gases are given in metric units.

$$\alpha(1+\beta)\frac{\partial C_A}{\partial t} = -V_z\frac{\partial C_A}{\partial z} + D_z\frac{\partial^2 C_A}{\partial z^2} + G \tag{11-4}$$

where α = total porosity, cm^3/cm^3 (ft^3/ft^3)
β = retardation factor accounting for sorption and phase change
C_A = concentration of compound A, g/cm^3 ($lb \cdot mole/ft^3$)
V_z = convective velocity in the vertical direction, cm/s (ft/d)
D_z = effective diffusion coefficient, cm^2/s (ft^2/d)
G = lumped parameter used to account for all generation terms, $g/cm^3 \cdot s$ ($lb \cdot mole/ft^3 \cdot d$)
z = depth, m (ft)

The convective velocity V_z in the vertical direction can be estimated using Darcy's law as follows:

$$V_z = -\frac{k}{\mu}\frac{dP}{dz} \tag{11-5}$$

where V_z = convective velocity, m/s (ft/d)
k = intrinsic permeability, m^2 (ft^2)
μ = gas-mixture viscosity, $N \cdot s/m^2$ ($lb \cdot d/ft^2$)
P = pressure, N/m^2 (lb/ft^2)
z = depth, m (ft)

Typical values for the convective velocity for the principal gases in landfills are on the order of 1 to 15 cm/d. Solutions of Eq. (11-4) are generally accomplished using finite difference or finite element numerical methods in conjunction with high-speed computers. The numerical solution of Eq. (11-4) in two and three dimensions is discussed in Ref. 26.

Simplified forms of Eq. (11-4) can be helpful in estimating emissions without having to resort to complex numerical computer-based solution techniques. For example, if sorptive and generative effects are neglected, then Eq. (11-4) reduces under steady-state conditions to

$$0 = -V_z \frac{dC_A}{dz} + D_z \frac{d^2 C_A}{dz^2} \tag{11-6}$$

If landfill gas is no longer being produced in significant quantities, only the diffusive portion of Eq. (11-6) remains, which can be integrated to yield the following expression:

$$N_A = -D_z \frac{dC_A}{dz} \tag{11-7}$$

where N_A = gas flux, g/cm^2 · s (lb · mol/ft^2 · d)

The effective diffusion coefficient is a function of both the molecular diffusion and the porosity of the soil. The following relationship was determined empirically for Lindane vapor movement through soil:

$$D_z = D \frac{(\alpha_{\text{gas}})^{10/3}}{\alpha^2} \tag{11-8}$$

where D_z = effective diffusion coefficient, cm^2/s (ft^2/d)
D = diffusion coefficient, cm^2/s (ft^2/d)
α_{gas} = gas-filled porosity, cm^3/cm^3 (ft^3/ft^3)
α = total porosity, cm^3/cm^3 (ft^3/ft^3)

Another approach used to determine the effective diffusion coefficient is as follows:

$$D_z = D\alpha\tau \tag{11-9}$$

where τ = tortuosity factor (typical value = 0.67)

Movement of Principal Landfill Gases. Although most of the methane escapes to the atmosphere, both methane and carbon dioxide have been found at concentrations up to 40 percent at lateral distances of up to 400 ft from the edges of unlined landfills. For unvented landfills, the extent of this lateral movement varies with the characteristics of the cover material and the surrounding soil. If methane is vented in an uncontrolled manner, it can accumulate (because its specific gravity is less than that of air) below buildings or in other enclosed spaces at, or close to, a sanitary landfill. With proper venting, methane should not pose a problem (except that it is a greenhouse gas). Carbon dioxide, on the other hand, is troublesome because of its density. As shown in Table 11-3, carbon dioxide is about 1.5 times as dense as air and 2.8 times as dense as methane; thus, it tends to move toward the bottom of the landfill. As a result, the concentration of carbon dioxide in the lower portions of a landfill may be high for years.

Upward migration of landfill gas. Methane and carbon dioxide can be released through the landfill cover into the atmosphere by convection and diffusion. The diffusive flow through the cover can be estimated using Eqs.(11-7) and (11-8) assuming the concentration gradient is linear and the soil is dry, thus $\alpha_{gas} = \alpha$. Assuming dry soil conditions introduces a safety factor in that any infiltration of water into the landfill cover will reduce the gas-filled porosity and thereby reduce the vapor flux from the landfill.

$$N_A = -\frac{D\alpha^{4/3}(C_{A_{atm}} - C_{A_{fill}})}{L} \tag{11-10}$$

where N_A = gas flux of compound A, g/cm$^2 \cdot$ s (lb $\cdot$ mol/ft$^2 \cdot$ d)
$C_{A_{atm}}$ = concentration of compound A at the surface of the landfill cover, g/cm^3 (lb $\cdot$ mol/ft^3)
$C_{A_{fill}}$ = concentration of compound A at bottom of the landfill cover, g/cm^3 (lb $\cdot$ mol/ft^3)
L = depth of the landfill cover, cm (ft)

Typical values for the coefficient of diffusion for methane and carbon dioxide are 0.20 cm^2/s (18.6 ft^2/d) and 0.13 cm^2/s (12.1 ft^2/d), respectively [26].

Downward migration of landfill gas. Ultimately, carbon dioxide, because of its density, can accumulate in the bottom of a landfill. If a soil liner is used, the carbon dioxide can move from there downward, primarily by diffusive transport through the liner, and through the underlying formation until it reaches the groundwater (note the movement of CO_2 can be limited with the use of a geomembrane liner). Carbon dioxide is readily soluble in water and can react with it to form carbonic acid, or

$$CO_2 + H_2O \rightarrow H_2CO_3 \tag{11-11}$$

This reaction lowers the pH, which in turn can increase the hardness and mineral content of the groundwater through solubilization. For example, if solid calcium carbonate is present in the soil structure, the carbonic acid will react with it to form soluble calcium bicarbonate, according to the following reaction:

$$CaCO_3 + H_2CO_3 \rightarrow Ca^{2+} + 2HCO_3^- \tag{11-12}$$

Similar reactions occur with magnesium carbonates. For a given carbon dioxide gas concentration, the reaction shown in Eq. (11-11) will proceed until equilibrium is reached, as described in Eq. (11-13).

$$\begin{array}{c} H_2O + CO_2 \\ \uparrow\downarrow \\ CaCO_3 + H_2CO_3 \leftrightarrow Ca^{2+} + 2HCO_3^- \end{array} \tag{11-13}$$

Thus, any process that increases the free carbon dioxide available to the solution will cause more calcium carbonate to dissolve. The resulting increase in hardness is the principal effect of the presence of carbon dioxide in groundwater. The solubility in water of the principal gases found in landfills can be computed using Henry's law as given in Appendix F. The effect of carbon dioxide on the the pH of leachate can be estimated using the first dissociation constant for carbonic acid (see Example 11-4 in Section 11-5).

Movement of Trace Gases. For the boundary conditions shown in Fig. 11-15, Eq. (11-10) can be modified for the trace gases found in landfills as follows [19]:

$$N_i = -\frac{D\alpha^{4/3}(C_{i_{\text{atm}}} - C_{i_s}W_i)}{L} \tag{11-14}$$

where N_i = vapor flux of compound i, g/cm$^2 \cdot$ s
D = diffusion coefficient, cm^2/s
α = dry soil porosity, cm^3/cm^3 (ft^3/ft^3)
$C_{i_{\text{atm}}}$ = concentration of compound i at the surface of the landfill cover, g/cm^3
C_{i_s} = saturation vapor concentration of compound i, g/cm^3
W_i = scaling factor to account for the actual fraction of trace compound i in the waste
$C_{i_s}W_i$ = concentration of compound i at bottom of the landfill cover, g/cm^3
L = depth of the landfill cover, cm (ft)

Equation (11-14) can be simplified by assuming that $C_{i_{\text{atm}}}$ is zero; this assumption is reasonable because the concentration of the trace constituent reaching the surface

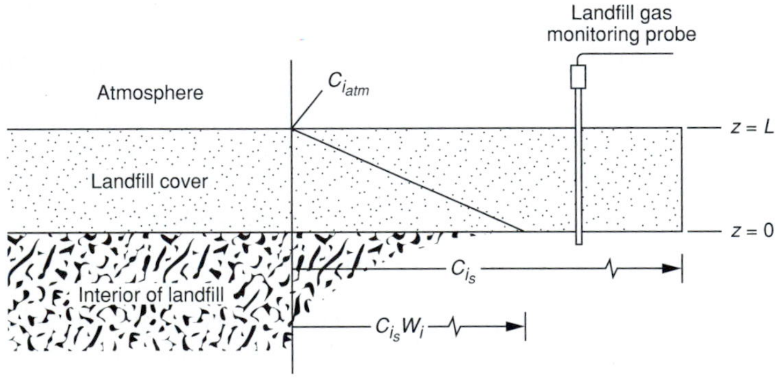

FIGURE 11-15
Definition sketch for the movement of trace landfill gases through a landfill cover [19].

TABLE 11-9

Selected physical properties for twelve trace compounds found in landfills[a]

	0°C			10°C			20°C			30°C			40°C			50°C		
Compound	D[b]	vp[c]	C_s[d]	D	vp	C_s	D	vp	C_s	D	vp	C_s	D	vp	C_s	D	vp	C_s
Ethyl benzene	.052	2.0	12.48	.055	3.9	23.47	.059	7.3	42.44	.062	13	73.08	.066	22	119.7	.069	36	189.9
Toluene	.056	6.7	36.26	.060	12	62.65	.064	22	110.9	.068	37	180.4	.073	59	278.5	.077	92	420.9
Tetrachloro-ethene	.053	4.1	39.95	.057	7.9	74.27	.061	15.6	127.1	.065	24	210.7	.069	40	340.0	.073	63	581.9
Benzene	.066	27	123.9	.070	47	208.1	.075	76	325.0	.081	122	504.6	.086	185	740.7	.091	274	1063
1,2-Dichloro-ethane	.063	24	139.6	.068	41	230.0	.072	62	363.0	.077	107	560.7	.082	164	831.9	.088	243	1194
Trichloro-ethene	.059	20	154.5	.063	36	268.4	.067	60	424.8	.072	94	654.5	.077	146	984.1	.082	217	1417
1,1,1-Trichloro-ethane	.058	36	282.2	.062	61	461.3	.067	100	715.9	.071	153	1081	.076	231	1580	.081	338	2240
Carbon tetra-chloride	.058	32	289.3	.062	54	470.9	.066	90	741.2	.071	138	1124	.075	209	1648	.080	308	2353
Chloroform	.065	61	427.9	.070	100	676.7	.075	160	1026	.080	240	1517	.085	354	2166	.090	508	3012
1,2-Dichloro-ethene	.077	110	626.7	.082	175	961.8	.087	269	1428	.092	399	2048	.097	576	2862	.102	810	3901
Dichloro-methane	.074	155	773.6	.080	242	1165	.085	349	1702	.091	536	2410	.097	763	3322	.103	1060	4472
Vinyl chloride	.080	1280	4701	.085	1810	6413	.091	2548	8521	.098	3350	11090	.104	4410	14130	.110	5690	17660

[a] From Ref. 19.

[b] Diffusion coefficient, cm^2/s.

[c] Vapor pressure, mm Hg.

[d] Saturation vapor concentration, g/m^3.

of the landfill will be quickly diminished by both wind dispersal and diffusion into the air. By making this assumption, the estimate for the mass flux of the gas will be conservative; any increase in $C_{i_{\text{atm}}}$ will result in a decrease in the mass flux. The simplified form of Eq. (11-14) is

$$N_i = \frac{D\alpha^{4/3}(C_{i_s}W_i)}{L} \tag{11-15}$$

Estimated values of the diffusion coefficient D for twelve trace compounds are reported in Table 11-9 for temperatures varying from 0 to 50°C. Porosity values typically vary from 0.010 to 0.30 for different types of clay. The term $C_{i_s}W_i$ corresponds to the concentration of the compound in question at the top of the landfill just below the cover. If field measurements are not available, the value of the term $C_{i_s}W_i$ can be estimated using the data given in Table 11-10 for C_{i_s} and W_i for the reported trace compounds. The values for the term W_i shown in Table 11-10 were derived from measurements made at 44 municipal waste landfills in California. If a compound of interest is not listed in Table 11-10, one can use a value of 0.001 as an estimate for W_i. The saturation concentration, C_{i_s}, for other trace organic compounds may be found in Appendix H. If the value of the term $C_{i_s}W_i$ is to be estimated in the field, measurements should be taken by inserting a gas probe through the landfill cover, to a point just beyond the bottom of the cover, and recording both the concentration of the compound and the temperature at this point in the landfill. By obtaining actual field measurements, one can estimate the average emission rate very quickly. The movement of trace gases by diffusion is considered in Example 11-3.

TABLE 11-10
Measured and saturation gas phase concentrations of 10 trace compounds

Compounds	Concentration, mg/m^3		Scaling factor, W_i
	Maximum measured[a]	Saturation value	
Benzene	135.9	319,000	0.0004
Chlorobenzene	6.8	54,000	0.0001
Ethylbenzene	414.5	40,000	0.01
1,1,1-Trichlorethane	86.3	715,900	0.0001
Chloroethene	89.2	8,521,000	0.00001
Tetrachloroethene	1331.7	126,000	0.01
Trichloroethene	85.1	415,000	0.0002
Dichloromethane	871.5	1,702,000	0.0005
Trichloromethane	63.9	1,027,000	0.00001
Toluene	1150.5	110,000	0.01

[a] Measurements taken from 44 California landfills (adapted from Ref. 5).

Example 11-3 Movement of trace gases. Estimate the emission of toluene, 1,1,1-trichloroethane, and vinyl chloride from the surface of a landfill due to diffusion. Assume the following conditions apply:

1. Temperature = 30°C
2. Landfill cover material = clay-loam mixture
3. Porosity of landfill cover material = 0.20
4. Landfill cover thickness = 2 ft (0.6 m)
5. Scaling factor to account for the actual fraction of trace compound present below landfill cover = 0.001
6. *Note:* $(\text{g/cm}^2 \cdot \text{s}) \times 0.864 \times 10^9 = \text{g/m}^2 \cdot \text{d}$

Solution

1. Estimate the concentration of the compounds just below the landfill cover.
   (*a*) From Table 11-9, the saturation concentrations for these compounds are:

Toluene:	180.4 g/m^3	$= 180.4 \times 10^{-6}$ g/cm^3
1,1,1-Trichloroethane:	1081 g/m^3	$= 1081 \times 10^{-6}$ g/cm^3
Vinyl chloride:	11,090 g/m^3	$= 11{,}090 \times 10^{-6}$ g/cm^3

   (*b*) Estimate the concentration of the compounds just below the landfill cover, $C_{i_s} W_i$, by multiplying the saturation concentration values by the scaling factor (0.001).

Toluene:	180.4×10^{-9} g/cm^3
1,1,1-Trichloroethane:	1081×10^{-9} g/cm^3
Vinyl chloride:	$11{,}090 \times 10^{-9}$ g/cm^3

2. Estimate the mass emission rate using Eq. (11-15) and the diffusion coefficients given in Table 11-9.
   (*a*) For toluene

$$N_i = \frac{D\alpha^{4/3}(C_{i_s} W_i)}{L}$$

$$N_i = \frac{(0.068 \text{ cm}^2/\text{s})(0.20)^{4/3}(180.4 \times 10^{-9} \text{ g/cm}^3)}{60 \text{ cm}}$$

$$= 2.39 \times 10^{-11} \text{ g/cm}^2 \cdot \text{s}$$

   (*b*) For 1,1,1-trichloroethane

$$N_i = \frac{(0.071 \text{ cm}^2/\text{s})(0.20)^{4/3}(1081 \times 10^{-9} \text{ g/cm}^3)}{60 \text{ cm}} = 1.5 \times 10^{-10} \text{ g/cm}^2 \cdot \text{s}$$

   (*c*) For vinyl chloride

$$N_i = \frac{(0.098 \text{ cm}^2/\text{s})(0.20)^{4/3}(11{,}090 \times 10^{-9} \text{ g/cm}^3)}{60 \text{ cm}} = 2.12 \times 10^{-9} \text{ g/cm}^2 \cdot \text{s}$$

3. Convert the mass emission rates to units of $g/m^2 \cdot d$ using the conversion factor given above.

   (*a*) For toluene

$$N_i = (2.39 \times 10^{-11} \text{ g/cm}^2 \cdot \text{s}) \times (0.864 \times 10^9) = 0.02 \text{ g/m}^2 \cdot \text{d}$$

   (*b*) For 1,1,1-trichloroethane

$$N_i = (1.5 \times 10^{-10} \text{ g/cm}^2 \cdot \text{s}) \times (0.864 \times 10^9) = 0.13 \text{ g/m}^2 \cdot \text{d}$$

   (*c*) For vinyl chloride

$$N_i = (2.39 \times 10^{-9} \text{ g/cm}^2 \cdot \text{s}) \times (0.864 \times 10^9) = 2.06 \text{ g/m}^2 \cdot \text{d}$$

Comment. In general, landfill covers composed of soil(s) offer little resistance to the movement of trace organic compounds found in landfills. It is interesting to compare the mass emissions that would occur for the trace compounds in this example based on convective flow. Typical convective velocity values for the principal gases range from 1 to 15 cm/d. The corresponding convective release of toluene would then range from (1 to 15 cm/d) $\times$ 180.4 $\times 10^{-9}$ g/cm^3 $\times$ (d/86,400 s) = 0.2 to 3.1 $\times 10^{-11}$ g/cm$^2 \cdot$ s. The conclusion that can be drawn from this example is that the convective transport of the trace compounds is often of less importance than their diffusive transport. To limit the release of these trace compounds, many landfill operating agencies have chosen to cover completed landfills with a flexible membrane liner.

Passive Control of Landfill Gases

The movement of landfill gases is controlled to reduce atmospheric emissions, to minimize the release of odorous emissions, to minimize subsurface gas migration, and to allow for the recovery of energy from methane. Control systems can be classified as passive or active. In passive gas control systems, the pressure of the gas that is generated within the landfill serves as the driving force for the movement of the gas. In active gas control systems, energy in the form of an induced vacuum is used to control the flow of gas. For both the principal and trace gases, passive control can be achieved during times when the principal gases are being produced at a high rate by providing paths of higher permeability to guide the gas flow in the desired direction. A gravel-packed trench, for example, can serve to channel the gas to a flared vent system. When the production of the principal gases is limited, passive controls are not very effective because molecular diffusion will be the predominant transport mechanism. However, at this stage in the life of the landfill it may not be so important to control the residual emission of the methane in the landfill gas. Control of VOC emissions may necessitate the use of both passive and active gas control facilities.

Pressure Relief Vents/Flares in Landfill Cover. One of the most common passive methods for the control of landfill gases is based on the fact that the lateral migration of landfill gas can be reduced by relieving gas pressure within the landfill interior. For this purpose, vents are installed through the final landfill cover extending down into the solid waste mass (see Fig. 11-16). If the methane

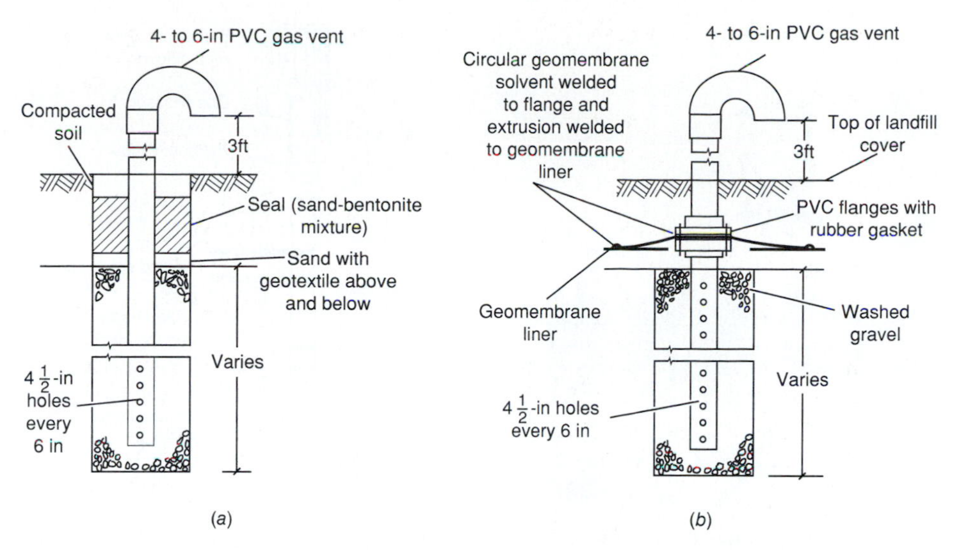

FIGURE 11-16
Typical gas vents used in the surface of a landfill for the passive control of landfill gas: (*a*) gas vent for landfill with a cover that does not contain a geomembrane liner and (*b*) gas vent for a landfill with a cover that contains a synthetic membrane liner.

in the venting gas is of sufficient concentration, several vents can be connected together and equipped with a gas burner (see Fig. 11-17*a*). Where waste gas burners are used the well should penetrate into the upper waste cells. The height of the waste burner can vary from 10 to 20 ft above the completed fill. The burner can be ignited either by hand or by a continuous pilot flame. To derive maximum benefit from the installation of a waste gas burner, a pilot flame should be used (see Fig. 11-17*b*). It should be noted, however, that passive vents with burners may not achieve the VOC and odor destruction efficiencies that are required by many urban air quality control agencies, and, thus, their use is not considered good practice. Gas burners are considered later in this section.

Perimeter Interceptor Trenches. A perimeter trench system, consisting of gravel-filled interceptor trenches containing horizontal perforated plastic pipe (typically polyvinyl chloride, PVC, or polyethylene, PE), can be used to intercept the lateral movement of landfill gases (see Fig. 11-18*a*). The perforated pipe is connected to vertical risers through which the landfill gas that collects in the trench backfill can be vented to the atmosphere. To facilitate gas collection in the trench, a membrane liner is often installed on the trench wall facing away from the landfill.

Perimeter Barrier Trench or Slurry Wall. Barrier trenches (see Fig. 11-18*b*) are usually filled with relatively impermeable materials such as bentonite or clay slurries. In this case, the trench becomes a physical barrier to lateral subsurface

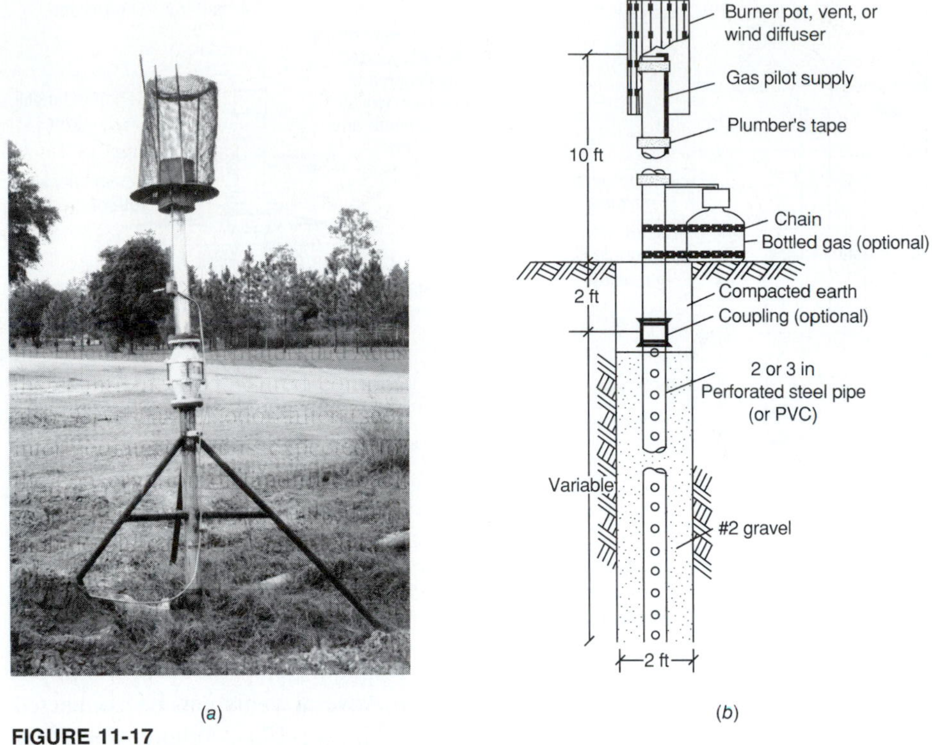

FIGURE 11-17
Typical candlestick type waste gas burner used to flare landfill gas from a well vent or several interconnected well vents: (*a*) without pilot flame and (*b*) with pilot flame.

gas movement. Landfill gas is removed from the inside face of the barrier with gas extraction wells or gravel-filled trenches. However, slurry trenches may be subject to desiccation cracking when allowed to dry out, and hence are more commonly used in groundwater interception projects. The long-term effectiveness of barrier trenches for the control of the migration of landfill gases is uncertain.

Impermeable Barriers within Landfills. In modern landfills, the movement of landfill gases through adjacent soil formations is controlled by constructing barriers of materials that are more impermeable than the soil before filling operations start (see Fig. 11-18*c*). Some of the landfill sealants available for this use are identified in Table 11-11. In connection with the control of leachate, the use of compacted clays and geomembranes of various types singly and in multilayer configurations is most common. Because the principal gases as well as the trace gases will diffuse through clay liners, many agencies now require the use of geomembranes to limit the movement of landfill gases.

Use of Sorptive Barriers within Landfills for Trace Gases. Based on results from sampling programs such as that performed by the California Integrated Waste Management Board, it is apparent that trace gases are present in landfills in widely

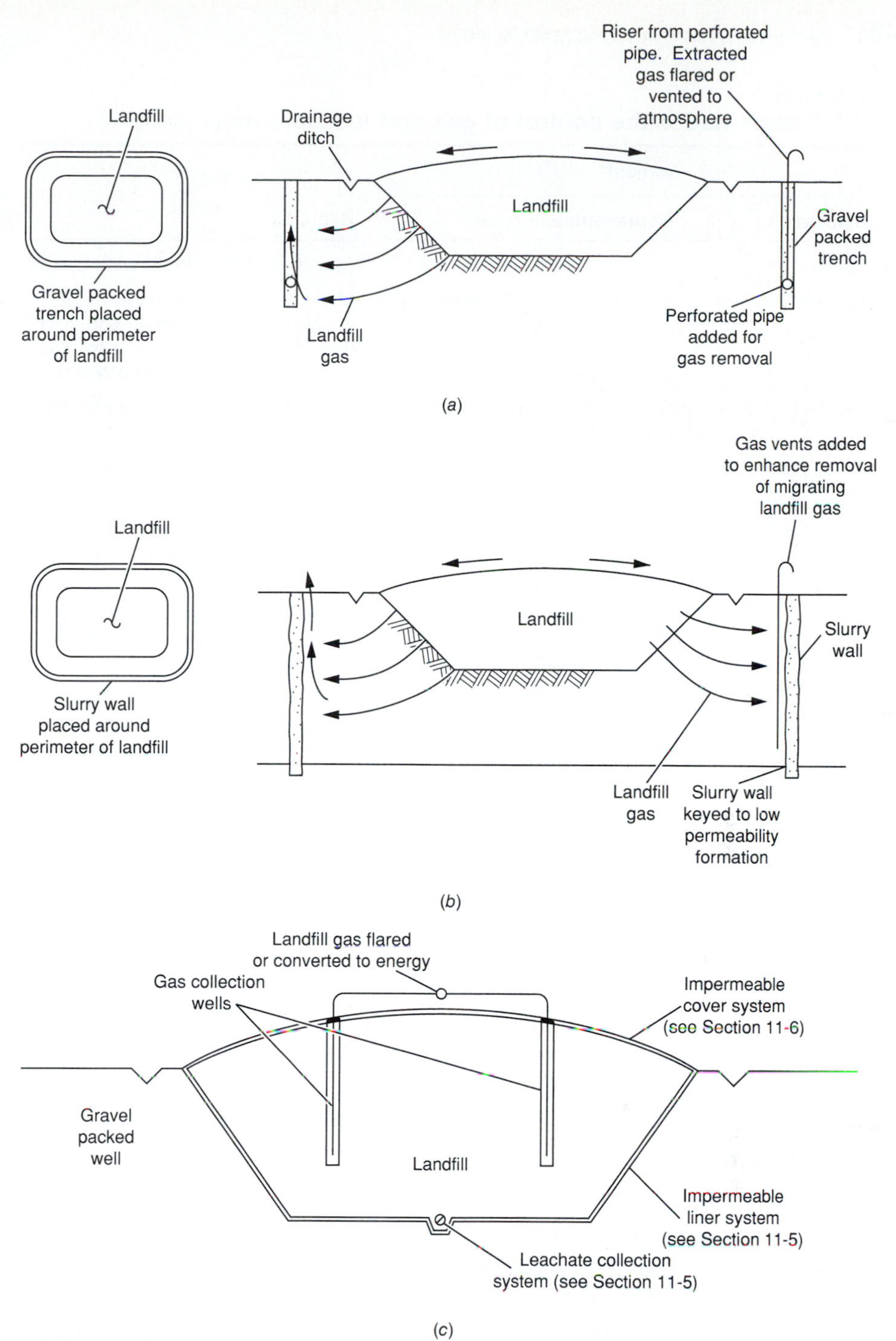

FIGURE 11-18

Passive facilities used for the control of landfill gas: (*a*) interceptor trench filled with gravel and perforated pipe, (*b*) perimeter barrier trench, and (*c*) use of impermeable liner in landfill. Note interceptor barrier perimeter trenches are used to control the off-site migration of landfill gas from existing unlined landfills.

TABLE 11-11
Landfill sealants for the control of gas and leachate movement

Sealant		
Classification	**Representative types**	**Remarks**
Compacted soil		Should contain some clay or fine silt
Compacted clay	Bentonites, illites, kaolinites	Most commonly used sealant for landfills; layer thickness varies from 6 to 48 in; layer must be continuous and not allowed to dry out and crack
Inorganic chemicals	Sodium carbonate, silicate, or pyrophosphate	Use depends on local soil characteristics
Synthetic chemicals	Polymers, rubber latex	Experimental; use in field not well established
Synthetic membrane liners	Polyvinyl chloride, butyl rubber, hypalon, polyethylene, nylon-reinforced liners	Commonly used for leachate control; increased usage for control of landfill gas
Asphalt	Modified asphalt, rubber-impregnated asphalt, asphalt-covered polyethylene fabric, asphalt concrete	Layer must be thick enough to maintain continuity under differential settling conditions
Others	Gunite concrete, soil cement, plastic soil cement	Not commonly used for control of leachate and gas movement because of shrinkage cracks after construction

varying concentrations. High concentration gradients result in a large diffusive component of the flow of trace gases, even during times when very little transport by convection from the flowing principal gas mixture is occurring. The use of sorptive material such as compost can be used to retard the release of trace gases. In turn, biotic and/or abiotic transformation mechanisms can have more time to degrade the sorbed trace compounds.

Active Control of Landfill Gas with Perimeter Facilities

The lateral movement of landfill gas can be controlled by using perimeter gas extraction wells and trenches and by creating a partial vacuum, which induces a pressure gradient toward the extraction well. The extracted gas is either flared to control the emission of methane and VOCs or used for the production of energy. The use of air injection wells is also considered in the following discussion.

Perimeter Gas Extraction and Odor Control Wells. Perimeter extraction wells (see Fig. 11-19*a*) are typically used in landfills with solid waste depths of at least 25 ft, where the distance between the landfill and off-site development is relatively

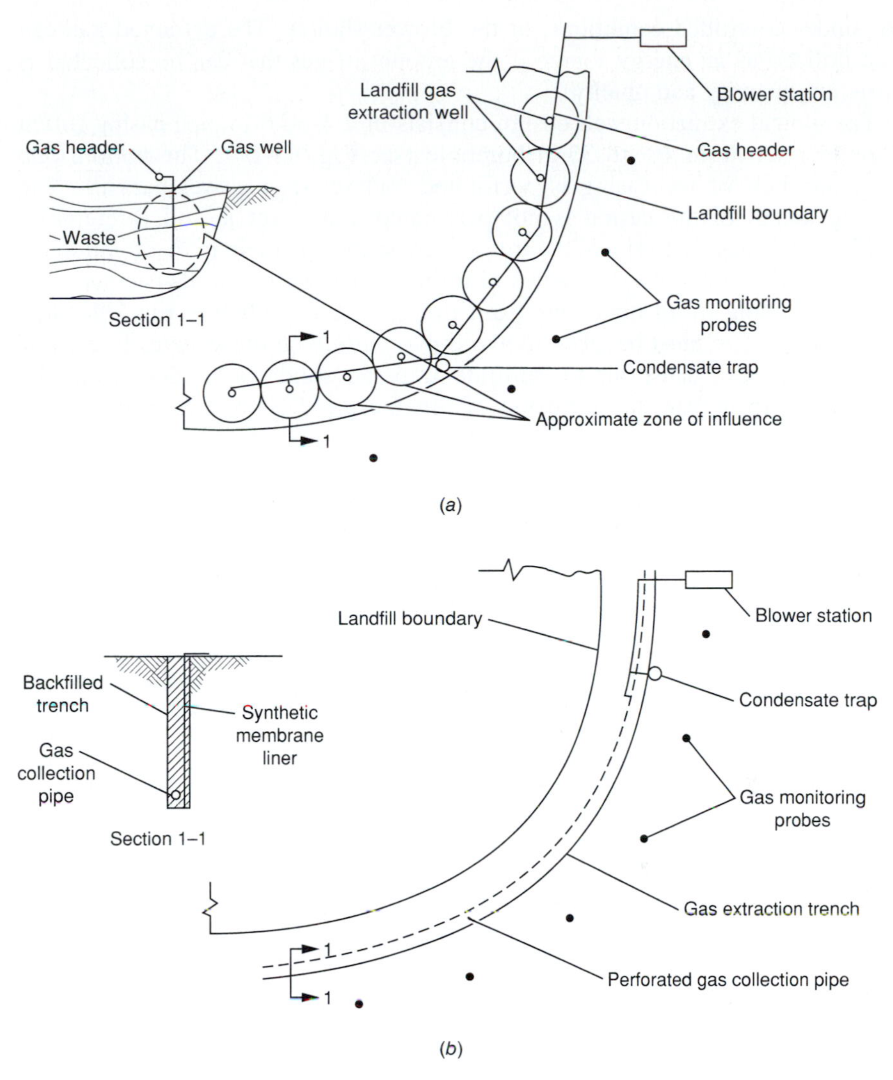

FIGURE 11-19
Active facilities used for the subsurface control of landfill gas migration: (*a*) perimeter landfill gas extraction wells and (*b*) perimeter landfill gas extraction trench.

small. They consist of a series of vertical wells installed either within the landfill along its edge or in the area between the edge of the landfill and the site boundary. The individual landfill gas extraction wells are connected by a common header pipe that in turn is connected to an electrically driven centrifugal blower, which induces a vacuum (negative pressure) in the collection header and the individual wells. When a vacuum is applied, a *zone* or *radius of influence* is created that extends into the solid waste mass surrounding each well and within which the gas that is generated is drawn to the well. Extracted landfill gas is usually vented or flared, under controlled conditions, at the blower station. The extracted gas can also be utilized as an energy source if the amount of gas that can be collected is of sufficient quantity and quality.

The typical extraction well design consists of a 4- to 6-in pipe casing (often PVC or PE) set in an 18- to 36-in borehole (see Fig. 11-20). The bottom one third to one half of the casing is perforated and set in a gravel backfill. The remaining length of the casing is not perforated and is set in soil (preferable) or solid waste backfill [44]. Wells are spaced such that their radii of influence overlap. Unlike water wells, the radius of influence for vertical wells is essentially a sphere extending in all directions from the extraction well (see Fig. 11-19*a*). For this reason, care must be taken to avoid *overpulling* on the system. Excessive extraction rates can cause air to infiltrate into the solid waste mass from the adjacent soil. To prevent the intrusion of air, the gas flow rate from each well

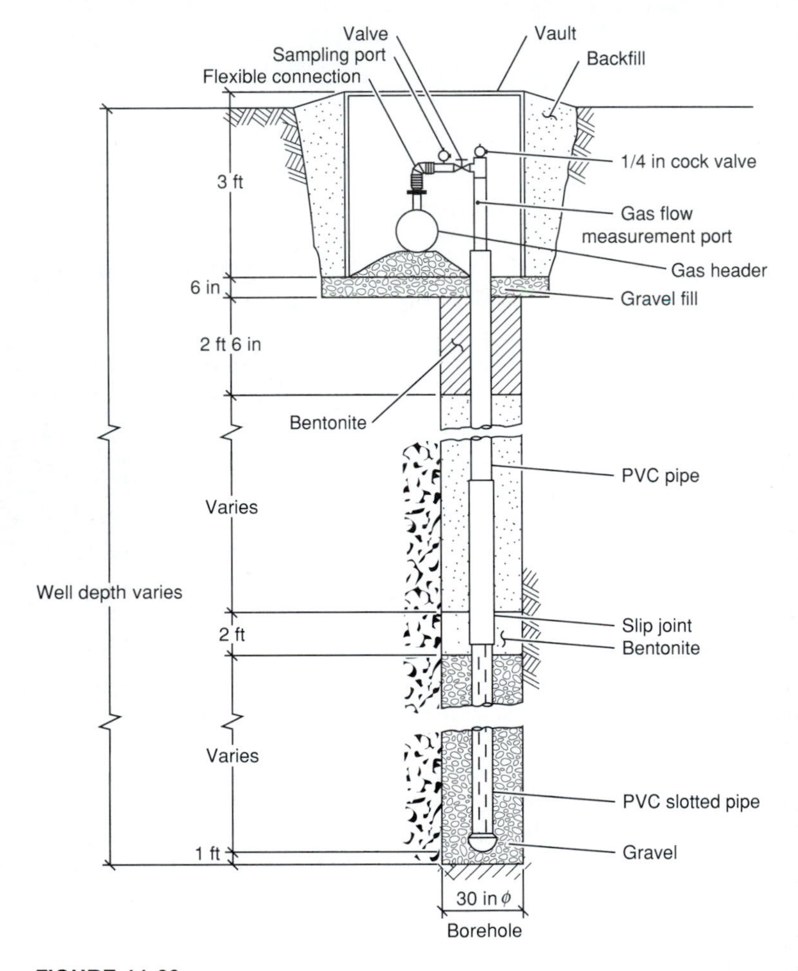

FIGURE 11-20
Representative detail of a landfill gas extraction well. (Courtesy of California Integrated Waste Management Board.)

must be controlled carefully. For this purpose, extraction wells are equipped with gas sampling ports and flow control valves. Depending on the depth of the landfill and other local conditions, well spacing for perimeter gas extraction wells will vary from 25 to 50 ft, although larger spacings have been used.

In large landfills, vertical perimeter wells are also used in conjunction with larger horizontal and vertical gas extraction wells located in the interior of the landfill. The vertical perimeter wells are used to control the off-site migration of landfill gases from the edges and faces of the landfill. Where the perimeter wells are used for the control of odorous emissions from the surfaces of the landfill, the surfaces of the landfill are maintained at a slight vacuum.

Perimeter Gas Extraction Trenches. Perimeter extraction trenches (see Fig. 11-19*b*) are usually installed in native soil adjacent to the landfill perimeter. They are typically used for shallow landfill disposal sites with depths of 25 feet or less. The trenches are gravel-filled and contain perforated plastic pipes that are connected through laterals to a collection header and centrifugal suction blower. Extraction trenches can extend vertically down from the landfill surface to the depth of the solid waste or to groundwater and can be further sealed on the surface with a membrane liner. The suction blower creates a zone of negative pressure in each trench, which extends toward the solid waste. Landfill gas migrating into this zone is drawn into the perforated pipe and collection header, and subsequently vented or flared at the blower station. Flow adjustments can be made via control valves at each trench.

Perimeter Air Injection Wells (Air Curtain System). Perimeter air injection wells consist of a series of vertical wells installed in natural soils between the limits of the solid waste landfill and the facilities to be protected against the intrusion of landfill gas. Air injection wells are typically installed near landfills with solid waste depths of 20 ft or more in areas of undisturbed soil between the landfill and the potentially affected properties.

Active Control of Landfill Gas with Vertical and Horizontal Gas Extraction Wells

Both vertical and horizontal gas wells have been used for the extraction of landfill gas from within landfills. In some installations both types of wells have been used. The management of the condensate that forms when landfill gas is extracted is also an important element in the design of gas recovery systems.

Vertical Gas Extraction Wells. A typical gas recovery system using vertical gas extraction wells is illustrated in Fig. 11-21. The wells are spaced so that their radii of influence overlap (see Fig. 11-22). For completed landfills without gas recovery facilities, the radius of influence for gas wells is sometimes determined by conducting gas drawdown tests in the field. Typically, an extraction well is installed along with gas probes at regular distances from the well, and the vacuum within the landfill is measured as a vacuum is applied to the extraction well. Both

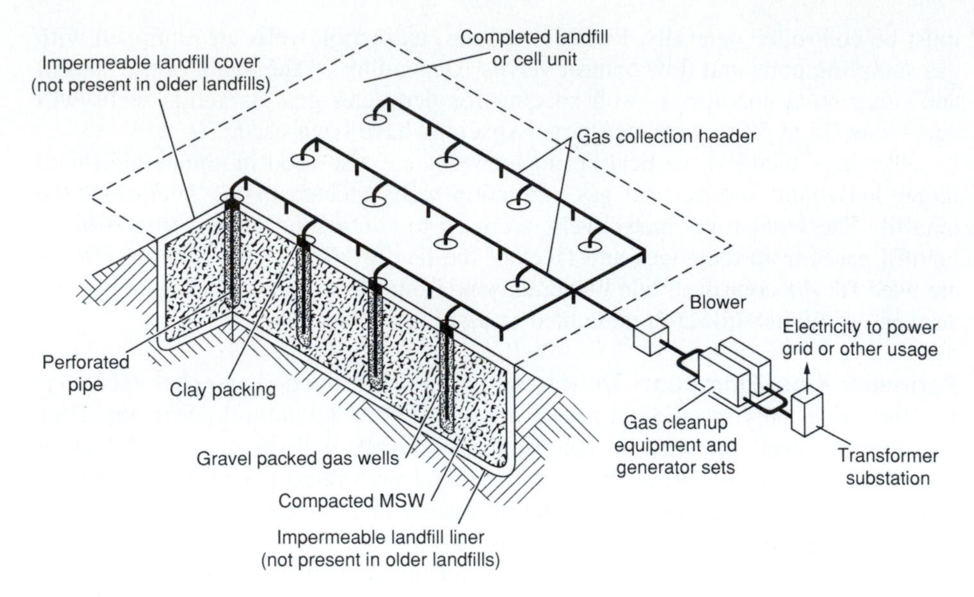

FIGURE 11-21
Landfill gas recovery system using vertical wells.

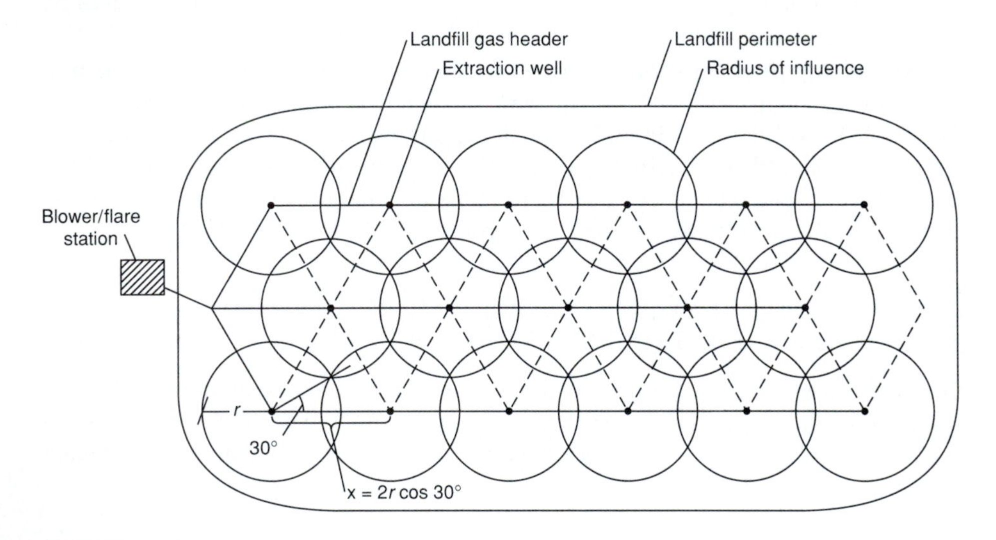

FIGURE 11-22
Equilateral triangular distribution for vertical gas extraction wells. (Courtesy of California Integrated Waste Management Board.)

short-term and long-term extraction tests can be conducted. Because the volume of gas produced will diminish with time, some designers prefer to use a uniform well spacing and to control the radius of influence of the well by adjusting the vacuum at the wellhead. Since the radius of influence of a vertical gas extraction well is essentially a sphere, the radius of influence will also depend on the depth of the landfill and on the design of the landfill cover. For deep landfills with a composite cover containing a geomembrane (see Section 11-6) a 150- to 200-ft spacing is common for landfill gas extraction wells. In landfills with clay and/or soil covers, a closer spacing (e.g., 100 ft) may be required to avoid pulling atmospheric gases into the gas recovery system.

Vertical gas extraction wells are usually installed after the landfill or portions of the landfill have been completed. In older landfills, vertical wells are installed both to recover energy and to control the movement of gases to adjacent properties. The typical extraction well design consists of 4- to 6-in pipe casing (usually PVC or PE) set in an 18- to 36-in borehole (see Fig. 11-20). The bottom third to one half of the casing is perforated and set in a gravel backfill. The remaining length of the casing is not perforated and is backfilled with soil and sealed with a clay [44]. Landfill gas recovery wells are typically designed to penetrate to 80 percent of the depth of the waste in the landfill, because their radii of influence will extend to the bottom of the landfill. However, to allay the public's fear concerning the escape of landfill gas, some designers now place gas recovery wells all the way to the bottom of the landfill. The available vacuum in the collection manifold at the well head is typically 10 in of water. The design of gas recovery facilities used in conjunction with the gas recovery wells is considered in Example 11-9 in Section 11-12.

Horizontal Gas Extraction Wells. An alternative to the use of vertical gas recovery wells is the use of horizontal wells. This usage was pioneered and developed by the County Sanitation Districts of Los Angeles County (see Figs. 11-23 and 11-24). The use of vertical perimeter wells in conjunction with horizontal gas extraction wells is also illustrated in Figs. 11-23 and 11-24. Horizontal wells are installed after two or more lifts have been completed (see Fig. 11-4). The horizontal gas extraction trench is excavated in the solid waste using a backhoe. The trench is then backfilled halfway with gravel and a perforated pipe with open joints is installed (see Fig. 11-25). The trench is then filled with gravel and capped with solid waste. By using a gravel-filled trench and a perforated pipe with open joints, the gas extraction trench remains functional even with the differential settling that will occur in the landfill with the passage of time. The horizontal trenches are installed at approximately 80 ft vertical intervals and at 200 ft horizontal intervals [45].

Management of Condensate in Gas Recovery Systems. Condensate forms when the warm landfill gas is cooled as it is transported in the header leading to the blower. Gas collection headers are usually installed with a minimum slope of 3 percent to allow for differential settlement. Because headers are constructed in

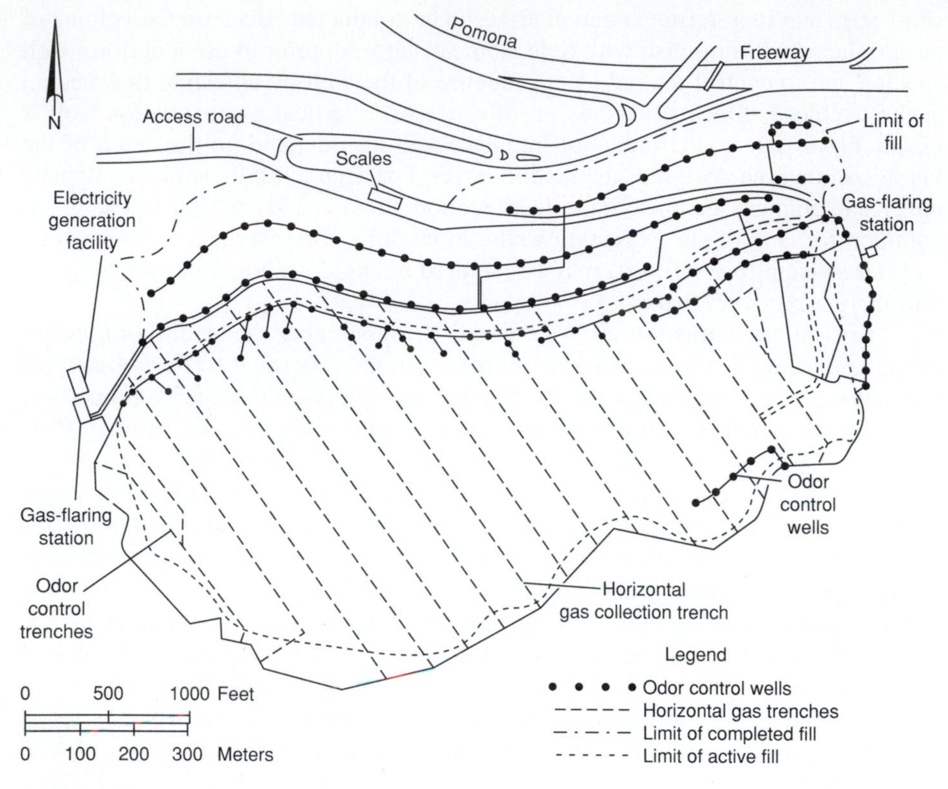

FIGURE 11-23
Plan view of gas collection facilities Puente Hills landfill. (Courtesy of County Sanitation Districts of Los Angeles County.)

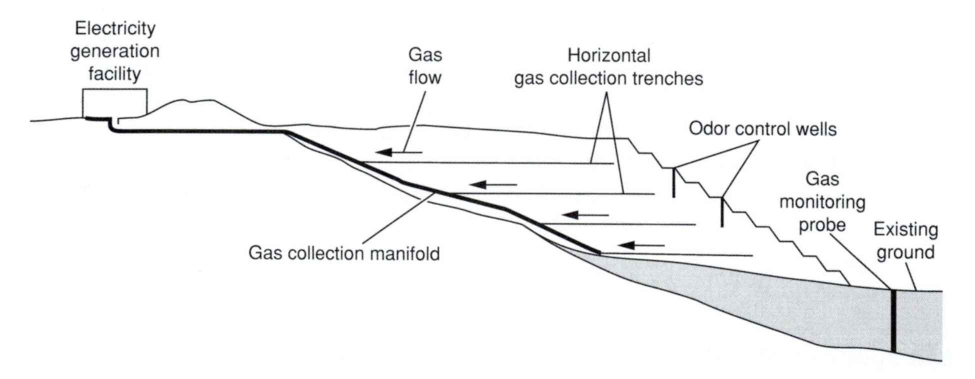

FIGURE 11-24
Sectional view through Puente Hills landfill showing horizontal gas collection trenches. (Courtesy of County Sanitation Districts of Los Angeles County.)

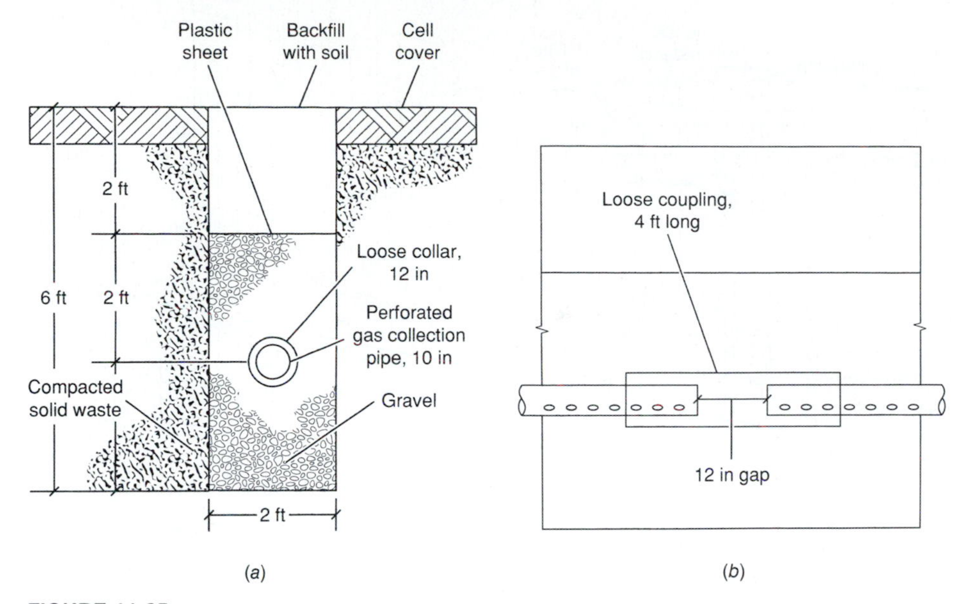

FIGURE 11-25
Details of horizontal gas extraction trench: (*a*) section through trench and (*b*) side view. (Courtesy of County Sanitation Districts of Los Angeles County.)

sections that slope up and down throughout the extent of the landfill, condensate traps are installed at low spots in the line (see Fig. 11-19*a*). A typical condensate trap in which the condensate is returned to the landfill is shown in Fig. 11-26*a*. In states where returning the condensate to the landfill is not allowed, the condensate traps are connected to holding tanks (see Fig. 11-26*b*). Condensate from the holding tanks is pumped out periodically and either transported to an authorized disposal facility or treated on-site before disposal or discharge to a local sewer. Computation of the volume of condensate formed is illustrated in Example 11-10 in Section 11-12.

Management of Landfill Gas

Typically, landfill gases that have been recovered from an active landfill are either flared or used for the recovery of energy in the form of electricity, or both. More recently, the separation of the carbon dioxide from the methane in landfill gas has been suggested as an alternative to the production of heat and electricity.

Flaring of Landfill Gases. A common method of treatment for landfill gases is thermal destruction; that is, methane and any other trace gases (including VOCs) are combusted in the presence of oxygen (contained in air) to carbon dioxide (CO_2), sulfur dioxide (SO_2), oxides of nitrogen, and other related gases. The thermal destruction of landfill gases is usually accomplished in a specially designed flaring facility (see Figs. 11-27 and 11-28). Because of concerns over air

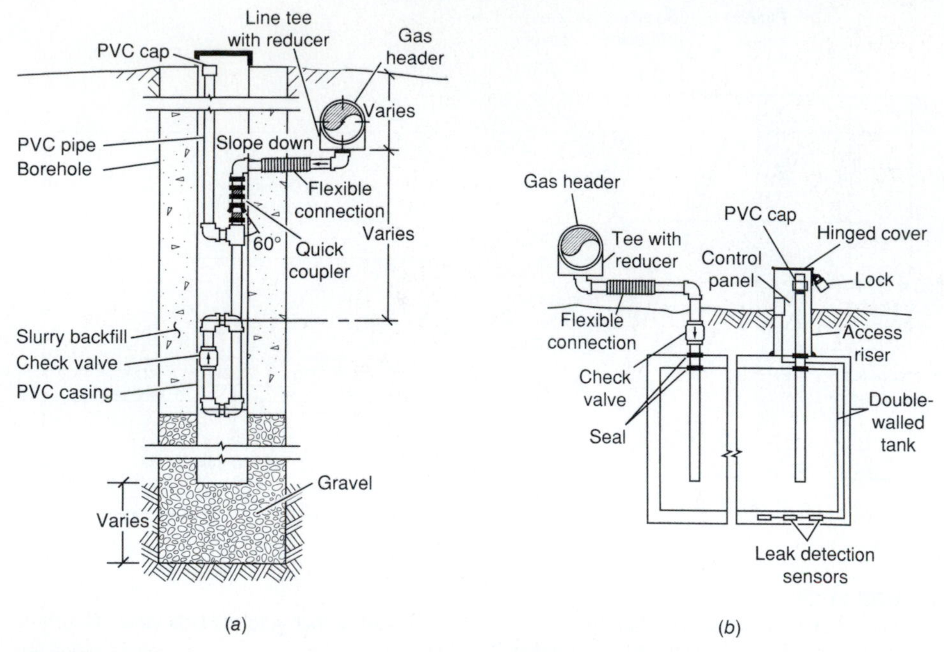

FIGURE 11-26
Typical condensate traps: (*a*) liquid is returned to landfill (Courtesy of California Integrated Waste Management Board) and (*b*) liquid is stored in holding tank.

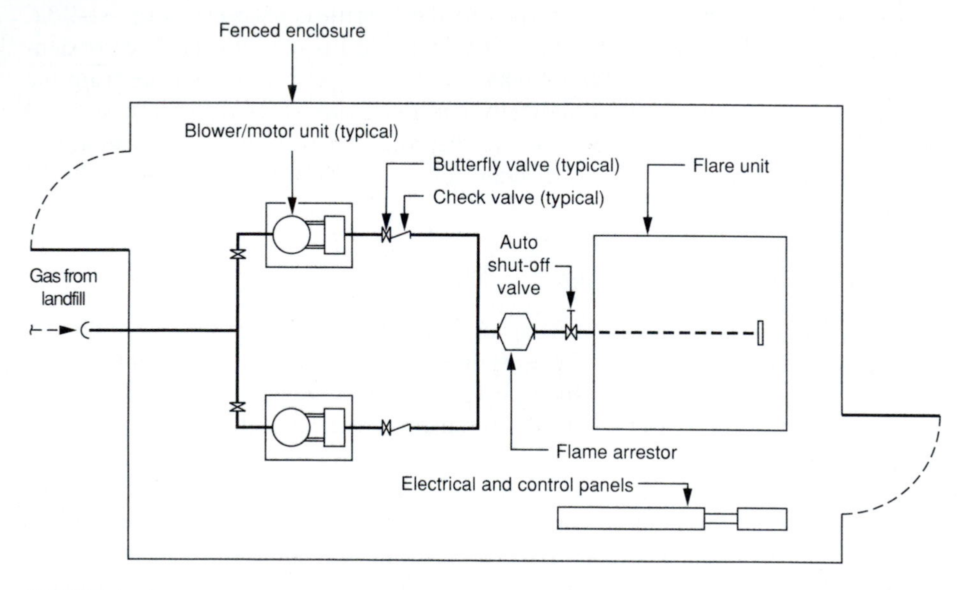

FIGURE 11-27
Schematic layout of blower/flare station for the flaring of landfill gas. (Courtesy of California Integrated Waste Management Board.)

FIGURE 11-28
View of large array of ground effects flares used to flare landfill gas.

pollution, modern flaring facilities are designed to meet rigorous operating specifications to ensure effective destruction of VOCs and other similar compounds that may be present in the landfill gas. For example, a typical requirement might be a minimum combustion temperature of 1500°F and a residence time of 0.3 to 0.5 s, along with a variety of controls and instrumentation in the flaring station. Typical requirements for a modern flaring facility are summarized in Table 11-12.

TABLE 11-12
Important design elements for enclosed ground-level landfill gas flares[a]

Item	Comments
Temperature indicator and recorder	Used to measure and record gas temperature in the flare stack. Whenever the flare is in operation, a temperature of 1500°F or greater must be maintained in the stack as measured by the temperature indicator 0.3 s after passing through the burner.
Automatic pilot restart system	To ensure continuous operation
Failure alarm with an automatic isolation system	The alarm and isolation system are used to isolate the flare from the landfill gas supply line, shut off the blower, and notify a responsible party of the shutdown.
Automatically controlled combustion air louvers	Used to control the amount of combustion air and the temperature of the flame
Source test ports with adequate and safe access provided	Test ports used for monitoring the combustion process and for sampling air emissions.
View ports	A sufficient number of view ports must be available to allow visual inspection of the temperature sensor location within the flare.
Heat shield	A heat shield should be provided around the top of the flare shroud for use during source testing.

[a] Adapted from Ref. 44.

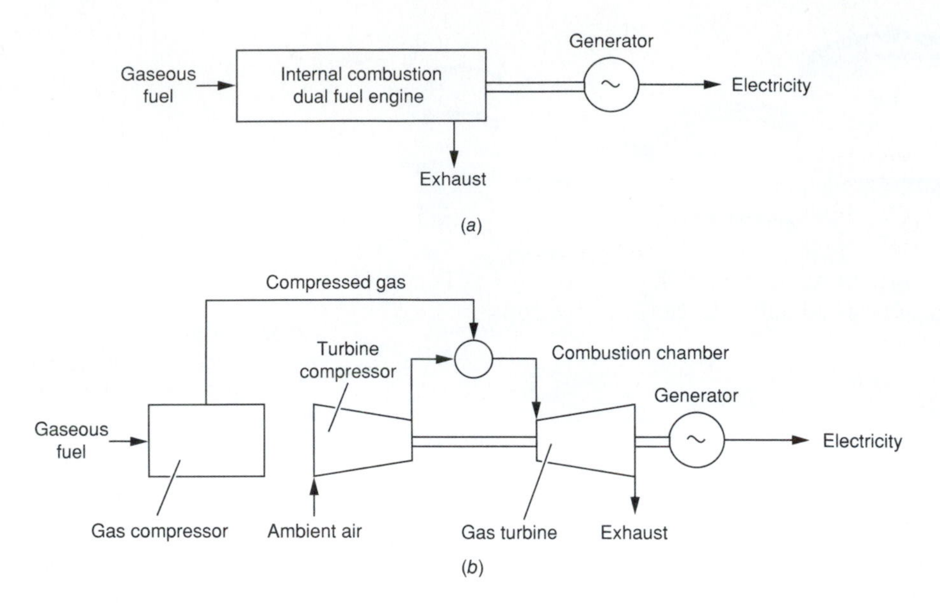

FIGURE 11-29
Schematic flow diagrams for the recovery of energy from gaseous fuels: (*a*) using internal combustion engine and (*b*) using a gas turbine.

Landfill Gas Energy Recovery Systems. Landfill gas is usually converted to electricity (see Figs. 11-29 and 11-30). In smaller installations (up to 5 MW), it is common to use dual fuel internal combustion piston engines (see Figs. 11-29*a* and 11-30*a*) or gas turbines (see Fig 11-29*b*). In larger installations, the use of steam turbines is common (see Fig. 11-30*b*). Where piston-type engines are used, the landfill gas must be processed to remove as much moisture as possible so as

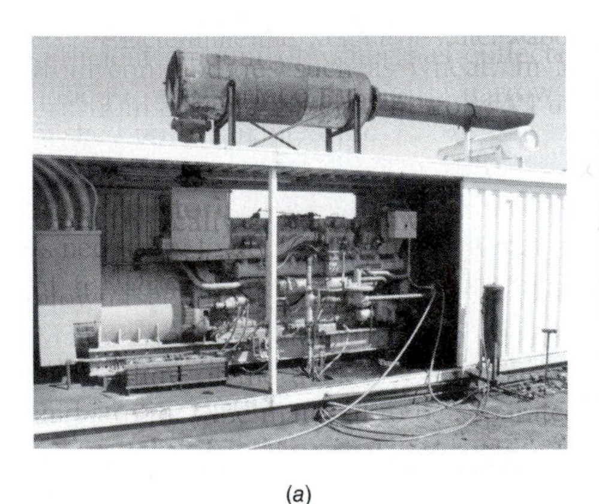

(*a*) (*b*)

FIGURE 11-30
Views of gas conversion facilities: (*a*) using dual fuel internal combustion piston engines and (*b*) using boilers and a steam turbine (see also Fig. 13-25*a*).

to limit damage to the cylinder heads. If the gas contains H_2S, the combustion temperature must be controlled carefully to avoid corrosion problems. Alternatively, the landfill gas can be passed through a scrubber containing iron shavings, or through other proprietary scrubbing devices, to remove the H_2S before the gas is combusted.

Combustion temperatures will also be critical where the landfill gas contains VOCs released from wastes placed in the landfill before the disposal of hazardous waste in municipal landfills was banned. The typical service cycle for dual fuel engines running on landfill gas varies from 3000 to 10,000 hours before the engine must be overhauled. In Fig. 11-30*a*, low-Btu landfill gas is compressed under high pressure so that it can be used more effectively in the gas turbine. The typical service cycle for gas turbines running on landfill gas is approximately 10,000 hours.

Gas Purification and Recovery. Where there is a potential use for the CO_2 contained in the landfill gas, the CH_4 and CO_2 in landfill gas can be separated. The separation of the CO_2 from the CH_4 can be accomplished by physical adsorption, chemical adsorption, and by membrane separation. In physical and chemical adsorption, one component is adsorbed preferentially using a suitable solvent. Membrane separation involves the use of a semipermeable membrane to remove the CO_2 from the CH_4. Semipermeable membranes have been developed that allow CO_2, H_2S, and H_2O to pass while CH_4 is retained. Membranes are available as flat sheets or as hollow fibers. To increase efficiency of separation, the flat sheets are spiral wound on a support medium while the hollow fibers are grouped together in bundles.

11-5 COMPOSITION, FORMATION, MOVEMENT, AND CONTROL OF LEACHATE IN LANDFILLS

Leachate may be defined as liquid that has percolated through solid waste and has extracted dissolved or suspended materials. In most landfills leachate is composed of the liquid that has entered the landfill from external sources, such as surface drainage, rainfall, groundwater, and water from underground springs and the liquid produced from the decomposition of the wastes, if any. The composition, formation, movement, and control of leachate are considered in this section.

Composition of Leachate

When water percolates through solid wastes that are undergoing decomposition, both biological materials and chemical constituents are leached into solution. Representative data on the characteristics of leachate are reported in Table 11-13 for both new and mature landfills. Because the range of the observed concentration values for the various constituents reported in Table 11-13 is rather large, especially for new landfills, great care should be exercised in using the typical values that are given. Typical physical, chemical, and biological monitoring parameters that are used to characterize leachate are summarized in Table 11-14.

TABLE 11-13
Typical data on the composition of leachate from new and mature landfills[a]

Constituent	Value, mg/L[b] New landfill (less than 2 years) Range[c]	Typical[d]	Mature landfill (greater than 10 years)
BOD_5 (5-day biochemical oxygen demand)	2,000–30,000	10,000	100–200
TOC (total organic carbon)	1,500–20,000	6,000	80–160
COD (chemical oxygen demand)	3,000–60,000	18,000	100–500
Total suspended solids	200–2,000	500	100–400
Organic nitrogen	10–800	200	80–120
Ammonia nitrogen	10–800	200	20–40
Nitrate	5–40	25	5–10
Total phosphorus	5–100	30	5–10
Ortho phosphorus	4–80	20	4–8
Alkalinity as $CaCO_3$	1,000–10,000	3,000	200–1,000
pH	4.5–7.5	6	6.6–7.5
Total hardness as $CaCO_3$	300–10,000	3,500	200–500
Calcium	200–3,000	1,000	100–400
Magnesium	50–1,500	250	50–200
Potassium	200–1,000	300	50–400
Sodium	200–2,500	500	100–200
Chloride	200–3,000	500	100–400
Sulfate	50–1,000	300	20–50
Total iron	50–1,200	60	20–200

[a] Developed from Refs. 2, 8, 9, 11, 39, 46.

[b] Except pH, which has no units.

[c] Representative range of values. Higher maximum values have been reported in the literature for some of the constituents.

[d] Typical values for new landfills will vary with the metabolic state of the landfill.

Variations in Leachate Composition. Note that the chemical composition of leachate will vary greatly depending on the age of landfill and the events preceding the time of sampling. For example, if a leachate sample is collected during the acid phase of decomposition (see Fig. 11-11), the pH value will be low and the concentrations of BOD_5, TOC, COD, nutrients, and heavy metals will be high. If, on the other hand, a leachate sample is collected during the methane fermentation phase (see Fig. 11-11), the pH will be in the range from 6.5 to 7.5, and the BOD_5, TOC, COD, and nutrient concentration values will be significantly lower. Similarly the concentrations of heavy metals will be lower because most metals are less soluble at neutral pH values. The pH of the leachate will depend not only on the concentration of the acids that are present but also on the partial pressure of the CO_2 in the landfill gas that is in contact with the leachate. The effect of the CO_2 in the landfill gas is illustrated in Example 11-4 below.

The biodegradability of the leachate will vary with time. Changes in the biodegradability of the leachate can be monitored by checking the BOD_5/COD

TABLE 11-14
Leachate sampling parameters[a]

Physical	Organic constituents	Inorganic constituents	Biological
Appearance	Organic chemicals	Suspended solids (SS), total dissolved solids (TDS)	Biochemical oxygen demand (BOD)
pH	Phenols	Volatile suspended solids (VSS), volatile dissolved solids (VDS)	Coliform bacteria (total; fecal; fecal streptococci)
Oxidation-reduction potential	Chemical oxygen demand (COD)	Chloride	Standard plate count
Conductivity	Total organic carbon (TOC)	Sulfate	
Color	Volatile acids	Phosphate	
Turbidity	Tannins, lignins	Alkalinity and acidity	
Temperature	Organic-N	Nitrate-N	
Odor	Ether soluble (oil and grease)	Nitrite-N	
	Methylene blue active substances (MBAS)	Ammonia-N	
	Organic functional groups as required	Sodium	
	Chlorinated hydrocarbons	Potassium	
		Calcium	
		Magnesium	
		Hardness	
		Heavy metals (Pb, Cu, Ni, Cr, Zn, Cd, Fe, Mn, Hg, Ba, Ag)	
		Arsenic	
		Cyanide	
		Fluoride	
		Selenium	

[a] Adapted from Ref. 44.

ratio. Initially, the ratios will be in the range of 0.5 or greater. Ratios in the range of 0.4 to 0.6 are taken as an indication that the organic matter in the leachate is readily biodegradable. In mature landfills, the BOD_5/COD ratio is often in the range of 0.05 to 0.2. The ratio drops because leachate from mature landfills typically contains humic and fulvic acids, which are not readily biodegradable.

As a result of the variability in leachate characteristics, the design of leachate treatment systems is complicated. For example, a treatment plant designed to treat a leachate with the characteristics reported for a new landfill would be quite different from one designed to treat the leachate from a mature landfill. The problem of interpreting the analytical results is complicated further by the fact that the leachate that is being generated at any point in time is a mixture of leachate derived from solid waste of different ages.

Trace Compounds. The presence of trace compounds (some of which may pose health risks) in leachate will depend on the concentration of these compounds in the gas phase within the landfill. The expected concentrations can be estimated using Henry's law as given in Appendix F and the Henry's law constants given in Table 5-8. As more communities and operators of landfills institute programs to limit the disposal of hazardous wastes with MSW, the quality of the leachate from new landfills is improving with respect to the presence of trace constituents.

Example 11-4 Estimate the pH of the leachate in contact with landfill gas. Assume the composition of the landfill gas in contact with the leachate is 50 percent carbon dioxide and 50 percent methane, the landfill gas is saturated with water vapor at a temperature of 50°C (122°F), and the pressure within the landfill is atmospheric. The alkalinity of the leachate is 500 mg/L as $CaCO_3$.

Solution

1. From Appendix F, the saturation concentration of carbon dioxide for the stated conditions is given as 379 mg/L.
2. Determine the pH of the leachate using the first dissociation constant for carbonic acid as given in Appendix G.

$$\frac{[H^+][HCO_3^-]}{[H_2CO_3^*]} = K_1$$

where $[H^+]$ = molar concentration of the hydrogen ion, mol/L
$[HCO_3^-]$ = molar concentration of the bicarbonate ion, mol/L
$[H_2CO_3^*]$ = molar concentration of carbonic acid, mol/L
$[H_2CO_3^*] = [CO_{2,aq}] + [H_2CO_3]$

For all practical purposes it can be assumed that the computed concentration value of $CO_{2,aq}$ is equal to the term $[H_2CO_3^*]$ and that at the pH values encountered in landfills all of the alkalinity is due to the bicarbonate ion, thus,

(*a*) The molar concentrations of HCO_3^- and $H_2CO_3^*$ are

$$[HCO_3^-] = \frac{500 \text{ mg/L}}{50 \text{ mg/meq}} \times \frac{61 \text{ mg/meq}}{61{,}000 \text{ mg/mol}} = 0.01 \text{ mol/L}$$

$$[H_2CO_3^*] \approx [CO_2] = \frac{379 \text{ mg/L}}{44{,}000 \text{ mg/mol}} = 0.00861 \text{ mol/L}$$

(*b*) Compute the pH of the leachate. The value of first dissociation constant, K_1, at 50°C as given in Appendix G is 5.07×10^{-7}

$$\frac{[H^+][0.01]}{[0.00861]} = 5.07 \times 10^{-7}$$

$$[H^+] = 4.37 \times 10^{-7}$$

$$pH = 6.36$$

Water Balance and Leachate Generation in Landfills

The potential for the formation of leachate can be assessed by preparing a water balance on the landfill [14]. The water balance involves summing the amounts of water entering the landfill and subtracting the amounts of water consumed in chemical reactions and the quantity leaving as water vapor. The potential leachate quantity is the quantity of water in excess of the moisture-holding capacity of the landfill material.

Description of Water Balance Components for a Landfill Cell. The components that make up the water balance for a landfill cell are identified in Fig. 11-31. The principal sources include the water entering the landfill cell from above, the moisture in the solid waste, the moisture in the cover material, and the moisture in the sludge, if the disposal of sludge is allowed. The principal sinks are the water leaving the landfill as part of the landfill gas (i.e., water used in the formation of the gas), as saturated water vapor in the landfill gas, and as leachate. Each of these components is considered below.

Water entering from above. For the upper layer of the landfill, the water from above corresponds to the precipitation that has percolated through the cover material. For the layers below the upper layer, water from above corresponds to the water that has percolated through the solid waste above the layer in question. One of the most critical aspects in the preparation of a water balance for a landfill

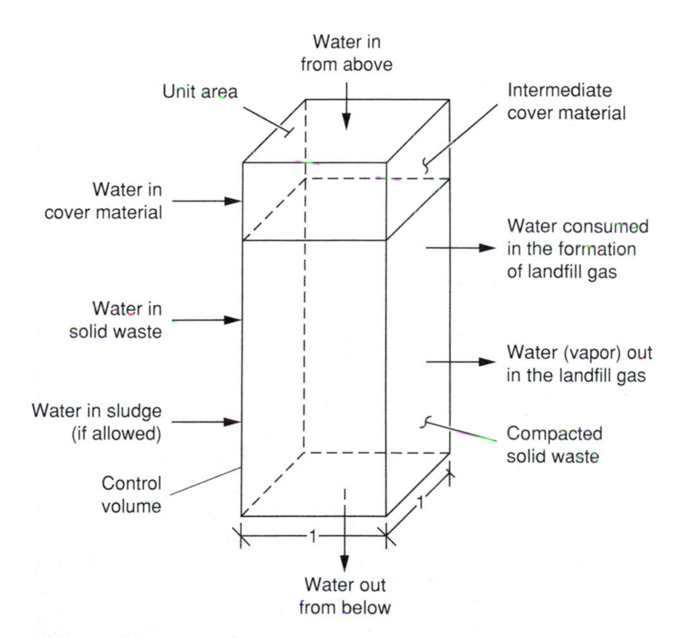

FIGURE 11-31
Definition sketch for water balance used to assess leachate formation in a landfill.

is to determine the amount of the rainfall that actually percolates through the landfill cover layer. Where a geomembrane is not used, the amount of rainfall that percolates through the landfill cover can be determined using the Hydrologic Evaluation of Landfill Performance (HELP) model [41, 42]. A simplified method for estimating the amount of percolation that can be expected is presented in Section 11-6.

Water entering in solid waste. Water entering the landfill with the waste materials is that moisture inherent in the waste material as well as moisture that has been absorbed from the atmosphere or from rainfall (where the storage containers are not sealed properly). In dry climates, some of the inherent moisture contained in the waste can be lost, depending on the conditions of the storage. The moisture content of residential and commercial MSW is about 20 percent, as reported in Table 4-1. However, because of the variability of the moisture content during the wet and dry seasons, it may be necessary to conduct a series of tests during the wet and dry periods.

Water entering in cover material. The amount of water entering with the cover material will depend on the type and source of the cover material and the season of the year. The maximum amount of moisture that can be contained in the cover material is defined by the field capacity (FC) of the material, that is, the liquid which remains in the pore space subject to the pull of gravity. Typical values for soils range from 6–12 percent for sand to 23–31 percent for clay loams. The FC of soils is considered further in Section 11-6 in connection with the storage of water in landfill covers.

Water leaving from below. Water leaving from the bottom of the *first* cell of the landfill is termed *leachate*. As noted previously, water leaving the bottom of the second and subsequent cells corresponds to the water entering from above for the cell below the cell in question. In landfills where intermediate leachate collection systems are used, water leaving from the bottom of the cell placed directly over the intermediate leachate collection system is also termed leachate.

Water consumed in the formation of landfill gas. Water is consumed during the anaerobic decomposition of the organic constituents in MSW. The amount of water consumed by the decomposition reactions can be estimated using the formula for the rapidly decomposable material developed in Example 11-2. The mass of water taken up per pound of dry organic waste consumed can be estimated as follows:

$$\underset{1741.0}{C_{68}H_{111}O_{50}N} + \underset{288.0}{16H_2O} \rightarrow \underset{560.0}{35CH_4} + \underset{1452.0}{33CO_2} + \underset{17}{NH_3}$$

The mass of water consumed per pound of dry rapidly biodegradable volatile solids (RBVS) destroyed is

$$\text{Water consumed} = \frac{288.0}{1741.0} = 0.165 \text{ lb } H_2O/\text{lb RBVS destroyed}$$

Using a gas production value of 13.9 ft^3/lb RBVS destroyed (from Example 11-2), the corresponding value for the amount of water consumed per cubic foot of gas produced is

$$\text{Water consumed} = \frac{(0.165 \text{ lb } H_2O/\text{lb RBVS destroyed})}{(13.9 \text{ ft}^3/\text{lb RBVS destroyed})} = 0.0119 \text{ lb } H_2O/\text{ft}^3$$

Water lost as water vapor. Landfill gas usually is saturated in water vapor. The quantity of water vapor escaping the landfill is determined by assuming the landfill gas is saturated with water vapor and applying the perfect gas law as follows:

$$p_v V = nRT \tag{11-16}$$

where p_v = vapor pressure of H_2O at temperature T, lb/in^2 (see Appendix C)
V = volume, ft^3
n = number of pound moles
R = universal gas constant = 1543 ft · lb/(lb · mole) · °R
T = temperature, degrees Rankine = (460 + T, °F) = °R

The numerical value for the mass of water vapor contained per cubic foot of landfill gas at 90°F is obtained as follows:

$$p_v = 0.70 \text{ lb/in}^2 = 100.8 \text{ lb/ft}^2 \text{ (see Appendix C)}$$

$$V = 1.0 \text{ ft}^3$$

$$n = \text{number of pound moles}$$

$$R = \text{universal gas constant} = 1543 \text{ ft} \cdot \text{lb/(lb} \cdot \text{mole)} \cdot {}^\circ\text{R}$$

$$T = (460 + 90) = 550^\circ\text{R}$$

$$n = \frac{(p_v)(V)}{RT} = \frac{(100.8)(1.0)}{(1543)(550)} = 0.00012 \text{ lb} \cdot \text{moles}$$

$$= (0.00012 \text{ lb} \cdot \text{mole})(18 \text{ lb/lb} \cdot \text{mole}) = 0.0022 \text{ lb } H_2O/\text{ft}^3 \text{ landfill gas}$$

Other water losses and gains. There will be some loss of moisture to evaporation as the waste is being landfilled. The amounts are not large and are often ignored. The decision to include these variables in the water balance analysis will depend on local conditions.

Landfill Field Capacity. Water entering the landfill that is not consumed and does not exit as water vapor may be held within the landfill or may appear as leachate. Both the waste material and the cover material are capable of holding water against the pull of gravity. The quantity of water that can be held against the pull of gravity is referred to as field capacity. The potential quantity of leachate is the amount of moisture within the landfill in excess of the landfill FC. The FC, which varies with the overburden weight, can be estimated using the following equation [21, 22]:

$$\mathrm{FC} = 0.6 - 0.55\left(\frac{\mathrm{W}}{10{,}000 + \mathrm{W}}\right) \tag{11-17}$$

where FC = field capacity (i.e., the fraction of water in the waste based on the dry weight of the waste)

W = overburden weight calculated at the midheight of the waste in the lift in question

The application of Eq. (11-17) is illustrated in Example 11-11 in Section 11-12.

Preparation of Landfill Water Balance. The terms that compose the water balance can be put into equation form as follows:

$$\Delta S_{\mathrm{SW}} = W_{\mathrm{SW}} + W_{\mathrm{TS}} + W_{\mathrm{CM}} + W_{\mathrm{A(R)}} - W_{\mathrm{LG}} - W_{\mathrm{WV}} - W_{\mathrm{E}} - W_{\mathrm{B(L)}} \tag{11-18}$$

where ΔS_{SW} = change in the amount of water stored in solid waste in landfill, lb/yd^3

W_{SW} = water (moisture) in incoming solid waste, lb/yd^3

W_{TS} = water (moisture) in incoming treatment plant sludge, lb/yd^3

W_{CM} = water (moisture) in cover material, lb/yd^3

$W_{\mathrm{A(R)}}$ = water from above (for upper landfill layer, water from above corresponds to rainfall or water from snowmelt), lb/yd^2

W_{LG} = water lost in the formation of landfill gas, lb/yd^3

W_{WV} = water lost as saturated water vapor with landfill gas, lb/yd^3

W_{E} = water lost due to surface evaporation, lb/yd^2

$W_{\mathrm{B(L)}}$ = water leaving from bottom of element (for the cell placed directly above a leachate collection system, water from bottom corresponds to leachate), lb/yd^2

The landfill water balance is prepared by adding the mass of water entering a unit area of a particular layer of the landfill during a given time increment to the moisture content of that layer at the end of the previous time increment, and subtracting the mass of water lost from the layer during the current time increment. The result is referred to as the available water in the current time increment for the particular layer of the landfill. To determine whether any leachate will form, the field capacity of landfill is compared with the amount of water that is present. If the field capacity is less than the amount of water present, then leachate will be formed.

In general, the quantity of leachate is a direct function of the amount of external water entering the landfill. In fact, if a landfill is constructed properly, the production of measurable quantities of leachate can be eliminated. When wastewater treatment plant sludge is added to solid wastes to increase the amount of methane produced, leachate control facilities must be provided. In some cases leachate treatment facilities may also be required.

Movement of Leachate in Unlined Landfills

Under normal conditions, leachate is found in the bottom of landfills. From there its movement in unlined landfills is downward through the underlying strata, although some lateral movement may also occur, depending on the characteristics of the surrounding material. Because of the importance of vertical seepage in the contamination of groundwater, this subject is considered further in the following discussion.

Darcy's Law. The rate of seepage of leachate from the bottom of a landfill can be estimated using Darcy's law, which can be expressed as follows [10]:

$$Q = -KA\frac{dh}{dl} \tag{11-19}$$

where Q = leachate discharge per unit time, gal/yr
K = coefficient of permeability, gal/ft^2 · yr
A = cross-sectional area through which the leachate flows, ft^2
dh/dl = hydraulic gradient, ft/ft
h = head loss, ft
l = length of flow path, ft

The minus sign in Darcy's law arises from the fact that the head loss, dh, is always negative. The coefficient of permeability is also known as the hydraulic conductivity, the effective permeability, or the seepage coefficient. In U.S. customary units, the coefficient of permeability is expressed in gallons per day per square foot, or feet per day. The conversion between these factors is accomplished by noting that 7.48 gal/ft^2 · yr = 1 ft/yr. Typical values for the permeability coefficient for various soils are given in Table 11-15.

Estimation of Vertical Seepage of Leachate. Before Darcy's law is applied to the estimation of seepage rates from a landfill, it is helpful to review the physical

TABLE 11-15
Typical permeability coefficients for various soils[a]

	Coefficient of permeability, K	
Material	**ft/d**	**gal/ft^2 · d**
Uniform coarse sand	1333	9970
Uniform medium sand	333	2490
Clean, well-graded sand and gravel	333	2490
Uniform fine sand	13.3	100
Well-graded silty sand and gravel	1.3	9.7
Silty sand	0.3	2.2
Uniform silt	0.16	1.2
Sandy clay	0.016	0.12
Silty clay	0.003	0.022
Clay (30 to 50 percent clay sizes)	0.0003	0.0022
Colloidal clay	0.000003	0.000022

[a] Adapted from Refs. 10 and 40 and based on laminar flow.

Note: ft/d × 0.3048 = m/d
gal/ft^2 · d × 0.0408 = m^3/m^2 · d

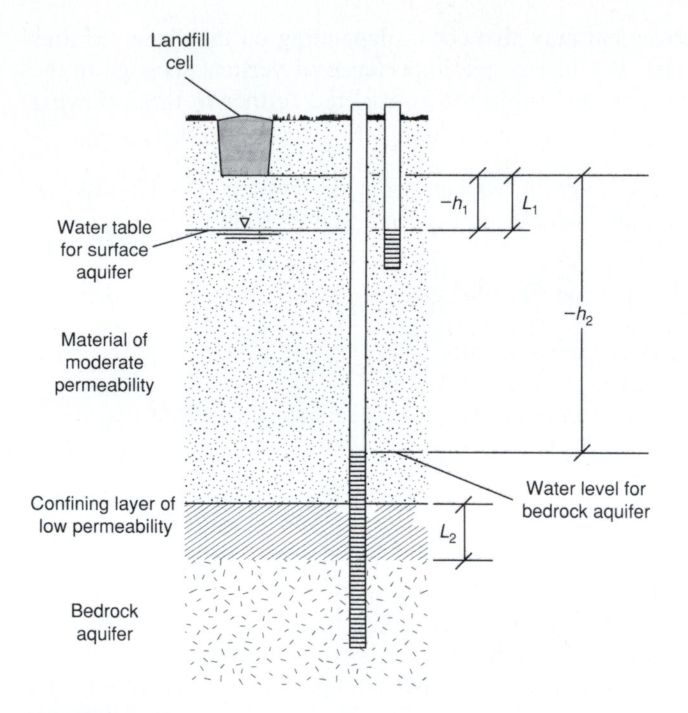

FIGURE 11-32
Definition sketch for determination of seepage from landfills and from surface to subsurface aquifers.

conditions of the problem by referring to Fig. 11-32. There, a landfill cell has been placed in a surface aquifer, composed of material of moderate permeability, that overlies a bedrock aquifer. In this situation, it is possible to have two different piezometric water surfaces if wells are placed in the surface and bedrock aquifers. With respect to the movement of leachate, two problems are of interest. The first is the rate at which leachate seeps from the bottom of the landfill into the groundwater in the surface aquifer. The second is the rate at which groundwater from the surface aquifer moves into the bedrock aquifer. These two problems are considered in the following analysis, but the question of how the mixing of the leachate and groundwater occurs in the surface aquifer is beyond the scope of this text.

In the first problem, the leachate flow rate from the landfill to the upper groundwater is computed by assuming that the material below the landfill to the top of the water table is saturated and that a small layer of leachate exists at the bottom of the fill. Under these conditions the application of Darcy's equation is as follows:

$$Q(\text{gal/yr}) = -K(\text{gal/ft}^2 \cdot \text{yr}) \times A(\text{ft}^2)\frac{-h_1(\text{ft})}{L_1(\text{ft})} \tag{11-20}$$

but because $h_1 = L_1$,

$$Q(\text{gal/yr}) = K(\text{gal/ft}^2 \cdot \text{yr}) \times A(\text{ft}^2)$$

If one assumes that flow occurs through 1.0 ft^2, then

$$Q(\text{gal/yr}) = K(\text{gal/ft}^2 \cdot \text{yr})(\text{ft}^2) \tag{11-21}$$

Thus, the leachate discharge rate per unit area is equal to the value of K multiplied by square feet. For example, if the upper stratum of material in Fig. 11-32 were sandy clay, the corresponding seepage rate would be 0.12 gal/ft^2 · d (see Table 11-15). The computed value represents the maximum amount of seepage that would be expected, and this value should be used for design purposes. Under normal conditions, the actual rate would be less than this value because the soil column below the landfill would not be saturated. Also, most of the leachate reaching the bottom of the landfill would have been removed in the leachate collection system.

In the second problem, the rate of movement of water from the upper aquifer to the lower aquifer would be given by Eq. (11-20). In this case, the thickness of the confining layer is used to determine the hydraulic gradient.

Hydraulic Equivalency. In some states the concept of hydraulic equivalency is used to assess alternative liner designs. Three equivalent liner configurations are illustrated in Fig. 11-33. If Darcy's law is applied to the first configuration, the flow rate per unit area is equal to 2.67 K. Applying Darcy's law to the remaining two liner configurations yields the same result. From this analysis one can see that the water level maintained within the landfill is an important design consideration.

Breakthrough Time. The breakthrough time in years for leachate to penetrate a clay liner of a given thickness can be estimated using the following equation:

$$t = \frac{d^2 \alpha}{K(d + h)} \tag{11-22}$$

where t = breakthrough time, yr
d = thickness of clay liner, ft

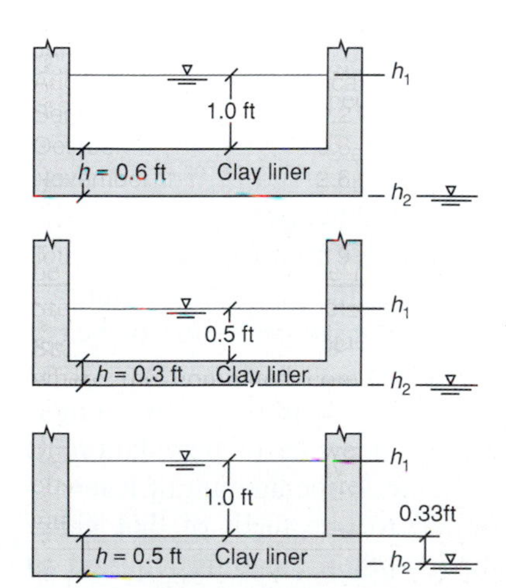

FIGURE 11-33
Definition sketch for assessing the equivalency of landfill liners. (Note that the discharge through each liner configuration is the same.)

α = effective porosity
K = coefficient of permeability, ft/yr
h = hydraulic head, ft

Typical effective porosity values for clays with a coefficient of permeability in the range from 10^{-6} to 10^{-8} cm/s will vary from 0.1 to 0.3, depending on the specific type of clay.

Fate of Constituents in Leachate in Subsurface Migration

The major concern with the movement of leachate into the subsurface aquifer below unlined and lined landfills is the fate of the constituents found in leachate. Mechanisms that are operative in the attenuation of the constituents found in leachate as the leachate migrates through the subsurface soil include mechanical filtration, precipitation and coprecipitation, sorption (including ion exchange), gaseous exchange, dilution and dispersion, and microbial activity [2, 29, 36]. The fate of heavy metals and trace organics, the two constituents of greatest interest, is considered in the following discussion.

Heavy Metals. In general, heavy metals are removed by ion exchange reactions as leachate travels through the soil while trace organics are removed primarily by adsorption. The ability of a soil to retain the heavy metals found in leachate is a function of the cation exchange capacity (CEC) of the soil. The uptake and release of positively charged ions by a soil is referred to as cation, or base, exchange. The total CEC of a soil is defined as the number of milliequivalents (meq) of cations that 100 grams of soil will adsorb. The CEC of a soil depends on the amount of mineral and organic colloidal matter present in the soil matrix. Typical CEC values, at a pH value of 7, are 100 to 200 meq/100 g for organic colloids, 40 to 80 meq/100 g for 2:1 clays (montmorillonite minerals), and 5 to 20 meq/100 g for 1:1 clays (kaolinite minerals). The reported CEC values are affected by the pH of the solution; they drop to about 10 percent of the given values at a pH value of 4. As noted previously, the presence of carbon dioxide in the bottom of a landfill will tend to lower the pH of the leachate [36].

The capacity of a clay landfill liner to take up heavy metals can be estimated as follows. Assume the CEC of the liner material is 100 meq/100 g. If the density of the clay material used in the liner is 137 lb/ft^3 (specific gravity equals 2.2), then about 3000 meq of cations can be adsorbed per cubic foot of liner material. Using a typical value of 20 mg/meq for the heavy metals, the amount of metal that could be adsorbed per cubic foot is equal to 60 g. If the concentration of heavy metals in the leachate was 100 mg/L, the heavy metals could be removed from about 600 L of leachate. If the permeability of the clay is equal to 1×10^{-7} cm/s, then 2.83 L would pass through 1 ft^2 each year. At this rate of percolation, it would take 212 years to saturate the original ft^3 of clay. If the amount of leachate allowed to percolate through the liner were limited to one tenth of that value by designing the leachate collection system properly, then the time required to

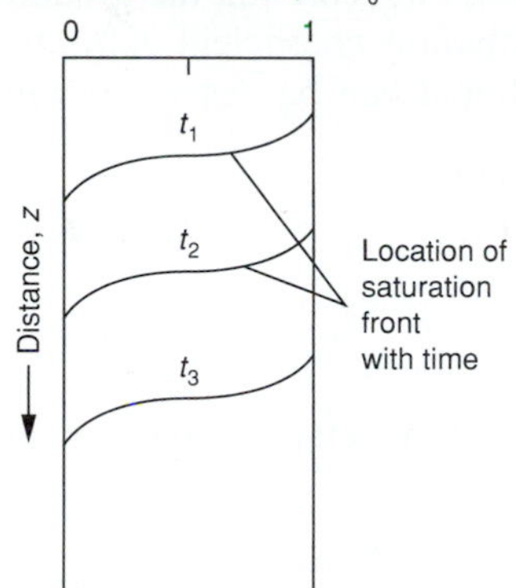

FIGURE 11-34
Typical movement of heavy metals saturation front in clay liner.

saturate the ft^3 of clay would be approximately 2000 years. Even with all of the simplifying assumptions that went into the above analysis, it can be concluded that with a properly designed landfill cover and clay liner, heavy metals should not pose a problem. The saturation front for a typical heavy metal with time can be depicted as shown in Fig. 11-34.

Trace Organics. Adsorption is the most common way in which the organic constituents in leachate are removed as it moves through a porous medium. If hydrodynamic dispersion is neglected, the materials balance for a contaminant subject to adsorption in a groundwater aquifer is given by the following modified form of Eq. (11-4):

$$\frac{\partial S}{\partial t}\frac{\rho_b}{\alpha} + \frac{\partial C}{\partial t} = -v_z\frac{\partial C}{\partial z} \tag{11-23}$$

where S = mass of solute sorbed per unit mass of dry soil, g/g
ρ_b = bulk density of soil, g/m^3
α = porosity
C = concentration of contaminant in the liquid phase, g/m^3
v_z = average fluid velocity in z direction, m/s

The mass of material sorbed per unit mass of dry soil is related to the concentration of the contaminant in the liquid phase and the soil distribution coefficient, as described in the following equation:

$$S = K_{SD} \times C \tag{11-24}$$

where K_{SD} = soil distribution coefficient, m^3/g

Note that Eq. (11-24) describes linear sorption. For some of the organic compounds found in landfills, the sorption may be nonlinear. Differentiating Eq. (11-24) with respect to time and substituting $(K_{SD})\partial C/\partial t$ for $\partial s/\partial t$ in Eq. (11-23) yields

$$-v_z\frac{\partial C}{\partial z} = \left(1 + \frac{\rho_b}{\alpha}K_{SD}\right)\frac{\partial C}{\partial t} \tag{11-25}$$

Where the partitioning of the contaminant between the soil and the groundwater can be described adequately by the soil distribution coefficient K_{SD}, the retardation of the contaminant front relative to the liquid can be described with the following relationship:

$$R = \frac{v_z}{v_{zc}} = \left(1 + \frac{\rho_b}{\alpha} K_{SD}\right) \tag{11-26}$$

where R = retardation factor, unitless
v_z = average velocity of groundwater, m/s
v_{zc} = average velocity of the $C/C_o = 0.5$ point of the retarded contaminant concentration profile, m/s

If it is assumed that α for most soils varies from 0.2 to 0.4 and that the corresponding values for ρ_b are approximately 1.6 to 2.1 × 10^6 g/m^3, then Eq. (11-26) can be written as follows:

$$R = \frac{v_z}{v_{zc}} = (1 + 4 \times 10^6 K_{SD}) \text{ to } (1 + 10 \times 10^6 K_{SD}) \tag{11-27}$$

If K_{SD} equals zero, the contaminant is nonreactive and no retardation occurs (see Fig. 11-35). If K_{SD} is greater than about 10^{-4} the contaminant is essentially immobile. The value of K_{SD} can be estimated by using the following expression:

$$K_{SD} = 6.3 \times 10^{-7} f_{OC}(K_{OW}) \tag{11-28}$$

where f_{OC} = fraction of organic carbon in the soil, g/g
K_{OW} = octanol:water distribution coefficient

Values for K_{OW} for various organic compounds are given in Appendix H.

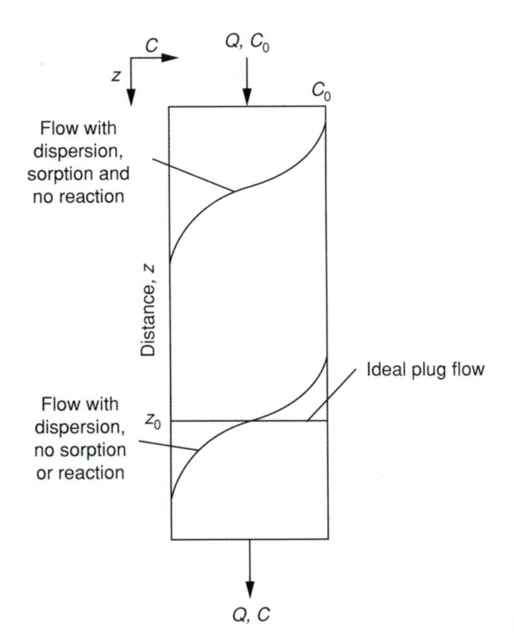

FIGURE 11-35
Typical retardation of trace organic compounds in subsurface movement.

Retardation of the organic constituents found in leachate is important because the retained material can be subjected to biological and chemical conversion reactions, in some cases rendering the retained material harmless.

Control of Leachate in Landfills

As leachate percolates through the underlying strata, many of the chemical and biological constituents originally contained in it will be removed by the filtering and adsorptive action of the material composing the strata. In general, the extent of this action depends on the characteristics of the soil, especially the clay content. Because of the potential risk involved in allowing leachate to percolate to the groundwater, best practice calls for its elimination or containment.

Landfill liners are now commonly used to limit or eliminate the movement of leachate and landfill gases from the landfill site. To date (1992), the use of clay as a liner material has been the favored method of reducing or eliminating the seepage (percolation) of leachate from landfills (see Table 11-11). Clay is favored for its ability to adsorb and retain many of the chemical constituents found in leachate and for its resistance to the flow of leachate. However, the use of combination composite geomembrane and clay liners is gaining in popularity, especially because of the resistance afforded by geomembranes to the movement of both leachate and landfill gases. The characteristics, advantages, and disadvantages of the geomembrane liners (also known as flexible membrane liners, FMLs) that have been used for MSW landfills are summarized in Table 11-16. Typical specifications for geomembrane liners are given in Table 11-17.

Liner Systems for MSW. The objective in the design of landfill liners is to minimize the infiltration of leachate into the subsurface soils below the landfill thus eliminating the potential for groundwater contamination. A number of liner designs have been developed to minimize the movement of leachate into the subsurface below the landfill. Some of the many types of liner designs that have been used are illustrated in Fig. 11-36. In the multilayer landfill liner designs illustrated in Fig. 11-36, each of the various layers has a specific function. For example, in Fig. 11-36*a* the clay layer and the geomembrane serve as a composite barrier to the movement of leachate and landfill gas. The sand or gravel layer serves as a collection and drainage layer for any leachate that may be generated within the landfill. The geotextile layer is used to minimize the intermixing of the soil and sand or gravel layers. The final soil layer is used to protect the drainage and barrier layers. A modification of the liner design shown in Fig. 11-36*a* involves the installation of leachate collection pipes in the leachate collection layer. Composite liner designs employing a geomembrane and clay layer provide more protection and are hydraulically more effective than either type of liner alone.

In Fig. 11-36*b*, a specifically designed open weave plastic mesh (geonet) and geotextile filter cloth (see Fig. 11-37*a*) are placed over the geomembrane which, in turn, is placed over compacted clay layer. A protective soil layer is placed above the geotextile. The geonet and the geotextile function together as the drainage

TABLE 11-16
Guidelines for leachate control facilities

Item	Comments
Synthetic flexible membrane liners (FMLs)	Liners must be designed and constructed to contain fluids, which include wastes and leachates. For MSW waste management units, synthetic liners are not required. However, if this alternative is selected, synthetic liners must have a minimum thickness of 40 mils. These liners must be installed to cover all natural geologic materials that are likely to be in contact with waste or leachate at a waste management unit.
Bottom seals	No specific regulations exist governing the application of bottom seals at MSW waste management units. Design, construction, and installation of bottom seals are subject to the approval of the local enforcement agencies.
Artificial earthen liners	Clay liners are optional for MSW landfills. If required by site conditions, clay liners for MSW waste management units must be a minimum of 1 ft thick and must be installed at a relative compaction of at least 90 percent. A clay liner must exhibit a maximum permeability of 1×10^{-6} cm/s. Clay liners, if installed, must cover all natural geologic materials that are likely to be in contact with waste or leachate at a waste management unit.
Subsurface barriers	A subsurface barrier is intended to be used in conjunction with natural geologic materials to assure that lateral permeability standards are satisified. Barriers may be required by regional agencies at MSW waste management units where there is potential for lateral movement of fluid, including waste and leachate, and the permeability of natural geologic materials is used for waste containment in lieu of a liner. Barriers must be a minimum of 2 ft thick for clay material or a minimum of 40 mils for synthetic materials. These structures are required to be keyed a minimum of 5 ft into natural geologic materials that satisfy permeability requirements of 1×10^{-6} to 1×10^{-7} cm/s. If cutoff walls are used, excavations for waste management units must also be keyed into natural geologic materials exhibiting permeabilities of no greater than 1×10^{-6} cm/s. Barriers are required to have fluid collection systems upgradient of the structure. The systems must be designed, constructed, operated, and maintained to prevent the buildup of hydraulic head against the structure. The collection system must be inspected regularly and accumulated fluid removed.

layer to convey leachate to the leachate collection system. The permeability of the liner system that is composed of a drainage layer and a filter layer is equivalent to that of coarse sand (see Table 11-15). Because of the potential for the geotextile filter cloth to clog, many designers favor the use of a sand or gravel layer as the drainage layer.

In the liner system shown in Fig. 11-36*c,* two composite liners, commonly identified as the primary and secondary composite liners, are used. The

TABLE 11-17
Performance tests used to measure properties of synthetic geomembrane liners and typical values for these properties[a]

Test	Test method	Typical values
Strength category		
Tensile properties	ASTM D638, Type IV; dumbbell 2 in/min	
Tensile strength at yield		2400 lb/in^2
Tensile strength at break		4000 lb/in^2
Elongation at yield		15%
Elongation at break		700%
Toughness		
Tear resistance initiation	ASTM D1004 die C	45 lb
Puncture resistance	FTMS 101B, method 2031	230 lb
Low temperature brittleness	ASTM D746, procedure B	−94°F
Durability		
Carbon black percent	ASTM D1603	2%
Carbon black dispersion	ASTM D3015	A-1
Accelerated heat aging	ASTM D 573, D1349	Negligible strength change after 1 month at 110°C
Chemical resistance		
Resistance to chemical waste mixtures	EPA method 9090	10% tensile strength change over 120 days
Resistance to pure chemical reagents	ASTM D543	10% tensile strength change over 7 days
Stress cracking resistance		
Environmental stress crack resistance	ASTM D1693, condition C	1500 h

[a] Adapted from Refs. 2, 52.

primary composite liner is used for the collection of leachate, whereas the secondary composite liner serves as a leak-detection system and a backup for the primary composite liner. A modification of the liner system shown in Fig. 11-36*c*, involves replacing the sand drainage layer with a geonet drainage system as shown in Fig. 11-35*b*. The two-layer composite liner shown in Fig. 11-35*d* is the same as the liner shown in Fig. 11-36*c*, with the exception that the clay layer below the first geomembrane liner is replaced with a geosynthetic clay liner (GCL). A manufactured product, the GCL is made from a high-quality bentonite clay (from Wyoming) and an appropriate binding material (see Fig. 11-37*b*). The bentonite clay is essentially a sodium montmorillonite mineral that has the capacity to absorb as much as 10 times its weight in water. As the clay absorbs water, it becomes putty-like and very resistant to the movement of water. Permeabilities as low as 10^{-10} cm/s have been observed. Available in large sheets (12 to 14 by

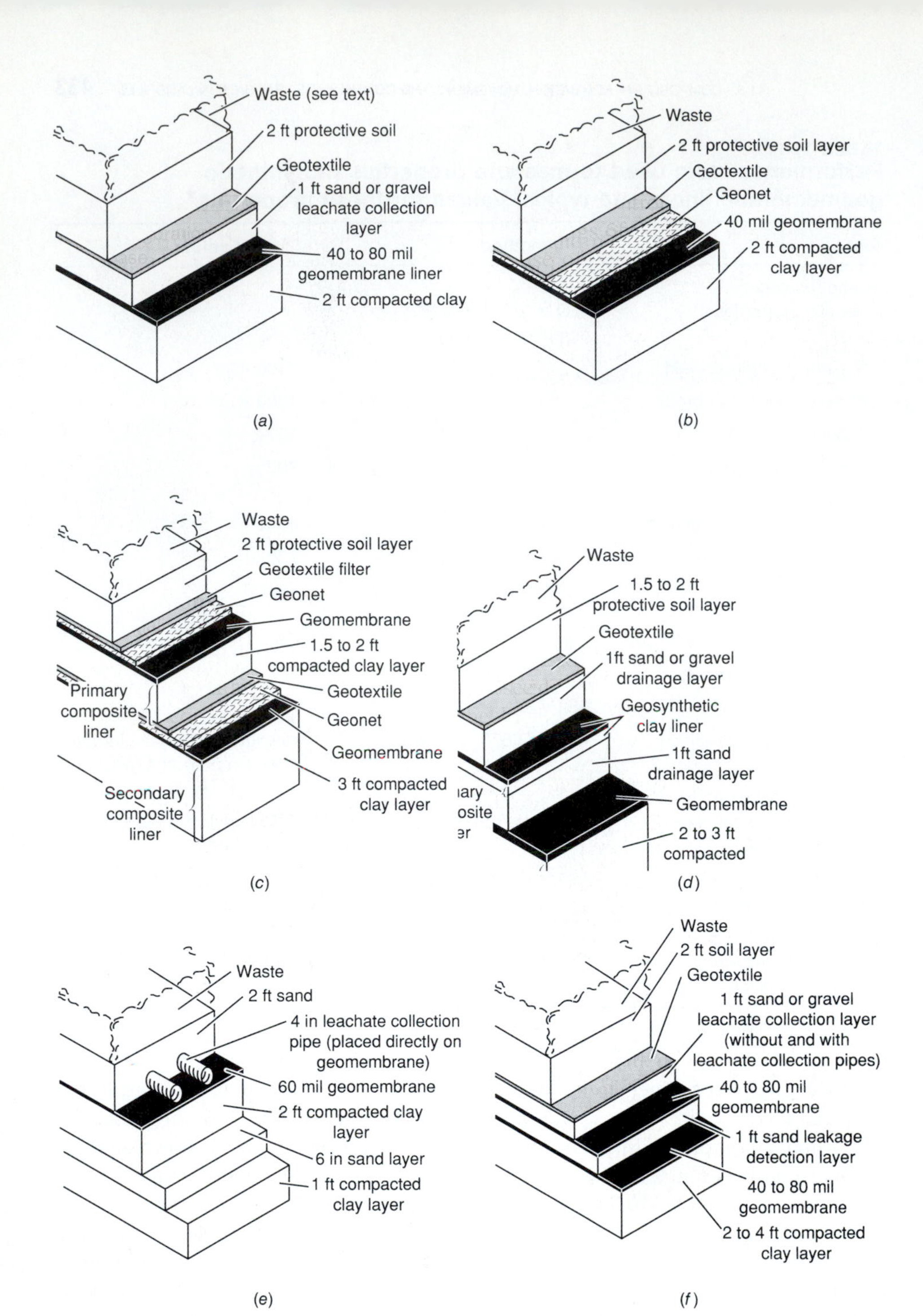

FIGURE 11-36

Typical landfill liners: (*a*, *b*) single-composite barrier types and (*c*–*f*) double-composite barrier types. Note in the double-liner systems the first composite liner is often identified as the primary liner or as the leachate collection system, while the second composite liner is identified as the leachate detection layer. Leachate detection probes are normally placed between the first and second liners.

(a) (b) (c) (d)

FIGURE 11-37
Manufactured materials used in the construction of landfill liners: (*a*) geonet used as a drainage layer is placed over a geomembrane; geotextile (shown folded back) is used to separate materials; (*b*) geosynthetic clay liner; bentonite clay at about 1 lb/ft^2 (gray side) is bonded to geomembrane; (*c*) geomembrane being installed on compacted clay layer; and (*d*) geosynthetic clay liner being installed with clay side up. ((*b*) and (*d*) courtesy of Gundle Lining Systems, Inc.)

100 ft), GCLs are overlapped in the construction of a liner system. Two additional two-layer liner systems are shown in Figs. 11-36*e* and 11-36*f*. In the two-layer composite layer landfill systems shown in Figs. 11-36*c* through *f*, leak-detection sensors are usually placed between the two liners (see Fig. 11-57, p. 461).

Liner Systems for Monofills. Liner systems for monofills usually comprise two geomembranes, each provided with a drainage layer and a leachate collection system (see Figs. 11-36*c* and 11-36*d*). A leachate detection system is placed between the first and second liners as well as below the lower liner. In many installations, a thick (3 to 5 ft) clay layer is used below the two geomembranes for added protection [7].

Construction of Clay Liners. In all of the liner designs illustrated in Fig. 11-36, great care must be exercised in the construction of the clay layer. Perhaps the most serious problem with the use of clay is its tendency to form cracks due to desiccation. It is critical that the clay not be allowed to dry out as it is being placed. To insure that the clay liner performs as designed, the clay liner should be laid in 4- to 6-in layers with adequate compaction between the placement of succeeding layers (see Fig. 11-38). Laying the clay in thin layers avoids the possibility of leaks due to the alignment of clods that could occur if the clay layer is applied in a single pass. Another problem that has been encountered when clays of different types have been used is cracking due to differential swelling. To avoid differential swelling only one type of clay must be used in the construction of the liner.

Leachate Collection Systems

The design of a leachate collection system involves (1) the selection of the type of liner system to be used, (2) the development of the grading plan including the placement of the leachate collection and drainage channels and pipelines for the removal of leachate, and (3) the layout and design of the leachate removal, collection, and holding facilities.

Selection of Liner System. The type of liner system selected will depend to a large extent on the local geology and environmental requirements of the landfill site. For example, in locations where there is no groundwater, a single compacted clay liner may be sufficient. In locations where both leachate and gas migration must be controlled, a combined liner comprising a clay liner and a geomembrane liner with an appropriate drainage and soil protection layer will be necessary.

FIGURE 11-38
Preparation of compacted clay layer before geomembrane liner is placed.

Design of Leachate Collection Facilities. A variety of liner designs have been used for the removal of leachate from landfills. The sloped terrace and piped bottom designs are discussed below.

Sloped terraces. To avoid the accumulation of leachate in the bottom of a landfill, the bottom area is graded into a series of sloped terraces. As shown in Fig. 11-39*a*, the terraces are shaped so that the leachate that accumulates on the surface of the terraces will drain to leachate collection channels. Perforated pipe placed in each leachate collection channel (see Fig. 11-39*b*) is used to convey the collected leachate to a central location, from which it is removed for treatment or reapplication to the surface of the landfill.

The cross-slope of the terraces is usually 1 to 5 percent, and the slope of the drainage channels is 0.5 to 1.0 percent. The slope and maximum length of the drainage channel is selected based on the capacity of the drainage facilities. The flow rate capacity of the drainage facilities is estimated using Manning's equation. The design objective is not to allow the leachate to pond in the bottom of the landfill so as to create a significant hydraulic head on the landfill liner (less than 1 ft at the highest point as specified in the new federal Subtitle D landfill regulations). The depth of flow in the perforated drainage pipe increases continually from the

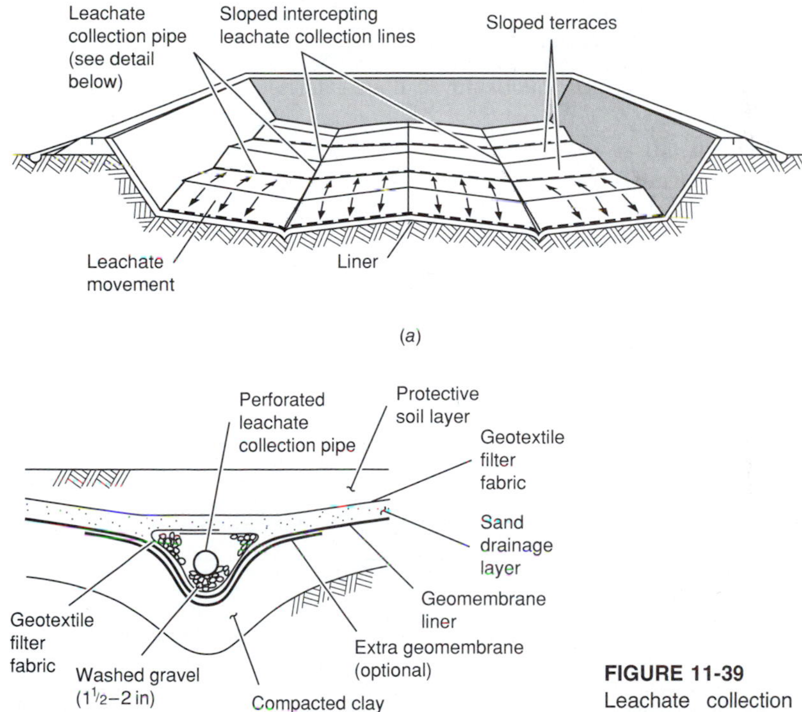

FIGURE 11-39
Leachate collection system with graded terraces: (*a*) pictorial view and (*b*) detail of typical leachate collection pipe.

upper reaches of the drainage channel to the lower reaches. In very large landfills, the drainage channels will be connected to a larger cross-collection system.

Piped bottom. An alternative plan for the collection of leachate is shown in Fig. 11-40. As shown, the bottom area is then divided into a series of rectangular strips by clay barriers placed at appropriate distances (see Fig. 11-40*a*). The barrier's spacing corresponds to the width of a landfill cell. Leachate collection pipes are then placed lengthwise directly on the geomembrane. The 4-in leachate collection pipes have laser-cut perforations, similar to a well screen, over one-half of the circumference. The laser-cuts are spaced 0.25 in apart and the size of the laser cut is 0.0001 in, corresponding to the smallest sand size. To promote effective drainage, the bottom is sloped from 1.2 to 1.8 percent. The leachate collection pipes, spaced every 20 ft, are covered with a two-foot layer of sand (see Fig. 11-40*b*) before landfilling commences. The use of a multiple-pipe leachate collection system will ensure the rapid removal of leachate from the bottom of the landfill. Further, the use of a 2-ft sand layer serves to filter the leachate before it is collected for treatment. The first 3-ft layer of solid waste, placed directly on the sand layer, is not compacted [33].

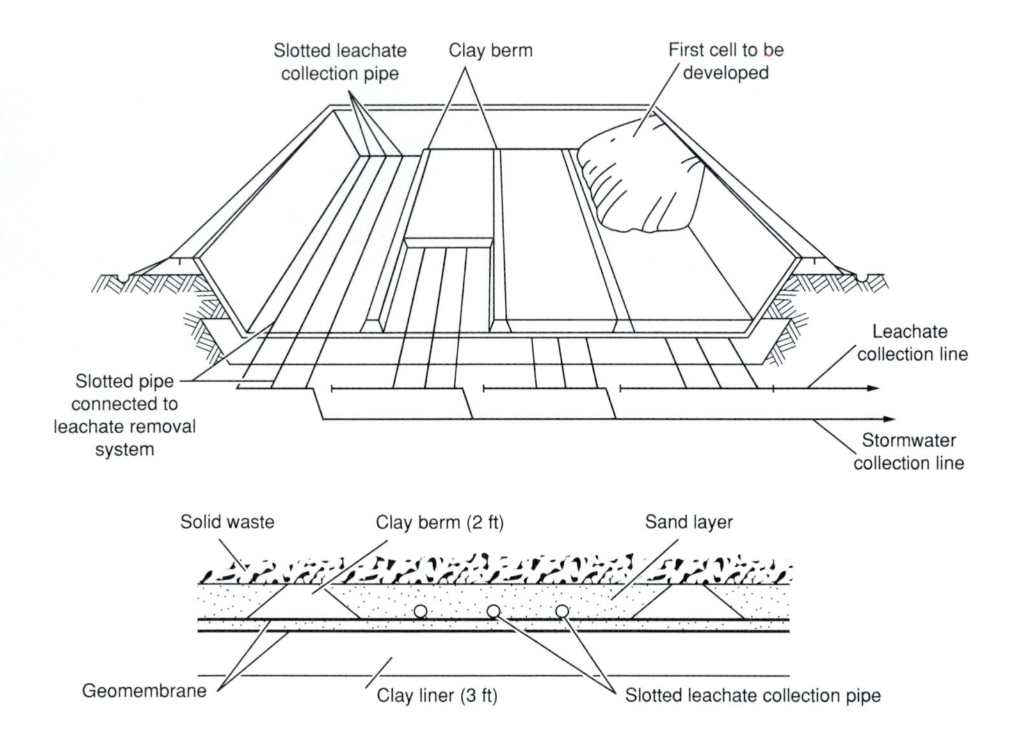

FIGURE 11-40
Typical leachate collection system using multiple leachate collection pipes: (*a*) pictorial view and (*b*) detail of typical leachate collection pipes (adapted from Ref. 33).

A unique feature of the design shown in Fig. 11-40 is the method used to remove the stormwater from the unused portion of the landfill. The method is detailed in Fig. 11-41. In the unused portion of the landfill, stormwater is collected in the lines that will ultimately be used for the collection of leachate. When the next landfill cell is to be placed in service, the leachate piping is reconnected to the leachate collection system, and the leachate collection pipe which extends into the next diked strip is capped [33].

Leachate Removal, Collection, and Holding Facilities. Two methods have been used for the removal of leachate that accumulates within a landfill. In Fig. 11-42*a*, the leachate collection pipe is passed through the side of the landfill. Where this method is used, great care must be taken to ensure that the seal where the pipe penetrates the landfill liner is sound. An alternative method used for the removal of leachate from landfills involves the use of an inclined collection pipe located within the landfill (see Fig. 11-42*b*). Leachate collection facilities are used where the leachate is to be recycled from or treated at a central location. A typical leachate collection access vault is shown in Fig. 11-43*a*. In some locations, the leachate removed from the landfill is collected in a holding tank such as shown in Fig. 11-43*b*. The capacity of the holding tank will depend on the type of treatment facilities that are available and the maximum allowable discharge rate to the treatment facility. Typically, leachate holding tanks are designed to hold from 1 to 3 days of leachate production during the peak leachate production period. Both double- and single-walled tanks have been used, but the double-

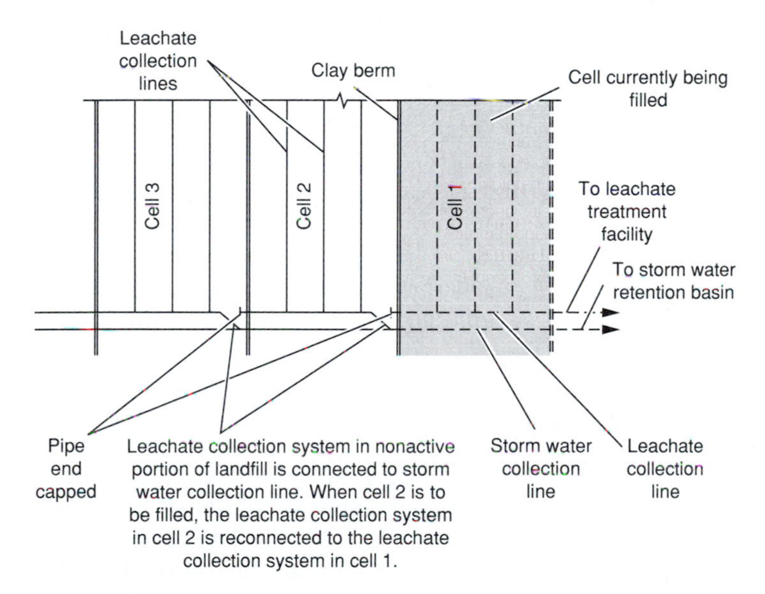

FIGURE 11-41
Storm water management in area-type landfill. (Courtesy of C. C. Miller, see also Ref. 33.)

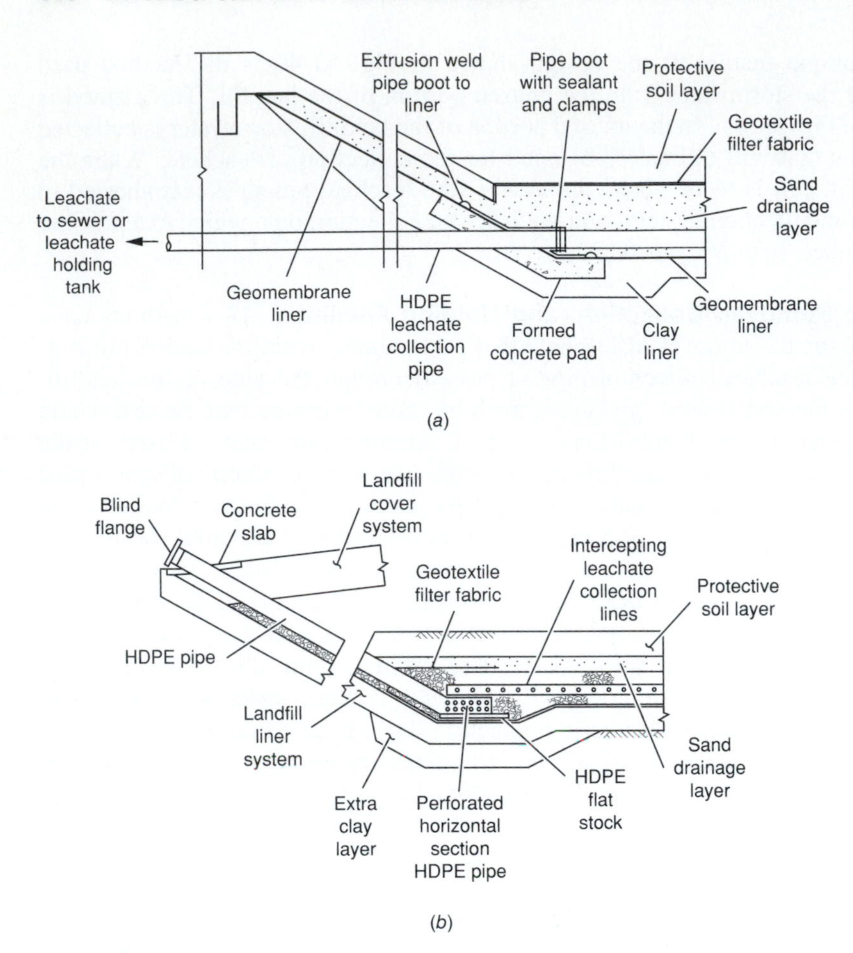

FIGURE 11-42
Typical systems used to remove leachate from landfills: (*a*) leachate collection pipe passed through side of landfill and (*b*) inclined leachate collection pipe located within landfill. Leachate is removed with a pump.

walled tanks are preferred over single-walled tanks because of the added safety afforded. Although both plastic and metallic tanks have been used, plastic tanks are more corrosion resistant.

Leachate Management Options

The management of leachate, when and if it forms, is key to the elimination of the potential for a landfill to pollute underground aquifers. A number of alternatives have been used to manage the leachate collected from landfills including: (1) leachate recycling, (2) leachate evaporation, (3) treatment followed by disposal, and (4) discharge to municipal wastewater collection systems. These options are discussed briefly below.

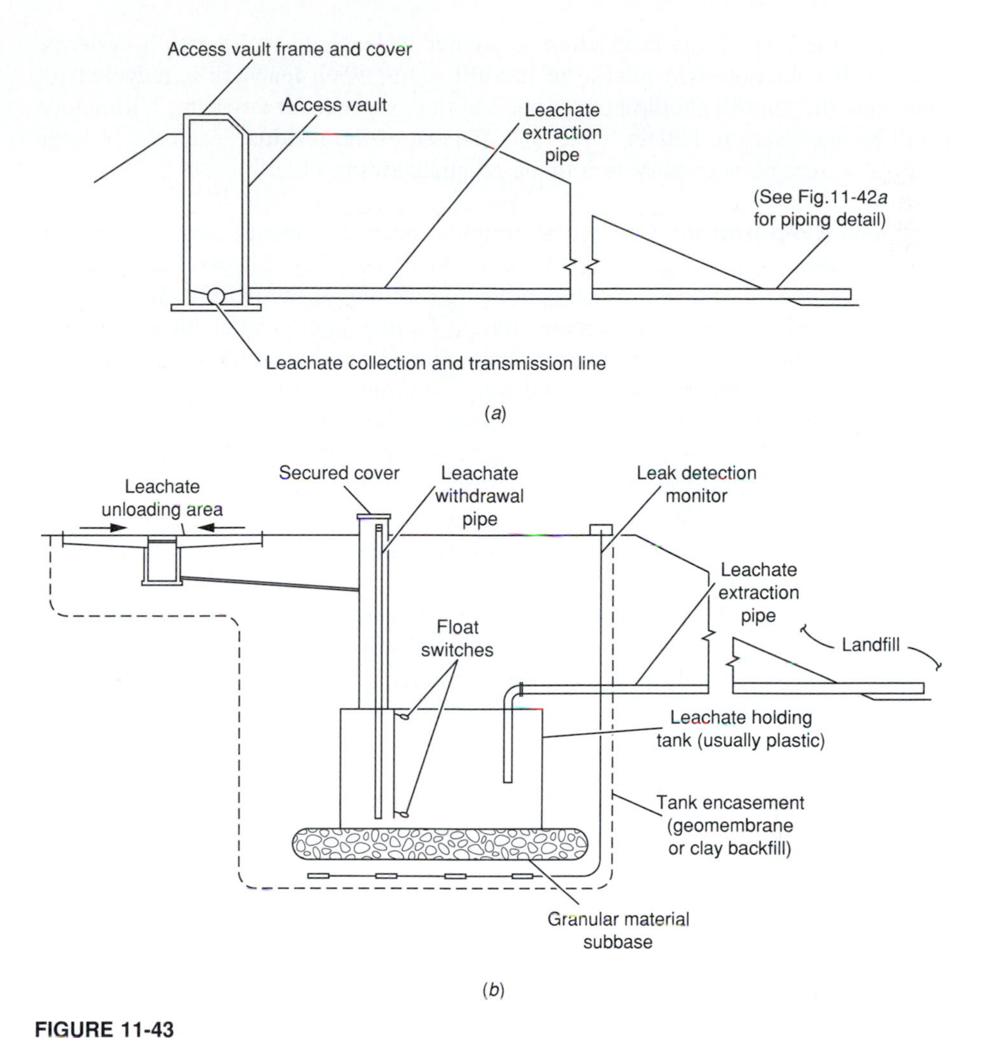

FIGURE 11-43
Examples of leachate collection facilities: (*a*) leachate collection and transmission vault and (*b*) leachate holding tank.

Leachate Recycling. An effective method for the treatment of leachate is to collect (see Fig. 11-43) and recirculate the leachate through the landfill. During the early stages of landfill operation the leachate will contain significant amounts of TDS, BOD_5, COD, nutrients, and heavy metals (see Table 11-13). When the leachate is recirculated, the constituents are attenuated by the biological activity and by other chemical and physical reactions occurring within the landfill. For example, the simple organic acids present in the leachate will be converted to CH_4 and CO_2. Because of the rise in pH within the landfill when CH_4 is produced, metals will be precipitated and retained within the landfill. An additional benefit of leachate recycling is the recovery of landfill gas that contains CH_4.

Typically, the rate of gas production is greater in leachate recirculation systems. To avoid the uncontrolled release of landfill gases when leachate is recycled for treatment, the landfill should be equipped with a gas recovery system. Ultimately, it will be necessary to collect, treat, and dispose of the residual leachate. In large landfills it may be necessary to provide leachate storage facilities.

Leachate Evaporation. One of the simplest leachate management systems involves the use of lined leachate evaporation ponds (see Fig. 11-44). Leachate that is not evaporated is sprayed on the completed portions of the landfill. In locations with high rainfall, the lined leachate storage facility is covered with a geomembrane during the winter season to exclude rainfall. The accumulated leachate is disposed of by evaporation during the warm summer months, by uncovering the storage facility, and by spraying the leachate on the surface of the operating and completed landfill. Odorous gases that may accumulate under the surface cover are vented to a compost or soil filter (see Fig. 11-45) [3, 51]. Soil beds are typically 2 to 3 ft deep, with organic loading rates of about 0.1 to 0.25 lb/ft^3 of soil. During the summer when the pond is uncovered, surface aeration may be required to control odors. If the storage pond is not large it can be left covered year round. Another example involves treatment of the leachate (usually biologically) with winter storage and spray disposal of the treated effluent on nearby lands during the summer. If enough land is available, spraying of effluent can be carried out on a continuous basis, even when it is raining.

Leachate Treatment. Where leachate recycling and evaporation is not used, and the direct disposal of leachate to a treatment facility is not possible, some form of pretreatment or complete treatment will be required. Because the characteristics of the collected leachate can vary so widely, a number of options have been used for

(a)

(b)

FIGURE 11-44
Views of lined evaporation ponds: (*a*) for leachate—see also Fig. 11-64 (liquid in pond is rainwater) and (*b*) for leachate and treatment plant sludges.

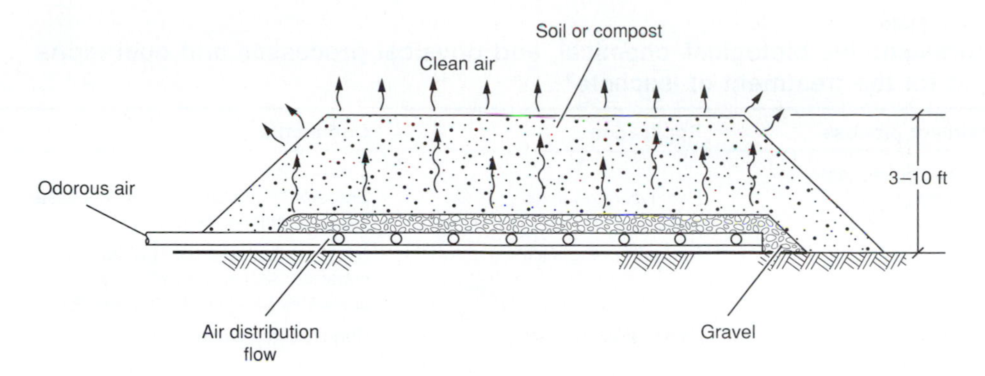

FIGURE 11-45
Typical compost or soil filter used to remove odors from gases [51].

the treatment of leachate. The principal biological and physical/chemical treatment operations and processes used for the treatment of leachate are summarized in Table 11-18. The treatment process or processes selected will depend to a large extent on the contaminant(s) to be removed. Typical examples of the types of aerobic and anaerobic biological processes that have been used for the treatment of leachate are shown in Fig. 11-46. Design details on the treatment options reported in Table 11-18 may be found in Ref. 49.

Selection of treatment facilities. The type of treatment facilities used will depend primarily on the characteristics of the leachate and secondarily on the geographic and physical location of the landfill. Leachate characteristics of concern include TDS, COD, SO_4^{2-}, heavy metals, and nonspecific toxic constituents. Leachate containing extremely high TDS concentrations (e.g., $> 50{,}000$ mg/L) may be difficult to treat biologically. High COD values favor anaerobic treatment processes because aerobic treatment is expensive. High sulfate concentrations may limit the use of anaerobic treatment processes because of the production of odors from the biological reduction of sulfate sulfide (see Eqs. 4-12 through 4-14). Heavy metal toxicity is also a problem with many biological treatment processes. Another important question is how large should the treatment facilities be? The capacity of the treatment facilities will depend on the size of the landfill and the expected useful life. The presence of nonspecific toxic constituents is often a problem with older landfills that received a variety of wastes, before environment regulations governing the operation of landfills were enacted.

Integrated leachate management system. An example of an integrated leachate management system is shown in Fig. 11-47. Liquid (leachate) that moves down through the solid waste is first filtered as it passes the sand layer in the landfill (see Fig. 11-40). The collected leachate is transported to a treatment lagoon where septage is also added. The liquid in the lagoon is aerated to reduce the organic content and to control odors. Liquid from the lagoon is then applied to

TABLE 11-18
Representative biological, chemical, and physical processes and operations used for the treatment of leachate[a]

Treatment process	Application	Comments
Biological processes		
Activated sludge	Removal of organics	Defoaming additives may be necessary; separate clarifier needed
Sequencing batch reactors	Removal of organics	Similar to activated sludge, but no separate clarifier needed; only applicable to relatively low flow rates
Aerated stabilization basins	Removal of organics	Requires large land area
Fixed film processes (trickling filters, rotating biological contactors)	Removal of organics	Commonly used on industrial effluents similar to leachates, but untested on actual landfill leachates
Anaerobic lagoons and contactors	Removal of organics	Lower power requirements and sludge production than aerobic systems; requires heating; greater potential for process instability; slower than aerobic systems
Nitrification/denitrification	Removal of nitrogen	Nitrification/denitrification can be accomplished simultaneously with the removal of organics
Chemical processes		
Neutralization	pH control	Of limited applicability to most leachates
Precipitation	Removal of metals and some anions	Produces a sludge, possibly requiring disposal as a hazardous waste
Oxidation	Removal of organics; detoxification of some inorganic species	Works best on dilute waste streams; use of chlorine can result in formation of chlorinated hydrocarbons
Wet air oxidation	Removal of organics	Costly; works well on refractory organics
Physical operations		
Sedimentation/flotation	Removal of suspended matter	Of limited applicability alone; may be used in conjunction with other treatment processes
Filtration	Removal of suspended matter	Useful only as a polishing step
Air stripping	Removal of ammonia or volatile organics	May require air pollution control equipment
Steam stripping	Removal of volatile organics	High energy costs; condensate steam requires further treatment
Adsorption	Removal of organics	Proven technology; variable costs depending on leachate
Ion exchange	Removal of dissolved inorganics	Useful only as a polishing step
Ultrafiltration	Removal of bacteria and high molecular weight organics	Subject to fouling; of limited applicability to leachate
Reverse osmosis	Dilute solutions of inorganics	Costly; extensive pretreatment necessary
Evaporation	Where leachate discharge is not permissible	Resulting sludge may be hazardous; can be costly except in arid regions

[a] Adapted from Ref. 43.

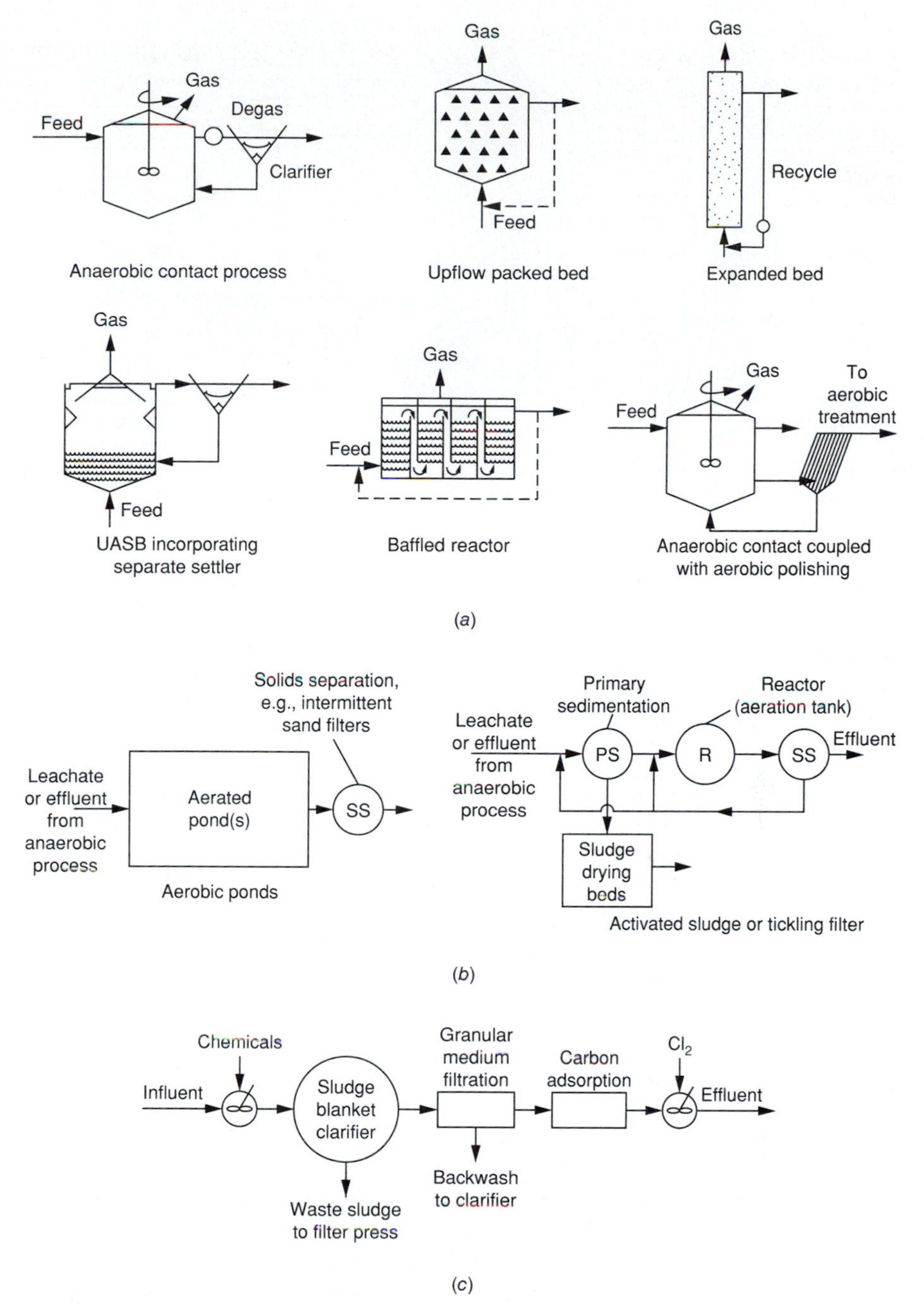

FIGURE 11-46
Typical processes used for the treatment of leachate: (*a*) anaerobic processes, (*b*) aerobic processes, and (*c*) chemical treatment process for the removal of heavy metals and selected organics.

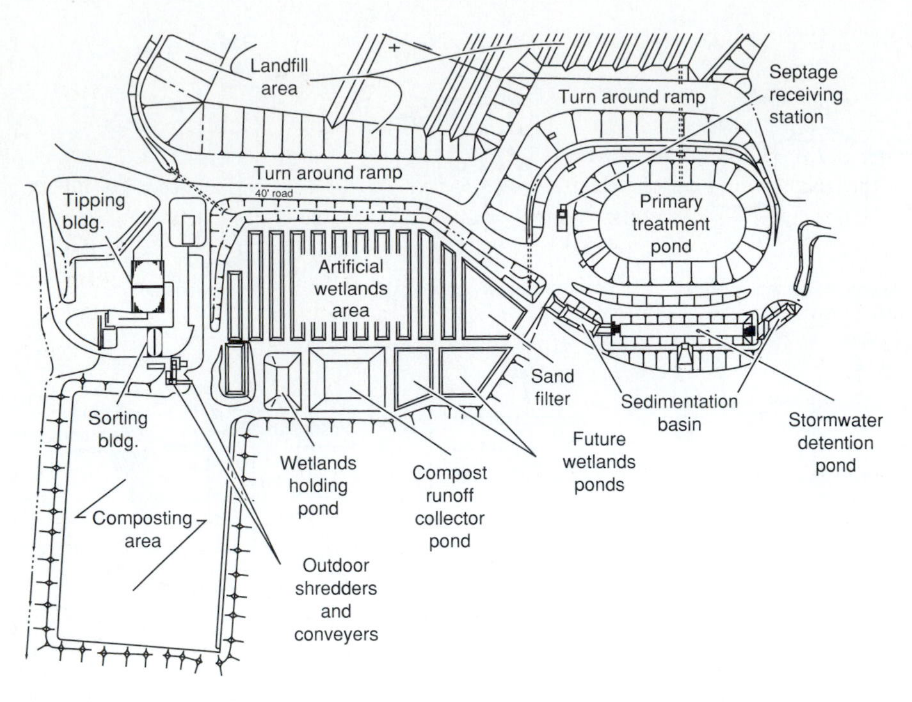

FIGURE 11-47
Integrated leachate treatment system employing constructed wetlands (from Ref. 33).

shredded MSW that is to be composted and used for intermediate cover material in the landfill (see Fig. 11-51 in Section 11-6). Recyclable materials and metals are removed before the MSW is shredded. Application of the leachate to the shredded MSW provides the moisture needed for optimum composting and reduces the volume of leachate through evaporation. The excess leachate is filtered as it passes through the shredded waste and the sand filter underdrain system. The collected leachate is piped to a series of constructed wetlands. The wetlands are used to remove organic material, nutrients, heavy metals, and other trace organics. The effluent from the constructed wetlands is passed through a slow sand filter and then used for spray irrigation on the grass-covered landscape at the landfill.

Discharge to Wastewater Treatment Plant. In those locations where a landfill is located near a wastewater collection system or where a pressure sewer can be used to connect the landfill leachate collection system to a wastewater collection system, leachate is often discharged to the wastewater collection system. In many cases pretreatment, using one or more of the methods reported in Table 11-18, may be required to reduce the organic content before the leachate can be discharged to the sewer. In locations where sewers are not available, and evaporation and spray disposal are not feasible, complete treatment followed by surface discharge may be required.

11-6 SURFACE WATER MANAGEMENT

Equally important in controlling the movement of leachate is the management of all surface waters including rainfall, stormwater runoff, intermittent streams, and artesian springs. The management of surface water is introduced in this section. With the use of a properly designed cover layer, an appropriate surface slope (3 to 5 percent), and adequate stormwater drainage, surface infiltration can be controlled effectively. With proper surface water controls, it may not be necessary to provide an impermeable surface barrier. Topics considered in this section include (1) surface water control systems, (2) design of intermediate cover layers, (3) design of final cover layers, and (4) determination of the percolation through the cover.

Surface Water Control Systems

Elimination or reduction of the amount of surface water that enters the landfill is of fundamental importance in the design of a sanitary landfill because surface water is the major contributor to the total volume of leachate. Stormwater runoff from the surrounding area must not be allowed to enter the landfill and surface water runoff (from rainfall) must not be allowed to accumulate on the surface of the landfill.

Surface Water Drainage Facilities. In those locations where stormwater runoff from the surrounding areas can enter the landfill (e.g., landfills located in canyons), the site must be graded appropriately and properly designed drainage facilities must be installed (see Fig. 11-48). The drainage facilities may be designed to remove the runoff from the surrounding area only, or from the surrounding area as well as the surface of the landfill. In locations where the entire landfill liner system is installed at one time, the design of the liner must allow for the diversion of stormwater not falling on the wastes being landfilled. The diversion of stormwater from the unused portion of a landfill is illustrated in Fig. 11-41.

In locations where only the surface water from the top of the landfill must be removed, the drainage facilities should be designed to limit the travel distance of the surface water. In many designs, a series of interceptor ditches are used. Flow from the interceptor ditches is routed to a larger main ditch for removal from the site. Examples of the types of drainage facilities used to protect landfills are illustrated in Fig. 11-49.

Stormwater Storage Basins. In many cases, it may be necessary to construct stormwater storage basins to contain the diverted stormwater flows so as to minimize downstream flooding. Typically stormwater must be collected from the completed portions of the landfill as well as from areas yet to be filled. An example of a stormwater retention/storage basin is illustrated in Fig. 11-50. Standard hydrological procedures are followed in sizing the stormwater basins [20, 27, 28].

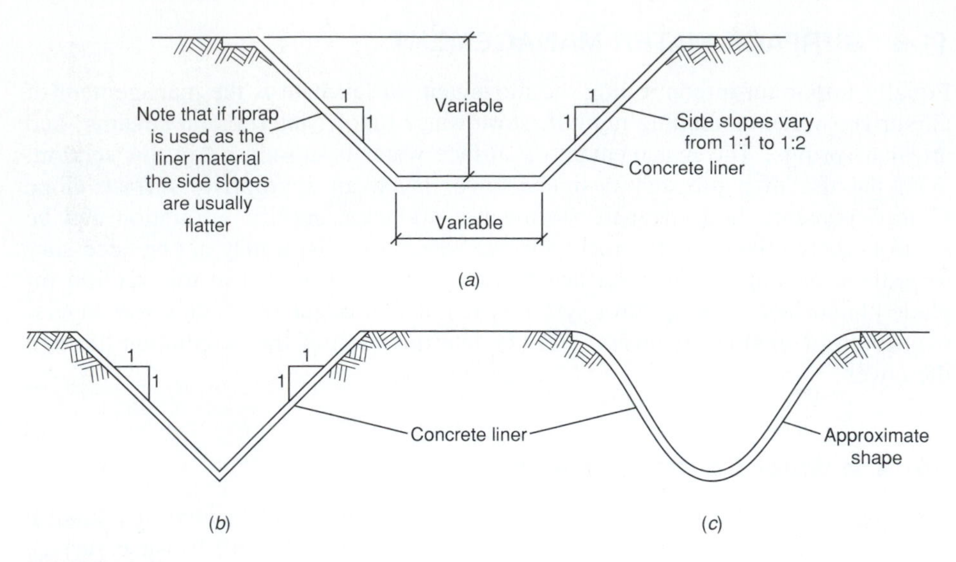

FIGURE 11-48
Examples of drainage facilities used at landfills: (*a*) trapezoidal lined ditch, (*b*) vee lined ditch, and (*c*) shaped vee lined ditch. Note that the trapezoidal ditch cross section is expandable to accomodate a wide range of flows.

Intermediate Cover Layers

Intermediate cover layers are used to cover the wastes placed each day to eliminate the harboring of disease vectors, to enhance the aesthetic appearance of the landfill site, and to limit the amount of surface infiltration. The greatest amount of water that enters a landfill and ultimately becomes leachate enters during the period when the landfill is being filled. Some of the water, in the form of rain and snow, enters while the wastes are being placed in the landfill. Water also enters the landfill by first infiltrating and subsequently percolating through the intermediate landfill cover. Thus, the materials and method of placement of the intermediate cover can limit the amount of surface water that enters the landfill.

Materials Used for Intermediate Cover Layers. Generalized ratings for the suitability of various types of materials that have been used as intermediate landfill cover are reported in Table 11-19. Of the materials listed, only compost produced from yard waste and MSW, the geosynthetic clay liner, and clay are effective in limiting the entry of surface water into the landfill. To be effective, the intermediate cover, using the materials cited above, must be sloped properly to enhance surface water runoff.

In some landfill operations, a very thick layer of soil (3 to 6 ft) is placed temporarily over the completed cell. Any rainfall that infiltrates the intermediate cover layer is retained by virtue of its field capacity. When a second lift is to be placed over the first lift, the soil is removed and stockpiled before filling

(*a*) (*b*) (*c*) (*d*) (*e*)

FIGURE 11-49
Views of drainage facilities used at landfills: (*a*) trapezoidal lined ditch; (*b*) trapezoidal lined ditch built in sections; (*c*) half-section corrugated pipe used to transport surface runoff from upper benches of landfill (note trapezoidal drainage channel in foreground); (*d*) typical vee type ditch used in upper portion of drainage area; and (*e*) shaped vee type ditch used to transfer runoff from upper portions of drainage area to stormwater retention basin.

FIGURE 11-50
View of large riprap lined stormwater retention/storage basin at a large landfill. The size of the basin can be estimated from the size of the vehicles parked in the bottom of the basin.

TABLE 11-19
Generalized ratings of the suitability of various materials for use as intermediate landfill cover[a]

	Generalized ratings[b]						
Function	**Yard waste mulch**	**Yard waste compost**	**MSW compost**	**Geosynthetic clay liner**	**Typical native soil**	**Clayey-silty sand**	**Clay**
Provides pleasing appearance and controls blowing paper	G–E	G–E	G	E	E	E	E[c]
Prevents rodents from burrowing or tunneling	P	P	P	G–E	P	F–G	P
Keeps flies from emerging	F	F–G	F	E	G	P	E[c]
Minimizes the entry of surface water into landfill	P	G–E	F–G	E	F–G[d]	P	E[c]
Retains rainfall and snowmelt	P	G–E	F–G	G	F–G[d]	P	G[c]
Minimizes landfill gas venting through cover	P	P	P	F–G	P	P	P–F[c]

[a] Adapted in part from Ref. 4.
[b] E = excellent; G = good; F = fair; P = poor.
[c] Except when allowed to dry out and cracks develop in the cover layer.
[d] When a thick layer of soil is used, the rating is G–E.

begins. The use of the operating technique of temporarily storing additional cover material over a completed cell can significantly limit the amount of water entering the landfill. Synthetic foam has also been used as an intermediate landfill cover material. In general, foam works well, except when it rains.

Intermediate Cover Layers Using Waste Materials. As noted in Section 11-4, where the amount of native soil available for use as intermediate cover material is limited, alternative waste materials have been used for the purpose. Suitable materials that can be used as a substitute for native soil include compost and mulch produced from yard wastes and compost produced from MSW (see Fig. 11-51). An important advantage of using compost and mulch produced from MSW is that the landfill volume that would have been occupied by the soil used for intermediate cover is now available for the disposal of waste materials. In locations where the

(*a*)

(*b*)

FIGURE 11-51
Composting of processed MSW for use as intermediate landfill cover: (*a*) shredding facility for commingled MSW from which selected recyclable materials and ferrous metals have been removed and (*b*) composting of the shredded MSW using the windrow method.

amount of cover material is limited, the use of composted MSW can increase the capacity of the landfill significantly.

In the composting operation shown in Fig. 11-51, approximately 40 percent of the waste from household and selected commercial solid waste is shredded after selected recyclable materials have been removed manually and ferrous metal has been removed using two stages of magnetic separation. The shredded material is placed in windrows for composting. Leachate from the landfill is sprayed on the shredded waste to increase the moisture content for optimum composting. The compost product is used as intermediate cover for the remaining 60 percent of the waste that was placed in the landfill directly. Where composted MSW is used as intermediate cover, the compost need not be cured fully before being used as intermediate cover material. Excess compost produced at the landfill site is stored there until needed. Cured compost placed on the MSW deposited in the landfill also serves as an odor filter (see Fig. 11-45). The use of composted MSW for intermediate cover is expected to increase significantly in the coming years, as the conservation of landfill capacity becomes a more important issue [33].

Other waste materials that have been used as intermediate cover material include old carpets, construction and demolition wastes, and agricultural residues. Old carpets can be stockpiled as they are received at the landfill and used as required. Carpets have also been used as part of the final cover design for landfills. The question of whether an intermediate cover layer is even needed or should be required is currently the subject of renewed debate [53].

Final Cover Layers

The primary purposes of the final landfill cover are (1) to minimize the infiltration of water from rainfall and snowfall after the landfill has been completed, (2) to limit the uncontrolled release of landfill gases, (3) to suppress the proliferation of vectors, (4) to limit the potential for fires, (5) to provide a suitable surface for the revegetation of the site, and (6) to serve as the central element in the reclamation of the site. To meet these purposes the landfill cover (1) must be able to withstand climatic extremes (e.g., hot/cold, wet/dry, and freeze/thaw cycles); (2) must be able to resist water and wind erosion; (3) must have stability against slumping, cracking and slope failure, and downslope slippage or creep; (4) must resist the effects of differential landfill settlement caused by the release of landfill gas and the compression of the waste and the foundation soil; (5) must resist failure due to landfilling operations such as surcharge loads due to stockpiling and the travel of collection vehicles across completed portions of the landfill; (6) must resist deformations caused by earthquakes; (7) must withstand alterations to cover materials caused by constituents in the landfill gas; and (8) must resist the disruptions caused by plants, burrowing animals, worms, and insects [18, 23]. It is important to note that under current legislation all of these purposes and attributes must continue to be satisfied far into the future. The general features of a landfill cover, some typical types of landfill cover designs, and the long-term performance requirements for landfill covers are considered below.

General Features of Landfill Covers. A modern landfill cover, as shown in Fig. 11-52, is made up of a series of layers, each of which has a special function. The subbase soil layer is used to contour the surface of the landfill and to serve as a subbase for the barrier layer. In some cover designs, a gas collection layer is placed below the soil layer to transport landfill gas to gas management facilities. The barrier layer is used to restrict the movement of liquids into the landfill and the release of landfill gas through the cover. The drainage layer is used to transport rainwater and snowmelt that percolates through the cover material away from the barrier layer and to reduce the water pressure on the barrier layer. The protective layer is used to protect the drainage and barrier layers. The surface layer is used to contour the surface of the landfill and to support the plants that will be used in the long-term closure design of the landfill.

It should be noted that not all of the layers will be required in each location. For example, a gas collection layer may not be required where an active gas recovery system is in place. Sometimes the subbase layer can also be used as the gas collection layer. Of the layers identified in Fig. 11-52, the barrier layer is the most critical for the reasons cited above [18, 23]. Although clay has been used in many existing landfills as the barrier layer, a number of problems are inherent with its use. For example, clay is difficult to compact on a soft foundation, compacted clay can develop cracks due to desiccation, clay can be damaged by freezing, clay will crack due to differential settling, the clay layer in a landfill cover is difficult to repair once damaged, and finally, the clay layer does not restrict the movement of landfill gas to any significant extent. As a consequence, the use of one or more geomembranes is recommended over the use of clay as a barrier layer in landfill covers. Geosynthetic clay liners (see Fig. 11-37*b*) have also been used for the barrier layer.

Typical Cover Designs. Some of the many types of cover designs that have been proposed and used are illustrated in Fig. 11-53. In Fig. 11-53*a*, the geotextile filter cloth is used to limit the intermixing of the soil with the sand layer. If the available topsoil at the landfill site is not suitable for plant growth, a suitable topsoil must be brought to the site or the available topsoil should be amended to improve its characteristics for plant growth. The modification of a soil through the addition of suitable amendments is discussed in Chapter 16. The use of a composite barrier

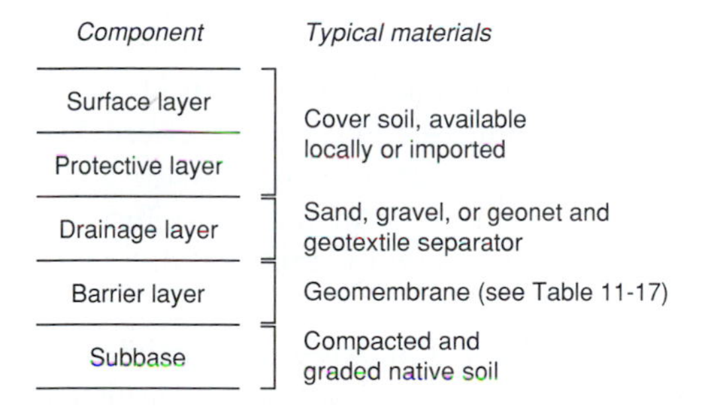

FIGURE 11-52
Typical components that constitute a landfill cover.

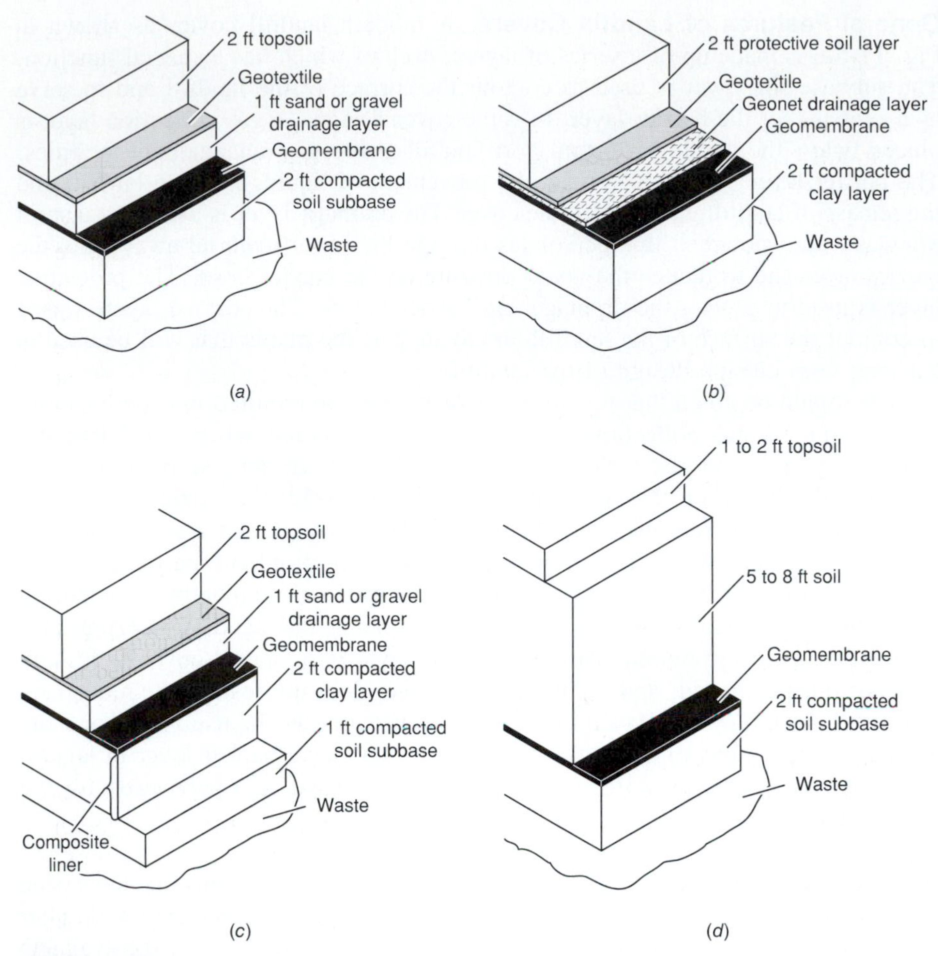

FIGURE 11-53
Typical landfill final cover configurations.

design composed of a geomembrane and clay layer is illustrated in Fig. 11-53*b*. In the design illustrated in Fig. 11-53*c*, a sand or gravel layer is substituted for the geonet drainage layer in Fig 11-53*b*. In the cover design illustrated in Fig. 11-53*d,* a 6- to 10-ft thick layer of soil is used as the cover layer. Functionally, the soil layer is sloped adequately to maximize surface runoff. The depth of soil is used to retain rainfall that does not run off and infiltrates into the soil cover. The flexible membrane liner is used to limit the release of landfill gases. Astro Turf™ has also been placed over a flexible membrane liner. Use of the Astro Turf™ is advantageous because the amount of maintenance required is minimized.

Long-Term Performance and Maintenance of Landfill Covers. Regardless of the design of the final landfill cover, the following question must be considered.

How will the integrity and performance of the landfill cover be maintained as the landfill settles, owing to the loss of weight resulting from the production of landfill gas and to long-term consolidation? For example, how will a composite liner be repaired to maintain adequate drainage? Typically, if settlement occurs, the landfill cover material is stripped back, soil or composted waste is added to adjust the grade, and the various layers are replaced. Where a thick soil cover is used, proper surface drainage may be restored by regrading the cover layer. Where vegetation is planted on the soil cover layer, a sprinkler system will be required to sustain the vegetation during the summer. In landfills where Astro Turf™ is used, when the turf starts to fall apart, the landfill cover is opened, the used turf is placed in the landfill, the flexible membrane is repaired, and a new Astro Turf™ layer is added to the top. Landscaping and the long-term maintenance of closed landfills are considered in Chapter 16.

Determination of Percolation Rate through Intermediate and Final Cover Layers

If one assumes (1) that the cover material is saturated, (2) that a thin layer of water is maintained on the surface, and (3) that there is no resistance to flow below the cover layer, then the theoretical amount of water, expressed in gallons, that could enter the landfill per unit area in a 24-h period for various cover materials is given in Table 11-15 in column 3. Clearly, these data are only theoretical values, but they can be used in assessing the worst possible situation. In actual practice, the amount of water entering the landfill will depend on local hydrological conditions, the design of the landfill cover, the final slope of the cover, and whether vegetation has been planted. In general landfill cover designs employing a flexible membrane liner are constructed to eliminate the percolation of rainwater or snowmelt into the waste below the landfill cover.

Estimation of the percolation of rainwater or snowmelt through the soil layer above the drainage layer (see Fig. 11-54*a*) or through a cover layer composed of soil only (see Fig. 11-54*b*) is usually accomplished using one of the many available hydrologic simulation programs. Perhaps the best known is the Hydrologic Evaluation of Landfill Performance (HELP) model [41, 42]. Percolation through the landfill cover layer can also be estimated using a standard hydrological water balance. Referring to Fig. 11-54, one can calculate the water balance for a soil landfill cover by the following expression:

$$\Delta S_{LC} = P - R - ET - PER_{SW} \tag{11-29}$$

where ΔS_{LC} = change in the amount of water held in storage in a unit volume of landfill cover, in

P = amount of precipitation per unit area, in

R = amount of runoff per unit area, in

ET = amount of water lost through evapotranspiration per unit area, in

PER_{SW} = amount of water percolating through unit area of landfill cover into compacted solid waste, in

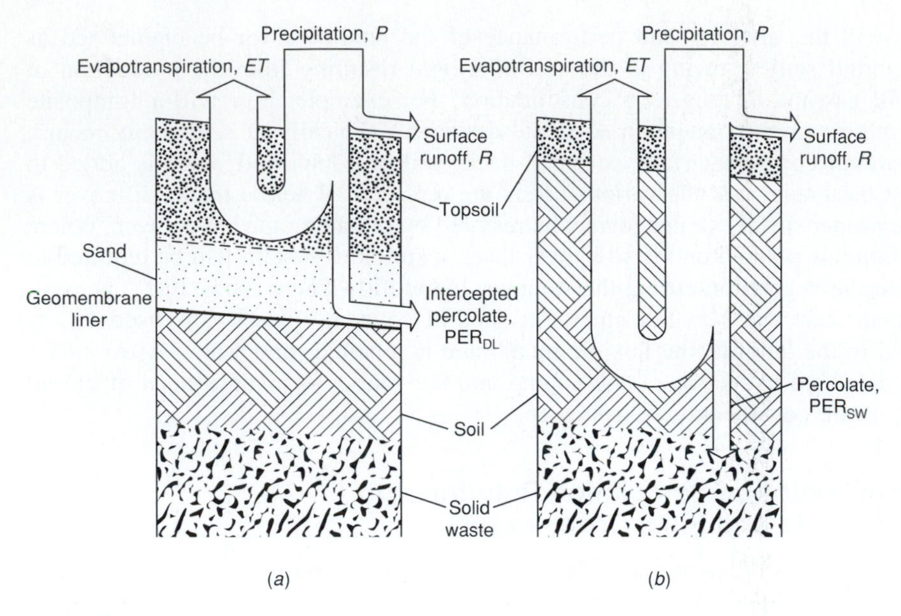

FIGURE 11-54
Definition sketch for moisture balance for landfill: (*a*) for landfill cover containing a drainage layer and geomembrane liner and (*b*) for landfill with no drainage layer (or geomembrane liner).

TABLE 11-20
Typical field capacity (FC) and permanent wilting point (PWP) values for various soil classifications[a]

Soil classification	Value, %			
	Field capacity		Permanent wilting point	
	Range	Typical	Range	Typical
Sand	6–12	6	2–4	4
Fine sand	8–16	8	3–6	5
Sandy loam	10–18	14	4–8	6
Fine sandy loam	14–22	18	6–10	8
Loam	18–26	22	8–12	10
Silty loam	19–28	24	9–14	10
Light clay loam	20–30	26	10–15	11
Clay loam	23–31	27	11–15	12
Silty clay	27–35	31	12–17	15
Heavy clay loam	29–36	32	14–18	16
Clay	31–39	35	15–19	17

[a] Adapted from Refs. 17, 27, 50.

TABLE 11-21
Typical runoff coefficients for storms of 5- to 10-year frequency[a]

		Runoff coefficient			
		With grass		Without grass	
Type of cover	Slope, %	Range	Typical	Range	Typical
Sandy loam	2	0.05–0.10	0.06	0.06–0.14	0.10
	3–6	0.10–0.15	0.12	0.14–0.24	0.18
	7	0.15–0.20	0.17	0.20–0.30	0.24
Silt loam	2	0.12–0.17	0.14	0.25–0.35	0.30
	3–6	0.17–0.25	0.22	0.35–0.45	0.40
	7	0.25–0.36	0.30	0.45–0.55	0.50
Tight clay	2	0.22–0.33	0.25	0.45–0.55	0.50
	3–6	0.30–0.40	0.35	0.55–0.65	0.60
	7	0.40–0.50	0.45	0.65–0.75	0.70

[a] Developed in part from Refs. 15, 27, 51.

The total amount of water that can be stored in a unit volume of soil will depend on the field capacity (FC) and the permanent wilting percentage (PWP). Soil moisture tension at FC is typically between 1/10 and 1/3 atm [17]. The PWP is defined as the amount of water left in a soil when plants are no longer able to extract any more. Soil moisture tension at PWP is approximately 15 atm [17]. The difference between the FC and PWP represents the amount of water that can be stored in a soil. Typical FC and PWP values for representative soils are given in Table 11-20. If a layered landfill cover is used, the field capacity of each layer must be considered in the analysis. Typical runoff coefficients for completed landfill covers are given in Table 11-21. Monthly precipitation and evapotranspiration data are site-specific, but local weather bureau data are usually acceptable. The application of Eq. (11-29) is illustrated in Example 11-12 in Section 11-12.

11-7 STRUCTURAL AND SETTLEMENT CHARACTERISTICS OF LANDFILLS

The structural characteristics and settlement of the landfill must be considered in the design of gas collection facilities, during filling operations, and before a decision is reached on the final use to be made of a completed landfill.

Structural Characteristics

When solid waste is initially placed in a landfill it behaves in a manner that is quite similar to other fill material. The nominal angle of repose for waste material placed in a landfill is approximately 1.5 to 1. Because solid waste has a tendency to slip when the slope angle is too steep, the slopes used for the completed portions of a landfill will vary from 2.5:1 to 4:1, with 3:1 being the most common. Because of the problems encountered with slippage due to settlement, many landfills where the height of the landfill will exceed 50 ft are benched (see Fig. 11-2). Benches help maintain slope stability and are also used for the placement of surface water drainage channels and for the location of landfill gas recovery piping.

In general, the construction of permanent facilities on completed landfills is not recommended because of the uneven settlement characteristics, variable bearing capacity of the upper layers of the landfill, and the potential problems that can result from gas migration, even with the use of gas collection facilities. When the final use of the landfill is known before waste placement begins, it is possible to control the deposition of certain materials during the operation of the landfill. For example, relatively inert materials such as construction and demolition wastes can be placed in those locations where buildings and/or other physical facilities are to be placed in the future.

Settlement of Landfills

As the organic material in landfill decomposes and weight is lost as landfill gas and leachate components, the landfill settles. Settlement also occurs as a result of increasing overburden mass as landfill lifts are added and as water percolates into and out of the landfill. Landfill settlement results in ruptures of the landfill surface and cover and breaks and misalignments of gas recovery facilities. It also interferes with subsequent use of the landfill after closure.

Effect of Waste Decomposition. Once placed in a landfill, the organic components of the waste will decompose, resulting in loss of as much as 30 to 40 percent of the total original mass. The loss of mass results in a loss of volume, which becomes available for refilling with new waste. The volume that is lost is usually filled in when the second lift is placed over the first lift. Weight and volume will also be lost after a landfill is closed. Evaluation of the effect of waste decomposition on settlement is considered in Example 11-13 in Section 11-12.

Effect of Overburden Pressure (Height). The specific weight of the material placed in the landfill will increase with the weight of the material placed above it, so that the average specific weight of waste in a lift depends on the depth of the lift. The maximum specific weight of solid waste residue in a landfill under overburden pressure will vary from 1750 to 2150 lb/yd^3 [21, 22]. The following relationship can be used to estimate the increase in the specific weight of the waste as a function of the overburden pressure:

$$SW_p = SW_i + \frac{p}{a + bp} \tag{11-30}$$

where SW_p = specific weight of the waste material at pressure p, lb/yd^3
SW_i = initial compacted specific weight of waste, lb/yd^3

p = overburden pressure, lb/in^2
a = empirical constant, $(\text{yd}^3/\text{lb})(\text{lb/in}^2)$
b = empirical constant, yd^3/lb

Typical specific weight versus applied pressure curves for compacted solid waste for several initial specific weights are shown in Fig. 11-55. For an initial specific weight of 1000 lb/yd^3 and a maximum specific weight of 2000 lb/yd^3, Eq. (11-30) can be written as follows:

$$D_{W_p} = 1000\ \text{lb/yd}^3 + \frac{p,\ \text{lb/in}^2}{0.0133\ (\text{yd}^3/\text{lb})(\text{lb/in}^2) + (0.001\ \text{yd}^3/\text{lb})(p,\ \text{lb/in}^2)}$$

The increase in the specific weight of the waste material in the landfill is important (1) in determining the actual amount of waste that can be placed in a landfill up to a given grade limitation and (2) in determining the degree of settlement that can be expected in a completed landfill after closure. Both of these issues are addressed in Example 11-13 in Section 11-12.

Extent of Settlement. The extent of settlement depends on the initial compaction, the characteristics of the wastes, the degree of decomposition, the effects of consolidation when water and air are forced out of the compacted solid waste, and the height of the completed fill. Representative data on the degree of settlement to be expected in a landfill as a function of the initial compaction are shown in Fig. 11-56. It has been found in various studies that about 90 percent of the ultimate settlement occurs within the first five years. In dry climates the settling rate is usually less. The settlement of landfills is modeled in Example 11-13 in Section 11-12.

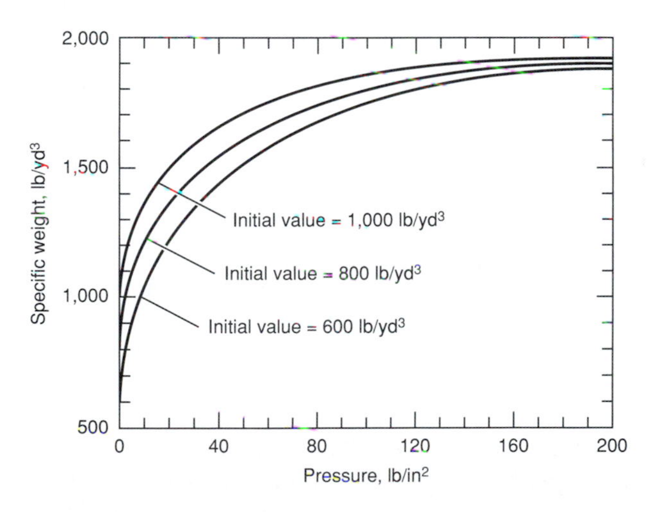

FIGURE 11-55
Specific weight of solid waste placed in landfill as function of the initial compacted specific weight of the waste and the overburden pressure.

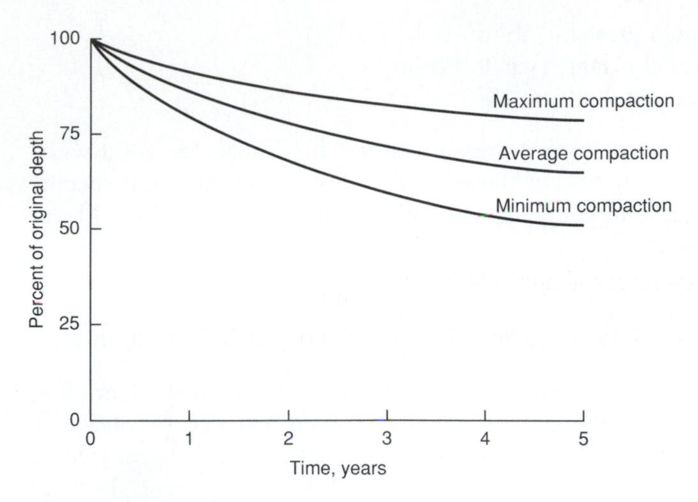

FIGURE 11-56
Surface settlement of compacted landfills.

11-8 ENVIRONMENTAL QUALITY MONITORING AT LANDFILLS

Environmental monitoring is conducted at sanitary landfills to ensure that no contaminants that may affect public health and the surrounding environment are released from the landfill. The monitoring required may be divided into three general categories: (1) vadose zone monitoring for gases and liquids, (2) groundwater monitoring, and (3) air quality monitoring. Environmental monitoring involves the use of both sampling and nonsampling methods. Sampling methods involve the collection of a sample for analysis, usually at an offsite laboratory. The typical instrumentation of a landfill for environmental monitoring is illustrated in Fig. 11-57. Nonsampling methods are used to detect chemical and physical changes in the environment as a function of an indirect measurement such as a change in electrical current. Representative devices that have been used to monitor landfill sites are listed in Table 11-22.

Vadose Zone Monitoring

The vadose zone is defined as that zone from the ground surface to where the permanent groundwater is found (see Fig. 11-58). An important characteristic of the vadose zone is that the pore spaces are not filled with water, and that the small amounts of water that are present coexist with air. Vadose zone monitoring at landfills involves both liquids and gases.

Liquid Monitoring in the Vadose Zone. Monitoring for liquids in the vadose zone is necessary to detect any leakage of leachate from the bottom of a landfill. In the vadose zone, moisture held in the interstices of the soil particles or within

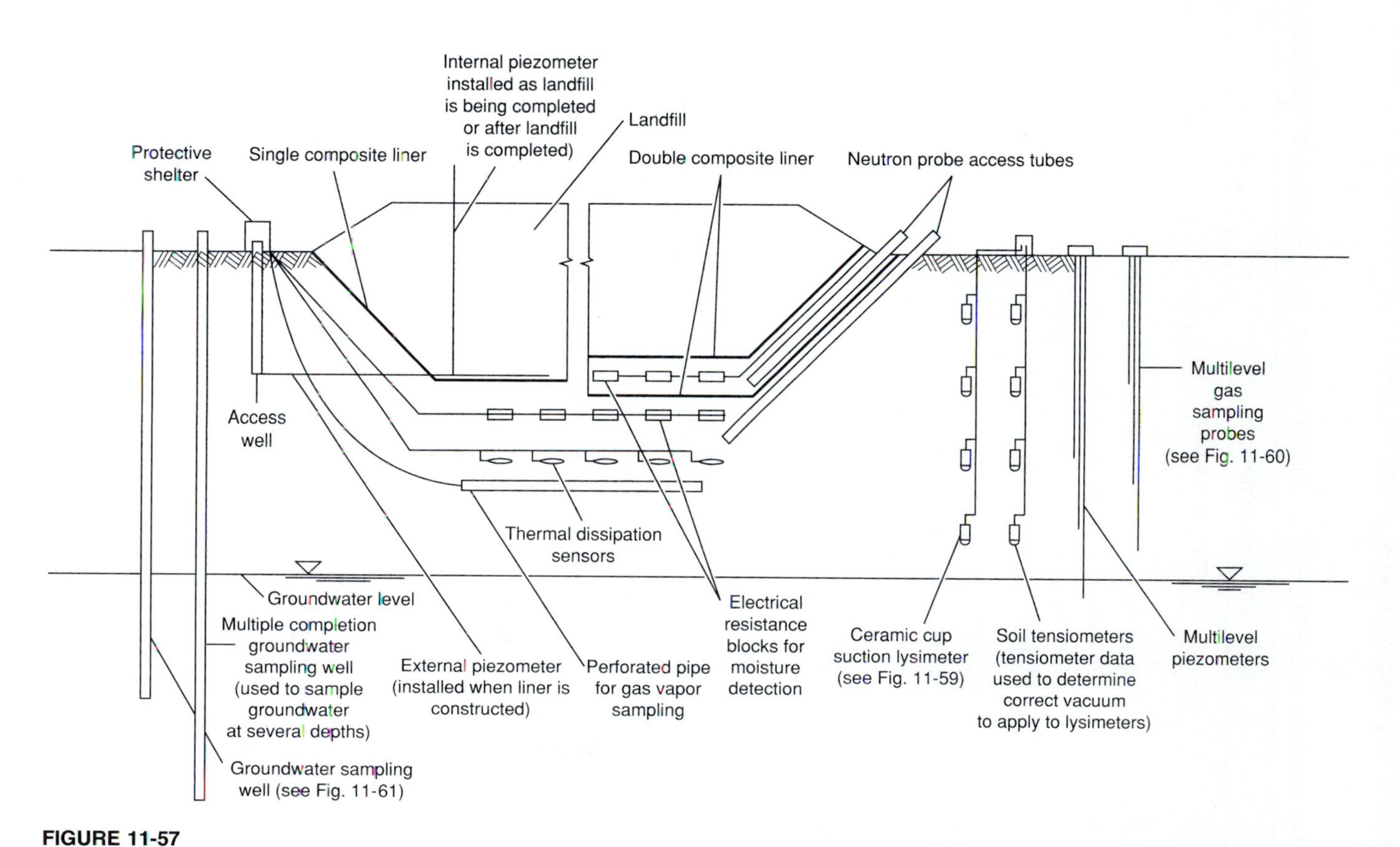

FIGURE 11-57
Instrumentation of a landfill for the collection of environmental monitoring data. Note not all of the instrumentation shown would be used at an individual landfill.

TABLE 11-22
Representative devices used to monitor landfill gases and leachate at landfills

Type	Application/description
	Sampling methods[a]
Air quality	
Active air sampler	Continuous collection and analysis of gas samples
Air collection bag	Collection of air grab samples for analysis
Evacuated flask	Collection of air grab samples for analysis
Gas syringe	Collection of air grab samples for analysis
Groundwater	
Monitoring wells; single and multiple depth	Used to collect groundwater samples. Multiple extraction wells are used to collect samples from different depths
Piezometers	Used to collect groundwater samples
In landfills	
Piezometers	Used to collect leachate samples. Piezometers can be installed before filling of the landfill is initiated or after the landfill has been completed.
Vadose zone	
Collection lysimeter	Used to collect liquid samples below landfill liners
Soil gas probes; single and multiple depth	Used to monitor landfill gases and volatile organic compounds (VOC) in the soil. The gas may be analyzed *in situ* using a portable gas chromatograph or tested in a laboratory after it has been absorbed in charcoal.
Suction cup lysimeter	Used to obtain liquid samples from the vadose zone
	Nonsampling methods[b]
Groundwater	
Conductivity cells	Used to monitor changes in groundwater conductivity. Conductivity cells are ofter located in or near monitoring wells
In landfills	
Piezometers	Used to measure the depth of leachate in landfills
Temperature blocks	Used to measure temperature
Temperature probes	Used to measure temperature
Vadose zone	
Electrical probes	Used to determine the salinity of the vadose zone. A four-probe array is installed so that conductivity of the soil can be measured.

(*continued*)

porous rock is always held at pressures below atmospheric pressure. To remove the moisture it is necessary to develop a negative pressure or vacuum to pull the moisture away from the soil particles. Because suction must be applied to draw moisture out of the soil in the vadose zone, conventional wells or other open cavities cannot be used to collect samples in this zone. The sampling devices used for sample extraction in the unsaturated zone are called suction lysimeters.

TABLE 11-22 (*continued*)

Type	Application/description
	Nonsampling methods (*continued*)
Vadose zone (*cont.*)	
Electrical resistance blocks	Used to measure changes in water content of the vadose zone. Electrode blocks embedded in porous material are installed in the soil. Electrical properties of the blocks change with the changing water content of the vadose zone.
Gamma ray attenuation probes	Used for detecting changes in moisture content of the vadose zone. Based on the attenuation of gamma rays transmission and scattering. In the transmission method, two wells are installed at a known distance apart. A single well is used in the scattering method. Usually limited to shallow depth because of difficulties in installing parallel wells.
Heat dissipation sensors	Used to monitor water content of the vadose zone by measuring the rate of heat dissipation from the block to the surrounding soil.
Neutron moisture meter	Used to obtain a profile of the moisture content of the soil below the landfill. Meter can be installed below a landfill or moved through a borehole next to the landfill.
Salinity sensors	Used to monitor soil salinity. Electrodes attached to a porous ceramic cup are installed in the soil.
Tensiometers	Used to measure the matric potential of soil. Tensiometers measure the negative pressure (capillary pressure) that exists in unsaturated soil.
Thermocouple psychrometers	Used to detect changes in moisture content. Operation is based on cooling of a thermocouple junction by the peltier effect. Wet bulb and dew point. The dew point method is used more commonly in landfill monitoring.
Time domain reflectrometry (TDR)	Based on the difference in dielectric properties of water and soil. Wide-frequency bandwidth and short pulse length that are sensitive to the high-frequency electrical properties of the material are measured.
Wave sensing devices	Use of both seismic and acoustic wave propagation properties for leak detection. In the seismic wave technique, the difference in travel time of Rayleigh waves between the source and geophones is used to detect leaks. In the acoustic emmission monitoring (AEM) technique, sound waves generated by flowing water from a leak are utilized in leak detection.

[a] Methods involving the collection of samples for subsequent laboratory analysis.

[b] Methods involving physical and electrical measurements.

Three commonly used classes of lysimeters are (1) the ceramic cup, (2) the hollow fiber, and (3) the membrane filter [2, 43].

The most commonly used device for obtaining samples of moisture in the vadose zone is the ceramic cup sampler (see Fig. 11-59), which consists of a porous cup or ring made of ceramic material that is attached to a short section of nonporous tubing (e.g., PVC). When placed in the soil, because of its pores

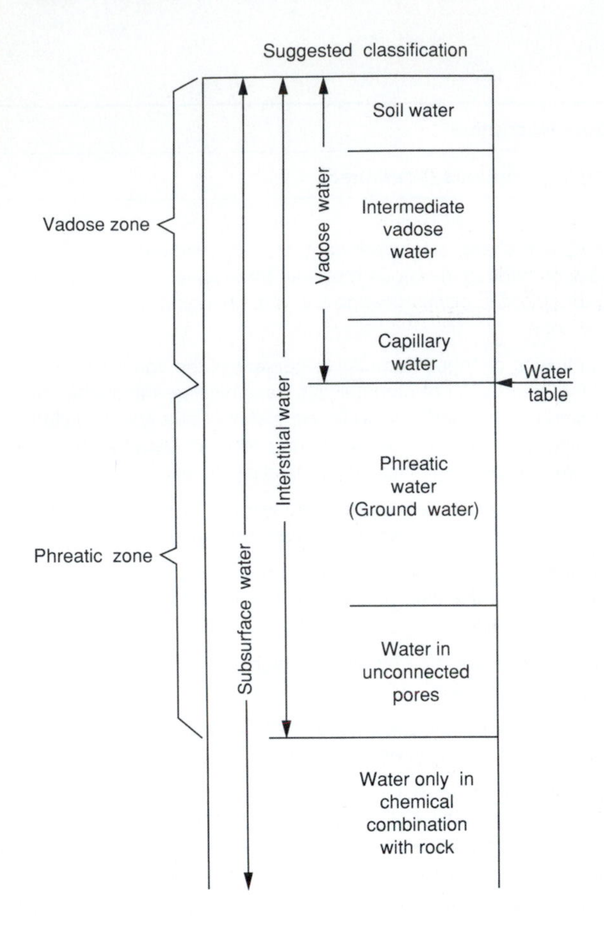

FIGURE 11-58
Classification of subsurface water. (Courtesy of California Integrated Waste Management Board.)

it becomes an extension of the pore space of the soil. Soil moisture is drawn in through the porous ceramic element by the application of a vacuum. When a sufficient amount of water has collected in the sampler, the collected sample is pulled to the surface through a narrow tube by the application of a vacuum or is pushed up by air pressure.

Gas Monitoring in the Vadose Zone. Monitoring for gases in the vadose zone is necessary to detect the lateral movement of any landfill gases. A typical example of a vadose zone gas monitoring probe is illustrated in Fig. 11-60. In many monitoring systems, gas samples are collected from multiple depths in the vadose zone.

Groundwater Monitoring

Monitoring of the groundwater is necessary to detect changes in water quality that may be caused by the escape of leachate and landfill gases. Both down- and up-

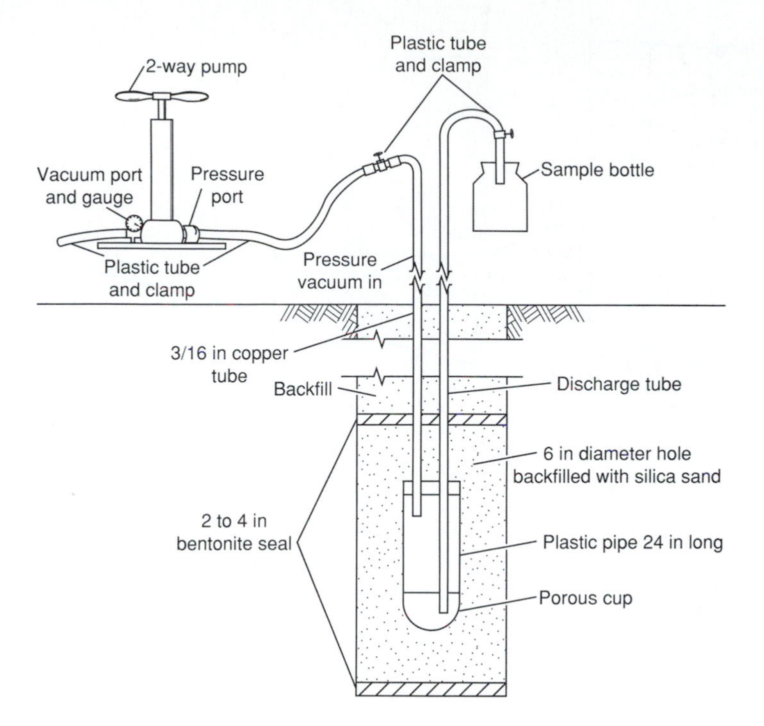

FIGURE 11-59
Porous cup suction lysimeter for the collection of liquid samples from the vadose zone. (Courtesy of California Integrated Waste Management Board.)

gradient wells are required to detect any contamination of the underground aquifer by leachate from the landfill. An example of a well used for the monitoring of groundwater is illustrated in Fig. 11-61. To obtain a representative sample, the liquid in permanent sample collection tubing, where used, must be purged before the sample is collected.

Landfill Air Quality Monitoring

Air quality monitoring at landfills involves (1) the monitoring of ambient air quality at and around the landfill site, (2) the monitoring of landfill gases extracted from the landfill, and (3) the monitoring of the off gases from any gas processing or treatment facilities.

Monitoring Ambient Air Quality. Ambient air quality is monitored at landfill sites to detect the possible movement of gaseous contaminants from the boundaries of the landfill site. Gas sampling devices can be divided into three categories: (1) passive, (2) grab, and (3) active. Passive sampling involves the collection of a gas sample by passing a stream of gas through a collection device in which the contaminants contained in the gas stream are removed for subsequent analysis.

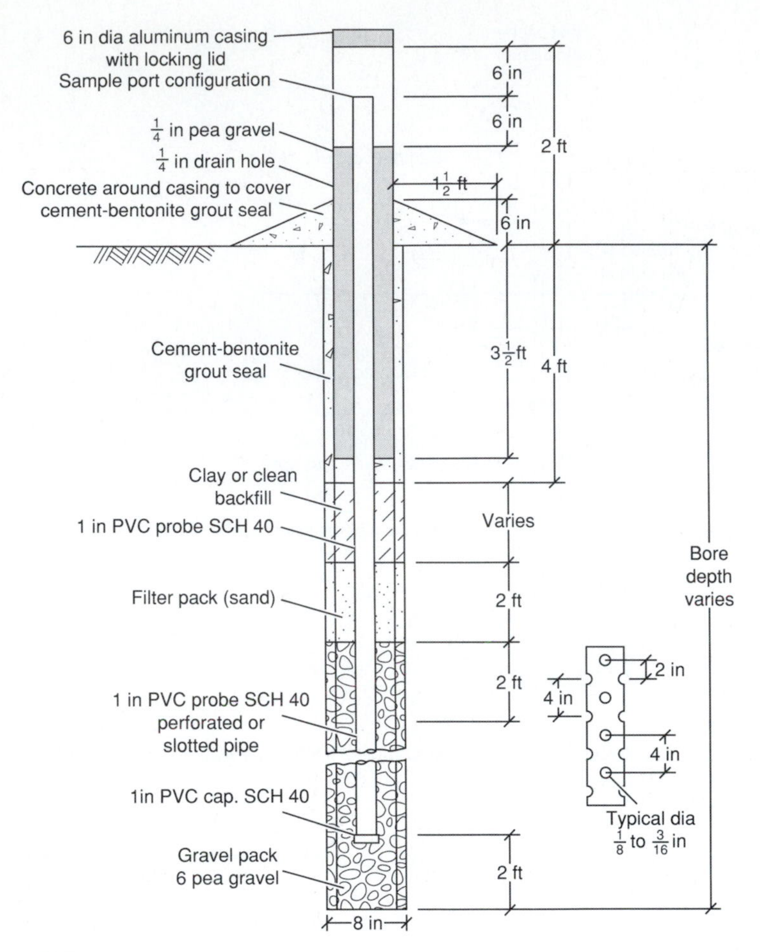

FIGURE 11-60
Vadose zone gas monitoring probe. (Courtesy of Waste Management, Inc.)

Commonly used in the past, passive sampling is seldom used today. Grab samples are collected using an evacuated flask, gas syringe, or an air collection bag made of a synthetic material (see Fig. 11-62). An active sampler involves the collection and analysis of a continuous stream of gas.

Monitoring Extracted Landfill Gas. Landfill gas is monitored to assess the composition of the gas, and to determine the presence of trace constituents that may pose a health or environmental risk.

Monitoring Off-Gases. Monitoring off-gases from treatment and energy recovery facilities is done to determine compliance with local air pollution control requirements. Both grab and continuous sampling have been used for this purpose.

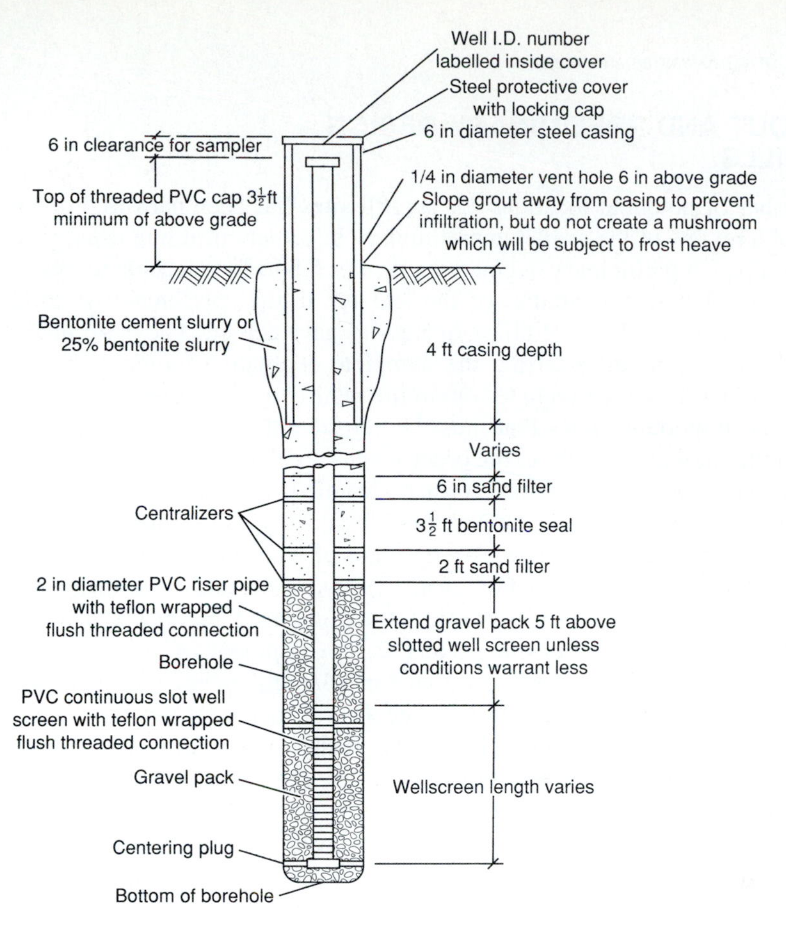

FIGURE 11-61
Typical groundwater monitoring well. (Courtesy of Waste Management, Inc.; see also Fig. 16-5.)

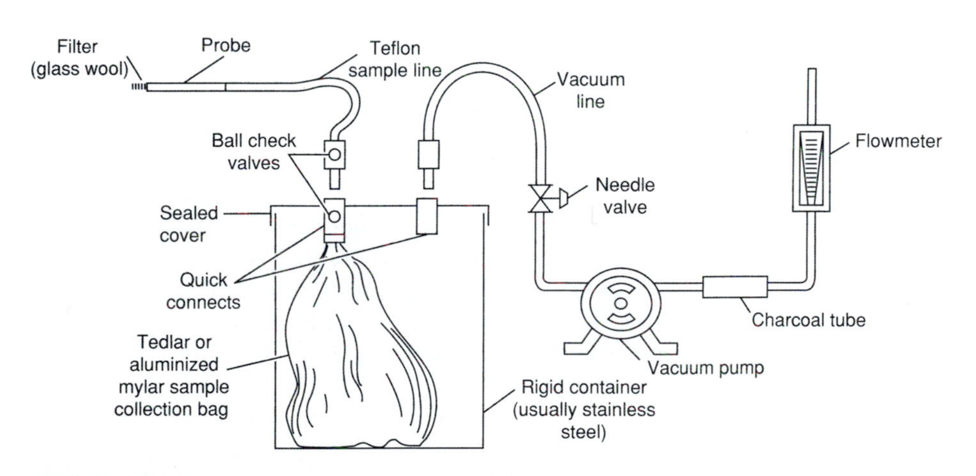

FIGURE 11-62
Sampling apparatus for the collection of air grab samples at landfills.

11-9 LAYOUT AND PRELIMINARY DESIGN OF LANDFILLS

Once the number of potential locations for landfill sites has been narrowed down on the basis of a review of the available preliminary information, it will usually be necessary to prepare a preliminary engineering design report for each site to assess the costs associated with preparation of the site for filling, placement of solid wastes, and closing of the site once filling operations have ceased. The preliminary engineering design report differs from the complete evaluation required for the final selection of a site, which includes environmental considerations.

Among the important topics that must be considered in an engineering design report, though not necessarily in the order given, are the following: (1) layout of landfill site, (2) types of wastes that must be handled, (3) the need for a convenience transfer station, (4) estimation of landfill capacity, (5) evaluation of the geology and hydrogeology of the site, (6) selection of leachate control facilities, (7) selection of landfill gas control facilities, (8) layout of surface drainage facilities, (9) aesthetic design considerations, (10) monitoring facilities, (11) determination of equipment requirements, and (12) development of an operations plan. The development of an operations plan for a landfill is considered in the following section. Closure and postclosure care is considered in Section 11-11. Important factors that must be considered in the design of landfills are reported in Table 11-23. Throughout the development of the engineering design report, careful consideration must be given to the final use or uses to be made of the completed site. Land reserved for administrative offices, buildings, and parking lots should be filled with dirt only and should be sealed against the entry of gases.

Layout of Landfill Sites

In planning the layout of a landfill site, the location of the following must be determined: (1) access roads; (2) equipment shelters; (3) scales, if used; (4) office space; (5) location of convenience transfer station, if used; (6) storage and/or disposal sites for special wastes; (7) areas to be used for waste processing (e.g., composting); (8) definition of the landfill areas and areas for stockpiling cover material; (9) drainage facilities; (10) location of landfill gas management facilities; (11) location of leachate treatment facilities, if required; (12) location of monitoring wells; and (13) plantings. A typical layout for a landfill disposal site is shown in Fig. 11-63. Because site layout is specific for each case, Fig. 11-63 is meant to serve only as a guide. However, the items identified on Fig. 11-63 can be used as a check list of the areas that must be addressed in the preliminary layout of a landfill. An aerial view of an operating landfill is shown in Fig. 11-64.

Types of Wastes

Knowledge of the types of wastes to be handled is important in the design and layout of a landfill, especially if special wastes are involved. It is usually best

TABLE 11-23
Important factors to consider in the design of landfills

Factors	Remarks
Access	Paved all-weather access roads to landfill site; temporary roads to unloading areas.
Land area	Area should be large enough to hold all community wastes for a minimum of 5 yr, but preferably 10 to 25 yr; area for buffer strips or zones must also be included.
Landfilling method	Landfilling method will vary with terrain and available cover; most common methods are excavated cell/trench, area, canyon (see Figs. 11-7, 11-8, and 11-9).
Completed landfill characteristics	Finished slopes of landfill, 3 to 1; height to bench, if used, 50 to 75 ft; slope of final landfill cover, 3 to 6%.
Surface drainage	Install drainage ditches to divert surface water runoff; maintain 3 to 6% grade on finished landfill cover to prevent ponding; develop plan to divert stormwater from lined but unused portions of landfill.
Intermediate cover material	Maximize use of onsite soil materials; other materials such as compost produced from yard waste and MSW can also be used to maximize the landfill capacity; typical waste to cover ratios vary from 5 to 1 to 10 to 1.
Final cover	Use multilayer design (see Fig. 11-53); slope of final landfill cover, 3–6%.
Landfill liner	Single clay layer (2 to 4 ft) or multilayer design incorporating the use of a geomembrane (see Figs. 11-36, 11-39, and 11-40). Cross slope for terrace type (see Fig. 11-39) leachate collection systems, 1 to 5%; maximum flow distance over terrace, 100 ft; slope of drainage channels, 0.5 to 1.0%. Slope for piped type (see Fig. 11-40) leachate collection system, 1 to 2%; size of perforated pipe, 4 in; pipe spacing, 20 ft.
Cell design and construction	Each day's wastes should form one cell; cover at end of day with 6 in of earth or other suitable material; typical cell width, 10 to 30 ft; typical lift height including intermediate cover, 10 to 14 ft; slope of working faces, 2:1 to 3:1.
Groundwater protection	Divert any underground springs; if required, install perimeter drains, well point system, or other control measures.
Landfill gas management	Develop landfill gas management plan including extraction wells (see Fig. 11-20), manifold collection system, condensate collection facilities (see Fig 11-26), the vacuum blower facilities, and flaring facilities (see Fig. 11-27) and/or energy production facilities (see Fig. 11-29). Operating vacuum at well head, 10 in of water.
Leachate collection	Determine maximum leachate flow rates and size leachate collection pipe and/or trenches; size leachate pumping facilities; select collection pipe materials to withstand static pressures corresponding to the maximum height of the landfill.
Leachate treatment	Based on expected quantities of leachate and local environmental conditions, select appropriate treatment process (see Table 11-18 and Fig. 11-46).
Environmental requirements	Install vadose zone gas and liquid monitoring facilities; install up- and downgradient groundwater monitoring facilities; locate ambient air monitoring stations.
Equipment requirements	Number and type of equipment will vary with the type of landfill (see Figs. 11-68 and 11-69) and the capacity of the landfill (see Table 11-26).
Fire prevention	Water onsite; if nonpotable, outlets must be marked clearly; proper cell separation prevents continuous burn-through if combustion occurs.

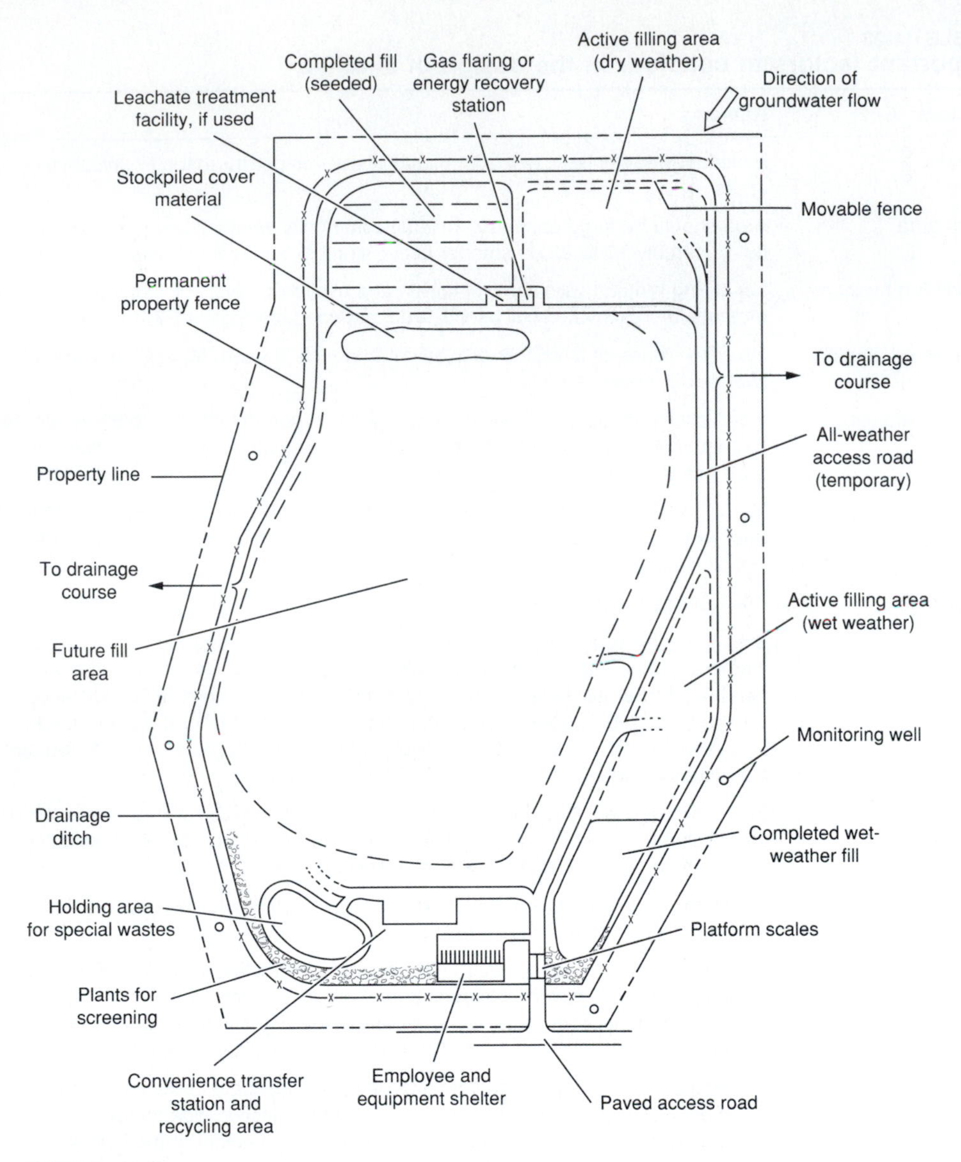

FIGURE 11-63
Typical layout of a landfill site.

to develop separate disposal sites or monofills for designated and special wastes such as asbestos because under most conditions special preparation of the site will be necessary before these wastes can be landfilled. The associated disposal costs are often significant, and it is wasteful to use this landfill capacity for wastes that do not require special precautions. If significant quantities of demolition wastes are to be handled, it may be possible to use them for embankment stabilization.

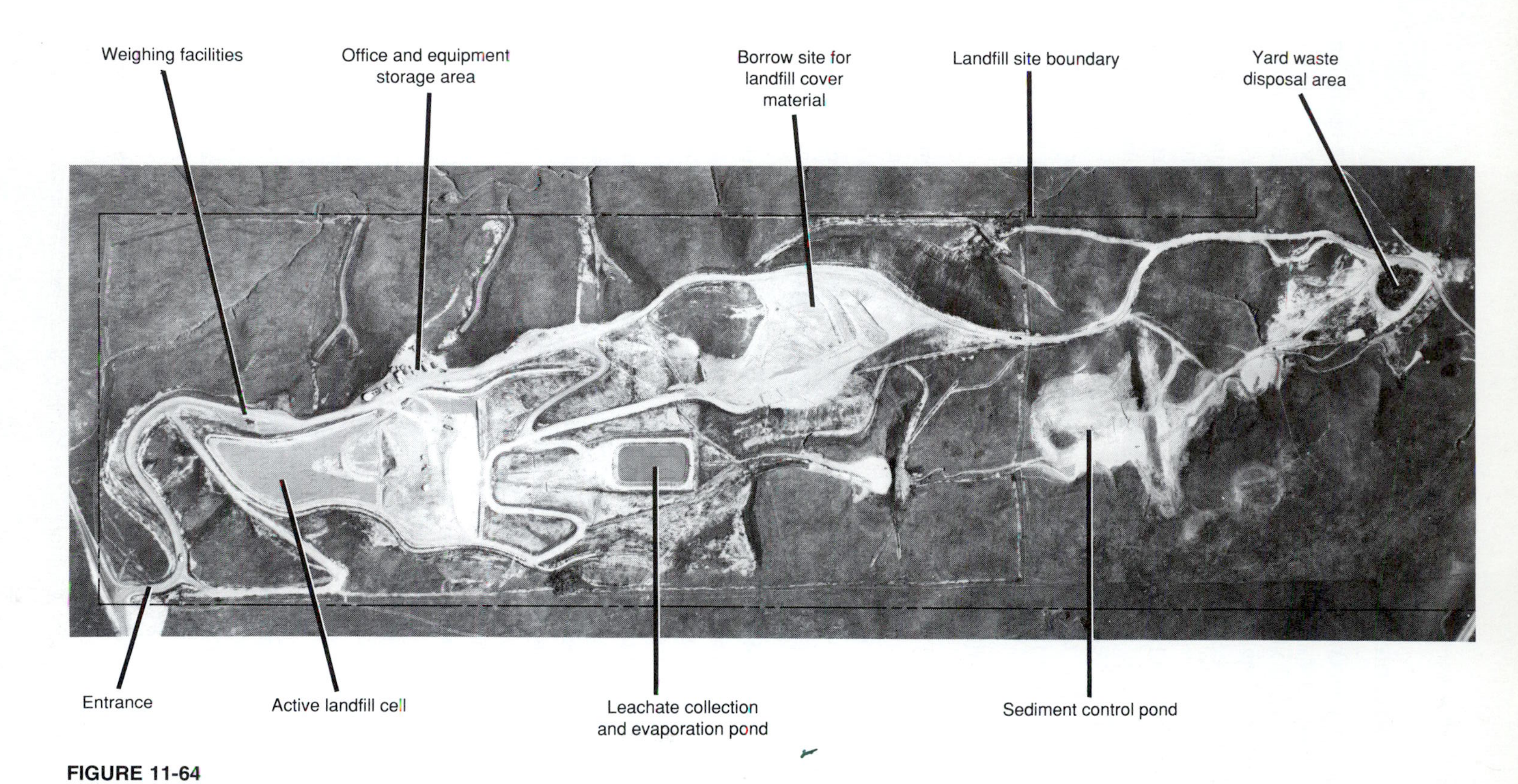

FIGURE 11-64
Aerial view of an operating landfill which receives about 80 ton/d of solid wastes. (Courtesy of Jack Scroggs, KASL Engineers.)

Need for a Convenience Transfer Station

Because of safety concerns and the many new restrictions governing the operation of landfills, many operators of landfills have constructed convenience transfer stations at the landfill site for the unloading of wastes brought to the site by individuals and small-quantity haulers (see Figs. 10-9 and 10-11). Having a separate transfer facility reduces the potential for accidents at the working face of the landfill significantly. The transfer facilities are also used for the recovery of recyclable materials. Waste materials are usually emptied into two large transfer trailers each of which is hauled to the disposal site, emptied, and returned to the transfer station. The need for a convenience transfer station will depend on the physical characteristics and the operation of the landfill and whether there is a separate location where the public can be allowed to dispose of waste safely.

Estimation of Landfill Capacity

Earlier in the chapter, an approximate method was given for determining the area requirements for landfill (see Example 11-1). In this section consideration is given to (1) the method used to estimate the nominal volume of the site, (2) the impact of the compactability of the individual solid waste components, (3) the impact of daily cover, and (4) the impact of waste decomposition and overburden height.

Determination of Nominal Landfill Volume. The nominal volumetric capacity of a proposed landfill site is determined by first laying out several different landfill configurations, taking into account appropriate design criteria (see Fig. 11-65). The next step is to determine the surface area for each lift. The nominal volume of the landfill is determined by multiplying the average area between two adjacent contours by the height of the lift and summing the volume of successive lifts. If the cover material will be excavated from the site, then the computed volume corresponds to the volume of solid waste that can be placed in the site. If the cover material has to be imported, then the computed capacity must be reduced by a factor to account for the volume occupied by the cover material. For example, if a cover to waste ratio of 1 to 5 is adopted, then the capacity reported must be multiplied by a factor of 0.833 (5/6). The determination of the nominal volume of a landfill site is considered in Example 11-7 in Section 11-12.

The nominal volumetric capacity of the landfill is used as a preliminary estimate of landfill capacity. The actual total capacity of the landfill to accept waste on a weight basis will depend on the initial specific weight at which the residual solid waste is placed in the landfill, on the subsequent compaction of the waste material due to overburden pressure, and on loss of mass as a result of biological decomposition. The impacts of these factors on the capacity of the landfill are considered in the following discussion.

Impact of Compactability of Solid Waste Components. The initial density of solid wastes placed in a landfill varies with the mode of operation of the

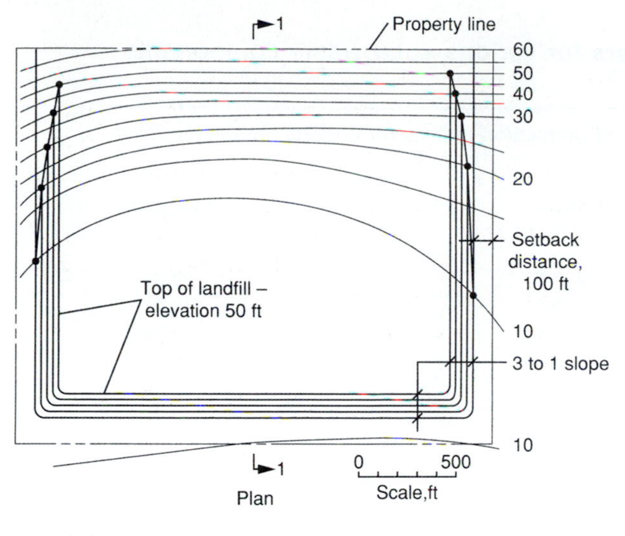

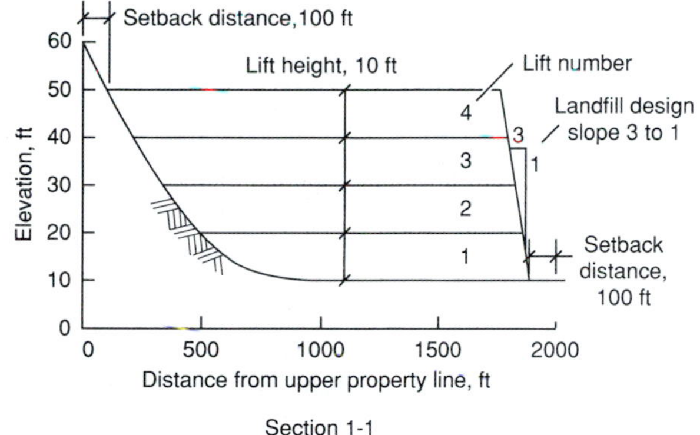

FIGURE 11-65
Layout of typical landfill for the purpose of estimating the volumetric capacity of a proposed site taking into account setback distances (100 ft), finished landfill slopes (3 to 1), and lift height (10 ft).

landfill, the compactability of the individual solid waste components, and the percentage distribution of the components. If the waste placed in the landfill is spread out in thin layers and compacted against an inclined surface, a high degree of compaction can be achieved. With minimal compaction, the initial specific weight will be somewhat less than the compacted specific weight in a collection vehicle. In general, the initial specific weight of solid waste placed in a landfill will vary from 550 to 1200 lb/yd^3, depending on the degree of initial compaction given to the waste. The diversion of waste materials before disposal will not only reduce the landfill volume requirements but will also affect the overall compactability of the remaining waste materials. Typical compactability data for the components found in MSW are reported in Table 11-24. Volume-reduction factors are given for both normally compacted and well-compacted landfills. The use of the data presented in Table 11-24 is illustrated in Example 11-5.

TABLE 11-24
Typical compaction factors for various solid waste components placed in landfills

Component	Compaction factors for components in landfills[a]		
	Range	Normal compaction	Well compacted
Organic			
Food wastes	0.2–0.5	0.35	0.33
Paper	0.1–0.4	0.2	0.15
Cardboard	0.1–0.4	0.25	0.18
Plastics	0.1–0.2	0.15	0.10
Textiles	0.1–0.4	0.18	0.15
Rubber	0.2–0.4	0.3	0.3
Leather	0.2–0.4	0.3	0.3
Garden trimmings	0.1–0.5	0.25	0.2
Wood	0.2–0.4	0.3	0.3
Inorganic			
Glass	0.3–0.9	0.6	0.4
Tin cans	0.1–0.3	0.18	0.15
Nonferrous metals	0.1–0.3	0.18	0.15
Ferrous metals	0.2–0.6	0.35	0.3
Dirt, ashes, brick, etc.	0.6–1.0	0.85	0.75

[a] Compaction factor = V_f / V_i where V_f = final volume of solid waste after compaction and V_i = initial volume of solid waste before compaction.

Example 11-5 Determination of density of compacted solid wastes without and with waste diversion. Determine the specific weight in a well-compacted landfill for solid wastes with the characteristics given in Table 3-4. Also determine the impact of a resource recovery program on landfill area requirements in which 50 percent of the paper and 80 percent of the glass and tin cans are recovered. Assume that the wastes have the characteristics reported in Table 3-4.

Solution

1. Set up a computation table with separate columns for (1) the weight of the individual solid waste components, (2) the volume of the wastes as discarded, (3) the compaction factors for well-compacted solid wastes, and (4) the compacted volume in the landfill. The required table, based on a total weight of 1000 lb, is given on page 475.
2. Compute the compacted specific weight of the solid wastes.

$$\text{Compacted specific weight} = \frac{1000 \text{ lb} \times 27 \text{ ft}^3/\text{yd}^3}{28.95 \text{ ft}^3}$$

$$= 933 \text{ lb/yd}^3 \ (554 \text{ kg/m}^3)$$

3. Determine the compacted specific weight of the wastes in the landfill in which 50 percent of the paper and 80 percent of the glass and tin cans are recovered.

Component	Weight of solid waste,[a] lb	Volume as discarded,[b] ft^3	Compaction factor[c]	Compacted volume in landfill, ft^3
Organic				
Food wastes	90	4.96	0.33	1.64
Paper	340	61.2	0.15	9.18
Cardboard	60	19.06	0.18	3.53
Plastics	70	17.18	0.10	1.72
Textiles	20	4.91	0.15	0.74
Rubber	5	0.61	0.3	0.18
Leather	5	0.50	0.3	0.15
Yard wastes	185	29.38	0.2	5.88
Wood	20	1.35	0.3	0.41
Inorganic				
Glass	80	6.55	0.4	2.62
Tin cans	60	10.80	0.15	1.62
Aluminum	5	0.50	0.15	0.08
Other metal	30	1.50	0.3	0.45
Dirt, ashes, brick, etc.	30	1.00	0.75	0.75
Total	1000			28.95

[a] See Table 3-4.
[b] See Table 4-1.
[c] See Table 11-24.

(*a*) Determine the weight of waste after resource recovery.

$$\text{Weight remaining} = 1000\text{ lb} - (340\text{ lb} \times 0.5 + 80\text{ lb} \times 0.80 + 60\text{ lb} \times 0.80)$$
$$= 718\text{ lb}$$

(*b*) Determine the volume and compacted specific weight of waste after resource recovery.

$$\text{Volume remaining} = 28.95\text{ ft}^3 - (9.18\text{ ft}^3 \times 0.5 + 2.62\text{ ft}^3 \times 0.80 + 1.62\text{ ft}^3 \times 0.80) = 20.97\text{ ft}^3$$

$$\text{Compacted specific weight} = \frac{718\text{ lb} \times 27\text{ ft}^3/\text{yd}^3}{20.97\text{ ft}^3}$$
$$= 924\text{ lb/yd}^3\ (548\text{ kg/m}^3)$$

Comment. The specific weight value of 933 lb/yd^3 (computed in Step 2) would then be used to determine the required landfill area. Because the specific weight computed in Step 2 is essentially the same as that computed in Step 3, the impact of the materials recovery program can be assessed on the basis of the weight reduction alone. In cases where the computed compacted specific weight changes significantly as a result of a materials recovery program, the required landfill area can also be reduced by the ratio of compacted specific weights. Large changes in the specific weight value will not be observed with materials recovery where a sizable fraction of the wastes are composed of garden trimmings.

Impact of Cover Material. Cover material, typically soil, is incorporated into a landfill at each stage of its construction. Daily cover, consisting of 6 in to 1 ft of soil, is applied to the working faces of the landfill at the close of operation each day to control disease vectors such as insects and rats, and to stop material from blowing from the working face. Interim cover is a thicker layer of daily cover material applied to areas of the landfill that will not be worked for some time. Final covers usually are 3 to 6 ft thick and include a layer of compacted clay, with other layers to enhance drainage and support surface vegetation. The quantity of cover material necessary for operation of the landfill is an important factor in determining the capacity of a landfill site. Usually, daily and interim cover needs are expressed as a waste:soil ratio, defined as the volume of waste deposited per unit volume of cover provided. Typically, waste:soil ratios range from 4:1 to 10:1.

The waste:soil ratio can be estimated by considering the geometry of a landfill cell. Cells usually are roughly parallelepipeds, with cover material on three of the six sides. The surface area of those faces depends on the slope of the working faces of the landfill, the cell volume, the lift height, and the width of the bench in which the waste is placed. Working face slopes are usually in the range of 2:1 to 3:1. The volume of the cell can be calculated by dividing the average mass of material deposited per day by the average density of the lift. Lift height and cell width should be selected to provide the lowest acceptable waste:soil ratio. The volume of daily cover should be calculated for different lift heights and bench widths, and for the minimum and maximum waste deposition rates. Calculation of waste:soil ratio is illustrated in Example 11-6.

Example 11-6 Determination of waste to soil ratio. Determine the ratio of waste to cover material (volume basis) as a function of the initial compacted specific weight for a solid waste stream of 70 tons per day to be placed in 10 ft lifts with a cell width of 15 ft. The slope of the working faces is 3:1. Assume that the waste is compacted initially to an average specific weight of 600, 800, and 1000 lb/yd^3. The daily cover thickness is 6 in.

Solution

1. Determine the daily volume of the deposited solid waste.

   (*a*) For 600 lb/yd^3

$$V_d = 70 \text{ ton/d} \times 2000 \text{ lb/ton} \times \frac{1 \text{ yd}^3}{600 \text{ lb}}$$

$$V_d = 233.3 \text{ yd}^3$$

   (*b*) For 800 lb/yd^3

$$V_d = 175.0 \text{ yd}^3$$

   (*c*) For 1000 lb/yd^3

$$V_d = 140.0 \text{ yd}^3$$

2. Determine the length of each daily cell.
   (*a*) For 600 lb/yd³

$$L = \frac{233.3 \text{ yd}^3 \times 27 \text{ ft}^3/\text{yd}^3}{10 \text{ ft} \times 15 \text{ ft}} = 41.9 \text{ ft}$$

   (*b*) For 800 lb/yd³

$$L = 31.5 \text{ ft}$$

   (*c*) For 1000 lb/yd³

$$L = 25.2 \text{ ft}$$

3. Determine cell surface areas.
   (*a*) For the top of the cell

$$A_{T_{600}} = 41.9 \text{ ft} \times 15 \text{ ft} = 628.5 \text{ ft}^2$$

$$A_{T_{800}} = 31.5 \text{ ft} \times 15 \text{ ft} = 472.5 \text{ ft}^2$$

$$A_{T_{1000}} = 25.2 \text{ ft} \times 15 \text{ ft} = 378.0 \text{ ft}^2$$

   (*b*) For the face of the cell

$$A_{F_{600}} = 41.9 \text{ ft} \times \sqrt{(10 \text{ ft})^2 + (3 \times 10 \text{ ft})^2} = 1325 \text{ ft}^2$$

$$A_{F_{800}} = 31.5 \text{ ft} \times \sqrt{(10 \text{ ft})^2 + (3 \times 10 \text{ ft})^2} = 996 \text{ ft}^2$$

$$A_{F_{1000}} = 25.2 \text{ ft} \times \sqrt{(10 \text{ ft})^2 + (3 \times 10 \text{ ft})^2} = 797 \text{ ft}^2$$

   (*c*) For the side of the cell

$$A_S = 15 \text{ ft} \times \sqrt{(10 \text{ ft})^2 + (3 \times 10 \text{ ft})^2} = 474 \text{ ft}^2$$

4. Determine volume of soil for daily cover.

$$V_C = 6 \text{ in} \times \frac{1 \text{ ft}}{12 \text{ in}} \times \left(A_T + A_F + A_S\right)$$

$$V_{C_{600}} = 6 \text{ in} \times \frac{1 \text{ ft}}{12 \text{ in}} \times \left(628.5 \text{ ft}^2 + 1325 \text{ ft}^2 + 474 \text{ ft}^2\right) = 1214 \text{ ft}^3$$

$$V_{C_{800}} = 6 \text{ in} \times \frac{1 \text{ ft}}{12 \text{ in}} \times \left(472.5\text{ft}^2 + 996 \text{ ft}^2 + 474 \text{ ft}^2\right) = 971 \text{ ft}^3$$

$$V_{C_{1000}} = 6 \text{ in} \times \frac{1 \text{ ft}}{12 \text{ in}} \times \left(378 \text{ ft}^2 + 797 \text{ ft}^2 + 474 \text{ ft}^2\right) = 825 \text{ ft}^3$$

5. Determine ratio of waste to cover soil.
   (*a*) For 600 lb/yd³

$$R_{W:C} = \frac{233.3 \text{ yd}^3 \times 27 \text{ ft}^3/\text{yd}^3}{1214 \text{ ft}^3} = 5.19{:}1$$

(*b*) For 800 lb/yd^3

$$R_{W:C} = \frac{175\ yd^3 \times 27\ ft^3/yd^3}{971\ ft^3} = 4.87:1$$

(*c*) For 1000 lb/yd^3

$$R_{W:C} = \frac{140\ yd^3 \times 27\ ft^3/yd^3}{825\ ft^3} = 4.58:1$$

Comment. Note that as the initial compacted specific weight of the waste placed in the landfill increases, the ratio of the waste to cover material decreases. However, the total volume occupied by the waste that has been compacted to an initial specific weight of 1000 lb/yd^3 is 0.6 times the volume occupied by the waste compacted to an initial specific weight of 600 lb/yd^3.

Impact of Waste Decomposition and Overburden Height. The loss of mass through biological decomposition results in a loss of volume, which becomes available for refilling with new waste. In the preliminary assessment of site capacity, only compaction due to overburden is considered. At later stages of landfill design, the loss of landfill material to decomposition should be considered. The specific weight of the landfilled material can be estimated using Eq. (11-30). Evaluation of the impact of waste decomposition on settlement is illustrated in Example 11-13 in Section 11-12.

Evaluation of Local Geology and Geohydrology

To evaluate the geologic and hydrogeological characteristics of a site that is being considered for a landfill, core samples must be obtained. Sufficient borings should be made so that the geologic formations under the proposed site can be established from the surface to (and including) the upper portions of the bedrock or other confining layers (see Fig. 11-66). At the same time, the depth to the surface water table should be determined along with the piezometric water levels in any bedrock or confined aquifers that may be found. The resulting information is then used (1) to determine the general direction of groundwater movement under the site, (2) to determine whether any unconsolidated or bedrock aquifers are in direct hydraulic connection with the proposed landfill site, and (3) to determine the type of liner system that will be required.

Selection of Leachate Management Facilities

The principal leachate management facilities required in the design of a landfill include the landfill liner and leachate collection system and the leachate treatment facilities.

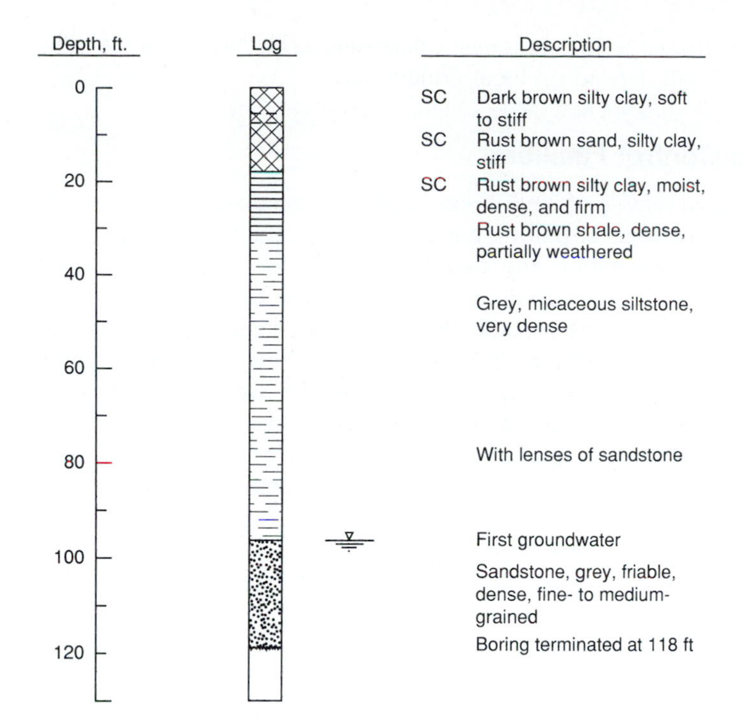

FIGURE 11-66
Typical boring log from well drilled at proposed landfill site.

Landfill Liner and Leachate Collection Facilities. The type of landfill liner used will depend on the local geology and hydrogeology. In general, landfill sites should be located where there is little or no possibility of contaminating potable water supplies. To provide assurances to the public that leachate will not contaminate underground waters, most states now require some type of liner for all landfills. Commonly used landfill liner designs are illustrated in Fig. 11-36. Typical leachate collection facilities are illustrated in Fig. 11-41 through 11-43. The current trend is toward the use of composite liners including a geomembrane and clay layer. In extremely arid areas where no possibility exists of contaminating the groundwater, it may be possible to develop a landfill without a liner. Nevertheless, the use of a liner system is a critical factor in siting new landfills. Further, the relative cost of a liner system is not great considering the potential environmental benefits. To determine the size of the leachate collection and treatment facilities required, the quantity of leachate must be estimated using the methods outlined in Section 11-5 and illustrated in Example 11-11 in Section 11-12. The selection of a liner system is illustrated in Example 11-14 in Section 11-12.

Leachate Treatment Facilities. As noted in Section 11-5, the most common alternatives that have been used to manage the leachate collected from landfills include (1) leachate recycling, (2) leachate evaporation, (3) treatment followed by disposal, and (4) discharge to municipal wastewater collection systems. The particular option used will depend on local conditions.

Selection of Gas Control Facilities

Because the uncontrolled release of landfill gas, especially methane, contributes to the greenhouse effect, and because landfill gas can migrate laterally and potentially cause explosions or kill vegetation and trees, most new landfills are equipped with gas collection and treatment facilities. To determine the size of the gas collection and processing facilities needed, the quantity of landfill gas must first be estimated using the methods outlined in Section 11-4 and illustrated in Example 11-8 in Section 11-12. Because the rate of gas production varies depending on the operating procedures (e.g., without or with leachate recycle) several rates should be analyzed. The decision to use horizontal or vertical gas recovery wells depends on the design and capacity of the landfill. The decision to flare or to recover energy from the landfill gas is determined by the capacity of the landfill site and the opportunity to sell power produced from the conversion of landfill gas to energy. In many small landfills located in remote areas, gas collection equipment is not used routinely.

Selection of Landfill Cover Configuration

As discussed previously, a landfill cover is usually composed of several layers, each with a specific function (see Fig. 11-53). The use of a geomembrane liner as a barrier layer is favored by most landfill designers to limit the entry of surface water and to control the release of landfill gases. The specific cover configuration selected will depend on the location of the landfill and the climatalogical conditions. For example, to allow for regrading, some designers favor the use of a deep layer of soil. To ensure the rapid removal of rainfall from the completed landfill and to avoid the formation of puddles, the final cover should have a slope of about 3 to 5 percent. The selection of a landfill cover is illustrated in Example 11-14 in Section 11-12.

Surface Water Drainage Facilities

An important step in the design of a landfill is to develop an overall drainage plan for the area that shows the location of storm drains, culverts, ditches, and subsurface drains as the filling operation proceeds. Depending on the location and configuration of the landfill and the capacity of the natural drainage courses, it may be necessary to install a stormwater retention basin.

Environmental Monitoring Facilities

Monitoring facilities are required at new landfills for (1) gases and liquids in the vadose zone, (2) for groundwater quality both upstream and downstream of the landfill site, and (3) for air quality at the boundary of the landfill and from any processing facilities (e.g., flares). The specific number of monitoring stations will depend on the configuration and size of the landfill and the requirements of the local air and water pollution control agencies.

Aesthetic Design Considerations

Aesthetic design considerations relate to minimizing the impact of the landfilling operation on nearby residents as well as the public that may be passing by the landfill.

Screening of Landfilling Areas. Screening of the daily landfilling operations from nearby roads and residents with berms, plantings, and other landscaping measures is one of the most important examples of an aesthetic design consideration (see Fig. 11-67*a*). Screening of the active areas in the landfill must be incorporated in the preliminary design and layout of the landfill.

(*a*) (*b*) (*c*) (*d*)

FIGURE 11-67
Aesthetic considerations in landfill design: (*a*) view of landscaped landfill in which filling operations are not visible from nearby freeway, (*b*) overhead wire system used to control sea gulls at landfills, (*c*) wire screen used to control blowing papers and plastic, and (*d*) daily cover used to control vectors at landfills.

The Control of Birds. The presence of birds at the landfill site is not only a nuisance, but also they can cause serious problems if the landfill site is located near an airport. Techniques that have been used to control birds at landfill sites include the use of noise makers, the use of recordings of the sounds made by birds of prey, and the use of overhead wires. The use of overhead wires to keep birds out of reservoirs and fishponds dates back to the early 1930s [1, 31]. The County Sanitation Districts of Los Angeles County pioneered the use of overhead wires to control sea gulls at landfills in the early 1970s (see Fig. 11-67*b*). Because sea gulls descend in a circular pattern when landing, it appears that the wires may interfere with the birds' guidance mechanism. The poles are usually spaced 50 to 75 ft apart, with line spans from 500 to 1200 ft [30]. Crisscrossing improves the effectiveness of the wire system. Typically, 100 lb test monofilament fish line is used, although stainless steel wire has also been used.

The Control of Blowing Materials and Dust. Depending on the location, windblown paper, plastics, and other debris can be a problem at some landfills. The most common solution is to use portable screens near the operating face of the landfill (see Fig. 11-67*c*). To avoid problems with vectors, the material accumulated on the screens must be removed daily. Dust is controlled by spraying water on the approach and internal access roads (see Fig. 11-69*e*).

The Control of Pests and Vectors. The principal vectors of concern in the design and operation of landfills are pests including mosquitos and flies and rodents such as rats and other burrowing animals. Flies and mosquitos are controlled by the placement of daily cover and by the elimination of standing water. The latter can be a problem in areas where white goods and used tires are stored for recycling. The use of covered facilities for the storage of these materials will eliminate most problems. Rats and other burrowing animals are controlled by the use of daily cover (see Fig. 11-67*d*).

Equipment Requirements

The type, size, and amount of equipment required will depend on the size of the landfill and the method of operation. The types of equipment that have been used at sanitary landfills include crawler tractors, scrapers, compactors, draglines, and motorgraders (see Figs. 11-68 and 11-69). Of these, crawler tractors are most commonly used. Properly equipped tractors can be used to perform all the necessary operations at a sanitary landfill, including spreading, compacting, covering, trenching, and even hauling cover materials [4]. Some generalized information on the performance of landfill equipment is summarized in Table 11-25. Typical cost information for landfill equipment may be found in Appendix E. The size and amount of equipment will depend primarily on the size of the landfill operation. Local site conditions will also influence the size of the equipment. Equipment requirements that may be used as a guide for landfill operations are reported in Table 11-26.

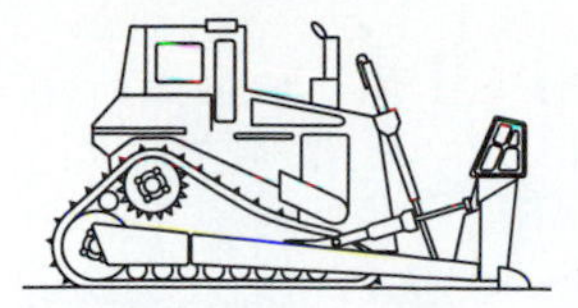

High track compactor with trash blade

Steel-wheeled compactor with trash blade

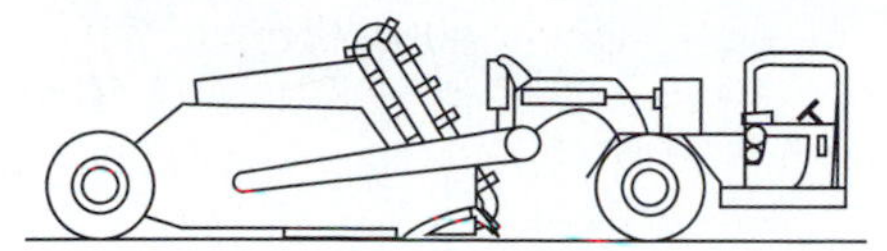

Self-loading earth moving scraper

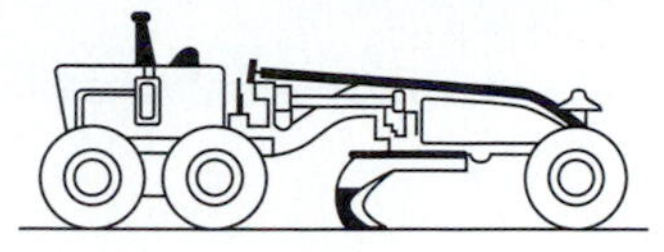

Motor grader

Drag line (for excavation of landfill cells and trenches)

Rubber-tired front end loader

FIGURE 11-68
Typical equipment used at landfills for the placement and covering of solid waste.

TABLE 11-25
Perfomance characteristics of landfill equipment[a,b]

	Solid waste		Cover material			
Equipment	**Spreading**	**Compacting**	**Excavating**	**Spreading**	**Compacting**	**Hauling**
Crawler tractor	E[c]	G	E	E	G	NA
Wheeled compactor	E	E	P	F–G	E	NA
Scraper	NA	NA	G	E	NA	E

[a] From Ref. 4.

[b] Basis of evaluation: easily workable soil and cover material haul distance greater than 1000 ft.

[c] Rating key: E, excellent; G, good; F, fair; P, poor; NA, not applicable.

FIGURE 11-69
Views of equipment used at landfills: (*a*) crawler tractor with dozer blade, (*b*) high track crawler tractor with trash blade, (*c*) steel wheel compactor with trash blade—engine in this unit is air cooled, (*d*) self-loading scraper, (*e*) water wagon used for dust control, and (*f*) drag line.

TABLE 11-26
Typical equipment requirements for sanitary landfills

Approximate population	Daily wastes, tons	Equipment: Number	Type	Equipment weight, lb	Accessory[a]
0–20,000	0–50	1	Tractor, crawler	10,000–30,000	Dozer blade Front-end loader (1 to $2\ yd^3$) Trash blade
20,000–50,000	50–150	1	Tractor, crawler	30,000–60,000	Dozer blade Front-end loader (2 to $4\ yd^3$) Bullclam Trash blade
		1	Scraper or dragline		
		1	Water truck		
50,000–100,000	150–300	1–2	Tractor, crawler	30,000+	Dozer blade Front-end loader (2 to $5\ yd^3$) Bullclam Trash blade
		1	Scraper or dragline[b]		
		1	Water truck		
>100,000	300[c]	1–2	Tractor, crawler	45,000+	Dozer blade Front-end loader (2 to $5\ yd^3$) Bullclam Trash blade
		1	Steel wheel compactor		
		1	Scraper or dragline[b]		
		1	Water truck		
		—[a]	Road grader		

[a] Optional, depends on individual needs.

[b] The choice between a scraper or dragline will depend on local conditions.

[c] For each 500-ton increase add one more of each piece of equipment.

11-10 LANDFILL OPERATION

The development of a workable operating schedule, a filling plan for the placement of solid wastes, landfill operating records and billing information, a load inspection plan for hazardous wastes, and site safety and security plans are important elements of a landfill operation plan. Other factors that must be considered in the operation of a landfill are reported in Table 11-27.

Landfill Operating Schedules

Factors that must be considered in developing operating schedules include (1) arrival sequences for collection vehicles, (2) traffic patterns at the site, (3) the time

TABLE 11-27
Important factors that must be considered in the operation of landfills

Factors	Remarks
Days and hours of operation	Usual practice is 5 to 6 d/wk and 8 to 10 h/d
Communications	Telephone for emergencies
Employee facilities	Restrooms and drinking water should be provided
Equipment maintenance	A covered shed should be provided for field maintenance of equipment
Litter control	Use movable fences at unloading areas; crews should pick up litter at least once per month or as required
Operation plan	With or without the codisposal of treatment plant sludges and the recovery of gas
Operational records	Tonnage, transactions, and billing if a disposal fee is charged
Salvage	No scavenging; salvage should occur away from the unloading area
Scales	Essential for record keeping if collection trucks deliver wastes; capacity to 100,000 lb
Security	Provide locked gates and fencing; lighting of sensitive areas
Spread and compaction	Spread and compact waste in layers less than 2 ft thick to achieve optimum compaction
Unloading area	Keep small, generally under 100 ft on a side; operate separate unloading areas for automobiles and commercial trucks

sequence to be followed in the filling operations, (4) effects of wind and other climatic conditions, and (5) commercial and public access. For example, because of heavy truck traffic early in the morning, it may be necessary to restrict public access to the site until later in the morning.

Solid Waste Filling Plan

Once the general layout of the landfill site has been established, it will be necessary to select the placement method to be used and to lay out and design the individual solid waste cells. The specific method of filling will depend on the characteristics of the site, such as the amount of available cover material, the topography, and the local hydrology and geology. Details on the various filling methods were presented in Section 11-2. To assess future development plans, it will be necessary to prepare a detailed plan for the layout of the individual solid waste cells. The filling sequence should be established so that the landfill operations are not impeded by unusual weather or adverse winter conditions. A typical example of such a plan is shown in Fig. 11-70.

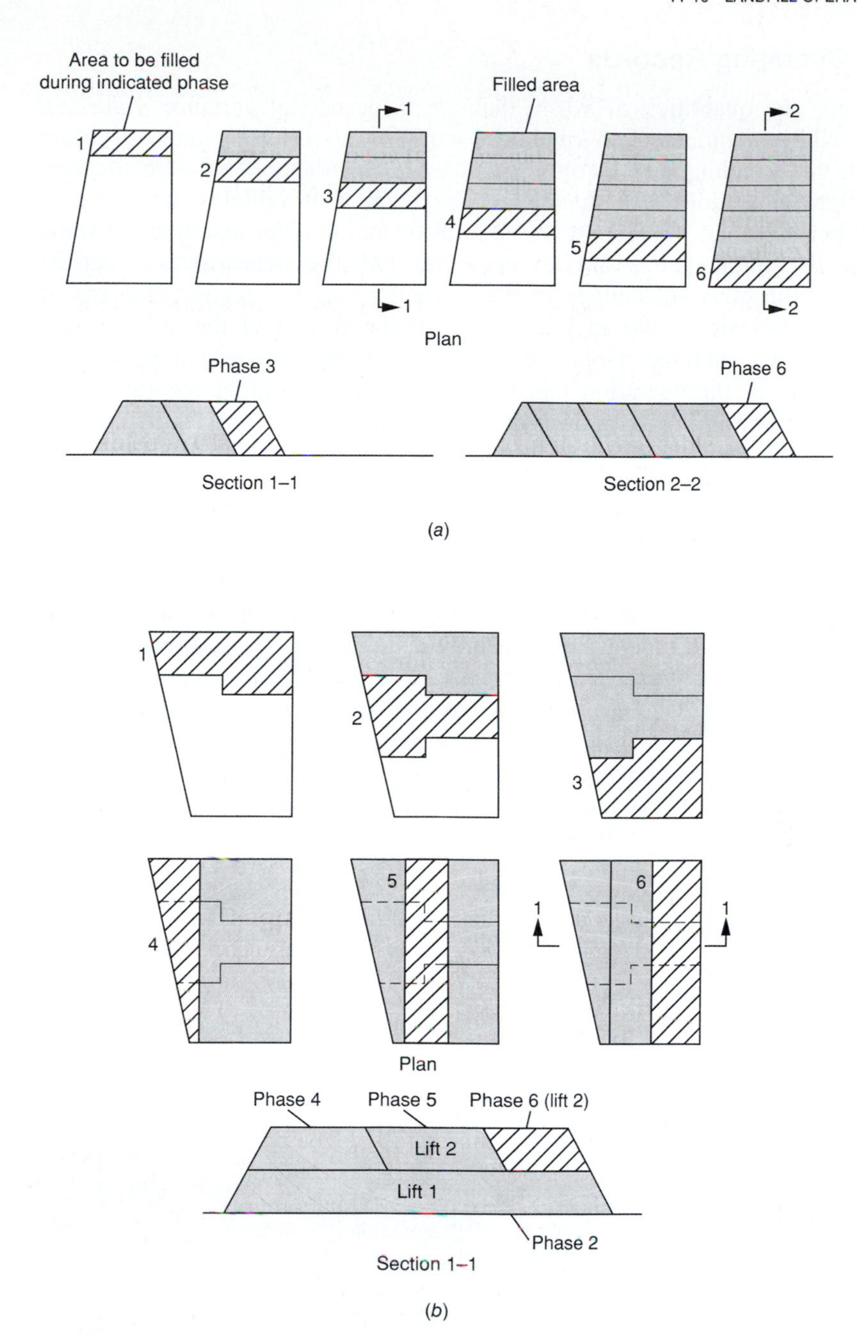

FIGURE 11-70
Typical examples of solid waste filling plans: (*a*) filling plan for single-lift landfill and (*b*) filling plan for a multilift landfill.

Landfill Operating Records

To determine the quantities of waste that are disposed, an entrance scale and gatehouse will be required. The gatehouse would be used by personnel who are responsible for weighing the incoming and outgoing trucks. The sophistication of the weighing facilities will depend on the number of vehicles that must be processed per hour and the size of the landfill operation. (For example, in some larger landfills, weigh stations are equipped with radiation detectors to detect the presence of radioactive substances in the incoming wastes.) Some examples of weighing facilities are shown in Fig. 11-71. If the weight of the solid wastes delivered is known, then the in-place density of the wastes can be determined and the performance of the operation can be monitored. The weight records would also be used as a basis for charging participating agencies and private haulers for their contributions.

Load Inspection for Hazardous Waste

Load inspection is the term used to describe the process of unloading the contents of a collection vehicle near the working face or in some designated area, spreading the wastes out in a thin layer, and visually inspecting the wastes to determine whether any hazardous wastes are present (see Fig. 11-72). The presence of radioactive wastes can be detected with a hand-held radiation measuring device or at the weigh station, as described above. If hazardous wastes are found, the waste collection company is responsible for removing them. In the operation of some landfills, if a company is caught bringing in hazardous wastes a second time, a high fine is levied. If caught a third time, the company is banned from discharging wastes at the landfill.

(*a*)

(*b*)

FIGURE 11-71
Typical truck weighing facilities: (*a*) at small landfill and (*b*) at large landfill.

(a)

(b)

FIGURE 11-72
Inspection of solid waste for the presence of hazardous wastes at the Frank R. Bowerman landfill in Orange Co., CA: (a) residential load and (b) commercial load.

Public Health and Safety

Public health and safety issues are related to worker health and safety and to the health and safety of the public.

Health and Safety of Workers. The health and safety of the workers at landfills is critical in the operation of a landfill. The federal government through OSHA regulations and states through OSHA-type programs have established requirements for a comprehensive health and safety program for the workers at landfill sites. Because the requirements for these programs change continually, the most recent regulations should be consulted in the development of worker health and safety programs. Attention must be given to the types of protective clothing and boots, air-filtering head gear, and punctureproof gloves supplied to the workers.

Safety of the Public. As noted previously, safety concerns and the many new restrictions governing the operation of landfills have forced landfill operators to reexamine past operational practices with respect to public safety and site security. As a result, the use of a convenience transfer station at the landfill site, to minimize the public contact with the working operations of the landfill, is gaining in popularity.

Site Safety and Security

The increasing number of law suits over accidents at landfill sites has caused landfill operators to improve security at landfill sites significantly. Most sites now have restricted access and are fenced and posted, with no trespassing and other warning signs. In some locations, television cameras are used to monitor landfill operations and landfill access.

11-11 LANDFILL CLOSURE AND POSTCLOSURE CARE

Landfill closure and *postclosure care* are the terms used to describe what is to happen to a completed landfill in the future. To ensure that completed landfills will be maintained 30 to 50 years into the future, many states have passed legislation that requires the operator of a landfill to put aside enough money so that when the landfill is completed the amount of money that has been set aside will be sufficient to maintain the closed site into perpetuity.

Development of Long-Term Closure Plan

Perhaps the most important element in the long-term maintenance of a completed landfill is the availability of a closure plan in which the requirements for closure are delineated clearly. A closure plan must include a design for the landfill cover and the landscaping of the completed site. Closure must also include long-term plans for the control of runoff, erosion control, gas and leachate collection and treatment, and environmental monitoring.

Cover and Landscape Design. The landfill cover must be designed to divert surface runoff and snowmelt from the landfill site and to support the landscaping design selected for the landfill. Increasingly, the final landscaping design is based on local plant and grass species as opposed to nonnative plant and grass species. In many arid locations in the Southwest, a desert type of landscaping is favored. The subject of landscaping is considered further in Chapter 16.

Control of Landfill Gases. The control of landfill gases is a major concern in the long-term maintenance of landfills. Because of the concern over the uncontrolled release of landfill gases, a gas control system is now installed before most modern landfills are completed. Older completed landfills without gas collection systems are being retrofitted with gas collection systems. The retrofitting of older landfills with gas collection facilities is considered in Chapter 17, along with the remedial actions that may be required at abandoned disposal sites.

Collection and Treatment of Leachate. As with the control of landfill gas, the control of leachate discharges is another major concern in the long-term maintenance of landfills. Again, most modern landfills have some sort of leachate control system as discussed above. Older completed landfills without leachate collection systems are being retrofitted with leachate collection systems (see Chapter 17).

Environmental Monitoring Systems. To be able to conduct long-term environmental monitoring after a landfill has been completed, monitoring facilities must be installed. The monitoring required at completed landfills usually involves (1) vadose zone monitoring for gases and liquids, (2) groundwater monitoring, and (3) air quality monitoring. The required facilities have been described previously.

Postclosure Care

Postclosure care involves the routine inspection of the completed landfill site, maintenance of the infrastructure, and environmental monitoring. These subjects are considered briefly below and in more detail in Chapter 16.

Routine Inspections. A routine inspection program must be established to monitor continually the condition of the completed landfill. Criteria must be established to determine when a corrective action(s) must be taken. For example, how much settlement will be allowed before regrading must be undertaken?

Infrastructure Maintenance. Infrastructure maintenance typically involves the continued maintenance of surface water diversion facilities; landfill surface grades; the condition of liners, where used; revegetation; and maintenance of landfill gas and leachate collection equipment. The amount of regrading that will be required will depend on the amount of settlement. In turn, the rate of settlement will depend on the rate of gas formation and the degree of initial compaction achieved in the placement of the waste materials in the landfill. The amount of equipment that must be available at the site will depend on the extent and capacity of the landfill and the nature of the facilities that must be maintained.

Environmental Monitoring Systems. Long-term environmental monitoring is conducted at completed landfills to ensure that there is no release of contaminants from the landfill that may affect health or the surrounding environment. The kinds of systems needed have already been enumerated. The number of samples collected for analysis and the frequency of collection will usually depend on the regulations of the local air pollution and water pollution control agencies. EPA has developed a baseline procedure for sampling of groundwater that should be reviewed (40 CFR 258).

11-12 LANDFILL PROCESS COMPUTATIONS

The general features of a sanitary landfill have been presented and described in this chapter. The purpose of this section is to illustrate the basic process computations involved in the development of a landfill site. Process computations are used to identify the quantities required for assessing the suitability of a site (e.g., volumetric capacity) and for sizing of the physical facilities (e.g., leachate collection pipes). The principal process and design computations to be considered in the following discussion include:

- 11-7 Determination of landfill capacity and useful life
- 11-8 Landfill gas generation
- 11-9 Analysis of landfill gas recovery system
- 11-10 Determination of the amount of water vapor collected in a landfill gas recovery system

11-11 Landfill leachate production
11-12 Estimation of water percolation rates through a landfill cover
11-13 Landfill compaction during operation and long-term compaction/consolidation
11-14 Selection of a landfill leachate collection system and cover configuration

These design computations have been grouped together so as not to break up the text discussion and to provide a more coherent presentation of the computations involved in designing a landfill. Wherever possible, the computations presented in the following examples have been set up to be solved using a spreadsheet. In addition, spreadsheet formats have been used for data presentation.

Example 11-7 Determination of landfill capacity and useful life. Determine the capacity and the expected useful life of the South Valley landfill site shown in the accompanying figure. The assumptions to be used in determining the capacity and useful life of the landfill are given below along with data on the expected population and daily waste quantities to be landfilled.

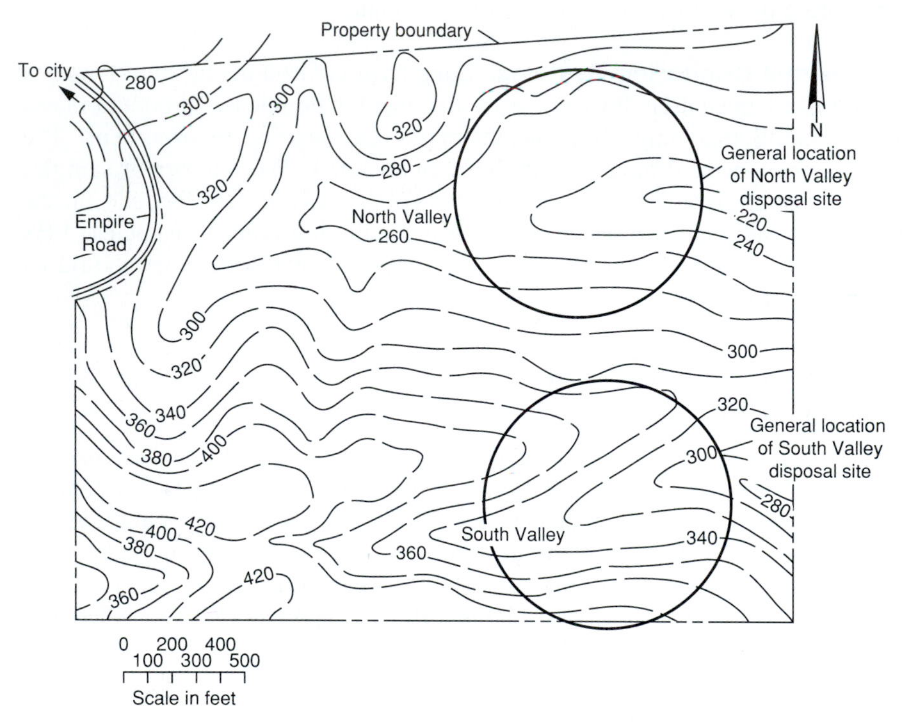

Height of individual landfill lift including the cover material = 10 ft
Slope of front face of landfill = 3 to 1 (see Table 11-23)
Specific weight of compacted solid waste in the landfill = 900 lb/yd^3
Maximum elevation of landfill = 400 ft

Year	Projected end-of-year population (× 1000)	Waste quantities, lb/capita · d
1995	38	4.8
1996	40	4.5
1997	42	4.2
1998	44	3.9
1999	46	3.7
2000	48	3.6
2001	50	3.5
2002	51	3.4
2003	52	3.3
2004	53	3.2
2005	54	3.1
2006	55	3.0
2007	56	3.0
2008	57	3.0
2009	58	3.0
2010	59	3.0

Solution

1. Develop a ground surface profile through the proposed landfill site.

   The profile of the south site, taken coincident with the flow line, is shown in the following figure. The ground surface profile is drawn by measuring, using the graphic scale provided, the distance along the flow line to the point where each contour line crosses the flow line. In preparing the profile through the landfill, an expanded vertical scale is used to allow for the superposition of alternative landfill configurations.

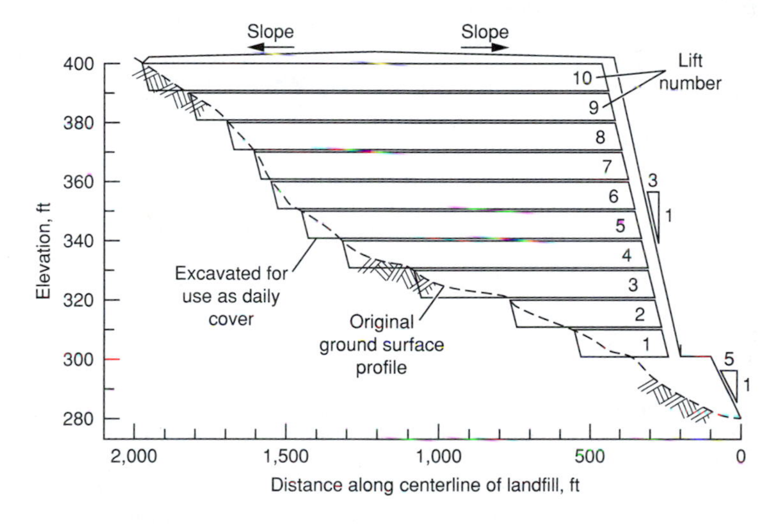

2. Develop a profile of the completed landfill.

   For operating lifts of 10 ft and a 3 to 1 slope for the front face of the landfill, a typical landfill profile is shown on the profile developed in Step 1 above. As shown, the maximum elevation of the landfill is 400 ft and a total of 10 lifts are to be used.

3. Develop a plan map of the completed landfill showing the 10 ft contours corresponding to the individual lifts.

   Developing a plan of the proposed landfill involves transferring information from the landfill profile developed in Step 2 to the location map. The plan view of the proposed landfill is shown in the following figure. From the profile map (see Step 2) there is a 10 ft vertical rise for each 30 ft of horizontal distance measured along the flow line. The locations where the 10 ft contours, corresponding to each lift, cross the flow line are marked off every 30 ft along the flow line. The method used to connect the contour intervals to the existing ground contours depends on the design of the front face of the landfill. For example, if the front face of the landfill is to be a flat inclined plane, then straight lines are passed through the contour intervals marked off along the flow line. Alternatively, if the front face of the completed landfill is to be curved, then the landfill might look as shown in the following figure.

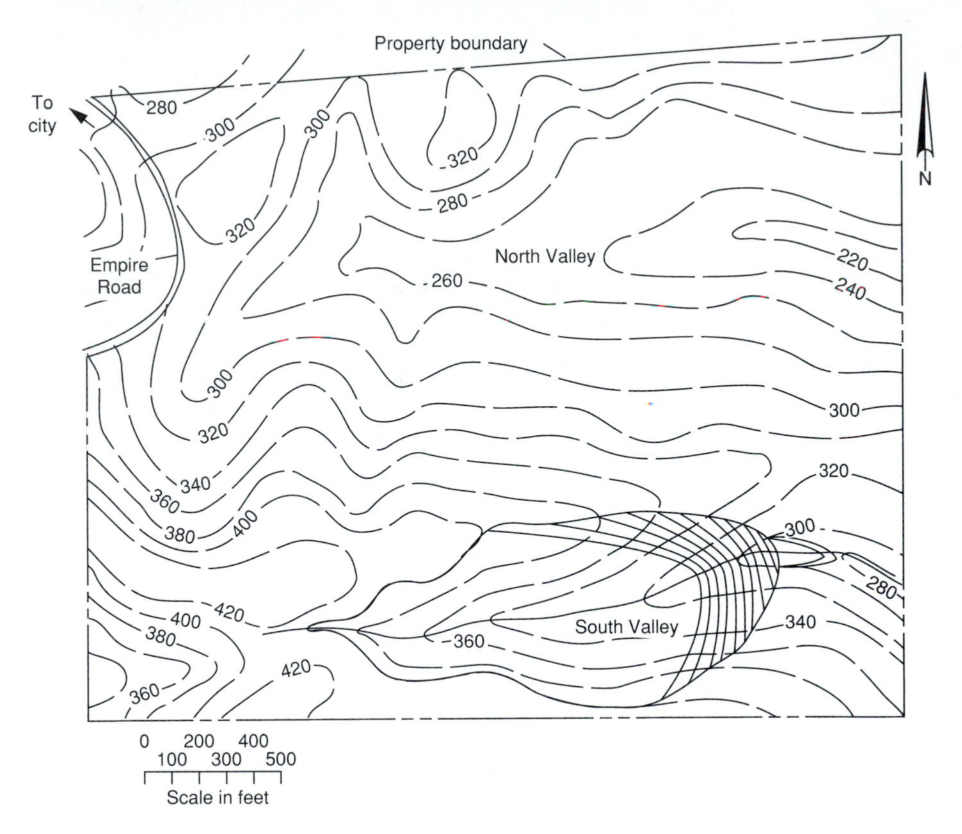

4. Determine the capacity of the proposed landfill.

   (*a*) The volumetric capacity of the South Valley landfill site in cubic yards is computed by determining the volume between contour intervals. The areas of the two adjacent contours are averaged, and the average value is multiplied by 10 ft (the lift height) and divided by 27 ft^3/yd^3 to convert to cubic yards. The necessary computations are presented in the table below. The area at each contour interval is obtained from the contour map developed in Step 3 using a planimeter. Alternatively, the surface area corresponding to each contour can be determined by tracing on a see-through grid the area enclosed by each contour and counting the squares. As computed, the total capacity of the landfill is 1,118,250 yd^3. Because the site will be excavated to obtain the necessary cover material, the capacity of the site is equal to the volume of the site.

Lift number	Elevation	Area, ft^2 At contour, interval[a]	Area, ft^2 Average between contours	Capacity between contours,[b] yd^3
	300	11,360		
1			28,405	10,520
	310	45,450		
2			79,540	29,460
	320	113,635		
3			136,360	50,500
	330	159,090		
4			193,180	71,550
	340	227,270		
5			255,680	94,700
	350	284,090		
6			321,500	115,740
	360	340,910		
7			423,865	156,990
	370	506,820		
8			526,135	194,860
	380	545,450		
9			537,495	199,070
	390	529,540		
10			526,135	194,860
	400	522,730		
Total capacity, yd^3				1,118,250

[a] From the figure given in Step 3.

[b] Volume = (average area, ft) × (10 ft)/(27 ft^3/yd^3)

(*b*) If cover material has to be brought to the site, then the volume of solid wastes determined in the above table must be multiplied by a factor to account for the cover material. For a cover to waste ratio of 1 to 5, the capacity of the proposed landfill is 931,875 yd^3.

5. Determine the useful life of the proposed landfill.

(*a*) Determine the expected daily, yearly, and cumulative yearly total waste quantities. These totals are summarized in the following table. The daily and yearly waste quantities were computed on the basis of the projected end-of-year population. This procedure is recommended even though it is on the conservative side. The volume was computed using an assumed value of 900 lb/yd^3 for the in-place compacted specific weight of the solid wastes. The computed values can be scaled for any other assumed specific weight values.

Year	Projected end-of-year population (× 1000)	Waste quantities, lb/capita · d	Waste quantities, yd^3		
			Daily volume	Yearly volume	Cumulative total
1995	38	4.8	202.7	73,986	73,986
1996	40	4.5	200.0	73,000	146,986
1997	42	4.2	196.0	71,540	218,526
1998	44	3.9	190.7	69,606	288,132
1999	46	3.7	189.1	69,022	357,154
2000	48	3.6	192.0	70,080	427,234
2001	50	3.5	194.4	70,956	498,190
2002	51	3.4	192.7	70,336	568,526
2003	52	3.3	190.7	69,606	638,132
2004	53	3.2	186.0	67,890	706,022
2005	54	3.1	186.0	67,890	773,912
2006	55	3.0	183.3	66,905	840,817
2007	56	3.0	186.7	68,146	908,963
2008	57	3.0	190.0	69,350	978,313
2009	58	3.0	193.3	70,555	1,048,868
2010	59	3.0	196.7	71,796	1,120,664

(*b*) When the waste quantities given in the above table are compared to the available capacity determined in Step 4, the useful life of the South Valley landfill site is found to be about 16 yr (1995 to 2010). At that time it would be necessary to develop the North Valley landfill site.

Comment. To start the landfill operation, the topsoil would be stripped away in the lower portions of the South Valley and stockpiled at the eastern end of the landfill site. The stockpile serves as a dam to capture and divert stormwater runoff as well as a site for topsoil storage. The computations performed in this example could also be performed using computer-aided design (CAD) software on a microcomputer or an engineering workstation.

Example 11-8 Landfill gas generation. Determine the distribution of gas production over time for a landfill with a useful life of five years based on the following data and assumptions:

1. Landfill life = 5 yr
2. The composition of the waste is as described in Table 3-4 for residential and commercial MSW, of which 79.5 percent is organic and 20.5 percent is inert.
3. The organic fraction (79.5 percent) is composed of 7 percent plastic (considered to be inert), 60.1 percent rapidly biodegradable material, and 12.4 percent slowly biodegradable material (see Example 11-2). The corresponding values for rapidly and slowly biodegradable material based on dry weight are 44.8 and 7.3 percent, respectively.
4. Of the rapidly biodegradable organic waste, 75 percent is available for degradation (i.e., some organic waste materials in plastic bags will not be degraded, some of the material will be too dry to support biological activity).
5. Of the slowly biodegradable organic waste, 50 percent is available for degradation (for the same reasons cited above).

6. The total amount of landfill gas produced from the biodegradable fraction of the rapidly and slowly biodegradable organic materials deposited each year is 14 and 16 ft^3/lb dry solids, respectively (see Example 11-2).
7. Time period for total decomposition of rapidly decomposable organic material is 5 yr.
8. Time period for total decomposition of slowly decomposable organic material is 15 yr.

Assume the yearly rate of decomposition for rapidly and slowly decomposable material is based on a triangular gas production model in which the peak rate of gas production occurs 1 and 5 years, respectively, after gas production starts. Gas production is assumed to start at the end of the first full year of operation.

Solution

1. Determine the amount of gas that has been produced at the end of each year from one pound of the rapidly and slowly biodegradable organic waste material as these materials decompose over a 5- and 15-year period, respectively.
   (*a*) Rapidly biodegradable waste (RBW):
      i. If one uses a triangular gas production model, the gas production over the five-year period can be illustrated graphically as shown in the following figure.

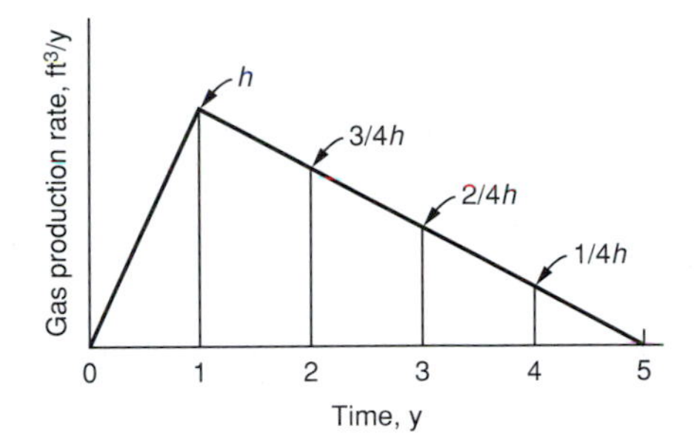

      ii. Because the area of the triangle is equal to one half the base times the altitude, the total amount of gas produced is equal to

Total gas produced, ft^3

$$= 1/2 \text{ (base, yr)} \times \text{(altitude, peak rate of gas production, ft}^3\text{/yr)}$$

      iii. If the total amount of gas produced from one pound of RBW is equal to 14.0 ft^3, then the peak rate of gas production, which occurs at the end of the first year that gas is produced, is equal to

$$\text{Peak rate of gas production, ft}^3\text{/yr} = 14.0 \text{ ft}^3 \times (2/5 \text{ yr}) = 5.6 \text{ ft}^3\text{/yr}$$

      iv. The amount of gas produced during the first year that gas is produced is equal to

$$\text{Gas produced during the first year, ft}^3 = 1/2\ (1.0 \text{ yr}) \times (5.6 \text{ ft}^3\text{/yr}) = 2.8 \text{ ft}^3$$

      v. The rate of gas production during the second year that gas is produced is

$$\text{Rate of gas production, ft}^3\text{/yr} = (5.6 \text{ ft}^3\text{/yr} + (3/4)\ 5.6 \text{ ft}^3\text{/yr})/2 = 4.9 \text{ ft}^3\text{/yr}$$

vi. The amount of gas produced during the second year that gas is produced is

$$\text{Gas produced during the second year, ft}^3 = [(5.6\ \text{ft}^3/\text{yr} + 5.6\ \text{ft}^3/\text{yr} \times 3/4) \times 1.0\ \text{yr}]/2 = 4.9\ \text{ft}^3$$

vii. The rate and amount of gas produced during the third, fourth, and fifth years are determined in a similar manner.

viii. Summarize the yearly gas production quantities.

End of year	Rate of gas production, ft³/yr	Gas production, ft³
1	0.0	
		2.8
2	5.6	
		4.9
3	4.2	
		3.5
4	2.8	
		2.1
5	1.4	
		0.7
6	0.0	
Total		14.0

(*b*) Slowly biodegradable waste (SBW):

Determine the amount of gas produced at the end of each year from one pound of the slowly decomposable biodegradable organic material as it decomposes during the 15-year period.

i. Using a triangular gas production model, the gas production over the 15-year period can be shown graphically in the following figure.

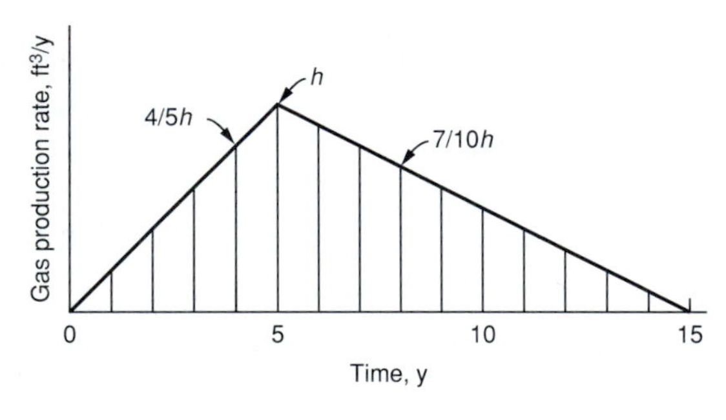

ii. If the total amount of gas produced from one pound of SBW is equal to 16.0 ft³, then the peak rate of gas production is equal to

$$\text{Peak rate of gas production, ft}^3/\text{yr} = 16.0\ \text{ft}^3 \times (2/15\ \text{yr}) = 2.133\ \text{ft}^3/\text{yr}$$

iii. The rate of gas production at end of the first year that gas is produced is

$$\text{Rate of gas production, ft}^3/\text{yr} = 1/5 \times 2.133\ \text{ft}^3/\text{yr} = 0.427\ \text{ft}^3/\text{yr}$$

iv. The amount of gas produced during the first year that gas is produced is

$$\text{Gas produced during the first year, ft}^3 = 1/2\,(1.0\text{ yr}) \times (0.427\text{ ft}^3\text{/yr}) = 0.213\text{ ft}^3$$

v. The amount of gas produced during the second year that gas is produced is

Gas produced during the second year, ft^3

$$= \left\{\left[\left(2.13\text{ ft}^3\text{/yr} \times 1/5\right) + \left(2.13\text{ ft}^3\text{/yr} \times 2/5\right)\right] \times 1.0\text{ yr}\right\}/2$$

$$= 0.64\text{ ft}^3$$

vi. The amount of gas produced during the remaining thirteen years is determined in a similar manner.

vii. Summarize the yearly gas production quantities.

End of year	Rate of gas production, ft^3/yr	Gas production, ft^3	End of year	Rate of gas production, ft^3/yr	Gas production, ft^3
1	0.000		9	1.493	
		0.213			1.387
2	0.427		10	1.280	
		0.640			1.173
3	0.853		11	1.066	
		1.067			0.960
4	1.280		12	0.853	
		1.493			0.747
5	1.706		13	0.640	
		1.920			0.534
6	2.133		14	0.427	
		2.027			0.320
7	1.920		15	0.213	
		1.813			0.107
8	1.706		16	0.000	
		1.600			
			Total		16.001

2. Determine the yearly gas production rates from the rapidly and slowly biodegradable organic material per pound of total waste. The computed values will be used to prepare a spreadsheet computation table to determine total quantity of gas produced per pound of total waste deposited in the landfill.

(*a*) Determine the distribution of gas produced from the rapidly and slowly biodegradable organic material per pound of total waste deposited.

i. Determine the fraction of the total waste that is rapidly biodegradable, based on dry weight.

$$(0.448)(0.75) = 0.336\text{ lb RBW/lb total waste}$$

ii. Determine the fraction of the total waste that is slowly biodegradable, based on dry weight.

$$(0.073)(0.50) = 0.0365\text{ lb SBW/lb total waste}$$

iii. Determine the total amount of gas produced per pound of RBW.

$$\text{Gas}_{\text{RB}} = 0.336\text{ lb RBW/lb waste} \times 14\text{ ft}^3\text{/lb RBW} = 4.7\text{ ft}^3\text{/lb waste}$$

iv. Determine the total amount of gas produced per pound of SBW.

$$\text{Gas}_{\text{SB}} = 0.0365 \text{ lb SBW/lb waste} \times 16 \text{ ft}^3/\text{lb SBW}$$
$$= 0.584 \text{ ft}^3/\text{lb waste}$$

(*b*) Determine the rapidly and slowly biodegradable waste gas generated based on total waste.

Determine the amount of gas produced at the end of each year from one pound of total waste as it decomposes during the five-year period. For rapidly decomposable waste, multiply the gas production per year values determined in Part 1 by 0.336 lb/lb; for slowly decomposable waste, multiply the gas productions per year determined in Step 1 by 0.0365 lb/lb (see Step 2*a*). The yearly gas production quantities are summarized as follows.

	Rapidly biodegradable		Slowly biodegradable		Total (rapid + slow)	
End of year	Rate of generation, ft^3/yr	Volume of gas, ft^3	Rate of generation, ft^3/yr	Volume of gas, ft^3	Rate of generation, ft^3/yr	Volume of gas, ft^3
0	0.000		0.000		0.000	
		0.000		0.000		0.000
1	0.000		0.000		0.000	
		0.941		0.008		0.949
2	1.882		0.016		1.898	
		1.646		0.023		1.669
3	1.411		0.031		1.442	
		1.176		0.039		1.215
4	0.941		0.047		0.988	
		0.706		0.055		0.761
5	0.470		0.062		0.532	
		0.235		0.070		0.305
6	0.000		0.078		0.078	
		0.000		0.074		0.074
7	0.000		0.070		0.070	
				0.066		0.066
8			0.062		0.062	
				0.058		0.058
9			0.055		0.055	
				0.051		0.051
10			0.047		0.047	
				0.043		0.043
11			0.039		0.039	
				0.035		0.035
12			0.031		0.031	
				0.027		0.027
13			0.023		0.023	
				0.019		0.019
14			0.016		0.016	
				0.012		0.012
15			0.008		0.008	
				0.004		0.004
16			0.000		0.000	
Total		4.704		0.584		5.288

3. Using the gas production data determined in Step 2, prepare a spread sheet computation table to determine total quantity of gas produced. Assume that equal amounts of waste will be deposited each of the five years that the landfill is used. For illustration purposes, in the following spreadsheet computation table 1 lb of waste is assumed to be deposited each year. Column 1 is the time since wastes were first accepted at the landfill. The yearly columns correspond to the total rate of gas production from the waste material deposited in the indicated year.

	Landfill gas as produced from waste deposited over a period of five years							
	Rate of landfill gas generation from waste deposited in indicated year, ft^3/yr[a]							
End of year	Year 1	Year 2	Year 3	Year 4	Year 5	Total	Gas, ft^3	Cumulative production, ft^3
0	0.000					0.000		
							0.000	0.000
1	0.000	0.000				0.000		
							0.949	0.949
2	1.897	0.000	0.000			1.897		
							2.618	3.567
3	1.442	1.897	0.000	0.000		3.340		
							3.833	7.400
4	0.988	1.442	1.897	0.000	0.000	4.327		
							4.593	11.993
5	0.533	0.988	1.442	1.897	0.000	4.860		
							4.899	16.892
6	0.078	0.533	0.988	1.442	1.897	4.938		
							4.024	20.916
7	0.070	0.078	0.533	0.988	1.442	3.111		
							2.420	23.336
8	0.062	0.070	0.078	0.533	0.988	1.730		
							1.264	24.600
9	0.055	0.062	0.070	0.078	0.533	0.797		
							0.544	25.154
10	0.047	0.055	0.062	0.070	0.078	0.311		
							0.292	25.446
11	0.039	0.047	0.055	0.062	0.070	0.273		
							0.253	25.699
12	0.031	0.039	0.047	0.055	0.062	0.234		
							0.214	25.913
13	0.023	0.031	0.039	0.047	0.055	0.195		
							0.175	26.088
14	0.016	0.023	0.031	0.039	0.047	0.156		
							0.136	26.224
15	0.008	0.016	0.023	0.031	0.039	0.117		
							0.097	26.321
16	0.000	0.008	0.016	0.023	0.031	0.078		
							0.062	26.383
17		0.000	0.008	0.016	0.023	0.047		
							0.035	26.418
18			0.000	0.008	0.016	0.023		
							0.016	26.434
19				0.000	0.008	0.008		
							0.004	26.438
20					0.000	0.000		

[a] Total waste deposited = 1 lb/year for five years.

4. Prepare a plot of the total yearly gas production rates and the cumulative amount of gas produced from the RBW and SBW deposited in the landfill over a five-year period.

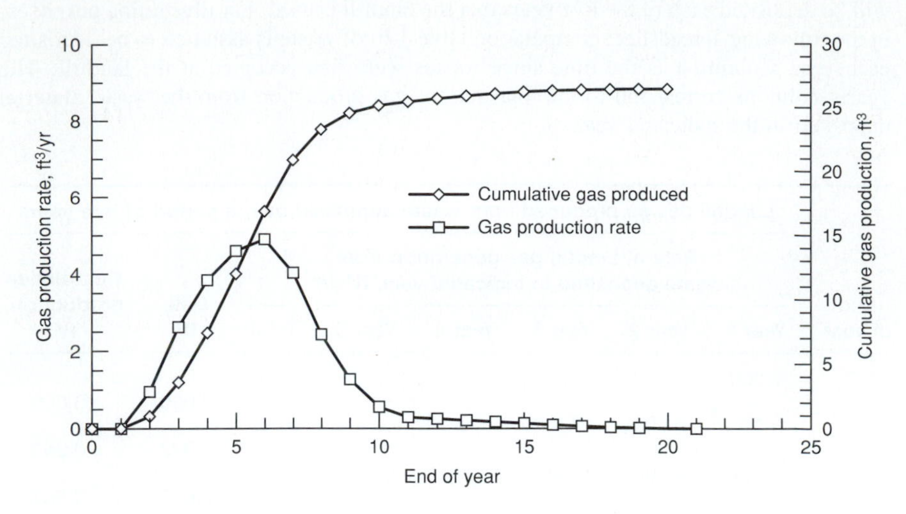

Comment. In developing the total gas production curve in this example, a triangular gas production function was used. Note that any type of gas production function can be used if better information is available.

Example 11-9 Analysis of landfill gas recovery system. Determine the head loss in the landfill gas recovery system shown in the accompanying figure. Also determine the required blower capacity. The analysis is to be based on the following data and assumptions:

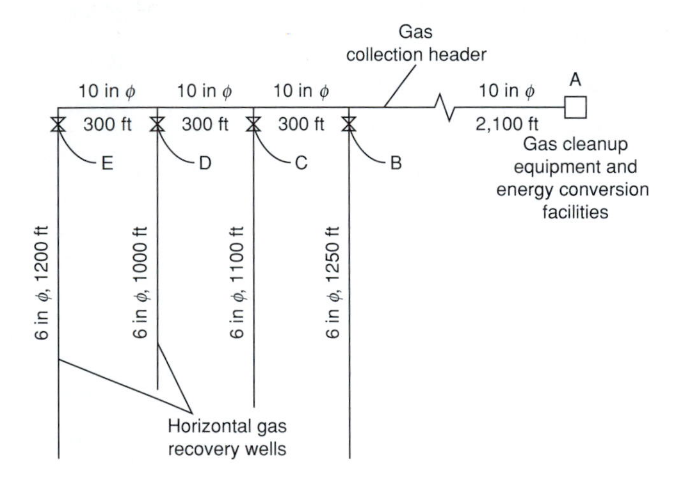

1. Diameter of horizontal gas extraction wells = 6 in
2. Diameter of header used to collect gas from the horizontal landfill gas recovery wells = to be determined

3. Absolute roughness for the plastic pipe used for the gas collection header, e = 0.00005 ft
4. Allowance for minor losses in header between extraction wells = 0.1 in H_2O
5. Allowance for minor losses in header between last extraction well and blower = 0.5 in H_2O
6. Estimated gas flow per horizontal gas extraction well = 200 ft^3/min (60°F, 14.7 lb/in^2)
7. Gas composition (by volume) = CH_4, 50%; CO_2, 50%
8. Temperature of landfill gas at the wellhead = 130°F
9. Temperature loss in manifold section between extraction wells = 5°F
10. Temperature of landfill gas at the blower station = 90°F
11. Landfill gas is saturated in water vapor at the wellhead.
12. Vacuum to be maintained at the wellhead of the farthermost horizontal gas extraction well (Point E) = 10 in H_2O
13. Vacuum at blower = to be determined, in H_2O

Solution

1. Determine the head loss in the header used to collect gas from the individual horizontal gas extraction wells starting at point E.
   Determine the headloss per 100 ft of header. Assume a 10 in header will be used. If the head loss using a 10 in header is too large, the size of the header is increased and the head loss computations are repeated. Friction losses in the gas piping can be calculated using the Darcy-Weisbach equation as given below (see Appendix I).

$$h_L = f \frac{L}{D} h_i$$

where h_L = friction loss, in of water
f = dimensionless friction factor obtained from Moody diagram (see Fig. I-1 in Appendix I)
L = length of pipe, ft
D = diameter of pipe, ft
h_i = velocity head of air, in of water

(*a*) Determine the velocity of flow of landfill gas in the 10 in header from point E to D. Using the perfect gas law, determine the volumetric flow rate at the average temperature in the header between the first and second extraction wells. Assume the temperature falls linearly with distance. Thus, the temperature at point D is 125°F. The volumetric flow rate of landfill gas at an average temperature of 127.5°F and 10 in H_2O vacuum can be determined as follows:

$$\left(\frac{PV}{T}\right)_1 = \left(\frac{PV}{T}\right)_2$$

$$P_1 = 14.7 \text{ lb/in}^2 = 2116.8 \text{ lb/ft}^2 = 33.9 \text{ ft of } H_2O$$

$$V_1 = 200 \text{ ft}^3\text{/min}$$

$$T_1 = 460 + 60 = 520°R$$

$$P_2 = 2116.8 \text{ lb/ft}^2 - [(10 \text{ in}/12 \text{ in/ft}) \times 61.60 \text{ lb/ft}^3] = 2065.5 \text{ lb/ft}^2$$

$$V_2 = ?\ \text{ft}^3/\text{min}$$

$$T_2 = 460 + 127.5 = 587.5°\text{R}$$

$$V_2 = \frac{2116.8 \times 200}{520} \times \frac{587.5}{2065.5} = 231.6\ \text{ft}^3/\text{min}$$

The velocity of flow is given by

$$v = q/A$$

v = velocity of flow, ft/min

q = volumetric landfill gas flow rate, ft^3/min

A = cross-sectional area of 10 in diameter pipe, $\text{ft}^2 = 0.545\ \text{ft}^2$

Thus:

$$v = (231.6\ \text{ft}^3/\text{min})/0.545\ \text{ft}^2 = 425.0\ \text{ft/min}$$

(*b*) Determine the value of the friction factor f in the Darcy-Weisbach equation using the Moody diagram given in Appendix I. The Reynolds number, N_R, may be computed using the following relationship:

$$N_R = \frac{dv\rho_{\text{gas}}}{\mu_{\text{gas}}} = \frac{dv\gamma_{\text{gas}}}{g\,\mu_{\text{gas}}}$$

where d = inside diameter of pipe, ft
v = velocity of gas flow in collection pipe = ft/s
ρ_{gas} = density of gas, slug/ft^3
μ_{gas} = viscosity of air, $\text{lb} \cdot \text{s/ft}^2$
γ_{gas} = specific weight of landfill gas at the operating temperature, lb/ft^3
g = acceleration due to gravity, ft/s^2

The specific weight of the landfill gas at a temperature of 127.5°F and a pressure of 2065.5 lb/ft^2 can be computed using the perfect gas law as given below (note that the specific volume is inversely proportional to the specific weight):

$$\gamma_{\text{gas}} = \frac{1}{V} = \frac{P}{RT}$$

where R = gas constant for the landfill gas, ft · lb/(lb-landfill gas) · °R
P = pressure at operating temperature, lb/ft^2

The gas constant for landfill gas is obtained by dividing the universal gas constant [1543 ft · lb/(lb · mole) · °R] by the number of lb/lb · mole in the landfill gas. The lb/lb · mole of landfill gas, based on the given composition of the landfill gas, is calculated as

$$\text{lb/lb} \cdot \text{mole} = (0.50\ CH_4 \times 16) + (0.50\ CO_2 \times 44) = 30.0$$

The gas constant for the landfill gas is

$$\text{R}_{\text{landfill gas}} = [1543\ \text{ft} \cdot \text{lb/(lb} \cdot \text{mole)} \cdot °\text{R}]/(30.0\ \text{lb/lb} \cdot \text{mole landfill gas})$$

$$= 51.43\ \text{ft} \cdot \text{lb/(lb-landfill gas)} \cdot °\text{R}$$

Thus, the specific weight of the landfill gas is equal to

$$\gamma_{gas} = \frac{2065.5 \text{ lb/ft}^2}{\left[\dfrac{51.43 \text{ ft} \cdot \text{lb}}{(\text{lb-landfill gas}) \cdot {}^\circ\text{R}}\right][(460 + 127.5)^\circ\text{R}]} = 0.068 \text{ lb/ft}^3$$

The viscosity of the landfill gas, μ_{gas}, at 127.5°F can be approximated, with sufficient accuracy for most practical purposes, using the following relationship:

$$\mu_{gas} = (0.0137) \times \mu_{\text{water at } 68^\circ\text{F}} \text{ (see Appendix I)}$$

$$\mu_{\text{water at } 68^\circ\text{F}} = 1.009 \text{ centipoise} = 2.11 \times 10^{-5} \text{ lb} \cdot \text{s/ft}^2$$

The Reynolds number at a temperature of 127.5°F is

$$N_R = \frac{dv\gamma_{gas}}{g\mu_{gas}} = \frac{(10/12)(425.3/60)(0.068)}{(32.17)(0.0137 \times 2.11 \times 10^{-5})} = 0.432 \times 10^5$$

Using an e/D value of 0.00006 (e = 0.00005 ft) and a Reynolds number of 0.432×10^5, the friction factor f from Fig. I-1 is found to be be equal to 0.020.

(*c*) The velocity head h_i in inches of water at a temperature of 127.5°F and a pressure of 2065.5 lb/ft² can be computed as follows:

$$h_i = \frac{(v, \text{ ft/min})^2}{2(32.17 \text{ ft/s}^2)}\left(\frac{1}{(60 \text{ s/min})^2}\right)\left(\gamma_{gas}, \frac{\text{lb-landfill gas}}{\text{ft}^3}\right)\left(\frac{1}{\gamma_w, \text{ lb/ft}^3}\right)\left(\frac{12 \text{ in}}{\text{ft}}\right)$$

where v = gas velocity, ft/min

γ_{gas} = specific weight of landfill gas at the operating temperature and pressure, lb/ft³

γ_w = specific weight of water, at the operating temperature, lb/ft³

The velocity head is

$$h_i = \frac{(425.0 \text{ ft/min})^2}{2(32.17 \text{ ft/s}^2)}\left(\frac{1}{(60 \text{ s/min})^2}\right)(0.068 \text{ lb/ft}^3)\left(\frac{1}{61.60 \text{ lb/ft}^3}\right)\left(\frac{12 \text{ in}}{\text{ft}}\right)$$

$$= 0.010 \text{ in of water}$$

(*d*) The head loss per 100 ft of 10 in pipe is

$$h_L = 0.020\left(\frac{100 \text{ ft}}{10 \text{ in}/(12 \text{ in/ft})}\right)0.010 \text{ in } H_2O = 0.024 \text{ in } H_2O$$

2. Set up a computation table to determine the head loss in the remaining portions of the manifold system. The computations for each section of the manifold are completed as outlined above. A new gas temperature must be computed at the point where each extraction well joins the manifold. The summary computation table is presented below.

Section	Pipe diameter, in	Pipe length, ft	Gas velocity, ft/min	Average gas temp., °F	Velocity head, h_i, in of H_2O	Friction factor, f	Friction loss, in/100 ft
E–D	10	300	425	127.5	0.010	0.020	0.024
D–C	10	300	850	125.0	0.041	0.018	0.089
C–B	10	300	1275	122.5	0.093	0.017	0.190
B–A	10	2100	1700	106.3	0.164	0.016	0.315

Section	Total friction loss, in of H_2O	Minor head loss, in of H_2O	Total head loss, in of H_2O
E–D	0.072[a]	0.1	0.172
D–C	0.267	0.1	0.367
C–B	0.570	0.1	0.670
B–A	6.615	0.5	7.115
Pipe loss in inches of H_2O			8.320
Vacuum at point E in inches of H_2O			10.000
Total			18.320

[a]0.024 in × (300 ft/100 ft) = 0.072 in

In the above computations the change in the gas volume due to the increased vacuum has not been considered. If the change in vacuum is significant, then successive computations must be conducted to account for the change in vacuum. In most cases, the change in temperature will offset the difference in vacuum. The total vacuum that must be supplied at the inlet of the vacuum blower, as computed above, is 18.32 in of H_2O at a gas flow of 893 ft^3/min. Typical vacuum levels at the blower inlet for landfill gas recovery systems vary from about 18 to 60 in of H_2O. The total pressure that the vacuum blower must overcome depends on the nature of the discharge facilities including meters, silencers, and check valves.

Comment. For the purposes of this example, the given values for the minor head losses were used. If assumed values for minor head losses are not used, the loss of head due to the presence of elbows, tees, valves, and so forth can be computed as a fraction of velocity head using the K values given in Appendix I of this text or in standard fluid mechanics texts. As noted in Appendix I, the minor losses due to fittings can also be expressed in terms of equivalent diameters of straight pipe that would result in the same loss of head. Meter losses can be estimated as a fraction of the differential head, depending on the type of meter. Losses owing to vacuum blower silencers and check valves should be obtained from equipment manufacturers.

Example 11-10 Determination of the amount of water vapor collected in a landfill gas recovery system. Determine the amount of condensed water vapor that must be removed daily from a landfill gas recovery system based on the following data and assumptions:

1. Total gas flow $= 2.5 \times 10^6$ ft^3/d (60°F, 14.7 lb/in^2)
2. Temperature of landfill gas as it exits the landfill = 130°F
3. Temperature of landfill gas at the blower station = 90°F
4. Vacuum at well head = 10 in H_2O
5. Vacuum at blower = 75 in H_2O
6. Landfill gas is saturated in water vapor at the well head

Solution

1. Determine the total pounds of water present in the water vapor in the saturated landfill gas at the well head.

(*a*) Determine the volume of gas at the well head relative to the volume at standard conditions (60°F, 14.7 lb/in^2).

$$\left(\frac{PV}{T}\right)_1 = \left(\frac{PV}{T}\right)_2$$

$$P_1 = 14.7 \text{ lb/in}^2 = 33.9 \text{ ft H}_2\text{O}$$

$$V_1 = 2.5 \times 10^6 \text{ ft}^3\text{/d}$$

$$T_1 = 460 + 60 = 520°\text{R}$$

$$P_2 = 33.9 \text{ ft H}_2\text{O} - [10 \text{ in}/(12 \text{ in/1 ft})] = 33.07 \text{ ft H}_2\text{O}$$

$$V_2 = ? \text{ ft}^3\text{/d}$$

$$T_2 = 460 + 130 = 590°\text{R}$$

$$V_2 = \frac{34 \times (2.5 \times 10^6)}{520} \times \frac{590}{33.07} = 2.92 \times 10^6 \text{ ft}^3\text{/d}$$

(*b*) Determine the moles of water vapor present in the landfill gas at the well head using the universal gas law.

$$p_v V = nRT$$

$$p_v = \text{vapor pressure of H}_2\text{O at } 130°\text{F}$$

$$= 2.22 \text{ lb/in}^2 \text{ (see Appendix C)} = 319.7 \text{ lb/ft}^2$$

$$V = 2.92 \times 10^6 \text{ ft}^3\text{/d}$$

$$R = \text{universal gas constant} = 1543 \text{ ft} \cdot \text{lb/(lb} \cdot \text{mole)} \cdot °\text{R}$$

$$T = 460 + 130 = 590°\text{R}$$

$$n = \frac{p_v V}{RT} = \frac{319.7 \times (2.92 \times 10^6)}{1543 \times 590} = 1025.4 \text{ lb} \cdot \text{mole/d}$$

(*c*) Determine the pounds of water vapor present in the landfill gas at the well head.

$$\text{lb H}_2\text{O} = (1025.4 \text{ lb} \cdot \text{mole/d}) \times (18 \text{ lb H}_2\text{O/lb} \cdot \text{mole}) = 18{,}457.2 \text{ lb/d}$$

2. Determine the total pounds of water present as water vapor in the landfill gas at the blower.

(*a*) Determine the volume of landfill gas at the blower.

$$P_1 = 33.07 \text{ ft H}_2\text{O}$$

$$V_1 = 2.92 \times 10^6 \text{ ft}^3\text{/d}$$

$$T_1 = 460 + 130 = 590°\text{R}$$

$$P_2 = 33.9 \text{ ft H}_2\text{O} - [75 \text{ in}/(12 \text{ in/ft})] = 27.65 \text{ ft H}_2\text{O}$$

$$V_2 = ? \text{ ft}^3\text{/d}$$

$$T_2 = 460 + 90 = 550°\text{R}$$

$$V_2 = \frac{33.07 \times (2.92 \times 10^6)}{590} \times \frac{550}{27.65} = 3.26 \times 10^6 \text{ ft}^3\text{/d}$$

(*b*) Determine the moles of water vapor present in the landfill gas at the blower.

$$p_v = \text{vapor pressure of } H_2O \text{ at } 90^\circ\text{F}$$

$$= 0.70 \text{ lb/in}^2 \text{ (see Appendix C)} = 100.8 \text{ lb/ft}^2$$

$$V = 3.26 \times 10^6 \text{ ft}^3\text{/d}$$

$$R = 1543 \text{ ft} \cdot \text{lb/(lb} \cdot \text{mole)} \cdot {}^\circ\text{R}$$

$$T = 460 + 90 = 550^\circ\text{R}$$

$$n = \frac{p_v V}{RT} = \frac{100.8 \times (3.26 \times 10^6)}{1543 \times 550} = 387.2 \text{ lb} \cdot \text{mole/d}$$

(*c*) Determine the pounds of water vapor present in the landfill gas at the blower.

$$\text{lb } H_2O\text{/d} = (387.2 \text{ lb} \cdot \text{mole/d}) \times (18 \text{ lb } H_2O\text{/lb} \cdot \text{mole}) = 6969.6 \text{ lb/d}$$

3. Determine the amount of condensed water vapor that must be removed daily.

$$\begin{aligned}\text{Total water vapor condensed} &= \text{lb } H_2O \text{ at well head} - \text{lb } H_2O \text{ at vacuum blower}\\ &= 18{,}457.2 \text{ lb/d} - 6969.6 \text{ lb/d}\\ &= 11{,}487.6 \text{ lb/d}\\ &= 1377.4 \text{ gal/d}\end{aligned}$$

Comment. Because significant amounts of water can be removed with the landfill gas, the capacity of the condensate traps should be constructed to handle at least two days of condensate flow. Depending on the location of the landfill and treatment facilities, larger condensate traps may be required.

Example 11-11 Landfill leachate production. Given the following information, calculate the yearly quantity of leachate produced from a landfill that is to be operated for a period of five years. The calculations should continue until the landfill reaches equilibrium; that is, the amount of water that enters the landfill will equal the amount of water that leaches out. Plot a curve of the yearly leachate production for the landfill. To simplify the calculations, determine the quantity of leachate produced for a surface area of one square yard, then convert the solution to account for the total quantity of waste deposited in the landfill.

1. Waste quantities
   (*a*) Waste deposited per day = 1000 tons
   (*b*) Number of operating days = 300
   (*c*) Waste deposited per year = 6×10^8 lb
2. Waste characteristics
   (*a*) Compacted specific weight of the waste = 1000 lb/yd^3
   (*b*) Initial moisture content of the waste = 20% by mass
   (*c*) The distribution of rapidly and slowly decomposable organic materials in the waste stream is as in Examples 11-2 and 11-6.
   (*d*) Assume no sludge will be deposited with the waste.

3. Landfill characteristics
   (*a*) General
      i. Lift height = 10 ft
      ii. Waste to cover ratio = 5:1 by volume
      iii. Number of lifts = 5 (one corresponding to each year)
   (*b*) Cover material
      i. Soil specific weight = 3000 lb/yd^3 (including moisture)
      ii. Moisture content of the soil is assumed to be at field capacity
   (*c*) Gas production
      i. Gas production: Use the following gas production data to estimate the total quantity of gas produced per lb of total waste deposited from each lift.

	Gas production, ft^3/lb		
End of year	Rapidly decomp.	Slowly decomp.	Total[a]
1	0.000	0.000	0.000
2	0.941	0.008	0.949
3	1.646	0.023	1.669
4	1.176	0.039	1.215
5	0.706	0.055	0.761
6	0.235	0.070	0.305
7	0.000	0.074	0.074
8	0.000	0.066	0.066
9	0.000	0.058	0.058
10	0.000	0.051	0.051
11	0.000	0.043	0.043
12	0.000	0.035	0.035
13	0.000	0.027	0.027
14	0.000	0.019	0.019
15	0.000	0.012	0.012
16	0.000	0.004	0.004
17	0.000	0.000	0.000
Total	4.704	0.584	5.288

[a] Based on a distribution of rapidly and slowly decomposable materials in the waste stream as given in Example 11-8.

      ii. Water consumed in the formation of landfill gas = 0.01 lb/ft^3 of gas produced
      iii. Water present as water vapor in landfill gas = 0.001 lb/ft^3 of gas produced
      iv. Specific weight of landfill gas = 0.0836 lb/ft^3
   (*d*) Field capacity

   Field capacity as a function of the overburden weight is expressed as

$$\mathrm{FC} = 0.6 - 0.55\left[\frac{W}{10{,}000 + W}\right]$$

FC = fraction of water in the waste based on dry weight

W = overburden weight calculated at the midheight of the waste in the lift in question, lb

4. Rainfall quantities
   (*a*) Rainfall that infiltrates the daily cover during the first five years of operation = 4 in/yr
   (*b*) Rainfall that infiltrates the final cover after five years = 1 in/yr

Solution—Part 1: for years 1 through 5

1. Define the elements of the water balance for the first lift. The pertinent definition sketch for the problem follows.

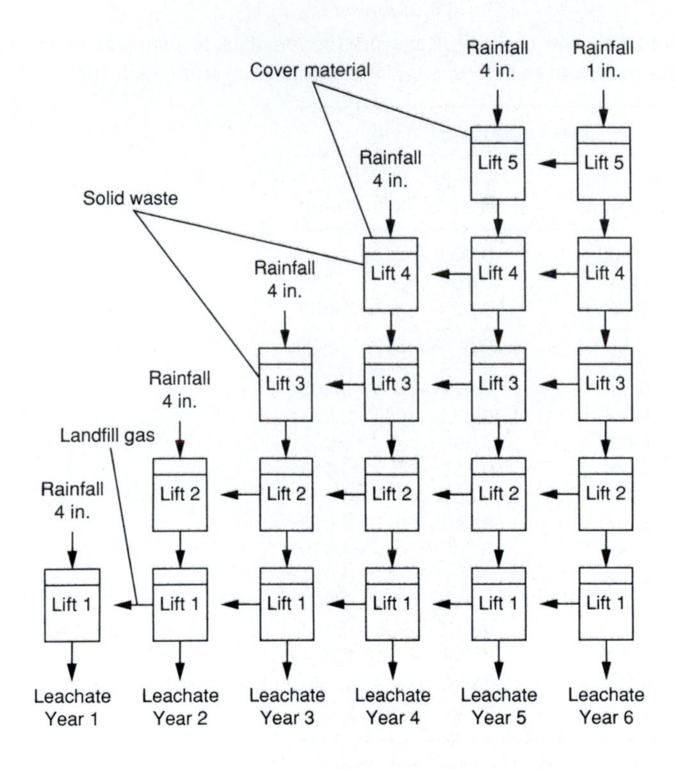

(*a*) Determine the weight of cover material and solid waste in each lift.

$$\text{Weight of cover material} = [3000 \text{ lb/yd}^3 \times (10 \text{ ft} \times 1/6) \times 1.0 \text{ yd}^2]/(3 \text{ ft/yd})$$
$$= 1666.7 \text{ lb}$$
$$\text{Weight of solid waste} = [1000 \text{ lb/yd}^3 \times (10 \text{ ft} \times 5/6) \times 1.0 \text{ yd}^2]/(3 \text{ ft/yd})$$
$$= 2777.8 \text{ lb}$$
$$\text{Total weight of lift} = (1666.7 \text{ lb} + 2777.8 \text{ lb})$$
$$= 4444.5 \text{ lb}$$

(*b*) Dry weight of solid waste = 2777.8 lb × 0.80 = 2222.2 lb
(*c*) Moisture content in solid waste = 2777.8 lb × 0.20 = 555.6 lb

(*d*) Weight of rainfall entering landfill during each of the first five years

$$\text{Rainfall weight} = [4 \text{ in}/(12 \text{ in/ft})] \times 1.0 \text{ yd}^2 \times (9 \text{ ft}^2/\text{yd}^2) \times (62.4 \text{ lb/ft}^3)$$
$$= 187.2 \text{ lb}$$

(*e*) Total weight of lift $= 2777.8 \text{ lb} + 1666.7 \text{ lb} + 187.2 \text{ lb} = 4631.7 \text{ lb}$

2. Prepare a water balance for lift 1 at the end of year 1 and determine the quantity of leachate to be expected from lift 1. The pertinent definition sketch for the first lift is shown above (see Step 1).

(*a*) Determine the amount and weight of gas produced from lift 1 during year 1. Note that gas production does not begin until the end of year 1, that is, it is assumed that no gas is produced in the first year.

$$\text{Gas produced} = 2777.8 \text{ lb} \times 0.0 \text{ ft}^3/\text{lb of waste deposited in lift 1}$$
$$= 0.0 \text{ ft}^3$$
$$\text{Weight of gas produced} = 0.0 \text{ ft}^3 \times 0.0836 \text{ lb/ft}^3$$
$$= 0.0 \text{ lb}$$

(*b*) Determine the weight of water consumed in the production of the landfill gas.

$$\text{Weight of water consumed} = 0.0 \text{ ft}^3 \times 0.01 \text{ lb/ft}^3 = 0.0 \text{ lb}$$

(*c*) Determine the weight of water vapor in the gas.

$$\text{Weight of water vapor} = 0.0 \text{ ft}^3 \times 0.001 \text{ lb/ft}^3 = 0.0 \text{ lb}$$

(*d*) Determine the weight of water in the solid waste in lift 1.

$$\text{Weight of water} = 555.6 \text{ lb} + 187.2 \text{ lb (from rainfall)} = 742.8 \text{ lb}$$

(*e*) Determine the dry weight of solid waste remaining in lift 1 at the end of year 1.

$$\text{Dry weight of solid waste} = 2222.2 \text{ lb} - [0.0 \text{ lb (landfill gas)} - 0.0 \text{ lb (water consumed in conversion reaction)}]$$
$$= 2222.2 \text{ lb}$$

(*f*) Determine the average weight on the waste placed in lift 1. (Note: the average weight in lift 1 will occur at the midpoint of the waste in lift 1.)

$$\text{Average weight} = 0.5 \times (2222.2 \text{ lb} + 742.8 \text{ lb}) + 1666.7 \text{ lb} = 3149.2 \text{ lb}$$

(*g*) Determine the field capacity factor using the following equation:

$$\text{FC} = 0.6 - 0.55 \frac{W}{10{,}000 + W}$$

$$\text{FC} = 0.6 - 0.55 \frac{3149.2}{10{,}000 + 3149.2} = 0.468$$

(*h*) Determine the amount of water that can be held in the solid waste.

$$\text{Water held in solid waste in lift 1} = 0.468 \times 2222.2 \text{ lb} = 1040 \text{ lb}$$

(*i*) Determine amount of leachate formed.

$$\text{Leachate formed} = \text{actual water in solid waste} - \text{field capacity of solid waste}$$

$$\text{Leachate formed} = 742.8 \text{ lb} - 1040 \text{ lb} = -297.2 \text{ lb}$$

Because the field capacity of the waste is greater than the actual amount of water present in the waste, no leachate will form.

(*j*) Determine the amount of water remaining in lift 1 at the end of year 1.

$$\text{Water remaining} = 742.8 - 0 = 742.8 \text{ lb}$$

(*k*) Determine the total weight of lift 1 at the end of year 1.

$$\text{Total weight of lift} = \text{dry waste} + \text{water remaining} + \text{cover}$$

$$= 2222.2 \text{ lb} + 742.8 \text{ lb} + 1666.7 \text{ lb} = 4631.7 \text{ lb}$$

3. Prepare a water balance for lifts 1 and 2 at the end of year 2 and determine the quantity of leachate to be expected from the first lift. The pertinent definition sketch for lifts 1 and 2 is shown above (see Step 1). Note that the computations for lift 2 in year 2 = the computations for lift 1 in year 1.

(*a*) Determine the amount and weight of gas produced from lift 1 during year 2.

$$\text{Gas produced} = 2777.8 \text{ lb} \times 0.949 \text{ ft}^3/\text{lb of waste deposited in lift 1}$$

$$= 2636.1 \text{ ft}^3$$

$$\text{Weight of gas produced} = 2636.1 \text{ ft}^3 \times 0.0836 \text{ lb/ft}^3 = 220.4 \text{ lb}$$

(*b*) Determine the weight of water consumed in the production of the landfill gas. Also note that the weight of solid waste that is consumed in the reaction is included in the weight of the gas determined in Step 3*a* above.

$$\text{Weight of water consumed} = 2636.1 \text{ ft}^3 \times 0.01 \text{ lb/ft}^3 = 26.4 \text{ lb}$$

(*c*) Determine the weight of water vapor in the gas.

$$\text{Weight of water vapor} = 2636.1 \text{ ft}^3 \times 0.001 \text{ lb/ft}^3 = 2.6 \text{ lb}$$

(*d*) Determine the weight of water in the solid waste in lift 1 at the end of year 2.

$$\text{Weight of water} = 742.8 \text{ lb} - 26.4 \text{ lb} - 2.6 \text{ lb} = 713.8 \text{ lb}$$

(*e*) Determine the dry weight of solid waste remaining in lift 1 at the end of year 2.

$$\text{Dry weight of solid waste} = 2222.2 \text{ lb} - (220.4 \text{ lb} - 26.4 \text{ lb})$$

$$= 2028.2 \text{ lb}$$

(*f*) Determine the average weight on the waste placed in lift 1.

$$\text{Average weight} = 4631.7 \text{ lb (lift 2)} + 0.5 \times (2028.2 \text{ lb} + 713.8 \text{ lb}) + 1666.7 \text{ lb}$$

$$= 7669.4 \text{ lb}$$

(*g*) Determine the field capacity factor.

$$\text{FC} = 0.6 - 0.55\frac{7669.4}{10{,}000 + 7669.4} = 0.361$$

(*h*) Determine the amount of water that can be held in the solid waste.

$$\text{Water held in solid waste in lift 1} = 0.361 \times 2028.3 \text{ lb} = 732.8 \text{ lb}$$

(*i*) Determine amount of leachate formed.

$$\text{Leachate formed} = 713.8 \text{ lb} - 732.8 \text{ lb} = -18.9 \text{ lb}$$

Because the field capacity of the waste is greater than the actual amount of water present in the waste, no leachate will form.

(*j*) Determine the amount of water remaining in lift 1 at the end of year 2.

$$\text{Water remaining} = 713.8 - 0 = 713.8 \text{ lb}$$

(*k*) Determine the total weight of lift 1 at the end of year 2.

$$\begin{aligned}\text{Total weight of lift} &= \text{dry waste} + \text{water remaining} + \text{cover}\\ &= 2028.2 \text{ lb} + 713.8 \text{ lb} + 1666.7 \text{ lb} = 4408.8 \text{ lb}\end{aligned}$$

4. Prepare a water balance for lifts 1, 2, and 3 at the end of year 3 and determine the quantity of leachate to be expected from lift 1. The pertinent definition sketch for lifts 1, 2, and 3 is shown above (see Step 1). Note that lift 3 = lift 2 and lift 2 = lift 1 in year 2.

(*a*) Determine the amount and weight of gas produced from lift 1 at the end of year 3.

$$\begin{aligned}\text{Gas produced} &= 2777.8 \text{ lb} \times 1.67 \text{ ft}^3/\text{lb of waste deposited in lift 1}\\ &= 4638.9 \text{ ft}^3\end{aligned}$$

$$\text{Weight of gas produced} = 4638.9 \text{ ft}^3 \times 0.0836 \text{ lb/ft}^3 = 387.8 \text{ lb}$$

(*b*) Determine the weight of water consumed in the production of the landfill gas. Also note that the weight of solid waste that is consumed in the reaction is included in the weight of the gas determined in Step 4*a* above.

$$\text{Weight of water consumed} = 4638.9 \text{ ft}^3 \times 0.01 \text{ lb/ft}^3 = 46.4 \text{ lb}$$

(*c*) Determine the weight of water vapor in the gas.

$$\text{Weight of water vapor} = 4638.9 \text{ ft}^3 \times 0.001 \text{ lb/ft}^3 = 4.6 \text{ lb}$$

(*d*) Determine the weight of water in the solid waste in lift 1 at the end of year 3.

$$\text{Weight of water} = 713.8 \text{ lb} - 46.4 \text{ lb} - 4.6 \text{ lb} = 662.8 \text{ lb}$$

(*e*) Determine the dry weight of solid waste remaining in lift 1 at the end of year 3.

$$\begin{aligned}\text{Dry weight of solid waste} &= 2028.3 \text{ lb} - (387.8 \text{ lb} - 46.4 \text{ lb})\\ &= 1686.9 \text{ lb}\end{aligned}$$

(*f*) Determine average weight on the waste placed in lift 1.

$$\begin{aligned}\text{Average weight} &= 4631.7 \text{ lb (lift 3)} + 4408.8 \text{ lb (lift 2)} \\ &\quad + 0.5 \times (1686.9 \text{ lb} + 662.8 \text{ lb}) + 1666.7 \text{ lb} \\ &= 11{,}882.0 \text{ lb}\end{aligned}$$

(*g*) Determine the field capacity factor.

$$FC = 0.6 - 0.55\frac{11{,}882}{10{,}000 + 11{,}882} = 0.301$$

(*h*) Determine the amount of water that can be held in the solid waste.

$$\text{Water held in solid waste in lift 1} = 0.301 \times 1686.9 \text{ lb} = 508.3 \text{ lb}$$

(*i*) Determine the amount of leachate formed.

$$\text{Leachate formed} = 662.8 - 508.3 \text{ lb} = 154.5 \text{ lb}$$

Because the field capacity of the waste is less than the actual amount of water present in the waste, leachate will be formed.

(*j*) Determine the amount of water remaining in lift 1 at the end of year 3.

$$\text{Water remaining} = (662.8 - 154.5) \text{ lb} = 508.3 \text{ lb}$$

(*k*) Determine the total weight of lift 1 at the end of year 3.

$$\begin{aligned}\text{Total weight of lift} &= \text{dry waste} + \text{water remaining} + \text{cover} \\ &= 1686.9 \text{ lb} + 508.3 \text{ lb} + 1666.7 \text{ lb} = 3861.9 \text{ lb}\end{aligned}$$

5. Prepare a water balance for lifts 1, 2, 3, and 4 at the end of year 4 and determine the quantity of leachate to be expected from the lift 1. The pertinent definition sketch for lift 4 is shown above (see Step 1). Note that lift 4 = lift 3, lift 3 = lift 2, and lift 2 = lift 1 in year 3. Also note the amount of water discharged from lift 3 to lift 4 = 154.5 lb.

(*a*) Determine the amount and weight of gas produced from lift 1 at the end of year 4.

$$\begin{aligned}\text{Gas produced} &= 2777.8 \text{ lb} \times 1.215 \text{ ft}^3\text{/lb of waste deposited in lift 1} \\ &= 3374.8 \text{ ft}^3\end{aligned}$$

$$\text{Weight of gas produced} = 3374.8 \text{ ft}^3 \times 0.0836 \text{ lb/ft}^3 = 282.1 \text{ lb}$$

(*b*) Determine the weight of water consumed in the production of the landfill gas. Also note that the weight of solid waste that is consumed in the reaction is included in the weight of the gas determined in Step 5*a* above.

$$\text{Weight of water consumed} = 3374.8 \text{ ft}^3 \times 0.01 \text{ lb/ft}^3 = 33.7 \text{ lb}$$

(*c*) Determine the weight of water vapor in the gas.

$$\text{Weight of water vapor} = 3374.8 \text{ ft}^3 \times 0.001 \text{ lb/ft}^3 = 3.3 \text{ lb}$$

(*d*) Determine the weight of water in the solid waste in lift 1 at the end of year 4. It should be noted that the initial amount of water remaining in lift 1 is equal to the field capacity determine in Step 4*h* above.

$$\text{Weight of water} = (508.3 \text{ lb} - 33.7 \text{ lb} - 3.3 \text{ lb}) + 154.5 \text{ lb (leachate from lift 3)}$$
$$= 625.7 \text{ lb}$$

(*e*) Determine the dry weight of solid waste remaining in lift 1 at the end of year 4.

$$\text{Dry weight of solid waste} = 1686.9 \text{ lb} - (282.1 \text{ lb} - 33.7 \text{ lb})$$
$$= 1438.5 \text{ lb}$$

(*f*) Determine the average weight on the waste placed in lift 1.

$$\text{Average weight} = 4631.7 \text{ lb (lift 4)} + 4408.8 \text{ lb (lift 3)} + 3861.9 \text{ lb (lift 2)}$$
$$+ [0.5 \times (1438.5 \text{ lb} + 625.7 \text{ lb}) + 1666.7 \text{ lb}] = 15{,}601.2 \text{ lb}$$

(*g*) Determine the field capacity factor.

$$\text{FC} = 0.6 - 0.55 \frac{15{,}601}{10{,}000 + 15{,}601} = 0.265$$

(*h*) Determine the amount of water that can be held in the solid waste.

$$\text{Water held in solid waste in lift 1} = 0.265 \times 1438.5 \text{ lb} = 381.0 \text{ lb}$$

(*i*) Determine amount of leachate formed.

$$\text{Leachate formed} = (625.7 - 381.0) \text{ lb} = 244.7 \text{ lb}$$

Because the field capacity of the waste is less than the actual amount of water present in the waste, leachate will be formed.

(*j*) Determine the amount of water remaining in lift 1 at the end of year 4.

$$\text{Water remaining} = (625.7 - 244.7) \text{ lb} = 381.0 \text{ lb}$$

(*k*) Determine the total weight of lift 1 at the end of year 4.

$$\text{Total weight of lift} = \text{dry waste} + \text{water remaining} + \text{cover}$$
$$= 1438.5 \text{ lb} + 381.0 \text{ lb} + 1666.7 \text{ lb}$$
$$= 3486.2 \text{ lb}$$

6. Prepare a water balance for lifts 1, 2, 3, 4, and 5 at the end of year 5 and determine the quantity of leachate to be expected from lift 1. The pertinent definition sketch for the lifts 1, 2, 3, 4, and 5 is shown above (see Step 1). Note that lift 5 = lift 4, lift 4 = lift 3, lift 3 = lift 2, and lift 2 = lift 1 in year 4. Also note the amount of water discharged from lift 3 to lift 2 = 244.7 lb and lift 2 to lift 1 = 154.5 lb.

(*a*) Determine the amount and weight of gas produced from lift 1 at the end of year 5.

$$\text{Gas produced} = 2777.8 \text{ lb} \times 0.760 \text{ ft}^3/\text{lb of waste deposited in lift 1}$$
$$= 2111.4 \text{ ft}^3$$
$$\text{Weight of gas produced} = 2111.4 \text{ ft}^3 \times 0.0836 \text{ lb/ft}^3 = 176.5 \text{ lb}$$

(*b*) Determine the weight of water consumed in the production of the landfill gas. Also note that the weight of solid waste that is consumed in the reaction is included in the weight of the gas determined in Step 6*a* above.

$$\text{Weight of water consumed} = 2111.4 \text{ ft}^3 \times 0.01 \text{ lb/ft}^3 = 21.1 \text{ lb}$$

(*c*) Determine the weight of water vapor in the gas.

$$\text{Weight of water vapor} = 2111.4\ \text{ft}^3 \times 0.001\ \text{lb/ft}^3 = 2.1\ \text{lb}$$

(*d*) Determine the weight of water in the solid waste in lift 1. Note that the initial amount of water remaining in lift 1 is equal to the field capacity determined in Step 5*h* above.

$$\begin{aligned}\text{Weight of water} &= (381.0 - 21.1 - 2.1)\ \text{lb} + 244.7\ \text{lb (from lift 2)}\\ &= 602.4\ \text{lb}\end{aligned}$$

(*e*) Determine the dry weight of solid waste remaining in lift 1 at the end of year 5.

$$\begin{aligned}\text{Dry weight of solid waste} &= 1438.5\ \text{lb} - (176.5 - 21.1)\ \text{lb}\\ &= 1283.1\ \text{lb}\end{aligned}$$

(*f*) Determine average weight on the waste placed in lift 1.

$$\begin{aligned}\text{Average weight} &= 4631.7\ \text{lb (lift 5)} + 4408.8\ \text{lb (lift 4)}\\ &\quad + 3861.9\ \text{lb (lift 3)} + 3486.2\ \text{lb (lift 2)}\\ &\quad + [0.5 \times (1283.1\ \text{lb} + 602.4\ \text{lb}) + 1666.7\ \text{lb}]\\ &= 18{,}998.0\ \text{lb}\end{aligned}$$

(*g*) Determine the field capacity factor.

$$\text{FC} = 0.6 - 0.55\frac{18{,}998}{10{,}000 + 18{,}998} = 0.240$$

(*h*) Determine the amount of water that can be held in the solid waste.

$$\text{Water held in solid waste in lift 1} = 0.240 \times 1283.1\ \text{lb} = 307.5\ \text{lb}$$

(*i*) Determine amount of leachate formed.

$$\text{Leachate formed} = (602.4 - 307.5)\ \text{lb} = 294.9\ \text{lb}$$

Because the field capacity of the waste is less than the actual amount of water present in the waste, leachate will be formed.

(*j*) Determine the amount of water remaining in lift 1 at the end of year 5.

$$\text{Water remaining} = (602.4 - 294.9)\ \text{lb} = 307.5\ \text{lb}$$

(*k*) Determine the total weight of lift 1 at the end of year 5.

$$\text{Total weight of lift} = \text{dry waste} + \text{water remaining} + \text{cover}$$

$$\text{Total weight of lift} = (1283.1 + 307.5 + 1666.7)\ \text{lb} = 3257.3\ \text{lb}$$

Solution—Part 2: for year 6 and following years. The weight of rainfall entering the landfill starting with year 6 is

$$\begin{aligned}\text{Rainfall weight} &= [1\ \text{in}/(12\ \text{in/ft})] \times 1.0\ \text{yd}^2 \times 9\ \text{ft}^2/\text{yd}^2 \times (62.4\ \text{lb/ft}^3)\\ &= 46.8\ \text{lb}\end{aligned}$$

To determine the leachate released from lift 1 each lift must be considered for each year. The analysis for year 6, which is the same for subsequent years, is illustrated below.

1. Determine the leachate from lift 5 in year 6.

(*a*) Determine the amount and weight of gas produced from lift 5 at the end of year 6 (see Part 1, Step 3*a*).

$$\text{Gas produced} = 2777.8 \text{ lb} \times 0.949 \text{ ft}^3/\text{lb of waste deposited in lift 5}$$
$$= 2635.0 \text{ ft}^3$$
$$\text{Weight of gas produced} = 2635.0 \text{ ft}^3 \times 0.0836 \text{ lb/ft}^3 = 220.3 \text{ lb}$$

(*b*) Determine the weight of water consumed in the production of the landfill gas. Note that the waste consumed is included in the weight of the gas determined in Step 3*a* above.

$$\text{Weight of water consumed} = 2635.0 \text{ ft}^3 \times 0.01 \text{ lb/ft}^3 = 26.3 \text{ lb}$$

(*c*) Determine the weight of water vapor in the gas.

$$\text{Weight of water vapor} = 2635.0 \text{ ft}^3 \times 0.001 \text{ lb/ft}^3 = 2.6 \text{ lb}$$

(*d*) Determine the weight of water in the solid waste in lift 5. Note that the calculations for lift 5 in year 5 correspond to the calculations for lift 1 in year 1 (See Part 1, Step 2*j*).

$$\text{Weight of water} = 742.8 \text{ lb} - 26.3 \text{ lb} - 2.6 \text{ lb} + 46.8 \text{ lb (from rainfall)}$$
$$= 760.6 \text{ lb}$$

(*e*) Determine the dry weight of solid waste remaining in lift 5.

$$\text{Dry weight of solid waste} = 2222.2 \text{ lb} - (220.3 - 26.3) \text{ lb}$$
$$= 2028.3 \text{ lb}$$

(*f*) Determine the average weight on the waste placed in lift 5.

$$\text{Average weight} = [0.5 \times (2028.3 \text{ lb} + 760.6 \text{ lb}) + 1666.7 \text{ lb}] = 3061.1 \text{ lb}$$

(*g*) Determine the field capacity factor.

$$\text{FC} = 0.6 - 0.55 \frac{3061.1}{10{,}000 + 3061.1} = 0.471$$

(*h*) Determine the amount of water that can be held in the solid waste.

$$\text{Water held in solid waste in lift 5} = 0.471 \times 2028.3 \text{ lb} = 955.5 \text{ lb}$$

(*i*) Determine the amount of leachate formed.

$$\text{Leachate formed} = (760.6 - 955.5) \text{ lb} = -194.9 \text{ lb}$$

Because the field capacity of the waste is greater than the actual amount of water present in the waste, no leachate will be formed.

(*j*) Determine the amount of water remaining in lift 5 at the end of year 6.

$$\text{Water remaining} = (760.6 - 0) \text{ lb} = 760.6 \text{ lb}$$

(*k*) Determine the total weight of lift 5 at the end of year 6.

$$\text{Total weight of lift 5} = \text{dry waste} + \text{water remaining} + \text{cover}$$
$$= 2028.3 \text{ lb} + 760 \text{ lb} + 1666.7 \text{ lb} = 4455.6 \text{ lb}$$

2. Determine the leachate from lift 4 in year 6.
   (*a*) Determine the amount and weight of gas produced from lift 4 at the end of year 6 (see Part 1, Step 4*a*).

$$\text{Gas produced} = 2777.8 \text{ lb} \times 1.67 \text{ ft}^3/\text{lb of waste deposited in lift 4}$$
$$= 4638.3 \text{ ft}^3$$

$$\text{Weight of gas produced} = 4638.3 \text{ ft}^3 \times 0.0836 \text{ lb/ft}^3 = 387.8 \text{ lb}$$

   (*b*) Determine the weight of water consumed in the production of the landfill gas.

$$\text{Weight of water consumed} = 4638.3 \text{ ft}^3 \times 0.01 \text{ lb/ft}^3 = 46.4 \text{ lb}$$

   (*c*) Determine the weight of water vapor in the gas.

$$\text{Weight of water vapor} = 4638.3 \text{ ft}^3 \times 0.001 \text{ lb/ft}^3 = 4.6 \text{ lb}$$

   (*d*) Determine the weight of water in the solid waste in lift 4 (see Part 1, Step 3*j*).

$$\text{Weight of water} = (713.8 - 46.4 - 4.6) \text{ lb} = 662.8 \text{ lb}$$

   (*e*) Determine the dry weight of solid waste remaining in lift 4.

$$\text{Dry weight of solid waste} = 2028.3 \text{ lb} - (387.8 - 46.4) \text{ lb}$$
$$= 1686.9 \text{ lb}$$

   (*f*) Determine the average weight on the waste placed in lift 4.

$$\text{Average weight} = 4455.6 \text{ lb (lift 5)} + [0.5 \times (1686.9 \text{ lb} + 662.8 \text{ lb}) + 1666.7 \text{ lb}]$$
$$= 7297.1 \text{ lb}$$

   (*g*) Determine the field capacity factor.

$$\text{FC} = 0.6 - 0.55 \frac{7297.1}{10{,}000 + 7297.1} = 0.368$$

   (*h*) Determine the amount of water that can be held in the solid waste.

$$\text{Water held in solid waste in lift 4} = 0.368 \times 1686.9 \text{ lb} = 620.7 \text{ lb}$$

   (*i*) Determine the amount of leachate formed.

$$\text{Leachate formed} = (662.8 - 620.7) \text{ lb} = 42.1 \text{ lb}$$

   Because the field capacity of the waste is less than the actual amount of water present in the waste, leachate will be formed.

   (*j*) Determine the amount of water remaining in lift 4 at the end of year 6.

$$\text{Water remaining} = (662.8 - 42.1) \text{ lb} = 620.7 \text{ lb}$$

   (*k*) Determine the total weight of lift 4 at the end of year 6.

$$\text{Total weight of lift 4} = \text{dry waste} + \text{water remaining} + \text{cover}$$
$$= 1686.9 \text{ lb} + 620.7 \text{ lb} + 1666.7 \text{ lb} = 3974.3 \text{ lb}$$

3. Determine the leachate from lift 3 in year 6.

(*a*) Determine the amount and weight of gas produced from lift 3 at the end of year 6 (see Part 1, Step 5*a*).

$$\text{Gas produced} = 2777.8 \text{ lb} \times 1.215 \text{ ft}^3/\text{lb of waste deposited in lift 3}$$
$$= 3374.8 \text{ ft}^3$$
$$\text{Weight of gas produced} = 3374.8 \text{ ft}^3 \times 0.0836 \text{ lb/ft}^3 = 282.1 \text{ lb}$$

(*b*) Determine the weight of water consumed in the production of the landfill gas.

$$\text{Weight of water consumed} = 3374.8 \text{ ft}^3 \times 0.01 \text{ lb/ft}^3 = 33.7 \text{ lb}$$

(*c*) Determine the weight of water vapor in the gas.

$$\text{Weight of water vapor} = 3374.8 \text{ ft}^3 \times 0.001 \text{ lb/ft}^3 = 3.4 \text{ lb}$$

(*d*) Determine the weight of water in the solid waste in lift 3 (see Part 1, Step 4*j*).

$$\text{Weight of water} = 508.3 \text{ lb} - 33.7 \text{ lb} - 3.7 \text{ lb} + 42.1 \text{ lb}$$
(leachate from above)
$$= 513.3 \text{ lb}$$

(*e*) Determine the dry weight of solid waste remaining in lift 4.

$$\text{Dry weight of solid waste} = 1686.9 \text{ lb} - (282.1 \text{ lb} - 33.7 \text{ lb})$$
$$= 1438.5 \text{ lb}$$

(*f*) Determine the average weight on the waste placed in lift 3.

$$\text{Average weight} = 4455.5 \text{ lb (lift 5)} + 3974.3 \text{ lb (lift 4)}$$
$$+ [0.5 \times (1438.5 \text{ lb} + 513.3 \text{ lb}) + 1666.7 \text{ lb}]$$
$$= 11{,}072.5 \text{ lb}$$

(*g*) Determine the field capacity factor.

$$\text{FC} = 0.6 - 0.55 \frac{11{,}072.5}{10{,}000 + 11{,}072.5} = 0.311$$

(*h*) Determine the amount of water that can be held in the solid waste.

$$\text{Water held in solid waste in lift 3} = 0.311 \times 1438.5 \text{ lb} = 447.4 \text{ lb}$$

(*i*) Determine the amount of leachate formed.

$$\text{Leachate formed} = (513.3 - 447.4) \text{ lb} = 65.9 \text{ lb}$$

Because the field capacity of the waste is less than the actual amount of water present in the waste, leachate will be formed.

(*j*) Determine the amount of water remaining in lift 3 at the end of year 6.

$$\text{Water remaining} = (513.3 - 65.9) \text{ lb} = 447.4 \text{ lb}$$

(*k*) Determine the total weight of lift 3 at the end of year 6.

$$\text{Total weight of lift 3} = \text{dry waste} + \text{water remaining} + \text{cover}$$
$$= 1438.5 \text{ lb} + 447.4 \text{ lb} + 1666.7 \text{ lb} = 3552.6 \text{ lb}$$

4. Determine the leachate from lift 2 in year 6.
   (*a*) Determine the amount and weight of gas produced from lift 2 at the end of year 6 (see Part 1, Step 6*a*).

$$\text{Gas produced} = 2777.8 \text{ lb} \times 0.760 \text{ ft}^3/\text{lb of waste deposited in lift 2}$$
$$= 2111.4 \text{ ft}^3$$
$$\text{Weight of gas produced} = 2111.4 \text{ ft}^3 \times 0.0836 \text{ lb/ft}^3 = 176.5 \text{ lb}$$

   (*b*) Determine the weight of water consumed in the production of the landfill gas.

$$\text{Weight of water consumed} = 2111.4 \text{ ft}^3 \times 0.01 \text{ lb/ft}^3 = 21.1 \text{ lb}$$

   (*c*) Determine the weight of water vapor in the gas.

$$\text{Weight of water vapor} = 2111.4 \text{ ft}^3 \times 0.001 \text{ lb/ft}^3 = 2.1 \text{ lb}$$

   (*d*) Determine the weight of water in the solid waste in lift 2 (see Part 1, Step 5*j*).

$$\text{Weight of water} = 381.0 \text{ lb} - 21.1 \text{ lb} - 2.1 \text{ lb} + \underset{\text{(leachate from above)}}{65.9 \text{ lb}}$$
$$= 423.6 \text{ lb}$$

   (*e*) Determine the dry weight of solid waste remaining in lift 4.

$$\text{Dry weight of solid waste} = 1438.5 \text{ lb} - (176.5 - 21.1) \text{ lb} = 1283.1 \text{ lb}$$

   (*f*) Determine the average weight on the waste placed in lift 2.

$$\text{Average weight} = 4455.5 \text{ lb (lift 5)} + 3974.3 \text{ lb (lift 4)} + 3552.6 \text{ lb (lift 3)}$$
$$+ [0.5 \times (1283.1 \text{ lb} + 423.6 \text{ lb}) + 1666.7 \text{ lb}] = 14{,}502.5 \text{ lb}$$

   (*g*) Determine the field capacity factor.

$$FC = 0.6 - 0.55 \frac{14{,}502.5}{10{,}000 + 14{,}502.5} = 0.274$$

   (*h*) Determine the amount of water that can be held in the solid waste.

$$\text{Water held in solid waste in lift 2} = 0.274 \times 1283.1 \text{ lb} = 352.2 \text{ lb}$$

   (*i*) Determine the amount of leachate formed.

$$\text{Leachate formed} = (423.6 - 352.2) \text{ lb} = 71.5 \text{ lb}$$

   Because the field capacity of the waste is less than the actual amount of water present in the waste, leachate will be formed.

   (*j*) Determine the amount of water remaining in lift 2 at the end of year 6.

$$\text{Water remaining} = (423.6 - 71.5) \text{ lb} = 352.2 \text{ lb}$$

   (*k*) Determine the total weight of lift 2 at the end of year 6.

$$\text{Total weight of lift 2} = \text{dry waste} + \text{water remaining} + \text{cover}$$
$$= 1283.1 \text{ lb} + 423.6 \text{ lb} + 1666.7 \text{ lb} = 3302.0 \text{ lb}$$

5. Determine the leachate from lift 1 in year 6.

(*a*) Determine the amount and weight of gas produced from lift 1 at the end of year 6.

$$\text{Gas produced} = 2777.8 \text{ lb} \times 0.305 \text{ ft}^3/\text{lb of waste deposited in lift 1}$$
$$= 848.0 \text{ ft}^3$$

$$\text{Weight of gas produced} = 848.0 \text{ ft}^3 \times 0.0836 \text{ lb/ft}^3 = 70.9 \text{ lb}$$

(*b*) Determine the weight of water consumed in the production of the landfill gas.

$$\text{Weight of water consumed} = 848.0 \text{ ft}^3 \times 0.01 \text{ lb/ft}^3 = 8.5 \text{ lb}$$

(*c*) Determine the weight of water vapor in the gas.

$$\text{Weight of water vapor} = 848.0 \text{ ft}^3 \times 0.001 \text{ lb/ft}^3 = 0.8 \text{ lb}$$

(*d*) Determine the weight of water in the solid waste in lift 3 (see Part 1, Step 6*j*).

$$\text{Weight of water} = 307.5 \text{ lb} - 8.5 \text{ lb} - 0.8 \text{ lb} + 71.5 \text{ lb (leachate from above)}$$
$$= 369.7 \text{ lb}$$

(*e*) Determine the dry weight of solid waste remaining in lift 1.

$$\text{Dry weight of solid waste} = 1283.1 \text{ lb} - (70.9 - 8.5) \text{ lb}$$
$$= 1220.7 \text{ lb}$$

(*f*) Determine average weight on the waste placed in lift 1.

$$\text{Average weight} = 4455.5 \text{ lb (lift 5)} + 3974.3 \text{ lb (lift 4)} + 3552.6 \text{ lb (lift 3)}$$
$$+ 3302.0 \text{ lb (lift 2)} + [0.5 \times (1220.7 \text{ lb} + 369.7 \text{ lb})$$
$$+ 1666.7 \text{ lb}] = 17{,}746.3 \text{ lb}$$

(*g*) Determine the field capacity factor.

$$FC = 0.6 - 0.55 \frac{17{,}746.3}{10{,}000 + 17{,}746.3} = 0.248$$

(*h*) Determine the amount of water that can be held in the solid waste.

$$\text{Water held in solid waste in lift 1} = 0.248 \times 1220.7 \text{ lb} = 303.0 \text{ lb}$$

(*i*) Determine the amount of leachate formed.

$$\text{Leachate formed} = (369.7 - 303.0) \text{ lb} = 66.7 \text{ lb}$$

Because the field capacity of the waste is less than the actual amount of water present in the waste, leachate will be formed.

(*j*) Determine the amount of water remaining in lift 1 at the end of year 6.

$$\text{Water remaining} = (368.7 - 66.7) \text{ lb} = 303.0 \text{ lb}$$

(*k*) Determine the total weight of lift 1 at the end of year 6.

$$\text{Total weight of lift 1} = \text{dry waste} + \text{water remaining} + \text{cover}$$
$$= 1220.7 \text{ lb} + 303.0 \text{ lb} + 1666.7 \text{ lb} = 3190.4 \text{ lb}$$

Solution—Part 3: estimate total leachate quantities

1. Determine the total number of square yards occupied by the landfill.
   (*a*) The total weight of solid waste placed in a landfill lift that is one yard square and 10 ft high = 2777.8 lb.
   (*b*) The total area occupied by each lift expressed in square yards is

$$\text{Total area} = (6 \times 10^8 \text{ lb/yr})/(2777.8 \text{ lb/yd}^2 \cdot \text{yr})$$
$$= 216{,}000 \text{ yd}^2$$

2. Determine the conversion factor to convert the lb of leachate obtained per square yard to gals/yr for the entire landfill.

$$\text{Conversion factor} = (\text{lb/yd}^2 \cdot \text{yr} \times 216{,}000 \text{ yd}^2)/(8.34 \text{ lb/gal})$$
$$= \text{lb/yd}^2 \cdot \text{yr} \times 25{,}900$$
$$= \text{gal/yr}$$

3. Prepare a summary table of the total leachate quantities to be expected with time and plot the results. The required data are summarized in the following table and illustrated in the figure presented below.

Leachate production

	Total	
Year	lb/yd²	10^6 gal
1	0.0	0.0
2	0.0	0.0
3	154.5	4.00
4	244.7	6.34
5	294.9	7.64
6	66.7	1.73
7	43.0	1.11
8	93.0	2.41
9	94.9	2.46
10	65.5	1.70
11	55.7	1.44
12	55.1	1.43
13	53.9	1.40
14	52.7	1.37
15	51.5	1.33
16	50.3	1.30
17	49.1	1.27
18	48.2	1.25
19	47.4	1.23
20	46.9	1.22
21	46.8	1.21
22	46.8	1.21

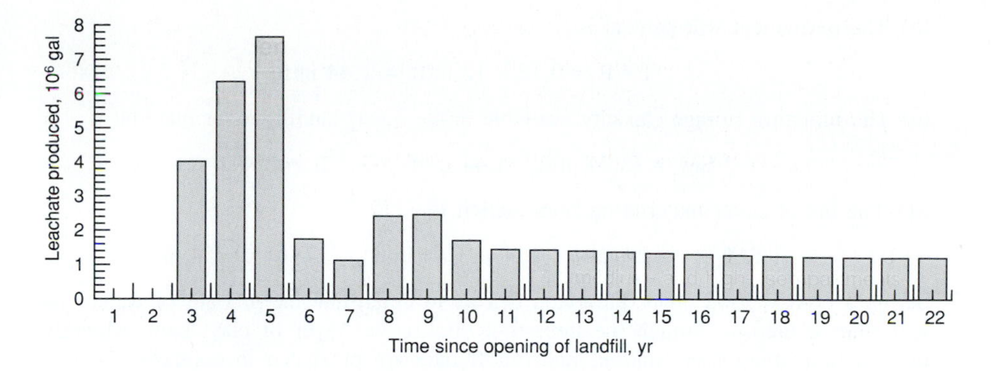

Comment. The spreadsheet solution developed above allows computation of the quantities of leachate for any expected quantity of surface infiltration and for varying rates of gas production.

Example 11-12 Estimation of water percolation rates through a landfill cover. Determine the amount of water that will enter a landfill if a three-foot thick layer of clay loam is used for the final cover. Make the following assumptions: (1) The following rainfall and evapotranspiration data are applicable to the landfill site. (2) The average monthly runoff coefficient is equal to 20 percent. (3) The cover material is a clay loam with the physical characteristics given in Table 11-20. (4) The moisture content of the cover material is 50 percent of field capacity.

Month	Precipitation, in	Evapotranspiration, in
January	4.5	0.7
February	3.5	1.5
March	3.0	3.1
April	2.4	3.9
May	1.6	5.2
June	0.5	6.5
July	0.1	7.0
August	Trace	6.5
September	0.2	4.4
October	0.6	3.9
November	2.6	1.5
December	3.9	0.8
Total annual	22.9	45.0

Solution

1. Determine the water storage capacity in the cover material using the data given in Table 11-20.

   (*a*) The field capacity of the cover material in inches is

$$\mathrm{FC} = 0.27 \times 12 \text{ in/ft} = 3.24 \text{ in/ft}$$

(*b*) The permanent wilt percent is

$$\text{PWP} = 0.12 \times 12 \text{ in/ft} = 1.44 \text{ in/ft}$$

(*c*) The moisture storage capacity available in the 3.0 ft landfill cover material is

$$\text{SM} = (3.24 \text{ in/ft} - 1.44 \text{ in/ft}) \times 3.0 \text{ ft} = 5.4 \text{ in}$$

(*d*) The initial cover material moisture deficit is

$$\text{SM}_d = (3.24 \text{ in/ft} \times 0.50 - 1.44 \text{ in/ft}) \times 3.0 \text{ ft} = 0.54 \text{ in}$$

2. Set up a computation table (presented below) to determine the amount of water that will enter a landfill through the three-foot thick cover layer of clay loam. Monthly precipitation, evapotranspiration, and runoff data are presented in columns (2), (3), and (4), respectively. The potential gain or loss of soil moisture from a unit volume of cover material is given in column (5). The cover material moisture deficit is given in column (6). The amount of water that potentially can percolate through the landfill cover is given in column (7).

	Value, in					
Month (1)	Precipitation (2)	Evapo-transpiration (3)	Runoff (4)	Moisture gain (+) or loss (−), (5)[a]	Cover material deficit (6)	Potential percolate through cover (7)
January	4.50	0.70	0.90	2.90	0.00	2.36[b]
February	3.50	1.50	0.70	1.30	0.00	1.30
March	3.00	3.10	0.60	−0.70	−0.70	0.00
April	2.40	3.90	0.48	−1.98	−2.68	0.00
May	1.60	5.20	0.36	−3.96	−5.40[c]	0.00
June	0.50	6.50	0.10	−6.10	−5.40	0.00
July	0.10	7.00	—	−6.90	−5.40	0.00
August	Trace	6.50	—	−6.50	−5.40	0.00
September	0.20	4.40	—	−4.20	−5.40	0.00
October	0.60	3.90	0.12	−3.42	−5.40	0.00
November	2.60	1.50	0.52	0.58	−4.82	0.00
December	3.90	0.80	0.78	2.32	−2.50	0.00
January	4.50	0.70	0.90	2.90	0.00	0.40
February	3.50	1.50	0.70	1.30	0.00	1.30
March	3.00	3.10	0.60	−0.70	−0.70	0.00
April	2.40	3.90	0.48	−1.98	−2.68	0.00
May	1.60	5.20	0.36	−3.96	−5.40	0.00

[a] (5) = (2) − (3) − (4)

[b] 2.36 = 2.90 − 0.54 (initial cover material deficit)

[c] 5.40 = maximum moisture storage capacity available in the cover material

Comment. The slope and depth of the landfill cover could be increased to limit the amount of rainfall that percolates through the landfill cover. Also, the runoff coefficient in this example was assumed to be constant. In practice the runoff coefficient will vary with antecedent conditions.

Example 11-13 Landfill compaction during operation and long-term compaction/consolidation. Given the following information, compute the additional capacity available in the landfill of Example 11-11 after five years as a result of compaction and gas production. Also estimate the actual height of the landfill at the end of year 5 and the long-term settlement of the landfill after closure. The calculations should continue until the landfill reaches equilibrium.

Use the same data as given in Example 11-11. Assume the initial compacted specific weight of the waste is 1000 lb/yd^3 and that the following relationship can be used to estimate the specific weight of the compacted waste as a function of the overburden pressure. Assume no compaction of the cover material.

$$SW_p = 1000 \text{ lb/yd}^3 + \frac{p, \text{ lb/in}^2}{0.0133 \text{ (yd}^3\text{/lb)(lb/in}^2) + (0.001 \text{ yd}^3\text{/lb)}(p, \text{ lb/in}^2)}$$

where SW_p = compacted specific weight of the waste at pressure p, lb/yd^3

Solution—Part 1: estimate the additional landfill capacity after five years

1. Calculate the height of each lift and of the cover material between the lifts. Use the specific weight and pressure at the middle of each lift to approximate the density and pressure for the whole lift.

   (*a*) Determine the height of the fifth lift.

   i. The total amount of waste in lift five at the end of year five is 4631.7 lb/yd^2 (waste = 2965.0 and cover material = 1666.7, see Example 11-11, Part 1, Step 1). The pressure at the midpoint of the lift can be calculated.

$$p = \left(1666.7 \text{ lb} + \frac{2965.0 \text{ lb}}{2}\right)\left(\frac{1}{\text{yd}^2}\right)\left(\frac{1 \text{ yd}}{3 \text{ ft}}\right)^2\left(\frac{1 \text{ ft}}{12 \text{ in}}\right)^2 = 2.43 \text{ lb/in}^2$$

   ii. The specific weight is related to the pressure by the equation given in the problem statement.

$$SW_p = 1000 \text{ lb/yd}^3 + \frac{p}{0.0133 + 0.001\ p}$$

$$SW_p = 1000 \text{ lb/yd}^3 + \frac{2.43 \text{ lb/in}^2}{0.0133 \text{ (yd}^3\text{/ft}^2\text{)(lb/in}^2) + 0.001 \text{ yd}^3\text{/lb (2.43 lb/in}^2)}$$

$$= 1154.5 \text{ lb/yd}^3$$

   iii. Estimate the height h of the waste material in lift 5 at the end of year 5. The height is related to the amount of the initial material remaining in the lift at the end of the year including water additions or losses and the average specific weight in the lift.

$$\text{Material remaining in lift, lb} = SW_p\left(\frac{\text{lb}}{\text{yd}^3}\right)\left(1 \text{ yd}^2 \times h \text{ ft} \times \frac{\text{yd}}{3 \text{ ft}}\right)$$

$$2965.0 \text{ lb} = 1154.4\left(\frac{\text{lb}}{\text{yd}^3}\right) h \text{ (ft)} \ \frac{\text{yd}^3}{3 \text{ ft}}$$

$$h = 7.70 \text{ ft}$$

iv. Estimate the total height of lift 5 at the end of year 5. Note that, because it is assumed that there is no decomposition or compression of the cover material over time, this value is the same for each lift each year.

$$\text{Height of cover material} = 10 \text{ ft } (1/6) = 1.67 \text{ ft}$$
$$h_{\text{total}} = 7.70 + 1.67 \text{ ft} = 9.37 \text{ ft}$$

(*b*) Determine the height of the fourth lift.

i. The total amount of waste in lift 4 at the end of year 5 is 4408.8 lb/yd^2 (2742.1 = waste, and 1666.7 = cover material). The pressure at the midpoint of the lift can be calculated.

$$p = \left(4631.7 + 1666.7 + \frac{2742.1}{2}\right)\left(\frac{\text{lb}}{\text{yd}^2}\right)\left(\frac{1 \text{ yd}}{3 \text{ ft}}\right)^2\left(\frac{1 \text{ ft}}{12 \text{ in}}\right)^2 = 5.92 \text{ lb/in}^2$$

ii. Determine the specific weight from the pressure.

$$SW_p = 1000 \text{ lb/yd}^3 + \frac{5.92 \text{ lb/in}^2}{0.0133 \text{ (yd}^3\text{/lb)(lb/in}^2) + 0.001 \text{ yd}^3\text{/lb } (5.92 \text{ lb/in}^2)}$$
$$= 1307.9 \text{ lb/yd}^3$$

iii. Estimate the height of the waste material in lift 4 at the end of year 5.

$$2742.1 \text{ lb} = 1307.9\left(\frac{\text{lb}}{\text{yd}^3}\right)\left(h \text{ (ft)} \frac{\text{yd}^3}{3 \text{ ft}}\right)$$
$$h = 6.29 \text{ ft}$$

iv. Estimate the total height of lift 4 at the end of year 5.

$$h_{\text{total}} = (6.29 + 1.67) \text{ ft} = 7.96 \text{ ft}$$

(*c*) Determine the height of the third lift.

i. The total amount of waste in lift 3 at the end of year 5 is 3861.9 lb/yd^2 (2195.2 = waste, and 1666.7 = cover material). The pressure at the midpoint of the lift can be calculated.

$$p = \left(4631.7 + 4408.8 + 1667.7 + \frac{2195.2}{2}\right)\left(\frac{\text{lb}}{\text{yd}^2}\right)\left(\frac{1 \text{ yd}}{3 \text{ ft}}\right)^2\left(\frac{1 \text{ ft}}{12 \text{ in}}\right)$$
$$= 9.11 \text{ lb/in}^2$$

ii. Determine the specific weight from the pressure.

$$SW_p = 1000 \text{ lb/yd}^3 + \frac{9.11 \text{ lb/in}^2}{0.0133 \text{ yd}^3\text{/ft}^2 + 0.001 \text{ yd}^3\text{/lb } (9.11 \text{ lb/in}^2)}$$
$$= 1406.5 \text{ lb/yd}^3$$

iii. Estimate the height of the waste material in lift 3 at the end of year 5.

$$2195.2 \text{ lb} = 1406.5\left(\frac{\text{lb}}{\text{yd}^3}\right)\left(h \text{ (ft)} \frac{\text{yd}^3}{3 \text{ ft}}\right)$$
$$h = 4.68 \text{ ft}$$

iv. Estimate the total height of lift 3 at the end of year 5.

$$h_{\text{total}} = (4.68 + 1.67)\ \text{ft} = 6.35\ \text{ft}$$

(*d*) Determine the height of the second lift.

i. The total amount of waste in lift 2 at the end of year 5 is 3486.2 lb/yd^2 (1819.5 = waste, and 1666.7 = cover material). The pressure at the midpoint of the lift can be calculated.

$$p = \left(4631.7 + 4408.8 + 3861.9 + 1666.7 + \frac{1819.5}{2}\right)\left(\frac{\text{lb}}{\text{yd}^2}\right)\left(\frac{1\ \text{yd}}{3\ \text{ft}}\right)^2\left(\frac{1\ \text{ft}}{12\ \text{in}}\right)^2$$
$$= 11.94\ \text{lb/in}^2$$

ii. Determine the specific weight from the pressure.

$$SW_p = 1000\ \text{lb/yd}^3 + \frac{11.9\ \text{lb/in}^2}{0.0133\ (\text{yd}^3/\text{lb})(\text{lb/in}^2) + 0.001\ \text{yd}^3/\text{lb}\ (11.9\ \text{lb/in}^2)}$$
$$= 1473.1\ \text{lb/yd}^3$$

iii. Estimate the height of the waste material in lift 3 at the end of year 5.

$$1819.5\ \text{lb} = 1473.1\ \frac{\text{lb}}{\text{yd}^3}\left(h\ (\text{ft})\ \frac{\text{yd}^3}{3\ \text{ft}}\right)$$
$$h = 3.71\ \text{ft}$$

iv. Estimate the total height of lift 2 at the end of year 5.

$$h_{\text{total}} = (3.71 + 1.67)\ \text{ft} = 5.38\ \text{ft}$$

(*e*) Determine the total height of the first lift.

i. The total amount of waste in lift 1 at the end of year 5 is 3257.3 lb/yd^2 (1590.6 = waste, and 1666.7 = cover material). The pressure at the midpoint of the lift can be calculated.

$$p = \left(4631.7 + 4408.8 + 3861.9 + 3486.2 + 1666.7 + \frac{1590.6}{2}\right)$$
$$\times\left(\frac{\text{lb}}{\text{yd}^2}\right)\left(\frac{1\ \text{yd}}{3\ \text{ft}}\right)^2\left(\frac{1\ \text{ft}}{12\ \text{in}}\right)^2 = 14.6\ \text{lb/in}^2$$

ii. Determine the specific weight from the pressure.

$$SW_p = 1000 + \frac{14.6\ \text{lb/in}^2}{0.0133\ (\text{yd}^3/\text{lb})(\text{lb/in}^2) + 0.001\ \text{yd}^3/\text{lb}\ (14.6\ \text{lb/in}^2)}$$
$$= 1522.4\ \text{lb/yd}^3$$

iii. Estimate the height of the waste material in lift 1 at the end of year 5.

$$1590.6\ \text{lb} = 1522.4\left(\frac{\text{lb}}{\text{yd}^3}\right)\left(h\ (\text{ft})\ \frac{\text{yd}^3}{3\ \text{ft}}\right)$$
$$h = 3.13\ \text{ft}$$

iv. Estimate the total height of lift 1 at the end of year 5.

$$h_{\text{total}} = (3.13 + 1.67)\ \text{ft} = 4.80\ \text{ft}$$

2. Estimate the additional capacity available in the landfill at the end of year 5.
   (*a*) Estimate the total height of the landfill at the end of year 5.

$$h_{total} = 4.80 \text{ ft} + 5.38 \text{ ft} + 6.35 \text{ ft} + 7.96 \text{ ft} + 9.37 \text{ ft} = 33.87 \text{ ft}$$

   (*b*) Estimate the additional capacity of the landfill. The landfill has 50.00 ft − 33.87 ft = 16.13 ft, or

$$\frac{16.13}{33.87} \times 100\% = 47.6\%$$

   more waste could be placed in the landfill.
   (*c*) Estimate the additional amount of waste that could be placed in the landfill. Note the amount of waste placed per lift is 2777.8 lb (Problem 11-11, Step 1).

$$0.476 \times 5(2777.8 \text{ lb/yd}^2) = 6611.2 \text{ lb/yd}^2$$

   For a landfill of 216,000 yd^2, the additional amount of waste which could be placed is

$$216{,}000 \text{ yd}^2 \times 6611.2 \text{ lb/yd}^2 = 1.43 \times 10^9 \text{ lb}$$

Solution—Part 2: long-term compaction of the landfill. To determine the long-term compaction of the landfill, each lift must be considered for each year. The analysis for year 6 and for subsequent years is the same as the analysis for year 5.

Prepare a summary table of the total height of each lift and of the total height of the landfill and plot the results. The required data are summarized in the following table and illustrated in the figure presented on page 529.

Year after opening of landfill	Height of lift, ft					
	Lift 1	Lift 2	Lift 3	Lift 4	Lift 5	h_{total}, ft
1						
2						
3						
4						
5	4.80	5.38	6.35	7.96	9.37	33.87
6	4.70	5.04	5.74	7.01	8.94	31.43
7	4.71	4.93	5.37	6.31	8.08	29.40
8	4.71	4.93	5.25	5.87	7.29	28.04
9	4.69	4.92	5.23	5.71	6.72	27.27
10	4.67	4.89	5.20	5.68	6.49	26.94
11	4.66	4.87	5.17	5.64	6.43	26.77
12	4.64	4.85	5.15	5.60	6.38	26.62
13	4.63	4.83	5.13	5.57	6.34	26.50
14	4.62	4.82	5.11	5.55	6.30	26.39
15	4.62	4.81	5.09	5.52	6.26	26.31
16	4.62	4.81	5.08	5.51	6.24	26.25
17	4.62	4.81	5.08	5.50	6.22	26.22
18	4.62	4.81	5.08	5.49	6.20	26.19
19	4.62	4.81	5.08	5.49	6.19	26.18
20	4.62	4.81	5.08	5.49	6.19	26.18
21	4.62	4.81	5.08	5.49	6.19	26.18
22	4.62	4.81	5.08	5.49	6.19	26.18

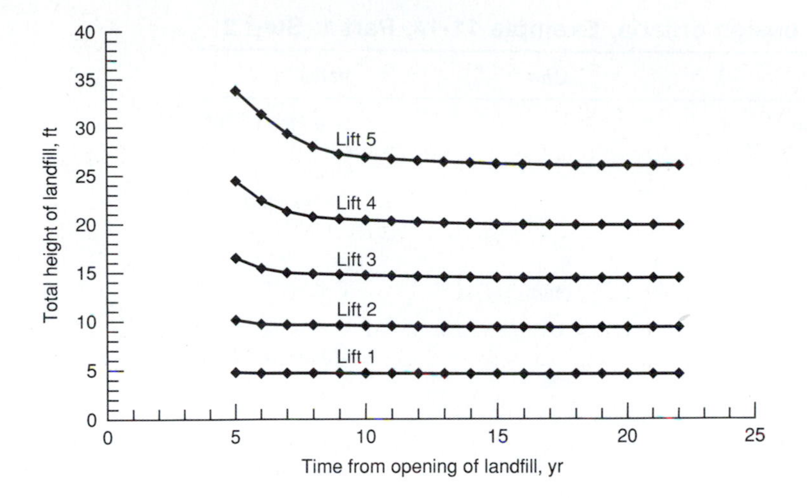

Example 11-14 Selection of a landfill leachate collection system and cover configuration. Select appropriate design criteria for a leachate collection system and landfill cover design for a municipal solid waste landfill. Assume the municipality has requested that a composite liner and cover design be used. The liner is to be composed of clay and a geomembrane with a drainage layer, and the cover is to incorporate the use of a drainage layer and a geomembrane. Also check to see if enough leachate can be transported through the drainage layer to the leachate collection channel to accommodate the leachate from the landfill in Example 11-9.

Solution—Part 1: leachate collection system

1. To meet the requirements specified by the municipality, a composite liner design of the type shown in Fig. 11-36*b* was selected.
2. Based on the information presented in Section 11-5, the appropriate design criteria for the liner configuration are reported in the table on page 530.
3. Check drainage capacity of liner cross slope.
   (*a*) Estimate drainage capacity of section of liner cross slope as shown below by considering a strip 3 ft wide by 200 ft in length with a slope of 1 percent.
   (*b*) Estimate the hydraulic capacity of the cross slope using Darcy's law. Assume the permeability of the combined drainage and filter layer is the same as coarse sand (1333 ft/d, see Table 11-15) and that the equivalent thickness of the combined drainage layer, composed of the drainage and filter layers, is 0.30 in.
   The quantity of leachate transmitted is determined using Darcy's law,

$$Q = Av = -AKi$$

where A = area, $\text{ft}^2 = [0.30 \text{ in}/(12 \text{ in/ft})] \times 3 \text{ ft} = 0.075 \text{ ft}^2$
K = permeability, ft/d = 1333 ft/d
$i = -dh/dl$ = slope = -0.01

$$Q = -0.075 \text{ ft}^2 \times 1333 \text{ ft/d} \times (-0.01)$$
$$= 1.0 \text{ ft}^3\text{/d} = 7.48 \text{ gal/d}$$

Landfill liner design criteria, Example 11-14, Part 1, Step 2

Item	Unit	Value
Liner configuration		See Fig. 11-36*b*
Subbase		
Material		Native soil
Treatment		Compacted
Clay layer		
Thickness	ft	2[a]
Permeability	cm/s	1×10^{-7}
Geomembrane		
Material		Polyethylene
Thickness	mil	80
Drainage layer		
Material		Polyethylene
Configuration		Cross weave
Thickness	in	0.25
Filter layer		
Material		Polyethylene
Thickness	in	0.25
Protective layer		
Material		Native soil
Thickness	ft	2
Liner design		
Length of cross slope	ft	200
Cross slope	%	1
Drainage channels	%	0.5

[a] To be applied in 6 in (150 mm) lifts.

(*c*) Compare the amount of leachate that can be transmitted with the actual quantity of leachate generated.

i. The maximum amount of leachate generated per yd^2 occurs at the end of year 5 and is equal to $282 \text{ lb/yd}^2 \cdot \text{yr} = 33.9 \text{ gal/yd}^2 \cdot \text{yr}$.

ii. The maximum quantity of leachate generated from the quantity of waste placed on an area of 66.7 yd^2 $[(3 \text{ ft} \times 200 \text{ ft})/(9 \text{ ft}^2/\text{yd}^2)]$ is equal to $33.9 \text{ gal/yd}^2 \cdot \text{yr} \times 66.7 \text{ yd}^2 = 2261 \text{ gal/yr} = 6.19 \text{ gal/d}$.

iii. Because the quantity of leachate that can be transmitted per day (7.48 gal/d) is greater than the amount of leachate generated per day (6.19 gal/d) the capacity of the cross slope is adequate. It should also be noted that after year 6 the quantity of leachate drops to 25 percent of the value used in these computations.

Solution—Part 2: cover design

1. Assume the cover will be a composite design as shown in Fig. 11-53*a*.
2. Based on the information presented in Sections 11-4 and 11-6, the appropriate design criteria for the landfill cover are reported in the following table.

Landfill cover design criteria, Example 11-14, Part 2, Step 2

Item	Unit	Value
Cover configuration		See Fig. 11-53*a*
Soil layer		
Material		Native soil
Thickness	ft	2
Permeability	cm/s	1×10^{-7}
Geomembrane		
Material		Polyethylene
Thickness	mil	80
Drainage layer		
Material		Sand
Thickness	ft	1.0
Separation layer		
Material		Geotextile filter fabric
Thickness (approx.)	in	0.125
Final earth cover		
Material		Native soil
Thickness	ft	2
Cover design		
Top slope	%	5
Maximum distance to drainage channels	ft	200

Comment. Although more sophisticated approaches are available for computing the hydraulic capacity of the drainage layer, the reliability of the computations is no better than the approach used in this example problem.

11-13 DISCUSSION TOPICS AND PROBLEMS

11-1. Estimate the theoretical amount of gas (methane and carbon dioxide) that could be produced under anaerobic conditions from wastes with the following chemical composition: (*a*) $C_{12}H_{22}O_{11}$ (sucrose), (*b*) $C_2H_5O_2N$ (glycine), and (*c*) $C_{60}H_{96}O_{38}N$.

11-2. Estimate the emission rates, expressed as $g/m^2 \cdot d$, for carbon dioxide and methane from the surface of a landfill due to diffusion alone. Assume the following conditions apply:

(*a*) Temperature = 30°C

(*b*) Landfill cover material = clay-loam mixture

(*c*) Porosity of landfill cover material = 0.23

(*d*) Landfill cover thickness = 2 ft

(*e*) Coefficient of diffusion for methane = $0.20\ cm^2/s$ ($18.6\ ft^2/d$)

(*f*) Coefficient of diffusion for carbon dioxide = $0.13\ cm^2/s$ ($12.1\ ft^2/d$)

(*g*) Note that $(g/cm^2 \cdot s) \times 0.864 \times 10^9 = g/m^2 \cdot d$

11-3. A 50-ft deep sanitary landfill in alluvial gravel has been completed for several years. The normal groundwater level is 150 ft below the surface, or 100 ft below the bottom of the fill. A special sampling well at the edge of the landfill shows that the atmosphere in the interstices of the soil 20 ft above the water table contains 48 percent CO_2, 28 percent CH_4, 20 percent N_2, 2 percent O_2, 1 percent H_2S, and 1 percent other gases, analyzed and calculated on a dry basis at 0°C and 760 mm pressure. On the basis of a long period of contact (i.e., equilibrium) at 10°C, compute the concentration in mg/L of each of these five gases to be expected in the upper layers of the groundwater under a total pressure of 1 atm at 10°C. Assume saturation with respect to vapor pressure. (Problem courtesy of Dr. Paul H. King.)

11-4. Using Henry's law (see Appendix F), estimate the maximum concentrations of methane and carbon dioxide that would be present in the leachate if the partial pressure of each gas within the landfill was equal to one atmosphere.

11-5. If the bicarbonate concentration of a leachate is 1000 mg/L and the pH measured in the field is found to be 5.8, estimate the partial pressure of the carbon dioxide within the landfill.

11-6. If the partial pressure of the carbon dioxide in the gas in contact with leachate within a landfill is one atmosphere, estimate the pH of the leachate. Assume the carbon dioxide in the landfill gas is the only factor affecting the pH of the leachate.

11-7. Determine the breakthrough time in years for leachate to penetrate a 4-foot thick clay liner. Assume the effective porosity is 0.20, the coefficient of permeability is 10^{-7} cm/s, and the hydraulic head is 6 ft.

11-8. If the breakthrough time for leachate to penetrate a 3-foot thick clay liner is 12 years, estimate the coefficient of permeability given that the effective porosity is 0.20 and the hydraulic head is 5.5 ft.

11-9. Determine the effect of a 10°C rise in temperature on the rate of percolation of leachate through a clay liner. Assume the coefficient of permeability is equal to 1×10^{-6} cm/s.

11-10. What thickness of clay liner would be required if the breakthrough time for leachate to penetrate the liner is to be 20 years? The coefficient of permeability is 5×10^{-8} cm/s and the effective porosity is 0.17. Assume the hydraulic head is 1 ft greater than the thickness of the liner.

11-11. If MSW with the composition given in Table 3-4 is to be mixed with wastewater treatment plant sludge containing 5 percent solids to achieve a final moisture content of 55 percent, estimate the ultimate amount of leachate that would be produced per cubic yard of compacted solid waste if no surface infiltration were allowed to enter the completed landfill. Assume that the following data and information are applicable:

(*a*) Initial moisture content of MSW = 20 percent

(*b*) In-place specific weight of compacted mixture of solid wastes and sludge = 1200 lb/yd^3

(*c*) Chemical formula for decomposable portion of the organic fraction of the MSW = $C_{60}H_{96}O_{38}N$

(*d*) Sixty-five percent of the organic fraction of the MSW is biodegradable

(*e*) Assume the biodegradable portion of the organic wastes will be converted according to Eq. (11-2)

(*f*) Final moisture content of wastes remaining in landfill = 35 percent

(*g*) Neglect surface evaporation

11-12. Review the current literature, and prepare a brief (two-page) assessment of the need for the use of soil as an intermediate cover material. In your review you should question whether any intermediate cover material is required.

11-13. Contact your local waste management agency and obtain the designs of the landfill liner and final cover used for landfills under their jurisdiction. Based on what you have read in this chapter and other literature sources, what is your assessment of the liner and cover designs that are being used?

11-14. Determine the compacted specific weight of the wastes in the landfill in Example 11-3 if 80 percent of the yard wastes are removed for composting.

11-15. Develop a spreadsheet program for the solution of Example 11-5 in Section 11-9 for wastes that are both normally and well compacted. Check your spreadsheet program using the results provided in the example. Once your spreadsheet program is working, determine the in-place specific weight assuming normal compaction of a waste with a composition given in the following table. Waste A, B, C, D, or E will be selected by your instructor.

	Percentage distribution by weight				
Component	**A**	**B**	**C**	**D**	**E**
Organic					
Food wastes	60	50	30	15	8
Paper	4	8	20	35	34
Cardboard	2	2	4	4	6
Plastics	2	4	6	6	9
Textiles	1	1	2	2	2
Rubber	—	0.5	0.5	0.5	0.5
Leather	—	0.5	0.5	0.5	0.5
Yard wastes	2	8	10	15	19.5
Wood	—	1	1	1	1.0
Misc. organics	—	—	—	—	—
Inorganic					
Glass	1	2	4	8	8
Tin cans	1	1	4	6	6.0
Aluminum	—	—	—	0	0.5
Other metal	—	—	1	1	2.0
Dirt, ash, etc.	27	22	17	6	3.0

11-16. Assuming that the asymptotic value for the compaction data given below is 2200 lb/yd^3, derive empirical equations to describe the degree of compaction that can be achieved as a function of the applied pressure, starting with initial specific weights of 750 and 1100 lb/yd^3.

Pressure, lb/in^2	Specific weight, lb/yd^3 1	2
0	750	1100
50	1400	1670
100	1740	1950
150	1950	2090
250	2120	2160

11-17. Given the following values, estimate the volumetric capacity of an excavated cell or area landfill using Fig. A or of a canyon landfill using Fig. B (see page 535 for both figures), expressed in cubic yards, that can be constructed within the property line of the figure (type of landfill to be selected by your instructor) subject to the following constraints and those listed on the figures:

(*a*) Slope of all landfill faces = 3 to 1

(*b*) Lift height = 10 to 12 ft

(*c*) Cover material will be excavated from the site

11-18. How many cubic yards of waste can be placed on a regulation soccer field subject to the following constraints?

(*a*) Slope of all landfill faces = 3 to 1

(*b*) Cover material will be excavated from the site

(*c*) Neglect cover material requirements

How many soccer fields would be required per year to dispose of the wastes from your community, assuming the in-place specific weight of the waste is 1000 lb/yd^3? State your assumptions clearly.

11-19. If the average final specific weight of the waste placed in the landfill of Problem 11-18 is equal to 1200 lb/yd^3, estimate the amount of as-delivered waste (expressed in cubic yards), with an average specific weight of 500 lb/yd^3, that can be placed in the landfill assuming that 30 percent of the original waste (dry basis) will be lost through the production of landfill gas.

11-20. If the average final specific weight of the waste placed in the landfill of Problem 11-16 is equal to 1400 lb/yd^3, estimate (1) the amount of as-delivered waste (expressed in cubic yards), with an average specific weight of 500 lb/yd^3, that can be placed in the landfill if 30 percent of the original waste (dry basis) will be lost through the production of landfill gas, and (2) the amount of as-delivered waste (expressed in cubic yards), with an average specific weight of 500 lb/yd^3, that can be placed in the landfill if 40 percent of the waste is to be composted and used as intermediate cover. Assume that 15 percent of the weight of the waste placed directly in the landfill will be lost as a result of the production of landfill gas. Also assume that the weight of the composted materials will be reduced by 45 percent as a result of the composting process.

11-21. Prepare a lift diagram and determine the volumetric capacity of the North Valley landfill disposal site in Example 11-7. Assume the final elevation of the landfill is to be 290 ft and that the required setback distance from the property line is 200 ft.

11-22. Using the data from Example 11-7, estimate the useful life of the South Valley landfill if 40 percent of the MSW is diverted to produce compost that is to be used as intermediate cover. Assume the waste that is diverted is primarily residential and commercial MSW and that the amount of the material removed from the diverted waste (both recyclable materials and contaminants) before the waste is shredded is equal to 22 percent by weight. Assume the weight reduction achieved in the composting process is 50 percent.

11-23. Using the results from Problem 11-21, estimate the useful life of the North Valley landfill if 50 percent of the MSW is diverted to produce compost that is to be used as intermediate cover. Assume the waste that is diverted is primarily residential and commercial MSW and that the amount of the material removed from the diverted waste (both recyclable materials and contaminants) before the waste is shredded

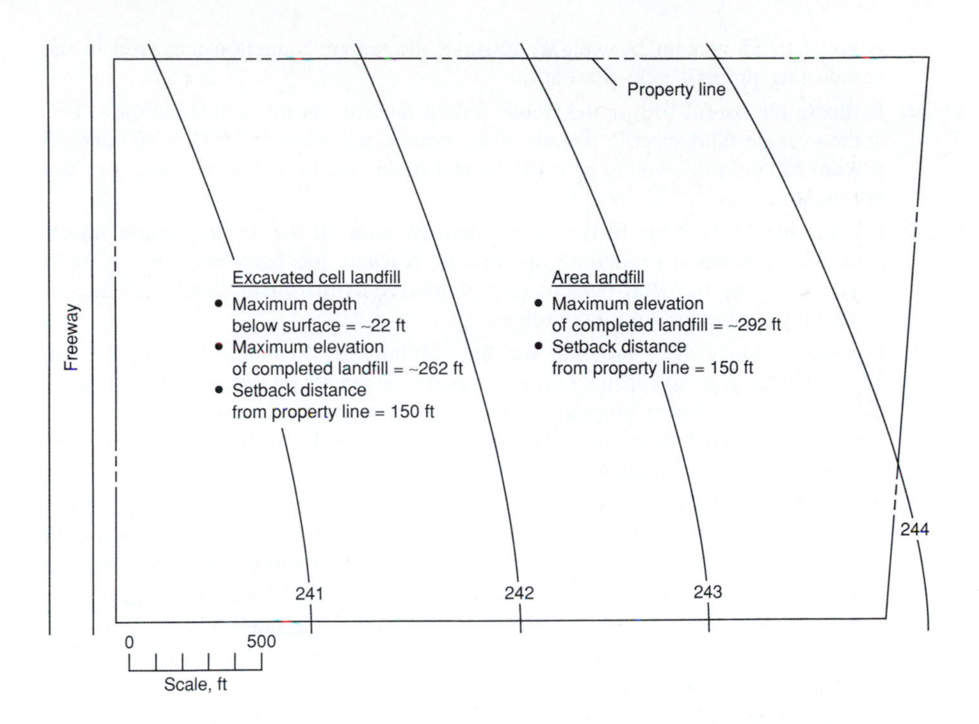

Property line
Excavated cell landfill
• Maximum depth below surface = ~22 ft
• Maximum elevation of completed landfill = ~262 ft
• Setback distance from property line = 150 ft
Area landfill
• Maximum elevation of completed landfill = ~292 ft
• Setback distance from property line = 150 ft
Freeway
241
242
243
244
0
500
Scale, ft

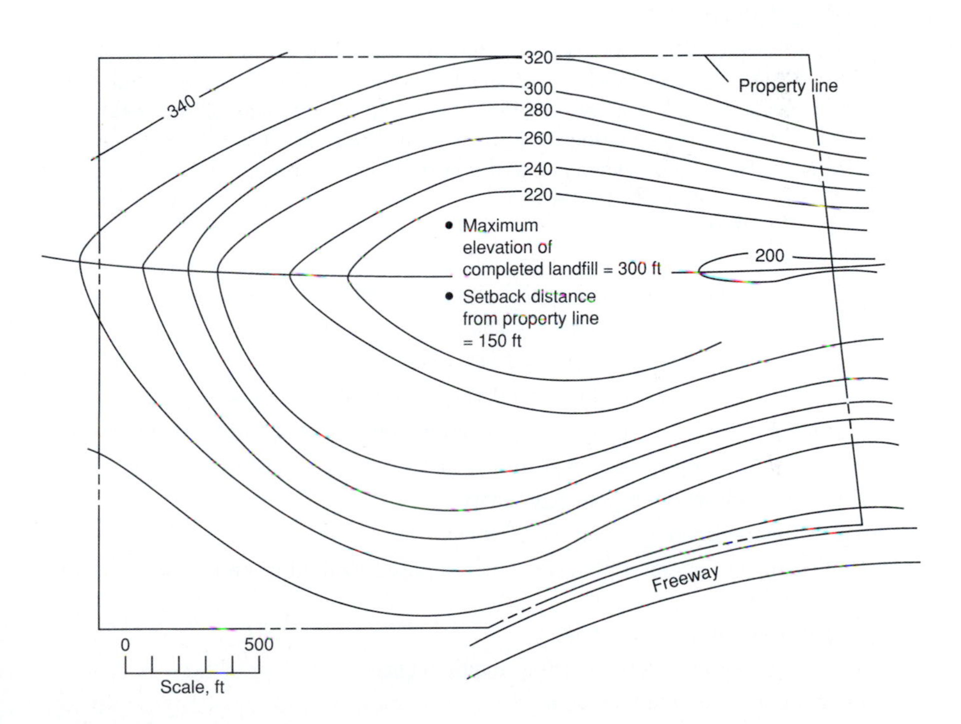

340
320
300
280
260
240
220
200
Property line
• Maximum elevation of completed landfill = 300 ft
• Setback distance from property line = 150 ft
Freeway
0
500
Scale, ft

is equal to 35 percent by weight. Assume the weight reduction achieved in the composting process is 50 percent.

11-24. Estimate the useful life of the South Valley landfill, as given in Example 11-7, if the average final specific weight of the compacted waste is 1500 lb/yd^3 and 35 percent of the initial weight of waste placed in the landfill is lost as a result of gas production.

11-25. In Example 11-7, if the final in-place density, after all the decomposable wastes have been converted to landfill gas and the leachate has been removed, is 1600 lb/yd^3, estimate the total percentage volume reduction. State clearly all the assumptions used in solving this problem.

11-26. Develop a spreadsheet program for the solution for Example 11-8 in Section 11-12. Check your spreadsheet program using the results provided in the example. Once your spreadsheet program is working, determine the amount of gas that would be expected from the waste A, B, C, D, or E (to be selected by your instructor) given in Problem 11-15.

11-27. Assuming that the curves shown in Fig. 11-55 can be approximated by a first-order equation, estimate the surface settlement after 10 yr in a well-compacted sanitary landfill (use maximum compaction curve). What will the maximum surface settlement be after 50 yr? Is the computed result realistic? Discuss.

11-28. Develop a spreadsheet program for the solution for Example 11-11 in Section 11-12. Check your spreadsheet program using the results provided in the example. Once your spreadsheet model is working, determine the amount of leachate that would be expected from waste A, B, C, D, or E (to be selected by your instructor) given in Problem 11-15.

11-29. Develop a spreadsheet program for the solution for Example 11-12 in Section 11-12. Check your spreadsheet using the results provided in the example. Once your spreadsheet program is working, determine the amount of infiltration in a landfill using the precipitation data given in the table on page 537 (A, B, or C to be selected by your instructor). Use the evapotranspiration data given in Example 11-12 with precipitation data A and B. Use the evapotranspiration data given on page 537 with precipitation data C.

11-30. Given the site plan for a parcel of land near the Fallen Oak River (see figure on page 537), prepare a sanitary landfill operation plan for the following conditions:

- Number of collection services = 2800 (average over 20 yr)
- Amount of solid wastes generated per service = 14.0 lb/d
- Compacted specific weight of solid wastes in landfill = 800 lb/yd^3
- Maximum allowable finish grade elevation above surrounding ground = 20 ft
- Slope of all landfill faces = 3 to 1

Include the following in your plan analysis:

(*a*) Required site preparation work, if any

(*b*) Placement operation plan (i.e., the proposed method to be followed in filling the site)

(*c*) Estimated useful life of site

(*d*) Equipment and storage facility requirements

(*e*) Work force and requirements

(*f*) Operational plan

Month	Precipitation, in/mo A	B	C	Evapotranspiration, in/mo
January	7.8	3.0	3.7	3.1
February	7.1	4.5	2.9	3.4
March	6.0	6.5	2.2	4.4
April	3.3	8.0	1.3	5.1
May	1.1	8.6	0.6	6.3
June	1.1	8.6	0.2	7.0
July	1.1	8.4	0.0	7.4
August	1.5	7.8	0.2	6.9
September	4.0	7.2	0.2	5.8
October	5.0	6.5	0.8	4.8
November	5.5	5.6	1.7	3.6
December	7.0	4.8	3.3	3.0
Total annual	50.5	79.5	17.1	60.8

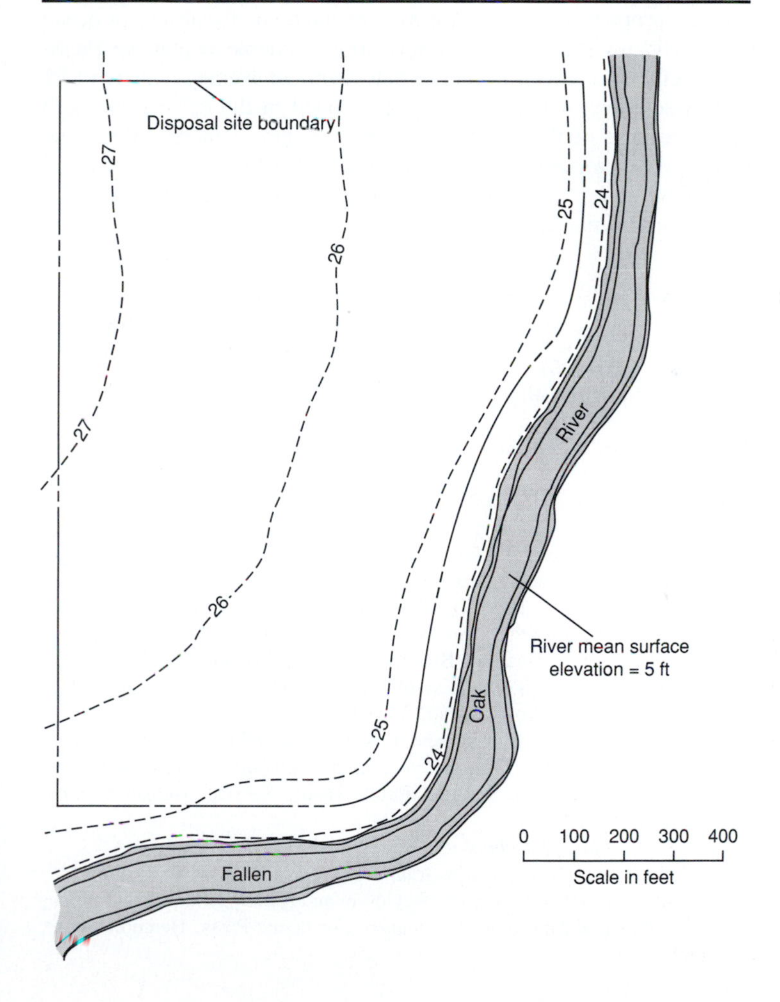

11-31. Given the site plan A or B (to be selected by your instructor) shown in Problem 11-17, prepare a sanitary landfill operation plan for the following conditions:

- Number of collection services = 1500 (average over 20 yr)
- Amount of solid wastes generated per service = 12.0 lb/d
- Compacted specific weight of solid wastes in landfill = 750 lb/yd^3
- Slope of all landfill faces = 3 to 1

Include the following in your plan analysis:

(*a*) Required site preparation work, if any

(*b*) Placement operation plan (i.e., the proposed method to be followed in filling the site)

(*c*) Estimated useful life of site

(*d*) Equipment and storage facility requirements

(*e*) Work force and requirements

(*f*) Operational plan

11-32. On your first day at work for a solid waste consulting organization, your supervisor asks you to prepare a proposal (in outline form) to evaluate the feasibility of ocean dumping of baled solid wastes. The only information available is that the Press-It-Tight Baling Co. claims that it can produce bales with an average density of 78 lb/ft^3 and that if these bales of solid wastes are dumped in the ocean, they will sink to the bottom because of their greater density and remain there, causing no problems. Structure your proposal by asking yourself what kinds of information, data, and criteria would be required to protect the environment and to formulate public policy concerning ocean dumping.

11-14 REFERENCES

1. Amling, W.: "Exclusion of Gulls from Reservoirs in Orange County, California," *Progressive Fish Culturist,* Vol. 43, No. 3, July 1981.
2. Bagchi, A.: *Design, Construction, and Monitoring of Sanitary Landfill,* John Wiley & Sons, New York, 1990.
3. Bohn, H. and R. Bohn: "Soil Beds Weed Out Air Pollutants," *Chemical Engineering,* Vol. 95, No. 6, 1988.
4. Brunner, D. R. and D. J. Keller: *Sanitary Landfill Design and Operation,* Publication SW-65ts, U.S. Environmental Protection Agency, Washington, DC, 1972.
5. California Waste Management Board: *Landfill Gas Characterization,* California Waste Management Board, State of California, Sacramento, October, 1988.
6. Christensen, T. H. and P. Kjeldsen: "2.1. Basic Biochemical Processes in Landfills," in T. H. Christensen, R. Cossu, and P. Stegmann (eds.), *Sanitary Landfilling: Process, Technology and Environmental Impact,* Academic Press, Harcourt Brace, Jovanavich, London, 1989.
7. Cope, F., G. Karpinski, J. Pacey, and L. Stein: "Use of Liners for Containment of Hazardous Waste Landfills," *Pollution Engineering,* Vol. 16, No. 3, 1984.
8. County of Los Angeles, Department of County Engineer, Los Angeles, and Engineering-Science, Inc.: *Development of Construction and Use Criteria for Sanitary Landfills, An Interim Report,* U.S. Department of Health, Education, and Welfare, Public Health Service, Bureau of Solid Waste Management, Cincinnati, OH, 1969.
9. Crawford, J. F. and P. G. Smith: *Landfill Technology,* Butterworth, London, 1985.
10. Davis, S. N. and R. J. M. DeWiest: *Hydrogeology,* John Wiley & Sons, New York, 1966.
11. Ehrig, H. J: "Leachate Quality," in T. H. Christensen, R. Cossu, and P. Stegmann (eds.): *Sanitary Landfilling: Process, Technology and Environmental Impact,* Academic Press, Harcourt Brace, Jovanavich, London, 1989.

12. Emcon Associates: *Methane Generation and Recovery from Landfills,* Ann Arbor Science, Ann Arbor, MI, 1980.
13. Farquhar, G. J. and F. A. Rovers: "Gas Production During Refuse Decomposition," *Water, Air and Soil Pollution,* Vol. 2, 1973.
14. Fenn, D. G., K. J. Hanley, and T. V. DeGeare: *Use of Water Balance Method for Predicting Leachate Generation from Solid Waste Disposal Sites,* EPA/530/SW-168, October 1975.
15. Frevert, R. K., G. O. Schwab, T. W. Edminster, and K. K. Barnes: *Soil and Water Conservation Engineering,* John Wiley & Sons, New York, 1955.
16. Ham, R. K., *et al.*: *Recovery, Processing and Utilization of Gas from Sanitary Landfills,* EPA-600/2-79-001, 1979.
17. Hansen, V. E., O. W. Israelsen, and G. E. Stringham: *Irrigation Principles and Practices,* John Wiley & Sons, New York, 1979.
18. Hatheway, A. W. and C. C. McAneny: "An In-Depth Look at Landfill Covers," *Waste Age,* Vol. 17, No. 8, 1987.
19. Herrera, T. A., R. Lang, and G. Tchobanoglous: "A Study of the Emissions of Volatile Organic Compounds Found in Landfills," *Proceedings of the 43rd Annual Purdue Industrial Waste Conference,* Lewis Publishers, Chelsea, MI, 1989.
20. Hjelmfelt, A. T., Jr., and J. J. Cassidy: *Hydrology for Engineers and Planners,* Iowa State University Press, Ames, IA, 1975.
21. Huitric, R. L., S. Raksit, and R. T. Haug: "In-Place Capacity of Refuse to Absorb Liquid Wastes," Presented at the Second National Conference on Hazardous Material Management, San Diego, CA, February 27–March 2, 1979.
22. Huitric, R. L., S. Raksit, and R. T. Haug: "Moisture Retention of Landfilled Solid Waste," County Sanitation Districts of Los Angeles County, Los Angeles, 1980.
23. Koerner, R. M. and D. E. Daniel: "Better Cover-Ups," *Civil Engineering,* Vol. 62, No. 5, May 1992.
24. Lang, R. J., T. A. Herrera, D. P. Y. Chang, G. Tchobanoglous, and R. G. Spicher: *Trace Organic Constituents in Landfill Gas,* Prepared for the California Waste Management Board, Department of Civil Engineering, University of California–Davis, Davis, CA, November 1987.
25. Lang, R. J., W. M. Stallard, L. Stiegler, T. A. Herrera, D. P. Y. Chang, and G. Tchobanoglous: *Summary Report: Movement of Gases in Municipal Solid Waste Landfills,* Prepared for the California Waste Management Board, Department of Civil Engineering, University of California–Davis, Davis, CA, February 1989.
26. Lang, R. J. and G. Tchobanoglous: *Movement of Gases in Municipal Solid Waste Landfills: Appendix A Modelling the Movement of Gases in Municipal Solid Waste Landfills,* Prepared for the California Waste Management Board, Department of Civil Engineering, University of California–Davis, Davis, CA, February 1989.
27. Linsley, R. K., J. B. Franzini, D. Fryberg, and G. Tchobanoglous: *Water Resources Engineering,* 4th ed., McGraw-Hill, New York, 1991.
28. Linsley, R. K., M. A. Kohler, and J. H. Paulhus: *Hydrology for Engineers,* McGraw-Hill, New York, 1958.
29. Mackay, K. M., P. V. Roberts, and J. A. Cherry: "Transport of Organic Contaminants in Groundwater," *Environmental Science and Technology,* Vol. 19, No. 5, pp. 384–392, May 1985.
30. Mathias, S. L.: "Discouraging Seagulls: The Los Angeles Approach," *Waste Age,* Vol. 15, No. 11, 1984.
31. McAtee, W. L.: "Excluding Birds from Reservoirs and Fishponds," Leaflet 120, U.S. Department of Agriculture, Washington, DC, September 1936.
32. Merz, R. C. and R. Stone: *Special Studies of a Sanitary Landfill,* U.S. Department of Health, Education, and Welfare, Washington, DC, 1970.
33. Moshiri, G. A. and C. C. Miller: "An Integrated Solid Waste Facility Design Involving Recycling, Volume Reduction, and Wetlands Leachate Treatment," in G. A. Moshiri (ed.), *Proceedings of the Constructed Wetlands for the Water Quality Improvement,* University of West Florida, Pensacola, FL, 1992.

34. Parker, A.: "Chapter 7. Behaviour of Wastes in Landfill–Leachate. Chapter 8. Behaviour of Wastes in Landfill–Methane Generation," in J. R. Holmes (ed.): *Practical Waste Management,* John Wiley & Sons, Chichester, England, 1983.
35. Perry, R. H., D. W. Green, and J. O. Maloney: Perry's (eds.): *Chemical Engineers' Handbook,* 6th ed., McGraw-Hill, New York, 1984.
36. Pfeffer, J. T.: *Solid Waste Management Engineering,* Prentice Hall, Englewood Cliffs, NJ, 1992.
37. Pohland, F. G.: *Critical Review and Summary of Leachate and Gas Production from Landfills,* EPA/600/S2-86/073, U.S. EPA Hazardous Waste Engineering Research Laboratory, Cincinnati, OH, 1987.
38. Pohland, F. G.: "Fundamental Principles and Management Strategies for Landfill Codisposal Practices," *Proceedings Sardinia 91 Third International Landfill Symposium,* Vol. II, pp. 1445–1460, Grafiche Galeati, Imola, Italy, 1991.
39. *Report on the Investigation of Leaching of a Sanitary Landfill,* California State Water Pollution Control Board, Publication 10, Sacramento, CA, 1954.
40. Salvato, J. A., W. G. Wilkie, and B. E. Mead: "Sanitary Landfill–Leaching Prevention and Control," *Journal Water Pollution Control Federation,* Vol. 43, No. 10, pp. 2084–2100, 1971.
41. Schroeder, P. R. *et al.*: "The Hydrologic Evaluation of Landfill Performance (HELP) Model," *User's Guide for Version I,* EPA/530/SW-84-009, 1, U.S. EPA Office of Solid Waste and Emergency Response, Washington, DC, 1984.
42. Schroeder, P. R. *et al.*: "The Hydrologic Evaluation of Landfill Performance (HELP) Model," *Documentation for Version I,* EPA/530/SW-84-010, 2, U.S. EPA Office of Solid Waste and Emergency Response, Washington, DC, 1984.
43. SCS Engineers, Inc.: *Procedural Guidance Manual for Sanitary Landfills: Volume I. Landfill Leachate Monitoring and Control Systems,* California Waste Management Board, Sacramento, CA, April 1989.
44. SCS Engineers, Inc.: *Procedural Guidance Manual for Sanitary Landfills: Volume II. Landfill Gas Monitoring and Control Systems,* California Waste Management Board, Sacramento, CA, April 1989.
45. Stahl, J. F., M. Moshiri, and R. Huitric: *Sanitary Landfill Gas Collection and Energy Recovery,* County Sanitation Districts of Los Angeles County, Los Angeles, 1982.
46. State Water Resources Control Board: *In-Situ Investigation of Movement of Gases Produced from Decomposing Refuse, Final Report,* The Resources Agency, Publication 35, State of California, Sacramento, 1967.
47. Stiegler, L., R. J. Lang, and G. Tchobanoglous: "Movement of Gases in Municipal Solid Waste Landfills: Appendix B Study of the Sorption of Trace Gases Found in Municipal Solid Waste Landfills," Prepared for the California Waste Management Board, Department of Civil Engineering, University of California–Davis, Davis, CA, February 1989.
48. Tchobanoglous, G. and E. D. Schroeder: *Water Quality: Characteristics, Modeling, Modifications,* Addison-Wesley, Reading, MA, 1985.
49. Tchobanoglous, G. and F. L. Burton: *Wastewater Engineering: Treatment, Disposal, Reuse,* 3rd ed., McGraw-Hill, New York, 1991.
50. U.S. Army Corp of Engineers: *Snow Hydrology,* Author, North Pacific Division, Portland, OR, June 1956.
51. Water Pollution Control Federation and American Society of Civil Engineers (Joint Committee): *Design and Construction of Sanitary and Storm Sewers,* WPCF Manual of Practice No. 9, Washington, DC, 1969.
52. World Wastes, *Equipment Catalog,* Communication Channels, Inc., Atlanta, GA, 1986.
53. Wright, T. D.: "To Cover or Not to Cover?," *Waste Age,* Vol. 17, No. 3, 1986.
54. Young, P. J. and L. A. Heasman: "An Assessment of the Odor and Toxicity of the Trace Components of Landfill Gas," *Proceedings of the GRCDA 8th International Landfill Gas Symposium,* Government Refuse Collection Disposal Association, Silver Spring, MD, April 1985.

PART IV

SEPARATION, TRANSFORMATION, AND RECYCLING OF WASTE MATERIALS

There is a growing awareness in the solid waste profession that managing solid wastes requires an integrated system that provides for source reduction, waste recycling, waste transformation, and disposal. The U.S. EPA has offered a strategy for solid waste management that is based on the concept of integrated solid waste management. To develop an effective integrated solid waste management system, information must be available on the options that can be used to manage solid waste. The focus of Part IV is on the options that are available for processing and transforming waste materials. The waste management options presented in Part IV are of importance because one or more of them will form the basis of most integrated waste management systems. Disposal, discussed in detail in Part III, is not considered in Part IV.

Part IV is intended to provide a sufficient understanding of the waste processing and transformation options so that informed choices of system components for managing solid wastes can be made. The separation, processing, and recovery of materials are considered in Chapter 12. Thermal transformation processes for volume reduction and the recovery of energy are presented in Chapter 13. Biological transformation processes are examined in Chapter 14. Recycling of source-separated materials is the primary focus of Chapter 15. Because of the importance of materials specifications in the recycling of waste materials, materials specifications are emphasized in Chapter 15.

CHAPTER 12

MATERIALS SEPARATION AND PROCESSING TECHNOLOGIES

In an "as collected," commingled state, municipal solid waste (MSW) is biologically unstable, can become odorous, and is essentially unusable. Materials recovery facilities (MRFs) are used to separate commingled MSW into usable materials. Waste components such as paper, plastics, glass, and metals can be recovered from MSW for remanufacturing into new products. The organic portions of solid waste can be recovered as a feedstock for composting or other biological processes or as refuse-derived fuel (RDF) for use in thermal processing for energy recovery. As defined in Chapter 9, the further separation and processing of wastes that have been separated at the source, as well as the separation of commingled wastes, usually occur at MRFs or at large integrated materials recovery/transfer facilities (MR/TFs). The purpose of this chapter is to discuss the theoretical aspects and applications of the unit operations used in materials recovery and to present and illustrate the computations involved in the design of MRFs.

12-1 UNIT OPERATIONS FOR THE SEPARATION AND PROCESSING OF WASTE MATERIALS

Unit operations used for the separation and processing of separated and commingled wastes are, as noted in Chapter 9, designed (1) to modify the physical characteristics of the waste so that waste components can be removed more easily and (2) to remove specific components and contaminants from the waste stream. The principal unit operations used for the recovery and processing of MSW are

TABLE 12-1
Methods used for the processing and the recovery of individual waste components from MSW

Processing options	Description
Size reduction	Unit operation used for the reduction of both commingled MSW and recovered materials. Typical applications include (1) hammer-mills for shredding commingled MSW; (2) shear shredders for use with commingled MSW and recycled materials such as aluminum, tires, and plastics; and (3) tub grinders used to process yard wastes.
Size separation	Unit operation in which materials are separated by size and shape characteristics, most commonly by the use of screens. Several types of screens are in common use, including (1) reciprocating screens for sizing shredded yard wastes; (2) trommel screens used for preparing commingled MSW prior to shredding; and (3) disc screens used for removing glass from shredded MSW.
Density separation	Unit operations in which materials are separated by density. Typical applications include (1) air classifiers for the preparation of RDF; (2) inertial separation for the processing of commingled MSW; and (3) flotation for the processing of construction debris.
Electric and magnetic field separation	Unit operations in which materials are separated by their electrostatic charge and magnetic permeability. Typical applications include (1) the separation of plastics from paper and (2) the separation of ferrous from nonferrous materials (e.g., "tin cans" from aluminum cans).
Densification (Compaction)	Densification and compaction are unit operations that are used to increase the density of recovered materials to reduce transportation costs and simplify storage. Typical applications include (1) the use of baling for cardboard, paper, plastics, and aluminum cans; and (2) the use of cubing and pelletizing for the production of densified RDF.
Materials handling	Unit operations used for the transport and storage of MSW and recovered materials. Typical applications include (1) conveyors for the transport of MSW and recovered materials; (2) storage bins for recovered materials; and (3) rolling stock such as forklifts, front-end loaders, and various types of trucks for the movement of MSW and recovered materials.

summarized in Table 12-1. Each of the processing methods listed in Table 12-1 is discussed in greater detail in the sections that follow.

12-2 SIZE REDUCTION

Size reduction is the unit operation used to reduce the size of the materials in MSW. Size reduction is used to process materials for direct use, such as mulch or compost, or as part of a materials recovery facility. The types, performance characteristics, design criteria, and selection of size reduction equipment are considered in the following discussion.

Types of Equipment

Several types of size reduction units are in common use: (1) the *hammermill,* which is very effective with brittle materials; (2) the *shear shredder*, which uses two opposing counterrotating blades to cut ductile materials in a scissorlike action; and (3) the *tub grinder,* which is widely used in the processing of yard wastes. Each type of shredder has unique properties that suit it to specific applications. These properties can be used to meet different size reduction goals, depending on the requirements of downstream processes such as air classifiers or screens.

Hammermill Shredders. Operationally, a hammermill shredder is an impact device in which a number of hammers are fastened flexibly to an inner shaft or disk that is rotated at high speed (700 to 1200 rev/min). Because of centrifugal force, the hammers extend radially from the center shaft. As solid wastes enter the hammermill, they are hit with sufficient force to crush or tear them and with such a velocity that they do not adhere to the hammers. Wastes are further reduced in size by being struck against breaker plates or cutting bars fixed around the inner periphery of the inner chamber. The cutting action continues until the material is of the size required and falls out of the bottom of the mill. Hammermills can be designed in either horizontal-shaft or vertical-shaft configurations (see Figs. 12-1*a* and 12-1*b*). Horizontal-shaft machines have proven to be the most reliable (see Fig. 12-2).

Shear Shredders. Shear shredders operate in scissorlike action (see Fig. 12-3), in which two counterrotating knives cut or shear the waste. Compared to hammermills, shear shredders are low-speed devices (60 to 190 rev/min). Most are driven by hydraulic motors that can be reversed automatically in the event of a jam.

Tub Grinders. A tub grinder is essentially a mobile hammermill shredder that can be taken to the source of wastes and used to process materials on site. Tub grinders have been used for a wide variety of materials, including tree prunings, tree stumps, and construction debris. As shown in Fig. 12-4, a typical tub grinder features a large-diameter feed throat, into which materials to be shredded are dropped. The rotary motion of this feed throat, or "tub," conveys the material into a horizontal hammermill shredder. Tub grinders are generally powered by diesel engines and are trailer-mounted for ease of moving. Some models feature a self-contained knuckle boom crane used to load the "tub."

Performance Characteristics

Raw (i.e., unprocessed), commingled MSW has a particle size distribution as illustrated in Fig. 9-5. After shredding, the particle distribution changes significantly, because hammermill shredders do not produce materials of uniform size. Brittle materials such as glass, sand, and rock form a higher proportion of fine particles than do ductile materials such as ferrous and nonferrous metals (see

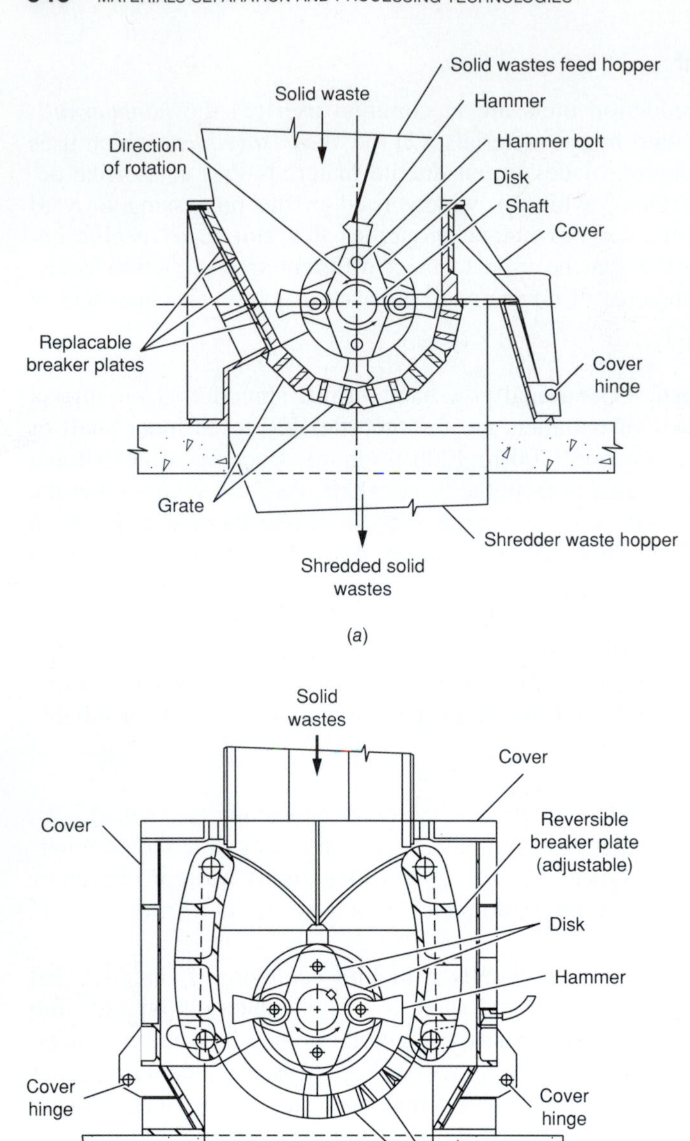

FIGURE 12-1
Cross section of hammermill shredders for commingled MSW: (*a*) one-way type and (*b*) reversible type. (Courtesy of Williams Crusher and Pulverizer Co.)

FIGURE 12-2
Horizontal shaft hammermill installation for commingled MSW. The shredder is loaded with an inclined conveyor. (Courtesy of Amadas Industries.)

FIGURE 12-3
Shear shredder for commingled MSW. The shredder shown is used for old and damaged pallets. (Courtesy of SSI Shredding Systems, Inc.)

FIGURE 12-4
Typical tub grinder equipped with boom crane to load wastes into the grinder and conveyor discharge system for yard and wood waste. (Courtesy of Morbark Industries, Inc.)

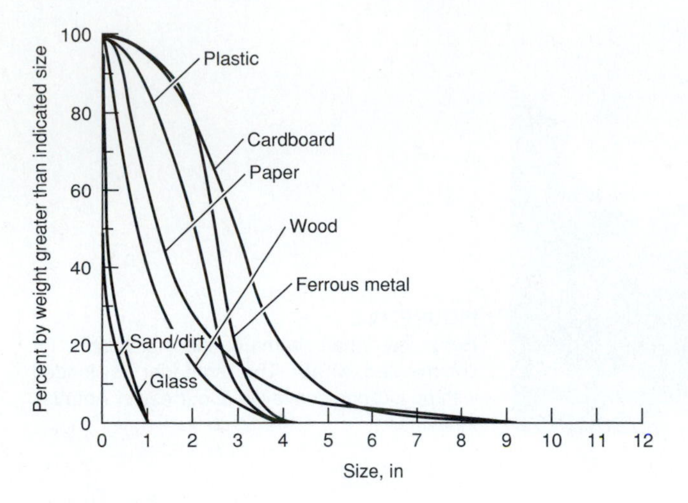

FIGURE 12-5
Representative size distribution by weight of hammermill-shredded MSW. (Adapted from Refs. 8 and 17.)

Fig. 12-5). This characteristic of hammermills can be used to advantage if screens are used following the hammermill to separate out glass, sand, and rocks from other materials. However, the high-speed rotation of the hammermill causes some of these finer particles of glass and sand to become embedded in softer materials such as paper and textiles. This contamination can affect the product specifications for these materials. Shear shredders tend to produce a more uniform product (see Table 12-2). They are particularly effective with MSW, some types of industrial waste, and tires.

TABLE 12-2
Size distribution characteristics of shear shredders[a,b]

	Percent retained		
Screen size, in	**Avg.**	**Min.**	**Max.**
+4	19.3	9.7	32.9
+2	30.7	23.0	35.0
+1	19.2	14.9	23.9
+1/2	13.2	8.9	17.4
+1/4	9.3	6.2	12.0
Pan	8.3	5.1	10.3

[a] Adapted from Ref. 10.

[b] Based on test results from 11 samples of unseparated MSW.

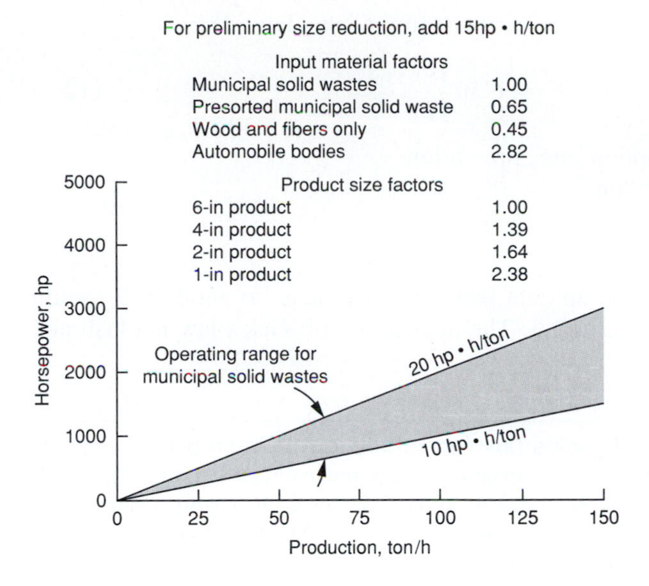

FIGURE 12-6
Horsepower requirements for hammermill shredding. (Adapted from Ref. 5.)

Design Criteria

Size reduction equipment is designed based on the mass loading of waste (ton/h) and the power consumed. Typical power requirements for shredding are given in Fig. 12-6. These data were derived from an analysis of information from equipment manufacturers and to a limited extent from actual installations [5]. The use of the data reported in Fig. 12-6 is illustrated in Example 12-1.

Example 12-1 Horsepower requirements for size reduction. Estimate the horsepower required to reduce commingled MSW to a final size of about 3 in for a plant with a capacity of 80 ton/h using the data presented in Fig. 12-6.

Solution

1. Using a conservative value of 20 hp · h/ton as given in Fig. 12-6 for the base horsepower, the required horsepower is

$$\text{Horsepower} = 80 \text{ ton/h} \times 20 \text{ hp} \cdot \text{h/ton} = 1600 \text{ hp } (1194 \text{ kW})$$

2. Using an estimated product size factor of 1.5 (see Fig. 12-6), the required horsepower is

$$\text{Horsepower} = 1600 \text{ hp} \times 1.5 = 2400 \text{ hp } (1790 \text{ kW})$$

Comment. The data presented in Fig. 12-6 or similar data should be used only for the preparation of preliminary estimates. Manufacturers' specifications, pilot testing, and operating experience from actual facilities should be used in arriving at a final design value.

The energy consumption can also be estimated by Kick's law [18]:

$$E = C \ln \frac{l_1}{l_2} \tag{12-1}$$

where E = energy consumption rate, hp · h/ton
C = constant, hp · h/ton
l_1 = initial size
l_2 = final size

Kick's law can be used to scale up data from one size range to another. It should be used in conjunction with Fig. 12-6. The application of Kick's law is illustrated in Example 12-2.

Example 12-2 Application of Kick's law for estimation of horsepower requirements for size reduction. Estimate the horsepower required to reduce commingled MSW with an average size of 12 in to a final size of about 2 in for a plant with a capacity of 80 ton/h, using Kick's law. Assume that a specific energy of 20 hp · h/ton is required to reduce MSW with an average size of 6 in to 2 in.

Solution

1. Find the constant C using Kick's law, Eq. (12-1):

$$E = C \ln \frac{l_1}{l_2}$$

$$C = \frac{E}{\ln \dfrac{l_1}{l_2}}$$

$$= \frac{20 \text{ hp} \cdot \text{h/ton}}{\ln \dfrac{6}{2}}$$

$$= 18.20 \text{ hp} \cdot \text{h/ton}$$

2. Find the energy consumption rate, E, for size reduction from 12 in to 2 in:

$$E = 18.20 \text{ hp} \cdot \text{h/ton} \ln \frac{12 \text{ in}}{2 \text{ in}}$$

$$= 32.58 \text{ hp} \cdot \text{h/ton}$$

3. Find the energy required.

$$\text{hp} = 32.58 \text{ hp} \cdot \text{h/ton} \times 80 \text{ ton/h}$$

$$= 2606 \text{ hp (1944 kW)}$$

Comment. Kick's law is an approximation of energy consumption that can be used to extrapolate experimental data from one size distribution to another.

TABLE 12-3
Factors to be considered in the selection of size reduction equipment

Factor	Comment
Materials to be shredded	Mechanical characteristics of the material must be known, such as shear strength and ductility.
Size requirements for shredded material by components	Hammermills tend to produce nonuniform product, whereas shear shredders produce a more uniformly shredded material.
Method of feeding shredder	Capacity of feed conveyors must be matched to the shredder.
Operational characteristics	Energy requirements (hp · h/ton), routine and specialized maintenance requirements, simplicity of operation, proven performance and reliability, noise output, and air and water pollution control requirements.
Site considerations	Floor space and height, access, and environmental considerations.
Materials storage and conveyance requirements	Shredded materials need to be stored and conveyed to downstream operations.

Selection of Size Reduction Equipment

Factors that must be considered in the selection of size reduction equipment are summarized in Table 12-3. Because shredders are available from several manufacturers, designers should develop performance-based specifications to choose the most cost-effective shredder that will meet the design goals.

Safety Issues. Size reduction devices have the potential to be safety hazards if proper design and operational procedures are not used. The most common danger is inadvertent shredding of containers containing volatile compounds, such as gasoline or solvents, thus releasing explosive vapors. Dust concentrations can also form explosive atmospheres. The hammermill shredder has the highest risk of explosion, caused by sparks from the high-speed impact of metal on metal.

Safety Systems. The first line of defense against such problems is a screening and inspection program to reduce the input of potentially explosive or combustible materials into the size reduction device. Shredders should be installed in structurally isolated rooms or buildings separate from other processing areas. Electrical controls, wiring, and lighting should be installed in explosion-proof housings and conduit to avoid sparking. Shredding rooms should be provided with explosion vents, which will direct explosions away from the structure.

Other safety systems include automatic fire and explosion suppression systems that can detect the presence of explosive vapors and the beginning of explosion or combustion. A few milliseconds after detection, the shredder and the feed and exit conveyors will be flooded with an inert gas such as nitrogen or carbon dioxide. Other extinguishing agents, such as Halon™, can also be used.

12-3 SIZE SEPARATION

Size separation, or screening, involves the separation of a mixture of materials into two or more portions by means of one or more screening surfaces, which are used as go or no-go gauges. Size separation can be accomplished dry or wet, with the former being most common in solid waste–processing systems. Screens have been used before and after shredding and after air classification in the processing of refuse-derived fuel (RDF). They are also used in the processing of compost and mulch to produce a more uniform product. The types of screens that are used, and the performance characteristics, design criteria, and selection of separation equipment are discussed in this section.

Types of Equipment

The most common types of screens used for the separation of solid waste are vibrating screens, trommel screens (also known as rotary drum screens), and disc screens. Each type of screen is best adapted to specific situations, as discussed in the following paragraphs.

Vibrating Screens. Vibrating screens are most often used to separate relative dry materials such as glass or metals. Vibrating screens have also been used to separate wood chips, used as a bulking agent in sludge composting, from the compost product and for the removal of broken pieces of concrete from construction debris. A typical vibrating screen is shown in Fig. 12-7.

Trommel Screens. The trommel screen is one of the most versatile types of screens for solid waste processing. It consists of a large-diameter (typically 10 ft)

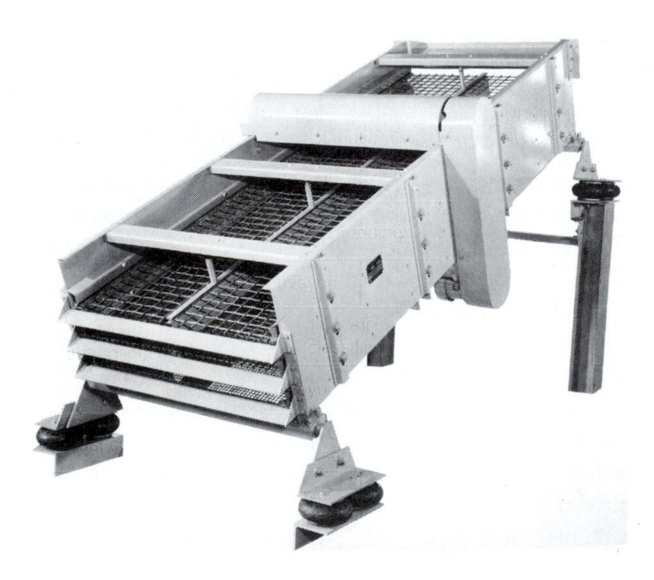

FIGURE 12-7
Typical vibrating screen used for separating commingled MSW by size. (Courtesy of Universal Vibrating Screen Company.)

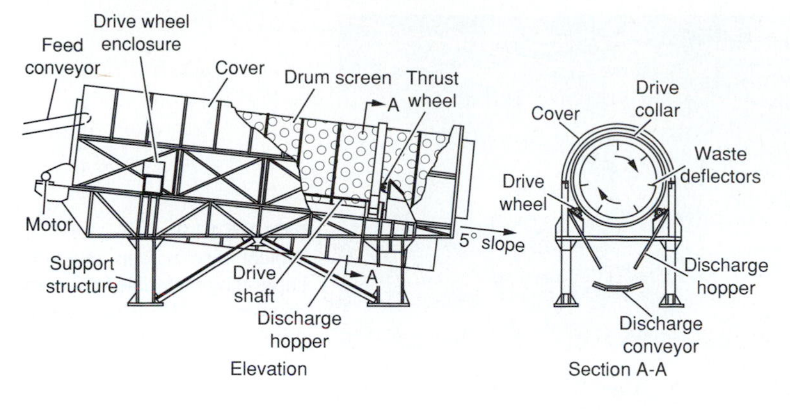

(*a*)

(*b*)

FIGURE 12-8
Typical trommel screen used for separating commingled MSW by size: (*a*) schematic views and (*b*) view of screen and appurtenant facilities. (Courtesy of Triple/S Dynamics Systems, Inc.)

screen, formed into a cylinder and rotating on a horizontal axis (see Fig. 12-8). Trommels have been used to protect shredders in RDF production facilities (by removing oversized material) and to separate cardboard and paper in materials recovery facilities.

Disc Screens. Disc screens are an alternative to reciprocating screens. As shown in Fig. 12-9, a disc screen consists of sets of parallel, interlocking, rotating discs. The materials to be separated fall between the spaces, and oversized materials are carried over the top of the discs as in a conveyor belt. Disc screens have several advantages over reciprocating screens, including self-cleaning and capability of adjustment by varying the spacing of the discs on the drive shafts.

FIGURE 12-9
Typical disc screen used for separating materials from commingled MSW. (Courtesy of Amadas Industries.)

Performance Characteristics

The efficiency of separation devices can be evaluated in terms of percentage recovery, purity, and efficiency by using binary separation theory [17]. Assume that a waste stream, composed of two components X and Y is to be separated. If the input to a binary separation device is denoted as $X_0 + Y_0$, then after separation, two waste streams result, 1 and 2, each containing a mixture of X and Y denoted as X_1, Y_1 and X_2, Y_2 respectively (see Fig. 12-10). The *recovery* percentage of the device for the two components in the waste stream can be defined as follows:

$$\text{Recovery of component } X = \text{Recovery } (X) = R_{(X_1)} = \left(\frac{X_1}{X_0}\right) \times 100\% \quad (12\text{-}2)$$

$$\text{Recovery of component } Y = \text{Recovery } (Y) = R_{(Y_2)} = \left(\frac{Y_2}{Y_0}\right) \times 100\% \quad (12\text{-}3)$$

Recovery is not a useful parameter if used by itself. For example, a hypothetical device could achieve 100 percent recovery of component X if all of the waste stream (both X and Y) went into side 1. In that case X_2 and Y_2 would both equal 0, but no separation would have occurred! A second parameter, *purity,* must also be used to describe the performance of a separation device. The purity function represents the quality of the material separated in terms of its contamination by the other material. The percentage of purity of the recovered material can be defined as follows:

$$\text{Purity for component } X = \text{Purity } (X) = P_{(X_1)} = \left[\frac{X_1}{(X_1 + Y_1)}\right] \times 100\% \quad (12\text{-}4)$$

$$\text{Purity for component } Y = \text{Purity } (Y) = P_{(Y_2)} = \left[\frac{Y_2}{(X_2 + Y_2)}\right] \times 100\% \quad (12\text{-}5)$$

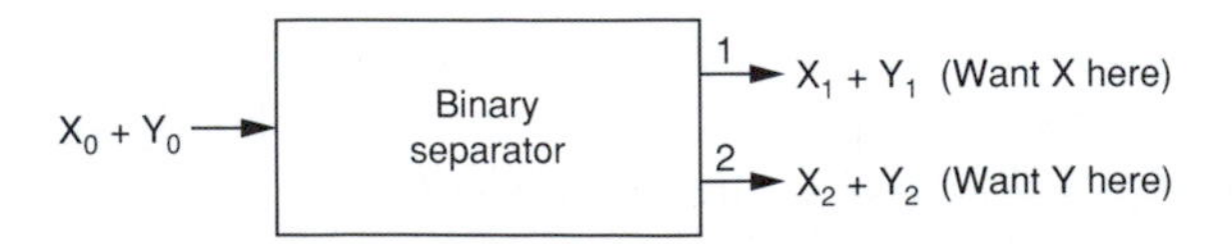

FIGURE 12-10
Definitional sketch for the binary separation of waste materials.

An ideal device would have both high recovery and high purity. A single parameter, *efficiency,* can be used to describe both recovery and purity:

$$\text{Efficiency} = E_{(X,Y)} = \left|\frac{X_1}{X_0} - \frac{Y_1}{Y_0}\right| \times 100\% = \left|\frac{X_2}{X_0} - \frac{Y_2}{Y_0}\right| \times 100\% \qquad (12\text{-}6)$$

The application of Eqs. (12-2) through (12-6) is demonstrated in Example 12-3.

Example 12-3 Determination of screen recovery, purity, and efficiency. Given that 100 ton/h of commingled MSW with the composition shown in Table 3-4 are applied to a trommel screen for the removal of glass prior to shredding, determine the recovery, purity, and efficiency of the screen based on the following experimental data.

1. Weight of underflow = 10 ton/h
2. Weight of glass in screen underflow = 7.2 ton/h

The MSW into the screen is defined as $X_0 + Y_0$, the overflow is denoted as $X_1 + Y_1$, and the underflow is $X_2 + Y_2$.

Solution

1. Find X_0 and Y_0 by determining the weight fraction of glass in the feed to the screen. From Table 3-4, the percentage of glass is given as 8 percent. Thus:

$$X_0 = 92 \text{ ton/h (mass flow rate of MSW} - \text{mass flow rate of glass)}$$

$$Y_0 = 8 \text{ ton/h (the mass flow rate of glass)}$$

2. Find X_1 and Y_1, the overflow of the trommel (also called the *reject*). The overflow is 90 ton/h (100 ton/h − 10 ton/h). Because 7.2 ton/h of glass are in the underflow, 0.8 ton/h must remain in the overflow. Thus:

$$X_1 = 90 - 0.8 = 89.2 \text{ ton/h “other”}$$

$$Y_1 = 0.8 \text{ ton/h glass}$$

3. Find X_2 and Y_2, the underflow in the trommel:

$$X_2 = 10 - 7.2 = 2.8 \text{ ton/h “other”}$$

$$Y_2 = 7.2 \text{ ton/h glass}$$

4. Find the recovery of glass in the underflow, using Eq. (12-3):

$$\text{Recovery } (Y) = R_{(Y_2)} = \left(\frac{Y_2}{Y_0}\right) \times 100\%$$

$$= \left(\frac{7.2}{8.0}\right) \times 100\% = 90\%$$

5. Find the purity of the overflow and of the recovered glass.
Using Eq. (12-4), find the purity of overflow:

$$\text{Purity } (X) = P_{(X_1)} = \left[\frac{X_1}{(X_1 + Y_1)}\right] \times 100\%$$

$$= P_{(X_1)} = \left[\frac{89.2}{89.2 + 0.8}\right] \times 100\% = 99\%$$

Using Eq. (12-5), find the purity of recovered glass:

$$\text{Purity }(Y) = P_{(Y_2)} = \left[\frac{Y_2}{(X_2 + Y_2)}\right] \times 100\%$$

$$= P_{(Y_2)} = \frac{7.2}{2.8 + 7.2} = 72\%$$

6. Find the efficiency of the trommel screen using Eq. (12-6):

$$\text{Efficiency} = E_{(X,Y)} = \left|\frac{X_1}{X_0} - \frac{Y_1}{Y_0}\right| \times 100\% = \left|\frac{X_2}{X_0} - \frac{Y_2}{Y_0}\right| \times 100\%$$

$$\text{Efficiency} = E_{(X,Y)} = \left|\frac{89.2}{92.0} - \frac{0.8}{8.0}\right| \times 100\% = 87\%$$

or

$$\text{Efficiency} = E_{(X,Y)} = \left|\frac{2.8}{92} - \frac{7.2}{8.0}\right| \times 100\% = 87\%$$

Comment. The trommel screen in this example appears to have a good recovery rate for glass (90 percent); however, note that while the purity of the overflow material is high (99 percent), the purity of the glass in the underflow is an unacceptable 72 percent. Without additional processing, it would be unmarketable as mixed glass. The overflow material would probably be acceptable, with further processing, as a feedstock for refuse-derived fuel (RDF).

Design Criteria

Size separation equipment is designed based on the mass loading of materials flow (ton/h) and the residence time in the device. Several empirical approaches have been developed for the design of screens, but the most accurate method of design is pilot testing of the actual material to be screened.

Reciprocating Screens. The following relationship has been suggested for finding the area of a reciprocating screen [12]:

$$A = \frac{Q}{C_u}\text{F} \tag{12-7}$$

where A = screen area, ft^2
Q = screen loading, ton/h
C_u = unit capacity (ton/h)/ft^2
F = product correction factor

The product correction factor F is a composite efficiency factor that is the product of several factors relating to size of particles, efficiency desired, number of decks (for multiple-deck screens), size of openings, percent open area, and specific weight of the material. For example, $F = 1.40$ for a single-deck screen processing crushed glass [17]. Additional details on this method may be found in Ref. 12.

The following alternative design equation has also been suggested [8]:

$$Q_{\text{nom}} = \frac{d\ D^{2/3}}{K} \tag{12-8}$$

where Q_{nom} = ton/h · ft^2 of open screen area
d = bulk specific weight, lb/ft^3
D = aperture width or diameter, in
K = an experimentally determined constant

Because K is determined experimentally, Eq. (12-8) can be used to extrapolate pilot data.

Trommel Screens. The design of a trommel is based on several parameters including diameter, length, rotational speed, angle of inclination, feedrate, and particle shape and size distribution.

The rotational speed is a function of the *critical speed:* the speed at which materials *centrifuge* or stick to the screening surface.

$$n_c = \frac{1}{2\pi}\sqrt{\frac{g}{r}} \tag{12-9}$$

where n_c = critical speed, rev/s
g = acceleration due to gravity, 32.2 ft/s^2
r = trommel radius, ft

The optimum speed occurs when the materials tumble in a *cataracting* motion; that is, they are partially carried up the interior wall of the drum and then fall back on themselves. Ideally the rotational speed should be 50 percent of the critical speed for a trommel with lifters, and up to 80 percent of the critical speed for a trommel without lifters (lifters are vertical plates attached to the inside of the trommel). For MSW, rotational speed in the range of 10 to 18 rev/min have been reported.

The angle of inclination affects the residence time of material in the trommel. Angles of 2 to 5° have been reported in operating MSW systems. Practical designs should allow for varying the slope of the trommel over this range to optimize performance.

The following relationship has been suggested for calculating trommel throughput [16]:

$$D = \left[\frac{11.36\, Q_m}{d_b\, F\, K_v\, g^{0.5} \tan\alpha}\right]^{0.4} \tag{12-10}$$

where D = trommel diameter, ft
Q_m = trommel throughput, lb/s
d_b = bulk specific weight of MSW, lb/ft^3
α = trommel inclination, degrees
K_v = velocity correction factor,
$K_v = 1.35$ when $\alpha = 3°$
$K_v = 1.85$ when $\alpha = 5°$
F = fillage factor (typically 0.25 to 0.33)
g = 32.2 ft/s^2

Operating characteristics of a typical trommel are summarized in Table 12-4.

TABLE 12-4
Operating characteristics of a typical trommel screen[a]

Parameter		Value
Diameter	m	3.5
Screen length	m	4.0
Screen size	mm	50
Screen open area	%	53
Inclination angle (variable)	degrees	3–7
Rotational speed (variable)	rev/min	11–13

[a] Adapted from Ref. 19.

Disc Screens. Disc screens are designed based on volumetric throughput in ft^3/h, disc spacing, disc rotation speed, and surface area. Because disc screens were developed originally for screening wood waste and wood chips, the mass throughput in ton/h of the screen will be less, because of the relatively low specific weight of shredded MSW (typically 2.5 to 5.0 lb/ft^3). Designers should rely on throughput data from operating plants to verify manufacturers' specifications.

Selection of Size Separation Equipment

Factors that should be considered in the selection of size reduction equipment are summarized in Table 12-5. Because screens are available from a number of manufacturers, designers should develop performance-based specifications for recovery, effectiveness, and purity to choose the most cost-effective screening system to meet their needs. Designs made with the methods suggested in this section, from the literature and from manufacturers' data, should be verified by measurements made at operating plants or by pilot testing.

TABLE 12-5
Factors to be considered in the selection of screening equipment

Factor	Comment
Waste characteristics	Particle size, shape, bulk specific weight, moisture content, particle size distribution, clumping tendency, rheological properties.
Materials specifications for screened components	Performance characteristics of screen should match required product.
Screen design parameters	Size of openings (in), percentage open space, total surface screening area (ft^2), oscillation rate for reciprocating screens (times/min), rotational speed for trommels (rev/min), elevation angle for trommels (degrees), loading rates (ton/$ft^2 \cdot$ h), and length (ft).
Separation efficiency	Recovery (%), efficiency (%), purity (%).
Operational characteristics	Energy requirements (hp), maintenance, complexity of operation, noise, and air and wastewater emissions.
Site factors	Floor space and vertical space availability, access.

12-4 DENSITY SEPARATION

Density separation is a technique widely used to separate materials based on their density and aerodynamic characteristics. Density separation has been applied to the separation of shredded MSW into two major components: (1) the *light fraction,* composed primarily of paper, plastics, and organics; and (2) the *heavy fraction,* which contains metals, wood, and other relatively dense inorganic materials.

Types of Equipment

Several technologies are used for density separation, including air classification, stoners, flotation, and heavy media separation. Air classification, the most widely used separation technology, is described in the most detail.

Air Classifiers. In the simplest types of air classifiers, shredded solid wastes are dropped into a vertical chute (see Fig. 12-11*a*). Air moving upward from the bottom of the chute is used to transport the lighter materials to the top of the chute. Because the upward airflow is insufficient to transport the heavier materials in the wastes, the heavier materials drop to the bottom. Control of the percentage split between the light and heavy fractions is accomplished by varying the waste-loading rate, the airflow rate, and the cross-sectional area of the chute. A rotary airlock mechanism is required to introduce the shredded wastes into the classifier. Another type of air classifier, known as the *zigzag air classifier,* consists of a continuous vertical column with internal zigzag deflectors, through which air is drawn up at a high rate (see Fig. 12-11*b*). Shredded wastes are introduced at the top of the column at a controlled rate, and air is introduced at the bottom of the column. In theory, the zigzag flow path creates turbulence in the airstream which, in turn, causes the wastes to tumble and allows bunched materials to be broken up [2]. Best

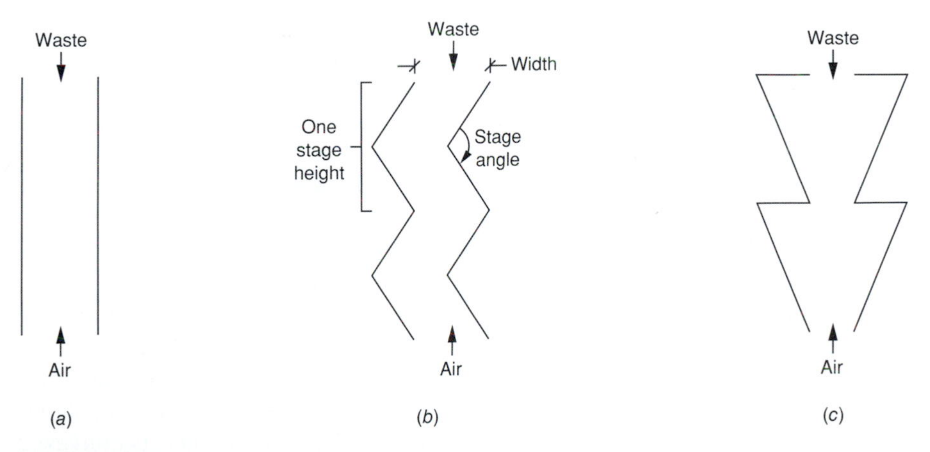

FIGURE 12-11
Typical functional designs for air classifiers used to separate the light and heavy fractions of solid waste: (*a*) straight (chute type) nonpulsed or active pulsed with louvered valve; (*b*) zigzag nonpulsed; and (*c*) stacked triangle passive pulsed.

separation is achieved through proper design of the separation chamber, airflow rate, and material feed rate.

The *pulsed air classifier* is a recent development that uses a varying airflow velocity instead of the constant airflow velocity used in the conventional and zigzag air classifiers discussed previously [7]. The pulsed air classifier achieves a greater discrimination between materials, because the velocity of a falling particle is a function of time until terminal velocity is reached. Varying the velocity of the airstream has the effect of keeping the falling particles in a velocity range such that particles with similar terminal velocities can be more completely separated. An *active pulsed* air classifier uses a simple straight throat as in a conventional air classifier (see Fig. 12-11*a*). However, the airflow to the unit is varied with a louver valve. An alternative design, the *passive pulsed* air classifier, uses a constant airflow in a "stacked triangle"–shaped throat cross section. This variable cross-sectional area results in an air velocity that varies with distance along the throat (see Fig. 12-11*c*).

A complete air classification system is comprised of one or more conveyors, the classifier, and a cyclone separator (see Fig. 9-10). Conveyors are required to transport processed wastes to the loading hopper and into the air classifier. Following the air classifier, a cyclone separator is used to separate the light fraction from the conveying air. Before being discharged into the atmosphere, the conveying air is passed through dust collection facilities, typically a baghouse (see Chapter 13). Alternatively, air from the cyclone separator can be recycled to the air classifier with or without dust removal. Air for the operation of the air classifier can be supplied by low-pressure blowers or fans. The heavy fraction that is removed from the air classifier is hauled either to a disposal site or to a subsequent resource recovery system. The light fraction may be stored in bins or transported or conveyed to another shredder for further size reduction before storage or utilization as a fuel or compost material. As noted in Chapter 9, conventional air classifiers are little used today because the shredding of commingled waste is no longer favored.

Stoner. The stoner was initially designed to remove stones and other heavy rejects from commodities such as wheat. In MRFs, stoners are used to separate heavy grit from organic material in trommel underflow (undersize) streams, and they are often called inerts separators. A recent trend in MRF design is to replace the use of shredders and air classifiers for unprocessed commingled MSW with one or more (typically two) trommels and a stoner.

The basic stoner consists of a vibrating porous deck through which air is blown (see Fig. 12-12). The deck vibrates in a straight line in the uphill direction. Material to be separated is fed onto the deck at a point between the center and the uphill third of the deck. Low-pressure air moving up through the deck fluidizes and stratifies the material to be separated according to the differences in the terminal velocity of the particles. The light material, lifted by the fluidizing air, flows downhill, while the heavier material remains on the deck surface and is conveyed uphill by the vibration action of the deck. Because the actual separating criterion is terminal velocity, not density or weight, stoners work as density separators only

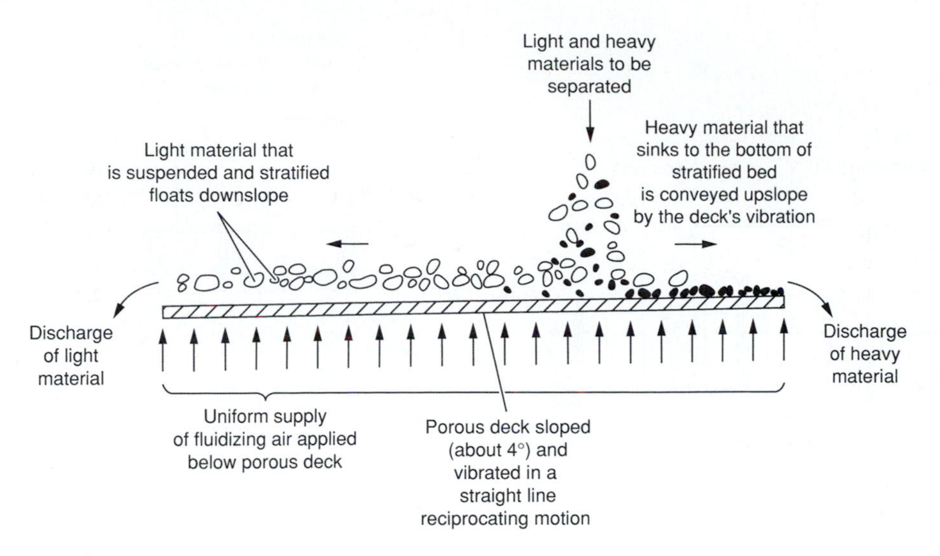

FIGURE 12-12
Definition sketch for the operation of a stoner used to separate the light and heavy fractions. (Courtesy of Triple/S Dynamics Systems, Inc.)

within a fairly narrow size distribution range. Important operational variables with a stoner are deck slope and air volumes.

A stoner is almost always equipped with a dust hood to collect the fluidizing air. The exhaust air is usually set to be a little higher than the fluidizing, or supply, air to keep dust from escaping through the product discharges at each end of the unit. Views of a stoner are shown in Fig. 12-13.

Flotation. Flotation is a unit operation which employs a fluid to separate two components of different densities. For example, glass-rich feedstocks, produced by screening the heavy fraction of the air-classified wastes after ferrous metal separation, can be separated from organic contaminants by immersion in water in a suitable tank. Glass chips, rocks, bricks, bones, and dense plastic materials that sink to the bottom are removed with belt scrapers for further processing. Light organics and other materials that float are skimmed from the surface. These materials may be hauled to a landfill for disposal or returned to the head end of the plant and passed through the operation with a new batch of solid wastes. Flotation can also be used to separate wood from mixed shredded construction debris and to separate plastics from organic contaminants.

Heavy Media Separation. Although the removal of aluminum can be accomplished in a number of different ways, heavy media separation is perhaps the process for which the greatest operating experience exists, principally in the automobile recovery industry. In this process a shredded feedstock that contains a high percentage of aluminum, such as air-classified shredded automobile bodies after ferrous metal and glass have been removed, is dumped into a liquid stream

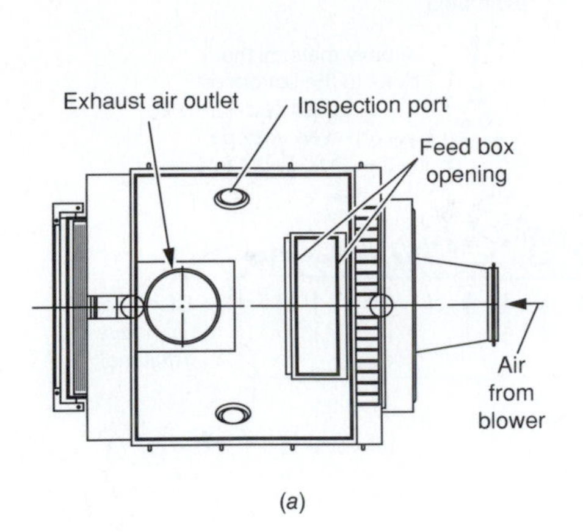

(*a*)

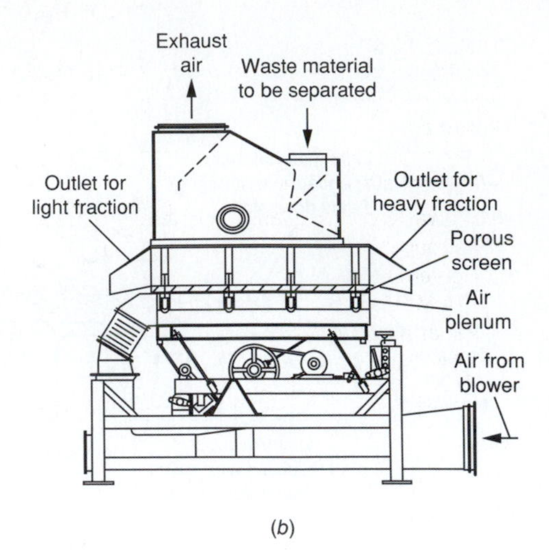

(*b*)

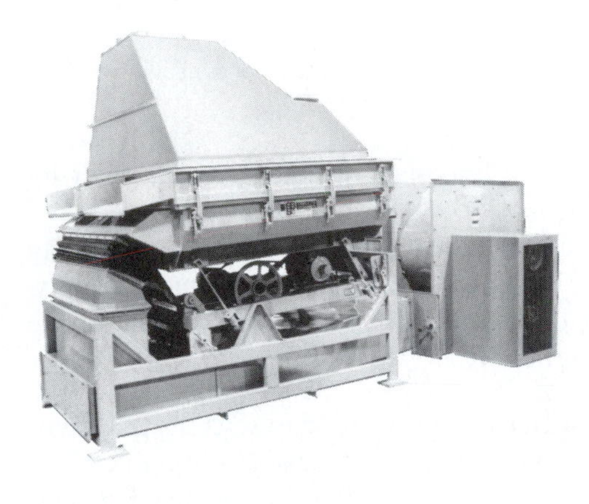

(*c*)

FIGURE 12-13
Views of a stoner used to separate materials from a trommel underflow stream: (*a*) schematic plan view, (*b*) schematic side view, and (*c*) view of commercial unit. (Courtesy of Triple/S Dynamics Systems, Inc.)

that has a high specific gravity. The specific gravity is maintained at a level that will permit the aluminum to float and the other materials to remain submerged. At present, the major disadvantage of this process is that the optimum-size plant requires about 2000 to 3000 ton/d of feedstock. Also, the increasing success of source separation systems is steadily decreasing the proportion of aluminum in the waste stream, decreasing the economic feasibility of aluminum separation systems.

Performance Characteristics

Typical performance characteristics for air classifiers operating with shredded MSW are summarized in Table 12-6. Notice that while the air-classified light fraction contains primarily paper and plastics, there is also an appreciable carry-over of nonferrous metals and fines. It has been suggested that the efficiency of

TABLE 12-6
Typical ranges of air classifier performance[a]

Parameter	Typical range
Critical air/solids ratio	2.0–7.0
Input waste composition to air classifier (percent)	
Ferrous metals	0.1–1.0
Nonferrous metals	0.2–1.0
–14 Mesh fines	15.0–30.0
Paper and plastic	55.0–80.0
Ash	10.0–35.0
Output light fraction from air classifier (percent)	
Ferrous metals	2.0–20.0
Nonferrous metals	45.0–65.0
–14 Mesh fines	80.0–99.0
Paper and plastic	85.0–99.0
Ash	45.0–85.0
Column loading (ton/h)/ft^2	5.0–40.0

[a] Adapted from Ref. 15.

an air classifier can be described by the following expression [7]:

$$\text{Efficiency} = \sqrt{\frac{X_e}{X_0} \times \frac{Y_r}{Y_0}} \times 100\% \qquad (12\text{-}11)$$

where X_e = mass of combustible light materials extracted from the air classifier
X_0 = mass of combustible light materials input to the air classifier
Y_r = mass of noncombustible materials exiting the air classifier as heavies
Y_0 = mass of noncombustible materials input to the air classifier as heavies

The efficiency of an air classifier is valid only for a specific air/solids ratio and fluidizing velocity, as discussed in the following sections.

Design Criteria

Air classification systems are designed primarily based on the air/solids ratio (lb air/lb waste material) and the required fluidizing velocity. An air/solids ratio of 2 to 7 is recommended for MSW air classification (see Table 12-6). Fluidizing velocity data for various materials are summarized in Table 12-7. It should be noted that the data reported in Table 12-7 were derived using small pilot-scale equipment and that data derived from full-scale units would be expected to vary with the geometry of the separator as well as with the loading rate. Fluidizing velocity data for pulsed air classifiers are summarized in Table 12-8.

Selection of Density Separation Equipment

Factors that should be considered in the selection of density separation equipment are summarized in Table 12-9. Characteristics of the waste stream, specifications of

TABLE 12-7
Fluidizing velocities for air separation of various solid waste components[a]

Component	Velocity, ft/min	
	Zigzag classifier with 2-in throat	Straight 6-in diameter pipe
Plastic wrapping (shirt bags)	Less than 400 (electrostatic)	—
Dry, shredded newspaper (25% moisture)	400–500	350
Dry, cut newspaper		
1-in rounds	500	350
3-in squares	—	350
Agglomerates of dry, shredded newspaper and cardboard	600	—
Moist, shredded newspaper (35% moisture)	750	—
Dry, shredded corrugated cardboard	700–750	450–500
Dry, cut corrugated cardboard		
1-in rounds	980	700
3-in squares	—	1000
Styrofoam, packing material	750–1000 (electrostatic)	—
Foam rubber (1/2-in squares)	2200	—
Ground glass, metal, and stone fragments (from automobile body trash)	2500–3000	—
Solid rubber (1/2-in squares)	3500	—

[a] Adapted from Ref. 2.

TABLE 12-8
Performance characteristics of pulsed air classifiers[a]

Classifier type	Air velocity range,[b] ft/min	
	High feed rate	Low feed rate
Zig-zag, nonpulsed (short classifier)	353.0	372.6
Zig-zag, nonpulsed (tall classifier)	313.8	431.4
Louvered valve, active pulsed (tall classifier)	392.2	588.3
Stacked triangle, passive pulsed (short classifier)	470.1	725.6
Stacked triangle, passive pulsed (tall classifier)	784.4	980.5

[a] Adapted from Ref. 7.

[b] Efficiency greater than 90%.

TABLE 12-9
Factors to be considered in the selection of density separation equipment

Factor	Comment
Characteristics of material produced by the shredder or other separation device (e.g., a trommel or disc separator)	Particle size, shape, bulk specific weight, moisture content, particle size distribution, clumping tendency, fiber content
Materials specifications for light fraction	Particle size, particle size distribution
Transfer method for materials to and from the separation unit	Conveyor characteristics and specifications
Air classifier design parameters	Air/solids ratio (lb air/lb solids), fluidizing velocities (ft/min), unit capacity (lb/h), total airflow (ft^3/min), and pressure drop (in of water)
Stoner design parameters	Bed slope, fluidizing air, exhaust air
Operational characteristics	Energy requirements (hp), maintenance, complexity of operation, noise, and air emissions
Site factors	Floor space and vertical space availability, access

the desired product, and the mechanical constraints of the installation are important selection criteria. Whenever possible, manufacturers data should be confirmed with pilot- or full-scale test results.

12-5 MAGNETIC AND ELECTRIC FIELD SEPARATION

Magnetic and electric field separation techniques use the electrical and magnetic properties of waste materials, such as *electrostatic charge* and *magnetic permeability,* to separate materials. *Magnetic separation* is the most commonly used technology for separating ferrous from nonferrous metals. *Electrostatic separation* can be be used to separate plastics from paper, based on the differing surface charge characteristics of the two materials. *Eddy current separation* is a technique in which varying magnetic fields are used to induce eddy currents in nonferrous metals such as aluminum, forming an "aluminum magnet."

Types of Equipment

The types of equipment that are used in magnetic and electric field separation are described in the following subsections.

Magnetic Separation. Either permanent magnets or electromagnets can be used in one of several configurations. In a typical multistage belt system designed to operate at the end of a conveyor (see Fig. 12-14*a*), three magnets are employed. The first magnet is used to attract the metal. The transfer magnet is used to

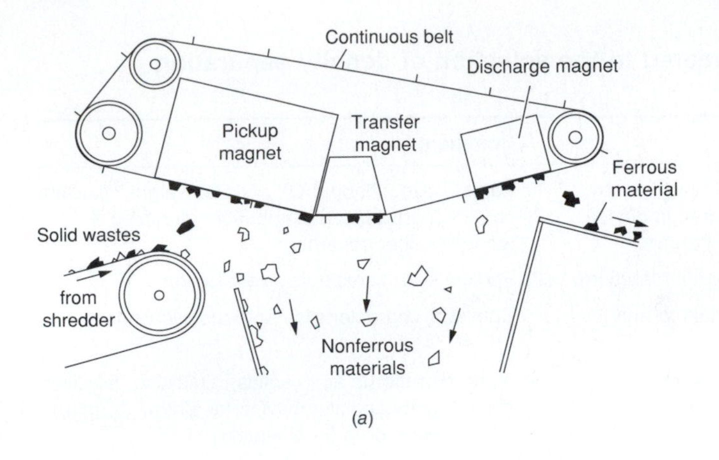

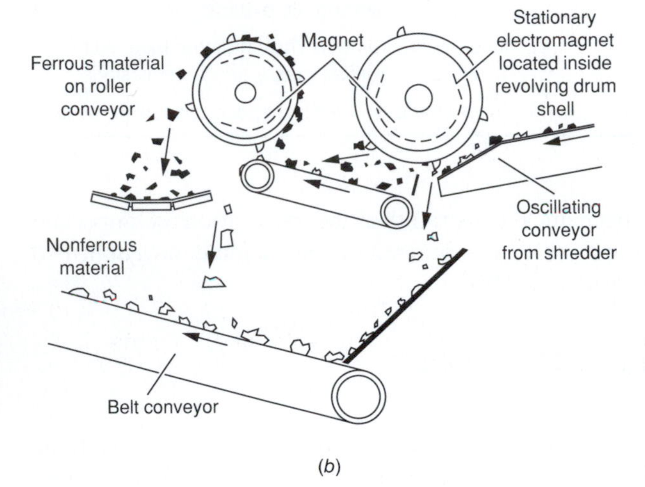

FIGURE 12-14
Typical magnetic separation systems used with shredded MSW: (*a*) belt-type magnetic separator and (*b*) two-drum magnetic separator.

convey the attracted material around a curve and to agitate it. When the attracted metal reaches an area where there is no magnetism, it falls away freely, and any nonferrous metal trapped by the ferrous metal against the belt also falls. The ferrous metal is then pulled back to the belt by the final magnet and is discharged to another conveyor or into storage containers (see Fig. 12-15).

To achieve the cleanest possible recovered material without secondary shredding or air classification, a two-drum installation such as the one shown in Fig. 12-14*b* may be used. The first magnetic drum is used to pick up ferrous material from the shredded waste and toss it forward to an intermediate conveyor. Most of the nonmagnetic material falls to a takeaway conveyor located below the primary separator. Because of the reduced burden on the intermediate conveyor, the second drum conveyor can be smaller and can be positioned closer to the conveyor. To ensure that no bridging or jamming occurs, the second drum rotates in a direction opposite to the flow of the material. Suspended magnets and magnetic pulleys, which have also been used, are shown in Figs. 9-11*a* and 9-11*b*, respectively.

FIGURE 12-15
View of magnetic separator installation used for the separation of ferrous metals from shredded MSW. An armored stainless steel self-cleaning belt is now used more commonly in solid waste separation applications to minimize belt damage. (Courtesy of Dings Co., Magnetic Group.)

Electrostatic Separation. High-voltage electrostatic fields can be used to separate nonconductors of electricity, such as glass, plastic, and paper, from conductors such as metals. It is also possible to separate nonconductors from each other based on differences in their *electrical permittivity,* ϵ, or ability to retain electrical charge. Thus, it is possible to separate paper from plastics, and different types of plastics from each other. Although this technology is not in widespread use at this time, it can be expected to be more fully developed as plastics recycling becomes more important.

Eddy Current Separation. Eddy current separation is based on Faraday's law of electromagnetic induction:

$$-\frac{dB}{dt} = \frac{V}{A} \tag{12-12}$$

where B = magnitude of magnetic flux density, T
V = voltage
A = cross-sectional area normal to magnetic field, m^2

If a conductor such as aluminum is placed in a time-varying magnetic field, a voltage will be generated in the material. This voltage will cause a current to flow and induce a magnetic field that is opposite in polarity to the applied time-varying field, thus producing a magnetic force, which will repel the conductor out of the magnetic field. This magnetic effect is used in a number of electromechanical devices, including the rotary induction motor and the linear induction motor, which may have application in future mass transit systems.

A time-varying field can be created either by rapidly reversing the voltage on an electromagnet (i.e., using alternating current) or by using strips of permanent magnets with alternating polarities. Several devices have been developed using this latter technique: the *ramp separator,* in which solid waste is rolled down a 45° inclined ramp in which the permanent magnets are embedded; and the *vertical eddy current separator*, in which solid waste is caused to fall freely between two plates with alternating permanent magnet strips. A recently developed variation of this principle is the *rotating disc separator,* in which a time-varying magnetic field is produced by a rotating disc that is constructed with sectors of alternating magnetic polarity [3].

Although these devices have demonstrated the ability to separate aluminum from shredded mixed solid waste streams, their use in an integrated solid waste management system is economically questionable. Source separation and buyback schemes for aluminum beverage cans—the predominant source of aluminum in MSW—have become so successful that the percentage of aluminum in the U.S. waste stream is decreasing. Such devices will probably have their widest use in the separation of automobile body shredder waste, which has an ever-increasing proportion of aluminum as automobiles are made lighter by the use of aluminum components.

Performance Characteristics

The performance of electrical and magnetic field separation devices can be evaluated by the purity, recovery, and efficiency expressions developed earlier, Eqs. (12-2) through (12-6). In general, magnetic separation devices have been found to have very high efficiencies (> 95 percent), while the purity and recovery of eddy current devices have been reported as high as 98 percent. Entrainment of nonconductive and nonmagnetic organic particles is the primary cause for impurities in the extract from these devices.

Design Criteria

Magnetic and electric field separation equipment is designed based on the mass loading of material flow (ton/h). Power consumption (kWh) is also an important consideration. The use of permanent magnets in both magnetic and eddy current devices is important for reducing their operating costs, although it increases the capital costs compared to devices using electromagnets. Because the use of magnetic separators is an established practice, ratings from established manufacturers can be used with confidence. Eddy current and electrostatic devices for waste separation are emerging technologies, therefore, pilot plant testing is recommended.

Selection of Magnetic and Electric Field Separation Equipment

Factors that should be considered in the selection of magnetic and electric field separation equipment are summarized in Table 12-10. The characteristics of the waste stream, the desired product, and the mechanical constraints of the installation are important selection criteria. The performance of a magnet drops off with increasing thickness of material on the conveyor belt. The effectiveness of a magnet as a function of the working distance is illustrated in Fig. 12-16.

TABLE 12-10

Factors to be considered in the selection of magnetic and electric field separation equipment

Factor	Comment
Characteristics of material to be separated	Particle size, shape, processed distribution, moisture content, material composition
Materials specifications for separated materials	Purity, recovery, and efficiency requirements
Transfer method for materials to and from the separation device	Conveyor characteristics and specifications
Device design parameters	Unit capacity (ton/h), power requirements (voltage, amperage), magnet strength, electrostatic field strength
Operational characteristics	Energy requirements (kWh), maintenance, complexity of operation, noise, and air emissions
Site factors	Floor space and vertical space availability, access

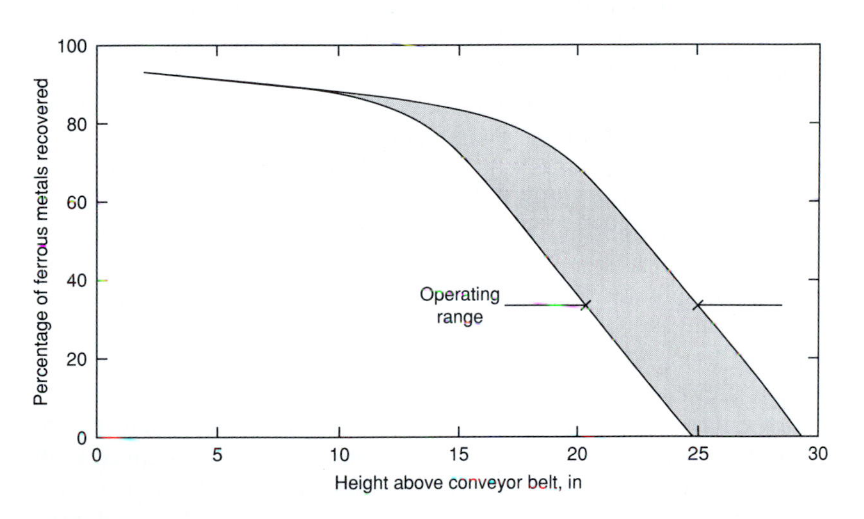

FIGURE 12-16

Typical ferrous metal recovery efficiency for magnetic separators. (Adapted from Refs. 8 and 17.)

12-6 DENSIFICATION (COMPACTION)

Densification (also known as compaction) is a unit operation that increases the density of waste materials so that they can be more efficiently stored and transported. Several technologies are available for the densification of solid wastes and recovered materials, including baling, cubing, and pelleting. Equipment for the densification of landfilled solid wastes is discussed in Chapter 11. Densification of solid waste is performed for several reasons, including reduction of storage requirements of recyclables, reduction of volume for shipping, and preparation of densified refuse-derived fuels (dRDF).

Types of Equipment

This section discusses the types of equipment that are available for the densification of solid wastes. This equipment includes stationary compactors; baling machines, which produce bales secured with wire or plastic ties; and cubing and pelleting machines, which produce woodlike cubes or pellets that are structurally stable because of chemical binding agents (such as Presto™ logs) or heat treatment during the densification process (as in cubing or pelleting machines).

Stationary Compactors. The types of compaction equipment used in solid waste operations may be classified as *stationary* or *movable*. Where wastes are brought to and loaded into the compactor either manually or mechanically, the compactor is stationary. Using this definition, the compaction mechanism used to compress wastes in a collection vehicle is, in fact, a stationary compactor. By contrast, the wheeled and tracked equipment used to place and compact solid wastes in a sanitary landfill is classified as movable. The types and applications of compaction equipment used routinely are reported in Table 12-11.

Typically, stationary compactors may be described according to their application as (1) light-duty, such as those used for residential and light commercial MSW; (2) commercial or light industrial; (3) heavy industrial; and (4) transfer station. Compactors used at transfer stations may be further divided according to the compaction pressure: low-pressure (less than 100 lb/in^2) and high-pressure (100 lb/in^2 or more). In general, all compactors in the other applications would also be classified as low-pressure units.

Where large stationary compactors are used, the wastes can be compressed (1) directly into the transport vehicle (see Fig. 10-7), (2) into steel containers that can be subsequently moved (see Fig. 12-17), (3) into containers equipped with internal compaction facilities (see Fig. 7-11*a*), (4) into specifically designed steel chambers in which the compressed block of solid wastes is banded or tied prior to being removed, or (5) into chambers where they are compressed into a block and then released and hauled away in an untied form (see Fig. 10-5). Application of such compaction equipment to transfer stations is discussed in Chapter 10.

Baling Equipment. Balers are an alternative to compaction equipment. Operating under high pressure, typically 100 to 200 lb/in^2, they produce relatively small,

TABLE 12-11
Compaction equipment used for volume reduction

Location or operation	Type of compactor	Remarks
Solid waste generation points	Stationary/residential	
	Vertical	Vertical compaction ram; may be mechanically or hydraulically operated; usually hand-fed; wastes compacted into corrugated box containers or paper or plastic bags; used in medium- and high-rise apartments.
	Rotary	Ram mechanism used to compact wastes into paper or plastic bags on rotating platform, platform rotates as containers are filled; used in medium- and high-rise buildings.
	Bag or extruder	Compactor can be chute-fed; either vertical or horizontal rams; single or continuous multibags; single bags must be replaced and continuous bags must be tied off and replaced; used in medium- or high-rise apartments.
	Undercounter	Small compactors used in individual residences and apartment units; wastes compacted into special paper bags; after wastes are dropped through a panel door into bag and door is closed, they are sprayed for odor control; button is pushed to activate compaction mechanism.
	Stationary/commercial	Compactor with vertical or horizontal ram; waste compressed into steel container; compressed wastes are manually tied and removed; used in low-, medium-, and high-rise apartments, commercial and industrial facilities.
Collection	Stationary/packer	Collection vehicles equipped with compaction mechanisms.
Transfer and/or processing station	Stationary/transfer trailer	Transport trailer, usually enclosed, equipped with self-contained compaction mechanism.
	Stationary	
	Low-pressure	Wastes are compacted into large containers.
	High-pressure	Wastes are compacted into dense bales or other forms.
Disposal site	Movable wheeled or tracked equipment	Specially designed equipment to achieve maximum compaction of wastes in situ.
	Stationary/track-mounted	High-pressure movable stationary compactors used for volume reduction at disposal sites.

FIGURE 12-17
Low-pressure compactors used at commercial facilities.

compact bales of solid waste or recovered materials. Typical bale sizes range from 48 × 30 × 42 in. up to 72 × 30 × 44 in. Weight of the bales depends on the material and ranges from 1150 to 1800 lb for the small and large bales of corrugated cardboard.

Although baled MSW has been landfilled in "balefills," the predominant use of baling is in the preparation of recovered materials for shipment to buyers of recycled materials. Virtually all of the most common recycled materials can be baled, including cardboard, newsprint, plastic, PETE bottles, and aluminum cans. Baled materials are easy to load with standard forklifts and can be economically shipped because of their high bulk density.

Cubing and Pelleting Equipment. Cubing and pelleting is a technology that can be used to produce densified refuse-derived fuels (dRDF) for combustion in incineration, gasification, or pyrolysis systems. The added expense of the process can be justified when the dRDF has to be shipped a long distance, stored for long periods, or burned in existing boilers designed for coal or wood. One of the first dRDF facilities in the United States was operated in the early 1970s by the City of Ft. Wayne, Indiana. A John Deere alfalfa-cubing machine, originally built to produce cattle feed, was used to densify the light fraction of MSW from a resource recovery plant. The cubes were burned in a 1:3 ratio with coal in a 40-MW electric generating plant [11]. Use of dRDF in a gasification system is described in Chapter 13. Densified RDF pellets made from sugar cane bagasse are being used in Hawaii to fuel a cogeneration plant at a sugar mill.

Cubing and pelleting machines operate on a similar principle. Waste paper or shredded RDF is extruded through extrusion dies with an eccentric rotating presswheel (see Fig. 12-18). A complete cubing or pelleting system requires a shredder, a conveyor, and a moisture control system (see Fig. 12-19). The cubes or pellets (see Fig. 12-20) are bonded together by heat caused by friction as the cubes or pellets are extruded. If kept dry, dRDF can be stored for months without decomposition. Typical cubes are approximately 3 in by 1 in^2. Pellets are cylindrical, typically $\frac{1}{2}$ to $\frac{3}{4}$ in diameter by $\frac{1}{2}$ to 1 in long.

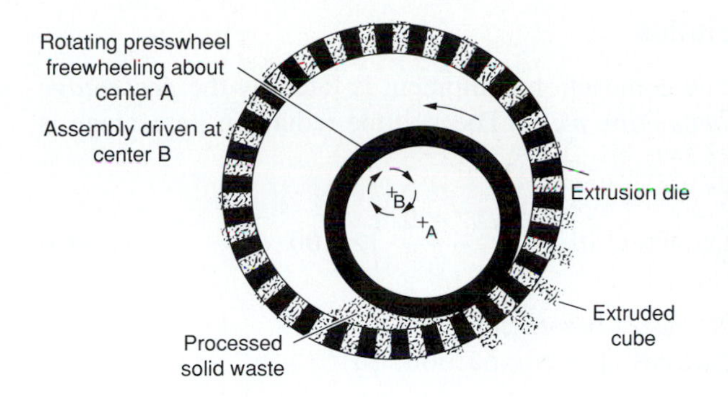

FIGURE 12-18
Cross section of extrusion dies used in a typical cubing machine for processed MSW. Note: Pelleting machines work on a similar principle, but the diameter of the dies is smaller.

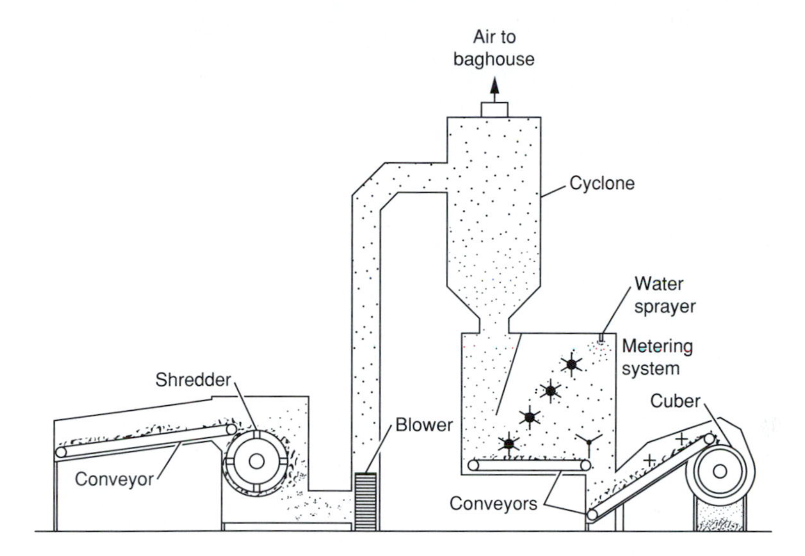

FIGURE 12-19
Schematic diagram of a complete cubing system for processed MSW.

FIGURE 12-20
Densified refuse-derived fuel (dRDF): cubes shown on top, pellets shown below ruler.

Performance Characteristics

The performance of baling and compaction equipment is rated by the *percentage volume reduction* and the *compaction ratio*. The volume reduction percentage is defined as follows:

$$\text{Volume reduction (\%)} = \left(\frac{V_i - V_f}{V_i}\right) \times 100 \qquad (12\text{-}13)$$

where V_i = initial volume of wastes before compaction, yd^3
V_f = final volume of wastes after compaction, yd^3

The compaction ratio is defined as

$$\text{Compaction ratio} = \frac{V_i}{V_f} \qquad (12\text{-}14)$$

where V_i, V_f = as defined in Eq. (12-13)

The relationship between the compaction ratio and the percent of volume reduction is shown graphically in Fig. 12-21. Because of the nature of the relationship, it can be seen that to achieve more than about 80 percent reduction requires a disproportionate increase in compaction ratio. For example, to achieve an increase from 80 to 90 percent requires an increase in compaction ratio from 5 to 10. This relationship is important in making a tradeoff analysis between compaction ratio and overall compaction cost.

Another important factor that must be considered is the final specific weight (density in metric units) of the wastes after compaction. Some typical curves for unprocessed municipal solid wastes are presented in Fig. 12-22. The asymptotic value used in developing these curves is 1800 lb/yd^3, which is consistent with values obtained by using high-pressure compactors. As a point of reference, the specific weight of water is 1685 lb/yd^3. When shredded wastes are compacted under the same conditions, the density may be up to 35 percent greater than that

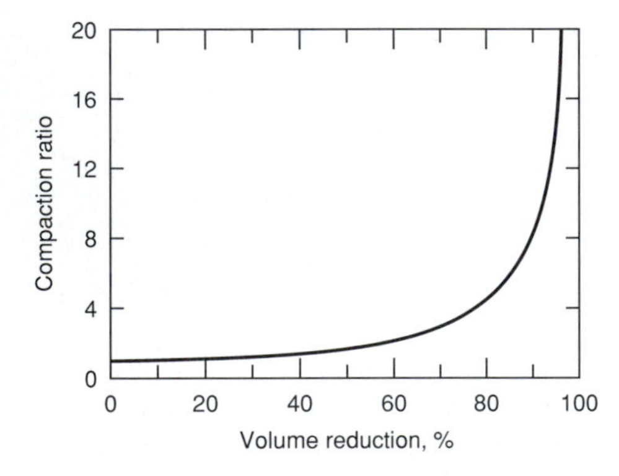

FIGURE 12-21
Compaction ratio versus percent volume reduction.

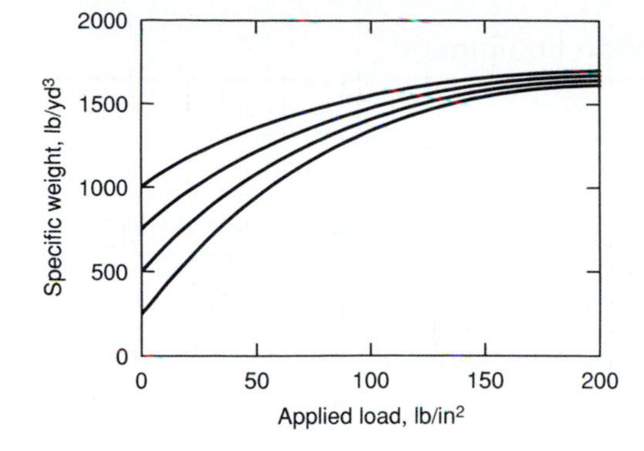

FIGURE 12-22
Specific weight of unprocessed solid wastes versus applied pressure.

of unprocessed wastes, up to an applied pressure of 100 lb/in^2. The maximum specific weight reached under the application of pressures greater than 100 lb/in^2 is not affected significantly by shredding.

Perhaps the most significant fact to be noted in Fig. 12-22 is that the initial increase in specific weight brought about by the application of pressure is highly dependent on the initial specific weight of the wastes to be compacted. This fact is important when claims made by manufacturers of compaction equipment are considered. The moisture content, which varies with location, is another variable that has a major effect on the degree of compaction achieved. In some stationary compactors, provision is made to add moisture, usually in the form of water, during the compaction process (see Fig. 12-19).

For cubing and pelleting equipment, the two most important performance characteristics are the *unit* and *bulk* specific weights of the cubes and pellets. The unit specific weight is the specific weight of an individual cube or pellet. The bulk specific weight is a measure of a fixed volume of cubes or pellets (i.e., 1 cubic foot) and is indicative of the specific weight of cubes or pellets in storage or during transport, because it includes the void space caused by randomly packing cubes in a container. For example, the cubing system shown in Fig. 12-19 has achieved unit specific weights of 1750 lb/yd^3 for shredded newsprint. However, the bulk specific weight of the cubes was only 814 lb/yd^3.

Design Criteria

For design purposes, densification equipment is rated on throughput in terms of ton/h. Attention must be paid to the bulk density of the incoming and outgoing materials in the design of conveyors and material-handling equipment. While some densified materials, such as plastic or aluminum, can tolerate moisture, most densified materials should be stored in a covered area protected against rain. Because the ultimate bulk specific weight of the material is highly dependent on the original specific weight and moisture content of the material, a witnessed field test should be performed to confirm performance, or a similar installation should

TABLE 12-12
Typical design factors for compaction equipment[a]

Factor	Value		Remarks
	Unit	Range	
Size of loading chamber	yd^3	<1–11	Fixes the maximum size of wastes that can be placed in the unit.
Cycle time	s	20–60	The time required for the face of the compaction ram, starting in the fully retracted position, to pack wastes in the loading chamber into the receiving container and return to the starting position.
Machine volume displacement	yd^3/h	30–1500	The volume of wastes that can be displaced by the ram in 1 h.
Compaction pressure	lb/in^2	15–50	The pressure on the face of the ram.
Ram penetration	in	4–26	The distance that the compaction ram penetrates into the receiving container during the compaction cycle. The further the distance, the less chance there is for wastes to fall back into the charging chamber and the greater the degree of compaction that can be achieved.
Compaction ratio		2:1–8:1	The initial volume divided by the final volume after compaction. Ratio varies significantly with waste compaction.
Physical dimensions of unit	variable	variable	Affects the design of service areas in new buildings and provision of service to existing facilities.

[a] Adapted from Ref. 2.

be visited. Design criteria for compaction and baling equipment are summarized in Tables 12-12 and 12-13.

Selection of Densification Equipment

Factors that should be considered in the selection of densification equipment are summarized in Table 12-14. Characteristics of the wastes to be densified, product throughput, and the mechanical constraints of the installation are important selection criteria. In general, compaction equipment is used to process solid waste prior to landfilling or combustion to reduce haul costs; baling is primarily used for the processing of recovered materials prior to sale; and cubing or pelleting is used in the preparation of densified RDF.

TABLE 12-13
Typical design factors for baling equipment[a]

Factor	Unit	Typical value	Comment
Small baler			
Bale size	in	42 × 30 × 30	Highly variable, depends on manufacturer
Motor size	hp	10	
Operating pressure	lb/in^2	100	
Bale weight	lb	500	Corrugated cardboard
Medium baler			
Bale size	in	62 × 45 × 30	Highly variable, depends on manufacturer
Motor size	hp	75	
Operating pressure	lb/in^2	225	
Hourly production	ton/h	5–9	Corrugated cardboard
		15–20	Commingled MSW
Bale weight	lb	to 1350	Corrugated cardboard
		to 1000	Aluminum cans
		to 2800	Commingled MSW
Large baler			
Bale size	in	62 × 45 × 30	Highly variable, depends on manufacturer
Motor size	hp	200	
Operating pressure	lb/in^2	225	
Hourly production	ton/h	12–20	Corrugated cardboard
		30–40	Commingled MSW
Bale weight	lb	to 1350	Corrugated cardboard
		to 1100	Aluminum cans
		to 2800	Commingled MSW

[a] Adapted from Ref. 1.

TABLE 12-14
Factors to be considered in the selection of densification equipment

Factor	Comment
Purpose of densification	Compaction equipment for processing of MSW, cubing and pelleting equipment for dRDF production, baling equipment for processing recyclables and MSW
Characteristics of material to be processed	Particle size, shape, particle size distribution, moisture content, material composition, specific weight
Transfer method for materials to and from the densification equipment	Conveyor characteristics and specifications, forklifts and other material-handling equipment
Equipment design parameters	Unit capacity (ton/h), power requirements (voltage, amperage, horsepower), compaction ratio, unit specific weight, bulk specific weight, bale weight, operating pressure
Operational characteristics	Energy requirements (kWh), maintenance, complexity of operation, noise, and air emissions
Site factors	Floor space and vertical space availability, access

12-7 SELECTION OF FACILITIES FOR HANDLING, MOVING, AND STORAGE OF WASTE MATERIALS

The effective movement of commingled MSW and recycled materials in a materials recovery facility (MRF) or transfer station is critical to the function of these facilities. Improperly designed materials-handling facilities can create bottlenecks, reducing the efficiency of the entire processing system. Belt and pneumatic conveyors and the use of conveyor belts for manual separation are discussed in this section.

Belt Conveyors

Belt conveyors, the most widely used materials-handling equipment, are used for conveying commingled MSW as well as recycled materials. Conveyors are also used to convey materials on manual sorting lines, where recycled materials can be handpicked. A conveyor is an endless belt, supported by antifriction idler bearings and driven from one end by a drive roller. Belts can be made from rubber, canvas, or synthetic materials for handling relatively lightweight recycled materials (see Fig. 12-23*a*). For heavy-duty applications, such as unseparated commingled MSW and recovered metals, hinged steel belts are used (see Fig. 12-23*b*).

Belt conveyors are designed on the basis of belt speed, mass throughput (lb/h or ton/h), horsepower, and thickness of material on the belt. Velocities of 10 to 100 ft/min have been used for metal belt conveyors and 100 to 450 ft/min for flexible belt conveyors. The following empirical equations can be used for determining the power requirements of a belt conveyor [9]:

$$\mathrm{HP}_{\mathrm{empty}} = \frac{(A + BL)}{100} S_{\mathrm{belt}} \qquad (12\text{-}15)$$

(*a*)

(*b*)

FIGURE 12-23
Typical belt conveyors used for the transport of waste materials at MRFs: (*a*) troughed belt used to transport shredded waste and (*b*) drag conveyor used for commingled MSW.

TABLE 12-15
Empirical constants for belt conveyors[a]

Conveyor belt width, in	Empirical constants[b] A	B
14	0.20	0.00140
16	0.25	0.00140
18	0.30	0.00162
20	0.30	0.00187
24	0.36	0.00224
30	0.48	0.00298
36	0.64	0.00396
42	0.72	0.00458
48	0.88	0.00538
54	1.00	0.00620
60	1.05	0.00765

[a] Adapted from Ref. 9.
[b] To be used with Eq. (12-15).

where HP_{empty} = horsepower to drive empty conveyor
A, B = empirical constants (see Table 12-15)
L = conveyor length, ft
S_{belt} = belt speed, ft/min

$$HP_{level} = M\frac{(0.48 + 0.00302L)}{100} \tag{12-16}$$

where HP_{level} = horsepower to convey material on level
M = throughput, ton/h

$$HP_{lift} = \frac{1.015HM}{1000} \tag{12-17}$$

where HP_{lift} = horsepower to lift material
H = lift height, ft

$$HP_{total} = HP_{empty} + HP_{level} + HP_{lift} \tag{12-18}$$

Conveyors Used for Manual Separation of Waste

Manual separation of materials is regaining popularity, because it can potentially produce materials with the highest quality. The use of conveyors on a picking line allows for partial automation of the process and increases worker efficiency and safety. A typical sorting line for separation of plastics, aluminum, and glass is illustrated in Fig. 12-24. Critical factors in the design of picking belts are the width of the belt, the speed of the belt, and the average thickness of the waste

FIGURE 12-24
Belt conveyor system for the manual sorting of recycled materials. (Courtesy of Mayfran International and Waste Management Inc.)

material on the belt (often referred to as the *average burden depth*). The maximum belt width, where separation is from either side of the belt, is about 4 ft. Belt speeds vary from about 15 to 90 ft/min. The thickness of the waste on the belt can be determined as follows:

$$\text{TW}_b,\ \text{in} = \frac{(\text{LR, ton/h}) \times (2000\ \text{lb/ton}) \times (12\ \text{in/ft})}{(60\ \text{min/h}) \times (S_{\text{belt}},\ \text{ft/min}) \times (W,\ \text{lb/ft}^3) \times (\text{BW}_{\text{eff}},\ \text{ft})} \tag{12-19}$$

where TW_b = thickness of waste on the conveyor belt, in
LR = solid waste loading rate, ton/h
S_{belt} = speed of the belt, ft/min
W = average waste specific weight, lb/ft^3
BW_{eff} = effective belt width, ft

For example, for a loading rate of 25 ton/h, an average belt speed of 60 ft/min, an average waste specific weight of 7 lb/ft^3, and an effective belt width of 4 ft, the average thickness of waste would be about 6.0 in.

Pneumatic Conveyors

Pneumatic conveying can be used to move shredded materials such as newsprint, plastic, or refuse-derived fuel. Pneumatic conveying systems consist of a fan, a feeding device, piping, and a discharge device, typically a cyclone. Systems can be operated as a vacuum system, operating below atmospheric pressure (see Fig. 9-17*a*), or as a pressure system, operating at up to 14 in of water pressure (see Fig. 9-17*b*). Air velocities for processed wastes are typically in the range of 4000 to 5000 ft/min, and the materials-to-air ratio is typically 0.1 lb material/1.0 lb air [17]. By comparison, the velocities needed to transport wood chips and shavings and fine coal are 3000 and 5000 ft/min, respectively [4]. Pneumatic conveyors offer considerable design flexibility, because the piping can be routed as required.

The following equations have been proposed for estimating the minimum carrying velocities for pneumatic transport of particulate material in horizontal and vertical ducts [4]. For horizontal ducts:

$$V = 6000\frac{S}{S+1}d^{0.4} \tag{12-20}$$

For vertical ducts:

$$V = 13{,}300\frac{S}{S+1}d^{0.6} \tag{12-21}$$

where V = air velocity, ft/min
S = specific gravity of material being transported
d = diameter of longest particle to be transported, in

Detailed design information on pneumatic conveyors may be found in Ref. 9.

Storage Containers and Buildings

Processed materials must be protected from the elements and stored prior to shipment. A variety of containers and buildings can be used for storing recycled materials, including (1) fully enclosed warehouse space, (2) open-sided roofed structures, (3) bulk shipping containers, and (4) custom bins for refuse-derived fuels.

Since recycled materials are generally of low value, storage must be low-cost. Storage in conventional warehouse space is generally not economically feasible unless special conditions exist. In mild climates, open-sided roofed storage is acceptable if rain can be kept from damaging sensitive materials such as baled paper. Other baled materials, such as plastics, aluminum, and cardboard can be safely stored outdoors.

One of the most economical storage systems is the use of shipping containers, particularly if they can be provided by the material buyer. This method

(*a*)

(*b*)

FIGURE 12-25
Loading and unloading facility for large containers used to ship baled paper overseas: (*a*) overall view of facility at the Port of Philadelphia's Packer Avenue Terminal and (*b*) empty container being loaded onto truck chassis to be returned to MRF. (Courtesy of Philadelphia Regional Port Authority.)

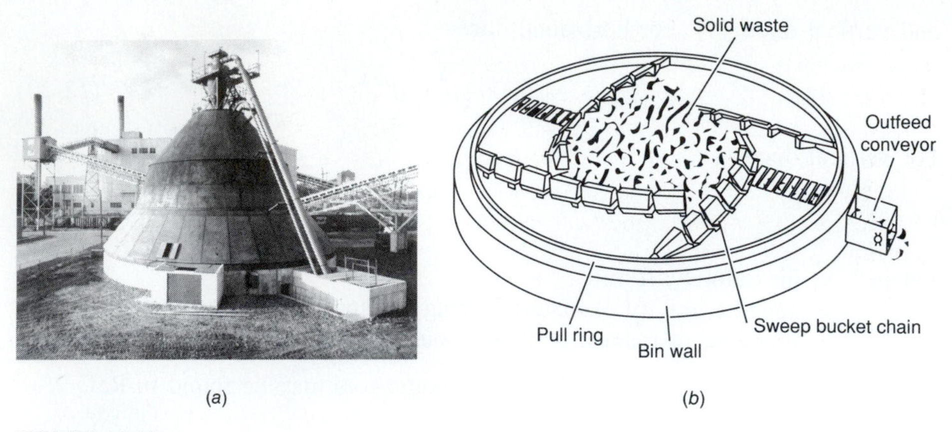

FIGURE 12-26
Atlas Bin™ for storage of RDF: (*a*) general view of bin and (*b*) diagram of the internal mechanism used to recover stored material. (Courtesy of Atlas Systems Corp.)

is widely used for the shipment of various grades of paper and cardboard to overseas markets on container ships (see Fig. 12-25). The shipboard containers are temporarily stored at the processing facility and filled as the paper is processed. They are then trucked to the buyer without reloading. A similar approach can also be used for domestic storage and shipment of glass and paper.

The storage of refuse-derived fuel (RDF) is more complex. Because RDF has low density and is potentially putrescible, it must be handled in special bins that operate on a first-in, first-out (FIFO) basis. This ensures that RDF will not be stored longer than necessary. Several designs have been developed that operate on a FIFO basis and also claim to avoid bridging or clogging of the low-density RDF. The Atlas Bin™ is a conical-shaped hopper with a rotary bottom feed system (see Fig. 12-26).

12-8 MOVABLE EQUIPMENT USED FOR MATERIALS HANDLING

In the design of materials-handling facilities, it is sometimes more economic to use mobile equipment such as front end loaders and forklifts to move materials. For example, in a typical application, the MSW would be dumped on a collection floor by collection vehicles and pushed by the front end loader onto belt conveyors for further processing (see Fig. 9-18*a*). Front end loaders can also be used to offload materials after processing, such as loading shredded wood waste into trucks for shipment to offsite customers. Forklifts are widely used to move baled materials from baling machines to storage areas and then to offload onto trucks for transport to market (see Fig. 9-18*b*).

The designer has the option of using fixed-belt and pneumatic conveyors or mobile equipment for many operations. The tradeoff is between the higher capital cost of the fixed systems and the higher labor costs of mobile equipment. In any case, some mobile equipment will always be required for loading and unloading of trucks. Because the mobile equipment is usually operated in tightly confined areas, operator training and safety is important. Front end loaders are generally powered by gasoline, LP gas, or diesel engines. Forklifts are available with either internal combustion engines or electric motors powered by batteries. For equipment operated indoors, proper ventilation of exhaust gases is critical.

12-9 DESIGN OF MATERIALS RECOVERY FACILITIES (MRFs)

As defined in Chapter 9, two basic MRF designs are possible: (1) MRFs for source-separated materials and (2) MRFs for commingled MSW. Design considerations, design examples, and equipment selection are presented and discussed in this section for both types of MRFs.

Design Considerations for MRFs

The design of a MRF involves three basic steps: (1) feasibility analysis, (2) preliminary design, and (3) final design. The first step was discussed in depth in Chapter 9. This section will focus on preliminary and final design.

The preliminary design process includes the development of process flow diagrams, the calculation of materials recovery rates, and the preparation of materials balances and loading rates. The final design process includes the preparation of detailed plans and specifications.

Process Flow Diagrams. In the feasibility study, the basic decision is made on how recyclable materials are to be recovered from MSW (e.g., source separation or separation from commingled MSW). The next step is to develop process flow diagrams. A *process flow diagram* for a MRF is defined as the assemblage of unit operations, facilities, and manual operations to achieve a specified set of design goals. Important factors that must be considered in the development of process flow diagrams include (1) characteristics of the waste materials to be processed, (2) specifications for recovered materials now and in the future, and (3) the available types of equipment and facilities. Typical process flow diagrams for source-separated and commingled MRFs have been presented in Chapter 9. Detailed preliminary design examples of MRFs for both source-separated materials and commingled MSW are presented in this chapter.

Materials Recovery Rates. A MRF is a major component of an integrated waste management system. A MRF developed to process source-separated waste can be thought of as an extension of the recycling program. To predict the materials flow to the MRF, it is necessary to estimate the effectiveness or performance of the recycling program. The performance of a recycling program is generally

TABLE 12-16
Recovery factors for source-separated recycled materials

	Percent recovery	
Material	**Range**	**Typical**
Mixed paper	40 to 60	50
Cardboard	25 to 40	30
Mixed plastics	30 to 70	50
Glass	50 to 80	65
Tin cans	70 to 85	80
Aluminum cans	85 to 95	90

reported as a *materials recovery rate* or *recycling rate,* which is the product of three factors, as shown in the following equation:

$$\begin{matrix}\text{Materials}\\ \text{recovery}\\ \text{rate}\end{matrix} = \left[\begin{matrix}\text{Composition}\\ \text{factor}\end{matrix} \times \begin{matrix}\text{Recovery}\\ \text{factor}\end{matrix} \times \begin{matrix}\text{Participation}\\ \text{factor}\end{matrix}\right] \qquad (12\text{-}22)$$

where Composition factor = fraction of waste component in total waste
Recovery factor = fraction of material recovered by a unit operation or recycling program
Participation factor = fraction of the public that participates in a recycling program

Composition factors are measured in waste composition studies. Typical data are given in Tables 3-4 and 3-7. The use of recovery factors or percentages, when applied to separation devices such as screens, was previously discussed in Section 12-3. Recovery factors can also be derived for source separation in the home. Recovery factors for the recyclable materials most commonly collected in source-separation recycling programs are reported in Table 12-16. The use of these factors to estimate the composition of source-separated materials is discussed in Example 12-4.

Example 12-4 Determination of the composition of recycled materials from a curbside recycling system. Estimate the composition of the recovered materials in a curbside recycling system in which mixed paper, cardboard, mixed plastics, glass, tin cans, and aluminum cans are to be collected. Also estimate the composition of recycled materials if 60 percent of the aluminum cans are removed prior to curbside recycling. The aluminum cans removed by the homeowners are returned to buyback centers. Assume that the following conditions apply:

1. Composition of the MSW is assumed to be the same as in column 4 of Table 3-7.
2. Recovery rates for recycled materials are as given in Table 12-16.

Solution

1. Set up a computation table to estimate the composition of recovered materials.

Component	MSW,[a] %	Weight before recycling,[b] lb	Recovery factor[c]	Weight recovered,[d] lb	Recyclables,[e] %
Organic					
Food wastes	8.0	8.0	—	—	—
Paper	35.8	35.8	0.50	17.9	52.2
Cardboard	6.4	6.4	0.30	1.9	5.5
Plastics	6.9	6.9	0.50	3.5	10.2
Textiles	1.8	1.8	—	—	—
Rubber	0.4	0.4	—	—	—
Leather	0.4	0.4	—	—	—
Yard wastes	17.3	17.3	—	—	—
Wood	1.8	1.8	—	—	—
Misc. organics	—	—	—	—	—
Inorganic					
Glass	9.1	9.1	0.65	5.9	17.2
Tin cans	5.8	5.8	0.80	4.6	13.4
Aluminum	0.6	0.6	0.90	0.5	1.5
Other metals	3.0	3.0	—	—	—
Dirt, ashes, etc.	2.7	2.7	—	—	—
Total	100.0	100.0		34.3	100.0

[a] Column 4 from Table 3-7.

[b] Based on 100 lb of MSW.

[c] From Table 12-16.

[d] Based on 100 lb of MSW.

[e] Based on the weight of material recovered (34.3 lb).

2. Set up a computation table to estimate the composition of recovered materials, taking into account the removal of 60 percent of the aluminum cans by the buyback center.

Component	Weight before aluminum removed, lb	Weight after aluminum removed, lb	Recovery factor	Weight recycled, lb	Recyclables, %
Paper	35.8	35.8	0.50	17.9	52.6
Cardboard	6.4	6.4	0.30	1.9	5.6
Plastics	6.9	6.9	0.50	3.5	10.3
Glass	9.1	9.1	0.65	5.9	17.4
Tin cans	5.8	5.8	0.80	4.6	13.5
Aluminum	0.6	0.24[a]	0.90	0.2	0.6
Total	64.6	64.24		34.0	100.0

[a] $0.24 = 0.6 - (0.6 \times 0.6)$

Comment. Although at first glance it would appear that 34.3 percent of the wastestream has been recycled, this computation represents the composition of the materials recycled by residents who actually participate in the recycling program. Thus, if the participation rate was 50 percent of the residents, the actual recycling rate would only be half of 34.3 percent, or 17.2 percent. Diversion of 60 percent of the aluminum cans has relatively little effect on the total amount of materials recovered by the recycling program, but the diversion does affect the distribution of the recycled materials.

Materials Balances and Loading Rates. One of the most critical elements in the design and selection of equipment for MRFs is the preparation of a *materials balance analysis* to determine the quantities of materials that can be recovered and the appropriate loading rates for the unit operations and processes used in the MRF. The steps involved in the preparation of a materials balance analysis and in determining the required process loading rates are as follows.

Step 1. The first step in performing a materials balance analysis is to define the *system boundary*. In most cases the system boundary is drawn around the MRF itself. This approach is illustrated in Example 12-5. In other cases, it is appropriate to draw the boundary around the community and to account for waste diversions that may occur prior to the MSW being delivered to the MRF. A buyback program for aluminum cans is an example of such a diversion. This latter approach is utilized in Examples 12-6 and 12-7.

Step 2. The second step is to identify all the waste or material flows that enter or leave the system boundary (e.g., a MRF) and the amount of material stored within the system boundary. Typically, for a MRF these can include MSW, source-separated materials, processed materials, nonrecyclable wastes to be landfilled, and crushed glass, as well as baled paper, cardboard, plastics, and aluminum and tin cans.

Step 3. The third step involves the application of the materials balance equations given previously in Chapter 6. From Chapter 6, the simplified word statement for a materials balance is

$$\text{Accumulation} = \text{Inflow} - \text{Outflow} + \text{Generation} \tag{6-2}$$

If waste transformation operations are not carried out at the MRF, the materials balance equation just given can be simplified to yield

$$\text{Accumulation} = \text{Inflow} - \text{Outflow} \tag{12-23}$$

Finally, for an individual unit operation or processing step where material is not accumulated, Eq. (12-23) becomes

$$\text{Inflow} = \text{Outflow} \tag{12-24}$$

Thus, the materials balance analysis for a unit operation or processing step is essentially an accounting process to ensure that all of the materials are completely tracked through the MRF. Computationally, a materials balance is most easily accomplished by the use of a spreadsheet, either on paper or, more commonly, with a spreadsheet program on a microcomputer. The application of the spreadsheet approach to materials balance analysis is illustrated in Examples 12-5, 12-6, and 12-7.

Step 4. The fourth and final step is to develop *materials loading rates* for the individual operations and processing steps in the MRF, using the data from the materials balance analysis. Generally, MSW or source-separated materials delivered to the MRF are expressed in terms of ton/d. Unit operations such as conveyors or screens must be specified in terms of ton/h, so the ton/d rate must be converted into ton/h by taking into account the effective working day. MRFs employing manual sorting lines are generally designed on a 7-h/d effective work day to allow for breaks. Mechanized MRFs are sometimes designed for 16-h/d effective operation to maximize the utilization of expensive equipment. If the total number of operating hours per year is assumed to be 1820, the hourly loading (or processing) rate is given by the following expression:

$$\text{Loading rate, ton/hr} = \frac{\text{Number of ton/yr (or ton/d)}}{\text{1820 processing hr/yr (or hr/d)}} \qquad (9\text{-}1)$$

To allow for scheduled and unscheduled equipment downtimes, some designers suggest that the base loading rate of the facility should be increased by about 10 to 15 percent.

Facility Layout and Design. *Facility layout* refers to the spatial arrangement of the components identified during the preliminary design. The overall MRF layout includes (1) sizing of the unloading areas for commingled MSW and source-separated materials, (2) sizing of presorting areas where oversize or undesirable materials are removed, (3) placement of conveyor lines, screens, magnets, shredders, and other unit operations, (4) sizing of storage and outloading areas for recovered materials, and (5) sizing and design of parking areas and traffic flow patterns in and out of the MRF. Many of these layout steps are also common to the layout and design of transfer stations.

Loading rates calculated during the materials balance analysis are used to select and size individual components. Designers must work with potential vendors to match loading rates with the capacity of available equipment. Particular care must be paid to the design of conveyor systems for manual waste separation, which are designed on the basis of material specific weight; see Eq. (12-19). Typical as-received specific weights for source-separated materials are given in Table 12-17.

Final design requires an interdisciplinary team of architects and environmental, civil, structural, electrical, and mechanical engineers to prepare construction drawings and specifications. At the present time there are no codes or standards developed specifically for MRF design, as there are in the case of water or wastewater treatment plants. Codes and standards that should be considered in the design of MRFs are listed in Table 12-18. Local and state building codes are the final authority.

TABLE 12-17
Typical specific weights of as-received source-separated materials[a]

Material	Typical specific weight, lb/yd^3
Paper	
Newspaper	475
Corrugated cardboard	350
High grades	300–400
Glass—whole bottles	
Clear	500
Green or amber	550
Glass—crushed	
Semi-crushed	1000
1-1/2" mechanically crushed	1800
1/4" furnace ready	2700
Aluminum cans	
Whole	50
Flattened	175
Tin-plated steel cans ("tin cans")	
Whole	150
Flattened	850
Plastics	
PETE, whole	34
PETE, flattened	75
HDPE (natural), whole	30
HDPE (natural), flattened	65
HDPE (colored), whole	45
HDPE (colored), flattened	90

[a] Adapted from Ref. 14.

TABLE 12-18
Codes and standards applicable to MRF design[a]

American Concrete Institute
American Institute of Steel Construction
American National Standards Institute
American Society of Heating, Refrigeration, and Air Conditioning Engineers
American Society of Mechanical Engineers
American Society for Testing and Materials
National Electric Code
Occupational Safety and Health Act
Underwriters Laboratories
Uniform Building Code
Uniform Fire Code
Uniform Mechanical Code
Uniform Plumbing Code
State and local building, safety, and construction codes

[a] This list is not all-inclusive; there may be other applicable codes and standards in addition to those listed.

MRFs for Source-Separated Materials

As discussed in Chapter 9, materials recovery from source-separated waste can be accomplished to recover newsprint, cardboard, several grades of plastics, aluminum, glass, and yard waste. A specialized MRF for the separation of various classes of paper was discussed in Chapter 9 (see Figs. 9-20*a* and 9-20*b*). The development of materials balances and loading rates for a similar MRF is illustrated in Example 12-5. The facility is designed to separate several grades of paper from residential and commercial sources, including old newsprint (ONP), old corrugated cardboard (OCC), and mixed paper. A similar analysis could be performed on a more complex MRF that processes paper, glass, plastics, and other types of source-separated materials.

Example 12-5 Determination of materials balances and loading rates for a MRF processing source-separated paper. Prepare a materials balance for the process flow diagram given in the accompanying figure. Assuming that the MRF is to be designed to

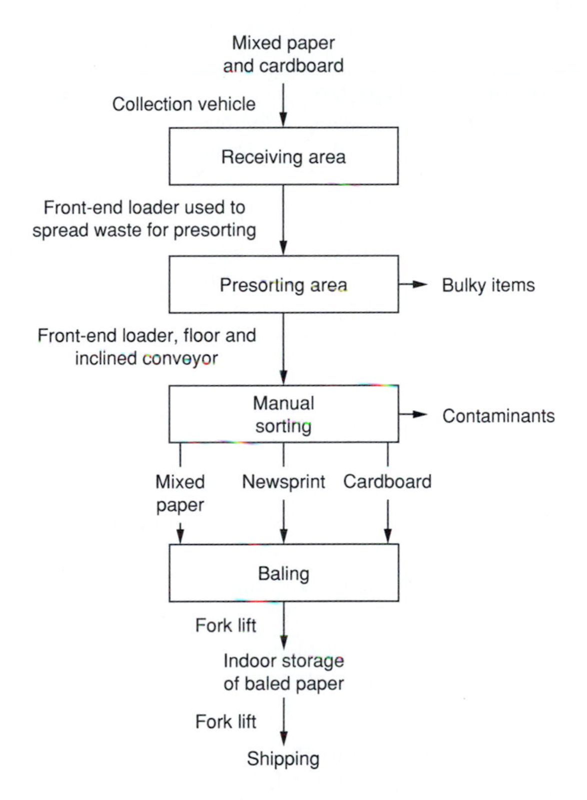

handle 150 ton/d, estimate the hourly loading rates for the various separation processes based on a 7-h/d operation. Determine the quantities recovered. Determine the number of workers required for manual sorting. Assume that the following conditions apply:

1. Composition of source-separated materials:

Old newsprint (ONP)	71.0%
Mixed paper	14.0%
Old corrugated cardboard (OCC)	13.0%
Contaminants	2.0%

2. The baler has a capacity of 16 ton/h. It is also used to process plastics from another process line, so it is fed in batches by a front end loader when a minimum of 4 tons of paper has been accumulated (about 15 minutes of operation).
3. An average worker can manually sort approximately 2.5 ton/h of paper.
4. The recovery factor for mixed paper, OCC, and contaminants is 95 percent. ONP does not carry over into mixed paper or OCC.

Solution

1. Determine the materials-mass balance quantities using the foregoing information; see the accompanying computation table.

Component	Material composition, %	Materials delivered, ton/d	Recovery factor	Materials recovered, ton/d
Old newsprint (ONP)	71.0	106.5		108.6[a]
Mixed paper	14.0	21.0	0.95	20.0
Old corrugated cardboard (OCC)	13.0	19.5	0.95	18.5
Contaminants	2.0	3.0	0.95	2.9
Total	100.0	150.0		150.0

[a]Old newsprint = total − [mixed paper + OCC + contaminants]

2. Percent impurities in ONP = [(108.6 − 106.5)/108.6] × 100% = 1.9% (Note: See Chapter 15 regarding purity requirements for ONP.)
3. Determine the number of workers needed on the sorting line: Workers = 150 ton/d × (1 d/7 h) × 1 worker/2.5 ton/h = 8.6 workers (9 workers will be required; thus, the estimated output per worker = 2.4 ton/h.)
4. Determine the materials sorting rate and time to accumulate 4 ton:

   (*a*) ONP

   i. Recovery rate = 108.6 ton/d × (1 d/7 h) = 15.5 ton/h
   ii. Time to accumulate 4 ton = 4 ton/(15.5 ton/h) = 0.26 h

   (*b*) Mixed paper

   i. Recovery rate = 20.0 ton/d × (1 d/7 h) = 2.9 ton/h
   ii. Time to accumulate 4 ton = 4 ton/(2.9 ton/h) = 1.4 h

   (*c*) OCC

   i. Recovery rate = 18.5 ton/d × (1 d/7 h) = 2.6 ton/h
   ii. Time to accumulate 4 ton = 4 ton/(2.6 ton/h) = 1.5 h

Comment. Note that the baler in this example would be required full-time for ONP baling and would not be available for baling other materials during the 7-hour shift. If space for stockpiling is available, the other materials could be baled on an overtime basis.

MRFs for Commingled Waste

Materials recovery from commingled wastes can be accomplished for the recovery of specific recyclable materials, such as ferrous metals, glass, and aluminum, or as a front-end process for combustion or composting. It has been found by experience that a MRF used to process commingled waste cannot produce recyclables of as high a quality as a MRF that is used for source-separated wastes. The designer must take into account the eventual market for the processed recyclables produced by the MRF and the quality specifications required for that market (see Chapter 15).

Two design approaches are possible: (1) MRFs in which a combination of manual and mechanical sorting operations are emphasized and (2) MRFs in which mechanized sorting is emphasized. A materials balance analysis for the first approach is illustrated in Example 12-6.

Example 12-6 Determination of materials balances and loading rates for a MRF used to process both commingled MSW and source-separated materials using manual separation. A community generates 1000 ton/d of MSW and has decided to build a MRF that will process commingled MSW, source-separated recyclables, and commingled recyclables. A portion of the wastestream, 15 percent, is being diverted to a curbside recycling program. An additional 4 percent diversion is achieved through a *commingled recycling program,* in which homeowners place recyclable materials in special clear plastic bags for later sorting at the MRF. Prepare a materials balance for the process flow diagram given in the accompanying figure and as described in Chapter 9 (see Fig. 9-28, page 284.) Determine the quantities of materials recovered. Estimate the hourly loading rates for the various separation processes based on a 7-h/d operation. Determine the number of workers required for manual sorting. Assume that the following conditions apply:

1. The wastestream delivered to the MRF contains 4 percent by weight of commingled recyclables separated in transparent bags. The bagged materials have the same composition as the materials collected by the curbside recycling program.
   (*a*) Bulky materials and other contaminants = 1.0 percent of total MSW generated in the community.
   (*b*) The presorting operation includes two front end loaders and six workers on the presorting floor.
   (*c*) First-stage presorting recovers 75 percent of cardboard and 100 percent of bagged commingled recyclables.
   (*d*) Second-stage presorting recovers 75 percent of the remaining cardboard.
2. Sixty percent of the aluminum cans are recovered and delivered to buyback centers by the residents of the community.
3. An average worker can manually sort approximately 2.5 ton/h of MSW on a conveyor belt sorting line.

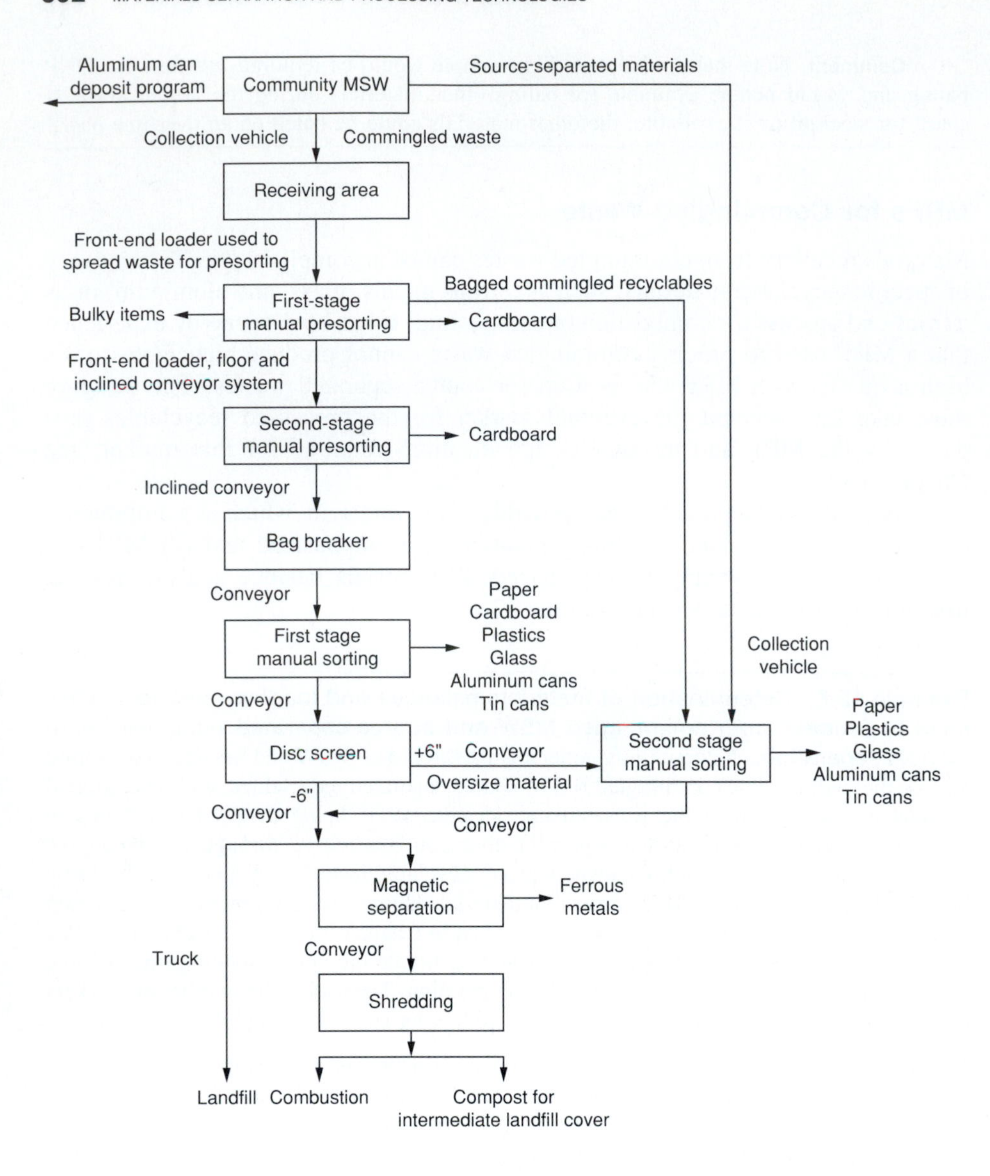

Solution

1. Determine the materials balance quantities for the presorting operation:

   (*a*) Amount of material received at MRF less bulky goods, white goods, and other contaminants removed during first-stage manual presort:

$$\text{Material remaining} = 1000 \text{ ton/d} \times (1 - 0.01) = 990 \text{ ton/d}$$

   (*b*) Set up a computation table to determine the quantities of each component delivered to the MRF, based on 990 ton/d. Also determine the total adjusted quantities, taking into account the aluminum that is returned to buyback programs. (See table, top of page 593.)

Component	MSW,[a] %	Total MSW, ton/d	Total adjusted MSW,[b] ton/d
Organic			
Food wastes	8.0	79.2	79.2
Paper	35.8	354.4	354.4
Cardboard	6.4	63.4	63.4
Plastics	6.9	68.3	68.3
Textiles	1.8	17.8	17.8
Rubber	0.4	4.0	4.0
Leather	0.4	4.0	4.0
Yard wastes	17.3	171.3	171.3
Wood	1.8	17.8	17.8
Misc. organics	—	—	—
Inorganic			
Glass	9.1	90.1	90.1
Tin cans	5.8	57.4	57.4
Aluminum	0.6	5.9	2.4
Other metals	3.0	29.7	29.7
Dirt, ashes, etc.	2.7	26.7	26.7
Total	100.0	990.0	986.5

[a] Distribution from Table 3-7, col. 4.

[b] Because of an aluminum can buyback program with a 60 percent recovery rate, aluminum cans = (1 − 0.60) × 5.9 = 2.4 ton/d.

(*c*) Set up a computation table to determine the quantities of each component in terms of commingled MSW, source-separated recyclables, and commingled, bagged recyclables.

Component	Total MSW,[a] ton/d	Recycled materials,[b] %	Source-separated recyclables,[c] ton/d	Commingled recyclables,[d] ton/d	Remaining MSW,[e] ton/d
Organic					
Food wastes	79.2	—	—	—	79.2
Paper	354.4	52.6	77.8	20.8	255.8
Cardboard	63.4	5.6	8.3	2.2	52.9
Plastics	68.3	10.3	15.2	4.1	49.0
Textiles	17.8	—	—	—	17.8
Rubber	4.0	—	—	—	4.0
Leather	4.0	—	—	—	4.0
Yard wastes	171.3	—	—	—	171.3
Wood	17.8	—	—	—	17.8
Misc. organics	—	—	—	—	—
Inorganic					
Glass	90.1	17.4	25.7	6.9	57.5
Tin cans	57.4	13.5	20.0	5.3	32.1
Aluminum	2.4	0.6	0.9	0.2	1.3
Other metals	29.7	—	—	—	29.7
Dirt, ashes, etc.	26.7	—	—	—	26.7
Total	986.5	100.0	147.9	39.5	799.1

[a] From Step 1*b*.

[b] Composition from Step 2, Example 12-4.

[c] Sample calculation: Paper = 986.5 × 0.15 × 0.526 = 77.8 ton/d.

[d] Sample calculation: Paper = 986.5 × 0.04 × 0.526 = 20.8 ton/d.

[e] Sample calculation: Paper = 354.4 − 77.8 − 20.8 = 255.8 ton/d.

(*d*) Set up a computation table to determine the quantities of each component recovered during presorting.

Component	Total MSW,[a] ton/d	Recovered cardboard 1st presort,[b] ton/d	Recovered cardboard 2nd presort,[c] ton/d	Remaining MSW,[d] ton/d
Organic				
Food wastes	79.2	—	—	79.2
Paper	255.8	—	—	255.8
Cardboard	52.9	39.7	9.9	3.3
Plastics	49.0	—	—	49.0
Textiles	17.8	—	—	17.8
Rubber	4.0	—	—	4.0
Leather	4.0	—	—	4.0
Yard wastes	171.3	—	—	171.3
Wood	17.8	—	—	17.8
Misc. organics	—	—	—	—
Inorganic				
Glass	57.5	—	—	57.5
Tin cans	32.1	—	—	32.1
Aluminum	1.3	—	—	1.3
Other metals	29.7	—	—	29.7
Dirt, ashes, etc.	26.7	—	—	26.7
Total	799.1	39.7	9.9	749.5

[a] From Step 1*c*.

[b] Cardboard recovered = 52.9 × 0.75 = 39.7 ton/d.

[c] Cardboard recovered = (52.9 − 39.7) × 0.75 = 9.9 ton/d.

[d] Cardboard remaining = 52.9 − 39.7 − 9.9 = 3.3 ton/d.

2. Set up a computation table to determine the quantities of each component recovered during first-stage manual sorting:

Component	Total MSW,[a] ton/d	Manual sorting recovery factor	Recovered materials,[b] ton/d	Remaining MSW,[c] ton/d
Organic				
Food wastes	79.2	—	—	79.2
Paper	255.8	0.80	204.6	51.2
Cardboard	3.3	0.75	2.5	0.8
Plastics	49.0	0.80	39.2	9.8
Textiles	17.8	—	—	17.8
Rubber	4.0	—	—	4.0
Leather	4.0	—	—	4.0
Yard wastes	171.3	—	—	171.3
Wood	17.8	—	—	17.8
Misc. organics	—	—	—	—

[a] From Step 1*d*.

[b] Sample calculation: Paper = 255.8 × 0.80 = 204.6 ton/d.

[c] Sample calculation: Paper = 255.8 − 204.6 = 51.2 ton/d.

(*continued*)

(continued)

Component	Total MSW,[a] ton/d	Manual sorting recovery factor	Recovered materials,[b] ton/d	Remaining MSW,[c] ton/d
Inorganic				
Glass	57.5	0.75	43.1	14.4
Tin cans	32.1	0.90	28.9	3.2
Aluminum	1.3	0.90	1.2	0.1
Other metals	29.7	—	—	29.7
Dirt, ashes, etc.	26.7	—	—	26.7
Total	749.5		319.5	430.0

[a] From Step 1*d*.

[b] Sample calculation: Paper = 255.8 × 0.80 = 204.6 ton/d.

[c] Sample calculation: Paper = 255.8 − 204.6 = 51.2 ton/d.

3. Set up a computation table to determine the quantities of each component as separated by the disc screen.

Component	Total MSW,[a] ton/d	Disc screen unders[b] (−6 in), recovery factor	Disc screen unders[c] (−6 in), ton/d	Disc screen overs[d] (+6 in), ton/d
Organic				
Food wastes	79.2	1.0	79.2	0
Paper	51.2	0.75	38.4	12.8
Cardboard	0.8	0.75	0.6	0.2
Plastics	9.8	0.75	7.4	2.4
Textiles	17.8	0.75	13.4	4.4
Rubber	4.0	0.75	3.0	1.0
Leather	4.0	0.75	3.0	1.0
Yard wastes	171.3	1.0	171.3	0
Wood	17.8	0.75	13.4	4.4
Misc. organics	—	—	—	—
Inorganic				
Glass	14.4	0.75	10.8	3.6
Tin cans	3.2	0.75	2.4	0.8
Aluminum	0.1	0.75	0.1	0
Other metals	29.7	0.75	22.3	7.4
Dirt, ashes, etc.	26.7	1.0	26.7	0
Total	430.0		392.0	38.0

[a] From Step 3.

[b] Actual recovery factors would need to be determined by pilot testing.

[c] Sample calculation: Paper = 0.75 × 51.2 = 38.4 ton/d.

[d] Sample calculation: Paper = 51.2 − 38.4 = 12.8 ton/d.

4. Set up a computation table to determine the quantities of each component recovered during second-stage manual sorting, assuming an 85 percent recovery factor for recyclables. (See table, top of page 596.)

Component	Disc screen overs (+6 in), ton/d	Source-separated recyclables, ton/d	Commingled recyclables, ton/d	Total materials 2nd stage, ton/d	Recovered materials, ton/d	Remaining materials, ton/d
Organic						
Food wastes	0	—	—	0	0	0
Paper	12.8	77.8	20.8	111.4	94.7	16.7
Cardboard	0.2	8.3	2.2	10.7	9.1	1.6
Plastics	2.4	15.2	4.1	21.7	18.5	3.2
Textiles	4.4	—	—	4.4	0	4.4
Rubber	1.0	—	—	1.0	0	1.0
Leather	1.0	—	—	1.0	0	1.0
Yard wastes	0	—	—	0	0	0
Wood	4.4	—	—	4.4	0	4.4
Misc. organics	—	—	—	0	0	0
Inorganic						
Glass	3.6	25.7	6.9	36.2	30.8	5.4
Tin cans	0.8	20.0	5.3	26.1	22.2	3.9
Aluminum	0	0.9	0.2	1.1	0.9	0.2
Other metals	7.4	—	—	7.4	0	7.4
Dirt, ashes, etc.	0	—	—	0	0	0
Total	38.0	147.9	39.5	225.4	176.2	49.2

5. Set up a computation table to determine the quantities of each component sent to the landfill or shredded for compost.

Component	Disc screen unders (−6 in), ton/d	Remaining materials from sorting, ton/d	Materials to landfill or composting, ton/d
Organic			
Food wastes	79.2	0	79.2
Paper	38.4	16.7	55.1
Cardboard	0.6	1.6	2.2
Plastics	7.4	3.2	10.6
Textiles	13.4	4.4	17.8
Rubber	3.0	1.0	4.0
Leather	3.0	1.0	4.0
Yard wastes	171.3	0	171.3
Wood	13.4	4.4	17.8
Misc. organics	0	0	0
Inorganic			
Glass	10.8	5.4	16.2
Tin cans	2.4	3.9	6.3
Aluminum	0.1	0.2	0.3
Other metals	22.3	7.4	29.7
Dirt, ashes, etc.	26.7	0	26.7
Total	392.0	49.2	441.2

6. The overall mass balance for the system is shown in the accompanying table and figure on page 597. Details on the quantities of specific recovered materials can be found in the preceding steps.

Materials in	Quantity, ton/d	Materials out	Quantity, ton/d
Community MSW	1000.0	Aluminum can buyback program	3.5
		Bulky items	10.0
		Cardboard (1st stage presort)	39.7
		Cardboard (2nd stage presort)	9.9
		Recovered materials (1st stage sort)	319.5
		Recovered materials (2nd stage sort)	176.2
		Waste to landfill or composting	441.2
Total	1000.0		1000.0

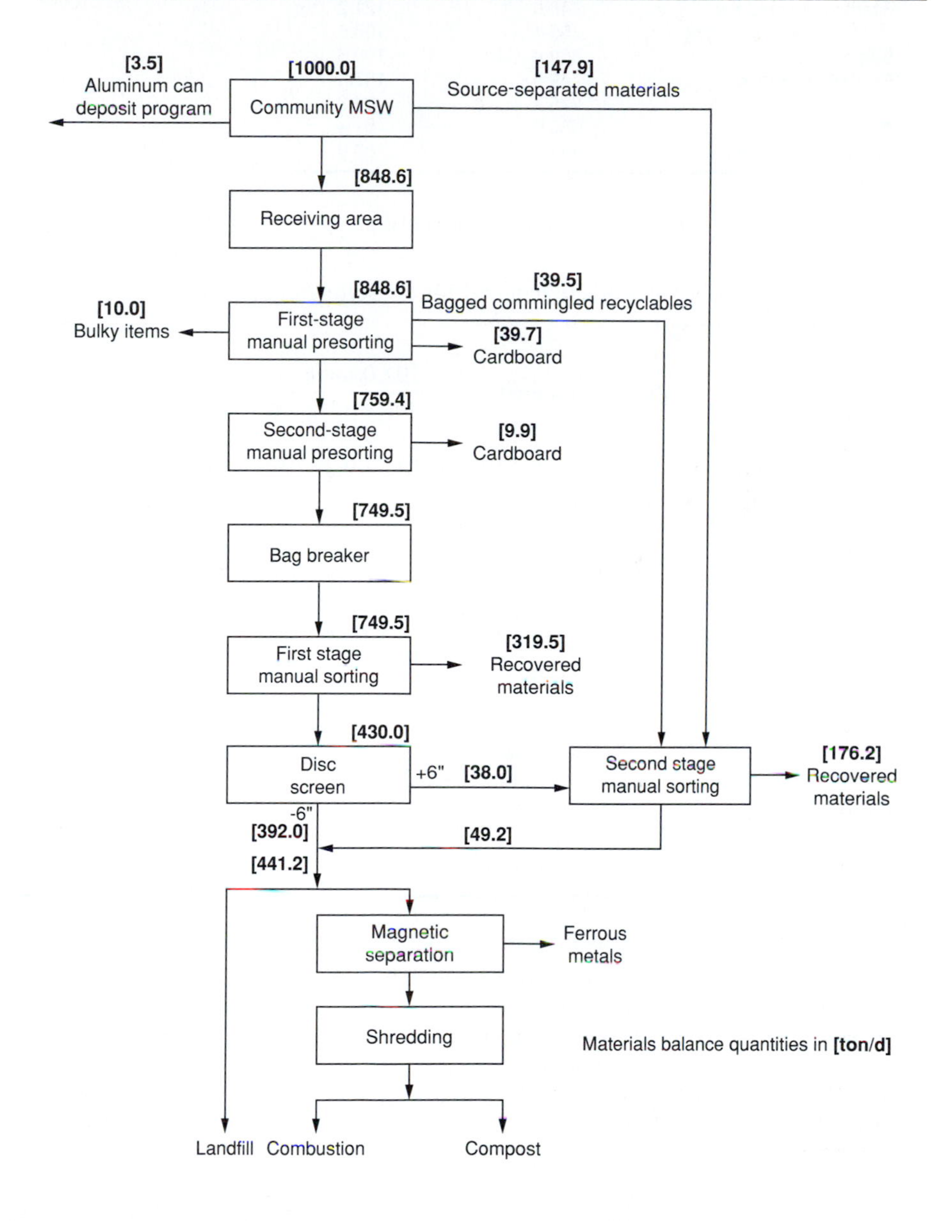

7. Determine the loading rates on the sorting lines and other processing operations, assuming 7-h/d operation. Loading rates in ton/d are summarized on the mass balance diagram. The results of the necessary computations are summarized in the accompanying table.

Operation	Process daily rate, ton/d	Process loading rate,[a,b] ton/h
Receiving area	848.6	121.2
First-stage presorting	848.6	121.2
Second-stage presorting	759.4	108.5
Bag breaker	759.4	108.5
First-stage sorting	749.5	107.0
Second-stage sorting	225.4	32.2
Disc screen	430.0	61.4
Shredder (optional)	441.2	63.0

[a] Based on 7 h/d.

[b] The actual loading on the sorting lines is determined in Step 9.

8. Calculate the number of workers required for the sorting lines.
   (*a*) First-stage sorting

$$\text{Workers required} = \frac{107.0 \text{ ton/h}}{2.5 \text{ ton/worker} \cdot \text{h}} = 42.8$$

   Round off to 40 workers, on 4 sorting lines with 10 workers each.

   (*b*) Second-stage sorting

$$\text{Workers required} = \frac{32.2 \text{ ton/h}}{2.5 \text{ ton/worker} \cdot \text{h}} = 12.9$$

   Round off to 12 workers, on 2 sorting lines with 6 workers each.

9. Calculate the line and worker loading rates.

Operation	Process daily rate, ton/d	Process loading rate, ton/h	Line loading rate, ton/h	Worker loading rate, ton/person · h
First-stage sorting (4 lines)	749.5	107.0	26.8	2.7
Second-stage sorting (2 lines)	225.4	32.2	16.1	2.7

Comment. The spreadsheet approach used in this example to develop a mass balance and to estimate loading rates can be applied to any MRF design. When this approach is used, recovery factors for the various unit operations should be chosen based on manufacturers' recommendations and pilot testing if available. Additional unit operations may be required to upgrade the purity of recovered materials as market-driven specifications change.

A MRF can also be designed with mechanical separation instead of the manual sorting used in the previous example. Two of the most common unit operations used in MRFs to separate commingled MSW are shredding and air classification. Data and information that can be used to estimate the required quantities are presented in Tables 12-19 and 12-20. The components that normally make up the light and heavy fractions after shredding and air classification are identified in Table 12-19. Waste composition data in this table are from Table 3-7. Recovery factors for ferrous metals, glass, and aluminum, along with information on the recovery of heavy materials from the light fraction, are reported in Table 12-20.

It should be noted that the moisture content that may be lost during shredding has not been considered in Table 12-19. The typical moisture content of solid wastes varies from 15 to 40 percent, depending on the geographic location and the season of the year. In the southwestern United States, the average moisture content value is about 20 percent. Under normal circumstances, from 5 to 25 percent of the initial moisture may be lost during shredding. If test data are not available, a value of 15 percent can be used for estimating this loss. To demonstrate the use of the data given in Tables 12-19 and 12-20, a materials balance analysis is performed in Example 12-7 on a MRF that incorporates a shredder and air classifier.

TABLE 12-19
Light and heavy fractions of solid waste components after shredding and air classification

Waste component	Typical percentage distribution of waste by weight[a,b]	Fraction by weight, percent: Light	Fraction by weight, percent: Heavy	Comment
Organic				
Food wastes	8.0	8.0	—	Components assumed to make up the light fraction after shredding. After air classification, the light fraction will contain from 2 to 8 percent of the components from the heavy fraction by weight.
Paper	35.8	35.8	—	
Cardboard	6.4	6.4	—	
Plastics	6.9	6.9	—	
Textiles	1.8	1.8	—	
Rubber	0.4	0.4	—	
Leather	0.4	0.4	—	
Yard wastes	17.3	17.3	—	
Wood	1.8	1.8	—	
Misc. organics	—	—	—	
Inorganic				
Glass	9.1	—	9.1	Components assumed to make up the heavy fraction after shredding. After air classification, the heavy fraction will contain from 5 to 20 percent of the components from the light fraction by weight.
Tin cans	5.8	—	5.8	
Aluminum	0.6	—	0.6	
Other metals	3.0	—	3.0	
Dirt, ashes, etc.	2.7	—	2.7	
Total	100.0	78.8	21.2	

[a] Moisture loss during shredding not considered.

[b] Composition data from Table 3-7, column 4.

TABLE 12-20
Estimated recovery factors for various components in MSW using mechanical separation equipment

Waste component	Recovery factor, % Range	Typical	Comment
Light fraction	80–95	90[a]	Recoverable portion will vary with the composition of the solid wastes and the characteristics of the wastes after shredding.
Heavy fraction	90–98	96[b]	
Ferrous metal	65–95	85	Varying amounts of light and heavy fraction material will also be removed with these components, depending on the specific process and equipment used.
Glass	50–90	80	
Aluminum	55–90	70	

[a] Varying amounts of the light fraction will be retained with the heavy fraction (see Table 12-19).
[b] Varying amounts of the heavy fraction will be carried over with the light fraction (see Table 12-19).

Example 12-7 Determination of materials balance and loading rates for a MRF processing commingled MSW using mechanical processing. A community generates 1000 ton/d of MSW and has decided to build a MRF which will process commingled MSW. Prepare a materials balance for the process flow diagram given in the accompanying figure and as described in Chapter 9. Determine the quantities recovered. Estimate

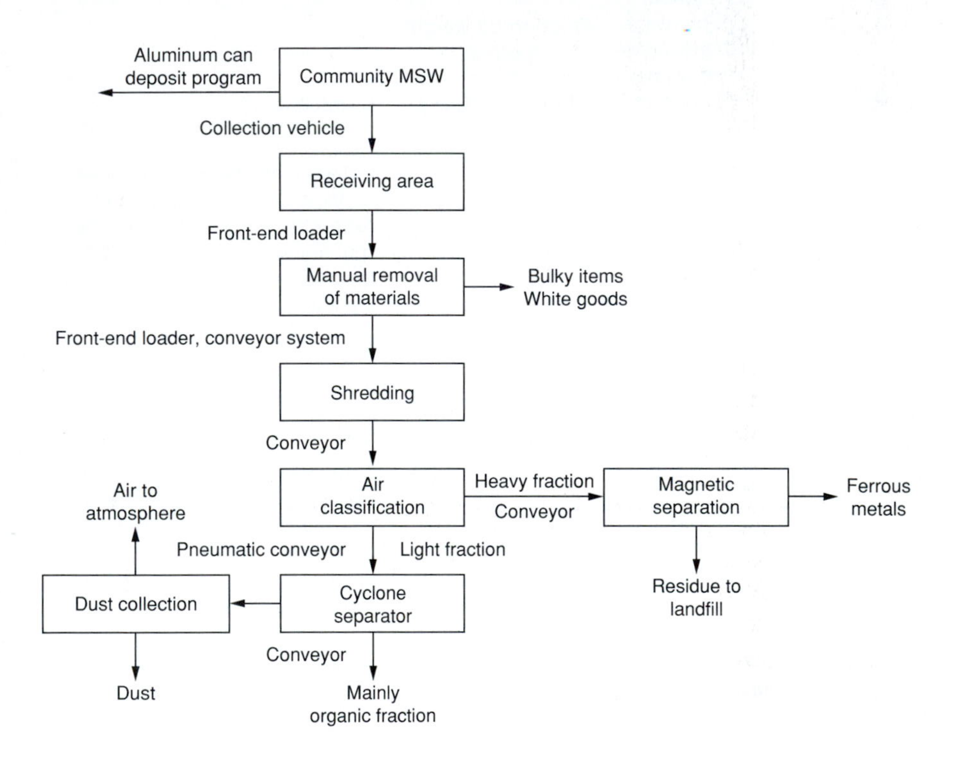

the hourly loading rates for the various separation processes based on 16-h/d operation. Because of the effectiveness of aluminum can recycling programs, it has been concluded that the mechanical recovery of aluminum cans is not justified. Assume that the following conditions apply:

1. Bulky materials and other contaminants = 1.0 percent of total MSW generated in the community.
2. 60 percent of the aluminum cans are recovered and delivered to buyback centers by the residents of the community.
3. Waste composition as given in column 4 of Table 3-7 and Table 12-16.
   (*a*) Initial moisture content = 20 percent.
   (*b*) Moisture lost during shredding = 20 percent of initial moisture content.
   (*c*) Moisture loss will be from moist light fraction components (food and yard wastes.)
   (*d*) Heavy fraction materials contained in light fraction = 6 percent of heavy fraction (based on weight).
   (*e*) Light fraction materials contained in heavy fraction = 10 percent of light fraction (based on weight after shredding).
   (*f*) The initial light fraction = 79.5 percent.
   (*g*) Initial heavy fraction = 20.5 percent.
   (*h*) Recovery factor for ferrous metals = 85 percent.

Solution

1. Determine the materials balance quantities for the presorting operation:
   (*a*) Amount of material received at MRF less bulky goods, white goods, and other contaminants removed during first-stage manual presort:

$$\text{Material remaining} = 1000 \text{ ton/d} \times (1 - 0.01) = 990 \text{ ton/d}$$

   (*b*) Set up a computation table to determine the quantities of each component delivered to the MRF based on 990 ton/d.

Component	MSW,[a] %	Total MSW, ton/d	Total adjusted MSW,[b] ton/d
Organic			
Food wastes	8.0	79.2	79.2
Paper	35.8	354.4	354.4
Cardboard	6.4	63.4	63.4
Plastics	6.9	68.3	68.3
Textiles	1.8	17.8	17.8
Rubber	0.4	4.0	4.0
Leather	0.4	4.0	4.0
Yard wastes	17.3	171.3	171.3
Wood	1.8	17.8	17.8
Misc. organics	—	—	—

[a] Distribution from Table 3-7, col. 4.

[b] Because of an aluminum can buyback program with a 60 percent recovery rate, aluminum cans = (1 − 0.60) × 5.9 = 2.4 ton/d.

(*continued*)

(continued)

Component	MSW,[a] %	Total MSW, ton/d	Total adjusted MSW,[b] ton/d
Inorganic			
Glass	9.1	90.1	90.1
Tin cans	5.8	57.4	57.4
Aluminum	0.6	5.9	2.4
Other metals	3.0	29.7	29.7
Dirt, ashes, etc.	2.7	26.7	26.7
Total	100.0	990.0	986.5

[a] Distribution from Table 3-7, col. 4.

[b] Because of an aluminum can buyback program with a 60 percent recovery rate, aluminum cans = (1 − 0.60) × 5.9 = 2.4 ton/d.

2. Determine the materials balance quantities for the shredder.

Component	Total adjusted MSW,[a] ton/d	Moisture loss from light fraction,[b] ton/d	MSW after shredding, ton/d
Light fraction			
Food wastes	79.2	15.8	63.4
Paper	354.4	—	354.4
Cardboard	63.4	—	63.4
Plastics	68.3	—	68.3
Textiles	17.8	—	17.8
Rubber	4.0	—	4.0
Leather	4.0	—	4.0
Yard wastes	171.3	34.3	137.0
Wood	17.8	—	17.8
Misc. organics	—	—	—
Heavy fraction			
Glass	90.1	—	90.1
Tin cans	57.4	—	57.4
Aluminum	2.4	—	2.4
Other metals	29.7	—	29.7
Dirt, ashes, etc.	26.7	—	26.7
Total	986.5	50.1	936.4

[a] From Step 1.

[b] 20% of food and yard wastes.

3. Determine the materials balance quantities for the air classifier. (See the accompanying table at the top of page 603.)

Component	MSW after shredding,[a] ton/d	Carryover from light to heavy,[b] ton/d	Carryover from heavy to light,[c] ton/d	Adjusted light fraction, ton/d	Adjusted heavy fraction, ton/d
Light fraction					
Food wastes	63.4	6.3	—	57.1	6.3
Paper	354.4	35.4	—	319.0	35.4
Cardboard	63.4	6.3	—	57.1	6.3
Plastics	68.3	6.8	—	61.5	6.8
Textiles	17.8	1.8	—	16.0	1.8
Rubber	4.0	0.4	—	3.6	0.4
Leather	4.0	0.4	—	3.6	0.4
Yard wastes	137.0	13.7	—	123.3	13.7
Wood	17.8	1.8	—	16.0	1.8
Misc. organics	—	—	—	—	—
Heavy fraction					
Glass	90.1	—	5.4	5.4	84.7
Tin cans	57.4	—	3.4	3.4	54.0
Aluminum	2.4	—	0.1	0.1	2.3
Other metals	29.7	—	1.8	1.8	27.9
Dirt, ashes, etc.	26.7	—	1.6	1.6	25.1
Total	936.4	72.9	12.3	669.5	266.9

[a] From Step 2.

[b] Carry over from light to heavy = 10 percent.

[c] Carry over from heavy to light = 6 percent.

4. Determine the weight of ferrous metals removed from the heavy fraction by magnetic separation (assuming that the tin can and other metals categories reported in Step 3 are included).

$$\text{Ferrous metals} = [54.0 + 27.9] \times 0.85 = 69.6 \text{ ton/d}$$

(Note: Weight of other material that might be removed along with the metal is not included).

5. The overall mass balance for the system is shown in the accompanying table and figure. Details on the quantities of specific recovered materials can be found in the preceding steps. (See page 602.)

6. Determine the loading rates on the individual unit operations. The results of the necessary computations are summarized in the accompanying table.

Operation	Total quantity, ton/d	Loading rate ton/h[a]
Shredder	986.5	61.7
Classifier	936.4	58.5
Magnetic separator	266.9	16.7

[a] Based on an operating time of 16 h/d.

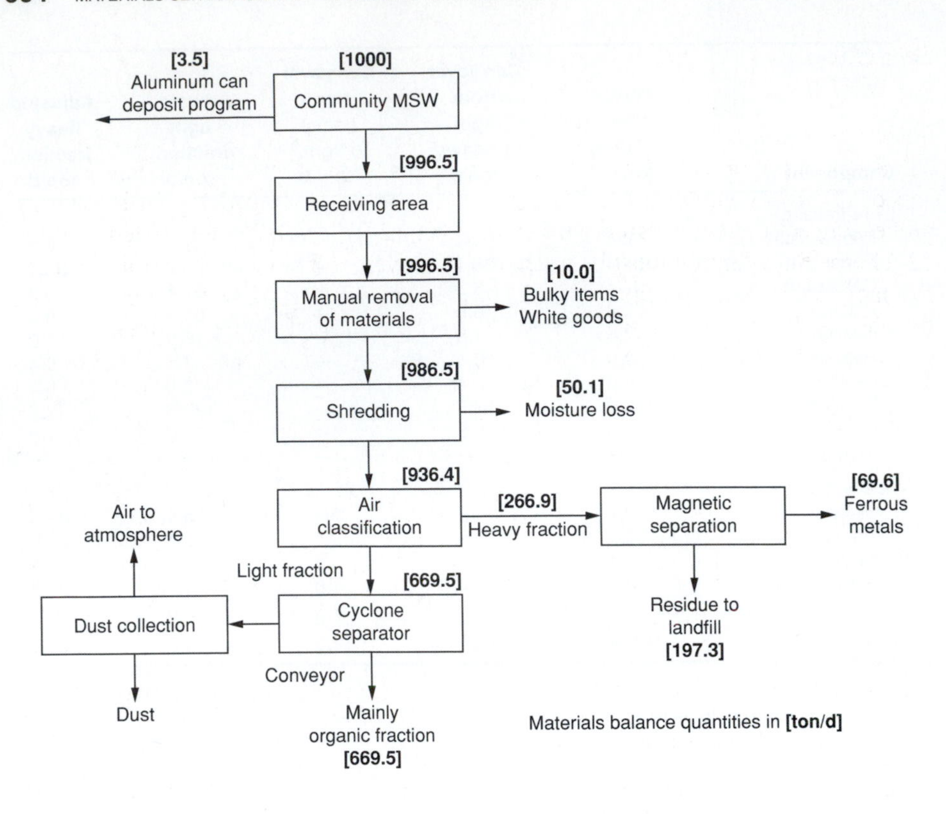

Materials in	Quantity, ton/d	Materials out	Quantity, ton/d
Community MSW	1000.0	Aluminum can buyback program	3.5
		Bulky items	10.0
		Moisture loss during shredding	50.1
		Light fraction	669.5
		Recovered ferrous metal	69.6
		Heavy fraction to landfill or further processing (heavy fraction − ferrous metals)	197.3
Total	1000.0		1000.0

Comment. The light fraction recovered in this example could be used for combustion or as a feedstock for composting. Although this example was simplified compared to the typical mechanized MRF for commingled MSW, the approach outlined is valid for any combination of separation operations. In applying such an analysis, it will be important to prepare a sensitivity analysis to assess how variable the relative loading rates may become in the future as the waste characteristics change and new regulations are implemented.

Selection of Equipment and Facilities for MRFs

Selection of the actual equipment and physical facilities that will be used in the MRF separation process flow diagram is perhaps the most challenging engineering aspect of implementing a MRF. Factors that should be considered in evaluating processing equipment are summarized in Table 12-21. Of the factors listed in Table 12-21, special attention must be given to (1) proven reliability and flexibility of the available equipment and facilities, (2) proven process performance efficiency, and (3) ease and economy of operation. Obtaining meaningful data on these factors is, at present, difficult, because the data base is essentially nonexistent. For this reason, it is recommended that visits be made to actual operating installations to obtain first-hand information on performance and maintenance requirements.

In general, it has been found that more equipment failures and operational problems will occur in processing operations than in other waste management operations. Because of the abrasive nature of many of the components found in solid wastes, the rate of wear on much of the processing equipment and the down time have been greater than anticipated. As a result of operational problems and other

TABLE 12-21
Factors that should be considered in evaluating processing equipment

Factor	Evaluation
Capabilities	What will the device or mechanism do? Will its use be an improvement over conventional practices?
Reliability	Will the equipment perform its designated functions with little attention beyond preventive maintenance? Has the effectiveness of the equipment been demonstrated in use over a reasonable period of time or merely predicted?
Service	Will servicing capabilities beyond those of the local maintenance staff be required occasionally? Are properly trained service personnel available through the equipment manufacturer or the local distributor?
Safety of operation	Is the proposed equipment reasonably foolproof so that it may be operated by personnel with limited mechanical knowledge or abilities? Does it have adequate safeguards to discourage careless use?
Efficiency	Does the equipment perform efficiently? What is the specific energy consumption (kWh/ton) relative to other equipment with similar capacity?
Environmental effects	Does the equipment pollute or contaminate the environment?
Health hazards	Does the device, mechanism, or equipment create or amplify health hazards?
Economics	What are the economics involved? Both first and annual costs must be considered. Future operation and maintenance costs must be assessed carefully. All factors being equal, equipment produced by well-established companies, having a proven history of satisfactory operation, should be given appropriate consideration.

equipment limitations, many system designers now recommend the installation of two or more independent process lines (also known as *trains*), especially where electric power is to be produced on a continuous basis. Further, because many of the firms in this developing field do not have a long history, it is recommended that the equipment selected be such that it can be repaired with standard parts and components that, if necessary, can be rebuilt or remade locally. The availability of a local distributor is also important.

12-10 DISCUSSION TOPICS AND PROBLEMS

12-1. Assume that the energy consumption required for size reduction for solid wastes can be estimated using Eq. (12-1). If it is found that 20 hp · h/ton is required to reduce the average size of solid wastes from about 6 to 2 in, estimate the energy required to reduce the average size of solid wastes from about 12 to 2 in at a loading rate of about 10 ton/h.

12-2. Estimate the horsepower requirement for the shredder used in the process flow diagram shown in Example 12-6. Use the power requirements given in Fig. 12-6, assuming that the (1) process loading rate is 63.0 ton/h for 7-h/d operation, (2) an input material factor for picked MSW will be used, and (3) the product size will be 2 in.

12-3. Calculate the annual cost to operate the shredder in Problem 12-2, assuming that electricity costs $0.10/kWh (or use your local rate).

12-4. A trommel screen is to be designed to replace the first-stage manual presorting in Example 12-6. It is assumed that the trommel will separate the cardboard and bagged commingled recyclables. Bulky items must be removed manually before the waste enters the trommel. Find the diameter of the trommel (ft) and the critical speed (rev/min). Assume the following conditions apply.

1. Bulk specific weight of MSW = 8 lb/ft^3.
2. Trommel inclination = 5 degrees.
3. Fillage factor = 0.25.

12-5. For the MRF described in Example 12-5, calculate the purity and efficiency for mixed paper and old corrugated cardboard if the recovery factor for both materials is 0.95.

12-6. If the recovery factor for paper is 0.80, calculate the purity and effectiveness of the first-stage manual sorting operation described in Example 12-6.

12-7. Assuming that an air-to-solids ratio (lb solids/lb air) of 3.0 is required for the separation of the light fraction of shredded solid wastes and that the head loss in a separation column is equal to 4 in of water, estimate the required blower horsepower to separate 50 ton/h of shredded solid wastes. Assume that the specific weight of air is 0.0750 lb/ft^3 and that the following equation can be used to calculate the blower horsepower (BHP):

$$\text{BHP} = 0.227Q\left[((14.7 + \rho)/14.7)^{0.283} - 1\right]$$

where BHP = blower horsepower, hp
Q = airflow rate, ft^3/min
p = pressure drop, lb/in^2

(Note: 1 in H_2O = 0.036 lb/in^2)

12-8. Use Eq. (12-11) to find the efficiency of the air classifier in Example 12-7.

12-9. Use binary separation theory to find the recovery, purity, and efficiency of the air classifier in Example 12-7.

12-10. What size magnetic separator (ton/h) would be required in Example 12-6? At 95 percent recovery, how much ferrous metal (primarily tin cans) could be recovered?

12-11. Assume that an eddy current "aluminum magnet" has been installed downstream of the magnetic separator discussed in Problem 12-10. At 98 percent recovery, how much aluminum could be recovered? Relative to the aluminum already recovered by manual separation, is the "aluminum magnet" cost-effective?

12-12. For the MRF shown in Example 12-6, calculate the purity of the organic fraction produced if the shredder and magnetic separator are used. Use the recovery assumptions given in Example 12-6 and in Problem 12-10. Comment on the acceptability of this material as a feedstock for composting.

12-13. You are designing a storage building for a new MRF that will produce 250 ton/d of RDF for a nearby coal-fired power plant. The storage building is to be sized to allow for 2 days of storage. Assume that the RDF will be piled no more than 20 feet high on a concrete floor and that it will be outloaded to trucks for shipping using front-end loaders.

(*a*) Compare the size of the buildings required depending on whether RDF or dRDF is used. The bulk specific weights are 200 and 800 lb/yd^3, respectively. List your assumptions.

(*b*) What are the other engineering and economic factors to be considered in choosing between the two alternatives?

12-14. If an air velocity of 2000 ft/min is to be used to transport finely ground material with a specific gravity of 0.75 in a horizontal duct, estimate the maximum particle size that can be transported.

12-15. For the MRF in Example 12-6, find HP_{total} and TW_b for the first-stage manual sorting lines. Assume that 4 lines are used, each 48 in wide, 50 ft long, and operated at 60 ft/min. The MSW on the lines has a specific weight of 8 lb/ft^3.

12-16. For the MRF in Example 12-6, find HP_{total} for the +6 in conveyor from the disc screen to the second-stage manual sorting lines. Assume that the conveyor is 24 in wide and 50 ft long, and has 10 ft of lift and a belt speed of 400 ft/min.

12-17. For the MRF in Example 12-7, use the BHP formula given in Problem 12-7 to find the airflow rate, ft^3/min, and blower horsepower to convey the light fraction pneumatically to an adjacent MSW combustion system. Assume that the pneumatic conveyor will operate at 10 in of water pressure and that a materials-to-air ratio of 0.1 will be used.

12-18. You have been appointed as recycling coordinator for a city of 100,000 population. The city council is concerned that 80 percent of potential aluminum is being diverted to a neighboring city that is across the state line. The other state offers a \$0.05/can deposit for returned aluminum cans; your state has no such program.

A wholesale buyer will pay the city $800/ton for aluminum cans. The estimated MSW generation at your city (without recycling or aluminum can diversion) is 4.5 lb/cap · d. If the MSW composition and assumptions of Example 12-4 apply, estimate the composition of recovered materials if a curbside recycling system were established (*a*) with the 80 percent diversion of aluminum cans and (*b*) without the diversion. What is the value of the diverted cans (21 aluminum cans/lb)? What is the value of the aluminum recovered through your program if sold to the buyer?

12-19. Analyze the MRF described in Example 12-6, assuming there are no source separation, aluminum can buyback, or bagged commingled recycling programs in operation. The second-stage manual sorting and disc screen are not used. Show your calculations in spreadsheet format and summarize your results in tables and on a flow diagram. Compare your results to the results obtained in Example 12-6.

12-20. Analyze the MRF described in Example 12-7, assuming that 15 percent of the wastestream has been diverted to a curbside recycling system as in Example 12-6. Sixty percent of the aluminum cans are diverted to buyback centers, as in Example 12-7. Assume that the source-separated materials will be processed at another newly constructed MRF. Show your calculations in spreadsheet format and summarize your results in tables and on a flow diagram. Compare your results to Example 12-7.

12-11 REFERENCES

1. Anonymous: "1991 Specification Catalog—Balers," *Management of World Wastes,* p. 76, January 1991.
2. Boettcher, R. A.: "Air Classification of Solid Wastes," *Solid Waste Management Program,* Publication SW-30c, U.S. Environmental Protection Agency, 1972.
3. Braam, B. C., H. J. L. Van der Valk, and W. L. Dalmijn: "Eddy-Current Separation by Permanent Magnets Part II: Rotating Disc Separators," *Resources, Conservation and Recycling,* 1(1), p. 3, March 1988.
4. Dallavale, J. M.: *The Industrial Environment and Its Control,* Pitman, New York, 1958.
5. Drobny, N. L., H. E. Hull, and R. F. Testin: *Recovery and Utilization of Municipal Solid Waste,* U.S. Public Health Service Publication 1908, 1971.
6. Endahl, R. B.: *Solid Waste Processing: A State-of the-Art Report on Unit Operations,* Publication SW-4c, Bureau of Solid Waste Management, US HEW, Washington, DC, 1969.
7. Everett, J. W. and J. J. Peirce: "The Development of Pulsed Flow Air Classification and Theory for Municipal Solid Waste Processing," *Resources, Conservation and Recycling,* 2(3), Sept. 1990.
8. Hasselriis, F.: *Refuse Derived Fuel,* An Ann Arbor Science Book, Butterworth Publishers, Boston, MA, 1984.
9. Henderson, S. M. and R. L. Perry: *Agricultural Process Engineering,* 3rd ed., The Avi Publishing Company, Westport, CT, 1976.
10. Hill, R. M.: "Three Types of Low Speed Shredder Design," *Proceedings of the 1986 National Waste Processing Conference,* American Society of Mechanical Engineers, New York, 1986, p. 265.
11. Hollender, H. I. and N. F. Cunningham: "Beneficiated Solid Waste Cubettes as Salvage Fuel for Steam Generation," *Proceedings of the 1972 National Incinerator Conference,* American Society of Mechanical Engineers, New York, 1972.
12. Matthews, C. W.: "Screening," *Chemical Engineering,* July 10, 1972.
13. Reinhardt, J. J. and R. K. Hamm: *Solid Waste Milling and Disposal on Land Without Cover,* NTIS Publication PB-234930, U.S. Environmental Protection Agency—NTIS, Springfield, VA, 1974.

14. Romeo, E. J.: "Material Recovery Design of Ocean County, NJ," *Proceedings of the 1992 Waste Processing Conference,* American Society of Mechanical Engineers, New York, 1992.
15. Savage, G. M. and L. F. Diaz: "Key Issues Concerning Waste Processing Design," *Proceedings of the 1986 National Waste Processing Conference,* American Society of Mechanical Engineers, New York, 1986, p. 361.
16. Sullivan, J. W., R. M. Hill, and J. F. Sullivan: "The Place of the Trommel in Resource Recovery," *Proceedings of the 1992 Waste Processing Conference,* American Society of Mechanical Engineers, New York, 1992.
17. Vesilind, P. A., and A. E. Rimer: *Unit Operations in Resource Recovery Engineering,* Prentice Hall, Englewood Cliffs, NJ, 1981.
18. Walker, W. H., *et al.*: *Principles of Chemical Engineering,* 3rd ed., McGraw-Hill, New York, 1937.
19. Wheeler, P. A., J. R. Barton, and R. New: "An Empirical Approach to the Design of Trommel Screens for Fine Screening of Domestic Refuse," *Resources, Conservation and Recycling,* 2(4), p. 261, September 1988.

CHAPTER 13

THERMAL CONVERSION TECHNOLOGIES

The thermal processing of solid waste, used both for volume reduction and energy recovery, is an important element in many integrated waste management systems. The purpose of this chapter is to introduce the fundamentals of thermal processing from the perspective of the environmental engineer and waste disposal system manager. An understanding of the function of thermal processing in an integrated solid waste management system is essential for selecting equipment and setting performance standards. Therefore, the focus of this chapter is on fundamentals of system analysis and not on design details. The chapter is organized into the following sections: (1) fundamentals of thermal processing, (2) combustion systems, (3) pyrolysis systems, (4) gasification systems, (5) environmental control systems, and (6) energy recovery systems.

13-1 FUNDAMENTALS OF THERMAL PROCESSING

Thermal processing of solid waste can be defined as the conversion of solid wastes into gaseous, liquid, and solid conversion products, with the concurrent or subsequent release of heat energy. Thermal processing systems can be categorized on the basis of their air requirements (see Fig. 13-1). Combustion with exactly the amount of oxygen (or air) needed for complete combustion is known as *stoichiometric combustion*. Combustion with oxygen in excess of the stoichiometric requirements is termed *excess-air combustion*. *Gasification* is the partial combustion of solid waste under substoichiometric conditions to generate a combustible gas

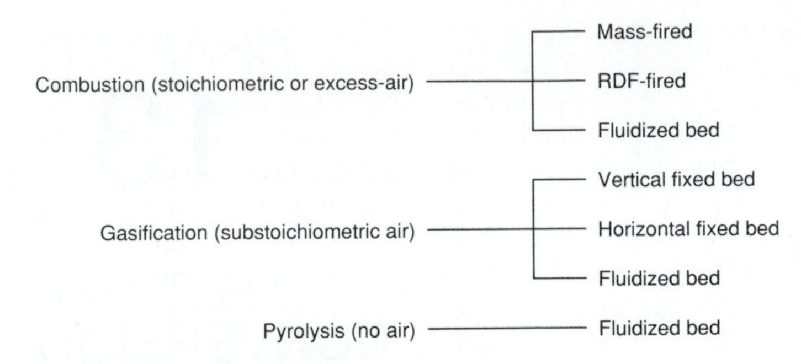

FIGURE 13-1
Representative thermal processing systems.

containing carbon monoxide, hydrogen, and gaseous hydrocarbons. *Pyrolysis* is the thermal processing of waste in the complete absence of oxygen. Gasification and pyrolysis are discussed in detail in subsequent sections.

Stoichiometric Combustion

As given in Chapter 9, the basic reactions for the stoichiometric combustion of the carbon, hydrogen, and sulfur in the organic fraction of MSW are as follows:

For carbon,

$$\underset{12}{C} + \underset{32}{O_2} \rightarrow CO_2 \tag{9-2}$$

For hydrogen,

$$\underset{4}{2H_2} + \underset{32}{O_2} \rightarrow 2H_2O \tag{9-3}$$

For sulfur,

$$\underset{32.1}{S} + \underset{32}{O_2} \rightarrow SO_2 \tag{9-4}$$

If it is assumed that dry air contains 23.15 percent oxygen by weight, then the amount of air required for the oxidation of 1 lb of carbon would be equal to 11.52 lb [(32/12)(1/0.2315)]. The corresponding amounts for hydrogen and sulfur are 34.56 and 4.31 lb, respectively. It should be noted that the amount of hydrogen must first be adjusted by subtracting one-eighth of the percent of oxygen from the total percent of hydrogen initially present in the waste (this subtraction accounts for the oxygen in the waste combining with hydrogen to form water). The necessary combustion computations to determine the amount of air required for stoichiometric combustion are illustrated in Example 9-2 in Chapter 9.

Excess-Air Combustion

Because of the inconsistent nature of solid waste, it is virtually impossible to combust solid waste with stoichiometric amounts of air. In practical combustion systems, excess air must be used to promote mixing and turbulence, thus ensuring that air can reach all parts of the waste. The use of excess air for combustion affects the temperature and composition of the combustion products (known as *flue gases*). As the percentage of excess air increases, the oxygen content of the flue gases increases and the temperature of combustion decreases; thus, the combustion air can be used to control combustion temperature. The temperature of flue gases is important from the viewpoint of odor control. When combustion temperatures are less than about 1450°F, the emission of odorous compounds may occur. It has also been found that combustion temperatures greater than 1800°F minimize the emission of dioxins, furans, volatile organic compounds (VOCs), and other potentially hazardous compounds in the flue gas. Dioxin and furan emissions and their control are discussed in Section 13-5. The computations required to assess the effects of excess-air combustion are illustrated in Example 13-1.

Example 13-1 Determination of the effects of excess air on temperature and composition of flue gases. Determine the composition and temperature of the flue gases for the solid wastes in Table 3-7 and Example 3-3. Show the effects of excess-air combustion on flue gas temperature. Assume that the following conditions apply:

1. All of the carbon initially present is converted to CO_2.
2. The energy content of the MSW is 5065 Btu/lb, as determined in Example 3-4.

Solution

1. Set up a computation table to determine the moles of oxygen and pounds of air required per 100 pounds of solid waste for stoichiometric combustion.

Component	Weight,[a] percent	Atomic weight	Atomic[b] weight units	Moles O_2 req.	Combustion reaction and products
Carbon	27.4	12.0	2.283	2.283	$C + O_2 \rightarrow CO_2$
Hydrogen	3.6	1.0	3.600	0.900	$2H_2 + O_2 \rightarrow 2H_2O$
Oxygen	23.0	16.0	1.438	−0.719	
Nitrogen	0.5	14.0	0.036		
Sulfur	0.1	32.1	0.003	0.003	$S + O_2 \rightarrow SO_2$
Water	21.4	18.0	1.189		
Inerts	24.0				
Total	100.0			2.467	

Moles of air[c] required per 100 pounds of solid wastes = 2.467/0.2069 = 11.92

Pounds of air[c] required per pound of solid waste = 11.92(28.7)/100 = 3.42

[a] Data from Example 3-3, Step 2.

[b] Sample calculation 27.4/12.0 = 2.283.

[c] Assumed air composition, in volume fractions: CO_2 = 0.0003; N_2 = 0.7802; O_2 = 0.2069; H_2O = 0.0126. Assuming ideal gases, the volume fractions may be taken as mole fractions. In the air composition just given, it is assumed that rare gases are included with the nitrogen and that the air moisture content is at 70 percent relative humidity at 60°F. Air of this composition has a weight of 28.7 lb/mol.

2. Set up a computation table to determine the moles of flue gases produced by the stoichiometric combustion of 100 pounds of solid waste.

Combustion product	Moles of flue gas			
	From combustion[a]	From air[b]	Total	Percent
CO_2	2.283	0.004[c]	2.287	15.5
H_2O	(1.800 + 1.189[d])	0.150	3.139	21.3
O_2	—	—	—	—
N_2	0.018[e]	9.30	9.318	63.2
SO_2	0.003	—	0.003	<0.1
Total			14.747	100.0
Moles of air per mole of flue gas = 11.92/14.75 = 0.81				

[a] Data derived from the computation table for Step 1.

[b] Moles air per mole 100 lb of solid waste = 11.92 (see Step 1).

[c] Sample calculation, 11.92(0.0003) = 0.004 (see the computation table for Step 1, footnote *c*).

[d] Moles of moisture in original sample.

[e] In Step 1, nitrogen was reported in atomic weight units N; in Step 2, nitrogen is reported as N_2; therefore, moles N_2 = 0.036/2 = 0.018.

3. Set up a computation table to determine the flue gas composition for various quantities of excess air assuming 100 moles of flue gas from stoichiometric combustion.

Percent excess air	Moles excess air[a]	Total moles of gas	Gas composition, percent				
			CO_2	O_2	N_2	H_2O	SO_2
0	0.0	100.0	15.5	0.0	63.2	21.3	<0.1
50	40.5[b]	140.5	11.0[c]	6.0[d]	67.5[e]	15.5[f]	<0.1
100	81.0	181.0	8.6	9.3	69.8	12.3	<0.1

[a] Moles excess air = percent excess air (moles of air/moles of flue gas).

[b] (50 percent excess air)(0.81) = 40.5.

[c] Percent CO_2 = {[15.3 + (40.5 × 0.0003)]/140.5}100 = 11.0.

[d] Percent O_2 = [(40.5 × 0.2069)/140.5]100 = 6.0.

[e] Percent N_2 = {[63.2 + 40.5(0.7802)]/140.5}100 = 67.5.

[f] Percent H_2O = {[21.3 = 40.5(0.0126)]/140.5}100 = 16.3.

4. Determine the enthalpy of the flue gas for the two percentages of excess air in Step 3 at four temperatures (1000, 1500, 2000, and 2500°F).

(*a*) Use the following equation with the enthalpy data given in the accompanying table.

$$\frac{\text{Btu in product gas}}{\text{lb of solid waste}} = \left[\left(\frac{\text{moles of flue gas}}{\text{lb of solid waste}}\right)\left(\frac{\text{total moles of gas}}{\text{moles of flue gas}}\right)\right]$$
$$\times\left[\sum(\text{mole fraction of gas component})\left(\frac{\text{Btu}}{\text{moles of gas component}}\right)\right]$$

where moles of flue gas are at stoichiometric conditions from Step 2 and total moles of gas include excess air from Step 3. (See table top of page 615.)

Temperature, °F	Btu/lb · mol over standard state[a]			
	CO_2	O_2	N_2	H_2O
1000	10,048	6,974	6,720	26,925
1500	16,214	11,008	10,556	31,743
2000	22,719	15,191	14,520	36,903
2500	29,539	19,517	18,609	42,405

From Ref. 38. SO_2 not included because it is <0.1 percent of flue gas volume.
[a] Gas, except liquid water, at 1 atm pressure, and 77°F.

(*b*) A sample enthalpy calculation follows for 1000°F and 50 percent excess air:

$$\begin{aligned}\text{Btu in flue gas/lb solid waste} &= (0.1475 \text{ moles flue gas/lb solid waste})(140.5/100)\\ &\quad\times [0.110(10{,}048 \text{ Btu/mole}) + 0.060(6974 \text{ Btu/mole})\\ &\quad + 0.675(6720 \text{ Btu/mole}) + 0.155(26{,}925 \text{ Btu/mole})]\\ &= 2121 \text{ Btu/lb solid waste}\end{aligned}$$

(*c*) Summarize the enthalpy calculations for the four temperatures:

Temperature, °F	Btu in flue gas/lb solid waste	
	Excess air, 50 percent	Excess air, 100 percent
1000	2121	2540
1500	3003	3655
2000	3923	4816
2500	4881	6023

5. Determine the temperature of the flue gas at 50 and 100 percent excess air.
   (*a*) If it is assumed that the energy content of the solid wastes is 5065 Btu/lb and that 15 percent of the energy is lost, then 4305 Btu/lb of solid waste must remain in the flue gas.
   (*b*) By interpolation from the summary table in Step 4*c*, the flue gas temperature is about 2200°F at 50 percent excess air and about 1780°F at 100 percent excess air.

Comment. The technique used in this example can be used to estimate the effects of heat losses and various amounts of excess air on the temperature of combustion. It is not entirely accurate, because the air distribution in an MSW combustor is not completely uniform. In some cases, the air distribution is deliberately modified to operate some portions of the combustion system at substoichiometric conditions and other parts at excess-air conditions. These combustion control techniques are discussed in Section 13-5.

Heat Released from Combustion

Heat released from the combustion process is partly stored in the combustion products and partly transferred by convection, conduction, and radiation to the walls of the combustion system, to the incoming fuel, and to the residue. If the elemental composition of the solid wastes is known, the energy content can be estimated by using the modified form of the Dulong equation, given in Chapter 4; see Eq. (4-10). Often the energy content of solid wastes is based on an analysis of the heating value of the individual waste components (see Table 4-5 and Example 4-3). The combustion computations necessary for estimating the heat that is available from the combustion process for conversion to steam and ultimately to electrical power are illustrated in Example 13-2.

Example 13-2 Materials and heat balance for the combustion of solid waste. Determine the heat available in the exhaust gases from the combustion of 125 ton/d of solid waste with the following characteristics:

Component	Percent of total	lb/day
Combustible	54.6	136,500
Noncombustible	24.0	60,000
Water	21.4	53,500

Element	Percent
Carbon	27.4
Hydrogen	3.6
Oxygen	23.0
Nitrogen	0.5
Sulfur	0.1
Water	21.4
Inerts	24.0

Assume that the following conditions are applicable:

1. The as-fired heating value of the solid wastes is 5065 Btu/lb.
2. The grate residue contains 5 percent unburned carbon.
3. Temperatures:

   Entering air, 80°F

   Grate residue, 800°F

4. Specific heat of residue = 0.25 Btu/lb-°F.
5. Latent heat of water = 1040 Btu/lb.
6. Radiation loss = 0.005 Btu/Btu of gross heat input.
7. All oxygen in waste is bound as water.
8. Theoretical air requirements based on stoichiometry (see Chapter 9):

Carbon:	$(C + O_2 \rightarrow CO_2)$	= 11.52 lb/lb
Hydrogen:	$(2H_2 + O_2 \rightarrow 2H_2O)$	= 34.56 lb/lb
Sulfur:	$(S + O_2 \rightarrow SO_2)$	= 4.31 lb/lb

9. The net hydrogen available for combustion is equal to percent hydrogen minus $\frac{1}{8}$ the percent oxygen. This accounts for the "bound water" in the dry combustible material.
10. The heating value of carbon is 14,000 Btu/lb.
11. Moisture in the combustion air is 1 percent.

Solution

1. Set up a computation table to compute the weights of the elements of the solid waste.

Element			lb/d
Carbon	= 0.274(250,000)	=	68,500
Hydrogen	= 0.036(250,000)	=	9,000
Oxygen	= 0.230(250,000)	=	57,500
Nitrogen	= 0.005(250,000)	=	1,250
Sulfur	= 0.001(250,000)	=	250
Water	= 0.214(250,000)	=	53,500
Inerts	= 0.240(250,000)	=	60,000
Total			250,000

2. Compute the amount of the residue:

$$\text{Inerts} = 60{,}000 \text{ lb/d}$$

$$\text{Total residue} = 60,000/0.95 = 63{,}158 \text{ lb/d}$$

$$\text{Carbon in residue} = 63{,}158 - 60{,}000 = 3158 \text{ lb/d}$$

3. Determine the available hydrogen and bound water:

$$\text{Net available hydrogen, \%} = (3.6\% - 23.0\%/8) = 0.725\% = 1812 \text{ lb/d}$$

$$\text{Hydrogen in bound water} = 3.6\% - 0.725\% = 2.875\% = 7188 \text{ lb/d}$$

$$\text{Bound water} = \text{oxygen} + \text{hydrogen in bound water}$$

$$= 57{,}500 + 7188 = 64{,}688 \text{ lb/d}$$

4. Set up a computation table to calculate the air required.

Element	Air requirement, lb/d
Carbon = (68,500 − 3,158)(11.52)	752,740
Hydrogen = 1,812(34.56)	62,623
Sulfur = 250(4.31)	1,078
Total dry theoretical air	816,441[a]
Total dry air including 100 percent excess	1,632,882
Moisture = 1,632,882(0.01)	16,329
Total air	1,649,211

[a] lb air/lb solid waste = 816,441 lb air/250,000 lb solid waste = 3.27. This result is virtually the same as the 3.42 lb air/lb solid waste found in Step 1 of Example 13-1, because the O_2 in the fuel in Example 13-1 was accounted for as a negative number and not in terms of bound water as in this example.

5. Determine the amount of water produced from the combustion of the available hydrogen:

$$H_2O = \frac{18 \text{ lb } H_2O}{2 \text{ lb H}}(1812 \text{ lb/d}) = 16{,}308 \text{ lb/d}$$

6. Prepare a heat balance for the combustion process.

Item	Value, 10^6 Btu/d
Gross heat input	
2.5×10^5 lb/d (5065 Btu/lb)	1266.3
Heat lost in unburned carbon	
3158 lb/d (14,000 Btu/lb)	−44.2
Radiation loss	
0.005 Btu/Btu (1266.3×10^6 Btu/d)	−6.3
Inherent moisture	
53,500 lb/d (1040 Btu/lb)	−55.6
Moisture in bound water	
64,688 lb/d (1040 Btu/lb)	−67.3
Moisture from the combustion of available hydrogen	
16,308 lb/d (1040 Btu/lb)	−17.0
Sensible heat in residue	
63,158 lb/d [0.25 Btu/lb-°F(800 − 80)°F]	−11.4
Total losses	
	−201.8
Net heat available in flue gases	
$(1266.3 - 201.8) \times 10^6$ Btu/d	1064.5
Combustion efficiency	
(1064.5×10^6 Btu/d/1266.3×10^6 Btu/d) $\times$ 100%	84.1%

Comment. If the boiler efficiency were 85 percent, then the overall efficiency would be equal to the combustion efficiency multiplied by the boiler efficiency (84.1% × 85%), about 71.5 percent. This value is consistent with values obtained in modern MSW combustion systems.

13-2 COMBUSTION SYSTEMS

Combustion can be defined as the thermal processing of solid waste by chemical oxidation with stoichiometric or excess amounts of air. End products include hot combustion gases, composed primarily of nitrogen, carbon dioxide, and water vapor (flue gas); and noncombustible residue (ash). Energy can be recovered by heat exchange from the hot combustion gases. The basic operations involved in the combustion of solid waste have been identified and described previously in Chapter 9 (see Fig. 9-31).

Types of Combustion Systems

Solid waste combustion systems can be designed to operate with two types of solid waste fuel: commingled solid waste (mass-fired) and processed solid waste refuse-derived fuel (RDF-fired). Mass-fired combustion systems are the predominant type. In 1987, 68 percent of the operational combustion capacity in the United States was provided by mass-fired units, versus 23 percent by RDF-fired units [52]. The remaining 9 percent of capacity was provided by mass-fired, modular combustion unit systems, described in a later section.

Mass-Fired Combustion Systems. In a mass-fired combustion system, minimal processing is given to solid waste before it is placed in the charging hopper of the system (see Fig. 13-2). The crane operator in charge of loading the charging hopper can manually reject obviously unsuitable items. However, it must be assumed that anything in the solid waste stream may ultimately enter the system, including bulky, oversize noncombustible objects (e.g., refrigerators) and even potentially hazardous wastes deliberately or inadvertently delivered to the system.

For these reasons, the system must be designed to handle these objectionable wastes without damage to equipment or injury to operational personnel. The energy content of mass-fired waste can be extremely variable, dependent on the climate, season, and source of waste. In spite of these potential disadvantages, mass-fired combustion has become the technology of choice for most existing and planned systems.

One of the most critical components of a mass-fired combustion system is the grate system. It serves several functions, including the movement of waste through the system, mixing of the waste, and injection of combustion air. Many variations of grates are possible, based on reciprocating, rocking, or rotating elements. Typical grate systems for mass-fired combustors are shown in Fig. 13-3.

RDF-Fired Combustion Systems. In RDF-fired combustors, RDF is typically burned on a traveling-grate stoker. The grate provides a platform on which the RDF can burn and provides for the introduction of underfire air to promote turbulence and uniform combustion. Best results have been obtained with combustion systems specifically designed for RDF, but some coal-fired boilers have been retrofitted to burn RDF or RDF/coal mixtures successfully (see Fig. 13-4).

The operation of front-end systems for the removal of metal, glass, and other noncombustible materials to produce refuse-derived fuel (RDF) was discussed in Chapters 9 and 12. Compared to the uncontrolled nature of unprocessed MSW, RDF can be produced with fair consistency to meet specifications for energy, moisture, and ash content. The RDF can be produced in shredded or fluff form, or as densified pellets or cubes. Densified RDF (dRDF) is more costly to produce but is easier to transport and store. Either form can be burned by itself or mixed with coal.

Because of the higher energy content of RDF compared to unprocessed MSW, RDF combustion systems can be physically smaller than comparatively rated mass-fired systems. An RDF-fired system can also be controlled more effectively than a mass-fired system because of the more homogeneous nature of RDF, allowing for better combustion control and better performance of air pollution control devices. Additionally, a properly designed front end system can remove significant portions of metals, plastics, and other materials that may contribute to harmful air emissions.

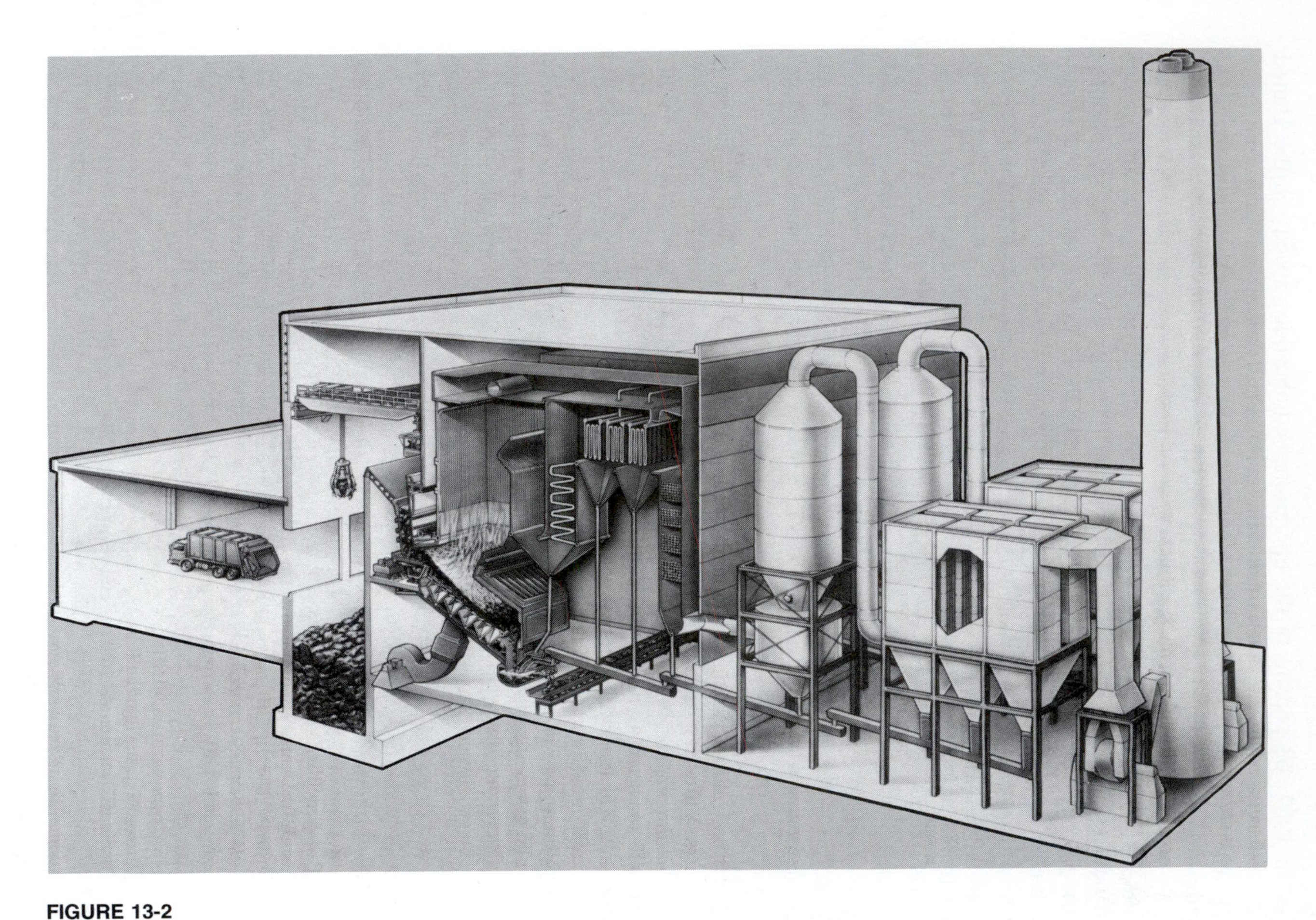

FIGURE 13-2
Section through modern mass-fired combustor for MSW. (Courtesy of Ogden Martin Systems, Inc.)

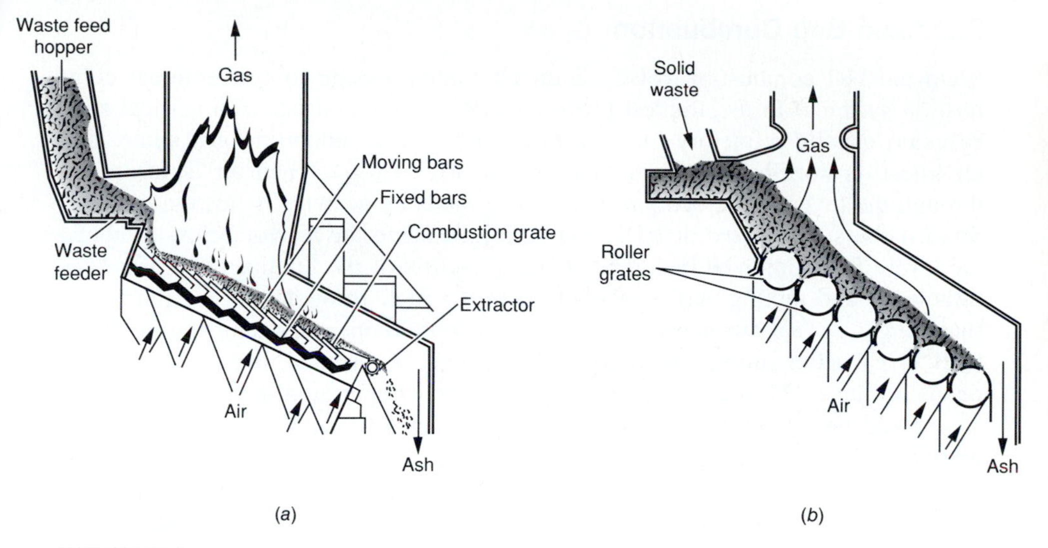

FIGURE 13-3
Representative grate systems used in mass-fired MSW combustors: (*a*) Martin grate (courtesy of Odgen Martin Systems, Inc.) and (*b*) Dusseldorf grate (courtesy of American Ref-Fuel, Inc.).

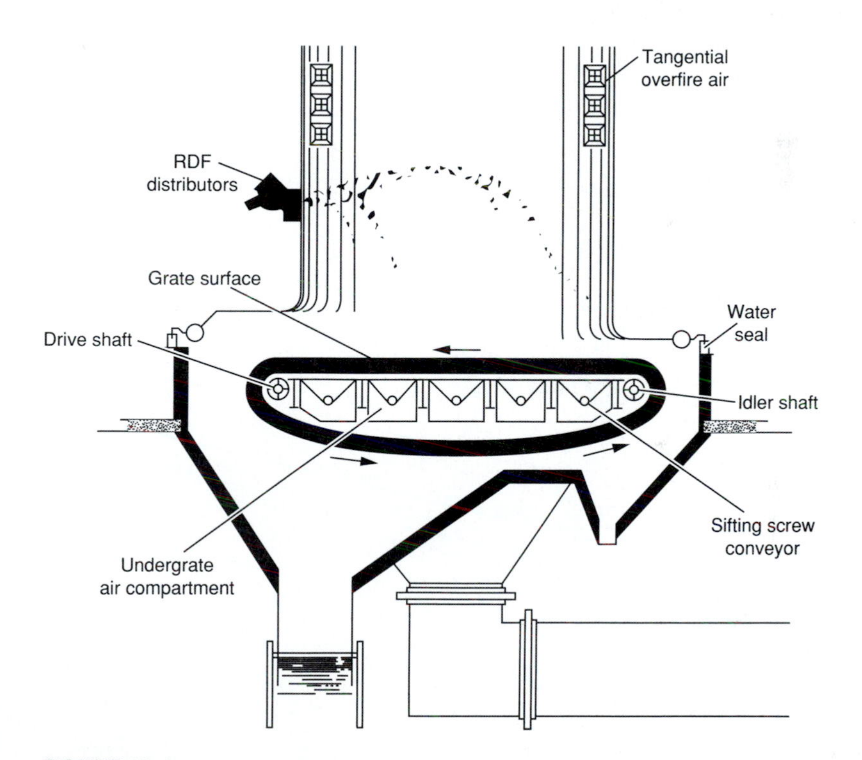

FIGURE 13-4
Section through modern RDF-fired combustor with traveling grate stoker. (Courtesy of ABB Resource Recovery Systems.)

Fluidized Bed Combustion

Fluidized bed combustion (FBC) is an alternative design to conventional combustion systems. In its simplest form, an FBC system consists of a vertical steel cylinder, usually refractory-lined, with a sand bed, a supporting grid plate, and air injection nozzles known as *tuyeres* (see Fig. 13-5). When air is forced up through the tuyeres, the bed *fluidizes* and expands up to twice its resting volume. Solid fuels, such as coal or RDF, can be injected into the reactor below or above the level of the fluidized bed. The "boiling" action of the fluidized bed promotes turbulence and mixing and transfers heat to the fuel. In operation, auxiliary fuel (natural gas or fuel oil) is used to bring the bed up to operating temperature (1450 to 1750°F). After startup, auxiliary fuel is usually not needed; in fact, the bed remains hot up to 24 hours, allowing rapid restart without auxiliary fuel.

Fluid bed combustion systems are quite versatile and can be operated on a wide variety of fuels, including MSW, sludge, coal, and numerous chemical wastes. The bed material can be plain sand or limestone ($CaCO_3$). When limestone is used, it reacts with oxygen and the sulfur dioxide (SO_2) formed by the combustion of sulfur-containing wastes to release carbon dioxide and form calcium sulfate ($CaSO_4$), a solid that can be removed with the ash. The use of limestone as the bed material allows the combustion of high-sulfur coal with minimum emissions of sulfur dioxide.

Several FBC systems are being used for solid waste combustion throughout the world. One of the first installations was a small (150-ton/d) fluidized bed unit

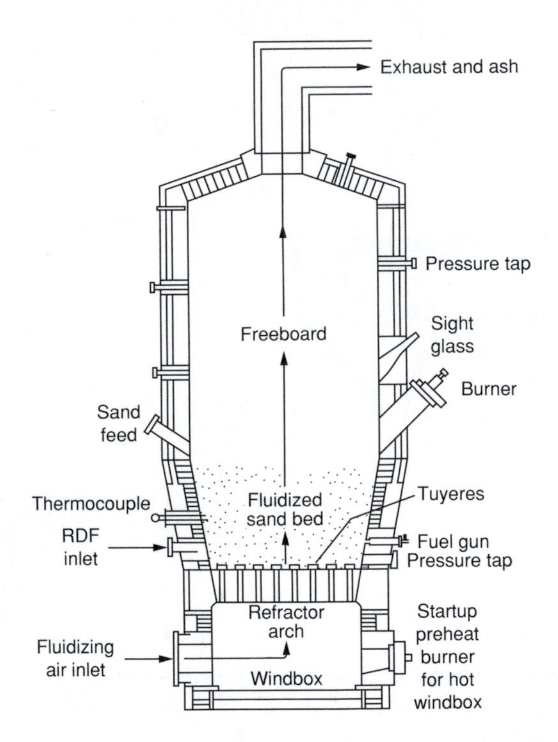

FIGURE 13-5
Typical fluidized bed combustion system for refuse-derived fuel.

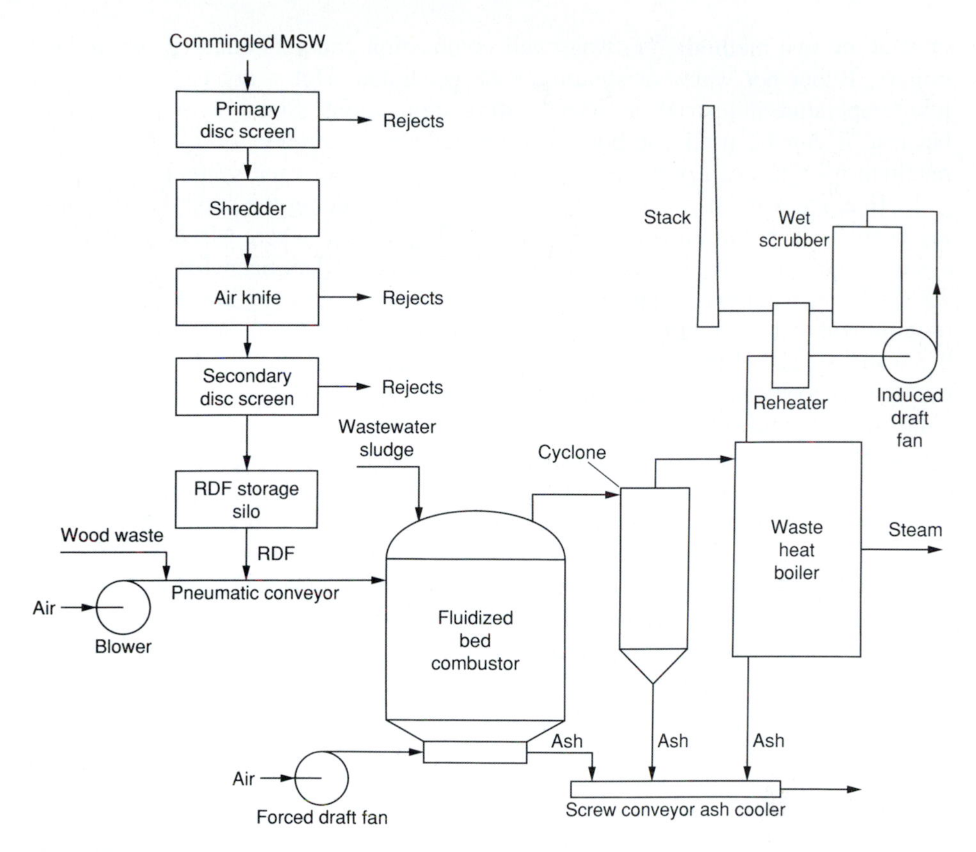

FIGURE 13-6
Schematic of Duluth, Minnesota fluidized bed system for the combustion of wastewater sludge, refuse-derived fuel, and wood waste [36].

in Lausanne, Switzerland [36]. It is used for the codisposal of MSW and dewatered wastewater treatment plant sludge. A waste heat boiler generates steam, which is used for heating and electricity generation. A larger scale plant (700 ton/d), was built in Duluth, Minnesota [36]. The plant is used to codispose of 300 ton/d of dewatered wastewater treatment plant sludge and 400 ton/d of MSW. The MSW is processed in a front-end system prior to combustion (see Fig. 13-6). A 390-ton/d fluidized bed combustion system is operating in Fujisawa, Japan. The system employs a proprietary fluidized bed–moving bed design, which allows mass firing of unprocessed MSW [20].

Heat Recovery Systems

Virtually all new solid waste combustion systems currently under construction in the United States and Europe employ some form of energy recovery to help offset operating costs and to reduce the capital costs of air pollution control equipment. Energy can be recovered from the hot flue gases generated by combusting MSW or RDF by two methods: (1) waterwall combustion chambers and (2) waste heat boilers. Either hot water or steam can be generated. Hot water can be used for low-temperature industrial or space-heating applications. Steam is more versatile, because it can be used for both heating and the generation of electricity. The resultant revenues can partially offset operational costs of the system.

Heat recovery also has a beneficial effect on reducing the capital and operating costs of air pollution control equipment. In practice, where MSW combustion systems without heat recovery equipment are used, it has been found that from 100 to 200 percent excess air must be supplied to meet combustion and turbulence requirements and to control slagging and the accumulation of other materials on the walls of the combustion system. The resultant large flue gas flow makes the use of such systems costly because of the extra capacity required for air pollution control equipment. In contrast, when heat recovery systems are used, it has been found that from 50 to 100 percent excess air is adequate, thus reducing the size of air pollution control devices. The cooling of flue gases that occurs during heat recovery also further reduces the volume of flue gases.

Waterwall Combustion Chamber. In this method, the walls of the combustion chamber are lined with boiler tubes that are arranged vertically and welded together in continuous sections. Water circulated through the tubes absorbs heat generated in the combustion chamber, generating steam. Usually the furnace wall areas adjacent to the grates are lined with refractory (heat-resistant) materials to protect the tubes from excessive temperatures and mechanical abrasion (see Fig. 13-7).

Waste Heat Boiler. In this method, the combustion chamber of the furnace is lined with insulating refractory materials to reduce heat losses through the furnace walls. The hot flue gases are passed through a separate waste heat boiler located

FIGURE 13-7
Section through a waterwall. (Courtesy of Combustion Engineering, Inc.)

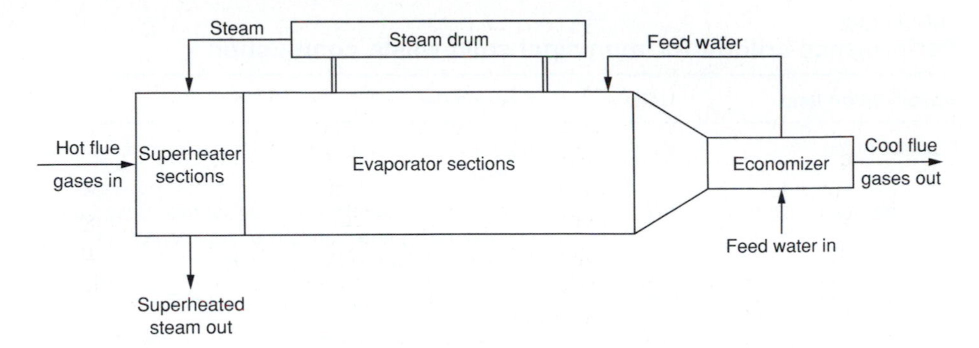

FIGURE 13-8
Waste heat boiler for heat recovery from combustion gases.

externally to the combustion chamber. This method of heat recovery is commonly used in modular combustion units, which are discussed later in this chapter. In some cases it is possible to retrofit a waste heat boiler on an existing refractory-lined furnace (see Fig. 13-8).

Analysis. A heat recovery system is essentially a heat exchanger. Heat from the burning waste is transferred into a working fluid (water) by means of a temperature gradient between the flue gases and the working fluid. Design and analysis of the heat transfer system is essentially the same as that for a coal or oil fired power plant. The reader is referred to Ref. 5 for details of heat recovery analysis.

A simplified analysis can be made by the use of steam production rates, which relate steam production to the energy content of the waste. As shown in Table 13-1, steam production rates range from 1.5 to 4.3 tons steam/ton waste, depending on the energy and moisture content of the waste.

Criteria for System Selection

Selection of a thermal processing system is a complex and expensive undertaking. Most systems are built on some form of a "turnkey" contract, where a single

TABLE 13-1
Steam production rates for MSW combustion[a]

	As-received energy content, HHV, Btu/lb				
Item	**6500**	**6000**	**5000**	**4000**	**3000**
MSW quality					
Moisture, %	15.0	18.0	25.0	32.0	39.0
Non-combustible, %	14.0	16.0	20.0	24.0	28.0
Combustible, %	71.0	66.0	55.0	44.0	33.0
Steam generated ton/ton MSW	4.3	3.9	3.2	2.3	1.5

[a] Adapted from Ref. 5.

TABLE 13-2
Performance criteria for municipal solid waste combustion

Specification item	Units	Comments
Nominal rating	ton/d	Based on commingled MSW for mass-fired systems or RDF.
Gross electrical output	kW	Does not include internal electrical use. Based on nominal rating and standard energy content of MSW or RDF, which must be specified.
Net electrical output	kW	Includes all internal uses and losses.
Availability	h/yr	Estimated time that system will be on-line, including allowances for all regularly scheduled maintenance.
Air emissions	lb/d[a]	As required to meet federal, state, and local air pollution control regulations.
Air pollution control equipment	—	Specifications of air pollution control equipment as required to meet air emission regulations.
Solid residue	ton/d	Estimate of residue of both bottom ash and fly ash, based on similar operating units and pilot testing.
Wastewater discharges	gal/d[b]	Estimate of wastewater quantity and quality.
Manpower	persons	Includes management, operations, and maintenance staff.
Capital cost	$	Includes engineering, construction, and capital cost of site work, structures, and all equipment.
Operating costs	$/yr	Includes manpower, routine maintenance and repairs, utilities, and disposal cost of solid residues and wastewater.

[a] May also be specified by permits and regulations in terms of concentration (see Section 13-5).

[b] Quality parameters include BOD, pH, heavy metals and others as required by local and state discharge permits.

contractor assumes complete responsibility for the design and construction of the system. Alternatively, some systems are built under a full-service contract, where the contractor designs, builds and operates the system for a fixed number of years.

Engineering Performance Criteria. Environmental engineers play an intermediary role in the system selection process by preparing a set of engineering performance specifications, which define the performance of the system instead of the details of specific technology such as grate type or ash-handling system. Performance specifications result in the most cost-effective system, because they broaden competition and encourage technical advancement by the industry. Typical performance criteria are listed in Table 13-2. They include throughput, reliability, volume and weight reduction, air emissions, energy output, space requirements, and utility requirements.

Economic Performance Criteria. The economic performance of a thermal processing system must also be evaluated to choose between competing systems.

The best way to compare alternatives is by the use of life cycle costing, which accounts for operating and maintenance costs over the lifetime of the system. The solid waste industry has developed a standardized approach to life cycle costing, known as the *pro forma income statement*. A net-present-worth approach is used to normalize capital and operating costs and revenues to a "zero year" time base, thus allowing costs to be evaluated on a dollar-per-ton basis. A computer program for calculating pro forma income statements is described in Ref. 54.

13-3 PYROLYSIS SYSTEMS

Pyrolysis, as previously defined, is the thermal processing of waste in the complete absence of oxygen. Unfortunately, there is quite a bit of confusion in the literature, and many so-called pyrolysis systems are actually gasification systems. Both pyrolysis and gasification systems are used to convert solid waste into gaseous, liquid, and solid fuels. The principal difference between the two systems is that pyrolysis systems use an external source of heat to drive the endothermic pyrolysis reactions in an oxygen-free environment, whereas gasification systems are self-sustaining and use air or oxygen for the partial combustion of solid waste.

Description of the Pyrolysis Process

Because most organic substances are thermally unstable, they can, upon heating in an oxygen-free atmosphere, be split through a combination of thermal cracking and condensation reactions into gaseous, liquid, and solid fractions. *Pyrolysis* is the term used to describe the process. In contrast to the combustion and gasification processes, which are highly exothermic, the pyrolytic process is highly endothermic, requiring an external heat source. For this reason, the term *destructive distillation* is often used as an alternative term for pyrolysis.

The three major component fractions resulting from the pyrolysis process are the following:

1. A gas stream, containing primarily hydrogen, methane, carbon monoxide, carbon dioxide, and various other gases, depending on the organic characteristics of the material being pyrolyzed.
2. A liquid fraction, consisting of a tar or oil stream containing acetic acid, acetone, methanol, and complex oxygenated hydrocarbons. With additional processing, the liquid fraction can be used as a synthetic fuel oil as a substitute for conventional No. 6 fuel oil.
3. A char, consisting of almost pure carbon plus any inert material originally present in the solid waste.

For cellulose, $C_6H_{10}O_5$, the following expression has been suggested as being representative of the pyrolysis reaction [19]:

$$3(C_6H_{10}O_5) \rightarrow 8H_2O + C_6H_8O + 2CO + 2CO_2 + CH_4 + H_2 + 7C \qquad (13\text{-}1)$$

In Eq. (13-1), the liquid tar or oil compounds normally obtained are represented by the expression C_6H_8O. It has been found that distribution of the product

TABLE 13-3
Materials balance for pyrolysis[a]

Temperature, °F	Wastes, lb	Gases, lb	Pyroligneous acids and tars, lb	Char, lb	Mass accounted for, lb
900	100	12.33	61.08	24.71	98.12
1200	100	18.64	59.18	21.80	99.62
1500	100	23.69	59.67	17.24	100.59
1700	100	24.36	58.70	17.67	100.73

[a] Adapted from Ref. 19.

fractions varies dramatically with the temperature at which the pyrolysis is carried out [19]. Representative data on the product as a function of the operating temperature are reported in Table 13-3. Typical analyses of pyrolysis gas as a function of temperature are given in Table 13-4. The energy content of pyrolytic oils has been estimated to be about 9000 Btu/lb. Under conditions of maximum gasification, it has been estimated that the energy contents of the resulting gas would be about 700 Btu/ft^3.

Operational Problems with MSW Pyrolysis Systems

Only one full-scale MSW pyrolysis system was built in the United States. Constructed in El Cajon, California, the Occidental Flash Pyrolysis System did not achieve its primary operational goal (production of a saleable pyrolysis oil) and was shut down after only two years of operation.

A simplified flow diagram of the system is shown in Fig. 13-9. The front-end system employed two stages of shredding, air classification, trommeling, and drying to produce a very finely ground organic fraction. Ferrous metals, aluminum, and glass were also recovered, by magnetic separation, eddy current separation, and froth flotation respectively. The pyrolysis portion of the system consisted of

TABLE 13-4
Gas composition for pyrolysis as a function of temperature[a]

	Percent by volume			
Gas	900°F	1200°F	1500°F	1700°F
H_2	5.56	16.58	28.55	32.48
CH_4	12.43	15.91	13.73	10.45
CO	33.50	30.49	34.12	35.25
CO_2	44.77	31.78	20.59	18.31
C_2H_4	0.45	2.18	2.24	2.43
C_2H_6	3.03	3.06	0.77	1.07
Accountability	99.74	100.00	100.00	99.99

[a] Adapted from Ref. 19.

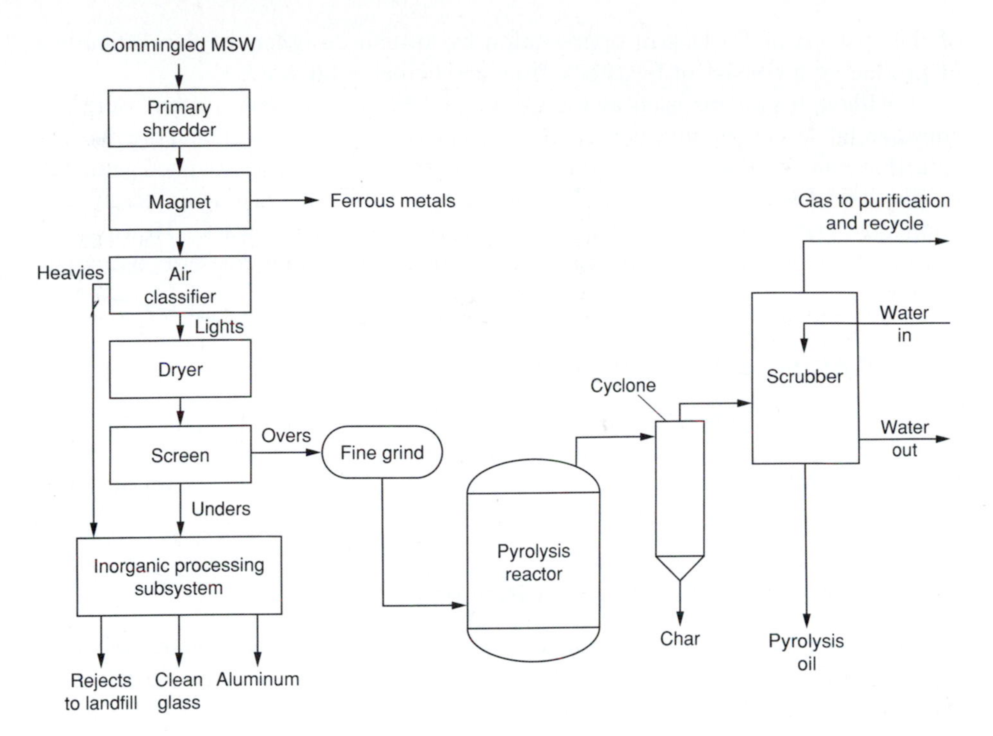

FIGURE 13-9
Schematic diagram of the Occidental flash pyrolysis system for the organic portion of MSW [30].

several interconnected process loops. The end products were pyrolytic oils, gases, char, and ash.

As might be expected with such a complex system, numerous operational problems were encountered. In an analysis of the system [30], the ultimate failure of the system was attributed to several factors, including the following:

1. Failure of the front-end system to meet purity specifications for aluminum and glass, which affected the economics of the system.
2. Failure of the system to produce a saleable pyrolysis oil. The oil produced had a moisture content of 52 percent, not the 14 percent predicted from the pilot plant results. The increased moisture in the oil decreased the energy content to 3600 Btu/lb, as compared to the 9100 Btu/lb predicted by the pilot plant tests.

Comment

Pyrolysis is still widely used as an industrial process for the production of charcoal from wood, coke and coke gas from coal, and fuel gas and pitch from heavy petroleum fractions. In spite of these industrial uses of pyrolysis, the pyrolysis of solid waste has not been as successful. The principal causes for the failure of pyrolysis technology in the past appeared to have been the inherent complexity of the systems and a lack of appreciation by system designers of the difficulties of producing a consistent feedstock from municipal solid waste.

Although systems such as the Occidental Flash Pyrolysis System were not commercial successes, they nevertheless produced valuable design and operational data that can be used by future designers. If the economics associated with the production of synthetic liquid fuels change, pyrolysis may once again be an economically viable process for the thermal processing of solid waste. However, if gaseous fuels are desired, gasification is a simpler and more cost-effective technology.

13-4 GASIFICATION SYSTEMS

Gasification is the general term used to describe the process of partial combustion in which a fuel is deliberately combusted with less than stoichiometric air. Although the process was discovered in the nineteenth century, it has only recently been applied to the processing of solid waste.

Description of the Gasification Process

Gasification is an energy-efficient technique for reducing the volume of solid waste and the recovery of energy. Essentially, the process involves partial combustion of a carbonaceous fuel to generate a combustible fuel gas rich in carbon monoxide, hydrogen, and some saturated hydrocarbons, principally methane. The combustible fuel gas can then be combusted in an internal combustion engine, gas turbine, or boiler under excess-air conditions. The historical development and basic theory of operation of the gasification process are discussed briefly in the following paragraphs.

Historical Development. Gasifiers have been used since the 19th century. The first coal gasifiers were built in Germany by Bischof, 1839, and by Siemens, 1861. Siemens's gasifiers were primarily used to fuel heavy industrial furnaces. The development of gas-cleaning and -cooling hardware by Dowson in England, 1881, extended the use of gasifiers to small furnaces and internal combustion engines [43].

By the early 1900s, gasifier technology had advanced to the point where virtually any type of cellulosic residue, such as olive pits, straw, and walnut shells could be gasified. These early gasifiers were used primarily to provide the fuel for stationary internal combustion engines for milling and other agricultural uses. Portable gasifiers also emerged in the 1900s. They were used for boats, automobiles, trucks, and tractors. The real impetus for the development of gasifier technology was the gasoline shortages of World War II. During the war years, France had over 60,000 charcoal-burning cars, while Sweden had about 75,000 wood-burning, gasifier-equipped buses, automobiles, trucks, and boats. With the return of relatively cheap and plentiful gasoline and diesel oil after the end of the World War II, gasifier technology was all but forgotten [48].

Gasification Theory. During the gasification process, five principal reactions occur:

$$C + O_2 \rightarrow CO_2 \quad \text{exothermic} \tag{13-2}$$

$$C + H_2O \rightarrow CO + H_2 \quad \text{endothermic} \tag{13-3}$$

$$C + CO_2 \rightarrow 2CO \quad \text{endothermic} \tag{13-4}$$

$$C + 2H_2 \rightarrow CH_4 \quad \text{exothermic} \tag{13-5}$$

$$CO + H_2O \rightarrow CO_2 + H_2 \quad \text{exothermic} \tag{13-6}$$

The heat to sustain the process is derived from the exothermic reactions, whereas the combustible components are primarily generated by the endothermic reactions. For a further discussion of gasification theory and reaction kinetics the reader is referred to Refs. 43 and 45.

When a gasifier is operated at atmospheric pressure with air as the oxidant, the end products of the gasification process are a low-Btu gas typically containing (by volume) 10 percent CO_2, 20 percent CO, 15 percent H_2, and 2 percent CH_4, with the balance being N_2; a char containing carbon and the inerts originally in the fuel; and condensible liquids resembling pyrolytic oil. Due to the diluting effect of nitrogen in the input air, the low-Btu gas has an energy content of about 150 Btu/ft^3. The operation of air-blown gasifiers is quite stable, with a fairly constant quality of gas being produced over a broad range of air input rates. The ability to function under different load conditions is known as the *turndown ratio*. When pure oxygen is used as an oxidant instead of air, as in the Purox® System (discussed later in this section), a medium-Btu gas can be produced with an energy content of about 300 Btu/ft^3 [26].

Gasifier Types

There are five basic types of gasifiers: (1) vertical fixed bed, (2) horizontal fixed bed, (3) fluidized bed, (4) multiple hearth, and (5) rotary kiln. Because the first three types have been the most widely used, they are described briefly in the following subsections.

Vertical Fixed Bed. The vertical fixed bed gasifier has a number of advantages over the other types of gasifiers, including simplicity and relatively low capital costs. However, this type of reactor is more sensitive to the mechanical characteristics of the fuel; it requires a uniform, homogeneous fuel, such as densified RDF. As shown in Fig. 13-10, fuel flow through the gasifier is by gravity, with air and fuel flowing concurrently through the reactor. The end products of the process are primarily low-Btu gas and char. It is also possible to operate a vertical fixed bed reactor in a countercurrent flow mode, with air and gas moving upwards through the reactor.

The operation of a vertical fixed bed gasifier using densified RDF as a fuel has been demonstrated at pilot scale (see Fig. 13-11) [2, 50, 53]. The gasifier was operated in a low-temperature (1200 to 1500°F) nonslagging mode, producing a

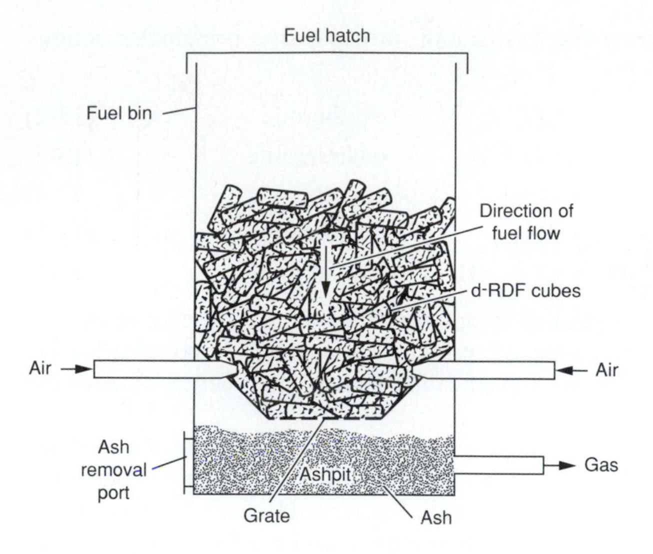

FIGURE 13-10
Schematic diagram of batch-fed vertical fixed-bed gasifier.

low-Btu gas (approximately 150 Btu/ft^3), a small amount of liquid condensate, and a dry, granular char and ash. The char was found to have adsorptive characteristics similar to commercial activated carbon and may be useful for the advanced treatment of wastewater [17].

The low-Btu gas produced by the system was tested in a Ford three-cylinder diesel engine [39]. The engine required only minor modifications to operate on a combination of low-Btu gas (80 percent of energy input) and diesel fuel (20

FIGURE 13-11
Experimental gasifier system for refuse-derived fuel MSW.

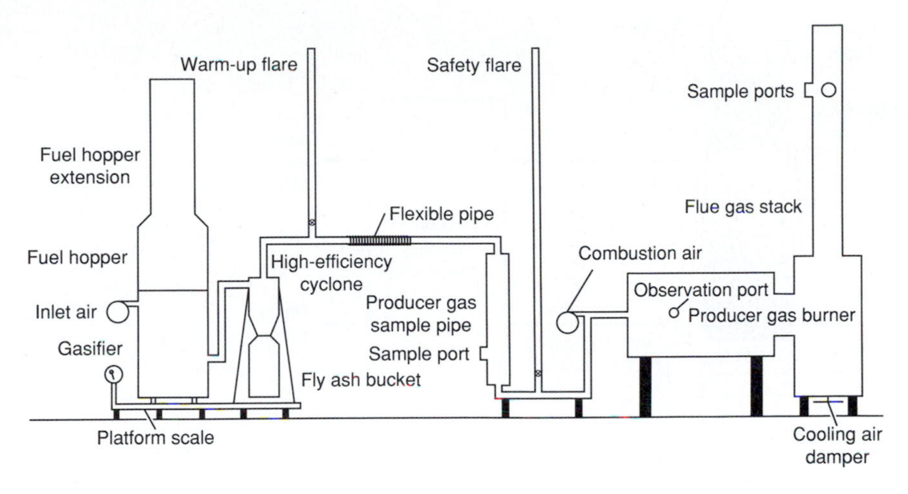

FIGURE 13-12
Schematic diagram of emission-testing system for experimental refuse-derived fuel gasifier system.

percent of energy input). From dynamometer tests it was found that the modified engine produced 76 percent of the output that the same engine produced on 100 percent diesel fuel.

Gasifiers have the potential to achieve low air pollution emissions with simplified air pollution control devices. The results of air emission testing, using the apparatus shown in Fig. 13-12, are reported in Table 13-5. The apparatus simulates the combustion of low-Btu gas in a boiler by burning the gas in an afterburner system. A simple high-efficiency cyclone was used for particulate control before the afterburner. The emissions are comparable to or less than the emissions from excess-air combustion systems employing far more complex emission control systems [49].

Vertical fixed bed gasifiers can also be operated with pure oxygen as an oxidant instead of air. Operation with pure oxygen results in the production of

TABLE 13-5
Gasifier air emissions[a]

Emission	Unit[b]	Values
NO_x	ppmv	60–115
SO_2	gr/dscm	0.091–0.227
Noncondensible hydrocarbons	ppmv	<1
Total particle emission rate (EPA Method 5)	gr/dscm	0.068–0.164
Particle cut diameter (from impactor tests)	μm	8

[a] Adapted from Ref. 49.

[b] ppmv = parts per million by volume; gr/dscm = grams per dry standard cubic meter.

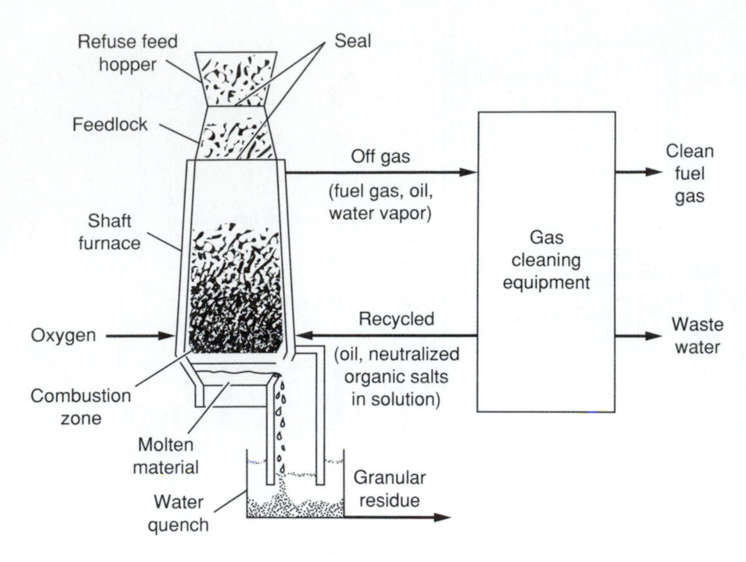

FIGURE 13-13
Schematic of Purox® oxygen-fed gasifier system for commingled MSW. (Courtesy of Union Carbide Corp.)

a medium-Btu gas with an energy content of 270 to 320 Btu/ft^3 and an average gas composition of 50 percent CO, 30 percent H_2, 14 percent CO_2, 4 percent CH_4, 1 percent hydrocarbons, and 1 percent N_2 [26, 28, 29]. Such a system was developed by the Union Carbide Corporation and marketed as the Purox® System. As shown in Fig. 13-13, the system consisted of the reactor, a minimal front-end system (shredding only), gas clean-up train (electrostatic precipitator, acid absorber, condenser, and water purifier), and an oxygen plant. The gasifier operated at relatively high temperatures (2600 to 3000°F), producing a molten slag as a by-product. Although a pilot plant was successfully tested on a variety of wastes, including MSW and sewage sludge, the Purox® System is no longer in commercial production.

Horizontal Fixed Bed. The horizontal fixed bed gasifier has become the most commercially available type. Ironically, it is not commonly referred to as a gasifier but rather by the terms *starved air combustor (incinerator), controlled air combustor,* or *pyrolytic combustor.* The terminology used in this section is *modular combustion unit* (MCU) as used in Ref. 5.

An MCU, as shown in Fig. 13-14, consists of two major components: a primary combustion chamber and a secondary combustion chamber. In the primary chamber, waste is gasified by partial combustion under substoichiometric conditions, producing a low-Btu gas, which then flows into the secondary combustion chamber, where it is combusted with excess air. The secondary combustion produces high-temperature (1200 to 1600°F) gases of complete combustion (CO_2, H_2O, N_2), which can be used to produce steam or hot water in an attached waste heat boiler [13]. Lower velocity and turbulence in the primary combustion cham-

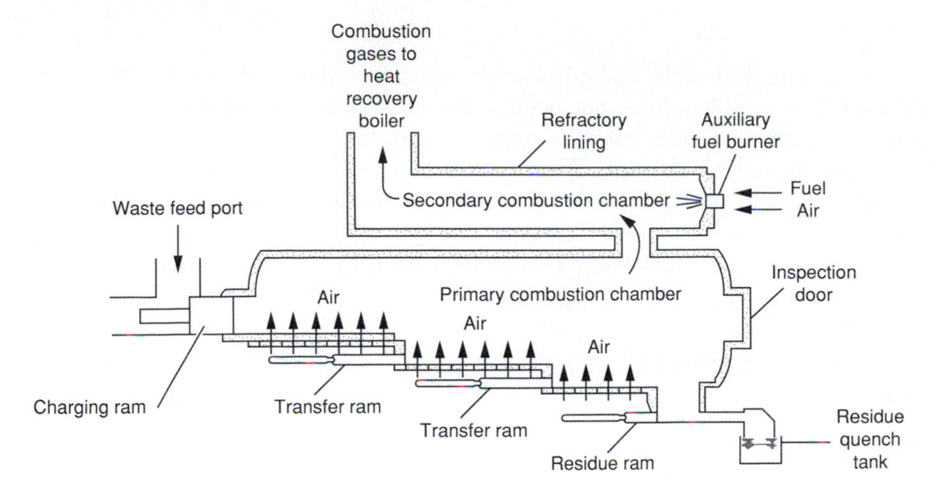

FIGURE 13-14
Modular combustion unit used for residential and commercial MSW and for selected industrial waste.

ber minimize the entrainment of particulates in the gas stream, leading to lower particulate emissions than in conventional excess-air combustors.

Modular combustion units are commercially available from several manufacturers in standard sizes ranging from 100 to 8400 lb/hr in capacity. The units are factory-preassembled and shipped by truck or rail to the project site, where they require a minimum of on-site labor for installation. The larger-sized MCUs feature continuous feeding and ash removal. The smaller-sized units are operated in batch mode and manually loaded. Typically, they are loaded during the working hours and fired unattended overnight.

When heat recovery boilers are installed, the steam or hot water produced must be matched carefully to a nearby market. Typical applications include the combustion of wood scraps at a plywood mill where the steam is used in the production process, and the combustion of MSW at a small industrial park where the steam is used for heating and cooling of nearby industrial and office buildings.

Fluidized Bed. The use of fluidized bed combustion for the excess-air combustion of MSW has been discussed previously. With minimal modifications, a fluidized bed combustion system can be operated in substoichiometric mode as a gasifier. Several pilot-scale tests have been conducted with municipal solid wastes as fuel. A 1 ton/h prototype fluidized bed gasifier fueled by RDF has been demonstrated in Kingston, Ontario [3]. A dual fluidized bed gasifier has been developed in Japan. The system employs two fluidized beds, one for fuel and one for char combustion, using the sand as a heat transfer medium between the two beds, producing medium-Btu gas [1]. A fluidized bed gasification system using dRDF has been constructed in Italy [18]. The system produces low-Btu gas, which is used in boilers for the production of steam and electricity.

Comment

Experience with full-scale and pilot-scale units has shown that reliable results with mass-fired gasifiers have not been achieved. Some form of RDF processing to remove metals and other inerts is required both to improve performance of the reactors and to reduce air emissions. Except for the modular combustion units, gasification systems cannot be considered a commercial technology at this time. Due to their lower air emissions, as compared to excess-air combustion systems, vertical fixed bed and fluid bed gasifiers may hold the most potential for future development and may again be "rediscovered."

13-5 ENVIRONMENTAL CONTROL SYSTEMS

The operation of thermal recovery systems produces several impacts on the environment, including gaseous and particulate emissions, solid residues, and liquid effluents. The proper design of control systems for these emissions is a critical part of the design of a thermal processing system. In some cases, the cost and complexity of the environmental control system is equivalent to or even greater than the cost of the thermal recovery system itself. An introduction to these environmental impacts, and to the control technologies used to manage them, is presented in this section.

Air Emissions

The Federal Clean Air Act required the Environmental Protection Agency (EPA) to identify pollutants of specific importance. Scientific data were collected on the relationships between various concentrations of air pollutants and their adverse effect on humans and the environment. This information was used to develop a list of *criteria pollutants* and acceptable ambient levels. These levels are known as the National Ambient Air Quality Standards. Criteria pollutants identified include carbon monoxide, sulfur dioxide, nitrogen dioxide, ozone, inhalable particulate matter (also known as PM10), and lead. In addition to these criteria pollutants, the EPA has identified other air pollutants of concern but has not yet established Ambient Air Standards for them. These pollutants are collectively known as *noncriteria pollutants*. Brief descriptions of the criteria pollutants and the most significant noncriteria pollutants are provided in this section.

Nitrogen Oxides (NO_x). The two most important nitrogen oxides are NO (nitric oxide) and NO_2 (nitrogen dioxide), collectively referred to as NO_x. There are two primary sources of NO_x in combustion. Thermal NO_x is formed by reactions between nitrogen and oxygen in the air used for combustion. Fuel NO_x is formed by reactions between oxygen and organic nitrogen in the fuel. Nitrogen oxides are precursors to the formation of ozone (O_3) and peroxyacetal nitrate (PAN), the photochemical oxidants known as *smog*. Nitrogen oxides also contribute to the formation of nitrate aerosols (liquid droplets), which can cause acid fog and rain.

Sulfur Dioxide (SO_2). Sulfur dioxide is formed by the combustion of fuels containing sulfur. Sulfur dioxide is an eye, nose, and throat irritant. In high concentrations, it can cause illness or death to persons already affected by lung problems such as asthma or bronchitis [12]. Sulfur dioxide is related to the production of acid rain and snow, which affect lakes, rivers, and forests in North America and Northern Europe.

Carbon Monoxide (CO). Carbon monoxide, formed during the combustion of carbonaceous materials when insufficient oxygen is present, reacts with the hemoglobin in the bloodstream to form carboxyhemoglobin (HbCO). The human body confuses HbCO with oxyhemoglobin (HbO_2), which normally transfers oxygen to the living tissues throughout the body. The lack of oxygen can cause headaches, nausea, and even death at extremely high concentrations.

Particulate Matter (PM). Particulate matter is formed during combustion by several processes, including incomplete combustion of fuel and the physical entrainment of noncombustibles. Particulate emissions cause visibility reductions and health effects. Particles smaller than 10 μm (10×10^{-6} m) are critical because they can be inhaled deeply into the lungs. The relative size of these particles is compared to other dusts in Fig. 13-15. Particle emissions from several facilities

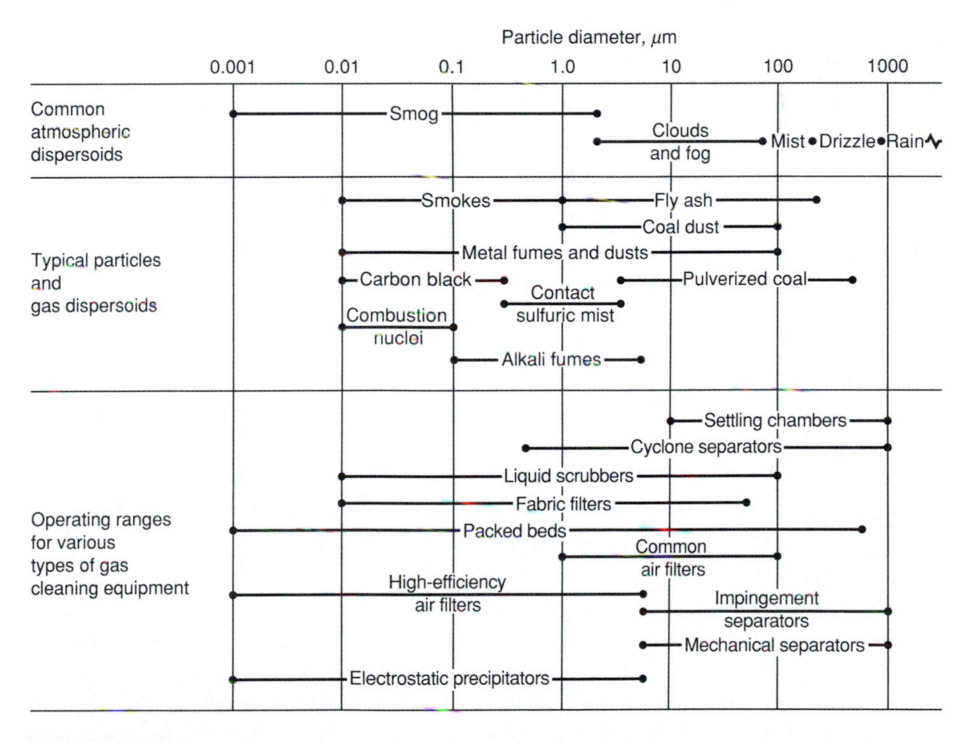

FIGURE 13-15
Particle size classification chart.

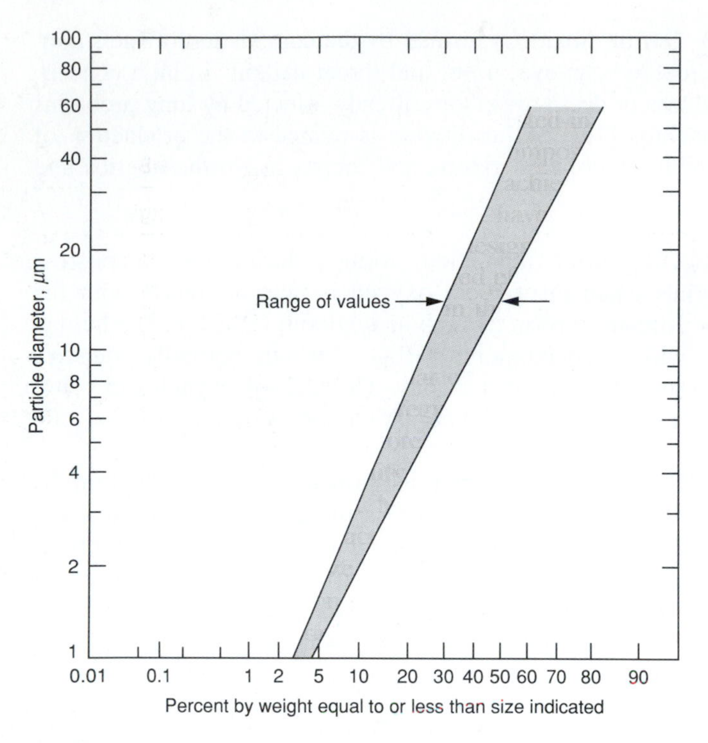

FIGURE 13-16
Particle size distribution of fly ash from MSW combustors. (Adapted from Ref. 25.)

are summarized in Fig. 13-16. Note that approximately 20 to 40 percent of the particulate emissions are less than 10 μm in diameter, and 7 to 10 percent are less than 2 μm in diameter. Several states, such as California, have enacted concentration limits for particulates smaller than 2 μm [7].

Metals. Municipal solid waste is a heterogeneous mixture. Many relatively innocuous items, such as plastics, glossy magazines, and flashlight batteries, contain metallic elements. Concentrations of fifteen metals in solid waste samples are compared with oil and U.S. coal in Table 13-6 [7]. For most metals, concentrations are higher in MSW and RDF than in coal or oil. Metals of particular concern from a public health viewpoint include cadmium (Cd), chromium (Cr), mercury (Hg), and lead (Pb). After combustion, metals are either emitted as particulate matter or vaporized into their gaseous form. Mercury is a particular problem in this respect, because it volatilizes at a relatively low temperature, 675°F.

Some metals are found in relatively few consumer products and could be removed from the solid waste stream prior to combustion. For example, virtually all of the mercury in MSW is due to the disposal of household dry cell batteries (mercury, alkaline, and carbon-zinc types). A smaller amount of mercury can be attributed to the disposal of broken home thermometers. Removal of these

TABLE 13-6
Metals in oil, U.S. coal, and MSW[a]

	Values in (μg/MJ) as fired								
	Oil			U.S. coal			MSW		
Metal	Low	High	Avg.	Low	High	Avg.	Low	High	Avg.
As	nr[b]	nr	nr	20.3	1501.0	331.0	23.0	1392	380
Be	nr	nr	nr	4.2	1550.0	171.0[d]	nd[c]	nd	nd
Cd	nr	nr	nr	1.9	1001.0	110.0	17.4	3538	702
Cr	0.26	7.4	3.2	186.0	3367.0	1017.0	280.0	125,623	15,102[d]
Cu	1.1	26.0	4.6	138.0	1218.0	510.0	1046.0	180,039	26,237
Hg	nr	nr	nr	2.3	96.7	18.7	<130.0	362	166[e]
Mn	0.69	9.4	3.6	47.0	4753.0	2081.0	1059.0	48,022	15,419
Mo	0.50	4.3	1.4	35.5	390.0	173.0	59.8	2982	938
Ni	9.7	371.0	251.0	104.0	2131.0	677.0	90.0	51,564	6380[d]
Pb	3.1	28.7	13.9	83.0	3084.0	738.0	877.0	136,663	39,290
Sb	nr	nr	nr	2.7	537.0	98.4	30.1	5404	2152
Se	nr	nr	nr	9.2	542.0	109.0	8.1	237	61.4
Sn	2.6	57.3	16.2	10.2	710.0	153.0	80.9	8179	2466
V	5.2	358.0	129.0	342.0	3341.0	1401.0	586.0	5988	2432
Zn	5.0	42.6	17.8	186.0	8825.0	1582.0	3018.0	303,716	60,943[d]

[a] Adapted from Ref. 7.
[b] Not reported.
[c] Not detected.
[d] Average strongly influenced by high value.
[e] Average of a very small number of measurements.

two products from the solid waste stream through a source separation or deposit program would substantially reduce the mercury load to the environment. The recycling of household batteries is discussed further in Chapter 15.

Acid Gases. The combustion of wastes containing fluorine and chlorine leads to the generation of the acid gases hydrogen fluoride (HF) and hydrogen chloride (HCl). Fluorine is found in trace amounts in many products, whereas chlorine is found primarily in plastics, chiefly polyvinyl chloride (PVC), polystyrene (PS), and polyethylene (PE).

The combustion of sulfur- and nitrogen-containing wastes can also lead to the formation of acid gases. The SO_2 may be oxidized to SO_3 and then react with water droplets in the atmosphere to form sulfuric acid (H_2SO_4). The emission of NO_2 can form nitric acid (HNO_3) in the atmosphere.

Dioxins and Furans. The emission of the organic compounds in the dioxin and furan families has become one of the most complex and controversial issues in the thermal processing of municipal solid waste. A dioxin is a member of a family of organic compounds known as polychlorinated dibenzodioxins (PCDD). A molecule of the PCDD family consists of a triple-ring structure in which two

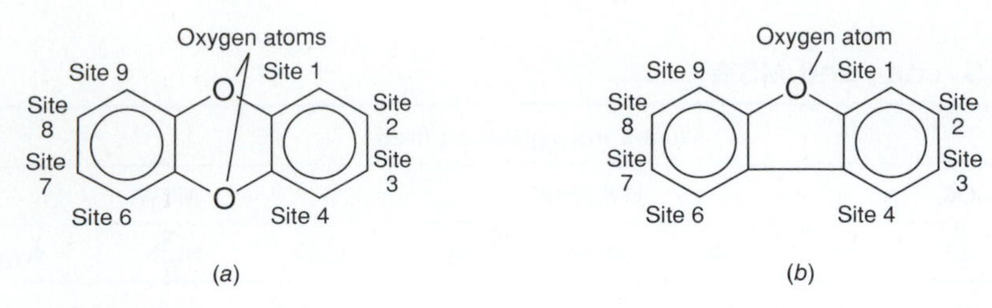

FIGURE 13-17
Structure of (*a*) polychlorinated dibenzodioxin (PCDD) and (*b*) polychlorinated dibenzofuran (PCDF) families.

benzene rings are interconnected by a pair of oxygen atoms (see Fig. 13-17*a*). The polychlorinated dibenzofuran (PCDF) family has a similar structure, except that only one oxygen atom links the two benzene rings together (see Fig. 13-17*b*). The numbered sites represent carbon atoms where bonds are free to attach to hydrogen or chlorine atoms. A total of 75 possible PCDD isomers and 135 PCDF isomers exist. In chemical notation these isomers are referred to by the site number of the chlorine atom. For example, 2,3,7,8-TCDD, or 2,3,7,8-tetrachloro-dibenzo-*p*-dioxin, is a PCDD with four chlorine atoms located at sites 2, 3, 7, and 8 (see Fig. 13-18).

The significance of the PCDD and PCDF families of organic compounds is that some of the isomers have been found to be among the most toxic substances in existence. For example, the LD_{50} of 2,3,7,8-TCDD for guinea pigs is less than 1 μg/kg of body weight [15]. Isomers of PCDDs and PCDFs have been analyzed as contaminants in various industrial chemicals. For example, the herbicide Agent Orange, used as a defoliant in the Vietnam War, was later found to have been contaminated with trace amounts of 2,3,7,8-TCDD [15].

Although evidence exists that PCDDs and PCDFs have carcinogenic (cancer-causing) properties in animals [16], their potential carcinogenicity in humans has recently been questioned. According to V. N. Houk of the U.S. Centers for Disease Control, dioxins are a low carcinogenic risk in high doses (such as in industrial accidents), but are not a carcinogenic risk in low doses, such as might be experienced in the ambient air near an MSW combustor [33]. In any case, both the EPA and state agencies have made control of dioxins and furans a high priority. Control strategies and emission limits for dioxins and furans are discussed later in this chapter.

It is known that PCDDs and PCDFs are emitted in low concentrations from combustion systems burning MSW and RDF [21, 40, 47]. There is considerable

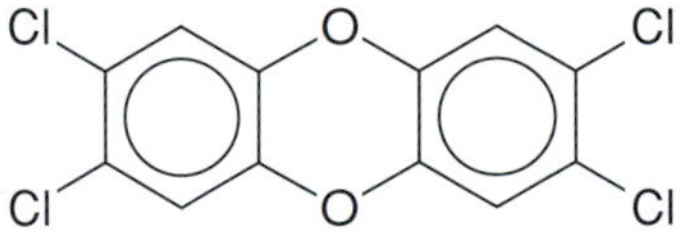

FIGURE 13-18
Structure of tetrachloro-dibenzo-*p*-dioxin (2,3,7,8-TCDD).

debate within the scientific and engineering communities as to the source of these emissions. There is some evidence that PCDDs and PCDFs are produced in all combustion processes, even home wood stoves and fireplaces [6].

Studies of sediments in the Great Lakes have shown that the input flux to the sediments from atmospheric fallout has been increasing since the 1940s [15]. Concentrations of PCDDs and PCDFs before this time period were very low, contradicting the argument that significant PCDD and PCDF emissions are caused by combustion of wood and coal. In the same study, a declining trend in PCDD and PCDF was found in the sediments since the mid-1970s. The decline was attributed to improved particulate controls on thermal processing systems since the passage of the Clean Air Act in 1970.

The exact mechanism of formation of PCDDs and PCDFs in the thermal processing of solid wastes has not yet been determined. Three sources of dioxin and furans in the emissions from MSW combustion have recently been proposed [32]: (1) the presence of dioxins and furans in MSW itself; (2) their formation during combustion due to chlorinated aromatic precursor compounds during combustion; and (3) their formation during combustion from simpler hydrocarbon and chlorine compounds.

Dioxins may find their way into the waste stream as contaminants in chemical compounds such as chlorophenols and chlorobenzenes, which are used in pesticides, paper, and wood preservatives. It has also been suggested that chlorophenol may act as a precursor compound in the formation of PCDDs and PCDFs [47]. Another hypothesis is that PCDDs and PCDFs are synthesized in the combustion system itself by reactions between lignins (a component of wood and paper) and chlorine compounds derived from polyvinyl chloride (PVC) or inorganic compounds such as NaCl [12].

Air Pollution Control Systems

Gaseous and particulate air emissions from resource recovery systems can be controlled with five classes of control equipment:

1. Electrostatic precipitators, fabric filters, electrostatic gravel bed filters (particulate control)
2. Source separation, combustion controls, flue gas treatment (NO_x control)
3. Source separation, wet or dry scrubbing (SO_2 and acid gas control)
4. Combustion controls (CO and HC control)
5. Source separation, combustion controls, particulate control (noncriteria pollutant controls)

Control devices are selected to achieve the required *removal efficiency,* which is defined as

$$E = \frac{W_{\text{inlet}} - W_{\text{outlet}}}{W_{\text{inlet}}} \times 100\% \tag{13-7}$$

where E = collection efficiency, percent
W_{inlet} = pollutant inlet weight
W_{outlet} = pollutant outlet weight

Control Equipment for Particulates. Fine particulates ($< 10\mu$m) are controlled with three technologies: (1) electrostatic precipitators (ESPs), which operate on the principle of electrostatic attraction; (2) fabric filters, which mechanically filter particulates from the flue gas stream; and (3) electrostatic gravel bed filters, which combine the operational characteristics of both ESPs and fabric filters.

Electrostatic precipitator. The electrostatic precipitator (ESP) was the first particle control device used on MSW combustors that was capable of removing fine (less than 10 μm) and very fine (less than 2 μm) particles. Electrostatic precipitators operate on the principle of electrostatic attraction (see Fig. 13-19). A high negative voltage, 20,000 to 100,000 volts, applied to the discharge electrodes, produces a strong electric field between the discharge and collector electrodes. Particles in the gas stream acquire a negative charge as they pass through the electrical field. Because of their charge, the particles are then attracted to the grounded collection electrode. After collection on the plates, particles are removed by mechanical vibration of the plates. The theory of operation and design characteristics of ESPs are discussed in greater detail in Refs. 12 and 51.

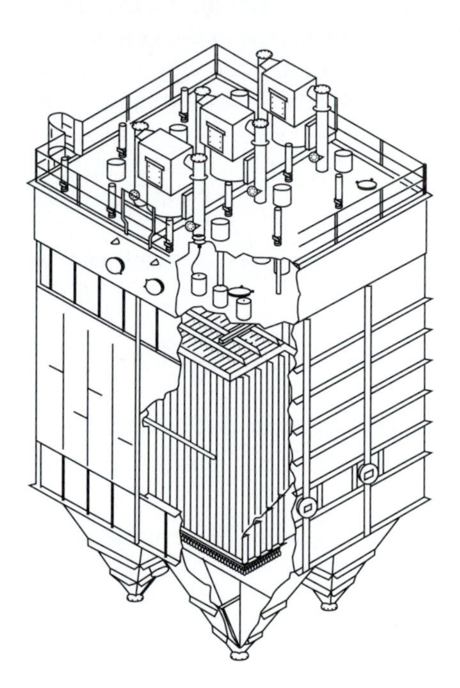

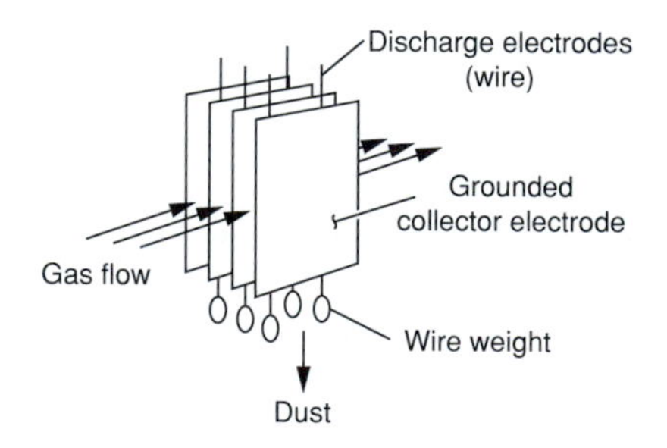

FIGURE 13-19
Electrostatic precipitator for particulate removal from MSW combustors. (Courtesy of Research-Cottrell, Inc.)

The efficiency of an ESP is a function of the flue gas characteristics (especially temperature and moisture) and the electrical resistivity of the particles. Typically, ESP efficiencies vary from about 93 percent for very fine particles (less than 2 μm) up to 99.8 percent for fine particles (up to 10 μm) [11]. While this performance is impressive, it does not meet the emission control requirements of some states, such as California, which have set requirements stricter than federal guidelines [7]. New ESP designs are being developed that are achieving particle removal efficiencies comparable to those of fabric filters [10].

Fabric filter. The fabric filter has become the technology of choice on most recently constructed MSW combustion systems in the United States. The fabric filter, or *baghouse* as it is sometimes referred to, is an intrinsically simple device (see Fig. 13-20). A number of filter bags are connected in parallel in a housing. Particles in the flue gas are trapped on a dust bed that gradually builds up on the surface of the fabric. The dust bed allows the fabric to filter particles as small as 0.1 μm, much smaller than the 50- to 75-μm open space between the fibers of the fabric. As particles build up on the surface of the fabric, the pressure drop across the fabric filter gradually increases. The particles are removed from the filter bags by several techniques, including mechanical shaking, reverse air flow, and pulse-jet [12]. A typical fabric filter installation is illustrated in Fig. 13-20.

The major design parameters for a fabric filter are filter area, material, and method of cleaning. Felted glass, woven glass, and Teflon™ have been used as fabric filters with some success in MSW combustion applications. The performance

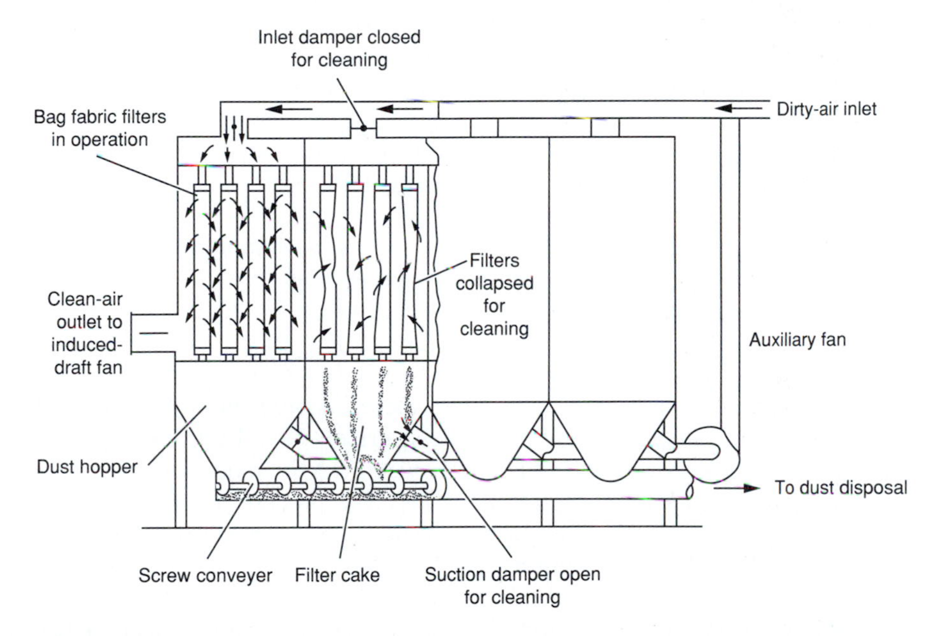

FIGURE 13-20
Baghouse with bag fabric filters for particulate removal from MSW combustors.

of fabric filters on recent installations has exceeded the most strict state guidelines. Fabric filters installed in 12 out of 13 MSW combustors that were tested during the period 1986 to 1987 achieved a particle emission rate of less than 0.01 grains/dscf as required by some states such as California [9, 11].

Electrostatic gravel bed filter. The electrostatic gravel bed filter is a hybrid device that employs both mechanical filtering and electrostatic attraction. The technology has been employed on wood-burning furnaces and more recently on the Pittsfield, Massachusetts, MSW combustor. A particle emission rate of 0.035 grains/dscf has been reported [7].

Control Equipment for NO_x. Fuel NO_x, formed by reactions between oxygen and organic nitrogen in the fuel, and thermal NO_x, formed by reactions between nitrogen and oxygen in the air used for combustion, are the two primary sources of NO_x in combustion. Source separation of MSW to remove organic nitrogen sources, such as food and yard wastes, from the waste stream prior to combustion could be used to control fuel NO_x [7].

Thermal NO_x control can be accomplished by both combustion controls and flue gas treatment. Combustion controls include (1) flue gas recirculation and (2) low-excess-air operation and staging of combustion. In flue gas recirculation, a portion of the exhaust gases from the combustor is recycled back into the combustion air side of the furnace. Low-excess-air operation and combustion staging involve carefully controlling the air input to the combustor by dividing the combustion air into primary and secondary flows. Thus, part of the furnace operates in a "starved air" or gasification mode and the rest of the furnace operates in excess-air mode, reducing the amount of thermal NO_x formed in the furnace. Because gasifiers are designed to operate in this mode, they have intrinsically low NO_x emissions.

Two technologies used for flue gas treatment include selective catalytic reduction (SCR) and selective noncatalytic reduction (SNCR).

Selective catalytic reduction. Selective catalytic reduction employs ammonia injection into the flue gas, followed by gas passage over a catalyst bed. The following reaction occurs in the temperature range of 530 to 800°F:

$$NO + NH_3 + 1/4\ O_2 \rightarrow N_2 + 3/2\ H_2O \tag{13-8}$$

Base metals, such as copper, iron, chromium, nickel, molybdenum, cobalt, and vanadium, in various shapes and forms (pellets, grids), have been employed as catalysts. The technology is quite efficient, achieving NO_x reductions up to 90 percent in coal- and oil-burning applications [7]. The process has not yet been applied to MSW combustion because of the sensitivity of the catalysts to contamination by particulates and poisoning by lead.

Selective noncatalytic reduction. Selective noncatalytic reduction has been commercially developed by the Exxon Research and Engineering Company, which licenses the process as the Thermal DeNO$_x$™ Process [34]. It is in use at over 60 installations. The process employs ammonia injection, but no catalyst is involved. Ammonia in gaseous form is directly injected into the furnace. When ammonia is injected into a furnace in the temperature range 1300 to 2200°F, the following approximate reaction predominates [34]:

$$NO + NH_3 + O_2 + H_2O + H_2 \rightarrow N_2 + H_2O \tag{13-9}$$

If H_2 is not added, the lower temperature limit of the reaction given by Eq. (13-9) is 1600°F. At temperatures above 2200°F, the approximate reaction given by Eq. (13-10) predominates [34]. Because producing extra NO would be counterproductive, the process must be controlled carefully to prevent NO formation.

$$NH_3 + O_2 + H_2O \rightarrow NO + H_2O \tag{13-10}$$

The process is mechanically simple, as shown in Fig. 13-21. Temperature control of the process is achieved by installing multiple wall injectors in the furnace, so that the NH_3 can be injected in a region of the furnace in the optimum temperature range. Removal efficiencies in the 50 to 80 percent range have been achieved.

Control Equipment for Acid Gases. Uncontrolled emissions from MSW combustion can contain hydrochloric acid (HCl), hydrofluoric acid (HF), nitrogen dioxide (NO_2), and sulfur dioxide (SO_2). Hydrochloric acid and hydrofluoric acid are emitted as fine aerosols, and nitrogen dioxide and sulfur dioxide are emitted as gases that combine with water droplets in the atmosphere to form nitric acid and sulfuric acid mists. These acid mists can result in reduced visibility, corrosion of metals, and the production of acid rain or fog. Several methods are available to control acid gases, including (1) source separation of chlorine- and sulfur-containing wastes; (2) wet scrubbing of flue gases, in which liquid solutions are used to scrub and neutralize acid gases; and (3) dry scrubbing, in which neutralizing slurries are injected directly into the flue gas stream.

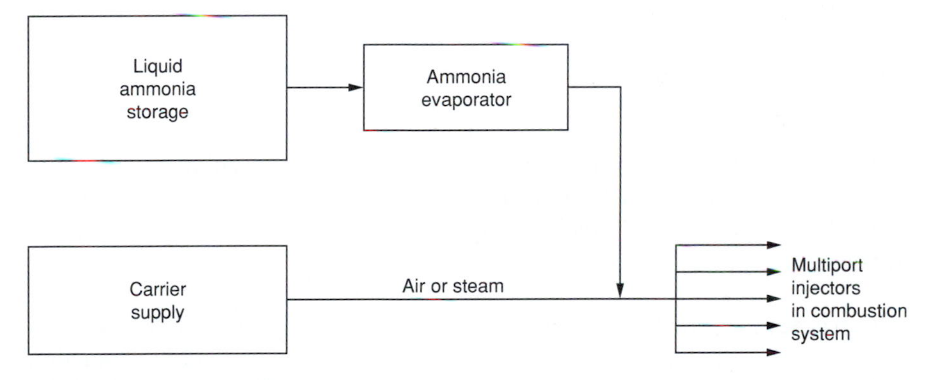

FIGURE 13-21
Schematic of Thermal DeNO$_x$™ process for control of NO$_x$ emissions [34].

Source separation. One method of emission control for HCl and SO_2 is source separation of waste components that contain large amounts of chlorine and sulfur, such as plastics. Separation of these wastes would tend to reduce the energy content of the remaining waste materials [7]. The chlorine and sulfur content of various waste components are listed in Table 13-7. The source separation of plastics, yard wastes, and miscellaneous organics would seem to offer the most potential for the reduction of HCl, SO_2, and the fuel portion of NO_2 [7]. The effectiveness of this approach has not yet been demonstrated.

Wet scrubbers. Wet scrubbers have been used widely in Europe and Japan. For example, wet scrubbing with a lime solution in a venturi scrubber is used in the MSW combustion system at Kiel, Germany [24]. The system, as shown in Fig. 13-22, is fairly complex, consisting of a venturi scrubber and demister, a lime-slaking system, and a filter press for dewatering the resultant scrubber sludge prior to disposal. A regenerative heat exchanger is also part of the system, first cooling the flue gases by about 90°F prior to the scrubber, then reheating the gases prior to discharge in the stack. The cooling step is required to enhance the efficiency of the scrubbing operation and the reheat step is required to enhance the buoyancy of the plume. The performance of the system averages HCl removals of 89 to 98 percent; HF removals of 84 to 96 percent; and SO_2 removals of 55 to 79 percent [24].

Dry scrubbers. Dry scrubbing systems are another approach to acid gas removal. Two techniques are used: (1) spray drying and (2) dry injection. A typical spray dryer system is shown in Fig. 13-23. Sodium carbonate and lime solutions are pumped into the spray dryer, where they react with the flue gas. The acid gases and SO_2 are adsorbed on the surface of the droplets, reacting to form neutral salts such as calcium sulfate ($CaSO_2$), calcium chloride ($CaCl_2$), sodium sulfate ($NaSO_2$), and sodium chloride (NaCl). The solid salt particles are removed in a downstream baghouse along with the fly ash still in the flue gas. Removal efficiencies of SO_2 for the combined dry scrubber/baghouse combination are quite high, approaching 98 percent [7].

The other approach to dry scrubbing is dry injection as in the Teller Dry Scrubbing System™. As shown in Fig. 13-24, a lime solution is sprayed into a quench chamber, neutralizing the acid gases. The water in the solution completely evaporates, so there is no liquid sludge to deal with. A proprietary agglomerating agent, Tesisorb™, is added to the flue gas stream after the quenching chamber to help coagulate very fine particles prior to collection in the downstream baghouse. Removal efficiencies for HCl and SO_2 up to 99 percent have been reported for the Teller system [11].

Control Equipment for CO and HC. The control of carbon monoxide (CO) and hydrocarbons (HC) is directly related to combustion efficiency and is a function of both design and operation. The formation of CO and HC are both caused by incomplete combustion of waste, due to fuel-rich burning (overloading of the furnace) and to insufficient temperature caused by high-moisture-content waste. The

TABLE 13-7
Sulfur and chlorine content of waste materials found in MSW[a]

Material	Sulfur content, % dry weight	Chlorine content, % dry weight
Cardboard	0.05	0.05
Newsprint	0.03	0.05
Yard wastes	0.07	0.34
Rubber, wood, textiles, etc.	0.18	0.14
Plastic	0.27	6.48
Mixed paper	0.04	0.12
Misc. organics	1.15	1.8

[a] Adapted from Ref. 7.

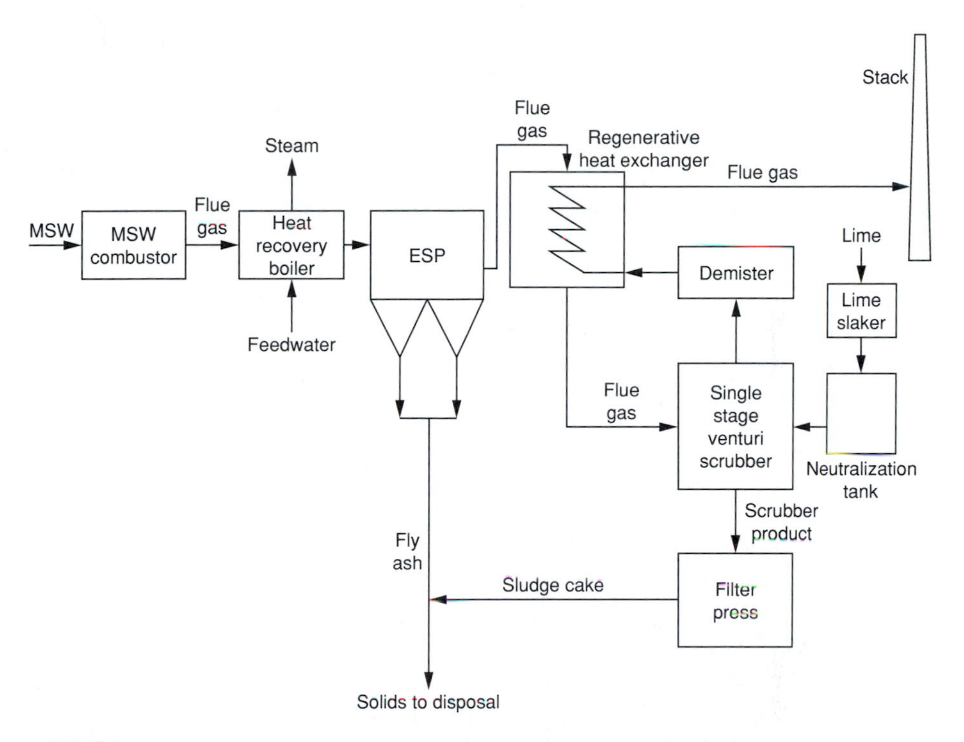

FIGURE 13-22
Schematic flow diagram of wet scrubber system used in conjunction with MSW combustor at Kiel, Germany, for the control of acid gases and SO_2 [24].

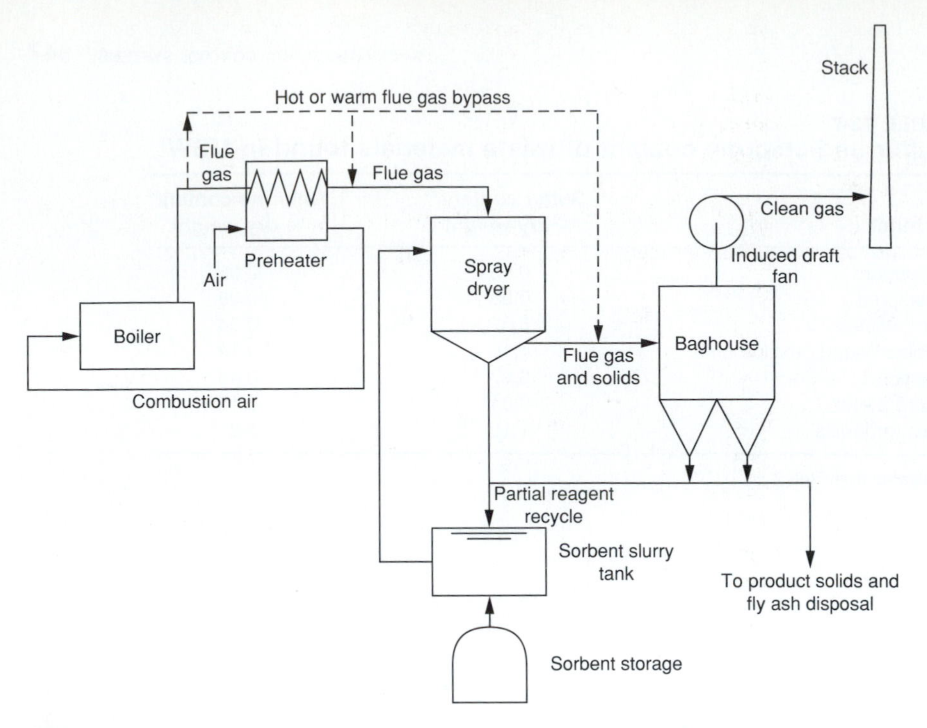

FIGURE 13-23
Schematic flow diagram of spray dryer for acid gas and SO_2 removal.

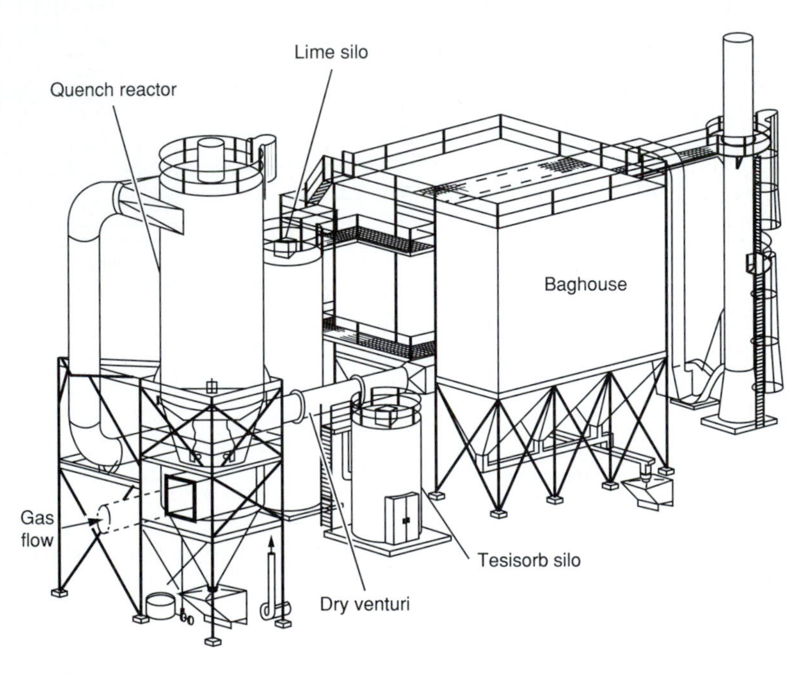

FIGURE 13-24
Teller dry scrubbing system for acid gas and SO_2. Tesisorb®, a crystalline powder of a specified size range, or another reagent, is injected and mixed in the dry venturi before the baghouse. (Courtesy of Research-Cottrell, Air Pollution Control Division.)

use of excess air is the primary tool to control CO and HC production. Excess air must be balanced to avoid burning too hot and generating excess NO_x emissions. Balancing of these factors is accomplished in modern MSW combustion systems by continuous emission monitoring (CEM) of flue gas constituents (CO, CO_2, NO_x, HC, and O_2). The CO and O_2 readings are used to balance excess air. Temperature readings in critical portions of the furnace are also used to assist in control. Sophisticated computer data displays and instrument control systems are used to assist operators in achieving efficient operation.

In mass-fired systems, the skill of the crane operator in loading the combustion chamber is critical to optimum combustion, because the crane operator must mix the waste prior to loading to minimize uneven fuel quality and must maintain as constant a flow of waste into the combustion unit as possible. RDF systems can be more automated because of the more uniform quality of the waste stream. Uniform fuel quality simplifies the role of the operator feeding waste into the system.

Control Equipment for Dioxins, Furans, and Metals. Some of the most important noncriteria pollutants in the combustion of MSW include dioxins, furans, and metals. Several technologies are available to facilitate their control, including source separation, combustion controls, and particulate control. As discussed previously, source separation can be an effective means of limiting heavy metal emissions from MSW combustion. Source separation of batteries is already being used in Japan and Sweden to control mercury and cadmium emissions [11].

The use of source separation for the control of dioxins and furans is more problematical. While removal of chlorine-containing wastes, primarily plastics, has been recommended as a control step [7], experimental evidence to date has not proven that it is effective. At recent tests at the Pittsfield, Massachusetts MSW combustor, test burns were conducted with "spiked" samples of MSW that contained extra polyvinyl chloride plastic (PVC). While extra HCl gas was generated, there was no correlation between PVC and HCl and dioxin and furan concentrations [32].

Combustion controls have become the principal control strategy to reduce dioxin and furan emissions. It has been found that a strong correlation exists between combustion temperature and residence time and dioxin and furan emissions [32]. The California Air Resources Board recommends minimum temperatures in thermal processing systems of 1800°F ± 190°F with a minimum residence time of 1 second [7]. It has also been found that conditions that minimize the generation of CO also minimize generation of dioxins and furans [31, 32]. Thus, CO can be used as a surrogate to monitor the emission of dioxins and furans. This monitoring technique is significant because CO can be measured in real time with continuous emission monitoring systems. Measurement of dioxins and furans cannot, at the present time, be done on a continuous basis.

Particulate control is also important in controlling emissions both of metals and of dioxins and furans. It has been found that metallic oxides and chlorides will tend to condense on submicron fly ash particles at temperatures below 500°F [7]. Properly designed fabric filters have the most efficient removal of submicron particles and are thus the most consistent and efficient method for control of metals. Mercury, because of its low volatilization temperature, is not removed efficiently by this mechanism. Source separation remains the most effective control technique for mercury control.

There is some evidence that dioxins and furans can be captured in the filter cake in fabric filters and in wet and dry scrubbers used for SO_2 control if temperatures are maintained below 284°F, allowing condensation of the dioxins and furans [32]. However, the results to date have not been conclusive enough to rely on this technique as a control strategy. Combustion controls remain the principal control technique for control of dioxins and furans.

Air Pollution Regulations

Under the authority of the Clean Air Act, the Environmental Protection Agency (EPA) issued Standards of Performance for new MSW combustors [23] and Emission Guidelines for existing MSW combustors [22] in February 1991. For regulatory purposes, MSW combustors (MWCs) that commenced construction on or before December 20, 1989 (the date when the Preliminary Regulations were published in the *Federal Register* for public comment) are considered to be *existing* MWCs.

New MWCs. The Standards of Performance for new MWCs are summarized in Table 13-8. The Standards currently apply only to MWCs larger than 250 ton/d. The Standards are divided into six parts: (1) definition of good combustion practices, (2) emission limits for organics (dioxins and furans), (3) emission limits for metals, (4) emission limits for acid gases, (5) emission limits for NO_x, and (6) monitoring requirements. The Standards also specify the best demonstrated technology (BDT) for meeting these emission limits.

Existing MWCs. The Emission Guidelines for existing MWCs are summarized in Table 13-9. The Guidelines contain only five parts, as NO_x emissions are currently not regulated [22]. However, the Clean Air Act Amendments of 1990 allow EPA to add NO_x regulation to existing MWCs at a later time.

The Guidelines for existing MWCs are similar to the Standards of Performance for new MWCs except that the emission limits are less stringent. For example, the emission limit for dioxins and furans on new MWCs is 30 ng/dscm (nanograms/dry standard cubic meter), whereas the emission limit for existing MWCs is 60 ng/dscm for very large MWCs ($>$ 1100 ton/d) and 125 ng/dscm for large MWCs ($>$ 250 ton/d and $\leq$1100 ton/d). The BDT is also different for existing MWCs. For example, BDT for acid gas controls on an existing very large MWC is a spray dryer and electrostatic precipitator, whereas BDT for a new MWC is a spray dryer and fabric filter.

State and Local Regulations. State and local air pollution control districts can issue emission limits that are stricter than federal regulations. For example, the California 1991 Guidelines for MSW combustion systems, summarized in

TABLE 13-8
Summary of EPA standards for new municipal waste combustors[a]

Applicability

The New Source Performance Standards (NSPS) apply to municipal waste combustors (MWC) with unit capacities above 225 Mg/day (250 ton/d) that combust residential, commercial, and/or institutional discards. Industrial discards are not covered by the NSPS.

Good Combustion Practices (GCP)

Maximum load level demonstrated during dioxin/furan performance test.
Cannot exceed 110 percent of maximum demonstrated load, 4-hour average.

Maximum particulate matter (PM) control device inlet temperature cannot exceed 17°C (30°F) above maximum demonstrated temperature during dioxin/furan performance test.

CO level[b] (block averaging time) as follows:

Modular starved and excess-air MWCs	50 ppmv (4 hour)
Mass burn waterwall and refractory MWCs	100 ppmv (4 hour)
MWCs using fluidized bed combustion	100 ppmv (4 hour)
Mass burn rotary waterwall MWCs	100 ppmv (24 hour)
RDF stokers	150 ppmv (24 hour)
Coal/RDF mixed fuel-fired MWCs	150 ppmv (4 hour)

ASME or State certification for MWC supervisors. Operator training and training manual for other MCW personnel.

MWC organic emissions (measured as total dioxins/furans)	
Dioxins/furans[b,c,d]	30 ng/dscm (12 gr/billion dscf)
Best Demonstrated Technology (BDT)	Good Combustion Practices (GCP), spray dryer, and fabric filter
MWC metal emissions (measured as PM)[b,d]	
Particulate matter (PM)	34 mg/dscm (0.015 gr/dscf)
Opacity	10% (6-minute average)
BDT	Fabric filter
MWC acid gas emissions (measured as SO_2 and HCl)[b]	
SO_2	80% or 30 ppmv (24 hour)
HCl[c]	95% reduction or 25 ppmv
BDT	Spray dryer and fabric filter
Nitrogen oxides emissions[b]	
NO_x	180 ppmv (24 hour daily [block])
BDT	Selective noncatalytic reduction
Monitoring requirements	
SO_2	Continous emission-monitoring system (CEMS), 24-hour geometric mean
NO_x	CEMS, 24-hour arithmetic average
Opacity	CEMS, 6-minute average
CO, load, temperature	CEMS, 4 or 24-hour average
PM, dioxins/furans[c] and HCl	Annual stack test

[a] Adapted from Ref. 27.

[b] All emission levels are at 7 percent O_2 dry basis.

[c] Dioxins/furans measured as total tetra- through octa-chlorinated dibenzo-*p*-dioxins and dibenzofurans, and not as toxic equivalents.

[d] Verified at annual stack compliance test.

TABLE 13-9
Summary of EPA emission guidelines for existing municipal waste combustors[a]

Applicability		
The guidelines apply to existing MWCs with unit capacities above 225 Mg/day (250 ton/d) that combust residential, commercial, and/or institutional discards. Industrial discards are not covered by the guidelines. MWCs with unit capacities above 250 ton/d at plants with aggregate capacity $>$ 250 ton/d but $\leq$ 1100 ton/d (large) and $>$ 1100 ton/d (very large) are subject to the provisions below.		
Good Combustion Practices (GCP)		
Maximum load level demonstrated during dioxin/furan performance test. Cannot exceed 110 percent of maximum demonstrated load, 4-hour average.		
Maximum particulate matter (PM) control device inlet temperature cannot exceed 17°C (30°F) above maximum demonstrated temperature during dioxin/furan performance test.		
CO level[b] (averaging time) as follows:		
Modular starved and excess air MWCs		50 ppmv (4 hour)
Mass burn waterwall and refractory MWCs		100 ppmv (4 hour)
MWCs using fluidized bed combustion		100 ppmv (4 hour)
Coal/RDF mixed fuel-fired MWCs		150 ppmv (4 hour)
RDF stokers		200 ppmv (24 hour)
Mass burn rotary waterwall MWCs		250 ppmv (24 hour)
ASME or State certification for MWC supervisors. Operator training and training manual for other MCW personnel.		
MWC organic emissions (measured as total dioxins/furans)		
Dioxins/furans [b,c,d]	Very large MWC plants	60 ng/dscm (25 gr/billion dscf)
	Large MWC plants[e]	125 ng/dscm (50 gr/billion dscf)
Best Demonstrated Technology (BDT)	Very large MCW plants	Good Combustion Practices (GCP), spray dryer, and electrostatic precipitator (ESP)
	Large MWC plants	GCP, dry sorbent injection, and ESP

(continued)

Table 13-10, are stricter than the 1991 federal regulations in several respects. Because federal, state, and local regulations are in a constant state of flux, the reader is cautioned to consult directly with regulatory officials for the most current standards and regulations. Publications and conferences sponsored by professional societies such as the Air and Waste Management Association are also a good source for up-to-date information.

Solid Residuals

Several solid residuals are produced by resource recovery facilities, including (1) bottom ash, (2) fly ash, and (3) scrubber product. Management of these solid residuals is an essential part of the design and operation of a resource recovery facility.

Bottom Ash. The unburned and nonburnable portion of MSW is known as *bottom ash*. In a mass-fired facility, bottom ash can contain considerable amounts of

TABLE 13-9 (*continued*)

MWC metal emissions (measured as PM)[b,c]		
Particulate matter (PM)	Very large MWC plants Large MWC plants	34 mg/dscm (0.015 gr/dscf) 69 mg/dscm (0.030 gr/dscf)
Opacity	Very large and large MWC plants	10% (6-minute average)
BDT		ESP
MCW acid gas emissions (measured as SO_2 and HCl)[b]		
SO_2	Very large MWC plants Large MWC plants	70% or 30 ppmv (24 hour) 50% or 30 ppmv (24 hour)
HCl[c]	Very large MWC plants Large MWC plants	90% reduction or 25 ppmv 50% reduction or 25 ppmv
BDT	Very large MWC plants Large MWC plants	Spray dryer and ESP Dry sorbent injection and ESP
Monitoring requirements		
SO_2		Continous emission-monitoring system (CEMS), 24-hour geometric mean
Opacity		CEMS, 6-minute average
CO, load, temperature		CEMS, 4- or 24-hour average
PM, dioxins/furans,[c] and HCl		Annual stack test

[a]Adapted from Ref. 27.

[b]All emission levels are at 7 percent O_2 dry basis.

[c]Dioxins/furans measured as total tetra- through octa-chlorinated dibenzo-*p*-dioxins and dibenzofurans.

[d]Verified at annual stack compliance test.

[e]Level for RDF units at large MWC plants is 250 ng/dscm. (100 gr/billion dscf)

metals and glass as well as unburned organics. Less metal and glass occur in the bottom ash from RDF-fired facilities, because most of this material has already been removed from the waste stream. The amount of unburned organic material in the ash is a measure of performance of the facility. It can be estimated using the Ash Burnout Index (ABI) [14]:

$$\mathrm{ABI} = \left[1 - \frac{(a - b)}{a}\right] \times 100\% \tag{13-11}$$

where a = original weight of ash sample
b = weight of ash sample after firing in muffle furnace

A well-operating MSW combustor should be able to achieve a 95 to 99 percent Ash Burnout Index.

Bottom ash from most MSW combustion systems in the United States is landfilled without processing. It is possible to recover metals and other materials from bottom ash by magnetic separation and screening [55]. The limiting factor is finding a market for the materials. For example, in the Netherlands it is common

TABLE 13-10
California 1991 guidelines for MSW combustion[a]

Item	Unit[b]	Value[c,d]
SO_2	ppmv	30
NO_x (as NO_2)	ppmv	30–50
CO	ppmv	50–100
THC (as CH_4)	ppmv	1–10
HCl	ppmv	25
Total suspended particulates	grains/dscf	0.01
Particulates ($<2\ \mu m$)	grains/dscf	0.008

[a] Adapted from Ref. 8.

[b] ppmv = parts per million by volume; grains/dscf = grains per dry standard cubic foot.

[c] At 12 percent CO_2 unless otherwise noted.

[d] 8-hour average.

practice to use bottom ash for dike maintenance. Bottom ash has also been used successfully for road base construction in the United States. Researchers have reported some success in using mixtures of bottom ash, fly ash, hydrated lime, and Portland cement to make building blocks. Although the blocks could be used for conventional construction, it has been proposed that they be used for artificial offshore reefs [46].

Fly Ash. As the efficiency of air pollution control systems increases, greater proportions of particulates, or *fly ash,* are removed from the flue gases. Particulate removal efficiencies exceeding 99 percent are common with modern ESP and fabric filter systems. The resulting fly ash is another solid residual, which must be managed.

Because fly ash is composed of the micron and submicron particulates that have been collected by the air pollution control system, it must be handled very carefully to avoid fugitive dust emissions, which may be harmful to workers and the surrounding environment. Fly ash should be removed from collection devices with pneumatic conveyors and transported in closed containers to an acceptable disposal site. When permitted by local regulations, fly ash can also be moistened and mixed with bottom ash prior to disposal.

Scrubber Product. Scrubber product is the sludge produced by a wet scrubber used for SO_2 and acid gas cleanup. Scrubber product consists of the calcium and sodium sulfate salts formed in the scrubbing reaction as well as trace organics and heavy metals. Management of scrubber product includes dewatering, to reduce volume, and subsequent disposal of the sludge as a solid residue and the supernatant as a wastewater.

Heavy Metals and Trace Organics. It is well known that ash from MSW combustion contains trace amounts of heavy metals and trace organics [21, 55]. Therefore solid residuals must be managed carefully to protect the public from contact with these materials. The primary concern is that when the ash is landfilled, it may under certain conditions leach into the groundwater. Elements of concern include arsenic (As), barium (Ba), cadmium (Cd), chromium (Cr), lead (Pb), mercury (Hg), selenium (Se), and silver (Ag).

Several tests have been used to assess the leaching potential of ash including (1) the EP Toxicity Test and (2) the Toxicity Characteristic Leaching Procedure (TLCP) Test. In the EP Toxicity Test, a mixture of 5 percent ash and 95 percent MSW is leached with an acidic solution in a test column. This test is intended to simulate leachate production in a mixed waste landfill in which organic acids from the biological decomposition of organics are present. In the TLCP test, ash samples are ground to $\leq \frac{3}{8}$ inch, mixed with a pH 5 acetate buffer, and mixed for 18 hours. The supernatant is then filtered and tested for heavy metals. The appropriateness of the tests is a matter of active research, and a number of studies are being undertaken to find a correlation, if any, between the test results and actual leachate from operational ash landfills.

Recommended Ash-Handling Procedures. Good engineering practice for the management of MSW combustion ash is summarized as follows [42]:

1. Handling: Ash should be properly wetted or covered so that there are no fugitive dust emissions.
2. Transport: Truck containers should be covered and leak-resistant.
3. Disposal of fly ash only: Disposal should normally be in a monofill (ash only) equipped with double liners and a leachate collection system.
4. Combined or bottom ash only: Disposal should normally be in a monofill equipped with a composite or clay liner, or by codisposal in an MSW landfill equipped with a double liner. Some designers also include gas recovery facilities to control the release of odors from the decomposition of partially combusted organic matter.

Disposal policies for solid residues are dynamic. The reader should consult the EPA and appropriate state and local agencies for current guidance and regulations.

Wastewater Discharges

Wastewater discharges arise from several sources in resource recovery plants, including (1) cooling and wash water from wet ash removal systems, (2) wet scrubber effluent from SO_2 and acid gas cleaning equipment, (3) wastewater from sealing, flushing, and housekeeping activities, (4) wastewater from boiler feedwater production, and (5) cooling tower blowdown. The last two sources relate to the electricity generation system and are common to any power plant using steam turbines. Compared to the leachate produced from a landfill, the quantities of wastewater produced are relatively minor, but they may require pretreatment before discharge into a municipal sewer system.

Ash Removal Wastewater. Ash can be handled dry or wet. Typical systems use water to quench and cool the ash prior to disposal and to control fugitive dust emissions from the ash. If the ash is disposed of off-site, the quench water is removed with the ash and no wastewater results. The water content of the ash is of concern when the ash is eventually landfilled. In some systems, the bottom ash is recycled into a road base or aggregate material. The bottom ash can be washed to remove readily soluble contaminates such as chlorides and sulfates.

Wet Scrubber Effluent. Wet scrubbing is an effective method of treatment for SO_2 and acid gas cleanup. However, two waste products are produced: a solid residue (scrubber product) and a wastewater (wet scrubber effluent). Treatment consists of neutralization, precipitation, and settling. At Bamberg, Germany, discharge of settled scrubber effluent exceeded German limiting values for mercury (0.05 mg/L). Precipitation with trimercaptotriazine, $(C_3N_3S_5)_5$, reduced effluent mercury levels to 0.045 mg/L. An alternate type of wet scrubbing treatment system is being installed at the Kiel, Germany, MSW combustion facility. Scrubber sludge will be dewatered in a filter press and dried in multi-effect evaporators using waste heat from combustion of the MSW, virtually eliminating wastewater from the system [24].

Wastewater from Sealing, Flushing, and Housekeeping Activities. Small amounts of wastewater are generated by the water used to seal and cool pumps and other equipment. Cooling and sealing water is sometimes contaminated with oils and greases. Washdown water, used to clean tipping areas, is also of concern, because it contains organics from the MSW. Typically, these wastewaters are settled prior to discharge to municipal sewers.

Wastewater from Boiler Feedwater Production. The water used to make steam in a steam turbine system must meet stringent water quality requirements for total dissolved solids (TDS), pH, and alkalinity. Meeting these requirements requires that boiler water treatment systems be used to treat water from municipal sources or onsite wells. Typical treatment systems may employ a combination of water softening, ion exchange, precipitation, and reverse osmosis units. Discharges from these systems may be regulated by local and state discharge orders.

Cooling Tower Blowdown. Another water discharge related to power production is cooling tower *blowdown*. Cooling towers are used to condense steam back into water after the steam passes through the steam turbines. They are essentially heat exchangers in which the steam in the closed circuit steam loop is cooled by a water droplet /air mixture in a separate loop. The water in the cooling loop is recirculated over wooden slats or other packing inside the cooling tower. Typically, chromium salts are used to retard algae growth inside the tower, which would reduce cooling efficiency. As the cooling water is recirculated, it gradually evaporates, increasing the content of total dissolved solids and chromium. Periodically, the water must be replaced, producing an effluent called *blowdown*. Because the water is high in dissolved solids and chromium salts, it may require pretreatment by reverse osmosis or precipitation prior to discharge to municipal sewers. Fortunately, the power systems used with MSW combustors are relatively small, 10 to 50 MW as compared to 500 to 1000 MW for a typical utility power plant, where cooling tower blowdown is a major problem.

13-6 ENERGY RECOVERY SYSTEMS

Once solid waste has been converted to thermal energy in the form of steam by combustion, or to chemical energy in the form of gases or liquids by pyrolysis or gasification, it can be converted to mechanical or electrical energy. Steam can be used directly for industrial processes or building heating. Steam can also be used to produce mechanical or electrical energy with a steam turbine. Gases and liquids produced from solid waste by both thermal and biological processes can be used to fuel boilers to produce steam. Gases and liquids can also be used directly to fuel reciprocating engines and gas turbines. The purpose of this section is threefold: (1) to present basic flow diagrams available for accomplishing these conversions, (2) to present data on the efficiency of the components used in the various conversion process flow diagrams, and (3) to illustrate the use of efficiency data in computing energy output.

Energy Recovery Flow Diagrams

The principal components used for energy recovery are boilers for steam production; steam turbines, gas turbines, and reciprocating engines as prime movers for mechanical energy; and electric generators for the conversion of mechanical energy into electricity. Steam turbines are used in larger systems (10 to 50 MW), and gas turbines and reciprocating engines are used in smaller systems. Typical flow diagrams for energy recovery are shown in Fig. 13-25.

Steam Turbine Systems. The most common energy recovery system for the production of electricity is the steam turbine system. As shown in Fig. 13-25*a*, steam is produced in a boiler by burning MSW or RDF (gaseous or liquid conversion products can also be used). The steam is used to drive a steam turbine and then condensed back into boiler feed water. The steam turbine drives an electrical generator, which supplies onsite power and excess power for export. The system is essentially a scaled-down version of a coal- or gas-fired electrical utility plant.

Gas Turbine Generator Systems. Gas turbines require gaseous or liquid fuels. These fuels can be supplied by biological processes, such as landfill gas or the anaerobic digestion of MSW, or by pyrolysis or gasification. As shown in Fig. 13-25*b*, a gas turbine is similar to a jet engine in that it consists of a compressor section to increase the density of the gas/air mixture, a combustor, and a turbine section to convert the hot combustion gases to mechanical energy. An electrical generator is connected directly to the output shaft of the gas turbine. Gas turbines are efficient and compact and widely used in landfill gas systems (see Chapter 11).

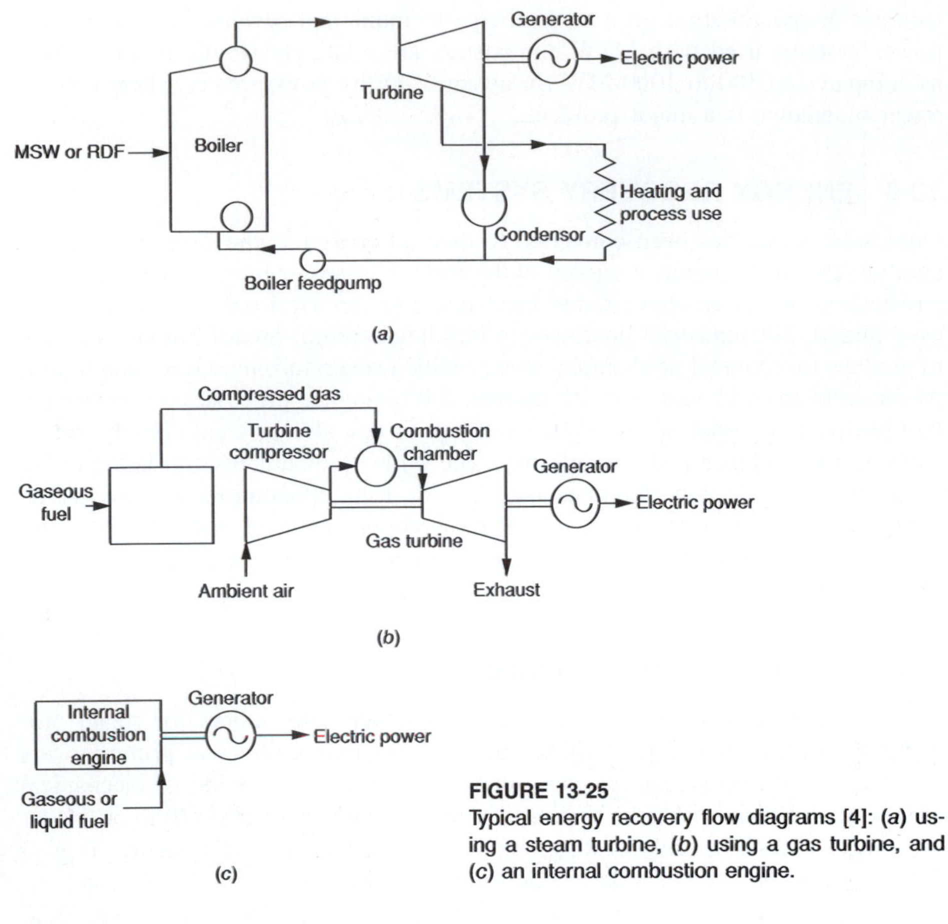

FIGURE 13-25
Typical energy recovery flow diagrams [4]: (*a*) using a steam turbine, (*b*) using a gas turbine, and (*c*) an internal combustion engine.

Internal Combustion Engine Systems. Internal combustion engines using pistons and a crankshaft are an alternative to gas turbines for gaseous or liquid fuels from the thermal or biological processing of solid wastes (see Fig. 13-25*c*). The engines are modified versions of industrial engines designed for natural gas or propane. Because natural gas has an energy content of about 1000 Btu/ft^3 compared to 400 to 500 Btu/ft^3 for landfill gas and 150 to 300 Btu/ft^3 for low-Btu gas, the engines use modified carburetors and intake manifolds to handle the lower-quality gas. Internal combustion engines are the most common prime movers used in landfill gas recovery systems (see Chapter 11).

Cogeneration Systems. Cogeneration is defined as the generation of both thermal and electrical power. Cogeneration systems are used widely in industry to generate electricity and process or building heat at the same time. Applications in energy recovery from solid waste are limited by the requirement that a use for the heat recovered must be located at the site with the power generation system. Several cogeneration flow diagrams are possible (see Fig. 13-26). In steam turbine systems, steam for heating is generated by extracting some of the steam from the

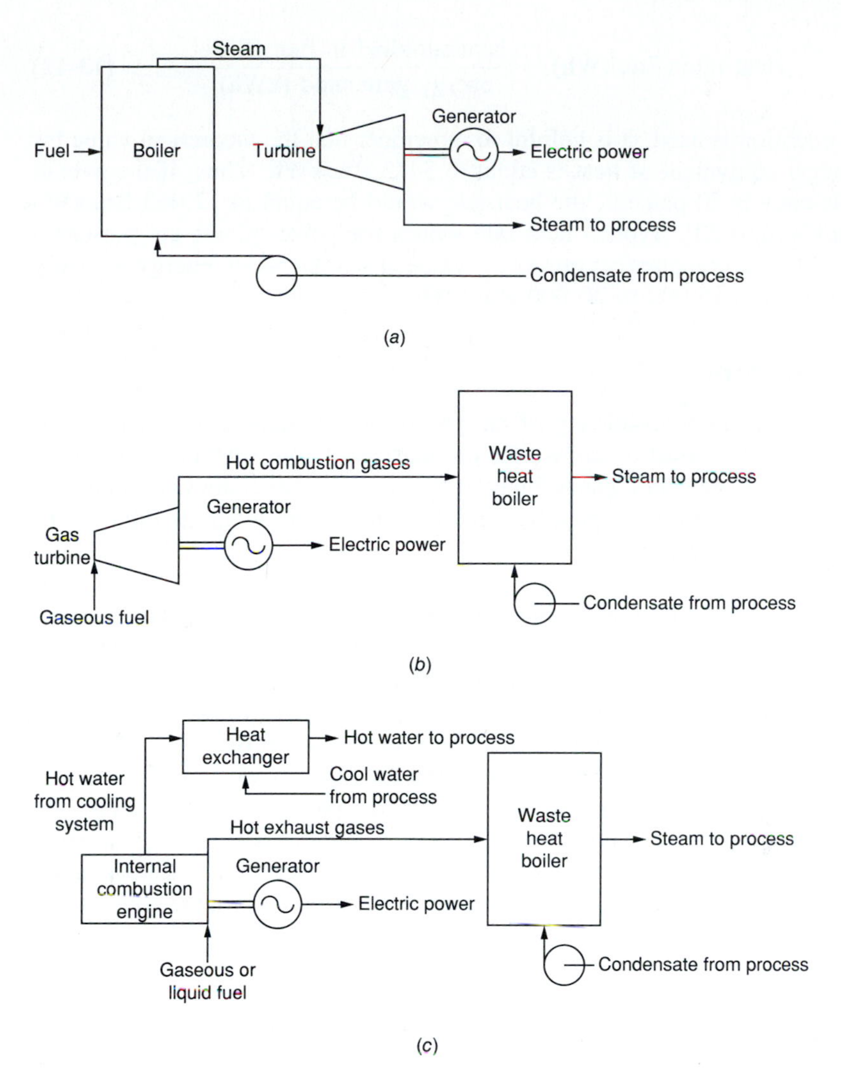

FIGURE 13-26
Typical flow diagrams for the cogeneration of electricity, steam, and/or hot water: (*a*) using a steam turbine with boiler back pressure topping, (*b*) using a gas turbine with waste heat boiler, and (*c*) using an internal combustion engine with waste heat boiler.

low-pressure stages of the steam turbine (see Fig. 13-26*a*). In gas turbine systems, a waste heat boiler can be added to generate steam from the hot gas turbine exhaust stream (see Fig. 13-26*b*). In internal combustion engine systems, heat can be recovered by special radiators and exhaust cooling jackets (see Fig. 13-26*c*).

Process Heat Rate

In the production of electrical power, it is common practice to consider the overall conversion efficiency in terms of the *heat rate,* as expressed in Eq. (13-12):

$$\text{Heat rate (Btu/kWh)} = \frac{\text{heat supplied in fuel (Btu)}}{\text{energy generated (kWh)}} \tag{13-12}$$

When this equation is used, it is helpful to remember that the theoretical value for the mechanical equivalent of heat is equal to 3413 Btu/kWh. Thus, if the overall system efficiency is 20 percent, the heat rate would be equal to 17,065 Btu/kWh [(3413 Btu/kWh)/0.20]. Typical heat rate values for power plants are presented in Table 13-11. For comparative purposes, values for solid waste energy recovery systems range from 15,000 to 30,000 Btu/kWh.

Efficiency Factors

To assess the conversion efficiency of the proposed flow diagrams given in Fig. 13-25, efficiency data must be known for the individual components. Representative data for boilers, gasifiers, gas turbines, steam turbine–generator combinations, electric generators, and related plant use and loss factors are given in Table 13-12 and discussed in this section.

Boilers. Boilers can be built as an integral part of the thermal processing system, as in a waterwall combustion chamber, or as a separate device, as in a waste heat boiler. It is important to be able to estimate the efficiency of a boiler, because it is typically the first component of an energy recovery system. Important variables that affect the efficiency of boilers include the energy content of the solid wastes, moisture content, flue gas exit temperature, and the physical design of the boiler water tubes. Although these variables are specific to a particular boiler design, the data presented in Fig. 13-27 and Table 13-12 can be used as a guide in estimating boiler efficiency. The curves presented in Fig. 13-27 are based on cellulosic solid wastes (e.g., paper or cardboard) with 50 percent excess air used in the combustion

TABLE 13-11
Typical heat rates of representative power plants[a]

Type of plant	Plant heat rate, Btu/kWh	Plant thermal efficiency
All stationary steam plants, average	25,000	0.14
Central-station steam plants, average	11,500	0.30
Best large central-station steam plant	8,500	0.40
Small noncondensing industrial steam plant	35,000	0.10
Small condensing industrial steam plant	20,000	0.17
"By-product" steam power plant	4,500–5,000	0.70–0.75
Diesel plant	11,500	0.30
Natural-gas-engine plant	14,000	0.24
Gasoline-engine plant	16,000	0.21
Low-Btu gas-engine plant	18,000	0.19

[a] Adapted from Ref. 41.

TABLE 13-12
Typical efficiency and loss factors for thermal processing systems

Component	Efficiency[a] Range	Efficiency[a] Typical	Comment
Incinerator-boiler	40–68	63	Mass-fired.
Boiler			
Solid fuel	65–72	70	Mass-fired.
Solid fuel	60–75	72	Processed MSW.
Low-Btu gas	60–80	75	Burners must be modified.
Oil-fired	65–85	80	Oils produced from MSW may have to be blended to reduce corrosiveness.
Pyrolysis reactor			
Conventional	65–75	70	
Purox	70–80	75	
Gasifiers			
Hot gas	70–80	75	Includes sensible heat of gas, when gas is used in a boiler.
Cold, filtered gas	60–70	65	Does not include sensible heat of gas, when gas is used in an engine or gas turbine.
Methanation process	80–90	85	Conversion of low-Btu gas to natural-gas quality.
Turbines			
Combustion gas			
Simple cycle	8–12	10	
Regenerative	20–26	24	Includes necessary appurtenances.
Expansion gas	30–50	40	
Steam turbine–generator system			
Less than 12.5 MW	24–30	29[b,c]	Includes condenser, heaters, and all other necessary appurtenances, but does not include boiler.
Over 25 MW	28–32	31.6[b,c]	
Electric generator			
Less than 10 MW	88–92	90	
Over 10 MW	94–98	96	
Plant use and loss factors			
Station service allowance			
Steam turbine–generator plant	4–8	6	
Purox process	18–24	21	
Methanation process	18–22	20	
Unaccounted heat losses	2–8	5	

[a] Theoretical value for mechanical equivalent of heat = 3413 Btu/kWh.

[b] Efficiency varies with exhaust pressure. Typical value given is based on an exhaust pressure in the range of 2 to 4 in Hg (absolute).

[c] Heat rate = 10,800 Btu/kWh [3413 Btu/kWh)/0.316].

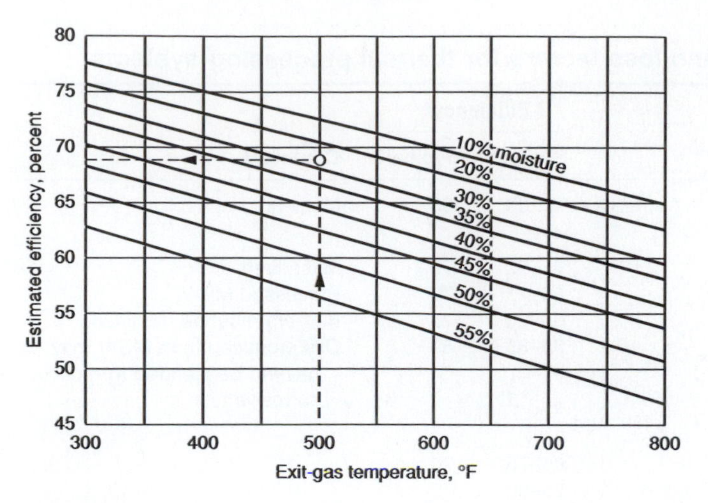

FIGURE 13-27
Estimated boiler efficiency versus exit-gas temperature for MSW combustion [37].

process. The reported boiler efficiencies are assumed to include latent heat losses and radiation, sensible heat, and unburned carbon losses. In comparison to boilers burning conventional fuels, where efficiencies up to 85 percent are common, solid waste boilers are typically in the 60 to 70 percent efficiency range. This lower efficiency is due primarily to the relatively high moisture content of solid wastes as compared to conventional solid fuels such as coal or wood.

Gasifiers. Data on the thermal efficiency of gasifiers are presented in Table 13-12. Note that several efficiency factors are given for gasifiers. Hot, raw gas is suitable for fueling boilers, whereas cold, cleaned gas is required for fueling of gas turbines or reciprocating engines. Energy balance computations for a gasifier engine system are shown in Example 13-3.

Example 13-3 Computation of energy output and efficiency for energy recovery system using a gasifier-engine-generator plant. Estimate the amount of energy produced from a solid waste energy conversion system with a capacity of 50 ton/d. The system consists of a fluidized bed gasifier–internal combustion engine–electric generator combination. Also estimate the heat rate and overall process efficiency. Assume that the energy value of the solid wastes is 5065 Btu/lb, as used in Example 13-1.

Solution

1. Set up a computation table to determine the energy output, using data reported in Table 13-12. (See the accompanying table shown on page 663.)
2. Determine the heat rate for the proposed plant by using Eq. (13-12).

$$\text{Heat rate} = \frac{21{,}100{,}000 \text{ Btu/h}}{976 \text{ kW}}$$

$$\text{Heat rate} = 21{,}619 \text{ Btu/kW}$$

Item	Value
Gross energy available in MSW, 10^6 Btu/h	
[(50 ton/d)(2000 lb/ton)(5065 Btu/lb)]/(24 h/d)	21.1
Chemical energy available from gasifier, cold cleaned gas, 10^6 Btu/h	
$(21.1 \times 10^6$ Btu/h)(0.7)	14.8
Mechanical energy available from engine, 10^6 Btu/h	
$(14.8 \times 10^6$ Btu/h)(0.25)	3.7
Net electrical energy generation, kW	
$(3.7 \times 10^6$ Btu/h)(0.90)/(3413 Btu/kWh)	976

3. Determine the overall efficiency.

$$\text{Efficiency} = \frac{(976\ \text{kW})(3413\ \text{Btu/kWh})}{(21{,}100{,}000\ \text{Btu/h})} \times 100\%$$

$$\text{Efficiency} = 16\%$$

Comment. Compare the heat rate and efficiency of this small-scale gasifier system (50 ton/d) with the large scale steam turbine system (1000 ton/d) of Example 13-4. Although the values are lower for the gasifier system, a steam turbine system would not be cost effective at this size range. Overall efficiency could be improved by using a more efficient prime mover, such as an expansion gas turbine (see Table 13-12).

Gas Turbines. Data on the thermal efficiency of various gas turbines are given in Table 13-12. The efficiency values include an allowance for the necessary appurtenances.

Steam Turbine–Generator Systems. The data reported in Table 13-12 for the steam turbogenerator are consistent with modern practice and reflect all the necessary allowances for condensers, heaters, and other appurtenances. Using the reported typical efficiency factor of 31.6 percent, the corresponding heat factor would be 10,800 Btu/kWh. If a boiler efficiency of 75 percent were achieved, the overall heat rate would be about 14,400 Btu/kWh. This value compares well with the values given in Table 13-11 for central-station steam plants.

Examples. Data from two operating MSW combustion systems that use steam turbine generators are summarized in Table 13-13. The Stanislaus County, California, system is a mass-fired system employing the Martin reverse-reciprocating grate system (see Fig. 13-28). The Biddeford, Maine, system is an RDF-fired semi-suspension system employing Babcock & Wilcox boilers (see Fig. 13-29). Although the systems represent two different design philosophies, mass burn versus RDF-fired, both facilities are successfully meeting performance standards and achieving the emission requirements of their respective localities.

Other Use and Loss Factors. In any installation where energy is being produced, allowance must be made for the station or process needs and for unac-

TABLE 13-13
Comparison of mass-fired and RDF-fired combustion systems[a]

Mass-fired—Stanislaus County, California	
Capacity	800 ton/d
Grate	Martin inclined, reverse-reciprocating
Steam pressure and temperature	865 lb/in^2 830°F
Net energy output	20 MW (electrical)
Air pollution controls	NO_x—Exxon Thermal $DeNO_X$™ System Acid Gas—dry flue gas scrubbing Particulates—fabric filter
Owner/Operator	Ogden Martin Systems, Inc.
Startup year	1989
RDF-fired—Biddeford, Maine	
Capacity[b]	600 ton/d
Boiler	Babcock & Wilcox Waterwall
Net energy output	20 MW (electrical)
Air pollution controls	NO_x—not required Acid Gas—dry flue gas scrubbing Particulates—fabric filter
Owner/Operator	Maine Energy Recovery Company
Startup year	1987

[a] Adapted from Ref. 35.

[b] Can also fire up to 10 percent energy input from wood chips, and up to 5 percent by natural gas.

FIGURE 13-28
Exterior view of Stanislaus County, California, mass-fired combustor. (Courtesy of Ogden Martin Systems, Inc.)

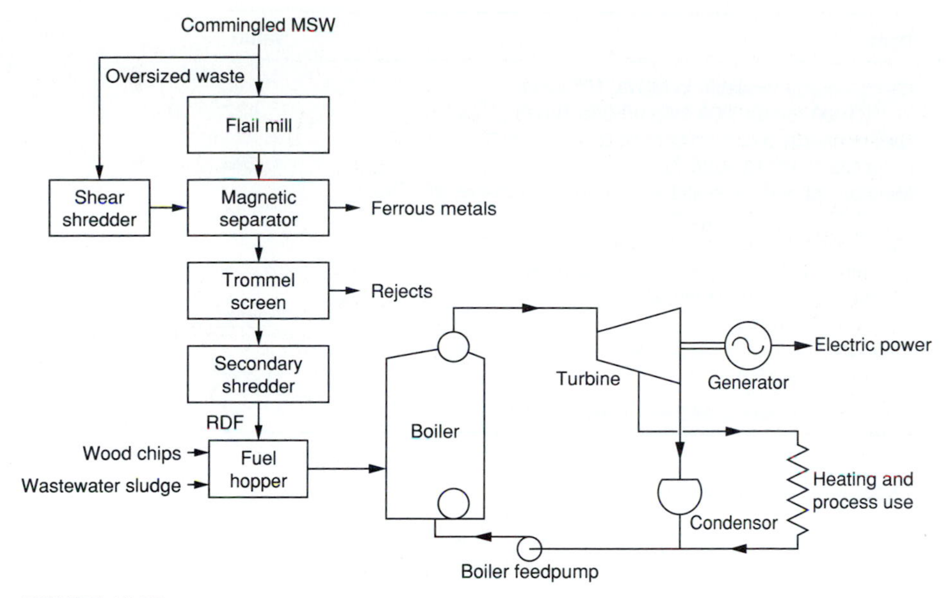

FIGURE 13-29
Schematic flow diagram for the Biddeford, Maine, RDF combustor.

counted process heat losses. Typically, the auxiliary power allowance varies from 4 to 8 percent. Both of these values must be considered in estimating the net heat rate when Eq. (13-12) is used. Efficiency computations for a typical steam-turbine generator system are given in Example 13-4.

Example 13-4 Computation of energy output and efficiency for energy recovery system using a steam boiler–turbine–generator plant. Estimate the amount of energy produced from a solid waste energy conversion system with a capacity of 1000 ton/d. The system consists of an MSW combustor–boiler–steam turbine–electric generator combination. Also estimate the heat rate and overall process efficiency, assuming that the station service allowance and the unaccounted heat losses are 6 and 5 percent, respectively, of the total power produced. Assume that the energy value of the solid wastes is 5065 Btu/lb, as used in Example 13-1.

Solution

1. Set up a computation table to determine the energy output, using data reported in Table 13-12. (See the accompanying table shown on page 666.)
2. Determine the heat rate for the proposed plant by using Eq. (13-12).

$$\text{Heat rate} = \frac{422{,}000{,}000 \text{ Btu/h}}{22{,}280 \text{ kW}}$$

$$\text{Heat rate} = 18{,}941 \text{ Btu/kWh}$$

Item	Value
Gross energy available in MSW, 10^6 Btu/h	
[(1000 ton/d)(2000 lb/ton)(5065 Btu/lb)]/(24 h/d)	422
Steam energy available from boiler, 10^6 Btu/h	
(422×10^6 Btu/h)(0.7)	295
Mechanical energy available from steam turbine, 10^6 Btu/h	
(295×10^6 Btu/h)(0.3)	89
Gross electrical energy generation, kW	
(89×10^6 Btu/h)(0.96)/(3413 Btu/kWh)	25,034
Station service allowance, kW	
(25,034)(0.06)	−1,502
Unaccounted heat losses, kW	
(25,034)(0.05)	−1,252
Net electric power available for export, kW	22,280

3. Determine the overall efficiency.

$$\text{Efficiency} = \frac{(22{,}280 \text{ kW})(3413 \text{ Btu/kWh})}{(422{,}000{,}000 \text{ Btu/h})} \times 100\%$$

$$\text{Efficiency} = 18\%$$

Comment. If it is assumed that 10 percent of the power generated is used for the front-end processing system (typical values vary from 8 to 14 percent), then the net power for export would be 20,052 kW and the overall efficiency would be 16 percent.

13-7 DISCUSSION TOPICS AND PROBLEMS

13-1. Using the data in Chapter 3 on moisture content and elemental composition (carbon, hydrogen, oxygen, nitrogen, sulfur, and ash), estimate the theoretical amount of air, in pounds, that would be required for the stoichiometric combustion of 1 ton of processed MSW having the composition given for the light fraction in Example 12-7.

13-2. The sludge from a wastewater treatment plant serving 500,000 persons is currently disposed of in a sanitary landfill. Because the capacity of the existing landfill will soon be exhausted, it has been proposed to combust the treatment plant sludge, using processed MSW as a fuel source. Two alternative modes of operation are to be evaluated. In the first, treatment plant sludge with a fuel value of 7500 Btu/lb (based on dry solids) and a solids content of 5 percent are to be mixed with processed MSW (5065 Btu/lb) and combusted. In the second alternative, dewatered sludge with a solids content of 20 percent and a fuel value of 6500 Btu/lb (based on dry solids) is to be mixed with the processed MSW before being combusted. It should be noted that the fuel value of the dewatered solids is lower because of the chemicals added to aid in dewatering.

Assuming that the specific gravity of the combined dry sludge with or without the addition of chemicals is 1.10, the per capita sludge production on a dry basis is 0.35 lb/day, and the moisture content of the processed MSW is 20 percent, determine the amount of MSW that must be added to the treatment plant sludge for both alternatives to achieve a final moisture content of 60 percent. Would the required quantities of processed MSW be available in the wastes from the community? State clearly all your assumptions.

13-3. Recalculate the materials and energy balances for Example 13-2 if the following data are applicable:

1. Combustibles = 80 percent of total
2. Noncombustibles = 10 percent of total
3. Water = 10 percent of total
4. All other conditions of Example 13-2 apply.

13-4. You have been retained as a consultant by a city in Southeast Asia. A vendor has proposed to the city that they purchase a 100-ton/d modular combustion unit (MCU). The following MSW composition data is available:

Component	Composition, % by weight	% Moisture	Energy content as-discarded, Btu/lb
Organics	95.6	70	2,000
Paper	1.3	6	7,200
Cardboard	0.3	5	7,000
Plastics	1.5	2	14,000
Glass	0.6	2	60
Iron	0.1	3	300
Tin cans	0.6	3	300
Total	100.0		

Calculate the as-fired moisture content, energy content, and chemical composition (see Chapter 4) of the wastes. Then calculate a materials and energy balance on the proposed MCU using the procedure of Example 13-2. Comment on the feasibility of using the MCU for this waste as compared to the results of Example 13-2, which are typical of MSW in the United States. Assume that the following data are applicable:

1. The dry portion of the waste is 90 percent combustible.
2. Other than physical and chemical composition and the as-fired energy content, conditions 2 through 11 of Example 13-2 apply.

13-5. Assuming that electricity will be produced, compare the use of a pyrolysis or gasification system to a conventional mass burn combustion system. Under what conditions would a pyrolysis or gasification system be used?

13-6. What are the emission standards for MSW combustors in your state? How do they compare with current federal standards for new and existing MSW combustors? *Note:* Be sure to research the current literature, as the standards and guidelines in Tables 13-8 and 13-9 may be out of date.

13-7. Prepare a review paper of the current (post-1985) literature on dioxin and furan research. Look at both health effects and control strategies. The journals cited in the reference section of this chapter are a good source of information on these topics.

13-8. What are the current regulations in your state for the disposal of MSW combustion ash? What are the responsible agencies for regulation and permitting of this waste material in your state?

13-9. Estimate the available energy for export from a 1000-ton/d fluidized bed gasification plant. Assume that the following data are applicable:

1. Energy content of solid wastes = 5065 Btu/lb
2. Energy efficiency of gasifier = 70 percent (for cold, filtered gas)
3. Low-Btu gas usage for building heat, and process maintenance based on the percentage of energy available in low-Btu gas = 8 percent
4. Gas turbine thermal efficiency = 24 percent
5. Electrical generator efficiency = 96 percent
6. In-plant electric power usage based on the percentage of total power generated = 21 percent

13-10. Electricity is sold by MSW combustion facilities to electric utilities on the basis of the *avoided cost,* which represents the average cost of the utility to generate electricity. The avoided cost is set by the state public utilities commission on a quarterly basis. If the avoided cost is $0.025/kWh, what is the annual value of electricity produced by the small-scale gasifier system of Example 13-3, assuming 24 h/d operation, 365 d/yr, and an 85 percent availability factor? *Note:* The avoided cost is much less than the retail cost of electricity, which is currently (1992) about $0.10/kWh in California. There is also an additional payment paid to MSW facilities, known as a *capacity payment,* which is a flat yearly payment based on the size of the plant (kW), the availability during peak use periods (percent), and reliability (percent).

13-8 REFERENCES

1. Andoh, N. Y., Y. Ishii, *et al.*: "Disposal of Municipal Refuse by the Two-Bed Pyrolysis System," in J. L. Jones and S. B. Radding, eds., *Thermal Conversion of Solid Wastes and Biomass (ACS Symposium Series 130),* American Chemical Society, Washington, DC, p. 541, 1980.
2. Bartley, D. A., S. A. Vigil, and G. Tchobanoglous: "The Use of Source-Separated Waste Paper as a Biomass Fuel," *Biotechnology and Bioengineering Symposium (10),* Wiley, New York, p. 67, 1980.
3. Black, J. W., K. G. Bircher, and K. A. Chisholm: "Fluidized-Bed Gasification of Solid Wastes and Biomass: The CIL Program," in J. L. Jones and S. B. Radding, eds., *Thermal Conversion of Solid Wastes and Biomass (ACS Symposium Series 130),* American Chemical Society, Washington, DC, 1980.
4. Brown and Caldwell Consulting Engineers: *Solid Waste Resource Recovery Study,* report prepared for the Central Contra Costa Sanitary District, Pleasant Hill, CA, 1974.
5. Brunner, C. L.: *Incinerator Systems Selection and Design,* Van Nostrand Reinhold, New York, 1984.
6. Bumb, R. R., *et al.*: "Trace Chemistries of Fire: A Source of Chlorinated Dioxins," *Science,* 210(4468), p. 385, October 1980.

7. California Air Resources Board: *Air Pollution Control at Resource Recovery Facilities,* Sacramento, CA, May 24, 1984.
8. California Air Resources Board: *Air Pollution Control at Resource Recovery Facilities 1991 Update,* California Air Resources Board, Sacramento, CA, May 1991.
9. Clarke, M. J.: "Debating the Virtues of ESP's," *Waste Age,* June 1988.
10. Clarke, M. J.: "Emission Control: A Never Ending Quest," *Waste Age,* January 1988.
11. Clarke, M. J.: "Issues, Options, and Choices for Control of Emissions from Resource Recovery Plants, 2nd Edition," presented at the Sixth Annual Resource Recovery Conference, sponsored by the U.S. Conference of Mayors and the National Resource Recovery Association, Washington DC, March 26–27, 1987.
12. Cooper, C. D., and F. C. Alley: *Air Pollution Control: A Design Approach,* PWS Publishers, Boston, 1986.
13. Cross, F. L., and H. E. Hesketh: *Controlled Air Incineration,* Technomic Press, Lancaster, PA, 1985.
14. Cross, F., P. O'Leary, and O. Walsh: "Operation and Maintenance Considerations for Waste-to-Energy Systems," *Waste Age,* August 1987.
15. Czuczwa, J. M., and R. A. Hites: "Airborne Dioxins and Dibenzofurans: Sources and Fates," *Environmental Science and Technology,* 20(2), p. 195, 1986.
16. Czuczwa, J. M., and R. A. Hites: "Environmental Fate of Combustion-Generated Polychlorinated Dioxins and Furans," *Environmental Science and Technology,* 18(6), p. 444, 1984.
17. Davis, D. A., S. A. Vigil, and G. Tchobanoglous: "Evaluation of Residual Char from the Gasification of Solid Wastes as a Substitute for Powdered Activated Carbon," *Biotechnology and Bioengineering Symposium (11),* Wiley, New York, p. 211, 1981.
18. Dhargalkar, P. H.: "An Integrated Waste Management Plant in Italy," *Proceedings of the 84th AWMA Meeting, Vancouver, BC,* Paper No. 91-43.6, Air and Waste Management Association, June 16–21, 1991.
19. Drobny, N. L., H. E. Hull, and R. F. Testin: *Recovery and Utilization of Municipal Solid Waste,* U.S. Environmental Protection Agency, Publication SW10c, 1971.
20. Ebara Corporation: "A New Type Fluidized Bed Incineration Plant," presented at the World Industry Conference on Environmental Management, Paris, November 14–16, 1984.
21. Eiceman, G. A., R. E. Clemant, and F. W. Karasek: "Analysis of Fly Ash from Municipal Incinerators for Trace Organic Compounds," *Analytical Chemistry,* 51(14), p. 2343, December 1979.
22. "Emission Guidelines: Municipal Waste Combustors," *Federal Register,* Vol. 56, No. 28, Rules and Regulations, p. 5519, February 11, 1991.
23. "Standards of Performance for New Stationary Sources: Municipal Waste Combustors," *Federal Register,* Vol. 56, No. 28, Rules and Regulations, p. 5493, February 11, 1991.
24. Feindler, K. S.: "Long Term Results of Operating TA LUFT Acid Gas Scrubbing Systems," *Proceedings of the 1986 National Waste Processing Conference,* American Society of Mechanical Engineers, New York, p. 17, 1986.
25. Fernandes, J. H.: "Incinerator Air Pollution Control," in *Proceedings of the 1970 National Waste Incinerator Conference,* American Society of Mechanical Engineers, New York, 1970.
26. Fisher, T. F., M. L. Kasbohmn, and J. R. Rivero, "The PUROX System," in *Proceedings of the 1976 National Waste Processing Conference,* American Society of Mechanical Engineers, New York, 1976.
27. Fricilli, P. W.: "Impact of EPA's Air Pollution Emission Standards and Guidelines on Municipal Waste Combustion Units," *Proceedings of the 84th AWMA Meeting, Vancouver, BC,* Paper No. 91-26.2, Air and Waste Management Association, June 16–21, 1991.
28. General Electric Company: *Solid Waste Management Technology Assessment,* Van Nostrand Reinhold, New York, 1975.
29. Gorman, P. G., M. Markus, *et al.*: *Project Summary—Environmental Assessment of Waste-to-Energy Process: Union Carbide's Purox Process,* EPA-600/S7-80-161, U.S. Environmental Protection Agency, Cincinnati, OH, December 1980.
30. Harrison, B., and P. A. Vesilind: *Design and Management for Resource Recovery, Volume 2: High Technology—A Failure Analysis,* Ann Arbor Science, Ann Arbor, MI, 1980.

31. Hasselriis, F.: "Optimization of Combustion Conditions to Minimize Dioxin Emissions," *Waste Management & Research,* 5, p. 311, 1987.
32. Hasselriis, F.: "Optimization of Combustion Conditions to Minimize Dioxin, Furan, and Combustion Gas Data from Test Programs at Three MSW Incinerators," *Journal APCA,* 37(12), p. 1451, December 1987.
33. Houk, V. N.: "Dioxin: Risk Assessment for Human Health," *The Diplomate,* Vol. 27, No. 4, American Academy of Environmental Engineers, Annapolis, MD, October 1991.
34. Hurst, B. E., and C. W. White: "Thermal $DeNO_X$: A Commercial Selective Noncatalytic NO_x Reduction Process for Waste-to-Energy Applications," *Proceedings of the 1986 National Waste Processing Conference,* American Society of Mechanical Engineers, New York, p. 119, 1986.
35. Kisker, J. V. L.: "A Comprehensive Report on the Status of Municipal Waste Combustion," *Waste Age,* November 1990.
36. Kleinan, J. H.: "Fluid Bed Combustion: Lessons Learned," *Waste Age,* p. 274, April 1988.
37. Meissner, H. G.: "Central Incineration of Community Wastes," in R. C. Corey, *Principles and Practices of Incineration,* Wiley-Interscience, New York, 1969.
38. Orning, A. A.: "Principles of Combustion," in R. C. Corey, *Principles and Practices of Incineration,* Wiley-Interscience, New York, 1969.
39. Ortiz-Canavate, J., S. A. Vigil, J. R. Goss, and G. Tchobanoglous: "Comparison of Operating Characteristics of a 34kW Diesel Engine Fueled with Low Energy Gas, Biogas, and Diesel Fuel," *Biotechnology and Bioengineering Symposium,* Vol. 11, Wiley, New York, p. 225, 1981.
40. Ozvacic, V., *et al.*: "Emissions of Chlorinated Organics from Two Municipal Incinerators in Ontario," *Journal APCA,* 35(8), p. 849, 1985.
41. Perry, R. H., C. H. Chilton, and S. D. Kirkpatrick: *Chemical Engineers Handbook,* 4th ed., McGraw-Hill, New York, 1963.
42. Porter, J. W.: "A Few Words About Incinerator Ash," *Waste Age,* April 1988.
43. Rambush, N. E.: *Modern Gas Producers,* Van Nostrand Company, New York, 1945.
44. Rankin, S.: "Plastic Wastes Are Not Harmful," *Waste Age,* May 1987.
45. Reed. T. B.: *A Summary of Biomass Gasification, Volume II—Principles of Gasification,* Report No. SERI/TR-33-239, Solar Energy Research Institute, Golden, CO, 1979.
46. Roethel, F. J.: "Ash Disposal Solution is 2000 Years Old," *Waste Age,* February 1987.
47. Shaub, W. M., and W. Tsang: "Dioxin Formation in Incinerators," *Environmental Science and Technology,* 17(12), p. 721, 1983.
48. Skov, N. A., and M. C. Papworth: *The PEGASUS Unit—Petroleum/Gasoline Substitute Systems,* Pegasus Publishers Inc., Olympia, WA, 1974.
49. Sorbo, N. W., G. Tchobanoglous, and S. A. Vigil: *Performance and Economic Feasibility of a Sludge/Wastepaper Gasifier System, A Project Summary,* EPA-600/S2-84-063, U.S. Environmental Protection Agency, Municipal Environmental Research Laboratory, Cincinnati, OH, 1984.
50. Tchobanoglous, G., N. W. Sorbo, and S. A. Vigil: "Gasification of Densified Sludge and Wastepaper in Down Draft Packed-Bed Gasifiers," *International Conference on Thermal Conversion of Municipal Sludge, Hartford, Connecticut, March 21–24, 1983,* U.S. Environmental Protection Agency, Washington, DC, 1983.
51. Turner, J. H., P. A. Lawless, *et al.*: "Sizing and Costing of Electrostatic Precipitators: Part 1—Sizing Considerations," *Journal APCA,* 38(4), p. 458, April 1988.
52. U.S. Environmental Protection Agency: *Report to Congress on Municipal Waste Combustion,* EPA/530-SW-87-021a, 1987.
53. Vigil, S. A., D. A. Bartley, R. Healy, and G. Tchobanoglous: "Operation of a Down Draft Gasifier Fueled with Source Separated Solid Waste," in J. L. Jones and S. B. Radding, eds., *Thermal Conversion of Solid Wastes and Biomass, ACS Symposium Series 130,* American Chemical Society, Washington DC, p. 257, 1980.
54. Vigil, S. A. and J. A. Zeveley: "Microcomputer Solid Waste Financial Model," *Proceedings of the 1986 National Waste Processing Conference,* American Society of Mechanical Engineers, New York, p. 463, 1986.
55. Walsh, P., P. O'Leary, and F. Cross: "Residue Disposal from Waste-to-Energy Facilities," *Waste Age,* April 1988.

CHAPTER

14

BIOLOGICAL AND CHEMICAL CONVERSION TECHNOLOGIES

The purpose of this chapter is to introduce and review the biological and chemical processes that can be used to transform the organic fraction of MSW into gaseous, liquid, and solid conversion products. The major focus of this chapter is on the biological processes, because they have been used most commonly for the transformation of organic waste materials. Biological processes considered in this chapter include aerobic composting, low-solids anaerobic digestion, high-solids anaerobic digestion, and high-solids anaerobic digestion/aerobic composting. Before the individual biological processes are considered, some basic biological principles, fundamental to all biological processes, must be introduced.

14-1 BIOLOGICAL PRINCIPLES

Before considering the specific processes employed for the biological conversion of wastes, it will be helpful to review (1) the general nutritional requirements of the microorganisms commonly encountered in solid waste conversion facilities, (2) the type of microbial metabolism based on the need for molecular oxygen, (3) the types of microorganisms of importance in the conversion of solid waste, (4) environmental requirements, (5) aerobic and anaerobic transformations, and (6) process selection.

Nutritional Requirements for Microbial Growth

To continue to reproduce and function properly, an organism must have a source of energy; carbon for the synthesis of new cell tissue, and inorganic elements (*nutrients*) such as nitrogen, phosphorus, sulfur, potassium, calcium, and magnesium. Organic nutrients (*growth factors*) may also be required for cell synthesis. Carbon and energy sources, usually referred to as *substrates*, and nutrient and growth factor requirements for various types of organisms are considered in the following discussion.

Carbon and Energy Sources. Two of the most common sources of carbon for cell tissue are organic carbon and carbon dioxide. Organisms that use organic carbon for the formation of cell tissue are called *heterotrophs*. Organisms that derive carbon from carbon dioxide are called *autotrophs*. The conversion of carbon dioxide to organic cell tissue is a reductive process, which requires a net input of energy. Autotrophic organisms must therefore spend more of their energy for synthesis than do heterotrophs, resulting in generally lower growth rates among the autotrophs.

The energy needed for cell synthesis may be supplied by light or by a chemical oxidation reaction. Those organisms that are able to use light as an energy source are called *phototrophs*. Phototrophic organisms may be either heterotrophic (certain sulfur bacteria) or autotrophic (algae and photosynthetic bacteria). Organisms that derive their energy from chemical reactions are known as *chemotrophs*. Like the phototrophs, chemotrophs may be either heterotrophic (protozoa, fungi, and most bacteria) or autotrophic (nitrifying bacteria). Chemoautotrophs obtain energy from the oxidation of reduced *inorganic* compounds, such as ammonia, nitrite, and sulfide. Chemoheterotrophs usually derive their energy from the oxidation of *organic* compounds. The classification of microorganisms by sources of energy and cell carbon is summarized in Table 14-1.

Nutrient and Growth Factor Requirements. Nutrients, rather than carbon or an energy source, may at times be the limiting material for microbial cell synthesis and growth. The principal inorganic nutrients needed by microorganisms

TABLE 14-1
General classification of microorganisms by sources of energy and carbon

Classification	Energy source	Carbon source
Autotrophic		
Photoautotrophic	Light	CO_2
Chemoautotrophic	Inorganic oxidation-reduction reaction	CO_2
Heterotrophic		
Chemoheterotrophic	Organic oxidation-reduction reaction	Organic carbon
Photoheterotrophic	Light	Organic carbon

[a] Adapted from Ref. 19.

are nitrogen (N), sulfur (S), phosphorus (P), potassium (K), magnesium (Mg), calcium (Ca), iron (Fe), sodium (Na), and chlorine (Cl). Minor nutrients of importance include zinc (Zn), manganese (Mn), molybdenum (Mo), selenium (Se), cobalt (Co), copper (Cu), nickel (Ni), and tungsten (W) [19].

In addition to the inorganic nutrients just cited, some organisms may also need organic nutrients. Required organic nutrients, known as growth factors, are compounds needed by an organism as precursors or constituents of organic cell material that cannot be synthesized from other carbon sources. Although growth factor requirements differ from one organism to another, the major growth factors fall into the following three classes: (1) amino acids, (2) purines and pyrimidines, and (3) vitamins [19].

Microbial Nutrition and Biological Conversion Processes. The major objective in most biological conversion processes is the conversion of the organic matter in the waste to a stable end product. In accomplishing this type of treatment, the chemoheterotrophic organisms are of primary importance because of their requirement for organic compounds as both carbon and energy source. The organic fraction of MSW typically contains adequate amounts of nutrients (both inorganic and organic) to support the biological conversion of the waste. With some commercial wastes, however, nutrients may not be present in sufficient quantities. In these cases, nutrient addition is necessary for the proper bacterial growth and for the subsequent degradation of the organic waste.

Types of Microbial Metabolism

Chemoheterotrophic organisms may be further grouped according to their metabolic type and their requirement for molecular oxygen. Organisms that generate energy by enzyme-mediated electron transport from an electron donor to an *external* electron acceptor (such as oxygen) are said to have a *respiratory metabolism*. In contrast, *fermentative metabolism* does not involve the participation of an external electron acceptor. Fermentation is a less efficient energy-yielding process than respiration; as a consequence, heterotrophic organisms that are strictly fermentative are characterized by lower growth rates and cell yields than respiratory heterotrophs.

When molecular oxygen is used as the electron acceptor in respiratory metabolism, the process is known as *aerobic respiration*. Organisms that are dependent on aerobic respiration to meet their energetic needs can exist only when there is a supply of molecular oxygen. These organisms are called *obligate aerobic*. Oxidized inorganic compounds such as nitrate and sulfate can function as electron acceptors for some respiratory organisms in the absence of molecular oxygen (see Table 14-2). In environmental engineering, processes that make use of these organisms are often referred to as *anoxic*.

Organisms that generate energy by fermentation and that can exist only in an environment that is devoid of oxygen are *obligate anaerobic*. There is another group of microorganisms, which has the ability to grow in either the presence or the absence of molecular oxygen. These organisms are called *facultative anaerobes*.

TABLE 14-2
Typical electron acceptors in bacterial reactions

Environment	Electron acceptor	Process
Aerobic	Oxygen, O_2	Aerobic metabolism
Anaerobic	Nitrate, NO_3^-	Denitrification
	Sulfate, SO_4^{2-}	Sulfate reduction
	Carbon dioxide, CO_2	Methanogenesis

The facultative organisms fall into two subgroups, based on their metabolic abilities. True facultative anaerobes can shift from fermentative to aerobic respiratory metabolism, depending upon the presence or absence of molecular oxygen. *Aerotolerant anaerobes* have a strictly fermentative metabolism but are relatively insensitive to the presence of molecular oxygen.

Types of Microorganisms

Microorganisms are commonly classified, on the basis of cell structure and function, as eucaryotes, eubacteria, and archaebacteria, as shown in Table 14-3. The procaryotic groups (eubacteria and archaebacteria) are of primary importance in biological conversion of the organic fraction of solid wastes and are generally referred to simply as *bacteria*. The eucaryotic group includes plants, animals, and protists. Eucaryotes important in biological conversion of organic wastes include (1) fungi, (2) yeasts, and (3) actinomycetes. Because of their importance in the biological conversion of organic wastes, these organisms are described briefly in the following paragraphs.

TABLE 14-3
Classification of microorganisms[a]

Group	Cell structure	Characterization	Representative members
Eucaryotes	Eucaryotic[b]	Multicellular with extensive differentiation of cells and tissue	Plants (seed plants, ferns, mosses)
			Animals (vertebrates, invertebrates)
		Unicellular or coenocytic or mycelial; little or no tissue differentiation	Protists (algae, fungi, protozoa)
Eubacteria	Procaryotic[c]	Cell chemistry similar to eucaryotes	Most bacteria
Archaebacteria	Procaryotic[c]	Distinctive cell chemistry	Methanogens, halophiles, thermacidophiles

[a] From Stanier, R. Y., J. L. Ingraham, M. L. Wheelis, and P. R. Painter, *The Microbial World,* 5th ed., ©1986. Reprinted by permission of Prentice-Hall, Inc.

[b] Contain true nucleus.

[c] Contain no nuclear membrane.

Bacteria. Typically, bacteria are single cells—spheres, rods, or spirals. Spherical forms (cocci) vary from 0.5 to 4 μm in diameter; rods (bacilli) are from 0.5 to 20 μm long and 0.5 to 4 μm wide; spirals (spirilla) may be more than 10 μm long and about 0.5 μm wide [19]. Bacteria are ubiquitous in nature and are found in aerobic (in the presence of oxygen) and anaerobic (in the absence of oxygen) environments. Because of the wide variety of inorganic and organic compounds that can be used by bacteria to sustain growth, bacteria are used extensively in a variety of industrial operations to accumulate intermediate and end products of metabolism. Tests on a number of different bacterial species indicate that they are about 80 percent water and 20 percent dry material, of which 90 percent is organic and 10 percent is inorganic. An approximate empirical formula for the organic fraction is $C_5H_7NO_2$ [10]. On the basis of this formula, about 53 percent by weight of the organic fraction is carbon. Compounds that make up the inorganic portion include P_2O_5 (50 percent), CaO (9 percent), Na_2O (11 percent), MgO (8 percent), K_2O (6 percent), and Fe_2O_3 (1 percent). Because all these elements and compounds must be derived from the environment, a shortage of these substances would limit, and in some cases alter, the growth of bacteria [10].

Fungi. Fungi are considered to be multicellular, nonphotosynthetic, heterotrophic protists. Most fungi have the ability to grow under low-moisture conditions, which do not favor the growth of bacteria. In addition, fungi can tolerate relatively low pH values. The optimum pH value for most fungal species appears to be about 5.6, but the viable range is from 2 to 9. The metabolism of these organisms is essentially aerobic, and they grow in long filaments, called *hyphae,* composed of nucleated cell units and varying in width from 4 to 20 μm. Because of their ability to degrade a wide variety of organic compounds over a broad range of environmental conditions, fungi have been used extensively in industry for the production of valuable compounds, such as organic acids (e.g., citric, gluconic), various antibiotics (e.g., penicillin, griseofulvin), and enzymes (e.g., cellulase, protease, amylase).

Yeasts. Yeasts are fungi that cannot form filaments (*mycelium*) and are therefore unicellular. Some yeasts form elliptical cells 8 to 15 μm by 3 to 5 μm, whereas others are spherical, varying in size from 8 to 12 μm in diameter. In terms of industrial processing operations, yeasts may be classified as "wild" and "cultured." In general, wild yeasts are of little value, but cultured yeasts are used extensively to ferment sugars to alcohol and carbon dioxide.

Actinomycetes. The actinomycetes are a group of organisms with intermediate properties between bacteria and fungi. They are similar in form to fungi, except that the width of the cell is only 0.5 to 1.4 μm. In industry, this group of microorganisms is used extensively for the production of antibiotics. Because their growth characteristics are similar, actinomycetes are often grouped with fungi for discussion purposes [4].

Environmental Requirements

Environmental conditions of temperature and pH have an important effect on the survival and growth of microorganisms. In general, optimal growth occurs within a fairly narrow range of temperature and pH values, although the microorganism may be able to survive within much broader limits. For instance, temperatures below the optimum typically have a more significant effect on the bacterial growth rate than temperatures above the optimum. It has been observed that growth rates double with approximately every 10°C increase in temperature until the optimum temperature is reached. According to the temperature range in which they function best, bacteria may be classified as *psychrophilic, mesophilic,* or *thermophilic*. Typical temperature ranges for bacteria in each of these categories are presented in Table 14-4.

The hydrogen ion concentration, expressed as pH, is not a significant factor in the growth of microorganisms, in and of itself, within the range from 6 to 9 (which represents a thousandfold difference in the hydrogen ion concentration). Generally, the optimum pH for bacterial growth lies between 6.5 and 7.5. However, when the pH goes above 9.0 or below 4.5, it appears that the undissociated molecules of weak acids or bases can enter the cell more easily than hydrogen and hydroxide ions and, by altering the internal pH, damage the cell.

Moisture content is another essential environmental requirement for the growth of microorganisms. The moisture content of the organic wastes to be converted must be known, especially if a dry process such as composting is to be used. In many composting operations, it has been necessary to add water to obtain optimum bacterial activity. The addition of water in anaerobic fermentation processes will depend on the characteristics of the organic waste and the type of anaerobic process that is used.

The biological conversion of an organic waste requires the biological system to be in a state of dynamic equilibrium. To establish and maintain dynamic equilibrium, the environment must be free of inhibitory concentrations of heavy metals, ammonia, sulfides, and other toxic constituents.

TABLE 14-4
Some typical temperature ranges for various bacteria

	Temperature, °C	
Type	**Range**	**Optimum**
Psychrophilic[a]	−10–30	15
Mesophilic	20–50	35
Thermophilic	45–75	55

[a] Also called cryophilic.

Note: 1.8 × (°C) + 32 = °F.

Aerobic Biological Transformations

The general aerobic transformation of solid waste can be described by means of the following equation.

$$\text{Organic matter} + O_2 + \text{nutrients} \xrightarrow{\text{Bacteria}} \text{new cells} + \begin{matrix}\text{resistant}\\ \text{organic}\\ \text{matter}\end{matrix}$$

$$+ CO_2 + H_2O + NH_3 + SO_4^{2-} + \cdots + \text{heat} \quad (4\text{-}20)$$

If the organic matter in solid waste is represented (on a molar basis) as $C_aH_bO_cN_d$, the production of new cells and sulfate is not considered, and the composition of the resistant material is represented (on a molar basis) as $C_wH_xO_yN_z$, then the amount of oxygen required for the aerobic stabilization of the biodegradable organic fraction of MSW can be estimated by using the following equation [17]:

$$C_aH_bO_cN_d + 0.5(ny + 2s + r - c)O_2 \rightarrow nC_wH_xO_yN_z + sCO_2$$
$$+ rH_2O + (d - nz)NH_3 \quad (14\text{-}1)$$

where $r = 0.5[b - nz - 3(d - nz)]$
$s = a - nw$

The terms $C_aH_bO_cN_d$ and $C_wH_xO_yN_z$ represent the empirical mole composition of the organic material initially present and at the conclusion of the process. If complete conversion is accomplished, the corresponding expression is

$$C_aH_bO_cN_d + \left(\frac{4a + b - 2c + 3d}{4}\right)O_2 \rightarrow aCO_2 + \left(\frac{b - 3d}{2}\right)H_2O + dNH_3 \quad (14\text{-}2)$$

In many cases the ammonia, NH_3, produced from the carbonaceous oxidation of organic matter is oxidized further to nitrate, NO_3^- (a process known as nitrification). The amount of oxygen required for the oxidation of ammonia to nitrate can be computed by the following equations:

$$NH_3 + \tfrac{3}{2}O_2 \rightarrow HNO_2 + H_2O \quad (14\text{-}3)$$

$$HNO_2 + \tfrac{1}{2}O_2 \rightarrow HNO_3 \quad (14\text{-}4)$$

$$NH_3 + 2O_2 \rightarrow H_2O + HNO_3 \quad (14\text{-}5)$$

Computation of the amount of oxygen required for the stabilization of prepared solid wastes is illustrated in Example 14-1.

Example 14-1 Oxygen requirements for the aerobic conversion of solid waste. Determine the amount of oxygen required to oxidize 1000 lb of an organic solid waste aerobically. Assume that the initial composition of the organic material to be decomposed is given by $[C_6H_7O_2(OH)_3]_5$, that the final composition of the residual organic matter is estimated to be $[C_6H_7O_2(OH)_3]_2$, and that 400 lb of material remains after the oxidation process.

Solution

1. Determine the moles of material present initially and at the end of the biological conversion process.

Moles present initially:

$$\frac{1000 \text{ lb}}{[(30 \times 12) + (50 \times 1) + (25 \times 16)]} = 1.23$$

Moles present at end:

$$\frac{400 \text{ lb}}{[(12 \times 12) + (20 \times 1) + (10 \times 16)]} = 1.23$$

2. Determine the moles of material leaving the process per mole of material entering the process.

$$n = \frac{1.23}{1.23} = 1.0$$

3. Determine the values for a, b, c, d, w, x, y, and z, and then determine the value of r and s in Eq. (14-1).

For the initial compound ($C_{30}H_{50}O_{25}$):

$$a = 30 \qquad b = 50 \qquad c = 25 \qquad d = 0$$

For the final compound ($C_{12}H_{20}O_{10}$):

$$w = 12 \qquad x = 20 \qquad y = 10 \qquad z = 0$$

The value for r is

$$\begin{aligned} r &= 0.5[b - nx - 3(d - nz)] \\ &= 0.5[50 - 1.0(20)] = 15.0 \end{aligned}$$

The value for s is

$$\begin{aligned} s &= a - nw \\ &= 30 - 1.0(12) = 18.0 \end{aligned}$$

4. Determine the amount of oxygen required.

$$\begin{aligned} O_2, \text{lb} &= 0.5(ny + 2s + r - c)O_2 \\ &= 0.5[1.0(10) + 2(18) + 15.0 - 25.0]1.23(32) \\ &= 708 \text{ lb } (321 \text{ kg}) \end{aligned}$$

5. Check the computations using a materials balance.

	lb	kg
Process input		
Organic material	1000	454
Oxygen	708	321
	1708	775
Process output		
Organic material	400	181
Carbon dioxide 1.23(18.0)(44.0)	974	442
Water 1.23(15.0)(18.0)	332	151
	1706	774

Anaerobic Biological Transformations

The production of methane from solid wastes by anaerobic digestion, or *anaerobic fermentation* as it is often called, is described in the following discussion.

Process Microbiology. The biological conversion of the organic fraction of municipal solid waste under anaerobic conditions is thought to occur in three steps (see Fig. 14-1). The first step in the process involves the enzyme-mediated transformation (hydrolysis) of higher-molecular-mass compounds into compounds suitable for use as a source of energy and cell tissue. The second step involves the bacterial conversion of the compounds resulting from the first step into identifiable lower-molecular-mass intermediate compounds. The third step involves the bacterial conversion of the intermediate compounds into simpler end products, principally methane and carbon dioxide [8, 9, 15, 18].

In the anaerobic decomposition of wastes, a number of anaerobic organisms work together to bring about the conversion of organic portion of the wastes to a stable end product. One group of organisms is responsible for hydrolyzing organic polymers and lipids to basic structural building blocks such as fatty acids, monosaccharides, amino acids, and related compounds (see Fig. 14-1). A second group of anaerobic bacteria ferments the breakdown products from the first group to simple organic acids, the most common of which in anaerobic digestion is acetic acid. This second group of microorganisms, described as nonmethanogenic, consists of facultative and obligate anaerobic bacteria that are often identified in the literature as "acidogens" or "acid formers."

A third group of microorganisms converts the hydrogen and acetic acid formed by the acid formers to methane gas and carbon dioxide. The bacteria responsible for this conversion are strict anaerobes, called methanogenic, and are identified in the literature as "methanogens" or "methane formers." Many of the methanogenic organisms identified in landfills and anaerobic digesters are similar to those found in the stomachs of ruminant animals and in organic sediments taken from lakes and river. The most important bacteria of the methanogenic

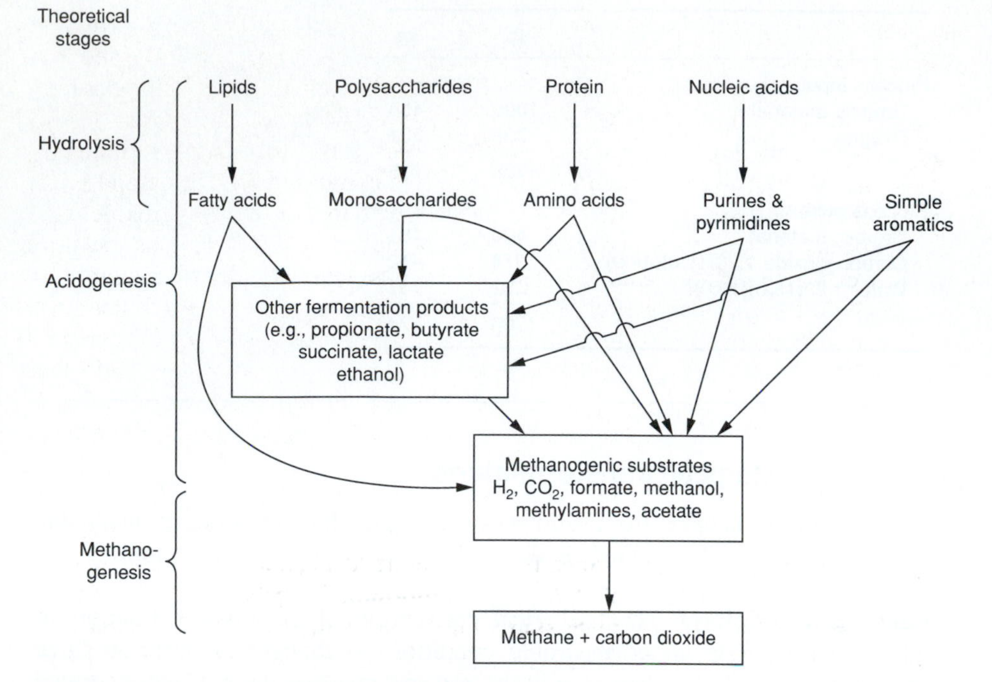

FIGURE 14-1
Pathways leading to the production of methane and carbon dioxide from the anaerobic digestion of the organic fraction of MSW [from Holland, *et al.*, *Anaerobic Bacteria*, ©1987, p. 184. Reprinted by permission of Blackie Academic & Professional (an imprint of Chapman and Hall)].

group are the ones that utilize hydrogen and acetic acid. They have very slow growth rates; as a result, their metabolism is usually considered rate-limiting in the anaerobic treatment of an organic waste. Waste stabilization in anaerobic digestion is accomplished when methane and carbon dioxide are produced. Methane gas is highly insoluble, and its departure from a landfill or solution represents actual waste stabilization.

Biochemical Pathways. It is important to note that methane bacteria can only use a limited number of substrates for the formation of methane. Currently, it is known that methanogens use the following substrates: $CO_2 + H_2$, formate, acetate, methanol, methylamines, and carbon monoxide. Typical energy-yielding conversion reactions involving these compounds are as follows [8, 9]:

$$4H_2 + CO_2 \rightarrow CH_4 + 2H_2O \tag{14-6}$$

$$4HCOOH \rightarrow CH_4 + 3CO_2 + 2H_2O \tag{14-7}$$

$$CH_3COOH \rightarrow CH_4 + CO_2 \tag{14-8}$$

$$4CH_3OH \rightarrow 3CH_4 + CO_2 + 2H_2O \tag{14-9}$$

$$4(CH_3)_3N + 6H_2O \rightarrow 9CH_4 + 3CO_2 + 4NH_3 \tag{14-10}$$

$$4CO + 2H_2O \rightarrow CH_4 + 3CO_2 \tag{14-11}$$

In anaerobic fermentation, the two principal pathways involved in the formation of methane (see Fig. 14-1) are (1) the conversion of carbon dioxide and hydrogen to methane and water, shown in Eq. (14-6), and (2) the conversion of formate and acetate to methane, carbon dioxide, and water, shown in Eqs. (14-7) and (14-8). The methanogens and the acidogens form a *syntrophic* (mutually beneficial) relationship in which the methanogens convert fermentation end products such as hydrogen, formate, and acetate to methane and carbon dioxide. The methanogens are able to utilize the hydrogen produced by the acidogens, because of their efficient hydrogenase. Because the methanogens are able to maintain an extremely low partial pressure of H_2, the equilibrium of the fermentation reactions is shifted towards the formation of more oxidized end products (e.g., formate and acetate). The utilization of the hydrogen, produced by the acidogens and other anaerobes, by the methanogens is termed *interspecies hydrogen transfer*. In effect, the methanogenic bacteria remove compounds that would inhibit the growth of acidogens.

Environmental Factors. To maintain an anaerobic treatment system that will stabilize an organic waste efficiently, the nonmethanogenic and methanogenic bacteria must be in a state of dynamic equilibrium. To establish and maintain such a state, the reactor contents should be void of dissolved oxygen and free from inhibitory concentrations of free ammonia and such constituents as heavy metals and sulfides. Also, the pH of the aqueous environment should range from 6.5 to 7.5. Sufficient alkalinity should be present to ensure that the pH will not drop below 6.2, because the methane bacteria cannot function below this point. When digestion is proceeding satisfactorily, the alkalinity will normally range from 1000 to 5000 mg/L and the volatile fatty acids will be less than 250 mg/L. Values for alkalinity and volatile fatty acids in the high-solids anaerobic digestion process (see Section 14-4) can be as high as 12,000 and 700 mg/L, respectively. A sufficient amount of nutrients, such as nitrogen and phosphorus, must also be available to ensure the proper growth of the biological community. Depending on the nature of the sludges or waste to be digested, growth factors may also be required. Temperature is another important environmental parameter. The optimum temperature ranges are the mesophilic, 30 to 38°C (85 to 100°F), and the thermophilic, 55 to 60°C (131 to 140°F).

Gas Production. The general anaerobic transformation of solid waste can be described by means of the following equation.

$$\text{Organic matter} + H_2O + \text{nutrients} \rightarrow \text{new cells} + \begin{matrix}\text{resistant}\\\text{organic}\\\text{matter}\end{matrix}$$

$$+ CO_2 + CH_4 + NH_3 + H_2S + \text{heat} \quad (4\text{-}21)$$

For practical purposes, the overall conversion of the organic fraction of solid waste to methane, carbon dioxide, and ammonia can be represented with the following equation [17].

$$C_aH_bO_cN_d \rightarrow nC_wH_xO_yN_z + mCH_4 + sCO_2 + rH_2O + (d - nx)NH_3 \quad (14\text{-}12)$$

where $s = a - nw - m$
$r = c - ny - 2s$

The terms $C_aH_bO_cN_d$ and $C_wH_xO_yN_z$ are used to represent (on a molar basis) the composition of the organic material present at the start and the end of the process, respectively. If it is assumed that the organic wastes are stabilized completely, as noted in Chapter 11, the corresponding expression is

$$C_aH_bO_cN_d + \left(\frac{4a - b - 2c + 3d}{4}\right)H_2O \rightarrow \left(\frac{4a + b - 2c - 3d}{8}\right)CH_4 + \left(\frac{4a - b + 2c + 3d}{8}\right)CO_2 + dNH_3 \quad (11\text{-}2)$$

In operations where solid wastes have been mixed with wastewater sludge, it has been found that the gas collected from the digesters contains between 50 and 60 percent methane. It has also been found that about 10 to 16 ft^3 of gas is produced per pound of biodegradable volatile solids destroyed. Computation of the amount of gas produced by the complete anaerobic stabilization of solid wastes is illustrated in Example 14-2.

Example 14-2 Estimation of the amount of gas produced from the organic fraction of MSW under anaerobic conditions. Estimate the total theoretical amount of gas that could be produced under anaerobic conditions in a sanitary landfill per unit weight of solid wastes. Assume that the wastes are of the composition shown in Table 3-4 and that the overall chemical formula for the organic constituents, as determined in Example 4-2, is $C_{60.0}H_{94.3}O_{37.8}N$.

Solution

1. The total weight of organic material in 100 lb of solid wastes is equal to 79.5 lb including moisture (see Table 3-4 and Example 4-2).
2. Determine the total amount of dry decomposable organic waste, assuming that 5 percent of the decomposable material will remain as an ash.

   Decomposable organic wastes (dry basis), lb = 58.1 lb (0.95) = 56.0 lb

3. Using the chemical formula $C_{60.0}H_{94.3}O_{37.8}N$ (determined in Example 4-2, Step 4), estimate the amount of methane and carbon dioxide that can be produced, using Eq. (11-2).

$$C_aH_bO_cN_d + \left(\frac{4a - b - 2c + 3d}{4}\right)H_2O \rightarrow \left(\frac{4a + b - 2c - 3d}{8}\right)CH_4 + \left(\frac{4a - b + 2c + 3d}{8}\right)CO_2 + dNH_3$$

For the given chemical formula,

$$a = 60.0 \qquad b = 94.3 \qquad c = 37.8 \qquad d = 1$$

The resulting equation is

$$\underset{1433.1}{C_{60.0}H_{94.3}O_{37.8}N} + \underset{329.0}{18.28H_2O} \rightarrow \underset{511.4}{31.96CH_4} + \underset{1233.8}{28.04CO_2} + \underset{17}{NH_3}$$

4. Determine the weight of methane and carbon dioxide from the equation derived in Step 3.

$$\text{Methane} = \frac{511.4}{1443.1}(56.0\text{ lb}) \qquad \text{(see Step 2)}$$

$$= 20.0\text{ lb }(9.1\text{ kg})$$

$$\text{Carbon dioxide} = \frac{1233.8}{1433.1}(56.0\text{ lb}) \qquad \text{(see Step 2)}$$

$$= 48.2\text{ lb }(21.9\text{ kg})$$

5. Convert the weight of gases, determined in Step 4, to volume, assuming that the specific weights of methane and carbon dioxide are 0.0448 and 0.1235 lb/ft^3, respectively (see Table 11-3).

$$\text{Methane} = \frac{20.0\text{ lb}}{0.0448\text{ lb/ft}^3}$$

$$= 446.4\text{ ft}^3\ (12.6\text{ m}^3)$$

$$\text{Carbon dioxide} = \frac{48.2\text{ lb}}{0.1235\text{ lb/ft}^3}$$

$$= 390.3\text{ ft}^3\ (11.0\text{ m}^3)$$

6. Determine the percentage composition of the resulting gas mixture.

$$\text{Methane (\%)} = \left(\frac{446.4\text{ ft}^3}{446.4\text{ ft}^3 + 390.3\text{ ft}^3}\right)$$

$$= 53.3\%$$

$$\text{Carbon dioxide (\%)} = 100\% - 53.3\% = 46.7\%$$

7. Determine the total theoretical amount of gas generated per unit weight of solid waste.

Based on the dry weight of organic material, ft^3/lb

$$\frac{446.4\text{ ft}^3 + 390.3\text{ ft}^3}{56.0\text{ lb}} = 14.9\text{ ft}^3\text{/lb }(0.93\text{ m}^3\text{/kg})$$

Based on 100 lb of solid waste, ft^3/lb:

$$\frac{446.4\text{ ft}^3 + 390.3\text{ ft}^3}{100.0\text{ lb}} = 8.4\text{ ft}^3\text{/lb }(0.52\text{ m}^3\text{/kg})$$

Comment. Total gas produced from the theoretical evaluation is equal to 14.9 ft^3/lb of dry decomposable organic material. This theoretical gas production is normally higher than those that can be achieved in practice.

Biological Process Selection

Aerobic and anaerobic processes both have a place in solid waste management. Each process offers different advantages. In general, the operation of anaerobic processes is more complex than that of aerobic processes. However, anaerobic processes offer the benefit of energy recovery in the form of methane gas and thus are net energy producers. Aerobic processes, on the other hand, are net energy users because oxygen must be supplied for waste conversion, but they offer the advantage of relatively simple operation and, if properly operated, can significantly reduce the volume of the organic portion of MSW. The relative advantages of aerobic and anaerobic processes are summarized in Table 14-5. The operational characteristics of both aerobic and anaerobic solid waste–processing systems are described in the following sections.

14-2 AEROBIC COMPOSTING

Aerobic composting is the most commonly used biological process for the conversion of the organic portion of MSW to a stable humus-like material known as *compost*. Applications of aerobic composting include (1) yard waste, (2) separated MSW, (3) commingled MSW, and (4) co-composting with wastewater sludge. Process descriptions and design guidelines for aerobic composting are presented in this section. Product specifications for compost produced from yard waste and MSW are given in Tables 15-8 and 15-9, respectively.

Process Description

All aerobic composting processes are similar in that they all incorporate three basic steps: (1) preprocessing of the MSW, (2) aerobic decomposition of the organic fraction of the MSW, and (3) product preparation and marketing. *Windrow, aerated*

TABLE 14-5
Comparison of aerobic composting and anaerobic digestion processes for processing the organic fraction of MSW

Characteristic	Aerobic processes	Anaerobic processes
Energy use	Net energy consumer	Net energy producer
End products	Humus, CO_2, H_2O	Sludge, CO_2, CH_4
Volume reduction	Up to 50%	Up to 50%
Processing time	20 to 30 days	20 to 40 days
Curing time	30 to 90 days	30 to 90 days
Primary goal	Volume reduction	Energy production
Secondary goal	Compost production	Volume reduction, waste stabilization

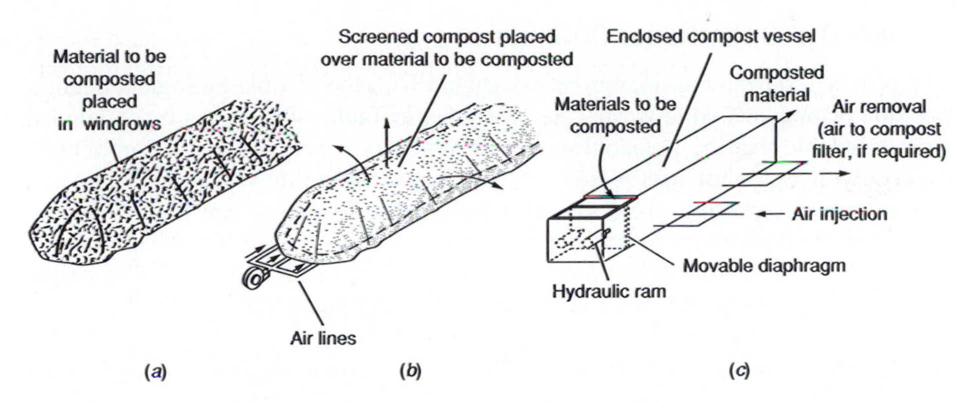

FIGURE 14-2
Commonly used composting methods: (*a*) windrow with periodic turning, (*b*) aerated static pile, and (*c*) in-vessel plug flow.

static pile, and *in-vessel* are the three principal methods used for the composting of the organic fraction of MSW (see Fig. 14-2). These methods are described in detail in Chapter 9. While these processes differ primarily in the method used to aerate the organic fraction of solid waste, the biological principles remain the same, and, when designed and operated properly, all produce a similar-quality compost in approximately the same time period.

Process Microbiology

During the aerobic composting process a succession of facultative and obligate aerobic microorganisms is active. In the beginning phases of the composting process, mesophilic bacteria are the most prevalent. After the temperatures in the compost rise, thermophilic bacteria predominate, leading to thermophilic fungi, which appear after 5 to 10 days. In the final stages, or *curing period* as it is sometimes known, actinomycetes and molds appear. Because significant concentrations of these microorganisms may not be present in some types of biodegradable waste (e.g., newspaper), it may be necessary to add them to the composting material as an additive or inoculum.

The microbiology of all aerobic composting processes is similar. Critical parameters in the control of aerobic composting processes include moisture content, C/N ratio, and temperature. For most biodegradable organic wastes, once the moisture content is brought to a suitable level (50 to 60 percent) and the mass aerated, microbial metabolism speeds up. The aerobic microorganisms, which utilize oxygen, feed upon the organic matter and develop cell tissue from nitrogen, phosphorus, some of the carbon, and other required nutrients. Much of the carbon serves as a source of energy for the organisms and is burned up and respired as carbon dioxide. Because organic carbon can serve both as a source of energy and cell carbon, more carbon is required than, for example, nitrogen.

Design and Operational Considerations

The principal design considerations associated with the aerobic biological decomposition of prepared solid wastes are presented in Table 14-6. It can be concluded from this table that the preparation of a composting process is not a simple task, especially if optimum results are to be achieved. For this reason, most of the commercial composting operations that have been developed are highly mechanized and are carried out in specially designed facilities, where the design factors reported in Table 14-6 can be controlled effectively (see Fig. 14-3). The principal design considerations are discussed in the following paragraphs.

Particle Size. Most materials that comprise the organic fraction of MSW tend to be irregular in shape. This irregularity can be reduced substantially by shredding the organic materials before they are composted (see Fig. 14-4). Particle size influences the bulk density, internal friction and flow characteristics, and drag forces of the materials. Most important of all, a reduced particle size increases the biochemical reaction rate during aerobic composting process. The most desirable particle size for composting is less than 2 inches (5 cm), but larger particles can be composted. The particle size of the material being composted is governed to some extent by the finished-product requirements and by economic considerations.

Carbon-to-Nitrogen Ratio. The most critical environmental factor for composting is the carbon-to-nitrogen ratio (C/N ratio). The optimum range for most organic wastes is from 20 to 25 to 1. As shown in Table 14-7, sludges have low C/N ratios, whereas yard wastes, such as leaves and newspaper, have relatively high C/N ratios. It should be noted that the C/N ratios given in Table 14-7 are based on the total dry weights of carbon and nitrogen, not on the dry weight of the biodegradable fraction of the organic material. In general, all of the organic nitrogen present in most organic compounds will become available, whereas not all of the organic carbon will be biodegradable. Depending on the particular waste material, the C/N ratio computed on the basis of total weights of carbon and nitrogen could be quite misleading, especially in those cases where all of the available nitrogen is biodegradable, but only a portion of organic carbon is biodegradable. (e.g., lignin in waste paper) [14]. For example, assuming that all of the nitrogen is available, the C/N ratio for the organic fraction of MSW can vary from about 34 to 60 depending on whether it is assumed that the available carbon is partially or totally biodegradable [14]. Blending of a waste high in carbon and low in nitrogen (e.g., newsprint) with a waste that is high in nitrogen (e.g., yard wastes) is used to achieve optimum C/N ratios for composting.

Blending and Seeding. Two design factors that may affect the blending of organic fraction of municipal solids waste for composting are C/N ratio and moisture content. Laboratory analyses are usually required to determine how the various

TABLE 14-6
Important design considerations for aerobic composting process

Item	Comment
Particle size	For optimum results the size of solid wastes should be between 25 and 75 mm (1 and 3 in).
Carbon-to-nitrogen (C/N) ratio	Initial carbon to nitrogen ratios (by mass) between 20 and 30 are optimum for aerobic composting. At lower ratios, ammonia is given off. Biological activity is also impeded at lower ratios. At higher ratios, nitrogen may be a limiting nutrient.
Blending and seeding	Composting time can be reduced by seeding with partially decomposed solid wastes to the extent of about 1 to 5 percent by weight. Sewage sludge can also be added to prepared solid wastes. Where sludge is added, the final moisture content is the controlling variable.
Moisture content	Moisture content should be in the range between 50 and 60 percent during the composting process. The optimum value appears to be about 55 percent.
Mixing/turning	To prevent drying, caking, and air channeling, material in the process of being composted should be mixed or turned on a regular schedule or as required. Frequency of mixing or turning will depend on the type of composting operation.
Temperature	For best results, temperature should be maintained between 122 and 131°F (50 and 55°C) for the first few days and between 131 and 140°F (55 and 60°C) for the remainder of the active composting period. If temperature goes beyond 151°F (66°C), biological activity is reduced significantly.
Control of pathogens	If properly conducted, it is possible to kill all the pathogens, weeds, and seeds during the composting process. To do this, the temperature must be maintained between 140 and 158°F (60 and 70°C) for 24 h.
Air requirements	The theoretical quantity of oxygen required can be estimated using Eq. (14-2). Air with at least 50 percent of the initial oxygen concentration remaining should reach all parts of the composting material for optimum results, especially in mechanical systems.
pH control	To achieve an optimum aerobic decomposition, pH should remain at 7 to 7.5 range. To minimize the loss of nitrogen in the form of ammonia gas, pH should not rise above about 8.5.
Degree of decomposition	The degree of decomposition can be estimated by measuring the final drop in temperature, degree of self heating capacity, amount of decomposable and resistant organic matter in the composted material, rise in the redox potential, oxygen uptake, growth of the fungus *Chaetomium gracilis,* and the starch-iodine test.
Land requirement	The land requirements for a plant with a capacity of 50 ton/d will be 1.5 to 2.0 acres. The land area required for a larger plant will be less on a ton/d basis.

Note: $1.8 \times (°C) + 32 = °F$.

FIGURE 14-3
Aerial view of large modern 660 ton/d materials recovery facility. Approximately 220 ton/d of compost are produced, using the windrow method of composting. The composting operation is carried out under a covered open-sided building. (Courtesy of Reuter Recycling of Florida.)

organic materials should be blended for aerobic composting. If the organic fraction of MSW contains significant amounts of paper or other substrates rich in carbon, other organic materials such as yard wastes, manure, or sludge from wastewater treatment plants can be blended to provide a near optimum C/N ratio. The blending of wastes to optimize the C/N ratio is illustrated in Example 14-3. Similarly, materials too wet and too dry for good composting can be blended in proper proportion to achieve an optimum moisture content. *Seeding* involves the addition of a volume of microbial culture sufficiently large to effect the decomposition of the receiving material at a faster rate.

FIGURE 14-4
View of large shredding facility used to reduce size of sorted MSW for effective composting. (Courtesy of SSI Shredding Systems, Inc.)

TABLE 14-7
Nitrogen content and nominal C/N ratios of selected compostable materials (dry basis)[a]

Material	Percent N	C/N ratio[b]
Food processing wastes		
Fruit wastes	1.52	34.8
Mixed slaughterhouse waste	7.0–10.0	2.0
Potato tops	1.5	25.0
Manures		
Cow manure	1.7	18.0
Horse manure	2.3	25.0
Pig manure	3.75	20.0
Poultry manure	6.3	15.0
Sheep manure	3.75	22.0
Sludges		
Digested activated sludge	1.88	15.7
Raw activated sludge	5.6	6.3
Wood and straw		
Lumber mill wastes	0.13	170.0
Oat straw	1.05	48.0
Sawdust	0.10	200.0–500.0
Wheat straw	0.3	128.0
Wood (pine)	0.07	723.0
Paper		
Mixed paper	0.25	173
Newsprint	0.05	983
Brown paper	0.01	4490
Trade magazines	0.07	470
Junk mail	0.17	223
Yard wastes		
Grass clippings	2.15	20.1
Leaves (freshly fallen)	0.5–1.0	40.0–80.0
Biomass		
Water hyacinth	1.96	20.9
Bermuda grass	1.96	24

[a] After Refs. 4, 6, and 21.

[b] C/N ratio based on total dry weights.

Example 14-3 Blending of wastes to achieve an optimum C/N ratio. Leaves, with a C/N ratio of 50, are to be blended with waste-activated sludge from a wastewater treatment plant, with a C/N ratio of 6.3. Determine the proportions of each component to achieve a blended C/N ratio of 25. Assume that the following conditions apply:

1. Moisture content of sludge = 75%
2. Moisture content of leaves = 50%
3. Nitrogen content of sludge = 5.6%
4. Nitrogen content of leaves = 0.7%

Solution

1. Determine the percentage composition for leaves and sludge.

   (*a*) For 1 lb of leaves:

$$\text{Water} = 1\text{ lb }(0.50) = 0.50\text{ lb}$$

$$\text{Dry matter} = 1\text{ lb} - 0.50\text{ lb} = 0.50\text{ lb}$$

$$\text{N} = 0.50\text{ lb }(0.007) = 0.0035\text{ lb}$$

$$\text{C} = 50\ (0.0035\text{ lb}) = 0.175\text{ lb}$$

   (*b*) For 1 lb of sludge:

$$\text{Water} = 1\text{ lb }(0.75) = 0.75\text{ lb}$$

$$\text{Dry matter} = 1\text{ lb} - 0.75\text{ lb} = 0.25\text{ lb}$$

$$\text{N} = 0.25\text{ lb }(0.056) = 0.014\text{ lb}$$

$$\text{C} = 6.3\ (0.014\text{ lb}) = 0.0882\text{ lb}$$

2. Determine the amount of sludge to be added to 1 lb of leaves to achieve a C/N ratio of 25:

$$\frac{\text{C}}{\text{N}} = 25 = \frac{\text{C in 1 lb of leaves} + x(\text{C in 1 lb of sludge})}{\text{N in 1 lb of leaves} + x(\text{N in 1 lb of sludge})}$$

   where x = weight of sludge required

$$25 = \frac{0.175 + x(0.0882)}{0.0035 + x(0.014)}$$

$$x = 0.33\text{ lb sludge/1 lb leaves}$$

3. Check the C/N ratio and moisture content of the resultant mixture.

   (*a*) For 0.33 lb sludge:

$$\text{Water} = 0.33\text{ lb }(0.75) = 0.25\text{ lb}$$

$$\text{Dry matter} = 0.33\text{ lb }(0.25) = 0.08\text{ lb}$$

$$\text{N} = 0.08\text{ lb }(0.056) = 0.004\text{ lb}$$

$$\text{C} = 6.3(0.004\text{lb}) = 0.03\text{ lb}$$

   (*b*) For 0.33 lb sludge + 1 lb leaves:

$$\text{Water} = 0.25\text{ lb} + 0.50\text{ lb} = 0.75\text{ lb}$$

$$\text{Dry matter} = 0.08\text{ lb} + 0.50\text{ lb} = 0.58\text{ lb}$$

$$\text{N} = 0.004\text{ lb} + 0.0035\text{ lb} = 0.008\text{ lb}$$

$$\text{C} = 0.03\text{ lb} + 0.175\text{ lb} = 0.205\text{ lb}$$

   (*c*) Find the C/N ratio:

$$\frac{\text{C}}{\text{N}} = \frac{0.205\text{ lb C}}{0.008\text{ lb N}} = 25.6 \qquad \text{ok}$$

(*d*) Find the moisture content:

$$\text{Moisture} = \frac{0.75\text{ lb water}}{0.75\text{ lb water } + 0.58\text{ lb dry matter}} = \frac{0.75\text{ lb}}{1.33\text{ lb}} = 56\% \qquad \text{ok}$$

Comment. Co-composting sludge with yard waste is a reasonable method of adding nitrogen to enhance composting. Because of the potential for pathogen and heavy metal contamination, the compost quality must be monitored carefully.

Moisture Content. The optimum moisture content for aerobic composting is in the range of 50 to 60 percent. Moisture can be adjusted by blending of components or by addition of water. When the moisture content of compost falls below 40 percent, the rate of composting will be slowed.

Mixing/Turning. Initial mixing of organic wastes is essential to increase or decrease the moisture content to an optimum level. Mixing can be used to achieve a more uniform distribution of nutrients and microorganisms. Turning of the organic material during the composting process is a very important operational factor in maintaining aerobic activity. Because turning can be specified by moisture content, waste characteristics, or air requirements, it is impossible to specify a minimum frequency of turning or number of turns in a general terms. For an organic waste having a maximum moisture of 55 to 60 percent and a composting period of 15 days, the first turn has been suggested at the third day. Thereafter, it should be turned every other day for a total of four to five turns.

Temperature. Aerobic composting systems can be operated in either the mesophilic, 30 to 38°C (85 to 100°F), or the thermophilic, 55 to 60°C (131 to 140°F), temperature regions. The temperature rise observed in actively composting wastes is caused by the exothermic reactions associated with respiratory metabolism; see Eq. (14-1). In aerated static pile and in-vessel composting systems, the temperature can be regulated by monitoring the temperature and controlling the airflow. In windrow composting, temperature can only be controlled indirectly, by varying the frequency of turning based on temperature measurements. A typical temperature profile range observed in windrow composting is illustrated in Fig. 14-5. In general, pile temperature will drop 5 to 10°C after turning, but will return to its previous level within several hours. Windrow temperatures decrease after 10 to 15 days as the readily biodegradable organic material is oxidized.

Control of Pathogens. The destruction of pathogenic organisms is an important design element in the compost process, as it will affect the temperature profile and aeration process. Data on the thermal death points for a number of pathogenic organisms are summarized in Table 14-8. As shown in Table 14-8, the die-off of pathogens is a function of time and temperature. For example, the *Salmonella* species of bacteria can be destroyed in 15 to 20 minutes when exposed to a temperature of 60°C, or in one hour at 55°C. From Table 14-8, it is apparent that

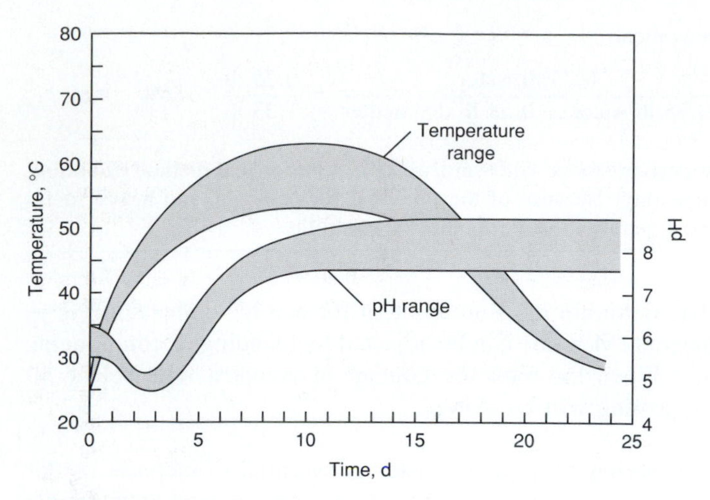

FIGURE 14-5
Typical temperature and pH ranges observed in windrow composting.

TABLE 14-8
Temperature and time of exposure required for destruction of some common pathogens and parasites[a]

Organism	Observations
Salmonella typhosa	No growth beyond 46°C; death within 30 minutes at 55–60°C and within 20 minutes at 60°C; destroyed in a short time in compost environment.
Salmonella sp.	Death within 1 hour at 55°C and within 15–20 minutes at 60°C.
Shigella sp.	Death within 1 hour at 55°C.
Escherichia coli	Most die within 1 hour at 55°C and within 15–20 minutes at 60°C.
Entamoeba histolytica cysts	Death within a few minutes at 45°C and within a few seconds at 55°C.
Taenia saginata	Death within a few minutes at 55°C.
Trichinella spiralis larvae	Quickly killed at 55°C; instantly killed at 60°C.
Brucella abortus or *Br. suis*	Death within 3 minutes at 62–63°C and within 1 hour at 55°C.
Micrococcus pyogenes var. *aureus*	Death within 10 minutes at 50°C.
Streptococcus pyogenes	Death within 10 minutes at 54°C.
Mycobacterium tuberculosis var. *hominis*	Death within 15–20 minutes at 66°C or after momentary heating at 67°C.
Corynebacterium diphtheriae	Death within 45 minutes at 55°C.
Necator americanus	Death within 50 minutes at 45°C.
Ascaris lumbricoides eggs	Death in less than 1 hour at temperatures over 50°C.

[a] From Ref. 6.
Note: 1.8 × (°C) + 32 = °F.

TABLE 14-9
EPA requirements for pathogen control in compost processes[a]

Requirement	Remarks
Processes to significantly reduce pathogens (PSRP)	Using the in-vessel, aerated static pile, or windrow composting methods, the solid waste is maintained at minimum operating conditions of 40°C for 5 days. For four hours during this period, the temperature exceeds 55°C.
Processes to further reduce pathogens (PFRP)	Using the in-vessel or aerated static pile composting methods, the solid waste is maintained at operating conditions of 55°C or greater for three days.
	Using the windrow composting method, the solid waste is maintained at operating conditions of 55°C or greater for at least 15 days during the composting period. Also, during the high-temperature period there will be a minimum of five turnings of the windrow.

[a] Based on "Criteria for Classification of Solid Waste Disposal Facilities and Practices," U.S. EPA, *Federal Register* 44:179 (1979).

most pathogens will be destroyed rapidly when all parts of the compost pile are subjected to a temperature of about 55°C. Only a few can survive at temperatures up to 67°C for a short period of time. Elimination of all pathogenic microorganisms can be accomplished by allowing the composting waste to reach a temperature of 70°C for 1 to 2 hours. The U.S. Environmental Protection Agency has required specific time-temperature standards for pathogen control in composting systems (see Table 14-9). These conditions are easily met in properly operating composting systems. The microbiology and kinetics of the heat deactivation of pathogens are discussed in greater detail in Refs. 5, 6, and 7.

Air Requirements. In processes with forced aeration, such as the aerated static pile and the in-vessel systems, the total air requirement and air flow rate are essential design parameters. Computation of the total air requirements and air flow rate for an in-vessel composting system is illustrated in Example 14-4. The computations for an aerated static pile system are similar.

Example 14-4 Air requirements for in-vessel composting. Determine the amount of air required to compost one ton of solid wastes using an in-vessel composting system with forced aeration. Assume that the composition of the organic fraction of the MSW to be composted is given by $C_{60.0}H_{94.3}O_{37.8}N$ (determined in Example 4-2, Step 4). Assume that the following conditions and data apply:

1. Moisture content of organic fraction of MSW = 25%.
2. Volatile solids, VS = 0.93 × TS (total solids).
3. Biodegradable volatile solids, BVS = 0.60 × VS.
4. Expected BVS conversion efficiency = 95%.
5. Composting time = 5 d.

6. Oxygen demand is 20, 35, 25, 15, 5% for the successive days of the 5-day composting period.
7. The ammonia produced during the aerobic decomposition of the waste is lost to the atmosphere.
8. Air contains 23 percent O_2 by mass, and the specific weight of air is equal 0.075 lb/ft^3.
9. A factor of 2 times the actual air supplied will be needed to be assured that the oxygen content of the air does not drop below 50 percent of its original value.

Solution

1. Determine the mass of biodegradable volatile solids in one ton of organic waste.

$$\text{Mass BVS} = 1.0 \text{ ton} \times 2000 \text{ lb/ton} \times 0.75 \text{ (dry matter content)} \times 0.93 \times 0.60 = 837.05 \text{ lb}$$

2. Determine the expected BVS mass conversion.

$$\text{Expected BVS mass conversion} = 837.05 \times 0.95 = 795.2 \text{ lb}$$

3. Determine the amount of oxygen required for the decomposition of one pound of the biodegradable volatile solids using Eq. (14-2), as given below:

$$C_aH_bO_cN_d + \left(\frac{4a + b - 2c - 3d}{4}\right)O_2 \rightarrow aCO_2 + \left(\frac{b - 3d}{2}\right)H_2O + dNH_3$$

For the given chemical composition, the coefficients are $a = 60.0$, $b = 94.3$, $c = 37.8$, and $d = 1$. The balanced equation is

$$\underset{1433.1}{C_{60.0}H_{94.3}O_{37.8}N} + \underset{2045.8}{63.93\ O_2} \rightarrow \underset{2640.0}{60.0\ CO_2} + \underset{822.6}{45.7\ H_2O} + \underset{17.0}{NH_3}$$

$$O_2 \text{ required} = \frac{2045.8 \text{ lb } O_2}{1433.1 \text{ lb BVS converted}} = 1.43 \frac{\text{lb } O_2}{\text{lb BVS converted}}$$

4. Determine the total amount of air required for one ton of MSW, containing 795.2 lb of BVS, as determined in Step 2:

$$\text{Air required} = \frac{\left(795.2 \text{ lb BVS} \times 1.43 \frac{\text{lb } O_2}{\text{lb BVS converted}}\right)}{\left(0.23 \frac{\text{lb } O_2}{\text{lb air}} \times 0.075 \frac{\text{lb air}}{\text{ft}^3 \text{ air}}\right)} = 65{,}921 \text{ ft}^3 \text{ air}$$

5. Determine the required capacity of the aeration equipment, expressed in ft^3/min.

$$\text{Air required} = \frac{\left(65{,}921 \text{ ft}^3 \text{ air} \times 2 \times \frac{0.35}{\text{d}}\right)}{\left(1440 \frac{\text{min}}{\text{d}}\right)} = 32.0 \frac{\text{ft}^3}{\text{min}}$$

Comment. The flow rate is computed using 35 percent of the total oxygen requirement, the most critical day. In an actual composting operation, some of the BVS would have been converted to cell tissue. However, because air is also required for the conversion of BVS to cell tissue, the computations presented in this example, which are based on the assumption that all of the BVS is converted, are reasonable.

pH Control. Control of pH is another important parameter in evaluating the microbial environment and waste stabilization. The pH value, like the temperature, of the compost varies with time during the composting process. The initial pH of the organic fraction of MSW is typically between 5 and 7. The pH of composting material will vary according to the pH-time profile shown in Fig. 14-5. In the first few days of composting, the pH drops to 5 or less. At this stage, the organic mass is at ambient temperature, the multiplication of indigenous mesophilic organisms begin, and the temperature rises rapidly. Among the products of this initial stage are simple organic acids, which cause the drop in pH. After about three days, the temperature reaches a thermophilic stage, and the pH begins to rise to approximately 8 or 8.5 for the remainder of the aerobic process. The pH value falls slightly during the cooling stage and reaches to a value in the range of 7 to 8 in the mature compost. If the degree of aeration is not adequate, anaerobic conditions will occur, the pH will drop to about 4.5, and the composting process will be retarded.

Degree of Decomposition. A suitable methodology for the measurement of the degree of decomposition is not available. However, several methodologies have been proposed [6, 7]. The proposed methods are (*a*) final drop in temperature, (*b*) degree of self-heating capacity, (*c*) amount of decomposable and resistant organic matter in the composted material, (*d*) rise in the redox potential, (*e*) oxygen uptake, (*f*) growth of the fungus *Chaetomium gracilis*, (*g*) the starch-iodine test, and (*h*) the respiratory quotient method. The laboratory analysis of chemical oxygen demand (COD) and the lignin test provide a quick check for determining the degree of decomposition. A low COD value and a high lignin content (greater than 30 percent) is indicative of a stable compost.

Control of Odor. The majority of the odor problems in aerobic composting processes are associated with the development of anaerobic conditions within the compost pile. In many large-scale aerobic composting systems, it is common to find pieces of magazines or books, plastics (especially plastic films), or similar materials in the organic material being composted. These materials normally cannot be decomposed in a relatively short time in a compost pile. Furthermore, because sufficient oxygen is often not available in the center of such materials, anaerobic conditions can develop. Under anaerobic conditions, organic acids will be produced, many of which are extremely odorous. To minimize the potential odor problems, it is important to reduce the particle size, remove plastics and other nonbiodegradable materials from the organic material to be composted, or use source-separated or uncontaminated feedstocks.

Land Requirements. Land area requirements are another important element which must be considered in the aerobic composting processes. For example, in windrow composting for a plant with a capacity of 50 ton/d, about 2.5 acres of land would be required (see Fig. 14-6). Of this total, 1.5 acres would be devoted to buildings, plant equipment, and roads. For each additional 50 tons, it is estimated that 1.0 acre would be required for the composting operation and that

(*a*)

(*b*)

FIGURE 14-6
Land area requirements for composting: (*a*) area required for storage of unprocessed waste materials and processing facilities (see also Figs. 9-37 and 14-7) and (*b*) area required for composting operation and storage of cured compost.

0.25 acre would be required for buildings and roads. The land requirement for highly mechanized systems varies with the process. An estimate of 1.5 to 2.0 acres for a plant with a capacity of 50 ton/d is not unreasonable; for larger plants, the unit area requirements would be less. For example, the Portland, Oregon, METRO compost plant, based on the DANO process, was designed to process 185,000 ton/yr of commingled MSW on an 18-acre site.

Processing Compost for Market. The economics of compost systems are greatly enhanced if the compost can be sold. To be marketable, compost must be of a consistent size; free from contaminants such as glass, plastic, and metals; and free of objectionable odors. The type of processing used to prepare the compost for marketing will depend on the specifications for the compost. Shredding and screening are commonly used to produce a more uniform product (see Fig. 14-7). In some cases, additives may be added to enhance the value of the final product.

FIGURE 14-7
Processing of cured compost with a trommel to produce a final product with a more uniform size. (Courtesy of Amadas Industries.)

Selection of Aerobic Composting Processes

Because the performance of properly operating windrow, aerated static pile, and in-vessel composting processes is essentially the same, the selection among alternative processes is based on capital and operating costs, land availability, operational complexity, and potential for nuisance problems. These factors are compared in Table 14-10.

An additional factor to consider is that the windrow and aerated static pile processes utilize standard components and have been designed and constructed successfully "in-house" by many municipalities and private firms. The in-vessel processes employ proprietary designs and custom-made equipment and are usually procured on a turnkey basis from a single vendor. Many in-vessel compost systems are procured on full-service contracts and are owned and operated by the vendor or a third party. The economics of these arrangements should be compared carefully to ownership and operation of a simpler windrow or aerated static pile system by the municipality itself. A successful composting operation is highly dependent on proper operation and maintenance as well as design.

14-3 LOW-SOLIDS ANAEROBIC DIGESTION

Low-solids anaerobic digestion is a biological process in which organic wastes are fermented at solids concentrations equal to or less than 4 to 8 percent. The low-solids anaerobic fermentation process is used in many parts of the world to generate methane gas from human, animal, and agricultural wastes, and from the organic fraction of MSW. One of the disadvantages of the low-solids anaerobic digestion process as applied to solid wastes is that considerable water must be added to wastes to bring the solids content to the required range of 4 to 8 percent. The addition of water results in a very dilute digested sludge, which must be dewatered prior to disposal. The disposal of the liquid stream resulting from the dewatering step is an important consideration in the selection of the low-solids digestion process.

Process Description

There are three basic steps involved whenever the low-solids anaerobic digestion process is used to produce methane from the organic fraction of MSW. As shown in Fig. 14-8, the first step involves the preparation of the organic fraction of the MSW. Typically, for commingled solid waste the first step involves receiving; sorting and separation; and size reduction. Size reduction is also required for source-separated materials.

The second step involves the addition of moisture and nutrients, blending, pH adjustment to about 6.8, and heating of the slurry to between 55 and 60°C, and the anaerobic digestion is carried out in a continuous-flow reactor whose contents are mixed completely (see Fig. 14-9). In some operations, a series of batch reactors have been used instead of one or more continuous-flow complete-mix reactors.

TABLE 14-10
Comparison of aerobic composting processes[a]

Item	Windrow	Aerated static pile	In-vessel, forced aeration: With agitation (dynamic)	In-vessel, forced aeration: No agitation (plug flow)
Capital costs	Generally low	Generally low in small systems, can become high in large systems	Generally high	Generally high
Operating costs	Generally low	High (in sludge systems where bulking agents are used)	Generally low	Generally low
Land requirements	High	High	Low, but can increase if windrow drying or curing required	Low, but can increase if windrow drying or curing required
Control of air	Limited unless forced aeration is used	Complete	Complete	Complete
Operational control	Turning frequency, amendment, or compost recycle addition	Airflow rate	Airflow rate, agitation, amendment, or compost recycle addition	Airflow rate, amendment, or compost recycle addition
Sensitivity to cold or wet weather	Sensitive unless in housing	Demonstrated in cold and wet climates	Demonstrated in cold and wet climates	Demonstrated in cold and wet climates
Control of odors	Depends on feedstock, potential large-area source	May be large-area source but can be controlled	Potentially good	Potentially good
Potential operating problems	Susceptible to adverse weather	Control of airflow rate is critical, potential for channeling or short-circuiting of air supply	High operational flexibility, system may be mechanically complex	Potential for channeling or short circuiting of air supply, system may be mechanically complex

[a] Adapted from Ref. 7.

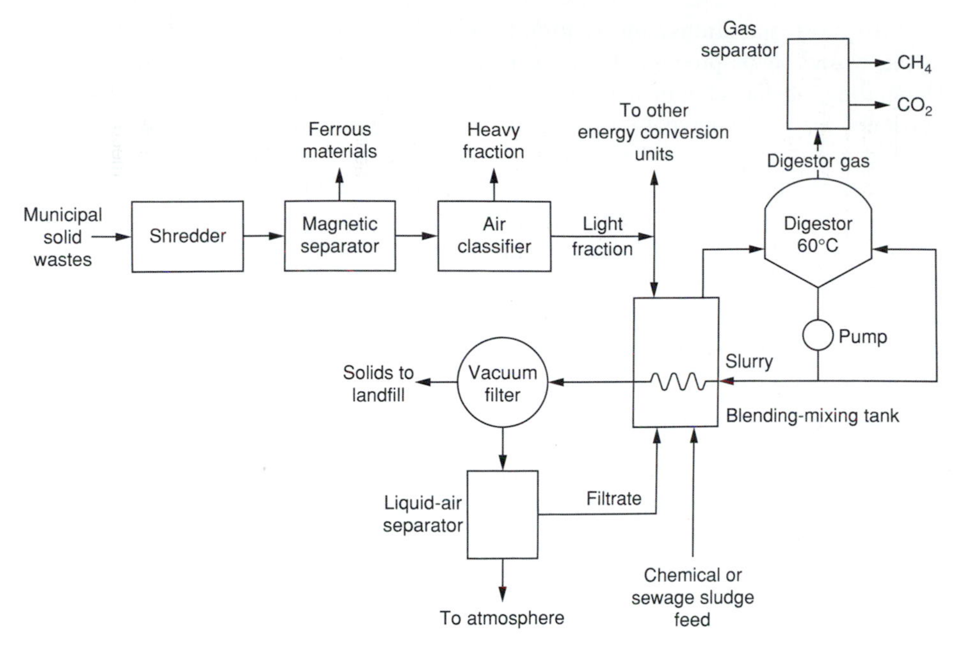

FIGURE 14-8
Flow diagram for the low-solids anaerobic digestion process for the organic fraction of MSW.

(*a*) (*b*)

FIGURE 14-9
Views of low-solids anaerobic digesters: (*a*) conventional circular type (both floating and fixed covers are used) and (*b*) egg-shaped type.

In most operations, the required moisture content and nutrients are added to the wastes to be processed, in the form of wastewater sludge or cow manure. Depending on the chemical characteristics of the sludge or manure, additional nutrients may also have to be added. Because foaming and the formation of surface crusts have caused problems in the digestion of solid wastes, adequate mixing is of fundamental importance in the design and operation of such systems.

The third step in the process involves the capture, storage, and, if necessary, separation of the gas components. The dewatering and disposal of the digested sludge is an additional task that must be accomplished. In general, the processing of the digested sludge produced from low-solids anaerobic digestion is so expensive that the process has seldom been used.

Process Microbiology

Carried out in the absence of oxygen, the anaerobic stabilization process or conversion of the organic materials in MSW occurs, as discussed in Section 14-1, in three steps. As discussed previously (see also Fig. 14-1), the first step in the process involves the enzyme-mediated transformation (hydrolysis) of higher-molecular-mass compounds into compounds suitable for use as a source of energy and cell tissue. The second step involves the bacterial conversion of the compounds resulting from the first step into identifiable lower-molecular-mass intermediate compounds. The third step involves the bacterial conversion of the intermediate compounds into simpler end products, principally methane and carbon dioxide [8, 9, 15, and 18].

Process Design Considerations

Although the process of anaerobic digestion of the organic fraction of MSW is not developed fully, some of the important design considerations are summarized in Table 14-11. In operations where solid wastes have been mixed with wastewater sludge, it has been found that the gas collected from the digesters contains between 50 and 60 percent methane. It has also been found that about 10 ft^3 of gas is produced per pound of biodegradable volatile solids destroyed. Because of the variability of the results reported in the literature, it is recommended that pilot plant studies be conducted if the digestion process is to be used for the conversion of MSW or other organic wastes.

Process Selection. Process selection between anaerobic processes is typically between the low-solids process and the high-solids process, described in the following section. Selection of equipment and facilities for the low-solids anaerobic digestion process usually involves the type of mixing equipment (internal mixers, internal gas mixing, and external pump mixing), the general shape of the digester (e.g., circular or egg-shaped), the control systems, and the ancillary facilities needed for mixing the incoming wastes and dewatering the digested sludge. Digester design is discussed in detail in Ref. 20.

TABLE 14-11
Important design considerations for the low-solids anaerobic digestion of the organic fraction of MSW

Waste component	Comment
Size of material	Wastes to be digested should be shredded to a size that will not interfere with the efficient functioning of pumping and mixing operations.
Mixing equipment	To achieve optimum results and to avoid scum buildup, mechanical mixing is recommended.
Percentage of solid wastes mixed with sludge	Although amounts of waste varying from 50 to 90+ percent have been used, 60 percent appears to be a reasonable compromise.
Hydraulic and mean cell-residence time	Washout time is in the range of 3 to 4 d. Use 10 to 20 d for design, or base design on results of pilot plant studies.
Loading rate	0.04 to 0.10 $lb/ft^3 \cdot d$ (0.6 to 1.6 $kg/m^3 \cdot d$). Not well defined at present time. Significantly higher rates have been reported.
Solids concentration	Equal to or less than 8 to 10% (4 to 8% typical).
Temperature	Between 85 and 100°F (30 to 38°C) for mesophilic and between 131 and 140°F (55 and 60°C) for thermophilic reactor.
Destruction of volatile solid wastes	Depends on the nature of the waste characteristics. Varies from about 60 to 80 percent; 70 percent can be used for estimating purposes.
Total solids destroyed	Varies from 40 to 60 percent, depending on amount of inert material present originally.
Gas production	8 to 12 ft^3/lb (0.5 to 0.75 m^3/kg) of volatile solids destroyed (CH_4 = 55 percent; CO_2 = 45 percent).

14-4 HIGH-SOLIDS ANAEROBIC DIGESTION

High-solids anaerobic digestion is a biological process in which the fermentation occur at a total solids content of about 22 percent or higher. The high-solids anaerobic digestion is a relatively new technology and its application for energy recovery from the organic fraction of MSW has not been developed fully [1, 3, 11, and 22]. Two important advantages of the high-solids anaerobic digestion process are lower water requirements and higher gas production per unit volume of the reactor size. The major disadvantage of this process is that at present (1992), limited full-scale operating experience is available.

Process Description

The three steps described for the low-solids anaerobic digestion are also applied in high-solids anaerobic digestion process. The principal difference is at the end of the process, where less effort is required to dewater and dispose of the digested sludge.

Process Microbiology

The process microbiology for the high-solids anaerobic digestion is as described previously for the low-solids anaerobic digestion process and in Section 14-1. However, because of the high solids concentration, the effects of many environmental parameters on microbial population are more severe. For example, ammonia toxicity can affect the methanogenic bacteria, which will have an adverse effect on system stability and methane production. In most cases, the ammonia toxicity can be prevented by a proper adjustment of the C/N ratio of the input feedstock.

Process Design Considerations

Although the high-solids anaerobic digestion process is not developed fully, some of the important design considerations are summarized in Table 14-12. In general, the high-solids anaerobic digestion process is capable of stabilizing more organic waste and producing more gas per unit of volume of reactor than is the low-solids process considered previously. The amount of methane that can be recovered from the organic fraction of MSW by means of the high-solids anaerobic digestion process is considered in Example 14-5.

TABLE 14-12
Important design considerations for the high-solids anaerobic digestion of the organic fraction of MSW

Item	Comment
Size of material	Wastes to be digested should be shredded to a size that will not interfere with the efficient functioning of feeding and discharging mechanisms.
Mixing equipment	The mixing equipment will depend on the type of reactor to be used.
Percentage of solid wastes mixed with sludge	Depends on the characteristics of the sludge.
Mass retention time	Use 20 to 30 d for design, or base design on results of pilot plant studies.
Loading rate based on biodegradable volatile solids, BVS	0.375 to 0.4 $lb/ft^3 \cdot d$ (6 to 7 $kg/m^3 \cdot d$). Not well defined at present time. Significantly higher rates have been reported.
Solids concentration	Between 20 and 35% (22 to 28% typical).
Temperature	Between 85 and 100°F (30 and 38°C) for mesophilic and between 131 and 140°F (55 and 60°C) for thermophilic reactor.
Destruction of BVS	Varies from about 90 to 98+ percent depending on the mass retention time and the BVS loading rate.
Total solids destroyed	Varies depending on the lignin content of the feedstocks.
Gas production	10 to 16 ft^3/lb of biodegradable volatile solids destroyed (0.625 to 1.0 m^3/kg), (CH_4 = 50 percent; CO_2 = 50 percent).

Example 14-5 Methane recovery from the high-solids anaerobic digestion of the organic fraction of MSW. Estimate the amount of methane that can be recovered from one ton of waste, consisting of the organic fraction of MSW, and its dollar value. Assume that the following conditions and data apply.

1. Moisture content of organic fraction of MSW = 20%.
2. Mass retention time = 30 d.
3. Volatile solids, VS = 0.93 × TS (total solids).
4. Biodegradable volatile solids, BVS = 0.70 × VS.
5. Expected BVS conversion efficiency = 85%.
6. Gas production = 12 ft^3/lb BVS destroyed.
7. Energy content of biogas = 500 Btu/ft^3.
8. One therm of power = 10^5 Btu/therm.
9. The value of one therm of power = \$0.60/therm.

Solution

1. Determine the mass of volatile solids in one ton of organic waste.

$$\text{Mass VS} = 1.0 \text{ ton} \times 2000 \text{ lb/ton} \times 0.80 \text{ (moisture content)} \times 0.93 = 1488 \text{ lb}$$

2. Determine the mass of BVS destroyed based on one ton of organic waste.

$$\text{Mass of BVS destroyed} = 1488 \text{ lb} \times 0.70 \times 0.85 = 885.4 \text{ lb}$$

3. Determine total volume of gas produced from one ton of organic waste.

$$\text{Gas produced} = 885.4 \text{ lb} \times 12 \text{ ft}^3/\text{lb BVS destroyed} = 10{,}625 \text{ ft}^3$$

4. Determine total Btu content of the gas.

$$\text{Btu} = 10{,}625 \text{ ft}^3 \times 500 \text{ Btu/ft}^3 = 5.31 \times 10^6 \text{ Btu}$$

5. Determine total value of the gas produced from one ton of organic waste.

$$\text{Value, \$/ton} = [(5.31 \times 10^6 \text{ Btu})/(10^5 \text{ Btu/therm})] \times \$0.60/\text{therm} = 31.86$$

Comment. The most effective use to be made of the organic fraction of MSW will depend on the market value of paper. For example, if the market value of mixed organic wastes is \$7.50/ton, the production of energy clearly makes economic sense. In assessing whether to market the mixed organic wastes, the cost of transporting the processed waste will be an important factor in the analysis.

Process Selection

At the present time, as described in the following section, there are no full scale high-solids digestion processes in operation in the United States. However, there are several different high-solids processes under development in the United States and Europe, and at least three are operational in Europe. As these processes become more fully developed and tested, it is anticipated that a number of anaerobic processes will be available commercially. For the purpose of comparison, the operational characteristics of the low-solids and the high-solids anaerobic digestion process are compared in Table 14-13.

TABLE 14-13
Comparative analysis of the low-solids and high-solids anaerobic digestion process consideration for the organic fraction of MSW

Design and/or operational parameter	Comment	
	Low-solids	High-solids
Reactor design	Complete-mix reactors have been used in large-scale systems for organic fraction of MSW. Plug-flow reactors are used widely other organic materials, especially cow manure.	Complete-mix, plug-flow, and batch reactors have been studied experimentally. None of these reactor types have been used commercially for processing MSW.
Solids content	4 to 8 percent.	22 to 32 percent.
Reactor volume	Large reactor volume is required per unit volume of organic waste.	A much smaller reactor volume is required for the same volume of organic waste than in the low-solids digestion process.
Water addition	A large volume of water is required to increase the moisture content of the organic fraction of MSW.	Water requirement is much less, because of the high solids concentration.
Organic loading rate	Relatively low organic loading rates per unit of reactor volume.	Relatively high organic loading rates per unit of reactor volume.
Gas production rate	Maximum gas production rates of up to 2 volumes per active reactor volume have been reported.	Maximum gas production rates of up to 6 volumes per active reactor volume have been achieved.
Mass removal rate	Mass removal rate is low due to higher water content.	A significantly higher mass removal rate can be achieved in the same retention time period as compared to low-solids digestion.
Mechanisms for feeding and discharging the effluent	Pumps of all types have been used.	Because this is a relatively new technology, appropriate mechanisms for feeding and discharging effluent from the anaerobic reactor are not well defined. High-solids pumps and screw conveyors have been used.
Toxicity problems	Toxicity problems in a low-solids anaerobic digester are less severe due to dilute nature of organic waste materials.	Salts and heavy metal toxicity are more common in high-solids anaerobic digestion due to large concentrations of these compounds and chemical elements. Ammonia toxicity is a major problem with low C/N ratios (less than 10 to 15).
Leachate problem	Due to high water content, the stabilized effluent can generate leachate problem.	Effluent from a high-solids digester normally contains 25 to 30 percent solids, which minimize the leachate generation potential.
Effluent dewatering	Large and expensive facilities are required to separate solids. For final disposal, the separated water should also be treated.	Inexpensive dewatering equipment is adequate.
Technology status	Not commercialized for energy recovery from organic fraction of MSW. The commercial use of the low-solids anaerobic digester for energy production from agricultural waste is worldwide.	Not commercialized for energy recovery from organic fraction of MSW.

14-5 DEVELOPMENT OF ANAEROBIC DIGESTION PROCESSES AND TECHNOLOGIES FOR TREATMENT OF THE ORGANIC FRACTION OF MSW

The purpose of this section is to introduce the reader to the emerging technologies involving the use of the anaerobic digestion process for the production of methane and a humus product from the organic fraction of MSW and other organic materials.

Anaerobic Digestion Technologies

In recent years, there has been a great interest in applying the anaerobic digestion process for the processing of the organic fraction of MSW because of the opportunity to recover methane and the fact that the digested material is similar to compost produced aerobically. The principal anaerobic digestion processes or technologies currently under investigation or in use are summarized in Table 14-14. As shown, most of the work with the anaerobic digestion process is going on in Europe. The combined high-solids anaerobic digestion/aerobic composting process under development in the United States is considered below.

Combined High-Solids Anaerobic Digestion/Aerobic Composting

The high-solids anaerobic digestion/aerobic composting process, developed by Professor Bill Jewell at Cornell University [11], combines the high-solids anaerobic digestion and aerobic composting processes. The major advantage of this process is the complete stabilization of the organic waste with a net energy recovery and without the need for major dewatering equipment. Other advantages include pathogen control and volume reduction.

Process Description. The high-solids anaerobic digestion/aerobic composting process is a two-stage process (see Fig. 14-10). The first stage of the two-stage process involves the high-solids (25 to 30 percent) anaerobic digestion of the organic fraction of MSW to produce a gas composed of methane and carbon dioxide. The anaerobic reactor operates under thermophilic conditions 129 to 133°F (54 to 56°C) with a nominal hydraulic retention time of 30 days (see Fig. 14-11).

The second stage involves the aerobic composting of the anaerobically digested solids to increase the solids content from 25 to 65 percent or more, depending on the final use. The output from the second stage is a fine humus-like material with a thermal content of about 6000 to 6400 Btu/lb (HHV) and a specific weight of about 35 lb/ft^3 (see Fig. 14-12). Because the final humus that is produced is combustible, it appears that it can be fired directly in a boiler when mixed with other fuels or pelletized for use as a fuel source. Alternatively, the humus-like material can be used as a soil amendment.

TABLE 14-14
Summary of anaerobic digestion processes and technologies for treatment of the organic fraction of MSW[a]

Anerobic digestion process	Country	Status	Description
Sequential batch anaerobic composting (SEBAC)	USA	Experimental stage	The SEBAC is a batch anaerobic three-stage process. In the first stage, a bed of coarsely-shredded feedstock is inoculated by recycling leachate from the third-stage reactor in the final stages of digestion. Volatile acids and other fermentation products generated during startup are removed from the first-stage reactor to the second-stage reactor for conversion to methane.
High-solids anaerobic digestion/aerobic composting process	USA	Under development	The high-solids anaerobic digestion/aerobic composting is a two-stage process. The first stage involves the dry digestion (solid content of 25 to 32 percent) to convert the organic fraction of MSW to methane. The second stage involves the aerobic composting of the anaerobically digested solids to produce a fine humus-like material that can be used as a fuel or soil amendment.
Semi-solid anaerobic digestion/aerobic composting process	Italy	Under development	The semi-solid anaerobic digestion/aerobic composting is a two-stage process. The first stage involves the semi-dry digestion (solids content of 15 to 22 percent) to convert the organic fraction of MSW to energy. The second stage involves the aerobic composting of the commingled anaerobically digested solids and the biodegradable fraction of organic MSW to produce humus-like material.
Leach-bed two-phase anaerobic digestion process	U.K.	Experimental stage	The leach-bed two-phase anaerobic digestion involves the rapid bio-leaching of organic matter from the putrescible material in MSW in specially engineered and prepared landfill. To accelerate the process, the resulting leachate is recirculated through the solid material.
Two-step anaerobic digestion	Germany	Experimental stage	The two-step anaerobic digestion process is used to treat organic substrates at low C/N ratios and high loading rates. The process is based on a sequential biochemical conversion of the organic solids which allows for better process control. This process is carried out in a semi-liquid phase in the mesophilic temperature range.
Biowaste process	Denmark	Under development	The biowaste anaerobic treatment system is designed to treat the source-separated household solid waste along with industrial and agricultural waste. The complete-mix digester is operated in the thermophilic temperature range.

(*continued*)

TABLE 14-14 (*continued*)

Anerobic digestion process	Country	Status	Description
KAMPOGAS process	Switzerland	Under development	The KAMPOGAS is a new anaerobic digestion process to treat fruit, yard waste, and vegetable waste. The digester is cylindrical and positioned horizontally. The digester, equipped with a hydraulically driven stirrer, is operated at high solids concentrations in the thermophilic temperature range.
DRANCO process	Belgium	Developed	The DRANCO process is used for the conversion of the organic fraction of MSW to produce energy and humus-like product, called humutex. The digestion process is carried out in a vertical plug-flow reactor with no mechanical mixing, but leachate from the bottom of the reactor is recirculated. The DRANCO digester operated at high-solids concentrations and in the mesophilic temperature range.
BTA process	Germany	Developed	The BTA process is developed especially to treat the organic fraction of MSW. The BTA treatment process includes: (1) pretreatment of incoming waste by mechanical, thermal, and chemical means; (2) separation of dissolved and undissolved biogenous solids; (3) anaerobic hydrolysis of biodegradable solids; and (4) methanization of dissolved biogenous materials. The methanization occurs at low solids and mesophilic temperature ranges. After dewatering, the non-degraded solids, with a total solids concentration of 35 percent, are used as a compost-like material.
VALORGA process	France	Developed	The VALORGA process is comprised of sorting unit, methane producing unit, and refining unit. The anaerobic fermenter operates at high-solids concentrations and in the mesophilic temperature range. The mixing of the organic matter in the reactor is achieved by recirculation of biogas under pressure at the bottom of the digester.
BIOCELL process	Netherlands	Under development	The BIOCELL process is a batch system developed to treat source-separated MSW (fruit, yard wastes, and vegetable wastes) and agricultural wastes. The digester used was circular in shape, 11.25 m in diameter and 4.5 m in height. The digester feedstock, at a total solids concentration of 30 percent, was obtained by mixing the incoming source-separated organic MSW with digested solids from previous digestion run.

[a] Abstracted from Ref. 2.

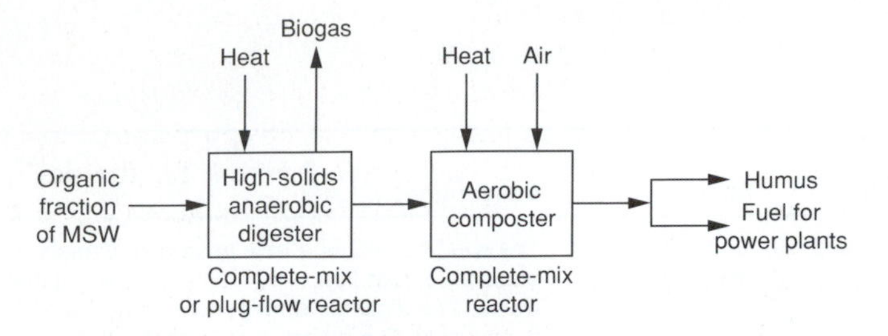

FIGURE 14-10
Flow diagram for high-solids anaerobic digestion/aerobic composting process.

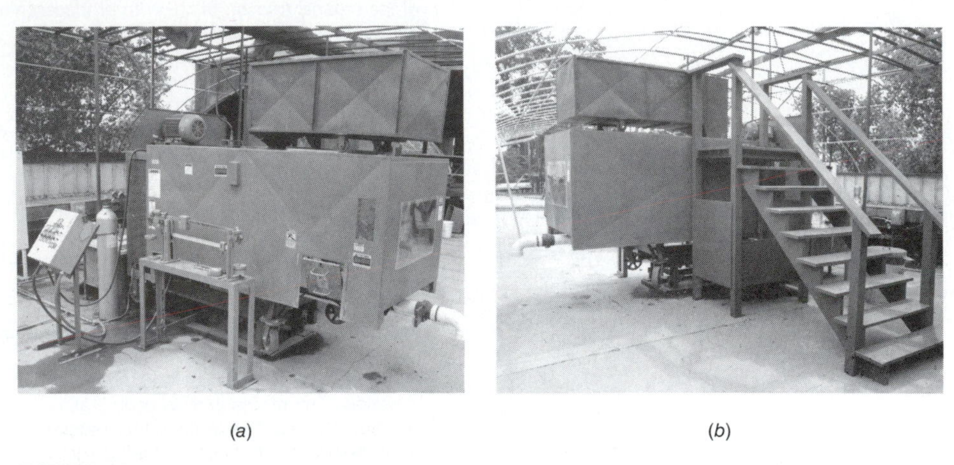

(*a*) (*b*)

FIGURE 14-11
Views of pilot scale high-solids anaerobic digester at UC, Davis: (*a*) front view (note that the digester is set on platform scales so that the weight lost each day in the form of biogas can be monitored) and (*b*) rear view showing platform leading to the reactor feed port located at the top of the unit.

FIGURE 14-12
Compost product produced using high-solids anaerobic digestion/aerobic composting process.

Process Applications. The combined high-solids anaerobic digestion/aerobic composting process is in the early stages of development [11, 12]. Design considerations for the first stage of this process are the same as for the high-solids anaerobic digestion process (see Table 14-12). The two major design parameters for the second stage are the aeration and thermal energy requirement for the destruction of pathogens. More detailed design and operational parameters can be found in Refs. 12 and 13.

As shown in Fig. 14-13*a*, this process can be used to process the combined organic fraction of MSW and wastewater treatment plant sludge. Because the high-solids anaerobic digestion/aerobic compostion process can be used to process both the organic fraction of MSW along with wastewater treatment plant sludge, the use of costly sludge dewatering facilities and the need to treat the liquid resulting from the dewatering of the sludge can be eliminated. Biogas produced from this process can be used for methanol production shown in Fig. 14-13*b*. The heat required for the anaerobic and aerobic reactors will be recovered from the thermal energy of the fluidized bed combustion reactor.

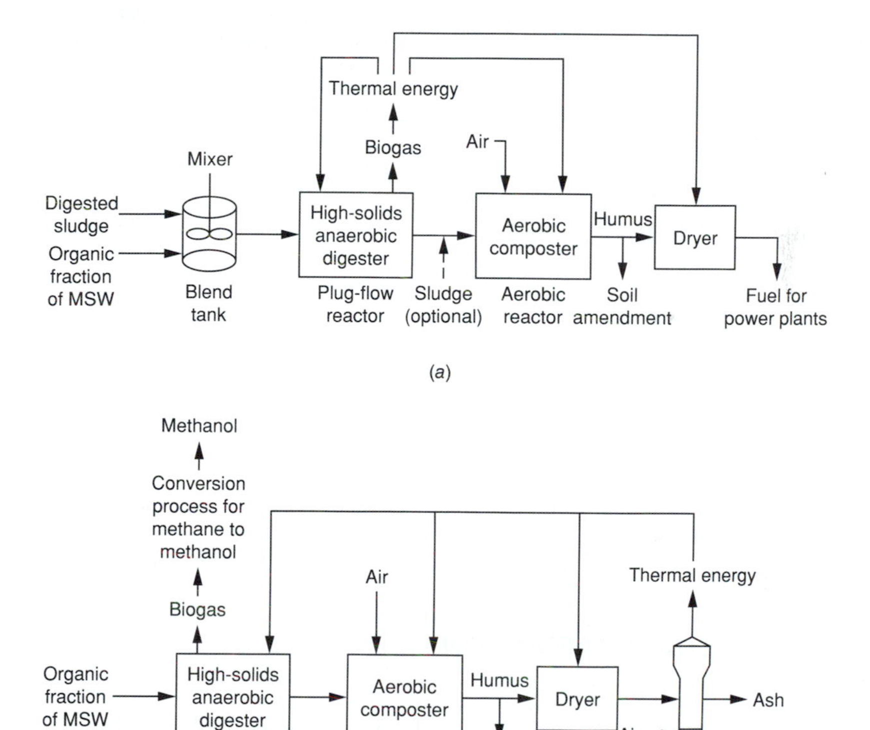

FIGURE 14-13
Flow diagrams for alternative applications of the high-solids anaerobic digestion/aerobic composting process: (*a*) codisposal of wastewater treatment plant sludge and (*b*) production of methanol.

TABLE 14-15
Biological processes for the recovery of conversion products from the organic fraction of MSW

Process	Conversion product	Preprocessing
Aerobic conversion	Compost (soil conditioner)	Separation of organic fraction, particle size reduction
Anaerobic digestion (in landfill)	Methane and carbon dioxide	None, other than placement in containment cells
Anaerobic digestion (low-solids, 4 to 8 percent solids)	Methane and carbon dioxide, digested solids	Separation of organic fraction, particle size reduction
Anaerobic digestion (high-solids, 22 to 35 percent solids)	Methane and carbon dioxide, digested solids	Separation of organic fraction, particle size reduction
Enzymatic hydrolysis	Glucose from cellulose	Separation of cellulose-containing materials
Fermentation (following acid or enzymatic hydrolysis)	Ethanol, single-cell protein	Separation of organic fraction, particle size reduction, acid or enzymatic hydrolysis to produce glucose

14-6 OTHER BIOLOGICAL TRANSFORMATION PROCESSES

The principal biological processes used for the transformation of the organic fraction of MSW are summarized in Table 14-15. Apart from the aerobic composting and the anaerobic digestion processes, enzymatic hydrolysis and fermentation following acid or enzymatic hydrolysis are the biological processes that have received most attention. A new process for the production of ethanol or other products is shown in Fig. 14-14. The novel features of the flow diagram shown in Fig. 14-14 are the recovery and recycling of enzymes and the use of new types of reactors (fermentors). With the national focus on solid waste management and increasing landfill disposal costs, it is anticipated that a number of new processes will become available in the future.

14-7 CHEMICAL TRANSFORMATION PROCESSES

Chemical transformation processes (see Table 14-16) include a number of hydrolysis processes, which are used to recover compounds such as glucose and furfural, and a variety of other chemical conversion processes used to recover compounds such as synthetic oil, gas, and cellulose acetate. Methanol, an alternative liquid fuel, can also be produced. These chemical processes are not used routinely for the transformation of the organic fraction of MSW, because these compounds can

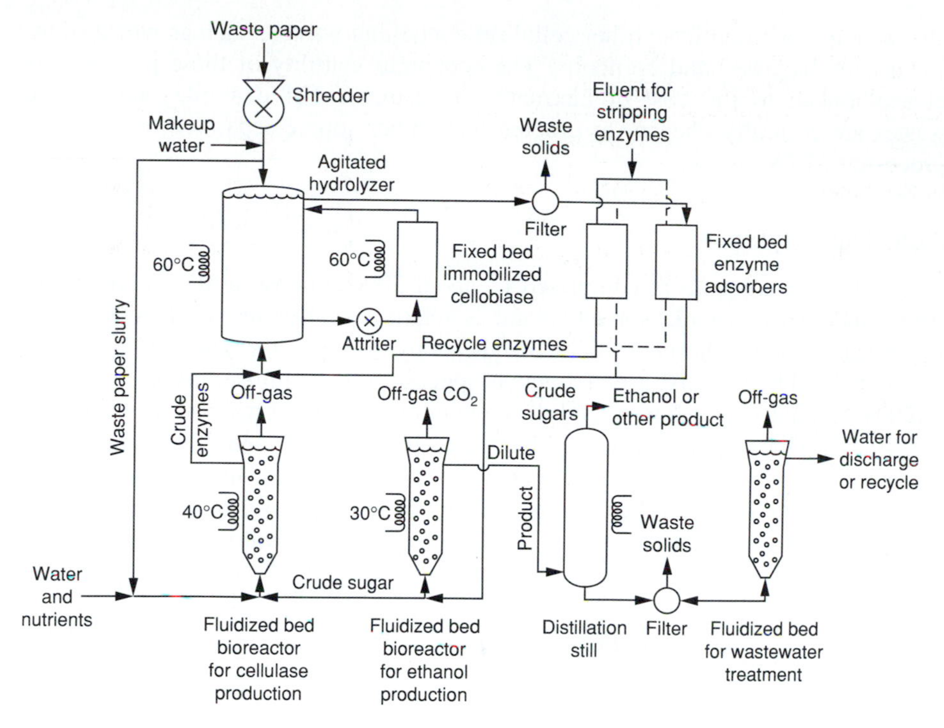

FIGURE 14-14
Major processing steps for the biological production of chemicals from waste paper using advanced bioreactor systems. (Personal information from Dr. Charles D. Scott, Oak Ridge National Laboratories.)

TABLE 14-16
Chemical processes for the recovery of conversion products from solid wastes

Process	Conversion product	Preprocessing
Acid hydrolysis	Organic acids	Separation of organic fraction, particle size reduction
Alkaline hydrolysis	Organic acids	Separation of organic fraction, particle size reduction
Various chemical conversion processes	Oil, gas, cellulose acetate	Separation of organic fraction, particle size reduction

also be manufactured from other cellulose-containing wastes, such as wheat straw, sugar cane bagasse, and corncobs. The economic viability of these processes is closely linked to the cost of alternative feedstocks. For example, agricultural wastes are currently cheaper to procure than either source-separated or machine-processed MSW.

Acid Hydrolysis

The cellulose molecule is comprised of about 3000 glucose units, is soluble in water and many organic solvents, and is relatively immune to attack by most microorganisms. If the cellulose molecule is hydrolyzed, the glucose can be recovered. Acid hydrolysis, used to recover glucose from cellulose, involves treating a finely divided suspension of cellulose-containing waste (e.g., newsprint) with a weak acid. The suspension is then heated to between 180 and 230°C and slight pressure is applied [4]. Under these conditions, the cellulose in the waste is converted into glucose and other sugars. The amount of glucose recovered depends on the characteristics of the waste. It is estimated that upwards of 80 percent of the weight of kraft paper may be recovered as sugar [4].

$$\underset{\text{cellulose}}{(C_6H_{10}O_5)n} + H_2O \xrightarrow{\text{acid}} \underset{\text{glucose}}{nC_6H_{12}O_6} \tag{14-13}$$

Lignin is not affected by the process. The sugar and glucose extracted from the cellulose can be converted by other chemical and biological processes into alcohols and other industrial chemicals.

Methanol Production from Methane

The methane produced by the anaerobic digestion of the organic portion of MSW can be converted to methanol, a liquid fuel. The conversion process involves the following two reactions, which are carried out in series.

$$CH_4 + H_2O \xrightarrow{\text{catalyst}} CO + 3H_2 \tag{14-14}$$

$$CO + 2H_2 \xrightarrow{\text{catalyst}} CH_3OH \tag{14-15}$$

In the first reaction, Eq. (14-14), which is endothermic, biogas, containing methane, is reacted with steam in a catalyst filled reactor to form carbon monoxide and hydrogen gas. In the second reaction, Eq. (14-15), which is exothermic, the products of the first reaction are converted catalytically to form methanol. The principal advantage of producing methanol from biogas that contains methane is that the resulting fuel is both storable and transportable. Methanol is currently manufactured from natural gas at a lower cost than it could be from biogas produced from the anaerobic digestion of MSW. Because the cost of fossil fuels is highly sensitive to political trends, the relative economics of methanol production from biogas could change in the future.

14-8 ENERGY PRODUCTION FROM BIOLOGICAL CONVERSION PRODUCTS

Once conversion products have been derived from solid wastes by either anaerobic digestion (methane), or chemical transformation (methanol), the next step involves their uses and/or storage. If energy is to be produced from these products, an additional conversion step is required. Biogas can be used directly with internal combustion (IC) engines and gas turbines to generate electricity (see Fig. 11-29).

Where IC engines are used, they should be started with propane or natural gas and operated until the engine is at its operating tempeature, at which point they can be switched to biogas operation. Before an engine is shut down, it again should be operated with propane or natural gas for about 20 minutes or so. By starting and stopping the engine with propane or natural gas, corrosion problems associated with hydrogen sulfide can be avoided. Biogas from landfills is being used to fuel IC engines in a wide range of power outputs from 50 kW to 5 MW. With gas turbines, the biogas is compressed under high pressure so that it can be used more effectively in the turbine. Such systems are in widespread use fueled with biogas recovered from landfills. Gas turbine systems are generally used in the 1 to 5 MW power range.

In large installations, the most common flow diagram for the production of electric energy involves the use of a steam turbine-generator combination (see Fig. 9-35). Steam can be produced in a boiler fired with either biogas (methane) or MSW derived liquid fuel (methanol). A number of steam turbine-generator installations are in operation thoughout the United States, fueled with biogas recovered from landfills. The largest such installation in the United States generates 50 MW of electricity at the Puente Hills landfill near Whittier, California.

14-9 DISCUSSION TOPICS AND PROBLEMS

14-1. Derive Eq. (14-3), assuming that the nitrogen in an organic compound goes to ammonia in the carbonaceous oxidation step.

14-2. Estimate the total theoretical amount of gas that could be produced under anaerobic conditions in a sanitary landfill per unit weight of solid wastes. Assume that the chemical formula for the biodegradable organic constituents is $C_{60}H_{100}O_{40}N$.

14-3. Determine the C/N ratio for the compound given in Problem 14-2.

14-4. Using the data given in Table 4-7 in Chapter 4 on the biodegradable fraction of several organic wastes, determine the C/N ratio for newspaper, mixed paper, and yard wastes, based on total and biodegadable carbon, assuming that all of the nitrogen is available.

14-5. Workers in New Zealand suggested in the 1950s that the following formula can be used to estimate the amount of carbon in waste materials to be composted [6]. Using the waste composition given in Table 3-4 in Chapter 3, determine whether you believe that the formula is acceptable for use as an approximate measure of the percentage carbon. If not, can you suggest some modifications to the equation?

$$\text{C, \%} = \frac{(100 - \%\ \text{ash})}{1.8}$$

14-6. Review the current issues of *BioCycle* and similar magazines with respect to the composting of yard wastes or of the organic fraction of MSW. Prepare a table of design parameters as reported in the literature. How do the reported values compare to the values given in Section 14-2? If there are differences, can you explain the differences? In your review also identify the sources of problems that have been encountered in the various facilities. Cite at least five references. Would you have known about the problem areas in composting from reading about composting in this and Chapter 9?

14-7. The importance of the C/N ratio in the aerobic composting process is well known. What is the effect of the C/N ratio on the anaerobic digestion process? Conduct a brief literature search and write a short review on your findings. Cite at least three references.

14-8. Identify at least 10 design parameters involved in the recovery of energy from the organic fraction of MSW, using the combined method of the high-solids anaerobic digestion and aerobic composting processes. Arrange them in order of their importance. State your reasons for the order you have selected.

14-9. A large city has developed a plan to use the combined high-solids anaerobic digestion/aerobic composting process for the management of the organic fraction of the MSW. The daily quantity of waste is 100 ton/d, based on a five-day week. The composition and results of the ultimate analysis for the organic fraction of the MSW from the city are given in the accompanying table. Using the data contained in the table, determine or estimate

(*a*) The ash-free empirical formula of the organic fraction.

(*b*) The theoretical biogas composition and total volume of methane gas.

(*c*) The total volume and mass reduction.

		Ultimate analysis					
Organic waste	**% wet basis**	**C**	**H**	**O**	**N**	**S**	**Residue**
Office paper	11	43.41	5.82	44.32	0.25	0.20	6.00
Newspaper	6	49.14	6.10	43.03	0.05	0.16	1.52
Cardboard	12	44.90	6.08	47.84	0.01	0.11	1.07
Food wastes	11	44.99	6.43	28.76	3.30	0.52	16.00
Yard wastes	20	43.33	6.04	41.68	2.15	0.05	6.75

14-10. Using the data given in Problem 14-9, determine the following:

(*a*) If the city plans to recycle all of the high-quality paper, what will be the effect on C/N ratio and total gas production?

(*b*) Assume that 5, 10, 20, 30, 40, 50, 60, 70, 80, 90 percent of high-quality paper is recycled. Calculate the C/N ratio and total biogas production using a spreadsheet.

14-11. Assume that an in-vessel aerobic composting process is to be used to transform the organic fraction of the MSW given in Problem 14-9. Determine the following items for the in-vessel composting process.

(*a*) Is it possible to achieve a 60 percent volume reduction?
(*b*) What is the total air requirement?
(*c*) What is the total land requirement?

14-12. Biogas naturally is not pure and contains small quantities of corrosive gases, which must be cleaned prior to use in internal combustion engines.
(*a*) Identify at least five available gas purification technologies and prepare a table to compare the various technologies.
(*b*) Select the most suitable techniques for purification of biogas and estimate the cost per ft^3 of biogas.
(*c*) What is the relationship between gas impurity and thermal energy?

14-10 REFERENCES

1. Cecchi, F., P. G. Traverso, J. Mata-Alvarez, J. Clancy, and C. Zaror: "State of the Art of Research and Development in the Anaerobic Digestion Process of Municipal Solid Waste in Europe," *Biomass*, Vol. 16, No. 4, 1988.
2. Cecchi, F., J. Mata-Alvarez, and F. G. Pohland (eds.): *Proceedings of the International Symposium on Anaerobic Digestion of Solid Waste,* Venice, Italy, April 14–17, 1992.
3. Chynoweth, D. P., F. K. Earl, G. Bosch, and R. Lagrand: "Biogasification of Processed MSW," *BioCycle*, Vol. 31, No. 10, 1990.
4. Golueke, C. G.: *Biological Reclamation of Solid Wastes,* Rodale Press, Emmaus, PA, 1977.
5. Golueke, C. G.: "When is Compost 'Safe?'" In *The Biocycle Guide to Composting Municipal Wastes*, The J. G. Press, Inc., Emmaus, PA, 1989.
6. Gotaas, H. B.: *Composting—Sanitary Disposal and Reclamation of Solid Wastes*, World Health Organization, Geneva, 1956.
7. Haug, R. T.: *Compost Engineering: Principles and Practices*, Ann Arbor Science Publishers Inc., Ann Arbor, MI, 1980.
8. Higgins, I. J., and R. G. Burns: *The Chemistry and Microbiology of Pollution*, Academic, London, 1975.
9. Holland, K. T., J. S. Knapp, and J. G. Shoesmith: *Anaerobic Bacteria*, Chapman and Hall, New York, 1987.
10. Hoover, S. R., and N. Porges: "Assimilation of Dairy Wastes by Activated Sludge, II: The Equation of Synthesis and Oxygen Utilization," *Sewage and Industrial Wastes*, Vol. 24, 1952.
11. Jewell, W. J.: "Future Trend in Digester Design," in D. A. Stafford, B. I. Wheatly, and D. E. Hughes, eds., *Proceedings of the First International Symposium on Anaerobic Digestion*, Cardiff, Wales, Applied Science Publishers Ltd, London, pp. 17–21, 1979.
12. Kayhanian, M., K. Lindenauer, S. Hardy, and G. Tchobanoglous: "Two-Stage Process Combines Anaerobic and Aerobic Method," *BioCycle*, Vol. 32, No. 3, pp. 48–52, 1991.
13. Kayhanian, M. and G. Tchobanoglous: "Pilot Investigation of an Innovative Two-Stage Anaerobic Digester and Aerobic Composting Process for the Recovery of Energy and Compost from the Organic Fraction of MSW," in F. Cecchi, J. Mata-Alvarez, and F. G. Pohland (eds): *Proceedings of the International Symposium on Anaerobic Digestion of Solid Waste,* Venice, Italy, April 14–17, 1992.
14. Kayhanian, M., and G. Tchobanoglous: "Computation of C/N Ratios for Various Organic Fractions," *BioCycle*, Vol. 33, No. 5, 1992.
15. McCarty, P. L.: "Anaerobic Waste Treatment Fundamentals," *Public Works*, Vol. 95, Nos. 8–12, 1964.
16. Ricci, L. J.: "Garbage Routes to Methane," *Chemical Engineering*, Vol. 81, No. 10, 1974.
17. Rich, L. G.: *Unit Processes of Environmental Engineering*, Wiley, New York, 1963.

18. Speece, R. E., "Anaerobic Biotechnology for Industrial Wastewater Treatment," *Environmental Science and Technology*, Vol. 17, No. 9, 1983.
19. Stanier, R. Y., J. L. Ingraham, M. L. Wheelis, and P. R. Painter: *The Microbial World*, 5th ed., Prentice Hall, Englewood Cliffs, NJ, 1986.
20. Tchobanoglous, G., and F. L. Burton: *Wastewater Engineering: Treatment, Disposal, and Reuse*, 3rd ed., Metcalf & Eddy, Inc., McGraw-Hill, Inc., New York, 1991.
21. Wilson, G. B.: "Combining Raw Materials for Composting," *The Biocycle Guide to Yard Waste Composting*, J. G. Press Inc., Emmaus, PA,1989.
22. Wujcik, J. W., and W. J. Jewell: "Dry Anaerobic Fermentation," *Biotechnology and Bioengineering Symposium No. 10*, Wiley, New York, 1980.

CHAPTER 15

RECYCLING OF MATERIALS FOUND IN MUNICIPAL SOLID WASTE

Recycling of postconsumer materials found in MSW involves (1) the recovery of materials from the waste stream, (2) intermediate processing such as sorting and compaction, (3) transportation, and (4) final processing, to provide a raw material for manufacturers or an end product. The primary benefits of recycling are conservation of natural resources and landfill space; however, the collection and transport of materials requires substantial amounts of energy and labor, and historically, most recycling programs are subsidized economically. The requirements for a successful program are that a strong demand exist for recovered materials and that the market value of the materials be sufficient to pay for collection and transportation costs. The purpose of this chapter is to introduce the reader to the key issues involved in the recovery and processing of waste materials and to discuss both the individual materials that are now recovered from MSW for recycling and the remaining materials that should be recovered. Emphasis is given to the reuse and recycling opportunities for waste materials recovered from MSW and to the applicable specifications for these materials. The actual separation and recovery of waste materials was considered in Chapters 9 and 12.

15-1 KEY ISSUES IN MATERIALS RECYCLING

Fundamental issues in materials recycling include identification of (1) the materials that are to be diverted from the waste stream, (2) reuse and recycling opportunities, and (3) specifications of buyers of recovered materials.

Identification of Materials to Be Diverted

Solid waste managers attempt to maximize landfill life and minimize operating costs, often within a framework of legislation that requires a certain percentage of the solid waste collected be diverted from landfills, or that imposes an outright ban on the landfill disposal of certain materials, such as yard wastes. Managers must decide what materials should be pulled from the waste stream to meet diversion goals, and the decision is complicated by the fact that many materials (e.g., glass) have weak markets or cannot be transported economically. Another problem is that materials with a high market value (e.g., aluminum) are often recovered by consumers and comprise only a small amount of the material that enters the waste management system, thus reducing the potential for income.

Identification of Reuse and Recycling Opportunities

Officials charged with developing a recycling program must consider the markets for recovered materials, the collection infrastructure, and the overall cost. Markets for recovered materials exist only when manufacturers or processors need those materials or can use them as economical substitutes for raw materials; therefore, the market depends on the quality of materials, overall industry capacity, and the cost of competing raw materials. In most cases, recovered materials are inferior in quality to virgin materials, so the market price must be attractive to buyers. Markets are also created by legislation that develops a long-term demand and by advances in technology.

Markets for Plastics: An Example. The following discussion of markets for plastics illustrates some of the problems encountered in marketing recovered materials, and why recovered materials are rarely as competitive as virgin materials.

Low value of recovered plastics. Scrap plastic has a low value because virgin materials are relatively inexpensive. There is little financial incentive for collection, and thus recycling must be legislated. Approximately 90 percent of the two-liter beverage bottles recovered by processors come from the 10 states with bottle deposit laws [8].

Lack of infrastructure. The collection and processing infrastructure for plastics is not nationwide (as it is with aluminum), but is generally limited to local areas. As a result, many consumers who wish to recycle find there is no outlet. Another consequence is that a dependable and consistent source of recovered material is not available to processors and manufacturers.

Low specific weight. The volume-to-weight ratio for plastics is very high, especially for polystyrene (PS) foam products. Isolated communities cannot afford to collect and ship plastics, and nobody is willing to come and pick them up. Attempts at on-vehicle compaction have been unsuccessful to date, and granulation is not acceptable until all plastics are separated. The relatively low weight also forces communities with landfill diversion programs to shift emphasis to other materials.

Potential contamination. Plastic bottles brought to processors are often contaminated by foreign material or undesired plastics. Foreign materials such as food and product residues cause premature wear on granulators and other equipment; noncompatible plastics degrade the quality of the "regrind" produced and must be removed.

Collection Infrastructure. The recovery of aluminum beverage containers is unique in that a nationwide network of regional transportation and processing centers has been established. Ideally, the development of a collection infrastructure should follow market demand, that is, the value of the recovered material should be sufficient to support the cost of collection, processing, and transportation. Processors of recovered materials usually establish processing plants in highly populated areas with large quantities of recoverable materials. Recyclers must pay the cost of transportation to these centralized facilities. The cost of collection and transportation to buyers, compared with the price paid for recovered materials, is usually the reason that small communities have not been able to maintain recycling programs without subsidies.

Subsidies for Recycling Programs. Solid waste managers often have limited control over program economics. Although landfill diversion legislation has been passed in many states and recycling programs are becoming more common, few municipal programs are self-supporting; generally they are subsidized by taxpayers or subscribers to the local waste hauling service. The type of collection system (e.g., curbside, buy-back center), length of collection routes, terrain, degree of sorting required, and transportation system all influence the program costs. In rural areas and in the western United States, the distance to markets makes the transportation cost prohibitive for many materials. Successful programs usually exist only for those materials that are in high demand, such as aluminum cans or two-liter plastic beverage bottles. As a rule, the market for recovered materials is a buyer's market, and as the number of collection programs becomes greater and the supply of recovered material increases, the price offered for some materials will decline.

Meeting Specifications for Recovered Materials

Processors and end users of recovered materials require that the materials be homogeneous and free of contamination that will cause product defects or damage to machinery; many buyers also require that baled material be compacted to specified sizes and specific weights. Some industries adhere to strict standards and cannot tolerate even very low levels of contamination (e.g., glass container manufacturers); others process materials sufficiently to remove almost all foreign materials (e.g., aluminum and tin can buyers). In general, there is less contamination in source-separated material, but collection is more labor intensive, and many communities are choosing to sort all materials at a central materials recovery facility (MRF). In many regions, markets for materials are not keeping pace with the volume collected, and it is expected that buyers will tighten specifications; consequently, vendors will no longer have assured markets, and will be competing to sell materials. As the specifications for recovered materials become more restrictive, recovery program managers must consider buyer specifications carefully when choosing collection and sorting systems, especially where large capital expenditures are involved.

15-2 ALUMINUM CANS

In 1990, approximately 85 billion aluminum beverage containers were produced in the United States and more than 53.8 billion were returned, for a recycling rate of 63.6 percent [26]. Nationwide, aluminum cans constitute less than 1 percent of municipal solid waste; in communities having established recycling programs or container deposit laws, the percentage in the local waste stream has become negligible. The annual recycling rate and weight of material recovered is shown in Fig. 15-1. The increased rate in the latter half of the 1980s is attributable to additional collection programs and container deposit legislation.

Why has aluminum recycling been so successful compared with other common postconsumer waste materials such as newspaper, glass, and plastics? The

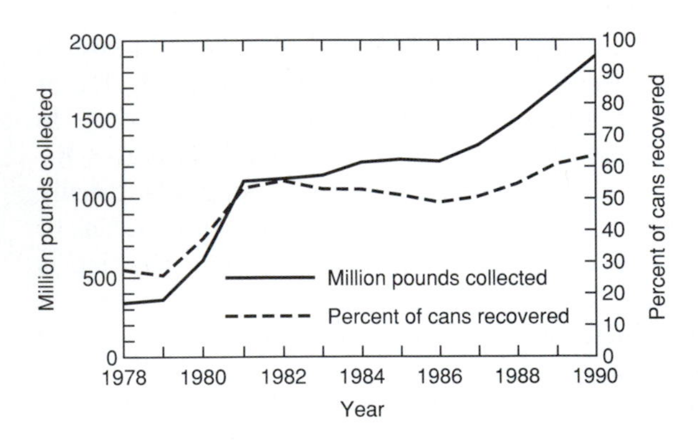

FIGURE 15-1
Aluminum can recycling, 1978 to 1990. Projected recovery for the year 2000 is 74 percent.

only other commodity with a recycling rate close to aluminum is corrugated containers (45 percent), primarily due to active industrial and commercial programs. The reason is that postconsumer newspaper, glass, and plastic must compete against the raw materials used for their manufacture, and these virgin materials are abundant and relatively cheap. By comparison, aluminum ore must be imported. Another reason is that the aluminum industry recognized the advantages of a domestic aluminum supply and established the necessary infrastructure for transportation and processing. A comparable infrastructure does not yet exist for other recyclable materials.

Aluminum producers and manufacturers such as Reynolds and Alcoa have actively promoted recycling since the mid-1960s. While other container manufacturers have resisted recycling schemes and mandatory container deposit legislation, the aluminum industry has developed collection and processing centers, a transportation network, and reclamation plants. Recycling makes economic sense to producers for several reasons: (*a*) Recycling provides a stable, domestic source of aluminum (approximately one-third of the industry's requirement). In contrast, most of the bauxite required to produce new aluminum must be imported (major producers are Jamaica, Australia, Surinam, Guyana, and Guinea), and four pounds of bauxite are required to produce each pound of new metal. (*b*) The energy required to produce a can from recycled aluminum is less than 5 percent of the energy needed to make a can from raw materials. (*c*) Recycled cans are of uniform and known composition, and the impurities are readily removed. (*d*) Recycling allows aluminum can manufacturers to compete favorably with glass and bi-metal container manufacturers. Virtually all metal beer containers and 93 percent of metal soft drink cans are aluminum [25].

Favorable economics have also changed public attitudes. As bottle bills are passed and recycling becomes institutionalized, people are beginning to regard aluminum as an inherently valuable material, rather than just a waste material that happens to be reusable.

Reuse and Recycling Opportunities

Aluminum cans are accepted in curbside pickup programs, at buy-back locations, at recycling collection centers, and by scrap metal dealers. A number of states have mandatory deposits for beverage containers and have established redemption centers at supermarkets. Scrap metal dealers also buy wrought and cast aluminum items such as lawn furniture, tubing, screen and storm doors, door sills, window frames, siding, gutters and downspouts, power tools, and pots and pans. Aluminum manufacturers complete the loop through contracts with independent recyclers, scrap dealers, aluminum fabricators, and auto dismantlers. Noncontainer scrap and aluminum alloys are not reclaimed with cans because aluminum cans are a particular alloy.

Cans brought to collection centers are crushed, baled (see Fig. 15-2), and shipped to regional mills or reclamation plants, where the cans are shredded to reduce volume. At the reclamation plant, the shredded cans are first heated in

FIGURE 15-2
Aluminum cans (left) and mixed plastics (right) that have been baled to reduce storage requirements and shipping costs.

a de-lacquering process to remove coatings and moisture and are then charged into a remelting furnace. Molten metal is formed into ingots of 30,000 pounds or more that are transferred to another mill and rolled into sheets. The sheets are sent to container manufacturing plants and cut into discs, from which the cans are formed. The cans are printed with the beverage maker's logo and are shipped (with tops separate) to the filling plant.

Specifications for Recovered Aluminum Cans

Collection centers and other buyers accept all cans that are free of gross contamination, such as dirt and food wastes. The buyers then compact and bale the material according to mill specifications regarding dimensions, weight, and number of bands; mills issue "report cards" to their suppliers, advising them of

TABLE 15-1
Specifications for purchased aluminum cans, foil, and foil products[a]

Source	Specifications
Beverage cans	Baled 3 ft × 4 ft × 5 ft, with a specific weight of 12 to 17 lb/ft^3 for unflattened containers, 12 to 20 lb/ft^3 for flattened containers, bales to be banded with four to seven 5/8 in by 0.020 in steel or aluminum bands, or 6 to 15 #13 gauge steel or aluminum wire. Bales must not contain excessive water or contamination, steel cans, or aluminum foil.
Foil and foil products	Some major producers of aluminum will accept clean foil products, but they are processed separately from beverage containers.

[a] Adapted from Ref. 5.

deficiencies. Most community recycling centers do not accept used aluminum foil because it is usually contaminated, but large buyers (e.g., Reynolds) accept foil if it is reasonably clean. Noncontainer aluminum products purchased by scrap dealers must simply be dry and free of contamination; the dealers consolidate and bale the material for shipment to end users. Typical specifications for aluminum beverage cans and foil products are presented in Table 15-1.

15-3 PAPER AND CARDBOARD

On a weight basis, paper constitutes the largest component of municipal solid waste. Including corrugated containers and boxboard, paper typically represents 25 to 40 percent of the total (see Table 3-4). Because this percentage is so large, one might expect that increased paper recycling represents a relatively easy opportunity to divert material from landfills, reuse available fibers, reduce impacts on forests, and reduce energy consumption. Unfortunately, only a portion of discarded paper can be reused, owing to both economic and logistical considerations: (1) virgin fiber is abundant and relatively cheap in the Northeast and Pacific Northwest, where there are extensive U.S. and Canadian forests; (2) many urban centers in the interior of the United States are located long distances from paper mills; and (3) mill capacity to de-ink and reuse postconsumer paper is limited.

Reuse and Recycling Opportunities

Paper mills have always recycled damaged product and scrap from converters (paper products manufacturing plants) because the material is of known composition, is usually unprinted, and often can be used as a direct pulp substitute. Paper mills purchase additional postconsumer waste paper based on fiber strength, fiber yield, and brightness, according to the type of product produced. In 1989, approximately 25 percent of the U.S. pulp supply came from recycled sources, including waste pulp, paper from converters, and postconsumer waste [9]. In the near future, the capacity for recycled and by-product materials is expected to grow twice as fast as that for virgin fiber [9].

Types of Paper Now Recycled. The principal types of paper now recycled are newspaper, corrugated cardboard, high-grade paper, and mixed paper (see Fig. 15-3*a*). Each of these types of paper is considered in the following discussion.

Newspaper. The Paper Stock Institute divides newspaper into four grades; the de-inking grade is used for newsprint, tissue, and higher-quality paper, whereas the remaining grades are mostly used to produce containerboard and construction products. In 1988, approximately 33 percent of the newspapers published were recycled [25]; of that amount, approximately 77 percent was used for recycled newsprint or used in bulk grades for conversion to containerboard and corrugated containers [10]. The remaining 23 percent was used for other grades of paper, cellulose insulation, and animal bedding [25].

(a)

(b)

FIGURE 15-3
Paper recovered for recycling: (*a*) mixed paper from residential sources and (*b*) baled cardboard material at a large supermarket.

Corrugated cardboard. Corrugated cardboard is the largest single source of waste paper for recycling; in 1988, approximately 9.7 million tons of old corrugated containers (OCC) were collected, representing a recycling rate of 45 percent [25]. Markets for good-quality baled cardboard have historically been steady, and many commercial generators, such as supermarkets and retail stores, handle enough containers to justify in-house balers (see Fig. 15-3*b*). Recycled corrugated containers are used primarily to make liner or medium for new containers (liner refers to the outside "skin" layers; medium is the internal rolled layer).

High-grade paper. High-grade postconsumer papers include computer paper, white and colored ledger paper (writing, typing, and other bond papers), guillotined books (i.e., covers and spine cut off), and reproduction paper. The market for this material has historically remained steady, as good quality paper (i.e., nontreated, noncoated paper containing a high percentage of long fibers) can be used as a direct substitute for wood pulp or can be de-inked to produce tissues or high-quality bond papers.

Mixed paper. According to standards of the Paper Stock Institute, mixed paper is not limited as to coatings or fiber content, but outthrows (such as carbon paper) are limited to 10 percent. In practice, the grading reflects market demand, and with the present oversupply of both mixed paper and newsprint, mixed paper may consist mostly of newspaper, magazines, and mixed long-fiber papers. Mixed paper is usually used to produce containerboard and miscellaneous pressed products. A higher grade, *super-mixed*, is limited to less than 10 percent groundwood, and is often used as a de-inking grade.

Major Uses of Recycled Paper. The four major grades of paper discussed above are often combined into three categories, depending on how they are processed or the type of end product.

Pulp substitutes. These are recycled papers that can be added directly to a paper pulper without treatment. In general, mills prefer clean industrial scrap from converters; the most commonly used postconsumer waste is non-groundwood computer printout (CPO).

De-ink grades. These are recycled papers that are pulped, chemically de-inked, washed, and bleached before introduction into the main pulp slurry. Typical grades are de-inking newsprint and higher-quality papers not suitable as direct pulp substitutes, such as colored ledger and printed white ledger. Most de-inking paper is used to produce newsprint, tissue, napkins, paper towels, and high-quality boxboard.

Bulk grades. These are recycled papers that are used without de-inking to produce containerboard, liner and medium for corrugated containers, egg cartons and pressboard, and building products such as felt paper and wallboard. Bulk grades include newspaper, OCC, and mixed paper. These papers make up the majority of the waste stream and theoretically can be diverted from landfills, but actual market demand and recycling potential is limited by mill capacity.

Other Uses for Recycled Paper. In addition to the uses cited above, paper collected for recycling can also be used to produce building products or refuse-derived fuel, or be exported.

Building products. Both newspaper and mixed paper are used to make gypsum wallboard, loose-fill and spray-on insulation, and saturated felt roofing paper. The manufacture of cellulosic insulation provides good potential for additional use of old newspaper; additional markets are important because the supply of newsprint is expected to increase as a result of mandated diversion programs.

Refuse-derived fuel (RDF). RDF has been produced from municipal solid waste for many years, and several firms now produce limited amounts of mixed-paper RDF in pellet form (see Chapter 12). The potential markets are existing biomass-fueled plants and other industrial users, depending on the proximity of pellet plants and transportation costs [7].

Exports. The United States is the world's largest exporter of waste paper; in 1989, 6.3 million tons were shipped abroad, an amount equal to 23 percent of the total waste paper, converter scrap, and waste pulp recovered [9]. Most export paper goes to Mexico, Japan, South Korea, and Taiwan, although Indonesia, Thailand, Hong Kong, and China are becoming larger consumers (see Fig. 12-25). Waste paper from the United States is desirable because a high proportion is made with long-fiber pulp.

Markets for Recycled Paper. Paper manufacturers acquire postconsumer waste paper by direct purchase or through independent brokers; both obtain supplies from government offices, businesses and corporations, materials collection centers,

FIGURE 15-4
Baled paper being loaded onto trailer for shipment.

and used materials dealers. Paper buyers usually require delivery to their premises, although some will make pickups if quantities are sufficient. Large buyers may provide containers to high-volume customers and make pickups on a regular schedule. To ensure a steady supply of high-grade paper, buyers encourage long-term contracts (typically three-year), often with flexible terms to account for changing market conditions.

The market for waste paper is strongly affected by the general economy, because a large portion of low-grade paper is used to make building products and containers for consumer goods. Because there is presently more waste paper than mill capacity, the closure of even one mill due to unfavorable economics, environmental restrictions, or labor disputes may affect market prices over a large region. For example, a mill shutdown will typically cause a decrease in waste paper prices because the supply available to other mills is increased. Paper companies continue to invest in new plant and equipment to use postconsumer waste as a source of fiber, but the consensus is that it will be several years before capacity and the increased supply are balanced. The growing export market to countries of the Pacific Rim has moderated the excess supply to some extent. Although normal market forces affecting supply and demand are now being disrupted by legislated minimum-recycled content requirements and landfill waste diversion laws, these changes may promote a more stable supply in the long term.

Specifications for Recovered Paper and Cardboard

The Paper Stock Institute of America, which represents buyers and processors of waste paper, has established standards for approximately 50 grades of paper and has listed 33 more specialty grades where specifications are mutually agreed upon by buyer and seller [17]. Most collection centers or buyers use six to eight grades that encompass almost all of the postconsumer paper collected: news, de-ink news, corrugated containers, sorted white ledger (includes office paper and

TABLE 15-2
Specifications for recycled paper and cardboard[a]

Grade number	Class	Description	Prohibitive materials,[b] %	Total outthrows,[c] %
1	Mixed paper	Consists of a mixture of various qualities of paper not limited as to type of packing or fiber content	2	10
6	News	Consists of baled newspapers containing less than 5 percent of other papers	0.5	2.0
7	Special news	Consists of baled, sorted, fresh dry newspapers, not sunburned, free from paper other than news, containing not more than the normal percentage of rotogravure and colored sections	None permitted	2.0
11	Corrugated	Consists of baled corrugated containers. Containers having liners of test liner, jute, or kraft	1.0	5.0
38	Sorted colored ledger	Consists of printed or unprinted sheets, colored shavings, and cuttings of colored or ledger white sulfite or sulfate ledger, bond, writing and other papers that have a similar fiber and filler content. This grade must be free of treated, coated, padded, or heavily printed stock.	None permitted	2.0
40	Sorted white ledger	Consists of printed or unprinted sheets, guillotined books, quire waste, and cuttings of white sulfite or sulfate ledger, bond, writing and other papers that have a similar fiber and filler content. This grade must be free of treated, coated, padded, or heavily printed stock.	None permitted	2.0
42	Computer printout	Consists of white sulfite or sulfate papers in forms manufactured for use in data processing machines. This grade may contain colored stripes and/or impact or nonimpact (e.g., laser) computer printing, and may contain not more than 5% of groundwood in the packing. All stock must be untreated and uncoated.	None permitted	2.0

[a] Adapted from the specifications of the Paper Stock Institute.

[b] Materials that could damage processing equipment.

[c] Papers unsuitable for consumption at the specified grade.

uncoated bond paper used for typing and reproduction), sorted colored ledger, computer printout, used brown kraft (grocery bags), and mixed (or super-mix). Larger buyers include additional grades: magazines, coated book stock, and solid fiber containers (boxboard, containerboard, carrier board).

Paper shipped to the mill must meet mill specifications regarding percentages of outthrows and contaminants. *Outthrows* are grades of a lesser quality than the specified grade. Sometimes mills will accept outthrow percentages higher than industry standards if the desired grade is in short supply. *Contaminants* are materials that are detrimental to the papermaking process or that cause damage to machinery. Examples of paper contaminants are sunburned newspaper, food containers, composites containing plastic or metal foil, waxed or treated paper, tissues or paper towels, bound catalogs or telephone directories, blueprints, Post-it™ notes, and FAX or carbonless carbon paper. (The adhesive backing on Post-it™ notes and that used for bound catalogs can be carried from the pulp to the rollers used in the papermaking process, where it forms deposits that cause damage to the continuous sheet of paper being formed. FAX and carbonless papers are considered contaminants because they are treated and are not chemically compatible with most pulps.)

Other contaminants are foreign materials such as dirt, metal, glass, food wastes, paper clips, and string. Most buyers prefer that newsprint not be presented in paper bags, and some mills insist that laser-printed computer paper be separated from other computer printout because the inks cannot be removed completely from the fibers during de-inking. Most brokers maintain quality control over the final shipment by inspecting and manually sorting all paper before it is baled, but some buyers require that paper or corrugated containers be baled to facilitate handling and reduce volume.

Specifications for the most common grades of postconsumer waste paper are reported in Table 15-2. The reader should be aware that the allowable percentage for outthrows may be modified, depending on availability of the grade and mill requirements.

15-4 PLASTICS

In 1973, twenty-nine billion pounds of plastics were produced in the United States; by 1990, the amount had nearly doubled to 54 billion pounds and was growing at an annual rate of 6 percent [26]. Although plastics have been used by consumers for nearly 50 years, their use in packaging has increased dramatically over the last 20 years and is expected to increase another 70 percent by the year 2000 [12]. Because most packaging is disposable, plastics in municipal solid waste have increased from 3 percent in the early 1970s to 7 percent (by weight) in 1990 (see Table 3-4). The growth in use of plastics in consumer products has occurred because plastics have largely replaced metals and glass as a container material and paper as a packaging material. Plastics have several advantages: They are light, and thus reduce shipping costs. They are durable and often provide a safer container (e.g., shampoo bottles). They can be formed into a variety of shapes and can be formulated to be flexible or rigid. They are good insulators. And they are well suited to wet foods and microwave oven use. The classifications, identification codes, and uses for the most widely used plastics are presented in Table 15-3.

Although plastic materials comprise only 7 percent of MSW by weight (see Table 3-4), they comprise a somewhat larger percentage on a volume basis. As landfills close and finding new sites becomes more difficult, the plastics and packaging industries have come under fire for contributing to the solid waste problem without a responsible attempt at solution. It is often suggested that plastics should be replaced with paper or biodegradable products, despite evidence that neither plastics nor paper will degrade rapidly in a well-managed landfill. Most consumers enjoy the benefits of plastics and recognize that additional recycling is a reasonable solution; nevertheless, only 2 percent of virgin production is recycled in the United States, compared with about 10 percent in the Netherlands and 6 percent in Germany [8].

Reuse and Recycling Opportunities

Most plastic container manufacturers now code their products with a number from 1 to 7, representing the most commonly produced resins, to facilitate separation and recycling (see Table 15-3). The reuse options for each of these resin types

TABLE 15-3
Classifications, identification codes, and uses for common plastics

Material	SPI[a] code	Original uses	Percent of total used for packaging[b]
Polyethylene terephthalate	1-PETE	Carbonated soft drink bottles, food containers	7
High-density polyethylene	2-HDPE	Milk bottles, detergent bottles, film products such as produce bags, etc.	31
Vinyl/polyvinyl chloride	3-PVC	Household and food product containers; pipe	5
Low-density polyethylene	4-LDPE	Thin-film packaging and wraps; other film materials	33
Polypropylene	5-PP	Crates, cases, closures and labels	10
Polystyrene	6-PS	Foamed cups and plates; injection molded items	10
All other resins and multilayered materials	7-other	Commingled plastics	4

[a] Society of the Plastics Industry.

[b] Adapted from Ref. 30.

is disussed below. In addition, the processing procedure used for the two major types of plastics that are recycled most widely now is reviewed.

Types of Plastics Now Recycled. The principal types of plastics now recycled are polyethylene terephthalate (PETE/1) and high-density polyethylene (HDPE/2) (see Fig. 15-5). These and other types of plastics are considered in the following discussion.

(1) Polyethylene terephthalate (PETE). PETE is recycled primarily to polyester fibers used in the manufacture of sleeping bags, pillows, quilts, and cold-weather clothing (green bottles are processed separately because green fibers may be used only in garments with a dark-colored outer shell). Postconsumer PETE is also used for carpet backing and fibers, molded products, polyisocyanurate insulation board, films, strapping, food and nonfood containers, and engineering-grade plastics for the automotive industry.

In a departure from conventional recycling technology, two large resin producers are now chemically depolymerizing postconsumer bottles to ethylene glycol and terephthalic acid, which are repolymerized to virgin-quality resins for new soft drink bottles. The use of postconsumer material in Coca-Cola™ bottles was approved by the Food and Drug Administration in January of 1991.

(2) High-density polyethylene (HDPE). The properties of HDPE vary widely depending on the product to be manufactured. Milk jugs are made from resin having a low melt index (roughly a measure of the viscosity, which determines the suitability for different manufacturing processes), which allows the resin to stretch while being expanded during blow-molding. Rigid HDPE is made from resin having a high melt index, which allows the resin to flow easily into a precision mold form. In turn, the properties of HDPE "regrind" depend on the feedstock material (*regrind* is the term used to describe granulated, cleaned plastic).

FIGURE 15-5
Mixed PETE and HDPE plastic containers to be separated at a materials recovery facility.

To control quality when producing a regrind, processors do not mix different types of resins, or mixtures of the same resin with different melt indexes. Instead, flakes or pellets are produced from homogeneous resins and blended by either the processor or end-use manufacturer to produce the melt index required.

The most common consumer items produced from postconsumer HDPE are detergent bottles and motor oil containers. The bottles are usually made in three layers, with the center layer containing the recycled material (see Fig. 15-6). The inner layer of virgin resin provides a dependable barrier, and the outer layer provides uniform color and appearance. Recycled HDPE is also used for protective wrap, grocery sacks, pipe, and molded products such as toys and pails.

(3) Polyvinyl chloride (PVC). PVC is widely used for food packaging, electrical wire and cable insulation, and plastic pipe. Although postconsumer PVC is a high-quality resin that needs little or no compounding, very little PVC is now recycled, because the cost of collection and sorting is prohibitive. Typical recycled products include nonfood containers, shower curtains, truck bed liners, laboratory mats, floor tiles, garden hose, flower pots, and toys. There is a huge potential

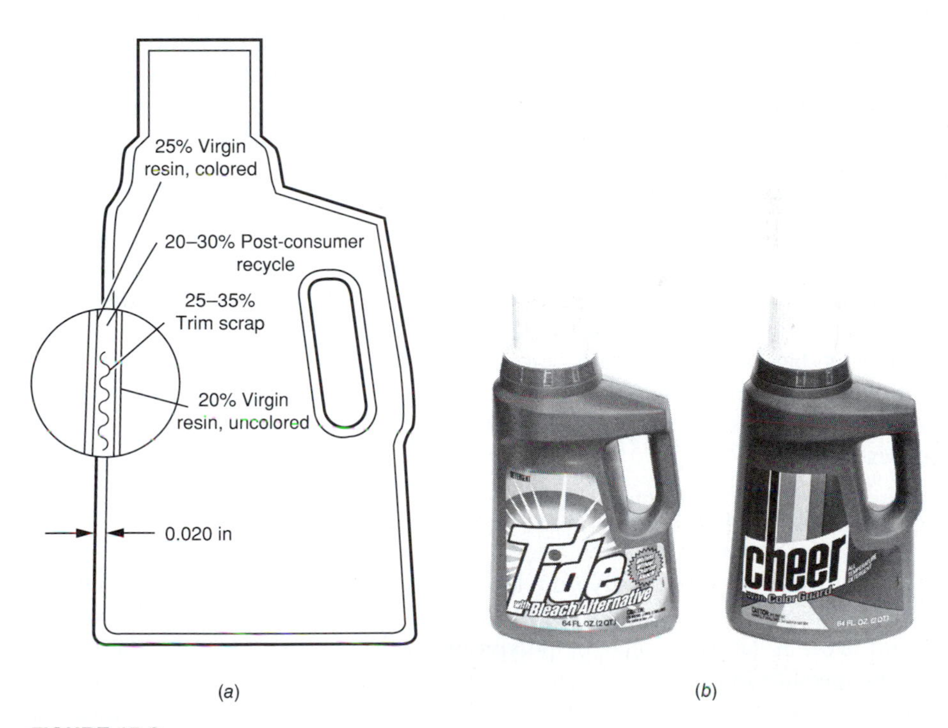

FIGURE 15-6
Multilayer construction of plastic container using recycled plastic: (*a*) schematic of three-layer bottle (courtesy of Procter & Gamble Co.) and (*b*) typical containers constructed with three layers. As more recycled plastic material becomes available, the percentage of recycled plastic used will increase, especially in dark-colored containers (e.g., Cheer™ container, which is blue).

market in the form of drainage pipe, fittings, moldings, sheet, and injection-molded parts, all of which could be made from recycled PVC.

The major impediment to PVC recycling is collection and sorting. To date, most sorting has been manual, based on either identifying codes or the characteristic "smile" line on the bottom of blow-molded PVC bottles. Both the EPA and resin producers have provided funds for sorting research; National Recovery Technologies has used electromagnetic processes to detect chlorine in plastics, and the Center for Plastics Recycling Research has used radiation technology, but neither process is yet cost-effective for full-scale operation [28].

(4) Low-density polyethylene (LDPE). In 1988, the production of polyethylene film was 3.6 billion pounds, including 1.2 billion pounds for food packaging and 2.1 billion pounds for trash bags, disposable diapers, agriculture, and construction [23]. Most film eventually ends up in the solid waste stream, and although it does not contribute much volume in a landfill, film accounts for about 16 percent by weight of discarded plastics. Several states have considered banning plastic bags and disposable diapers, and Florida has passed legislation requiring degradable bags. As a result, the industry is under pressure to collect and recycle both LDPE and HDPE film products. In addition, other polyethylene processors who normally use rigid feedstocks are increasing their effort to recycle film. Procter & Gamble even tried a disposable diaper recycling system in Seattle; the diapers were pulped, and the plastic was recovered and manufactured into plastic lumber. The demonstration program was never economical and has been terminated.

According to the Council on Plastics and Packaging in the Environment (COPPE), there are now more than 10,000 supermarkets in the U.S. that accept used film products. The bags are manually sorted for contaminants and are processed by granulating, washing, and pelletizing. The major problem is that printing inks in the original bags lead to a dark, inconsistently colored regrind; the solution has been to use dark colorants (as found in lawn and trash bags), or print over the mixed color. Other uses of LDPE are the plastic protectors used by truckers where ropes or cables touch cargo and mixed (HDPE, LDPE, and PP) plastics products.

(5) Polypropylene (PP). Polypropylene is commonly used for automotive battery cases, container closures (caps), bottle and jug labels, and, to a minor extent, for food containers. Polypropylene labels and caps are normally granulated with polyethylene products, and 10 to 13 percent can be left in bottle-grade HDPE regrind [28]. A larger amount of polypropylene is left in the mixed flake only for low-specification products such as plastic lumber, outdoor furniture, pilings, posts, and fencing. Lead-acid battery processors also recover polypropylene for use in new batteries.

(6) Polystyrene (PS). Approximately 5.2 billion pounds of polystyrene are produced annually in the United States, and about 25 percent is used for food packaging [24]. Familiar PS foamed products are clamshell fast-food containers, plates, meat trays, cups, and rigid packing material. Other common items are food utensils, clear drinking glasses, and pigmented cottage cheese and yogurt containers, which are produced by extrusion and injection molding.

According to the plastics industry, PS comprises only 0.26 percent of MSW by weight and only 1 percent by volume [24], and thus does not deserve the bad reputation it has acquired. Critics disagree, pointing out that much packaging is unnecessary, and have called for legislation to reduce or eliminate foam products; some bans have been enacted. As a result of public pressure, eight resin producers formed the National Polystyrene Recycling Company in the late 1980s. The goal of the NPRC is to recycle 25 percent of specific post-consumer products, or about 5 percent of all PS produced. The NPRC has set up five regional processing plants in the United States to reach this objective, and has provided funds to nonaffiliated PS recyclers who already have plants in operation.

The different types of PS packaging or food service containers can be reclaimed separately or together. A typical process includes semi-automatic sorting, granulation, washing, drying, and pelletizing. Solid foam board is processed differently; the foam is chopped without heat to form a taffy-like mixture, and then sprayed with water and chopped into pellets. Recycled polystyrene is used to produce foam foundation insulation board, office accessories, food service trays, trash receptacles, insulation, toys, and injection-molded products. Manufacturers are apparently satisfied with the reclaimed resin, but processors may require subsidies to cover collection, sorting, and shipping costs.

(7) Mixed and multilayer plastics (other). Manufacturers also use less common resins and multilayer containers to package products and foods with special requirements (such as salad dressings and ketchup). These containers have no value as a regrind product because there is no market. However, processors are using mixed streams of postconsumer plastics (especially polyethylenes and polypropylene) to produce resins for manufacturers of products that are massive and do not require strict resin specifications, such as outdoor benches, tables, car stops, fence posts, retaining timbers, pallets, and stakes. Because the plastics are not sorted, processors are usually able to obtain their feedstocks at very low cost. PETE is kept out of the regrind because it melts at a higher temperature than the other resins and forms inclusions in the final product.

Processing Plastics for Recycling. Processors receive postconsumer material from collection centers or brokers in bales varying in weight from 300 to 1700 pounds [28]. At a typical recovery plant, PETE bottles or HDPE jugs are transformed to clean flake through these processing steps (see Fig. 15-7).

Bale breaking and sorting. Pre-sorted bales are broken open and dumped onto a conveyor for final sorting; PETE bottles are manually sorted by color, and undesired plastics are removed. Sophisticated systems to sort by color and remove labels automatically are currently being developed and tested.

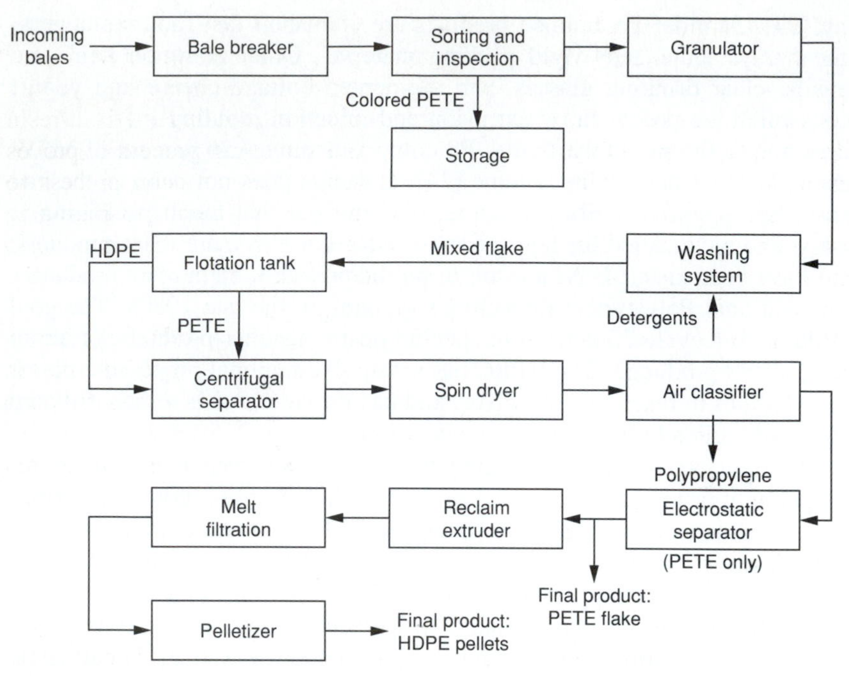

FIGURE 15-7
Typical flow diagram for the processing of recovered HDPE and PETE crushed for shipment.

Granulation and washing. Bottles are turned into small flakes by a granulator designed to cut clean chips without causing excessive heat, which would fuse them. The chips are washed using hot water, detergents, and agitation to remove labels, adhesives, and dirt; and a centrifugal separator is used to separate the flakes from the dirty water, paper, and debris.

Separation. After they are washed, the flakes proceed to a settling tank, where PETE sinks to the bottom and lighter plastics such as HDPE float. If the feedstock is nearly homogeneous, a single tank may be sufficient; if the mixture does not separate readily, a series of hydrocyclones (cyclones or centrifugal separators) may be required for both the light and heavy streams, with the process tailored to the mix of bottles.

Drying. After separation, the original stream of bottles has been converted to a stream of PETE and/or a stream of HDPE. A spin dryer is used to remove the free water, and the flakes are then dried with hot air to reduce moisture content to about 0.5 percent.

Air classification. Plants that granulate milk jugs (or other HDPE products with polypropylene caps or labels) use an air classification step between spin drying and forced-air drying to remove the light pieces of polypropylene.

Electrostatic separation. PETE bottles often contain aluminum caps when they are baled, and the granulated aluminum appears in the PETE flake. After drying, electrostatic separation is used to remove the aluminum.

Clean PETE is sold as flake, but most HDPE is pelletized. Pelletized resin is dust-free and flows smoothly, and the melting and screening process improves homogeneity and purity. Additives can be used during the extrusion process to change melt index or color. Processors and manufacturers attempt to minimize "heat history" (a measure of the number of times a resin is melted or the maximum temperature reached) because each heating degrades the resin.

Reclaim extrusion. Resin is fluidized by using an extruder, which in simplified terms is a tapered screw inside a long barrel. The flakes are fed into the extruder at the large diameter end of the screw, and are compressed as they are carried toward the extrusion die. The combined heat from flow friction and supplemental heating bands causes the resin to melt, and volatile contaminants are vented from the mixture. Immediately before the die, the melted mixture passes through a fine screen that removes remaining solid impurities; this step is known as melt filtration.

Pelletizing. The melt extruded through the die has the characteristics of a strand of spaghetti. As it passes through the orifice, a rotating knife chops the strand into short segments, which drop into a water bath where they are cooled. The pellets are dried in a centrifugal dryer to a moisture content of about 0.5 percent and packaged for shipment to the end user.

Specifications for Recovered Plastics

Trade groups representing manufacturers and processors have established specifications for recycled plastics. These standards are extensive and beyond the scope of this chapter. In general, though, buyers require postconsumer plastic to be well sorted, reasonably free of foreign material, free of excess moisture, and baled to within a specified range of size and weight. Poor separation by resin type is a major problem, because regrinders cannot afford to hire extra people to pull out unwanted materials. Some contamination is inevitable, but for badly separated or contaminated loads, buyers reduce the offering price. If poorly sorted loads continue from a supplier, processors may refuse delivery; as a rule, processors need materials and generally give vendors reasonable opportunity to resolve problems.

15-5 GLASS

Glass constitutes approximately 8 percent by weight of MSW (see Table 3-4). Ninety percent is flint (clear), green, or amber bottle and container glass; the remaining 10 percent is mostly glassware and plate glass. The benefits of recycling glass include reuse of the material, energy savings, reduced use of landfill space, and in some cases, cleaner compost or an improved refuse-derived fuel (RDF).

Reuse and Recycling Opportunities

Almost all recycled glass is used to produce new glass containers and bottles. Nationwide, new containers include approximately 30 percent of postconsumer glass (see Fig. 15-8) and recycled cullet (crushed glass) from manufacturing operations. A minor amount of glass is used to make glass wool or fiberglass insulation, paving material ("glasphalt"), and building products such as brick, ceramic and terrazzo tile, and lightweight foamed concrete.

Glass Bottles and Containers. Glass container manufacturers prefer to include cullet with the raw materials (sand, soda ash, and limestone) because furnace temperatures can be reduced significantly. Manufacturers are willing to pay a slightly higher price for cullet than for raw materials owing to the energy savings and extended furnace life. The disadvantage of using cullet from postconsumer glass is that it almost always contains contaminants that can alter product color or quality; in-house cullet from broken or "off-spec" products is always preferred because it is of known composition and free of contaminants.

Although the demand for clear glass cullet is high, the economics of recycling often vary with region of the country because of the cost of collecting, processing, and transporting used glass to manufacturing plants. The market for colored glass also varies with the capacity of plants that manufacture colored glass containers.

Fiberglass. The fiberglass industry uses cullet as an integral part of the manufacturing process, but because the specifications are very stringent, almost all cullet comes from in-house operations or from other glass manufacturers. To accomodate increased recycling, the major producers have expressed a willingness to use increased amounts of postconsumer cullet, if strict specifications are maintained.

FIGURE 15-8
Damaged glass Coca-Cola™ bottles are crushed to form cullet for the manufacture of new bottles (Bogotá, Colombia).

Other Uses. Glass that is not sorted by color is acceptable for the manufacture of glasphalt and masonry building materials, although contaminants such as ferrous metals, aluminum, and paper must first be removed by magnetic and vacuum processes. Interest in using glass as a paving material has fluctuated as a result of the high cost of processing and transporting the glass to asphalt plants, and the need for hydrated lime to promote adhesion. Further, the final product is not superior to paving material made with conventional materials [1]. There will likely be renewed interest in glasphalt as a consequence of legislation regarding mandatory waste diversion from landfills.

Specifications for Recovered Glass

Glass to be used for new bottles and containers usually must be sorted by color and must contain no contaminants such as dirt, rocks, ceramics, high-temperature glass cookware such as Pyrex™, or other glassware. These materials, known as refractory materials, have higher melting temperatures than container glass and form solid inclusions in the finished product. Laminated auto glass is prohibited because it contains a layer of plastic. Plate glass, although not a refractory material, affects the melting temperature of the mixture and is not usually accepted in cullet unless the amount is known reliably. Aluminum caps and paper labels are allowed if they are to be removed by further processsing before the cullet is added to the melting furnace. When cullet to be used for new containers is delivered to a manufacturing plant, the glass is sampled for contaminants and refractory materials; the presence of prohibited materials may be reason to reject the entire load. Typical materials specifications for recovered glass are presented in Tables 15-4 and 15-5.

The specifications for cullet to be used in the manufacture of fiberglass require clear glass with virtually no organics, metals, or refractory materials. Only a small amount of container glass can be used; plate glass is preferred because the chemical composition is closer to that of the raw materials and in-house cullet. Manufacturers set their own specifications, although industry-wide standards may be adopted someday to promote the use of postconsumer glass.

TABLE 15-4
Specifications for color-sorted glass[a,b]

	Permissible color mix levels, percent			
Color	**Flint**	**Amber**	**Green**	**Other**
Flint (clear)	97 to 100	0 to 3	0 to 1	0 to 3
Amber (brown)	0 to 5	95 to 100	0 to 5	0 to 5
Green	0 to 10	0 to 15	85 to 100	0 to 10[c]

[a] Adapted in part from Refs. 1 and 6.

[b] See Table 15-5 for other contaminants in sorted glass.

[c] "Georgia Green"—Lightly tinted green glass typically found in cola drink bottles.

TABLE 15-5
Specifications for contaminants in color-sorted glass[a]

Contaminant	Causes for rejection of loads
Ferrous (magnetic metal)	Any pieces larger than 6 in × 6 in × 12 in
	More than 1% of the load is smaller than 6 in × 6 in × 12 in and larger than 1/2 in pieces
	More than 0.05% of the load is smaller than 1/2 in pieces
Nonferrous metal (aluminum, lead, etc.)	+ 3/4 in glass packaging material (closures, aluminum foil) greater than normal amounts inherent to glass packaging
	+ 3/4 in nonglass packaging material (lead, copper, brass) greater than 0.5%
Organic material (labels, etc.)	Glass packaging materials (labels, Plasti-Shield™) greater than normal amounts inherent to glass packaging
	Nonglass packaging materials (paper, wood, rubber) greater than 5%
Refractory material (ceramics, tableware, tile, etc.)	Any particles in a 50 lb sample larger than U.S. 8 mesh
	More than one particle in a 50 lb sample smaller than U.S. 8 mesh, but larger than U.S. 20 mesh
	More than 40 particles in a 50 lb sample smaller than U.S. 20 mesh but larger than U.S. 40 mesh
Cullet sizing	More than 25% of the cullet is smaller than 3/4 in
Other contamination	Excessive amounts of dirt, gravel, asphalt, concrete, limestone, garbage, etc.
	Excessive amounts of moisture
	Contamination caused by burning glass containers
	Pyrex, ovenproof material, plate glass, automobile glass, light bulbs, fluorescent tubes, ceramics, etc.

[a] Adapted from Ref. 1.

15-6 FERROUS METAL (IRON AND STEEL)

Municipal solid waste typically contains about 6 percent "tin" cans and other steel products (see Table 3-4). The percentage has declined slightly over the past decade as steel beverage containers have been replaced by aluminum and plastic containers. Common durable goods that are not normally discarded in MSW but are available for recovery include household or commercial appliances (white goods), broken or used consumer electronics, and automobiles. Additional sources of steel are old and cut pieces of pipe, discarded building materials, industrial scrap and machine shop cuttings, construction debris, steel doors, desks, shelving, bicycle frames, and so on.

The demand for steel scrap is related to the general economy and to the demand for new autos, machine tools, and heavy construction equipment. Historically, the demand has been cyclical, but now appears stable owing to improved competitiveness of the U.S. steel industry and the increased number of minimills, which use almost 100 percent scrap. Foreign demand also provides a strong market; in 1990, exports of steel scrap were approximately 12.6 million tons [27].

Reuse and Recycling Opportunities

The principal categories of ferrous metals now recovered from MSW are tin cans and scrap metal. Reuse opportunities for these materials are reviewed below.

Steel Cans. Steel cans (also known as *tin cans* because of the tin plating used for corrosion control) are recovered from the consumer waste stream through curbside collection programs and at collection centers or MRFs. Cans are often mixed with nonferrous materials when they are delivered to collection centers and must be separated magnetically, compacted, and shipped to a detinning facility. Most detinning plants first shred the cans, this activity also serves to loosen food residues and paper labels. A vacuum system is used to remove these foreign materials. The shredded material is then sorted magnetically to remove aluminum (from bi-metal cans) and other nonferrous materials. The clean steel is then detinned, either by heating in a kiln to volatilize the tin, or by a chemical process, using sodium hydroxide and an oxidizing agent. The tin is recovered from the solution by electrolysis and formed into ingots (typically, about 5 to 6 pounds of tin is recovered per ton of cans).

Chemically detinned steel is used primarily for the production of new steel. Scrap that is detinned by heating is not suitable for steel production, because heat causes some tin to diffuse into the steel and appear eventually as an impurity in new steel. Instead, it is used to produce copper (see below), and a small amount is purchased by the paint industry as a source of iron oxides. Nationwide, about 96 percent of the steel cans recovered are used to produce new steel, and only 4 percent are used to produce copper [15].

In the copper extraction process, copper ores are treated with sulfuric acid to produce copper sulfate, and the solution is leached through steel scrap to produce ferrous sulfate and to precipitate metallic copper. Approximately 1.65 tons of steel scrap are needed to produce one ton of copper; steel scrap with a high ratio of surface area to weight is desirable, which makes cans an excellent material. The majority of steel cans collected in California and the southwestern United States are used for copper production at mines in Utah and Arizona.

The major impediment to recycling steel cans is the high cost of transportation, especially in the Western states. The market price for steel cans is limited by the price of new steel, so the value of cans is not expected to rise significantly. Typically, recycling centers recover their costs to collect, process, and deliver cans only if the buyer is close geographically, because the transportation cost to a detinning facility is usually the major expense. Despite the marginal economics, steel can recycling programs are expected to increase as a result of landfill-diversion legislation.

Appliances, Automobiles, and Miscellaneous Steel Scrap. White goods, automobiles, and miscellaneous postconsumer steel products are usually processed by scrap dealers and auto dismantlers, who consolidate and bale the material for brokers and end users (see Fig. 15-9). The first step in the recycling process is removal of useful or hazardous materials. Scrap dealers who process appliances must remove motors (which may contain polychlorinated biphenyls [PCBs], formerly used in motor-starting capacitors) and compressor units (which contain chlorofluorocarbon refrigerants). Auto dismantlers remove gas tanks, batteries, tires, and salable items such as windshields and radiators. If the engine and drive train are left intact, oils and transmission fluids must be drained completely.

Appliances, automobiles, and bulky items are compacted and sent to a shredder. Shredding and magnetic separation are used because it is not economical to recover steel piece by piece; shredding also increases bulk density for economical shipment. Industrial shredders reduce autos (including the engine block and drive train) to small chunks that are suitable for remelting in an electric furnace. There are approximately 220 shredders in the U.S., and approximately 9 million vehicles were processed in 1990 [18], as well as appliances and other bulky steel products. Not all scrap is shredded; industrial processors also prepare bundles of heavy steel scrap and bales of flattened sheet metal.

Specifications for Recovered Ferrous Materials

Specifications for tin cans and scrap metal are well defined. However, as an increased amount of material becomes available in the future, more rigorous specifications will probably be imposed.

FIGURE 15-9
Processing facility for ferrous metal using horizontal shaft hammermill. (Courtesy of Williams Crusher and Pulverizer Co.)

Steel Cans. It is generally accepted that most consumers will recycle steel cans only if it is very convenient, because the market price is too low to provide a financial incentive. Even with curbside programs, some consumer effort is required to prepare the cans for recovery. Requirements vary among states; for example, Washington State requires that labels be removed, while California does not (most cans from Washington are detinned chemically, while those from California are heated). The Steel Can Recycling Institute offers the following general guidelines [34]: Consumers are encouraged to remove paper labels, rinse the cans (with dishwater, if possible), and flatten them. Ends that have been removed may be reinserted before flattening. Bottle caps and jar lids may be recycled with steel cans. Aerosol cans are acceptable for recycling provided they are empty. Plastic caps should be removed, but spray nozzles may be left intact. Paint cans are recyclable if only a thin, dry coat of paint remains. Lids should be left loose.

Operators of collection centers and scrap dealers consolidate and bale cans as required by end users. The general specifications published by the Steel Can Recycling Institute are shown in Table 15-6.

Other Shredded and Compacted Scrap. Scrap dealers and processors consolidate and bale compacted and shredded materials according to buyer specifications and ISRI standards (Institute of Scrap Recycling Industries, Washington, DC), which are accepted throughout the industry. These specifications should always be verified by brokers or end users.

TABLE 15-6
General specifications for tin-coated and tin-free steel scrap[a]

Material	Requirements[b]
Baled can scrap for steel companies	Bales should be 2 ft × 2 ft × 2 ft (or 3 ft) in size, with a specific weight of 75 to 80 lb/ft^3. Cans may be baled without removal of paper labels, but must be free of water, plastic, wood, and other debris.
Densified biscuit scrap for steel companies	Scrap should be stacked and banded into bundles, with a density of 75 to 80 lb/ft^3. Bundle weight is subject to negotiation.
Baled can scrap for detinning	May be of varied dimensions. Specific weight should nominally be 30 lb/ft^3, subject to negotiation. Wire or other steel banding is acceptable.
Loose cans	Loose cans (whole or flattened) are acceptable, subject to negotiation.
Shredded can	Shredded cans (loose or baled) are acceptable, subject to negotiation.

[a] These specifications have been simplified; complete specifications are contained in Ref. 5.

[b] Steel may include bi-metal cans but must not contain all-aluminum cans or other nonmetallic materials such as wood, plastics, or water.

15-7 NONFERROUS METALS

Nonferrous metals compose about 3.5 percent of MSW, including commercial and industrial wastes. Recyclable materials are recovered from common household items (outdoor furniture, kitchen cookware and appliances, ladders, tools, hardware); from construction and demolition projects (copper wire, pipe and plumbing supplies, light fixtures, aluminum siding, gutters and downspouts, doors, windows); and from large consumer, commercial, and industrial products (appliances, automobiles, boats, trucks, aircraft, machinery). Virtually all nonferrous metals can be recycled if they are sorted and free of foreign materials such as plastics, fabrics, and rubber.

Reuse and Recycling Opportunities

Scrap metal dealers buy materials from the public, construction and demolition firms, repair shops, auto dismantlers, appliance dealers, metal fabricators, and

TABLE 15-7
Nonferrous metals: Common sources and end uses

Metal	Percentage supplied by recycling[a]	Typical sources	Products and uses
Aluminum	34	Containers, tubing, outdoor furniture, gutters, siding, doors, windows, cookware, cooling coils and fins, autos, boats, trucks, aircraft	Containers, tubing, outdoor furniture, gutters, siding, doors, windows, cookware, cooling coils and fins, autos, boats, trucks, aircraft
Copper (includes brass and bronze)	50	Wire, tubing, plumbing fixtures, valves, cooling coils and fins, radiators, bearings	Same as sources plus alloys, electronics, chemicals, electro-plating
Lead	61	Tire weights, batteries, cables, solders, wine bottle seals, bearings	Batteries, solder, bearings, shot, alloys
Nickel	27	High-strength and corrosion-resistant alloys, jet engines, industrial machinery	High-strength and corrosion-resistant alloys, stainless steel
Stainless steel[b]	NA	Commercial kitchen-ware, countertops, industrial scrap	Corrosion and heat-resistant alloys for a variety of products
Tin	18	Solders, bronze, bearing materials, tinplate	Solders, alloys, coatings, plating
Zinc	27	Alloy scrap, shredded automobiles and appliances, galvanizing wastes	Galvanized products, brasses, alloys

[a] From Ref. 19.

[b] Stainless steel is included here with nonferrous metals because it is a specialty alloy.

other suppliers, and sell to brokers or industrial buyers. Metals are sorted according to alloy type, if known, and manufacturing process (i.e., cast or wrought). Well-sorted items can be consolidated and baled directly. Complex scrap such as automobiles and appliances requires a combination of processes, including dismantling and sorting, compaction, shredding, magnetic separation, eddy current separation, and baling (see Chapter 12). Typical sources of nonferrous metals and some of the new products and uses are reported in Table 15-7.

Although there is a strong demand for scrap, dealers do not usually enter into long-term contracts because they have no control over market prices. The scrap market is global, and U.S. industries are in competition with foreign buyers, especially those from Japan, Taiwan, South Korea, Canada, and Mexico [27].

Specifications for Recovered Nonferrous Metals

Scrap dealers consolidate and bale material to meet the specifications required by mills and brokers; in turn, dealers have requirements for materials they purchase. For example, auto dismantlers are required to remove radiators and other nonferrous parts from vehicles, and vendors must remove compressors and motors from appliances. Most dealers buy scrap "as is" from the public; if extensive processing is required, the material is usually accepted but the offering price is reduced. For example, appliances requiring removal of the motors and nonmetallic materials might be accepted at half the value of those in clean condition.

15-8 YARD WASTES COLLECTED SEPARATELY

To reduce the amount of material going to a landfill, many communities now collect and process yard wastes separately. Typically, yard wastes are placed in containers or in the street for collection (see Figs. 9-25 and 9-26). The manner in which the wastes are placed in the street varies from city to city. Some communities require that grass clippings be placed in plastic bags for collection. The plastic bags are placed in the same pile with shrub and tree trimmings. In other communities, the bagging of grass clippings is not required and all yard wastes are commingled.

Reuse and Recycling Opportunities

The principal recycling opportunities for yard wastes are for (1) the production of compost, (2) the production of landscape mulch, (3) use as a biomass fuel, and (4) use as intermediate landfill cover material.

Production of Compost. Yard wastes are generally accepted as the best starting material for high-quality compost. Leaves are considered the easiest material to process; grass clippings, although high in nitrogen, produce odors when composted alone and are generally mixed with other wastes. Brush and woody materials are usually handled separately because they require chipping or shredding. Most yard waste composting operations are located at or near existing landfills, where diverted wastes serve as the major source of raw materials. In addition, roads to accommodate trucks are available, and haulers are familiar with the location. Composting methods for yard wastes were discussed in Chapter 9.

Compost operators sell compost to homeowners and landscapers, and some have contracts that require municipalities to take back a portion of the compost for public works projects. Some municipalities provide compost free of charge to the residents of the community, provided they pick it up. Although there is a relatively good market for clean yard compost, one obstacle to consumer acceptance is the presence of plastic, which results from incomplete removal of plastic bags before the yard wastes are ground up to produce a compostable material. The consensus of those in the industry is that quality should be the primary objective, especially as the supply of compost increases over the next few years.

Production of Mulch. Mulch production is a low-cost option to composting for some yard wastes. Brush and woody wastes such as tree prunings can be used to produce a mulch that can be used in residential and commercial landscaping projects as well as onsite at the landfill. For a landfill that is to be closed and revegetated, mulch provides a variety of benefits that enhance plant growth and development. Mulching of closed landfills is discussed in greater detail in Chapter 16. In dry climates, mulch retards evaporation and conserves water. It also controls weed propagation in unplanted areas. A rough mulch suitable for water conservation and public landscaping purposes can be produced simply by chipping or grinding wastes in a tub grinder. A more refined product, suitable for sale through retail nurseries, requires an additional screening step to produce a more uniform particle size. Mulch can also be produced by commercial tree-trimming companies, city landscape maintenance crews, and telephone and electric company maintenance crews. For example, in California the Pacific Gas and Electric Company collects and chips tree prunings during annual maintenance of vegetation near electric power lines. The mulch produced is given away free to the general public.

Biomass Fuel. Yard wastes can also be used as a biomass fuel. Two approaches to the production of biomass fuel are typically used. In the first approach, yard wastes are ground using a tub grinder (see Fig. 9-27). The material from the tub grinder is passed through a trommel screen to separate pieces of wood larger than $\frac{1}{2}$ inch. Wood chips larger than $\frac{1}{2}$ inch are sold as a biomass fuel. Green wastes and wood chips smaller than $\frac{1}{2}$ inch are composted. In the second approach, all of the ground-up yard wastes are sold as biomass fuel. Although the price paid for commingled biomass fuel (green waste and wood chips) is less than the amount paid for separated wood chips ($7/ton versus $35/ton, central California, September 1991), most waste managers prefer to sell the commingled waste and thus avoid the additional processing costs associated with the separation of wood chips and the production of compost.

(a)

(b)

FIGURE 15-10
Yard wastes used as intermediate landfill cover: (*a*) size of yard waste is reduced using a tub grinder and (*b*) ground up waste is applied to face of landfill.

Intermediate Landfill Cover. Yard wastes can also be used as intermediate landfill cover. Typically, the yard wastes are ground up and composted before being applied. Note that composted yard wastes usually are not cured completely before being placed on the landfill. In some operations, yard wastes are ground up to reduce their volume and are applied directly as intermediate cover (see Fig. 15-10).

Specifications for Yard Wastes

Specifications for yard wastes to be composted depend on the end use of the compost. The Environmental Protection Agency has developed guidelines for compost products, and some states are using these guidelines as interim standards. Other states have written their own specifications, both to satisfy legislative mandates and to provide guidance to agencies procuring composted materials for public uses, such as road landscaping and parks. Most regulations or guidelines classify compost products according to the intended use; for example, a Class 1 or Class AA compost might be suitable for unrestricted use by the public, whereas an *off-spec* or *restricted* compost would be allowed only for landfill cover or strip-mined land reclamation. Specifications are written to cover chemical and biological characteristics such as nutrient content, organic content, pH, texture, particle size, moisture content, moisture-holding capacity, amount of foreign matter, concentration of salts, residual odor, degree of stabilization or maturity, presence of pathogenic organisms, and concentration of heavy metals. Allowable limits are expected to become better defined as the industry grows and customer preferences determine the market. Typical specifications for general-use compost made from yard wastes are shown in Table 15-8.

Specifications for the use of yard wastes as a biomass fuel vary with the individual facility. Typical properties that are considered in assessing the suitability of yard wastes as a fuel source include composition of the yard wastes, particle size distribution, moisture content, and degree of contamination. As the available supply of yard wastes continues to increase, the specifications for biomass fuel will likely become more restrictive.

TABLE 15-8
Typical specifications for general-use compost produced from yard wastes[a]

Parameter	Units	Value: Range	Value: Typical	Remarks
Cation exchange capacity, CEC	meq/100g[b]	20–60	No specs	
Conductivity	mmho/cm	No limits, to $\leq$ 15	No specs	Parameter indicates soluble salts.
Foreign matter	%	0 to 2	No sharp, injurious material	Some states prohibit all foreign material.
Metals				
Arsenic	ppm[b] or mg/kg[b]	20–41	41	Typical metals values are the same as the values for sludge in the new EPA 503 regulations.
Cadmium		4–20	39	
Chromium		1000–1200	1200	
Copper		1000–1500	1500	
Lead		250–500	300	
Mercury		5–20	17	
Molybdenum		10–18	18	
Nickel		50–420	420	
Selenium		20–40	36	
Zinc		200–2800	2800	
Moisture	percent by dry weight	20–40	$\leq$40	
Nutrients	various	Nitrogen $\geq$ 1% NH_3:TKN $\leq$ 10	No typical value	Several states require testing.
Odor	none	No specs, or Not offensive	Not offensive	
Organic matter	%	No specs	No specs	Several states require testing.
Pathogens	none	No pathogens, to PFRP[c]	No typical value	See Table 14-9.
PCBs	ppm	No limits, to 1	No specs	
pH	unitless	5.5–7.0	No typical value	
Specific weight	lb/yd^3	No specs	No specs	
Stability				
Reheat	°C	No heat gain, to $\leq$ 20°C increase	No typical value	One state requires C/N ratio between 12 and 25 to ensure completion of biological activity.
Vol. reduct.	%	No specs	No specs	
Texture	in mm	1/8 to $\leq$1/2 3.7 to $\leq$ 13	No specs	Values given are for screened compost.

[a]Based, in part, on draft or final specifications from several states.

[b]Based on dry weight of compost.

[c]PFRP: Process to further reduce pathogens; required if co-composting with sludge.

15-9 ORGANIC FRACTION OF MSW

The components that constitute the organic fraction of MSW are food wastes, paper, cardboard, plastics, textiles, rubber, leather, yard wastes, and wood. All of these materials can be recycled, either separately or as commingled waste. The components can be recovered separately by source separation or at a MRF; they can also be recovered from MSW in commingled form by removal of inorganics. The choice of recovery method is dictated by the use of the material or end product. Source-separated materials generally have the least contamination and exhibit physical and chemical properties different from the commingled components. The reuse and recycling opportunities and the specifications for the commingled materials that compose the organic fraction of MSW are considered in the following discussion.

Reuse and Recycling Opportunities

The principal reuse and recycling opportunities for the materials that make up the organic fraction of MSW are the production of (1) compost, (2) methane, (3) organic compounds, and (4) refuse-derived fuel.

Production of Compost. Municipal solid waste typically contains 70 to 80 percent organic material (see Table 3-4), and composting is gaining popularity as a waste management option (see Chapter 9). Almost all MSW composting systems start with separation of recyclables, metals, and hazardous materials, followed by size reduction and additional separation. The end uses for MSW compost are usually limited to agricultural uses or land reclamation. Few operators sell the finished product, although some give it to public agencies, farmers, and nurseries. Owing to the poor separation of incoming materials, complaints regarding the presence of residual plastics and glass shards have been frequent. In some cases, the compost produced from MSW has been used for intermediate landfill cover. The intentional composting of partially sorted residential and commercial MSW for use as intermediate landfill cover material was considered in Chapter 11.

Production of Methane. The production of methane from the organic materials contained in commingled MSW is accomplished biologically under anaerobic conditions (see Chapter 14). Typically, methane is produced from the organic fraction of MSW under uncontrolled conditions in sanitary landfills (see Chapter 11), and under controlled conditions in either a low-solids (6 to 10 percent solids) or a high-solids (20 to 35 percent solids) anaerobic reactor (see Chapter 14). Methane can be used for the production of energy and heat, or for the conversion to methanol and/or other products. The production of methanol is of interest because methanol is a clean-burning, storable fuel. The digested solids from the low- and high-solids processes can be composted to produce a usable product or placed in a landfill.

Production of Organic Compounds. The organic materials contained in commingled MSW can also be used for the production of a variety of organic com-

TABLE 15-9
Typical specifications for general-use compost produced from MSW[a]

		Value		
Parameter	**Units**	**Range**	**Typical**	**Remarks**
Cation exchange capacity, CEC	meq/100g[b]	20–60	No specs	
Conductivity	mmho/cm	<2 to <10	Test only	Parameter indicates soluble salts.
Foreign matter	%	≤2 to ≤6	No sharp, injurious material	Most states allow no sharp materials.
Metals				
Arsenic	ppm[b]	20–41	41	Typical metals values are the same as the values for sludge in the new EPA 503 regulations.
Cadmium	or	4–20	39	
Chromium	mg/kg[b]	1000–1200	1200	
Copper		1000–1500	1500	
Lead		250–500	300	
Mercury		5–20	17	
Molybdenum		10–18	18	
Nickel		50–420	420	
Selenium		20–40	36	
Zinc		200–2800	2800	
Moisture	percent by dry weight	≤40 to 60	No specs	
Nutrients	various	No specs, to Nitrogen ≥ 1%	No specs	Several states require testing.
Odor	none	No specs, or Not offensive	Not offensive	Specs are not usually strictly defined.
Organic matter	%	No specs, to ≥3	No specs	
Pathogens	none	PFRP[c]	PFRP[c]	See Table 14-9.
PCBs	ppm	No specs, to 10	No specs	
pH	unitless	6.1–7.8	No typical value	
Phytotoxicity	% seed germination	No specs	No specs	
Specific weight	lb/yd^3	1000–1300	No specs	
Stability				
Reheat	°C	No heat gain, to ≤20°C increase	No typical value	Several states require that compost "not be a nuisance."
Vol. reduct.	%	No specs, to ≥ 60	No typical value	
Texture	in mm	1/8 to ≤1/2 3.7 to ≤13	No specs	Values given are for screened compost.

[a]Based, in part, on draft or final specifications from several states.

[b]Based on dry weight of compost.

[c]PFRP: Process to further reduce pathogens; may not be required if compost is used for landfill cover.

pounds including sugars, alcohols, solvents, organic acids, hydrocarbon gases, and aromatic compounds (see Chapter 14). For example, source-separated paper is comprised of approximately 61 percent cellulose; 16 percent hemicellulose; 21 percent lignin; and 2 percent protein, ash, and so on. With this composition, waste paper is ideally suited as a feedstock for the production of ethanol (see Chapter 14). Similarly, other organic materials in MSW can be used for the production of other organic compounds.

Production of Refuse-Derived Fuel. Refuse-derived fuel refers to solid waste that is processed to serve as a fuel for boilers used to produce steam or electricity [7]. RDF is frequently burned in utility boilers and in specially designed combustion systems (see Chapter 13). RDF has also been mixed and burned with coal. Although *as-discarded* waste is sometimes considered a refuse-derived fuel, the term usually refers to waste that has been sorted, reduced in size, and refined by removal of noncombustibles such as metals and glass. The production of RDF is discussed in Chapter 12.

Specifications for Organic Fraction of MSW

Specifications for the components that comprise the organic fraction of MSW depend on the individual applications. Compost specifications have been written by state regulatory agencies, both to provide guidelines for compost producers and to ensure minimum product quality. A summary of representative specifications for compost produced from MSW is provided in Table 15-9. These specifications were generally developed from sludge composting regulations and are likely to change as the market develops and the industry matures. The specifications for the organic materials used as a feedstock for the production of methane will vary depending on the final use to be made of the digested material.

There are no industry-wide specifications for RDF, but RDF has been classified by ASTM [5] and EPA [35], and the various forms of RDF are summarized in Table 15-10. Properties of RDF that should be considered and incorporated into supply contracts include proximate analysis (moisture content, ash content, volatiles, and fixed carbon); ultimate analysis (C, H, N, O, S, and ash percentages); higher heating value (HHV); and the content of chlorine, fluorine, lead, cadmium, and mercury.

15-10 CONSTRUCTION AND DEMOLITION WASTES

Construction and demolition (C/D) wastes result from the construction, renovation, and demolition of buildings; road repaving projects; bridge repair; and the cleanup associated with natural disasters (see Fig. 15-11). Typically C/D wastes are made up of about 40 to 50 percent rubbish (concrete, asphalt, bricks, blocks, and dirt), 20 to 30 percent wood and related products (pallets, stumps, branches, forming and framing lumber, treated lumber, and shingles), and 20 to 30 percent miscellaneous wastes (painted or contaminated lumber, metals, tar-based products, plaster, glass,

TABLE 15-10
Forms of refuse-derived fuel (RDF)[a]

Designation[b]	Description/uses
RDF-1	Wastes used as fuel in as-discarded form
RDF-2	Wastes processed to coarse particle size with or without ferrous metal separation
RDF-3	Shredded fuel derived from MSW that has been processed to remove metal, glass, and other inorganic materials (this material has a particle size such that 95 weight percent passes through a 50-mm square mesh screen)
RDF-4	Combustible waste processed into powder form, 95 weight percent passing 10 mesh (2 mm)
RDF-5	Combustible waste densified (compressed) into the form of pellets, slugs, cubettes, or briquettes
RDF-6	Combustible waste processed into liquid fuel
RDF-7	Combustible waste processed into gaseous fuel

[a] From Standard Test Method for Thermal Characteristics of Refuse-Derived Fuel Macrosamples, Designation E 955-88, Ref. 5.

[b] RDF-4, RDF-6, and RDF-7 are unavailable. RDF-5 is not widely available.

white goods, asbestos and other insulation materials, and plumbing, heating and electrical parts) [13].

Although a relatively small percentage of C/D wastes are now recovered, significantly greater amounts will probably be recycled in the future as a result of higher tipping fees, mandatory landfill diversion legislation, and the success of entrepreneurs in processing both source-separated and mixed wastes. Many landfills already use rubble for road building and daily cover, both of which may be considered as diversion by regulators. For those municipalities where C/D wastes are presently combined with household wastes, recycling programs afford an excellent opportunity to meet diversion goals and extend landfill life. The processing of C/D wastes was considered in Chapter 9.

FIGURE 15-11
Typical construction and demolition wastes received at a landfill.

Reuse and Recycling Opportunities

Reuse and recycling opportunities for C/D wastes depend on the markets for the individual materials comprising the wastes and the ability to process the commingled waste or separate the individual materials. The principal materials that are now recovered from C/D wastes include asphalt, concrete, wood, drywall, asphalt shingles, and metals [13].

Asphalt. Most asphalt waste comes from repaving projects (asphalt paving is composed of a mixture of about 5 percent heavy oil and 95 percent aggregate). Most old pavement that is reused is processed for road base, but up to 40 percent can be included in new pavement (a typical figure is 10 to 15 percent, because old material has already been somewhat degraded by weathering and sunlight). Old asphalt pavement is processed by itself or with concrete and other rubble; the mixture is crushed, ferrous metals are removed magnetically, and the crushed material is then screened to size. The graded material is supplemented with other crushed and screened rubble and is used as road base, or mixed with fresh asphalt binder to make new paving material.

Concrete. Most concrete is recovered from roads, bridges, and foundations; it is processed for road base, aggregate in asphalt pavement, and as a substitute for gravel aggregate in new concrete. Concrete chunks are crushed, ferrous materials such as bolts or reinforcing bar are removed, and the resultant aggregate is screened to sizes suitable for road building or new concrete (aggregate to be used in new concrete must meet standard specifications such as those of the American Society for Testing and Materials). Recovered aggregate must be competitive with new materials, and processors are able to keep prices low by charging tipping fees for concrete waste.

Wood. Construction and demolition wood wastes consist of framing and form lumber, treated wood, plywood and particle board, and wood contaminated by paint, asbestos, or insulation. Because most wood wastes are processed for fuel or landscaping cover, processors usually accept only clean wood. Wood wastes are typically shredded in a tub or other commercial wood grinders and passed through a classifier or trommel, where the oversized pieces are separated. Ferrous metals are removed magnetically, and the fines (undersized materials, which are often sold for mulch or soil amendments), are separated by screening.

Drywall. In most areas, drywall is landfilled with household wastes, but experience has shown that it is not entirely inert or benign. Anaerobic decomposition of drywall can result in the production of hydrogen sulfide gas. For this reason, the city of Seattle requires that old drywall be bagged or boxed before accepting it at transfer stations. There are only three plants in the United States and Canada that process drywall, although a fourth facility is planned. In the recycling process, the gypsum interior is pulverized and returned to drywall manufacturers; the reclamation company is now trying to process the paper waste so that it can be incorporated in the facing of new wallboard.

Asphalt Shingles. Asphalt shingles contain up to 30 percent asphalt, and several asphalt pavement manufacturers use shredded postconsumer shingles as a portion of their mix for both road base and paving. Used shingles are shredded and reduced in size with a hammer mill, and nails and other ferrous materials are removed magnetically. The material is screened to final size and is added to reclaimed aggregate mixtures.

Metals. Reinforcing steel used in foundations, slabs, and pavement is usually recovered and sold to scrap dealers. Processors also reclaim nonferrous scrap such as aluminum window frames, screen doors, gutters, and siding, and copper pipe and plumbing fixtures.

Specifications for Recovered C/D Wastes

There are no industry-wide specifications for C/D wastes. Specifications are negotiated individually with buyers of the separated materials.

15-11 WOOD

Wood wastes are a major component of yard wastes and account for more than 25 percent of construction and demolition (C/D) wastes [13]. Wood waste is commonly categorized according to the source of generation: harvested wood waste (generated from land-clearing and forest management activities); mill residue (waste from primary producers such as pulp and lumber mills, and secondary producers such as furniture manufacturers and cabinet shops); pallet and container waste; construction and demolition wastes; and other wood wastes (yard, orchard, nursery, and agricultural wastes) [14].

Wood reuse has increased throughout the last decade as a result of higher tipping fees, landfill diversion programs, and developing markets. The primary end uses are for boiler fuel and landscaping, with lesser amounts used for landfill cover, pulp and paper mill feedstocks, intermediate landfill cover, and wastewater treatment plant sludge composting [14]. Fine screenings are used for composting and soil amendments, and fine shavings and small, clean chips are highly desirable for use as animal bedding.

Successful wood processing operations require a steady market and a dependable supply of raw materials. Many wood processors are associated with MRFs, or sign agreements to process wood wastes at landfills with large supplies of material and convenient accommodations for trucks and haulers (see Fig. 15-12). The landfill operator may waive the normal tipping fee for wood wastes that are routed to the wood processing area, and processors charge a reduced fee to recover a portion of their operating costs. Tipping fees vary according to the desirability of the waste; for example, some processors may charge high rates for large stumps, but no fee for clean chipped wastes brought by tree service companies.

FIGURE 15-12
Wood processing facility located at a landfill site.

Reuse and Recycling Opportunities

Waste wood brought to a facility is inspected for contaminated wood (pressure-treated wood, painted wood, etc.) and unwanted materials such as dirt, rocks, or trash. Plants that process waste wood for both boiler fuel and landscape materials may separate construction wastes, demolition wastes, brush and limbs, and green waste. Because volume and type of material vary with the season, processors often operate on an intermittent schedule; for example, a processor might stay open during normal business hours, but simply stockpile material for several weeks until there is enough to keep a crew busy. The heart of the processing equipment is a tub or large commercial wood grinder, which is used to shred the wood waste.

After wastes have been shredded, a trommel is normally used to separate usable chips from fines, but oversized material is carried through with the chips and must be raked out manually. An alternative method is to pass all material from the tub grinder through a classifier, which is simply a conveyor that uses revolving discs to carry oversized material on top and let usable chips and fine material drop through. The combined stream of chips and fines is then screened, and the *overs* from the classifier are loaded back into the tub grinder. Because classifiers are expensive, operators may use two trommels in series, one of which removes oversized material. Some processors even sell material directly from the tub grinder, using manual labor to separate oversized material and trash.

The degree of allowable contamination depends on the end market. Ferrous metal contained in pallets and C/D waste is removed magnetically after the tub grinder or classifier. (Large plants that use conveyors to supply the tub grinder may also use magnetic separation in the feed system.) In some systems, a portion of the light materials, such as paper and plastic, is removed after the classifier by the use of compressed air. Fine screenings sold as soil amendments must be virtually free of plastics, paper, and other foreign matter; and the undesired materials must be removed before shredding because it is virtually impossible to remove them later.

Specifications for Recovered Wood

The specifications for recovered wood vary, depending on the markets available to wood processors. Processors accept a variety of wood wastes, depending on the supply available and the end market. Plants producing boiler fuel prefer clean construction and demolition waste, pallets and containers, and clean brush and tree trimmings; some will accept small, clean stumps. Processors do not want pressure-treated wood, telephone poles, or railroad ties (which are treated with tar or creosote), plywood, leaves, grass clippings, large tree trunks, or dirty stumps, because these materials affect boiler performance and may cause air pollution violations.

Waste-to-energy plants sign contracts with wood processors or brokers. Loads are purchased on a dry basis because the moisture content varies, depending on the season and the type of wood waste being processed. The moisture content of every load is sampled, and the percentage is deducted from the gross weight of chips delivered. Incoming loads are also sampled on a regular basis to verify the heating content of the fuel and ensure that the supplier is meeting contract specifications. Loads are visually checked for contaminants such as excessive metal, rocks, mud, or unwanted materials from demolition wastes. Loads containing excessive contaminants, water, or mud are rejected and suppliers are notified.

15-12 WASTE OIL

In 1991, approximately 1.35 billion gallons of petroleum-derived waste oil were produced in the United States [20]. An estimated 790 million gallons were associated with automotive uses and 560 million gallons were generated industrially [21]. Automotive oils include crankcase oil, diesel engine oil, and transmission, brake, and power steering fluids. Sources of automotive oil are do-it-yourself oil changers (see Fig. 15-13), auto garages, service stations, truck and taxi fleets,

FIGURE 15-13
Waste oil receiving facility located at a materials recovery facility.

TABLE 15-11
Disposition of waste oil produced in the United States[a]

	Managed oil[b]		Nonmanaged oil[c]	
End use	**gal ($\times 10^6$)/yr**	**%**	**gal ($\times 10^6$)/yr**	**%**
Burning	489.8	73.2	100.3[d]	18.7
Re-refined lube oil	62.7	9.4		
Road oiling	39.6	5.9	28.9	5.4
Disposal/dumping[e]	42.1	6.3	363.8	67.7
Nonfuel industrial	34.9	5.2		
Onsite recycling			44.0	8.2
Total	669.1	100.0	537.0	100.0

[a] (1988); Based on Ref. 21, "Background," Table 4, page 14.

[b] Managed oil = waste oil that is collected and reused and recycled.

[c] Nonmanaged oil = waste oil with no structured collection program.

[d] Most is mixed with home heating oil.

[e] Includes disposal in landfills and incineration.

military installations, and industrial and manufacturing facilities. Waste industrial oil includes metal working oils, hydraulic oils, process oils, lubricating oils, and engine crankcase oil.

Approximately 56 percent of waste oil from generators passes through a regulated management system of collectors, reclaimers, and marketers (primarily fuel oil dealers) [21]. The fate of both automotive and industrial waste oil is shown in Table 15-11. Approximately 34 percent of all waste oil is disposed of, either by landfilling, incineration, or dumping (of the 34 percent, about 59 percent is dumped). Approximately 49 percent of all waste oil is burned, primarily for cement manufacturing, space heating, and commercial, industrial, and marine boilers.

Waste oils often contain metals, chlorinated solvents, and miscellaneous organic compounds, including those listed as priority pollutants by the EPA. The presence of metals such as arsenic, barium, cadmium, chromium, and zinc is usually the result of engine or bearing wear, or the inclusion of these metals in oil additives. Lead from leaded gasoline is a common contaminant, although the concentration appears to be dropping significantly as the use of unleaded fuel increases. Chlorinated solvents such as PCBs are found in waste oil as a result of illegal or careless mixing. The presence of miscellaneous organic compounds, such as benzene and naphthalene, is usually associated with the base oil itself.

Reuse and Recycling Opportunities

From the standpoint of energy conservation, the recycling of used oil is an efficient use of resources. Most waste oil that passes through the regulated or managed system is treated by reprocessors or re-refiners, and most of the heavy process residue can be used with road asphalt products.

Reprocessing. As of 1989, there were 200 to 300 reprocessors in the United States [21]. Reprocessors use mild heating and cleaning to remove bottom sediments, water, suspended material, and ash; but the quantities of volatile organics and metals are not reduced appreciably, and the end product is suitable only for fuel. Reprocessing treatment systems typically involve settling, heating, vacuum filtration, and centrifugation (see Fig. 15-14). Untreated oil enters the settling tank, where larger particles are removed by sedimentation. The oil is then heated and vacuum filtered to remove water, volatile hydrocarbons, and suspended materials. After neutralization and demulsification, the oil is heated to 300°F (149°C) and centrifuged to remove particles that pass through the filtration process. Approximately 90 percent of the feed leaves as product, and the remainder is returned to the heater. A sludge of metals and sediments is produced, which is usually burned as in-plant fuel or incorporated into asphalt products.

Re-refining. In 1989, there were only 16 re-refiners in the United States [21]. Re-refiners use techniques similar to those of reprocessors to remove sediments and water, as well as advanced treatment to remove volatile contaminants and metals, allowing the oil to be used again as lubricating oil. A typical process includes heating, filtration, vacuum distillation, and solvent extraction, clay treatment, or treatment in a catalytic reactor. One treatment gaining popularity is the KTI Process, developed by Phillips Petroleum, shown in Fig. 15-15. Primary distillation removes water and light hydrocarbons; vacuum distillation produces a

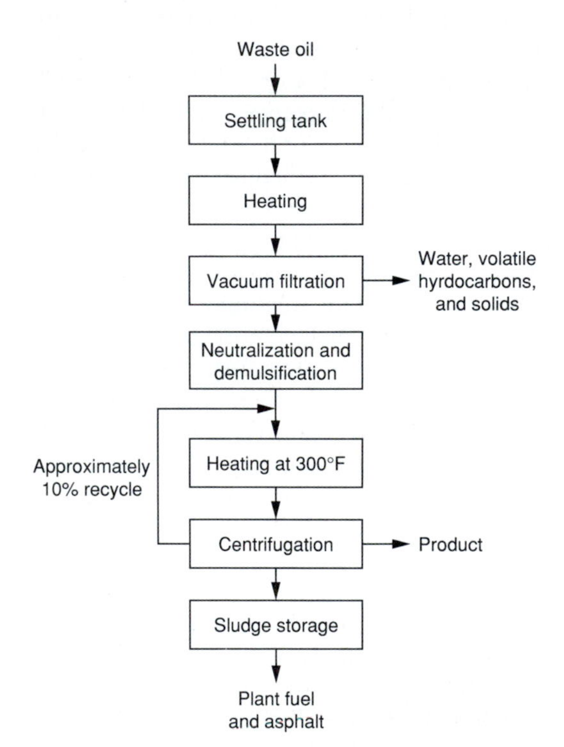

FIGURE 15-14
Centrifuge treatment of waste oil (adapted from Ref. 21).

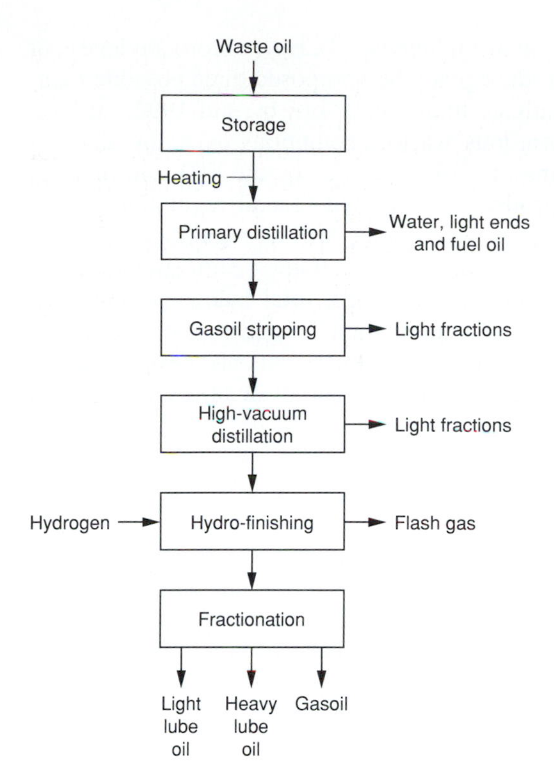

FIGURE 15-15
Kinetics Technology International (KTI) process for production of lubricating oil (adapted from Ref. 21).

major fraction in the lubricating oil range. The hydrotreating step consists of the addition of hydrogen gas and reaction over a catalyst. Hydrogenation removes the contaminants that boil in the same range as lube oil, including chlorine-, oxygen-, and nitrogen-containing compounds. The treated oil is flashed at low pressure to separate gaseous products, and is then fractionated into desired grades of lube oil. The KTI process is attractive because it provides good yield and a product quality equivalent to virgin lube oil. The process is able to treat oils contaminated by PCBs and other hazardous wastes.

In recent years, re-refiners have suffered financial difficulties, due in part to increasingly stringent regulations, but primarily due to very low prices for crude oil. Several refineries were constructed between 1980 and 1990, but were never put into operation. Low crude oil prices have forced down the value of re-refined oil, thus making collection, transportation, and refining economically marginal. Low crude oil prices have also been a disincentive to recycling; in many locations, waste oil generators must pay collection centers to accept used oil, and in turn, collection centers must pay transporters.

Regulations for Waste Oil

As of late 1991, the United States Environmental Protection Agency did not classify waste oil as a hazardous waste unless it had been contaminated with hazardous materials such as PCBs (polychlorinated biphenyls), TCE (trichloroethylene), or other chlorinated solvents. However, the agency had proposed three possible management plans and planned to promulgate final regulations by mid-1992. At least 10 states regulate used oil as a hazardous waste; exemptions exist for oils that meet recycled standards. The recyling of oil is regulated by the EPA and individual states, and burning is also controlled by state and federal regulations (e.g., Resource Conservation and Recovery Act and Clean Air Act regulations [21]).

Fuel oils produced by reprocessing must meet certain specifications regarding toxic metals (arsenic, cadmium, chromium, lead), total halogens, and flash point. Used oil containing less than 1000 ppm total halogens may be burned without federal restrictions. Oils that exceed allowable levels are designated *off-specification used oil* and may only be burned in industrial or utility boilers and furnaces.

15-13 USED TIRES

Approximately 281 million tires are replaced annually in the United States. An estimated 237 million tires are discarded, 10 million are reused, and about 33.5 million are retreaded [22]. Between two and three billion tires have accumulated in storage piles (see Fig. 15-16), and millions more have been dumped illegally [22]. Approximately 14 percent of the discarded tires are used as fuel; an estimated 5 percent are used for rubber-modified asphalt, crumb rubber, and miscellaneous uses; and 4 percent are exported [22].

Tire dealers, auto dismantlers, and even landfill operators pay independent contractors to pick up used tires. Whole tires are no longer buried at most landfills because they occupy a large volume and tend to rise to the surface. Landfills usually accept tires as a public service, but charge a fee to cover the collection costs. Tire collectors separate usable casings for retreading and deliver the remainder to a storage yard or shredder; after shredding, the scraps are landfilled. Owners of tire-burning waste-to-energy plants also operate their own collection services and recover reusable tires.

FIGURE 15-16
Portion of a remote canyon area used for the storage of discarded tires. (Courtesy of Bob Boughton, California Integrated Waste Management Board.)

Reuse and Recycling Opportunities

The principal reuse opportunities for rubber tires are for retreading and remanufacturing, tire-derived fuel, and rubber-modified asphalt. These uses and others are considered in the following discussion.

Retreading and Remanufacturing. The EPA has suggested that the number of tire discards could be reduced if consumers bought better-quality tires and purchased used or retreaded tires [22]. About 800 companies in the United States retread or remanufacture tires (retreaders replace only the tread, whereas remanufacturers replace both tread and sidewall rubber). Retreaded tires make up only 7.5 percent of passenger tire sales [31], in part because retreads still suffer a negative image, but also as a result of competition from cheaper import tires. The popularity of radial tires has also reduced retreading because most retreaders cannot afford the precision molds and equipment required. Truck tires, on the other hand, are made to be retreaded, and compose about 39 percent of all truck replacements. Replacement tires for very large vehicles (i.e., dump trucks, earth moving equipment, and so on) are so expensive that many tires are retreaded three times, and retreads make up 60 percent of all sales [32].

Tire-Derived Fuel. Currently, the largest use of discarded tires is for boiler fuel. Approximately 33 million tires are burned annually in waste-to-energy and manufacturing plants. The Oxford Energy Company plant in Westley, CA, uses 4.5 million tires per year; their plant in Sterling, CT, burns 10 million, and a new facility under construction near Las Vegas, NV, will have a capacity of 18 million tires per year (almost 50,000 per day). Cement producers burn about 6 million tires yearly, and pulp and paper mills in Oregon, Washington, and Wisconsin use about 12 million [22].

Rubber-Modified Asphalt. Rubber-modified asphalt has been used since the early 1960s. Approximately two million tires are used annually, accounting for less than 1 percent of the nation's discards. There are two general methods of preparing rubber-modified asphalt. In the *wet* process, crumb (finely ground) rubber is blended with asphalt at 400°F to form a chemical bond; in the *dry* process, the tire rubber is simply used as a substitute for aggregate.

Paving contractors have not been universally happy with the product, and state highway and city road departments, concerned about the increased cost, are worried that the use might be mandated as a solution to the growing supply of scrap tires. California and Arizona are pleased with results to date and will continue their programs; Minnesota and New York feel that the additional cost is not justified. The consensus of other states is that the increased cost will reduce their paving programs, and that the long-term performance remains to be evaluated.

Other Uses. Approximately 10 million tires are used annually for miscellaneous purposes. Whole tires have been used to create artificial reefs in New Jersey and

Florida, as erosion control structures in California and North Carolina, and for highway crash barriers. Split and punched tires are used to make miscellaneous products such as muffler hangers, belts, gaskets, and floor mats. A small amount of recovered crumb rubber is mixed with virgin rubber for new tire tread. Of these 10 million tires, 3 to 4 million are specially processed for rubber and plastic products. A new publicly-financed Tirecycle® plant in Minnesota produces polymer-treated rubber powder that is used to manufacture carpet backing, roofing, V-belts, and molded items [33]. Tires have also been experimentally treated by pyrolysis (see Chapter 13) to produce carbon black, an essential component of tires. Owing to the relatively low cost of natural gas, the traditional raw material for carbon black, this process has not been commercialized.

Specifications/Regulations for Used Tires

In general, there are no specifications for the recycling of used tires, as many of the applications are new and not fully developed. The specifications for reusable tires vary with individual tire retreaders and tire remanufacturers. The Federal Highway Administration is encouraging the use of rubberized asphalt, and is preparing guidelines and specifications for the use of waste tires in this application.

15-14 LEAD-ACID BATTERIES

Approximately 78 to 80 million automotive batteries are consumed and replaced annually in the United States, not including those used for large trucks or nonautomotive uses, such as lawn and garden machinery and emergency power [4]. A well-organized infrastructure exists to recycle lead-acid batteries (LABs), and the nationwide recycling rate is now about 90 percent. Because of the environmental hazards of lead, the goal of both the EPA and state governments is to maintain a high recycling rate despite market fluctuations in the price of lead or the loss of industry capacity. In an effort to create a recycling infrastructure that is stable when prices are depressed, the EPA has proposed fees on the use of virgin lead, mandatory take-back programs, and a minimum recycled lead content in new batteries.

Reuse and Recycling Opportunities

Battery manufacturers and secondary lead reclaimers carry out the recycling of lead-acid batteries in a cooperative effort. Approximately 1.4 million tons of lead are produced in the United States each year by primary producers and secondary smelters, and 80 percent of that production is used to make LABs (the remainder is used for leaded gasoline (still sold in some parts of the country), paint, solder, bearing material, and miscellaneous products such as tire weights and lead shot) [4]. In return, secondary smelters rely on the recovery of old batteries for more than 70 percent of their supply (the average battery contains about 18 pounds of recoverable lead) [4]. The primary lead industry is in decline, and more of the national requirement is being filled by secondary smelters using recycled materials.

Battery Processing. There are 22 active secondary processors in the United States; some are affiliated with battery manufacturers, but the majority are independent [2]. At a typical plant, truckloads of batteries are crushed and then the lead, plastic, and sulfuric acid are separated. All lead components are charged into a reverberatory furnace (a reverberatory furnace is one in which the flame is directed downward from the roof), where the oxides and sulfates are reduced to metallic lead. The molten lead is drained off from the furnace and the residue, still containing about 29 percent of the original lead, goes to a blast furnace, where silica, iron, and lime are added as fluxing and scavenging agents. Although nearly all of the remaining lead is recovered in the blast furnace, the slag still contains unrecovered lead, and each batch must be tested to determine if leaching is a danger. If the extraction procedure toxicity level exceeds 5 parts per million, the slag must be disposed of in a hazardous waste landfill [2].

New technologies are available that will meet environmental requirements and produce less slag. One is the use of a short rotary furnace in place of the reverberatory and blast furnaces. The resulting slag differs chemically from blast furnace slag and is not leached so easily. In the Engitec Impianti process developed in Italy in the late 1980s [27], batteries are crushed in a hammer mill and the components are separated on a vibrating screen. The acid/lead-paste slurry is neutralized, lead oxides are separated, and reusable sodium hydroxide and sulfuric acid are recovered from the solution by electrodialysis. The grids, poles, polyvinyl chloride separators, and polypropylene case fragments are separated by density in columns of liquid (see Chapter 12). Lead oxides are reduced by electrolysis and combined with metallic components, then melted at 400–500°C and cast into ingots. In addition to lead, polypropylene and battery-grade sulfuric acid are recovered for reuse, and it is claimed that the Engitec Impianti method produces no hazardous waste [27].

Recycling of Batteries by the Consumer. As far as the consumer is concerned, there are no special requirements for recycling a battery; it is simply turned in

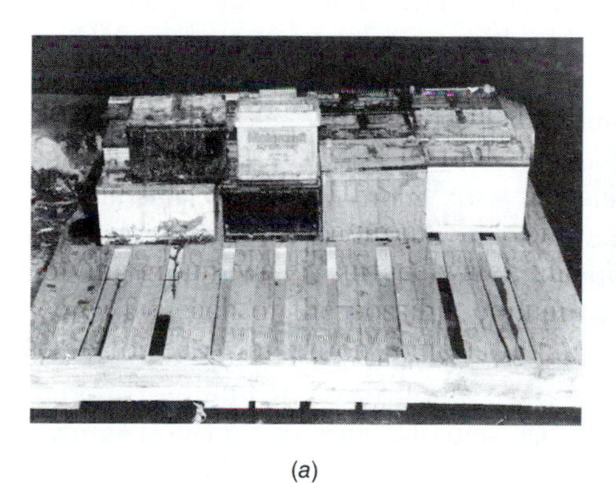

(*a*)

(*b*)

FIGURE 15-17
The recycling of batteries at a MRF: (*a*) lead-acid batteries stacked on a pallet for shipment to reprocessing facility and (*b*) the acceptance of small batteries for reprocessing and disposal.

to a dealer or retailer when a new one is purchased. (If proposed regulations by the EPA go into effect, vendors will be required to accept *all* used batteries, regardless of purchase.) Batteries are also accepted at MRFs (see Fig. 15-17*a*), scrap yards, auto dismantlers, and at some retail chain stores.

Regulations for Lead-Acid Batteries

Because approximately two thirds of all lead in municipal solid waste comes from auto batteries [3], the EPA declared in 1985 that LABs would be considered a hazardous waste, although storage and transportation regulations were written to allow consumers to return batteries to a regulated recovery network.

15-15 HOUSEHOLD BATTERIES

More than 2.5 billion household batteries are purchased annually in the United States, approximately 10 for each person. Most spent batteries are not separated by consumers but are simply discarded with household wastes. Batteries contain mercury, cadmium, lead, and other metals, which become toxic contaminants in landfill leachate or incinerator emissions. The Environmental Protection Agency has found that household batteries are the source of more than 50 percent of the mercury and cadmium in MSW [29]. Begining in 1993, the mercury content of alkaline and carbon-zinc batteries has essentially been eliminated through changes in the manufacturing process.

Reuse and Recycling Opportunities

Most consumers are not aware that household batteries are a potential source of toxic metals, and few states or municipalities attempt to recover them. In the few programs that exist, most batteries are collected at consumer electronics stores, at jewelers, and at some MRFs (see Fig. 15-17*b*). Recycling is difficult because very few companies have the technology to process household batteries, and there is no convenient collection infrastructure. In addition, mixed *button* batteries are difficult to sort and may present a storage hazard due to mercury vapor emissions. Another obstacle is that batteries may have to be separated individually to meet federal transportation requirements [29].

Consumers should not discard batteries with household waste but should instead turn them in during special collection drives, or contact public agencies that can dispose of them properly. Alkaline and carbon-zinc batteries containing mercury are not recyclable and must be disposed of in hazardous landfills. Nickel-cadmium cells or mercuric oxide and silver oxide button batteries are recyclable, although for a fee, one processor will deactivate and dispose of lithium batteries [29]. A summary of household battery types and the currently (1992) known processors is shown in Table 15-12.

Regulations for Household Batteries

Although the Environmental Protection Agency does not regulate household batteries as a hazardous waste, some states have passed legislation requiring

TABLE 15-12
Types of household batteries and U.S. recycling companies, as of 1992[a]

Battery type	Percentage of market	Metal of concern	Processors/recyclers[b]
Button cells			
Alkaline	5	Mercury	None[c]
Lithium	15	Lithium	MERECO (New York) Deactivation of lithium only
Mercuric oxide	20	Mercury	MERECO (New York)
Silver oxide	5	Silver	MERECO (New York) ECS Refining (California)
Zinc-air	60		None
Cylindrical, 6-volt, and 9-volt cells			
Alkaline	75	Mercury	None[c]
Carbon-zinc	15	Mercury	None[c]
Nickel-cadmium	1	Cadmium	Inmetco (Pennsylvania) MERECO
Other	4		None

[a] From Ref. 29.

[b] MERECO (Mercury Refining Company) will accept nickel-cadmium batteries for shipment to SNAM in France.

[c] Begining in 1993, mercury has essentially been eliminated in these batteries.

that manufacturers initiate recovery programs. New Jersey has considered legislation that would force manufacturers to take back batteries from retailers or municipal collection programs. California law prohibits disposal of household hazardous waste in landfills, but the regulations concerning batteries are not widely known or enforced, and the only official recovery mechanism is through periodic hazardous waste collections by counties and municipalities. The primary goal of public agencies is not to recycle batteries, but to dispose of them safely [29].

15-16 FUTURE RECYCLING OPPORTUNITIES

As existing landfill capacity is exhausted and replacement landfills are constructed, requiring expensive land purchases and extensive environmental protection, it is reasonable to expect that solid waste managers will attempt to reserve landfill space for materials that have no economic recycling potential. Because recycling opportunities are limited for many materials, source reduction will assume more importance for those waste materials that are now typically discarded. Approximately 30 percent of MSW consists of packaging material [31], so there is great potential for conservation of landfill space through source reduction. Packaging manufacturers can contribute to a solution by reducing the amount of material used per container or package, by using single materials rather than composites, by using recycled materials, and by clearly identifying the type of material to facilitate recycling. Consumers must assume responsibility for choosing products according to the recyclability of the packaging.

Because waste paper is such a large portion of the waste stream, additional capacity must be added to recycle newspapers, magazines, and lower grades of waste paper. There is a limit to the amount of used paper that can be incorporated into new paper, however, and additional uses for paper must be developed, such as fuel pellets for home or industrial heating, and built-up or laminated packaging to replace nonrecyclable plastic packaging. As consumers nationwide change habits to include the routine separation of materials, it is anticipated that new products will be developed.

In the future, plastics recycling may be achieved by chemical and refinery processes. Two resin producers are already depolymerizing postconsumer PETE bottles to produce ethylene glycol and terephthalic acid, which are used as feedstocks to synthesize virgin-quality PETE for new bottles. Major oil companies and resin producers are working on refinery processes that would transform plastics to combustible gases, hydrocarbon feedstocks, and oils. In addition, several laboratories are experimenting with auto shredder residue. One approach is to separate the polymers into hydrocarbon gases; another technique is to dissolve the thermoplastic components in different solvents to recover lubricants and usable polymers. Another approach that is being pursued by the plastics industry to minimize the impact of plastics in landfills is to produce biodegradable clones of the existing plastic materials identified in Section 15-4. The goal is to produce a biodegradable version of various plastics for single-use product applications such as fast-food containers and thin-film bags. Biodegradable clones are produced by incorporating photooxidation and biological and chemical triggers in the basic structure of the plastic material. Activation of these triggers will allow the plastics to be degraded biologically.

15-17 DISCUSSION TOPICS AND PROBLEMS

15-1. Prepare a list of the materials found in MSW that are now recovered for recycling in your community. How does your list compare with the materials listed in Table 3-10?

15-2. Obtain information on the infrastructure for the collection, processing, and marketing of plastics in your region. What might be done to improve the existing infrastructure?

15-3. Should government (state or federal) subsidies be provided to rural communities for the recycling of paper, plastics, glass, and tin cans? What are the advantages and disadvantages of such subsidies? Could the money that would be spent on such subsidies be spent more effectively? How?

15-4. Obtain the prices paid to your local waste management agency for waste paper and cardboard over the past five years. Are any trends discernible? What is your estimate of the total cost required to collect, process, and market these materials? What fraction of this cost is recovered by the sale of the materials?

15-5. Contact your local waste management agency and obtain a list of the specifications that must be met for aluminum cans, paper, and plastics that are to be recycled. How do the local specifications compare with the corresponding specifications given in Tables 15-1 and 15-2, and discussed in Section 15-4?

15-6. Develop a plan whereby the recycling of glass from a rural community can be accomplished without the need for economic subsidies, or with limited subsidies.

15-7. Describe the programs that are now in place in your community to deal with waste oil. Can you suggest some improvements to the existing programs?

15-8. Describe the programs that are now in place in your community to deal with used tires. Can you suggest some improvements to the existing programs?

15-9. Describe the programs that are now in place in your community to deal with household batteries. Can you suggest some improvements to the existing programs?

15-10. Describe the programs that are now in place in your community to deal with construction and demolition wastes. Can you suggest some improvements to the existing programs?

15-11. Review and summarize the regulations in your community that are applicable to household batteries. Do you believe that these regulations are adequate to protect the environment? Why?

15-12. Compare and contrast the life cycle of recovered aluminum and steel (tin) cans. Assume the recovered aluminum cans will be used to produce new containers and that the steel cans will be used to produce reinforcing bar.

15-13. Contact your local materials recovery facility or recycling center and obtain the price paid for one commonly recycled material. Estimate or obtain the cost of collecting, processing, and shipping this material to a buyer, and the price received. Does recycling show a net income? Under what conditions might the material be profitable to collect, process, and transport?

15-18 REFERENCES

1. Alpert, G.: (California Glass Recycling Corporation, 5709 Marconi Avenue, Carmichael, CA 95608); telephone communications, 1991.
2. Apotheker, S.: "Batteries power secondary lead smelter growth," *Resource Recycling*, Vol. IX, No. 2, pp. 46–47, 92, February 1990.
3. Apotheker, S.: "Does battery recycling need a jump?" *Resource Recycling*, Vol. IX, No. 2, pp. 21–23, 91, February 1990.
4. Apotheker, S.: "Get the lead out," *Resource Recycling*, Vol. X, No. 4, pp. 58–63, April 1991.
5. ASTM: *1989 Annual Book of ASTM Standards, Volume 11.04, Water and Environmental Technology*, American Society for Testing and Materials, Philadelphia, PA, 1989.
6. ASTM: "Standard Specifications for Waste Glass as a Raw Material for the Manufacture of Glass Containers," E 708-79 (Reapproved 1988), *1989 Annual Book of Standards, Volume 11.04*, American Society for Testing and Materials, pp. 299–300, Philadelphia, PA, 1989.
7. Bartley, D. A., S. A. Vigil, and G. Tchobanoglous: "Use of source separated waste paper as a refuse derived fuel," *Biotechnology and Bioengineering Symposium*, No. 10, pp. 67–79, 1980.
8. Basta, N., K. Fouhy, K. Gilges, A. Shanley, and S. Ushio: "Recycling everything, part 1: plastic recycling gains momentum," *Chemical Engineering*, Vol. 97, No. 11, pp. 37–43, 1990.
9. Basta, N., K. Gilges, and S. Ushio: "Recycling everything, part 3: paper recycling's new look," *Chemical Engineering*, Vol. 98, No. 3, pp. 45–48F, 1991.
10. Brown and Caldwell, Inc.: "Source Reduction and Recycling Element, and Household Hazardous Waste Element for Alameda County Jurisdictions," Appendix D, preliminary draft, August 1991; Brown and Caldwell, Inc., Pleasant Hill, CA, 1991.

11. Bye, J.: "Setting standards for compost utilization," *BioCycle*, Vol. 32, No. 5, pp. 66–70, May 1991.
12. COPPE: "Plastic packaging recycling," Council on Plastics and Packaging in the Environment, 1275 K St., NW, Washington, DC 20005, 1991.
13. Donovan, C. T.: "Construction and demolition waste processing: new solutions to an old problem," *Resource Recycling*, Vol. X, No. 8, pp. 146–155, August 1991.
14. Donovan, C. T.: "Wood waste recovery and processing," *Resource Recycling*, Vol. X, No. 3, pp. 84–92, March 1991.
15. Force, J.: (MRI Corporation/Proler International, 6000 Marginal Way SW, Seattle, WA 98106); telephone communications, April 1991 and August 21, 1991.
16. Hegberg, B. A., W. H. Hallenbeck, G. R. Brenniman, and R. A. Wadden: "Setting standards for yard waste compost," *BioCycle*, Vol. 32, No. 2, pp. 58–61, February 1991.
17. Hemphill, T.: (Paper Stock Institute of America, a division of the Institute of Scrap Recycling Industries, 1627 K St. NW, Suite 700, Washington, DC 20006); telephone communication, August 12, 1991.
18. Holusha, J.: "Recyclers wring new life from old cars," *The Sacramento Bee*, June 28, 1991.
19. Kaplan, R. S. and H. Ness: "Review article No. 13: recycling of metals," *Conservation and Recycling*, Vol. 10, No. 1, 1987.
20. Lohof, A.: "Used oil management in selected industrialized countries," *Resource Recycling*, Vol. X, No. 9, p. 62, September 1991.
21. Mueller Associates, Inc.: *Waste Oil: Reclaiming Technology, Utilization and Disposal*, Noyes Data Corporation, Park Ridge, NJ, 1989.
22. Pillsbury, H.: "Markets for scrap tires: an EPA assessment," *Resource Recycling*, Vol. X, No. 6, pp. 19–24, June 1991.
23. Plastics World: "Is it really practical to recycle baby diapers?" *Plastics World*, pp. 61–64, April 22, 1990.
24. Plastics World: "Producers plan to recycle 25% of food service plastics," *Plastics World*, pp. 45–53, April 22, 1990.
25. Powell, J.: "Recycling in the '80s: how are we doing?" *Resource Recycling*, Vol. 8, No. 2, May/June 1989.
26. Powell, J.: "How are we doing? the 1990 report," *Resource Recycling*, Vol. X, No. 4, April 1991.
27. "Recovery system for lead-acid batteries," *The Chemical Engineer*, No. 486, p. 15, November 29, 1990.
28. Rennie, C.: (Envirothene, Inc., 14312 Central Avenue, Chino, CA 91710); telephone communication, August 30, 1991.
29. Reutlinger, N. and D. de Grassi: "Household battery recycling: numerous obstacles, few solutions," *Resource Recycling*, Vol. X, No. 4, pp. 24–29, April 1991.
30. Selke, S.: *Packaging and the Environment: Alternatives, Trends, and Solutions*, Technomic Publishing Company, Lancaster, PA, 1990.
31. Sikora, M. B.: "A little retreading goes a lot of miles," *Resource Recycling*, Vol. IX, No. 12, pp. 50–58, December 1990.
32. Spencer, R.: "New Hampshire issues solid waste compost rule," *BioCycle*, Vol. 32, No. 11, pp. 71–74, November 1991.
33. Stark, F. J. III: "The Tirecycle solution: Minnesota's answer to the scrap tire disposal problem," *Journal of Resource Management and Technology*, Vol. 17, No. 3, October 1989.
34. Steel Can Recycling Institute, Foster Plaza 10, 680 Andersen Drive, Pittsburgh, PA 15220 (circa 1991).
35. U.S. Environmental Protection Agency: *Decision-Makers Guide to Solid Waste Management*, EPA 530-SW89-072, Washington, DC, November 1989.

PART V

CLOSURE, RESTORATION, AND REHABILITATION OF LANDFILLS

For as long as the wastes placed in landfills continue to produce gases and leachate—the byproducts of decomposition—these landfills will remain a legacy for the future. The management of closed landfills is part of any integrated solid waste management system. As new landfills are built, operated, and monitored, a data base is being developed that can be used as a basis for future engineering, social, and economic judgments about the continued use of landfills. To ensure that existing landfills, when closed, do not affect public health and the local environment, federal and state regulations have been developed that specify minimum standards for closure and postclosure maintenance.

The closure and postclosure care and rehabilitation of landfills is the subject of Part V. Landfill closure, restoration, and rehabilitation have been separated from the landfill design and operation details of Chapter 11, because the time that elapses between the original landfill design and final closure is significant. Many landfills are now owned by persons who had no responsibility for the original landfill design. Closure and postclosure standards and guidelines are presented and discussed in Chapter 16. Because of the special importance of the revegetation of closed landfill sites, special attention is focused on this subject. Remedial actions that must be taken to rehabilitate abandoned nonhazardous landfills are presented and discussed in Chapter 17.

CHAPTER 16

CLOSURE OF LANDFILLS

Landfill design and construction is a continuous activity that is completed only when all of the available or permitted capacity of the site has been filled with solid waste. Once that happens, the landfill must be closed, the final action of a facility that is to receive no more solid wastes. To ensure the functioning of environmental controls during closure and for a period of time after closure, a closure plan must be developed early in the life of a landfill, often at the design or site development phase. The goal of the plan is to define the steps that must be taken in closing the landfill and the elements of postclosure care required by federal or state laws.

The purpose of this chapter is to introduce the reader to the key elements involved in the development of closure plans for landfills and guidelines for the long-term care (operation and maintenance) of closed landfills (also known as *waste management units*). Note that landfill closure is a separate activity from the landfill design and operations described in Chapter 11. Such separation emphasizes the long period of time that elapses between the design and initiation of a landfill and the landfill closure.

16-1 DEVELOPMENT OF A CLOSURE PLAN

As a waste management unit, a landfill, when completed, must continue to function effectively as an environmental control unit for solid wastes for a long time into the future. As landfill regulations have become more prescriptive, many states have required the development of a landfill closure plan as a part of the site approval process, before construction and landfilling operations begin. The closure plan must show all features of the completed site (see Fig. 16-1) and identify the agencies responsible for implementing closure of facilities. Closure plans developed at the time a landfill is opened can be expected to change during the time the landfill is operational. Thus, it is important to update the closure plan periodically. In California, closure plans are updated every five years, or whenever

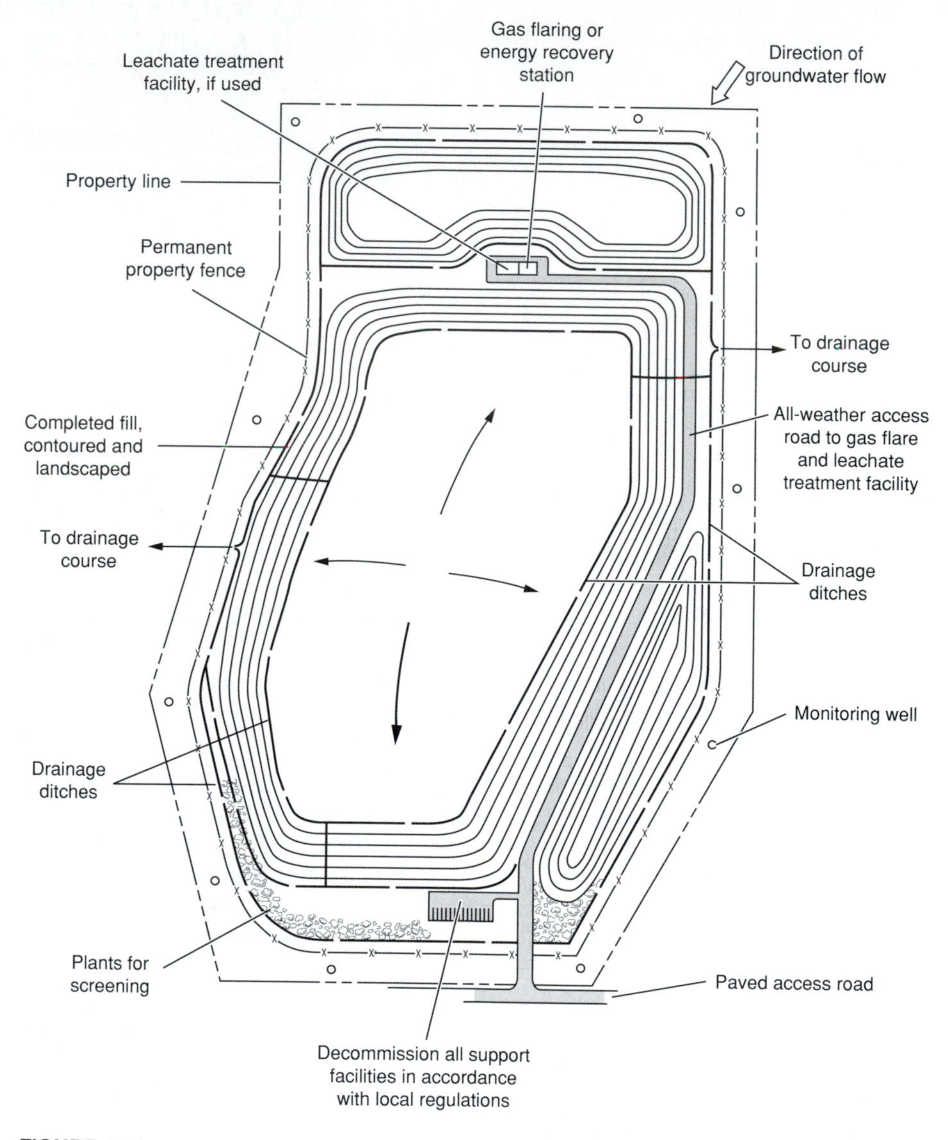

FIGURE 16-1
Plan view of completed landfill showing all of the elements involved in closure and postclosure care.

there are significant changes to landfill operations. A final closure plan is prepared and adopted just before the landfill is to stop receiving wastes. The elements of a closure plan are identified in Table 16-1. These elements must be addressed in a landfill closure plan:

- Final cover design
- Surface water and drainage control systems
- Control of landfill gases
- Control and treatment of leachate
- Environmental monitoring systems

Also, the natural biological processes occurring in the landfill will ultimately cause the landfill to stabilize and become usable for other community purposes. The potential uses for completed landfills should also be identified in the closure plan.

Final Cover Design

The final cover is the surface to be placed over a landfill after all wastes are received (see Fig. 16-2). Final cover design is an integral part of the site development plan. The design of the final cover must satisfy two functions for the site: (1) ensure the long-term postclosure integrity of the landfill with respect to any emissions to the environment, and (2) support the growth of vegetation or other site uses.

Cover Design Parameters. Typical cover design parameters include (1) design configuration, (2) final permeability, (3) surface slope, (4) landscape design,

TABLE 16-1
Typical elements of a landfill closure plan

Element	Typical activity
Postclosure land use	Designation and adoption
Final cover design	Select the infiltration barrier, final surface slopes and vegetation
Surface water and drainage control systems	Calculate stormwater quantities for runoff and select perimeter channel location and sizes to collect runoff and to prevent runon
Control of landfill gases	Select locations and frequency of gas monitoring and set the operations schedule for gas extraction wells and flare, if required
Control and treatment of leachate	Set the operation schedule for leachate removal and treatment, if required
Environmental monitoring systems	Select sampling locations and frequency of monitoring as well as constituents to be measured

FIGURE 16-2
Aerial view of completed Palos Verdes landfill in southern CA. The lower portion of the landfill is now a botanical garden. The main portion of the landfill site is scheduled to be developed as open space under the County Department of Parks and Recreation. The upper portion of the landfill, which received inert wastes, has been developed into Howlett Park, a multipurpose recreational complex. Note residential and commercial development surrounding the landfill. (Courtesy of County Sanitation Districts of Los Angeles County.)

(5) method of repair as landfill settles, and (6) slope stability under static and dynamic loadings. Typical landfill cover designs are presented in Fig. 11-53. The anticipated performance of the landfill cover can be evaluated by subjecting the final design to an engineering analysis for surface settlement, slope stability, and surface loadings at the site. Natural and synthetic materials used in the design of the final cover may have to be modified to resolve performance problems (e.g., stretching limits) identified in the evaluation.

Revegetation. Although a closed landfill provides a large land surface that can be used for many purposes with appropriate environmental control features, the most common use of closed landfills is for growth of plants. Where plants are used, planners must take special care in selecting plant species so that they survive the particular conditions at a closed landfill. Because the subject of selecting landscape vegetation for closed landfills is not widely covered in the literature, a separate section of this chapter is devoted to this topic.

Construction Quality Assurance Program. Because many regulations covering landfill closure are performance based, most regulatory agencies are requiring stringent controls and monitoring during construction of the final cover. A construction quality assurance (CQA) program provides the required details of how the final landfill cover is to be tested and monitored during construction. The CQA program should be coordinated closely with the final cover design, and should describe installation and testing procedures for soils, drainage grids, and synthetic membranes.

Soils. Each layer of soil in the final cover will have a test for density, in-place permeability, and thickness. Where soil is placed over geotextiles and synthetic membranes, there will be a specified control for the thickness of soil that can be placed in one pass when multiple passes will be required to achieve total soil layer thickness. The uppermost soil layer, often the layer that supports vegetation, will require testing to establish its ability to sustain plant growth. The analytical results for soil tests on the final cover are specific to each site and cannot be generalized.

Drainage grid. The drainage grid is the means by which water that penetrates the uppermost soil layer is routed off the landfill without penetrating the underlying wastes. The CQA program will focus on the integrity of the drainage layer to prevent fracture of the flow line or creation of ponding over the underlying synthetic membrane.

Synthetic membrane. In many landfill closure plans, synthetic membranes are used to prevent most of the water that penetrates the upper soil layer from entering the wastes. The membrane manufacturer provides guidance on construction of the membrane over a landfill. The CQA program will specify the responsibility of the construction contractor or city crews to ensure that the membrane manufacturer's recommendations are followed.

Surface Water and Drainage Control Systems

Artificial and natural features at the landfill site control surface water and groundwater (see Fig. 16-3). When integrated, the artificial and natural features must be effective in controlling run-on and runoff of surface waters as well as preventing groundwater from penetrating the landfill liner. When the landfill is closed, the drainage control system must be designed to function for the long-term use of the site. Rainfall and snowmelt must be removed from the final cover surface without soil erosion or excessive water infiltration. The greatest risk to the site is from ponding of surface waters in areas of land subsidence. The following features must be included in the design of drainage control facilities: (1) collection and routing of surface waters off the landfill surface in the shortest possible distance; (2) selection of channel and drainage ways that will carry waters at adequate velocities to avoid deposition; (3) use of sufficient surface slopes to maximize the removal of surface runoff and at the same time minimize surface scour; and (4) material specifications for the drainage features that allow repair and replacement as the landfill settles (see Fig 16-3*b*).

(a) (b) (c) (d)

FIGURE 16-3
Typical examples of drainage facilities at completed landfills: (*a*) permanent drainage facilities to divert storm water from landfill, the trapezoidal drainage ditch is cast in place in a continuous operation, (*b*) drainage facilities built in sections to allow for landfill settlement, (*c*) unlined ditch overgrown with natural grass, and (*d*) temporary drainage facilities installed to relieve surface flooding of landfill.

Control of Landfill Gases

Landfill gases must be controlled for as long as they are expected to be generated after the landfill is closed (see Fig. 11-13). As described in Chapter 11, typical landfill gas control facilities include extraction wells, collector and transmission piping, and gas flaring and/or combustion facilities (see Fig. 16-4). The landfill gas control system used while the landfill was active is also used for the control of the landfill gases after the landfill has closed. The most critical design steps are the selection of materials and the placement of wellheads, valves, and collection pipes in the final cover. The materials used in pipe manufacturing must be flexible, to withstand movement when the land settles, and strong enough to withstand the loadings of vehicles passing over the surface when maintaining landscape plants and the gas extraction and collection facilities.

(a)

(b)

(c)

FIGURE 16-4
Landfill gas management facilities: (*a*) extraction wells and gas piping, (*b*) gas combustion facilities (a gas turbine is shown), and (*c*) gas flaring facilities (a candlestick flare is shown).

An important consideration in the management of landfill gas is that the quantity of methane produced after landfill closure may not be sufficient to support combustion. When planners anticipate this situation, they must make provision for a supply of auxiliary fuel for the combustion of the extracted landfill gases, especially where the control of VOC emissions is an issue. The U.S. EPA is considering setting standards for nonmethane gas emissions from landfills. The standard would be similar to emission standards from combustors.

Control and Treatment of Leachate

In addition to contaminating the ground water, leachate can also transport dissolved organic substances that may be released in the unsaturated subsurface environment, by the change in the partial pressure of the constituents in the gas phase. To minimize the movement of leachate toward groundwater and the release of dissolved constituents, the liner must be constructed under strict quality control. The amount of leachate to be controlled and treated after the landfill is closed is a function of the final cover design, the types of waste placed in the landfill, and the climate, especially precipitation, of the region. With an effective landfill cover in place, the amount of leachate will decrease after closure until only the leachate generated from the decomposition of the waste will be collected.

Leachate collection and treatment facilities are designed and built when the landfill first starts operations. The same facilities are used after closure. As the closed landfill matures, the quantity of leachate generated normally decreases, as do the BOD_5 and COD concentrations (see Fig. 11-11). Owing to the reduced quantities of leachate and reduced waste strength, the treatment works may become underloaded and turn odorous. In many cases, the best treatment option is to discharge the leachate to a wastewater treatment plant, where the large quantities of wastewater buffer the smaller quantities of leachate.

Environmental Monitoring Systems

The final part of a closure plan involves the environmental monitoring facilities. Environmental monitoring is necessary to ensure that the integrity of the landfill is maintained with respect to the uncontrolled release of any contaminants to the environment. In most instances, the selection of facilities and procedures to be included in a closure plan will be a function of the environmental control facilities used during landfill operations before closure (see Table 16-2).

Selection of environmental monitoring methods and facilities for closed landfills will be most successful when done in accordance with the guidelines of the regulatory agency. Unfortunately, many state regulatory agencies have not yet developed landfill closure guidelines, so that solid waste management agencies are faced with the possibility of selecting environmental monitoring facilities that may be unacceptable under future guidelines. In the face of this uncertainty, designers should choose monitoring facilities that can be used to track the movement of any landfill emissions to the water, air, and soil environments.

Water. Monitoring of water quality and movement is done to identify leachate leakage from the landfill (Fig. 16-5). Monitoring facilities will be placed in soils under the landfill liner and in the uppermost groundwater aquifer. In dry climates, where moisture does not penetrate to soils beneath the landfill, the monitoring facilities must be capable of functioning in the vadose zone. The groundwater aquifer is monitored by wells. Equipment and facilities for vadose zone and groundwater monitoring were presented and discussed in Chapter 11.

Air. A landfill closure plan will show the manner in which methane and other gases are to be controlled and discharged to the atmosphere. Gas monitoring is also used to assess the degree of biological activity in the landfill. Typical gas monitoring

TABLE 16-2
Environmental monitoring facilities that are installed during landfill construction and operations and used after landfill closure

Monitoring facility	Function during operations	Function after closure
Groundwater monitoring wells		
Upgradient	Water sampling at location to get background water quality	Same functions as during operation
Downgradient	Water sampling at location to detect movement of leachate contaminants; if contaminants are present, stop operations and correct problem with liner; wells function as a control variable for operations	Water sampling at location to detect any leachate plume created by a leaking liner; a data reference location for defining the direction and rate of movement for a contaminant plume
Vadose zone lysimeters	Sampling location to detect liquids in soils above groundwater; if liquids are present, stop operations and determine the cause; correct problems before restarting operations	Sampling location to detect liquids in soils above groundwater; if liquids are present, complete additional investigations as to cause; correct any problems as required by regulatory agency
Gas vents	Sampling location for combustible gases	Sampling location for combustible gases; gas extraction wells for control and removal of methane gas after closure
Leachate treatment facilities	Leachate quantity measurement and quality sampling location	Same functions as during operations
Stormwater holding basins	Retain stormwater for regulated release of basins; measure quantity and sample for quality	Same functions as during operations

equipment used at closed landfills includes explosive gas meters, hydrogen sulfide meters, and sample collection equipment and containers for samples to be analyzed off-site.

Soil. In most landfill closure plans, cover soil is one of the most important features. It must be placed under strict construction supervision, and then maintained to prevent loss of soils. Environmental monitoring of soils includes measuring land surface settlement, soil slippage, and land surface erosion. Inspection of closed landfills requires training and good judgment in making visual observations and in the use of survey monuments to monitor cover layer movement.

(a)

(b)

FIGURE 16-5
Typical groundwater sampling well: (*a*) well with locked cap and (*b*) electrical connection and sampling hose stored in the top of the well casing.

16-2 REVEGETATION OF CLOSED LANDFILL SITES

In many areas population has increased so rapidly that former landfill sites, and even some currently used sites, are now surrounded by industrial, commercial, or residential developments. Because open space in urban areas is highly desirable, former landfill sites present a unique opportunity for land reclamation. Some possible uses for closed landfill sites include parks, recreational areas, nature preserves, botanic gardens, crop production, and even commercial development. Each of these uses has been implemented in parts of the country, but each presents a unique challenge.

Selection of an end use for a former landfill is dictated by the needs of the community and the funds available for the reclamation project. Parks with limited facilities and wildlife habitats, for example, would require less expenditure than golf courses and multi-use recreational areas. All of the end uses mentioned above have one thing in common—they require vegetation to realize their potential. The purpose of this section is to address the problems encountered and solutions that have been developed in seeking to revegetate closed landfills.

Factors That Limit Growth of Vegetation on Landfills

Factors that limit growth of plants on completed sanitary landfills include toxicity of landfill-generated gases (e.g., CO_2 and CH_4) to root systems, low soil oxygen, thin cover soils, limited cation exchange capacity, low nutrient status, low water-holding capacity, low soil moisture, high soil temperatures, high soil compaction, poor soil structure, and improper choice of plant species [5].

Root Toxicity. The main gases produced by anaerobic decomposition of wastes in a closed landfill are CO_2 and CH_4. High concentrations of CO_2 have been shown to be directly toxic to plants, although CH_4 is not phytotoxic (toxic to plants) by itself [2]. Its damaging action results from the displacement of oxygen, leading to anaerobic conditions detrimental to plants. These two gases compose approximately 95 percent of landfill gas by volume. Hydrogen sulfide, ammonia, hydrogen, mercaptans, and ethylene (among others) make up the remaining 5 percent. Of these gases, H_2S and C_2H_4 are known to be toxic to plants even in minute amounts.

From studies of soils at landfill sites, it has been concluded that the reduced state of soils saturated with landfill gas enhances the availability of trace elements, possibly at high enough levels to be phytotoxic. Although the potential for toxicity exists, it has not been proven that excessive levels of trace elements such as cadmium, copper, and zinc actually injure vegetation planted on landfill sites [2, 5].

Low Oxygen Supply. Pore spaces in the soil are alternately occupied by water and atmospheric gases. Following rain or irrigation, the pores fill up with water, displacing air. As gravity pulls the water out of the larger pores, air moves back in. Good plant growth depends on enough large pores to hold air and enough small pores to retain moisture between rains or irrigations. Because the oxygen supply to plant roots depends on the ability of a soil to hold air, any process that reduces pore space is detrimental to plant growth. Soil compaction by heavy machinery, in addition to the poor soil structure of most landfill cover soils, adds to the difficulty of growing plants.

Low Cation Exchange Capacity. Cation exchange capacity (CEC) relates to the ability of the soil to adsorb and retain nutrients. Colloidal organic matter and clays are the primary source of cation exchange sites in the soil. Positive ions (cations) are adsorbed at negatively charged surface locations of soil colloids. Adsorbed cations resist leaching from the cation exchange sites, but can be replaced by other cations through mass action. Cations found in large quantities on exchange sites are calcium, magnesium, hydrogen, sodium, potassium, and aluminum. The availability of many essential nutrients depends on the CEC of the soil. A soil with low organic matter content will be unable to hold on to nutrients and prevent their leaching out from the root zone. A typical range for organic matter in soils is 2 to 5 percent [9]. The generalized relationship between soil texture and CEC is presented in Table 16-3.

Low Nutrient Status. Soil fertility refers to the nutrients available in the soil necessary for plant growth. Sixteen plant nutrients are known to be essential. These are shown in Table 16-4 in elemental and plant-available ionic form. Hydrogen, carbon, and oxygen come from air and water; nitrogen is derived from both air and soil; and the rest originate from the soil [1]. Nitrogen, phosphorus, and potassium are called macronutrients and are absorbed in large quantities from soil and added

TABLE 16-3
Soil texture and cation exchange capacity[a]

Soil texture	Cation exchange capacity, meq/100 g dry soil
Sand	1–5
Fine sandy loams	5–10
Loams and silt loams	5–15
Clay loams	15–30
Clays	over 30

[a] Adapted from Ref. 1.

fertilizers. Micronutrients (trace elements) are taken up from the soil in smaller quantities; however, despite the small amounts needed by the plants, lack of these elements will adversely affect plant growth and development.

Soil used for the final cover on a landfill usually comes from the most readily available and least expensive source. As a result of economic considerations, landfill cover soils are often of poor quality both in texture and in nutrient content. Leone and Flower [8] found that landfill cover soils were "particularly low in organic matter and in ionic strength. They were generally at the low end of their respective ranges."

TABLE 16-4
Plant nutrients and their common forms in air, water, and soil[a]

Element and symbol	Ion or molecule
Carbon (C)	CO_2
Oxygen (O)	CO_2, OH^-, CO_3^{2-}
Hydrogen (H)	H_2O, H^+
Nitrogen (N)	NH_4^+ (ammonium), NO_3^- (nitrate), NO_2^- (nitrite)
Calcium (Ca)	Ca^{2+}
Potassium (K)	K^+
Magnesium (Mg)	Mg^{2+}
Phosphorus (P)	$H_2PO_4^-$, HPO_4^{2-} (phosphates)
Sulfur (S)	SO_4^{2-} (sulfates)
Chlorine (Cl)	Cl^- (chloride)
Iron (Fe)	Fe^{2+}, Fe^{3+} (ferrous, ferric) = Fe (II), Fe (III)
Boron (B)	H_3BO_3, $H_2BO_3^-$ (boric acid, borate)
Manganese (Mn)	Mn^{2+} = Mn (II)
Zinc (Zn)	Zn^{2+} = Zn(II)
Copper (Cu)	Cu^{2+} = Cu (II)
Molybdenum (Mo)	MoO_4^{2-} (molybdate)

[a] Adapted from Ref. 1.

Low Water-Holding Capacity. Water retention capacity of a soil depends on the physical properties of the soil. Soil texture and soil compaction are important considerations in the ability of a soil to retain water. During rain or irrigation the large pores in the soil fill with water. As this water percolates through the soil by gravity, air replaces water in the larger pores. Water is retained in the smaller soil pores by capillary force. The soils with the most plant available water are those of medium texture with an ideal ratio of large and small pores. Compaction by the use of heavy machinery reduces pore size and prevents the infiltration and retention of adequate amounts of water into the soil.

Low Soil Moisture. Low soil moisture and low water-holding capacity are related. Two factors may be responsible: compaction and soil discontinuity. Compaction is a necessary procedure in modern landfilling techniques, but it damages the soil structure by decreasing pore space in the soil. Without adequate pore space, the water retention capabilities of a soil are diminished and runoff is increased, resulting in drier soils and erosion. In general, similar soils adjacent to landfills have less runoff and greater water infiltration capacity than the soils on the landfill itself. Soil discontinuity results from the layering of waste and soil and may hinder water from rising through the soil as would happen in a normal soil profile.

High Soil Temperature. Soil temperatures in excess of 100°F have been found in closed landfills [2]. Although such extreme temperatures are not common, in conjunction with other soil-related problems, higher temperatures contribute to plant stress.

High Soil Compaction. The waste and soil layers placed on a landfill are compacted daily with heavy machinery, a practice that has many desirable effects from the engineering perspective. Soils placed as part of the final landfill cover are also compacted. Unfortunately, this activity results in a reduction of soil porosity and permeability. Water and air can no longer pass through the soil layers, and plant roots are denied the air and moisture needed for proper growth.

Poor Soil Structure. Soil structure refers to the aggregation of soil particles. Aggregation agents can be organic matter, iron oxides, carbonate clays, and silica [1]. Organic matter is the aggregation material most effective in improving soil structure for plant growth. Soils that are composed of mostly uniformly sized particles suffer from poor water-holding capacity in addition to other problems that affect plant growth. Addition of organic matter such as compost creates a granular soil that is most favorable for plant growth and development. A 2 to 5 percent range for organic matter is desirable in most soils.

Determination of Site Conditions

Before initiating a planting program one must determine existing soil conditions. A field survey is performed, then soil testing. After completion of these two steps, necessary soil improvements can be undertaken.

Field Survey of Soil Condition. The first step in a field survey is a visual inspection. If vegetation is already present, does it look unhealthy or is there dead vegetation? If so, is there a discernible pattern of dead or dying patches of vegetation? Such a visual inspection can be used to identify problem areas. These areas should be tested for landfill gases in addition to normal soil testing. The second step is done simultaneously with the visual inspection. Landfill gases from anaerobic decomposition have a putrid odor. Small surface cracks in the cover soil allow gases to escape, and the smell is noticeable when one walks on the site. Disturbing the soil surface releases trapped gases and is often an early sign of gas migration off-site. A more accurate method is the use of a portable methane detector and probe. A portable hydrogen sulfide meter is also available. The third step involves actual examination of the soil. Guidelines for the field evaluation of soils are presented in Table 16-5.

Soil Testing. Once the preliminary steps in determining soil conditions have been taken, a more thorough investigation of soil characteristics is needed. No former landfill site should ever be revegetated without first doing a detailed soil analysis. Time and money would be wasted in any attempt to grow plants without a thorough knowledge of site conditions and the need for certain soil improvements.

The soil tests that should be conducted as part of a site evaluation, which are summarized in Table 16-6, include the macronutrients, micronutrients, pH, conductivity, bulk density, and organic matter content. Nutrients can be supplied to the plants by carefully designed fertilizer programs based on soil test results. Amendments may be required to bring the soil pH within the desired range. Some trace elements become more readily available as pH decreases and can become phytotoxic. High concentrations of zinc, copper, magnesium, iron, cadmium, and lead are injurious to plants. Conductivity below 2 mmhos is necessary to preserve proper water balance in the soil. Bulk specific weights between 75 and 87 lb/ft^3 are ideal and bulk density should not exceed 106 lb/ft^3. The organic matter content should be on the order of 2 to 5 percent [1].

TABLE 16-5
Guide for field evaluation of soils[a]

	Soil environment	
Characteristic	**Aerobic**	**Anaerobic**
Odor	Pleasant	Septic
Color	Lighter	Darker
Moisture content	Lower	Higher
Friability	Good	Poor
Temperature	Lower	Higher

[a] Adapted from Ref. 3.

TABLE 16-6
Tests that should be conducted to assess the suitability of soils for plant growth

Test	Units	Elements
Macronutrients	mg/g dry soil	N, P, K, Ca, Mg, S
Micronutrients	mg/g dry soil	Cl, Cu, B, Fe, Mn, Mo, Zn
pH	unitless	
Conductivity	mmhos	
Bulk specific weight	lb/ft^3	
Organic matter content	mg/g dry soil	
Cation exchange capacity (CEC)	meq/100 g dry soil	

Improving Site Conditions

Proper planning produces the best possible results in landfill revegetation. Ideally, the end use of a landfill should be determined while the landfill itself is still in the planning stages. A variety of strategies exist that, when implemented, isolate the plants, particularly deeper-rooted trees and shrubs, from the waste.

Gas Removal. Active gas extraction systems remove the maximum amount of gases from the soil and, therefore, also from the root zones of plants. If active gas extraction is not possible, gas migration barriers and venting systems should be installed. In the immediate root zone of trees, the following methods have been found to be effective:

1. Soil mounds—a 3-ft high mound either with or without an underlying geomembrane barrier.
2. Geomembrane lined, excavated cavity (see Fig. 16-6). The depth of the excavation depends on the rooting characteristics of the shrub or tree to be planted.

Areas Left Free of Waste. If the areas for future landscaping are known, tree-planting sites can be left unfilled to allow for the planting of deeper-rooted trees and shrubs.

Waste Removal from Tree-Planting Sites. Waste can be removed from areas where trees will be planted. Although this method can be expensive, it may be the best solution for the planting of "islands" of trees where many trees will be planted in one area; otherwise each would require its own mound or gas barrier.

Segregating Biodegradable Materials in the Landfill. Anaerobic decomposition of putrescibles generates the gases that are toxic to plants. By locating biodegradable and nonbiodegradable wastes in separate areas of the landfill, zones that are relatively free of toxic gases can be created for tree growth.

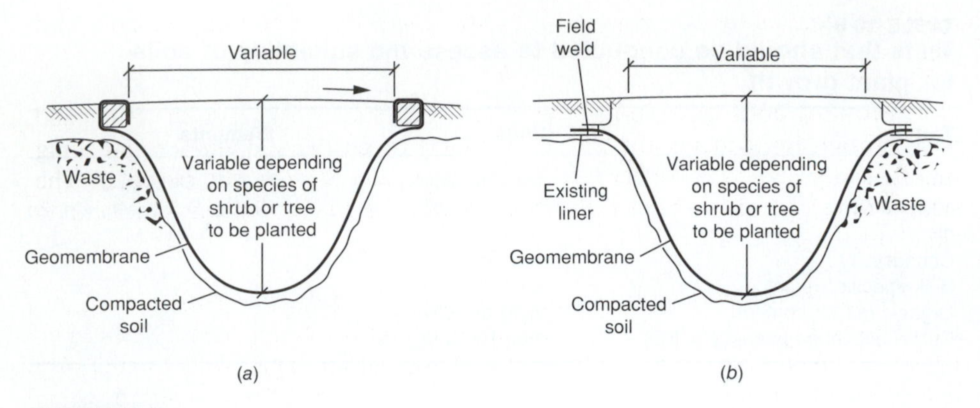

FIGURE 16-6
Methods used for planting shrubs and trees at landfills: (*a*) landfill without geomembrane liner in final cover and (*b*) landfill with geomembrane liner in final cover.

Stockpiling of Topsoil. Whenever feasible, the scraping and stockpiling of native topsoil for later use as the final cover for the closed landfill is recommended. Particularly when the end use is restoration of the site to its natural condition and native plants are to be used, the availability of local soil will greatly enhance the success of the plantings. Use of stockpiled soil will reduce one stress factor for plants growing under the inherently adverse conditions of a closed landfill site.

Preparing the Site

The final layer of cover soil should be optimized for plant growth. Cover soils of good quality should be used for the final 2- to 3-ft layer. The following procedures are recommended for site preparation:

1. Soil amendments should be mixed with the cover soil before spreading. Premixing of soil amendments produces a more uniform mix.
2. Cover soil should be spread when dry to avoid excessive compaction.
3. Earthmoving equipment other than earth scrapers should be used to apply and spread the cover soil to minimize compaction.

Use of Mulches on Closed Landfills

Mulching is the process of applying a top-dressing layer of organic or inorganic material to exposed soil surfaces. Consideration is given here only to organic mulches.

Benefits of Mulching. In addition to the benefits to the plants' growing environment, mulching provides a useful repository for chipped yard waste. Because mulching requires only chipping as a front end process, it provides economic as well as environmental benefits.

Erosion control. A 6-in layer of chipped yard waste reduces water and wind erosion by reducing the impact of raindrops on the soil surface, reducing runoff, and providing a barrier between the wind and erodible soil particles. The steeper the slope, the coarser the material should be so that the chips are able to "knit" together.

Addition of nutrients. As organic matter decomposes, nutrients are released back into the soil, reducing the need for fertilization over time. Organic matter supplies small quantities of all essential plant nutrients, which are slowly released as decomposition takes place, thereby allowing full utilization of these nutrients by the plants.

Retention of moisture. Because of the generally poor soils used as cover soils on closed landfills, water retention is an important benefit of mulch, particularly in arid areas such as the western United States. Mulch retards evaporation from the soil surface, allowing plants to draw on that water supply far longer than would otherwise be possible.

Soil temperature moderation. Mulch evens out soil temperatures, creating a more favorable environment for root growth. The beneficial effects of temperature moderation because of mulching are particularly noticeable in areas that have extremely hot or cold climates.

Weed growth inhibition. A 4- to 6-in layer of mulch is quite effective in inhibiting the germination of seeds. When seed germination is desired, the thickness of the mulch layer should be reduced to approximately 1 in. This thin layer of mulch will protect the emerging seedlings and still provide many of the benefits of the deeper layer.

Potential Disadvantages of Mulching. Despite the many advantages that mulching offers, a few words of caution apply. Potential disadvantages include these possibilities: (1) weed seeds may be introduced, (2) plant disease organisms could be introduced, (3) mulch must be reapplied yearly to maintain desired thickness until the plant canopies cover the soil surface, (4) some plants used in the production of mulch may generate allelopathic (toxic) compounds that could impair plant growth, particularly seed germination, and (5) mulching must be accompanied by changes in cultural methods, for example decreased irrigation to compensate for the moisture retentive characteristics of mulch.

Plant Selection

Until recently planners gave little thought to the end result of landfill revegetation and sought only the most economical solution, usually the planting of range grasses. However, with the increasing importance of landfill sites as potential recreation areas or urban open spaces, the selection of appropriate plant materials for revegetation has become far more important. Plant selection will largely depend on the chosen end use of the site. If restoration to a native habitat is the goal, then plants from the appropriate local plant communities must be used. Nonnative species used for specific landscaping situations such as golf courses and parks should be adapted to local climatic conditions.

Guidelines for the Selection of Plantings. It is not possible to make generalized statements regarding the proper selection of plant materials for landfill revegetation. Each region of the country has environmental conditions that suit certain plant species but not others. In the western United States alone, 24 distinct climate zones have been identified [7]. For example, plants adapted to desert conditions would not thrive in year-round high-rainfall areas such as the southeastern United States. Similarly, plants from coastal areas would not adapt to a high mountain environment. Therefore, plants used in revegetation must be adapted to the area where the landfill is located, particularly because landfills are already an adverse environment for plant growth. An additional concern when restoring a site to its natural state is the source of seed and plant materials. To preserve the local gene pool, it is desirable to collect seeds and cuttings from the immediate surrounding area and grow the plants used in the revegetation effort under contract.

Native versus Nonnative Species. Restoration of a closed landfill site to its native condition is often the desirable alternative. It is the least costly revegetation option in the long term and it provides much-desired open space and greenbelts in urban areas. If restoration is the goal, then the use of native plants is essential. Native plants are those that grow naturally in a geographic region (see Fig. 16-7). Some plants may also be endemic to an area, that is, their populations are

(*a*)

(*b*)

FIGURE 16-7
Typical plant vegetation at completed landfills: (*a*) native species and (*b*) combination of native and nonnative species.

restricted in their natural distribution to that geographical area. Endemic plants account for many of the rare and endangered plants found across the country. Native plants are the most highly adapted for the environmental conditions found in a particular region on native soils.

Nonnative plants may also be used to revegetate closed landfills (see Fig. 16-7). The most suitable plants in this case are those that naturally grow in very similar climatic zones throughout the world. For example, the eucalyptus is widely grown in California, but it is native to Australia. Both California and parts of Australia have Mediterranean climates, and plants from these regions are quite interchangeable.

Factors in the Selection of Woody Plants for Landfill Revegetation. Factors that should be considered when selecting woody plant material for landfill revegetation include growth rate, tree size, rooting depth, flood tolerance, mycorrhizal fungi, and disease resistance [4, 6].

1. Slower-growing trees seem to adapt more easily to landfill conditions than faster-growing trees. Slower-growing trees require less moisture, which is a limiting factor in landfill cover soils in general.
2. Smaller trees (under 1 meter tall) are able to grow roots close to the surface, thus avoiding contact with gas in lower soil layers. However, shallow-rooted trees require more frequent irrigation.
3. Trees with naturally shallow root systems are inherently better adapted to landfill conditions. Again, shallow roots require more frequent irrigation and are subject to wind toppling.
4. Similar changes in soil occur under both landfill gas and flooding conditions, with the exception of moisture content. Flood tolerant species show greater adaptability to landfill conditions than non–flood tolerant species, but their use requires adequate irrigation.
5. Mycorrhizal fungi have a symbiotic relationship with plant roots and enable the plant to take up more nutrients.
6. Plants that are particularly sensitive to specific disease or insect attacks should not be used under the marginal growing conditions of closed landfills.

Use of Grasses for Landfill Revegetation. In addition to woody plants, the use of grasses may be desirable in revegetating landfill sites (see Fig. 16-8). Emphasis has been given in this chapter to woody plants because of their special requirements under landfill conditions. Like other plants, grasses will be affected by poor soils and landfill gases, but they are easier to grow than woody plants. Whether native or non-native, the root systems of grasses are fibrous and shallow, allowing them to survive landfill conditions more readily than woody plants. Their life cycles also provide some advantages. Some grasses are annuals, which means they complete their life cycle in one year or less. Thus, annual grasses grow and set seed during the most favorable time of the year. For example, in the arid

(a)

(b)

FIGURE 16-8
Grasses used for the revegetation of closed landfills: (*a*) natural grass that has dried with the summer heat (note soil cracking in foreground) and (*b*) planted grass with irrigation system.

regions of the western United States, annual grasses take advantage of the rainy season. In the East, annual grasses grow during the warm season. Annual grasses are easily reseeded if necessary. Perennial grasses live longer than one year, but many of their other characteristics are similar to the annual grasses. Characteristics such as the type of root system, life cycle, and quick reproduction make grasses much easier to grow under adverse conditions.

Design Considerations for Revegetation

Determination of the end use of the closed landfill site should be an integral part of landfill design. Unless the closed landfill is used for golf courses or other intensive uses, designers should make every effort to blend the closed landfill into the natural surroundings. This requires the planting of native plants. Much more study is needed in the area of plant adaptability to landfill conditions and horticultural practices to overcome the inherent adverse conditions in landfills. Few species have been tested and those only in very select regions of the country. Given the diversity of environmental conditions across the United States, studies should be initiated to define native plant adaptability and specific horticultural practices under landfill conditions.

Successful design and implementation of landfill revegetation requires an interdisciplinary team of professionals including engineers, planners, landscape architects, soil scientists, botanists, and horticulturists. The goals of the end-use design include stabilization of the landfill surface and reduction of erosion, determination of specific end use, aesthetic restoration of the site, enhancement of soil fertility, selection of appropriate plant materials, and management of the installation and maintenance of the plantings. The steps involved in landfill revegetation are illustrated in Example 16-1.

Example 16-1 Preparing a landfill site for revegetation. Imagine you are working with a landfill that will close in two years. Located in southern California, it is surrounded by a mixed chaparral plant community. The end use selected is to restore the site to its pre-disturbed condition using plants found in the immediate surrounding area. Lay out and prioritize the steps required to accomplish revegetation.

Solution

1. Project coordination.

   Coordinate with engineers, planners, landscape architects and others who are involved in end-use planning for the site.

2. Identify vegetation types and sources of plant materials.

   Restoration being the goal for this site, prepare a list of plants to be used. Arrange with contract growers for the necessary plant materials. This step requires approximately two years of lead time. Ideally, as part of the restoration planning process, a plant testing program would have been established before this stage to gain information about the tolerances of the various species to landfill conditions and to study cultural methods aimed at improving survival rates. A list of plants typically found in the mixed chaparral plant community of southern California is presented below. Plant lists similar to that presented below are available for all parts of the United States from appropriate governmental agencies and others.

Common species of mixed chaparral plant community of southern California

Botanical name	Common name	Growth habit
Adenostoma fasciculatum	chamise	Shrub to 8 ft
Arctostaphylos spp.	manzanita	Ground cover to 30 ft tree
Ceanothus spp.	ceanothus	Ground cover to 20 ft tree
Cercocarpus betuloides	mountain-mahogany	Shrub to 12 ft
Eriodictyon spp.	yerba santa	Shrub to 4 ft
Garrya spp.	silk-tassel bush	Shrub to 10 ft
Heteromeles arbutifolia	toyon	Shrub-tree to 25 ft
Pickeringia montana	chaparral-pea	Shrub to 6 ft
Prunus ilicifolia	holly-leafed cherry	Shrub-tree to 30 ft
Quercus dumosa	scrub oak	Shrub-tree, height varies
Quercus wislizenii var. *frutescens*	shrub form of interior live oak	Shrub to 10 ft
Rhamnus californica	coffeeberry	Shrub to 12 ft
Rhamnus crocea	redberry	Shrub to 10 ft
Rhus ovata	sugarbush	Shrub to 12 ft
Ribes spp.	currants, gooseberries	Shrub to 8 ft
Toxicodendron diversilobum	poison oak	Vine to 6 ft shrub

3. Site inspection.

   Walk the site and perform a visual inspection. Note growth patterns of existing vegetation and any denuded areas. Sample for landfill gas, especially in areas denuded of vegetation or where abnormal dieback is evident. Has any adjacent natural vegetation invaded the site? Which species? How dense are the volunteer stands? Are certain species conspicuously absent? Check the soil. Is there a foul odor? Perform simple onsite soil tests.
4. Soil characterization.

   Collect soil samples from the cover soil and from surrounding undisturbed soil. Analyze both samples and compare the results. A soil scientist should be involved in this process to ensure collection of an adequate number of samples and to interpret the results and propose solutions. The goal is to approximate as closely as possible the natural surrounding soil conditions.
5. Preparation of site.

   Repair or install any additional environmental control devices prior to preparing the soil for planting. Necessary grading to improve drainage, particularly in areas of uneven settlement, should be done. Settlement cracks should be filled to prevent gas from escaping and oxygen and water from entering the landfill.
6. Soil modifications.

   Implement any soil modifications, including the addition of nutrients, as indicated by the soils tests (refer to Table 16-6). Add organic matter to match the levels in surrounding soils. The typical percentage range for soil organic matter across the United States is from 2 to 5 percent. In some parts of the United States where there is heavy rainfall and dense forests, organic matter levels can reach 10 to 15 percent. Adding and maintaining a layer of mulch, over time, will also add to the organic matter content of the soil.
7. Planting.

   Plant the site, using all of the knowledge and experience gained in the previous steps. Planting should be done during the most appropriate time of the year. In areas of the country with mild winters this might be in the fall, or just before the start of the rainy season. In cold winter areas, such as the northern and eastern United States, planting should be done in spring after the soil has warmed. In some situations, an irrigation system will be required to assist the plantings during the establishment period.
8. Monitoring.

   Monitor the plantings closely during the first three years of establishment, with less frequent checks in the later years. Migrating gas has the potential to create problems for the plantings as long as it is generated by the decomposing waste in the landfill.

Comment. Plant lists similar to that presented above in Step 2 are available in all parts of the United States from appropriate governmental agencies. Landfill revegetation is a complex and expensive undertaking involving both engineers and scientists. Botanists and horticulturists should be involved in Steps 2, 3, 7, and 8 and soil scientists in Steps 3, 4, and 6.

16-3 LONG-TERM POSTCLOSURE CARE

The facilities at a closed landfill must be maintained over the period of time that the landfill is producing products of decomposition. Because the waste placed in landfills will decompose at different rates depending on the design of the landfill, there can be extreme variations in the period of time that maintenance will be required at a closed landfill (see Fig. 11-13). For this reason, federal and state regulations have prescribed minimum time periods for the long-term care of closed landfills. In most regulations that maintenance period is 20 to 30 years.

At the time of site closure, or other time as specified by local regulations, a postclosure maintenance plan will be developed. Long-term postclosure care involves a series of continuing activities, beginning at the closed landfill with monitoring of environmental controls and finishing with written reports. Typical elements of a postclosure plan are identified in Table 16-7. The key issues that must be addressed in a landfill closure plan are as follows:

- Routine inspections
- Infrastructure maintenance
  - Grading and landscaping
  - Drainage control systems
  - Gas management systems
  - Leachate collection and treatment
- Environmental monitoring systems

The development and use of closure plans and long-term postclosure maintenance plans are recent activities for waste management agencies. These agencies must recognize that these plans will be tested over time as closed landfills are used as community assets, thereby causing conflicts over land use and uncertainty about any future environmental problems. In all future situations, it will be im-

TABLE 16-7
Typical elements of a landfill postclosure plan

Element	Typical activity
Postclosure land use	Designation and adoption
Routine inspection schedule	See Table 16-8
Infrastructure maintenance	Preparation of a description of the programs for final cover, vegetative cover, final grading, drainage systems, leachate collection and treatment systems, gas monitoring and control systems, and groundwater monitoring system
Environmental monitoring systems	Setting of specific tasks and schedules
System operation	Description of systems to be operated, frequency of operation, and responsible agency
Reporting	Identification of reports required, timing, and format
Facility changes	Preparation of as-built descriptions of current monitoring, collection, and treatment systems
Emergency response plan	Preparation and submittal

portant to manage the site as a closed landfill. The legal framework for long-term postclosure care of landfills, including the establishment of financing for the 30 years of care, is discussed below.

Routine Inspections

Routine inspections are conducted to characterize the condition of landfill closure facilities. Personnel from the solid waste management agency may be responsible for conducting the inspection, although a closed landfill may be used and the land area managed by a different department of the community. The responsible agency and its duties in regard to inspections will be identified and set forth in the postclosure maintenance plan. A listing of items to be inspected, the suggested frequency of inspection, and typical problems that might be observed is presented in Table 16-8. The suggested frequency of inspection is typical, and variations can be expected based on local climatic conditions.

Infrastructure Maintenance

The infrastructure of landfills includes grading and landscape features, drainage control systems, gas management systems, and leachate control systems. This infrastructure must be maintained systematically through a planned schedule of preventive maintenance to protect the integrity of the landfill cover and prevent contamination of the air, water, and soil environment adjacent to the landfill.

TABLE 16-8
Closed landfill inspection items, frequency of inspection, and potential problems to be observed

Inspection item	Frequency of inspection	Potential problems to be observed
Final cover	Once per year, and after each substantial rainfall	Erosion to expose the synthetic liner; landslides
Vegetative cover	Four times per year	Dead plants
Final grades	Twice per year	Standing ponds of water
Surface drainage	Four times per year, and after each substantial rainfall	Debris in drains; broken drain pipes
Gas monitoring	Continuous as required by the postclosure maintenance and management plan	Odors; compressor and flare equipment inoperable; high gas readings in monitoring probes; broken gas well pipes
Groundwater monitoring	As required by equipment and the postclosure maintenance and management plan	Damaged wells; inoperable sampling
Leachate management	As required by the postclosure maintenance and management plan	Inoperable leachate pumps; blockage in leachate collection pipes

Grading and Landscaping. Closed landfills will have significant settlement, which will affect land surfaces and the plants used for landscaping (see Section 11-17). The responsible site manager must have the equipment and funds necessary to maintain the required grades and plants. Each closed landfill will have a specific set of equipment and funding requirements based on the size and slope of the land area to be maintained and the type of vegetation.

In correcting a land settlement on the top of the closed landfill, particular attention must be given to the landfill cap. Repair of a landfill cap that includes a geomembrane liner is illustrated in Figs. 16-9, 16-10, and 16-11. The repair procedure is complicated because the geomembrane must be restored to its original watertight condition after the repair. Any of the three repair methods shown might be used at a closed site with a geomembrane in the cap, with the decision on a method of repair based on economics and the skill of the repair crew.

Drainage Control Systems. Drainage control at closed landfills includes both run-on and runoff of surface waters. The drainage control systems to be maintained will be the facilities identified and installed as a part of the landfill closure plan. Drainage facilities at closed landfills are subject to long-term settlement, which causes concerns for the preservation of gravity flow systems that discharge to offsite conveyance facilities (see Fig. 16-12). It may be necessary to install and operate stormwater pumps after many years of landfill settlement. Maintenance of drainage control systems must be coordinated with maintenance of land surfaces and revegetation of landscape plants.

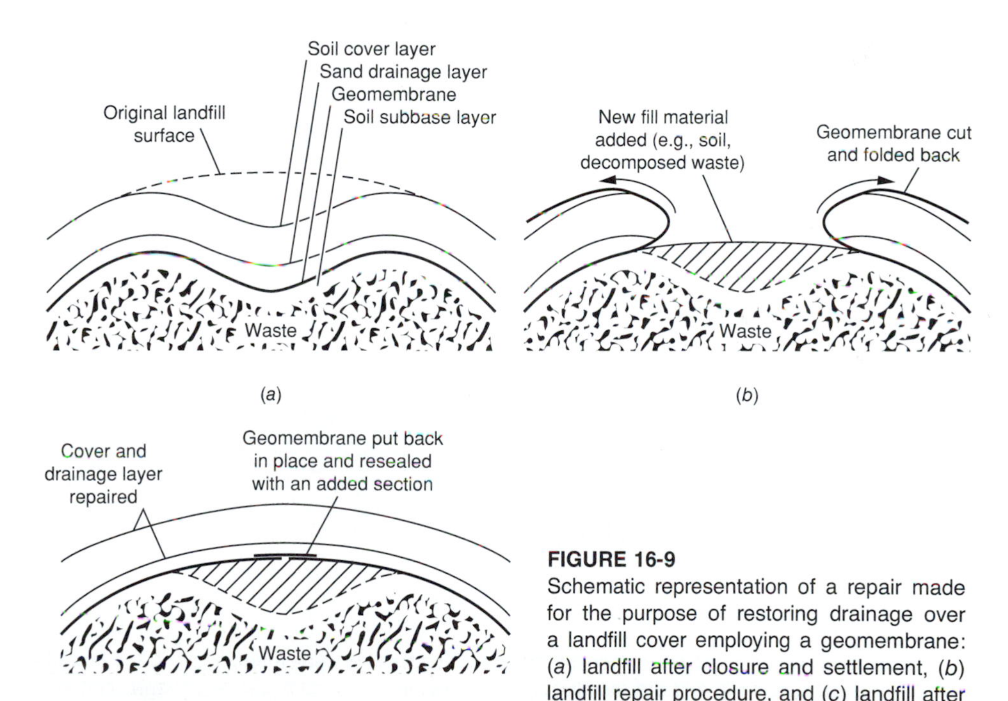

FIGURE 16-9
Schematic representation of a repair made for the purpose of restoring drainage over a landfill cover employing a geomembrane: (*a*) landfill after closure and settlement, (*b*) landfill repair procedure, and (*c*) landfill after repair to restore surface drainage.

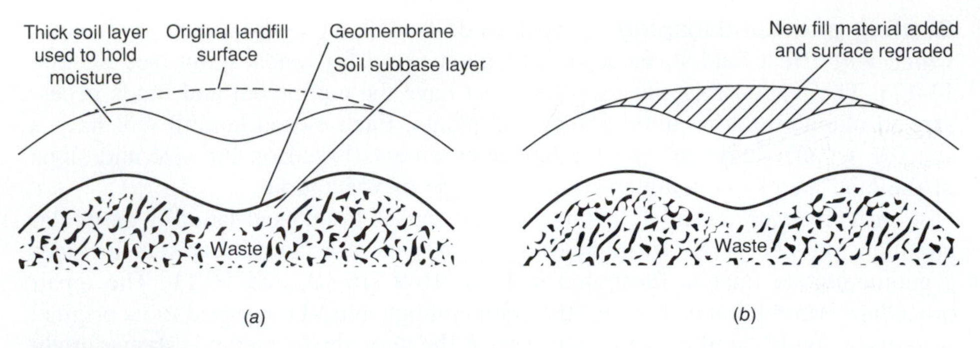

FIGURE 16-10
Schematic representation of a repair made for the purpose of restoring drainage over a landfill cover employing a thick soil cover and a geomembrane: (*a*) landfill after closure and settlement and (*b*) landfill after repair to restore surface drainage.

Gas Management Systems. Gas management will be required at closed landfills as long as landfill gas is produced. Gas management systems are installed as a landfill is operated before closure. (The types of gas management systems were identified in Chapter 11.) The gas extraction wells and gas collection pipes installed in the wastes or in the final cover over the wastes are a high-frequency maintenance item because of settlement of the wastes. Waste settlement will dislocate pipes and wells, and a significant dislocation will cause the pipe and well casings to break. If the breakage is near the land surface, the vacuum in the extraction system is lost and the system becomes inefficient or nonfunctional.

Gas management at a closed landfill is completed when the extracted gases are destroyed and discharged to the atmosphere. Gas destruction can be done in a flare, which does not recover heat energy, or in an energy recovery system. The gas destruction unit is a sensitive piece of equipment that requires maintenance by

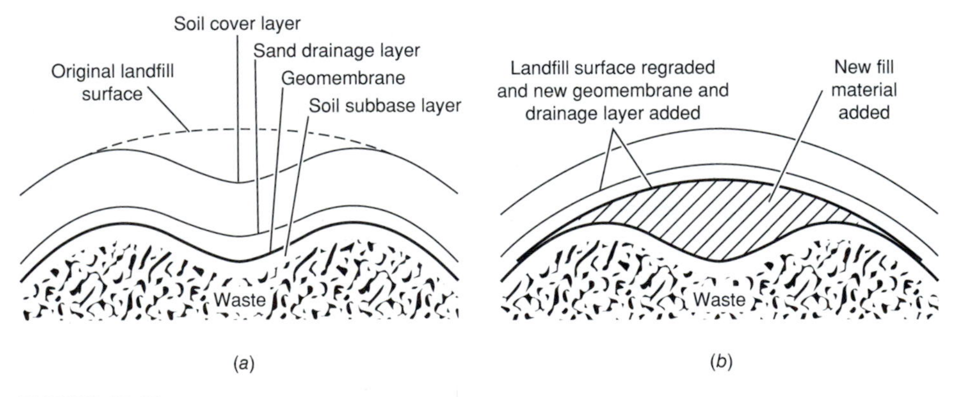

FIGURE 16-11
Schematic representation of the replacement of landfill cover after settlement involving the regrading of the landfill surface and the installation of a new geomembrane, drainage layer, and soil cover: (*a*) landfill after closure and settlement and (*b*) new landfill cover installed after settlement.

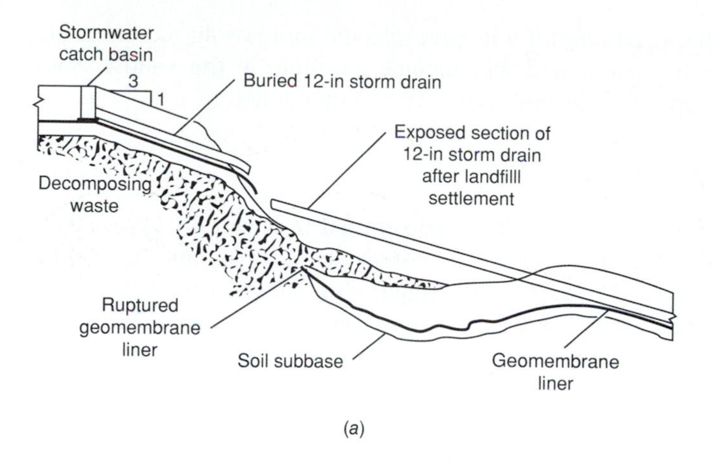

(a)

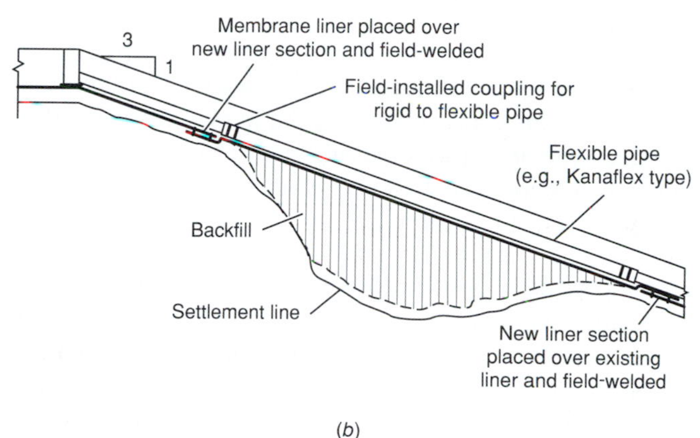

(b)

FIGURE 16-12
Schematic representation of the repair of a broken drainage pipe after landfill settlement: (a) damaged drainage pipe and (b) repair of broken section of drainage pipe.

trained and experienced technicians. The equipment must be tested periodically to ensure it is meeting air discharge standards, and its operation is the subject of annual review by regulatory agencies.

Leachate Collection and Treatment. Closed landfills with installed leachate collection systems must collect, remove and treat leachate as it appears over the postclosure period. The leachate management system will operate as long as is necessary to meet the applicable water discharge standards set for the closed landfill. The maintenance of leachate collection and treatment facilities requires system operators who are skilled in handling changing quantities of wastewater that varies in strength.

Environmental Monitoring Systems

The environmental monitoring systems to be maintained during the postclosure period are the systems designated in the closure plan and approved by the regulatory agency. Although each closed landfill will have specific monitoring requirements, the typical systems to be maintained will include monitors in the vadose zone, water wells and well caps, gas probes, and survey monuments.

16-4 LEGAL FRAMEWORK

Laws assign legal responsibilities for the waste placed in a landfill. These laws are currently being tested in the courts on a case-by-case basis, and no general guidelines are yet available to determine the impacts on solid waste management agencies. Within the framework of the law, it is possible to highlight legal issues, liability during postclosure maintenance, and financing of postclosure maintenance.

Legal Issues

The solid waste management agency or private landfill operator is affected by landfill closure laws in two areas: responsibilities as the site owner and liabilities for the wastes in the site. Where the landfill is owned by the public agency and its residents, the agency has a dual responsibility covering both areas. In situations where the landfill owner is independent of the waste generators who have delivered waste to that landfill, the responsibility is split.

Responsibilities of the Site Owner. Upon landfill closure, the law requires the site owner to have a postclosure maintenance plan and a means of paying for postclosure maintenance and monitoring. At the federal government level, the legal framework is defined in RCRA Subtitle D. In addition to the federal laws, many states have passed laws governing landfill closure and financial assurance. The laws require the site owner to close and maintain the site in accordance with an approved closure plan. That closure plan will have a financial assurance plan to provide sufficient funds to perform postclosure activities over a set period of time, usually the full period of time for postclosure maintenance.

Liabilities for Wastes at the Landfill. Federal Law has defined the liability for waste impacts at a closed landfill as staying with the waste generator. For large waste generators such as commercial establishments and industries, the assignment of liability is clear. However, the wastes from residential generators cannot be assigned to individual houses, as wastes from many houses are typically mixed in a single collection truck. Residential wastes remain the liability of the government jurisdiction within which the wastes are generated. The solid waste manager is faced with expanding legal issues as the number of landfills decreases in the future and many generators are delivering wastes to common use regional landfills. In each community, the solid waste manager is evaluating the costs today of landfill siting and construction within the community or the costs in the future if there is a problem at a regional landfill, and the community is assessed a liability cost by the courts. These issues are not yet cases in the courts and it is not possible to provide guidance on the options with the least risk to a community.

Liability for Postclosure Maintenance

The laws for postclosure maintenance become less clear as closed landfill sites are sold and resold. The land buyer is not interested in the property as a former landfill requiring maintenance to preserve environmental controls. In most instances, the buyer is interested in land improvements that increase land value. Former landfills become attractive property when surrounded by urban development. Liability for postclosure maintenance must transfer with the land deed when closed landfills are sold. The solid waste manager must prepare a clear statement of the postclosure maintenance responsibilities that go with the closed landfill, and register that statement with the land deed office.

Financing of Postclosure Maintenance

The law has prescribed financial assurance for postclosure maintenance so that sufficient funds will be available to pay for maintenance in the future. The basis of the fund is the postclosure maintenance cost estimate that is prepared as a part of the postclosure maintenance plan. A number of different methods have been developed for establishing and maintaining financial assurance for the closed landfill, including bonds, sinking funds, enterprise funds, and pledged revenues or assets of the site owner. Regardless of the method used, the assurance must be irrevocable. Where the sinking fund is to be accumulated from disposal fees paid at the operating landfill, the operator must set fees at a level that will accrue the entire fund within the remaining operating life of the landfill. Because financial assurance is new in 1992 federal regulations, there will be a period of trying methods other than those defined here. The solid waste manager should monitor the regulations and select the methods most appropriate to the community.

16-5 DISCUSSION TOPICS AND PROBLEMS

16-1. Locate one or more completed landfill sites in your community or region and identify the types of facilities that have been installed to control surface runoff from the landfill and to divert stormwater runoff from the land surrounding the landfill. Based on your observations, how effective do you feel the drainage facilities are in meeting the objective of diverting rainwater from the landfill site? Are they being maintained properly?

16-2. Locate one or more completed landfill sites in your community or region and identify what type of final landfill cover was used. Based on your observations, how effective do you believe the final cover is in limiting the uncontrolled release of landfill gases?

16-3. Locate one or more completed landfill sites in your community or region and identify the types of facilities that have been installed to control landfill gases. Are the facilities being maintained properly?

16-4. Locate one or more completed landfill sites in your community or region and identify the types of facilities that have been installed to capture and treat leachate.

16-5. Locate one or more completed landfill sites in your community or region and identify the types of facilities that have been installed to monitor groundwater quality.

16-6. Locate one or more completed landfill sites in your community or region and identify the types of plants that have been used to revegetate the landfill area. Classify the various plantings according to whether they are native or nonnative species.

16-7. Locate one or more completed landfill sites in your community or region. Prepare a list of native plants that could be used to revegetate a landfill.

16-8. In many landfill designs, the regulatory agency requires that the final cover have a permeability less than or equal to the permeability of the landfill liner. Do you agree with this requirement? If you agree, what are the implications for long-term maintenance of the site.

16-9. Refer to Fig. 16-12. Assume the settlement has occurred over five years, and the original liner installation contractor is no longer in business. You have been directed to repair all land surface settlements and to report the repair activity to the regulatory agency. As city engineer, describe the construction sequence you would use for repairs. Your choices are to use city work crews or a contractor. If you choose to use your own crews, how will you provide for construction quality assurance?

16-10. The U.S. EPA and many state agencies are requiring financial assurance for the long-term maintenance of landfills. What financial assurance methods would you use? Why?

16-6 REFERENCES

1. Donahue, R. L., R. W. Miller, and J. C. Schickluna: *Soils: An Introduction to Soils and Plant Growth*, 5th ed., Prentice-Hall, Englewood Cliffs, NJ, 1983.
2. Flower, F. B., E. F. Gilman, and I. A. Leone: "Landfill gas, what it does to trees and how its injurious effects may be prevented," *Journal of Arboriculture* Vol. 7, No. 2, February 1981.
3. Flower, F. B. and I. A. Leone: "Damage to vegetation by landfill gases," *The Shade Tree*, Vol. 50, Nos. 6 and 7, June and July 1977.
4. Gilman, E. F., F. B. Flower, and I. A. Leone: "Standardized procedures for planting vegetation on completed sanitary landfills," *Waste Management and Research*, Vol. 3, 65–80, 1985.
5. Gilman, E. F., I. A. Leone, and F. B. Flower: "Factors affecting tree growth on resource recovery residual landfills," *Proceedings of the 1980 National Waste Processing Conference*, The American Society of Mechanical Engineers, New York, 1980.
6. Gilman, E. F., I. A. Leone, and F. B. Flower: "Influence of soil gas contamination on tree root growth," *Plant and Soil*, Vol. 65, 3–10, 1982.
7. Hogan, E. L., *et al.*, (eds.): *Sunset Western Garden Book*, 5th ed., Lane Publishing, Menlo Park, CA, 1988.
8. Leone, I. A. and F. B. Flower: "Soil characteristics of landfill cover soils in nine U.S. climatological regions," presented at the International Symposium on Remote Sensing of the Environment, Third Thematic Conference, Remote Sensing for Exploration Geology, Colorado Springs, CO, April 16–17, 1984.
9. Lyon, T. L. and H. O. Buckman: *The Nature and Properties of Soils*, 4th ed., revised by H. O. Buckman, Macmillan, New York, 1949.
10. U.S. Environmental Protection Agency, *Remedial Action at Waste Disposal Sites*, revised edition, EPA Technology Transfer Publication, EPA/625/6-85/006, October 1985.

CHAPTER 17

REMEDIAL ACTIONS AT INACTIVE WASTE DISPOSAL SITES

Closed, inactive, or abandoned MSW disposal sites of various sizes can be found throughout the United States. Until the late 1970s, when the U.S. EPA mandated the identification and elimination of open dumps, every roadside ravine, every vacant land parcel, and every accessible wetland was a candidate site for waste disposal. Many sites were used for waste disposal and then abandoned without adequate long-term maintenance controls to protect the surrounding environment. The use of these lands or waters for waste disposal was an accepted practice until the occurrence of public health or environmental problems. Additional descriptions of the evolution of solid waste management are found in Chapter 1.

The closed, inactive, or abandoned landfills of the past remain as potential problems for communities in the future (see Fig. 17-1). The problems caused by these sites can be more severe and costly to solve than monitoring and engineering new landfills, because past landfill practices usually did not follow the standards and regulations that are now mandated by federal and state governments. Without standards, many landfills often accepted industrial wastes that today are classified as hazardous and excluded from MSW landfills. The process of identifying old landfills that have problems, and the actions that might be taken to remediate those problems, are described in this chapter. The topics discussed include (1) impact of inactive landfills, (2) determination of need for remediation, (3) hazardous waste landfill remediation, and (4) other designated waste landfill remediation.

(a)

(b)

FIGURE 17-1
Typical examples of abandoned landfills: (*a*) in a national forest (courtesy of Greg Haling, Metcalf & Eddy, Inc.) and (*b*) near a metropolitan area (courtesy of the State of New Jersey Department of Environmental Protection).

17-1 IMPACT OF INACTIVE LANDFILLS

Inactive landfills are a problem only when there is a health or environmental impact caused by the wastes or waste byproducts, such as the emission and migration of landfill gases and leachate. In many situations, the landfill remains a benign neighbor, undergoing the natural processes of decay that are undetected and not harmful. However, as population increases have caused changes in land use, inactive and abandoned landfills (see Fig. 17-1) may have an impact on human activity. In this situation, landfill impacts, whether real or perceived, are discussed widely in the community. People want to know what is in the inactive landfill and what impact the inactive landfill site will have on their homes. The steps that must be taken to answer such questions include (1) site identification, (2) identification and delineation of the pathways through which the contaminants might move to affect people and the environment, (3) analysis of the impacts of landfill settlement, (4) verification of visual impacts, and (5) measuring the public reactions to a problem.

Site Identification

The first indication of a problem with an inactive landfill often is a complaint from a community resident. If the complaint is based on visual sighting of waste materials (see Fig. 17-1) or of land movement, the inactive landfill site will have been identified. A complaint based on odors, fires, or bad drinking water from a well will require a more extensive investigation to identify the exact location of the site, because in many cases the contaminant causing the problem might have moved away from the landfill. Investigative steps that have been used to define the boundaries of the landfill are presented in Table 17-1. An aerial photograph of an operating landfill is shown in Fig. 17-2. Such photographs can be used to

TABLE 17-1
Identification methods and typical sources of information for inactive landfills

Identification method	Typical sources of information
Search records	Historical files of permit-issuing agencies; files of deed transfer; old facility plans
Contact current landfill operators	Franchise agreement; operating permit; copies of receipts from disposal revenues of previous operations; interviews with long-time employees
Contact current waste collection agencies	Franchise agreement; copies of receipts from disposal fees paid; interviews with long-time employees
Walk all areas within 1500 ft of the source of the complaint	Type of vegetation; land contours; interviews with long-time residents (see Fig. 17-1)
Aerial surveys	Files of companies that perform aerial photographic surveys and photogrammetry; landfill sites often are included in general surveys of the area (see Fig. 17-2)

FIGURE 17-2
An aerial photograph of an operating landfill. (Courtesy of SYUSA, Buenos Aires, Argentina.)

determine the extent of the landfilling operations. In most cases, the results of the field investigations are improved significantly when the field investigations are conducted by a person with working knowledge of landfills.

Identifying Contaminant Pathways

Contaminants might exist as gases in the air and soils or as leachate in surface waters and groundwaters. Unless the contamination is first detected in the landfill site, it is important to define the route of contaminant movement, the *pathway*, from the landfill site to the point of detection [2, 5, 7, 8]. For surface waters, the pathway is often a stream channel or an eroded or stained land surface. For groundwaters, the pathway is usually the uppermost groundwater aquifer. Gases in the soil will move from less permeable zones to more permeable zones until they enter the atmosphere. Once the pathway is established, it is common practice to identify all human and other activities along the pathway so that an assessment of the contaminant impacts can be completed.

Landfill Settlement

The land surface settles as the underlying wastes contained in abandoned landfills decompose. Many abandoned landfills are paved over as a part of urban development, and the land surface settlement causes depressions, which destroy the surface water flow contours. The resulting water accumulations cause a drainage problem for traffic as well as accelerating the cracking and failure of the pavement. Pavement failure or standing water may be a means of identifying an abandoned landfill. Such physical evidence can be supplemented with additional data from record searches and interviews to complete a site identification. The impacts of land settlement include destruction of surface water flow contours, failure of pavements, increased gas flow from underlying wastes, and damage to buried utility pipes and conduits.

Public Impressions and Perceptions

The impact of an abandoned landfill on the community residents is negative, causing concern and fear about water pollution and methane gas explosions. Public concern and fear is manifested in demands for action on elected officials who are perceived to be letting a significant problem go unsolved. In California, the state legislature responded to public concerns by passing a bill that mandated investigations of groundwater and the air at landfills to determine the impacts on the citizens of the state. These solid waste assessment tests, titled SWATs, are being conducted at all active and closed landfills, at a significant cost to the state's residents. In most cases, the costs of this investigation are being recovered from charges to the current users of active community solid waste disposal facilities. Public impressions can have a significant economic impact on the cost of a waste management system.

17-2 QUANTIFYING THE PROBLEM AND COMPLETING THE SITE DESIGNATION

The activities for remediation of a waste disposal site are greatly influenced by federal and state regulations regarding public health and the environment. In most situations, the regulations are structured to give requirements according to the type of contaminant, with the most toxic contaminants having the most restrictive standards for quantities allowed in the environment. The type and quantity of contaminants at an inactive site will be the basis for cleaning up site problems. Although it is not possible to define the multitude of problems that might develop from contaminants in an inactive site, it is possible to define a procedure to follow when responding to a reported site problem. The procedure followed, and suggestions for working with regulatory agencies on problem solving at an inactive waste disposal site, are discussed in this section.

Environmental Testing

Upon the determination that an inactive waste disposal site is part of a problem, it is important to conduct field tests to define the types and quantities of the

TABLE 17-2
Procedure for environmental testing at an inactive waste disposal site

Step	Objective
Meet with staff of regulatory agency	To open a line of communication; develop a preliminary understanding of which regulations might apply to the problem; obtain a list of data needs for the agency, based on preliminary assessments of the problem
Review document	To log all previous reports and field work done on the site
Lay out and conduct a first-stage field investigation	To better define the type and concentration of contaminants; develop a cost-effective pattern of soil borings and groundwater monitoring wells for problem definition; identify areal extent and concentrations of contaminants; using data from Chapter 5, data obtained from document review, and any interviews conducted with long-term employees, determine if contaminants are of the type found in solid waste landfills
Develop a model for contaminant movement	To predict the efficiency of remedial actions; meet the agency requirements for model accuracy for both hydraulics and solute transport; set cleanup standards for soils, groundwater, and surface waters
Lay out and conduct a second-stage field investigation	To verify the model and to set the boundaries of the area to be remediated; define the groundwater plume and horizontal and vertical dimensions of soil contamination to the limits of remedial actions by a construction contractor
Meet with agency staff	To report the results of testing

contamination causing the problem. The presentation here is one of responding to a problem in accordance with state or federal guidelines, the most common method of response provided by the law. Field investigations to identify the type and quantity of contaminants are costly and disruptive to current land use at the site or along the suspected pathway of contaminant movement. Quantifying the underground movement of contaminants from an inactive waste disposal site is costly, because usually little or no information is known about the underlying strata. Therefore, it is prudent to have a structured approach to field data gathering that takes the investigation through a series of steps, each of which has a stated objective and estimated cost to complete. The procedure is presented in Table 17-2. After each step, and before the next step is started, the investigator has the opportunity to modify the original scope of the total investigation to better match the data obtained from the field work. The procedure is illustrated in Example 17-1.

Example 17-1 Laying out a field investigation grid for an inactive waste disposal site. An abandoned landfill has been identified as the probable cause of a water quality problem. Using the accompanying site plan, develop a cost-effective field investigation procedure to quantify the problem and to verify the source of the contaminant. Both the source of the contaminant and the pathway of contaminant movement to the point where it is causing a problem must be identified. Assume that the following data apply.

1. The groundwater flow direction has been established on the basis of passing a plane through the water elevation in three wells in a triangular grid and on existing U.S. Geological Survey records.
2. The contaminant of concern is chlorides (Cl^-). The average chloride concentrations measured in water samples from the well, located as shown in the site plan, was 702 mg/L.
3. The maximum allowable Cl^- concentration for drinking water is 500 mg/L.

Solution. A cost-effective procedure for field investigations is to use multiple stages of field work, as described in Table 17-2.

1. Set up the first-stage grid as illustrated in the site plan. As shown, the grid starts at the point of an identified problem and moves outward in all directions. The soil borings and groundwater wells are to detect any contaminants and provide data that can be compared with data from the problem well and the inactive landfill.
2. Install two wells, one up-gradient of the inactive waste disposal site and one down-gradient of the contaminated well. The up-gradient well will be sampled to establish background water quality. The down-gradient well will be sampled to determine if the Cl^- contamination has moved past the problem well. The well locations should be a sufficient distance from the existing contaminated well and the inactive landfill so that no cross-contamination is possible. A review of local geology is needed to determine that distance; for this example, the appropriate distances are 500 ft for each well.
3. Place two soil borings to the depth of the landfill and in the locations shown on the site plan. The soil tailings should be classified in accordance with the Soil Conservation Service standards. A soil permeability test is run on the soils from the bore hole located between the contaminated well and the inactive landfill.

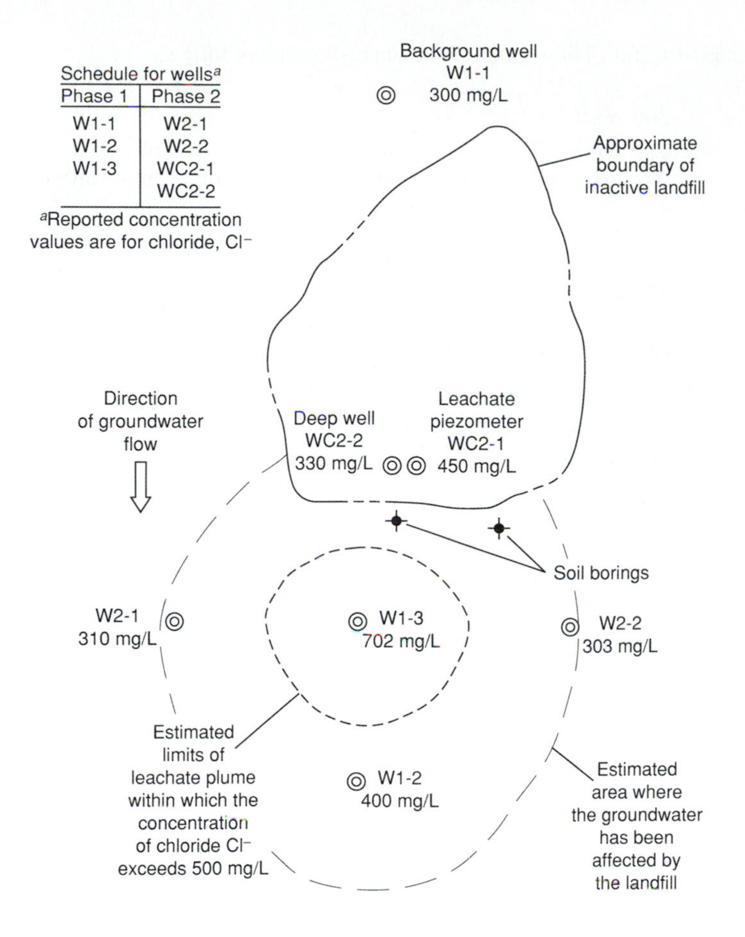

4. Compare the data from the wells and the soils testing. In this example, it can be seen from the data that the landfill was in contact with the groundwater aquifer. The up-gradient well had a Cl^- concentration of 300 mg/L.The new down-gradient well had a Cl^- concentration of 400 mg/L.
5. From the first-stage investigation, it was concluded that the inactive landfill may be the probable source of the high Cl^- concentrations observed in the well. The owner of the inactive landfill met with the regulatory agency to present the findings and to propose a remediation plan. Upon reviewing the data, the agency requested that the owner conduct additional field tests to better define the area affected by the leachate plume.
6. The second stage of field testing included placing two additional wells at right angles to the groundwater flow direction. The location is shown on the site plan. In addition, a well cluster was placed through the landfill, with the first well stopping just above the aquifer elevation. The second well in the cluster was drilled into the aquifer to a depth of 10 ft below the bottom of the wastes, and the well screen was set from the bottom up a space of 8 ft. All wells were numbered as shown on the site plan.

7. All wells were sampled again. The results of the analysis were as follows:

W1-1	300 mg/L
W1-2	398 mg/L
W1-3	698 mg/L
W2-1	310 mg/L
W2-2	303 mg/L
WC2-1 (leachate)	450 mg/L
WC2-2	330 mg/L

8. The field data from the 2nd stage investigation were interpreted as follows.
   (*a*) The leachate plume was localized around the contaminated well (W1-3).
   (*b*) The inactive landfill is no longer generating high concentrations of Cl^-, as the analysis results gave a concentration below drinking water standards for the landfill leachate.
   (*c*) The leachate was mixing with the groundwater under the inactive landfill, and the resulting groundwater Cl^- did not exceed drinking water standards.
   (*d*) Remediation of the Cl^- contamination in the problem well could be done at the well without creating a barrier in the aquifer.
9. The remediation selected had two parts:
   (*a*) The problem well was pumped to capacity until the Cl^- concentrations returned to background levels.
   (*b*) Bottled water was delivered to the home using the well until the water quality in the well met drinking water standards.

Comment. The remediation in this example is straightforward. The highest cost was for the cluster well put in as a part of the second-stage field investigation. However, the well cluster did demonstrate the lack of a strong leachate in the landfill, which resulted in a less costly remediation, involving the use of interceptor trenches or cutoff walls.

Legal Classification of Landfill Sites

The legal classification of a closed or abandoned landfill site [4, 6] is important because regulatory agencies set standards for remediation based on the site classification. Once a site is classified, the site owner can proceed with site remediation knowing that the standard of cleanup is legally binding. To aide in the discussion that follows, a glossary of terms that are used commonly when discussing MSW landfills that may be classified as hazardous waste sites is presented in Table 17-3. In the United States, the standards by which closed, inactive, and abandoned landfill sites are to be remediated is set forth in the Comprehensive Environmental Response, Compensation and Liability Act (CERCLA). Typically, the process of identifying the legal classification of an abandoned landfill site involves: (1) a preliminary assessment (PA), (2) a site inspection, and (3) a scoring of the site using the Hazard Ranking System (HRS) developed by the U.S. EPA. The process is also illustrated schematically in Fig. 17-3.

TABLE 17-3
Glossary of frequently used hazardous waste acronyms

Acronym	Meaning and common usage
CERCLA	The Comprehensive Environmental Response, Compensation, and Liability Act is the original superfund law.
CERCLIS	The Comprehensive Environmental Response, Compensation, and Liability Information System.
CRP	Community Relations Plan, required by CERCLA for a hazardous waste release.
FS	Feasibility Study, the phase of a CERCLA cleanup in which remedial action alternatives are developed, evaluated, and selected.
HRS	Hazard Ranking System, a scoring system the U.S. EPA uses to rank the relative risk of a site; an HRS score of 28.5 or more places the site on the NPL.
IRM	Interim Remedial Measures, taken at NPL sites to remove immediate health threats.
MCL	Maximum Contaminant Level, the maximum concentration of a contaminant in groundwater considered acceptable for potable water under the Safe Drinking Water Act.
NCP	The National Contingency Plan, describes the procedure for adding CERCLA sites to the NPL and the process for cleanup.
NPL	The National Priorities List, established by the U.S. EPA to rank the worst hazardous waste sites.
PA	Preliminary Assessment, the first step required by CERCLA when a hazardous substance has reportedly been released.
PRP	A Potentially Responsible Party, any entity that can be held responsible for contamination.
RA	Remedial Action, a CERCLA term meaning the implementation of the remedial design at the site.
RAP	Remedial Action Plan, required by CERCLA to detail the remedy selected from the feasibility study and the methods for implementation.
RD	Remedial Design, the phase of cleanup under CERCLA in which the remedy selected is developed into engineering plans and specifications for implementation.
RI	Remedial Investigation, the phase of cleanup under CERCLA during which data are collected and analyzed and the site is characterized.
ROD	Record of Decision, the official document, approved and signed by the U.S. EPA, of the remedial action selected for a cleanup under CERCLA.
SI	Site Investigation, the step following preliminary assessment in ranking the hazard potential of a site under CERCLA.

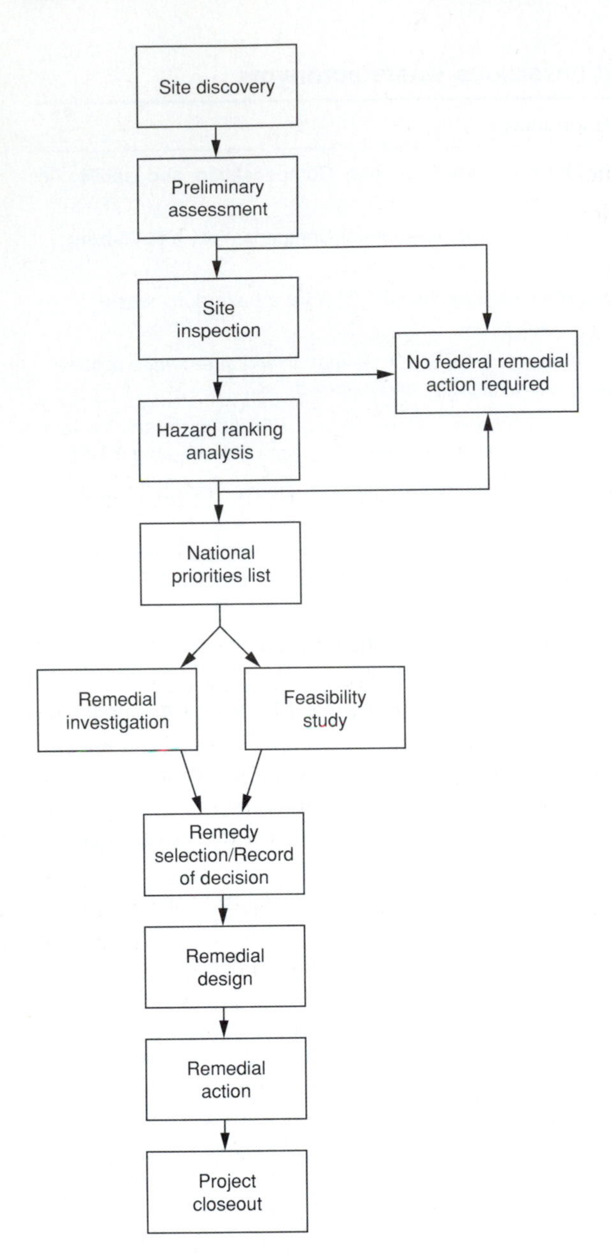

FIGURE 17-3
The CERCLA remedial response process for abandoned landfill [6].

Preliminary Assessment. The preliminary assessment step is used to identify the potential for hazards from a given site. The principal objectives of the PA are to (1) determine if there has been a release of contaminant to the environment from the site, (2) determine whether there is any immediate danger to persons living or working in the vicinity of the site, and (3) whether a site inspection is required. The preliminary assessment involves the following steps [6]:

- Review of existing information
- Site reconnaissance
- Development of preliminary and projected HRS scores (see Table 17-3)
- Application of qualitative criteria
- Prioritization for site inspection
- Report preparation
- Documentation
- Development of information for CERCLIS tracking

Based on the results of the preliminary site assessment, one of the following three actions may be taken:

1. If there is no threat to human health or the environment, no further action is needed.
2. Additional information is needed to complete the PA.
3. Inspection of the site is necessary.

Site Inspection. The site inspection involves sampling to determine the types of contaminants and to identify the extent of the contamination. Before the site is inspected, a detailed work plan and onsite safety plan must be prepared and approved. The objectives of the site investigation are to (1) determine which releases pose no threat to public health and the environment, (2) determine if there is any immediate threat to persons living or working in the vicinity of the site or point of contaminant release, and (3) to collect data to determine whether the site should be included on the National Priorities List (NPL). In general, if the contaminants are on the published list of hazardous and toxic materials, the site will be classified in accordance with legal definitions established by the U.S. EPA. If the contaminants are not on the published list of hazardous and toxic materials, the classification will be in accordance with state or local regulations.

Hazard Ranking Analysis. The final step in detemining whether a site should be placed on the NPL is to rank the site using the U.S. EPA Hazard Ranking System (HRS). The HRS score is based on the probability of harm to human populations and the environment from the migration of hazardous substances involving groundwater, surface water, and air. A composite score, based on separate scores for each of the possible contaminant migration routes, is developed. The score for each migration route is obtained by assigning a numerical value based on predetermined guidelines, as defined in the HRS, to a set of conditions or factors that can be used to characteize the potential of the release to cause harm. Sites investigated under CERCLA receiving a numeric rating greater than 28.5 are eligible for inclusion on the NPL. Approximately 25 percent of the 1200 or so sites on the NPL in 1992 are inactive MSW disposal sites.

17-3 HAZARDOUS WASTE LANDFILL REMEDIATION

The options for remediation of landfills that have been designated as hazardous waste sites are limited to those that are within the guidelines set by federal and state agencies. The reader should refer to the numerous U.S. EPA publications for the details on the conduct of a hazardous waste investigation and remediation. A broad overview of those regulations is presented in this section as background material for the reader who will be required to respond to those regulations if working with an inactive landfill that contains hazardous wastes.

The U.S. EPA has established a firm set of procedures to remediate hazardous waste sites that have been designated in accordance with the NCP (see Table 17-3). Where hazardous wastes are found at inactive waste disposal sites, the required remediation will usually include removal of the source of contamination. Complete removal of the hazardous materials is especially important when the inactive site is located adjacent to or under occupied structures. The public cannot understand nor accept a remedial action that leaves a hazardous material near homes. The high level of negative feelings by the public prevents the use of traditional landfill closure and postclosure actions on a hazardous waste problem.

The procedure for actions at Superfund sites is presented in various publications [2, 5, 7]. The significant actions are illustrated in Fig.17-3 and summarized in Table 17-4. The cost of remediation under Superfund is very high, and when solid waste landfills are not hazardous waste sites, it is not the remediation nor-

TABLE 17-4
Actions to be taken after a site has been designated as a Superfund site under the National Contingency Plan[a]

Action	Comments
Remedial Investigation/Feasibility Study (RI/FS)	The RI/FS is the second most costly action in a Superfund remediation because it addresses (1) the conditions at the site, including sources and extent of contamination, pathways, and potential exposure; (2) quality control plan for data sampling; and (3) health and safety procedures to be used during remediation.
Record of Decision (ROD)	The ROD is the official selection of the preferred alternative from the RI/FS. The ROD includes a conceptual design and preliminary cost estimate.
Remedial Action Design (RAD)	The RAD is the detailed engineering design that will be used to construct the alternative selected in the ROD. The RAD includes both plans and specifications and detailed operations requirements for alternatives that require long-term operation, such as groundwater remediation.
Remedial Action (RA)	The RA is the cleanup action. The RA includes construction and operation of the remedial action and will be the most costly activity in remediation.

[a] Adapted from Refs. 4, 6.

(a)

(b)

FIGURE 17-4
Views of abandoned landfill in which storage drums were disposed: (a) typical view before site was excavated and (b) excavated drums stored temporarily on landfill surface, for sampling, before disposal. (Courtesy of the State of New Jersey Department of Environmental Protection.)

mally chosen for an inactive landfill. However, the current list of some 1200 sites on the NPL includes about 300 former municipal solid waste landfills. Clearly, past landfill practices included accepting hazardous wastes in those landfills now identified as hazardous waste problems. The worst of the problem landfills are those that received and disposed of storage drums, many of which contained hazardous liquids (see Fig. 17-4).

The U.S. EPA has prepared design guidelines for remedial actions at waste disposal sites. In practice, these guidelines have become the prescriptive standards against which all other remediation designs are measured. The engineer preparing a remediation design must refer to the guidelines and use them as appropriate to the problem, recognizing that the guidelines are developed to be protective in a worst-case problem as defined by a regulatory agency.

17-4 OTHER DESIGNATED WASTE LANDFILL REMEDIATION

Remediation of inactive waste landfills other than hazardous waste sites will require actions similar to the actions taken to close existing landfills and to perform postclosure care. The unique nature of inactive landfill remediation is most evident when the site to be remediated has been built on or has other beneficial use. Depending on the severity of the contamination, the existing land uses can be severely affected or may even have to be abandoned. In severe cases, the cost of remediation can include a high cost for displacing an existing facility.

A complete remediation at an inactive landfill will in most situations have three parts: a *remediation part*, which eliminates all or a significant portion of the source of the problem; a *mitigation part*, which reduces the severity of the problem; and a *monitoring part*, used to test the remediation to ensure that the problem is eliminated. The regulatory agency and the inactive landfill site owner will determine the relative contribution of each part to the complete remediation.

Elimination of the Source of the Problem

Types of problems, and the actions that need to be taken to eliminate the problems at the source, are presented in Table 17-5. The problems listed in Table 17-5 and illustrated in Fig. 17-5 are the most common problems associated with nonhazardous waste landfills. The soil erosion problem is remediated at reasonable cost, because the inactive landfill site has not been improved for beneficial use. The methane and gas problems will be the most expensive to remediate if the inactive landfill site and surrounding lands have been improved for beneficial use, because drill rigs and trench-excavating machines will not have unrestricted access to the areas of work.

Mitigation to Reduce the Severity of the Problem

When the contamination from an inactive landfill site is difficult to remediate quickly, or an immediate action other than an action of remediation is required

TABLE 17-5
Types of problems at inactive landfills and the actions to eliminate the problems at the source

Type of problem	Action at the source to eliminate the problem
Land surface settlement	Bring additional soils onsite to fill in the areas of settlement; resurface paved areas as necessary; plant new vegetation as necessary to eliminate eyesores and to hold soils from erosion.
Soil erosion and exposure of wastes	Bring additional soils onsite to fill in the erosion channels; install metal underlying pipes or concrete channels to carry onsite stormwaters across the landfill to discharge points offsite; return all wastes to the site that have moved offsite when exposed, and bury wastes under the new soils filling the erosion channels.
Methane gas in onsite building	Install an active gas-venting system under the building and operate the system as necessary to remove gases.
Methane gas in offsite building	Install an active perimeter gas migration control system; operate the system as necessary to prevent gases from moving offsite; if there is sufficient gas being generated in the inactive landfill, install gas extraction wells in the landfill and use the gases in an energy generator.
Landfill leachate moving offsite in surface waters	Intercept the leachate at the point of surface breakthrough on the landfill; pump the leachate to an appropriate treatment facility for disposal.
Landfill leachate offsite	Intercept the leachate underground at or moving inside the landfill site boundary, using an interceptor trench or cutoff wall in groundwaters; install leachate-pumping facilities to remove leachate for treatment.

(a)

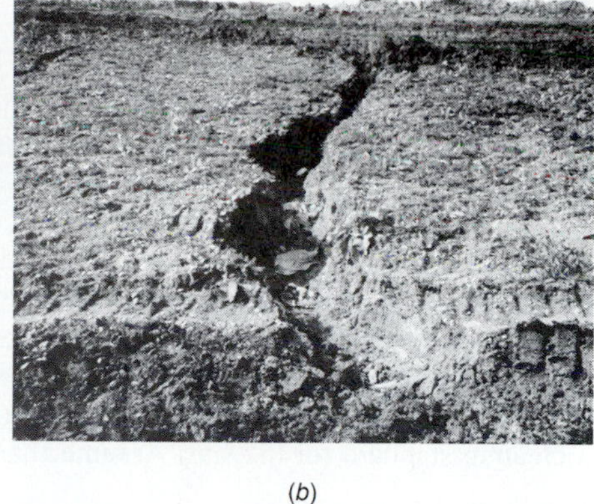

(b)

FIGURE 17-5
Conditions commonly encountered at abandoned landfills: (*a*) accumulation of leachate in unlined drainage ditch and (*b*) severe erosion of landfill cover, often exposing the buried waste. (Courtesy of the California Integrated Waste Management Board.)

to preserve public health, then there will be a mitigation action for the site. In some situations, mitigation actions will occur while the remediation action is in progress. In Example 17-1, the provision of bottled water is a mitigation action for the user of the contaminated well.

Mitigation actions for leachate can be classified into three types:

1. Sealing the site to prevent water from entering the wastes and thereby reducing the quantity of leachate in the landfill (a typical seal would be a new, impermeable cap placed on top of the existing cover)
2. Abandoning water wells that are contaminated and installing new water service lines to the homes served by the contaminated wells
3. Adding water treatment equipment at the well head of the contaminated well to remove the contamination to safe drinking water standards

Mitigation action for landfill gases requires the purchase and removal of buildings that are unsafe because of landfill gas migration. This extreme and costly mitigation would be used only if the gas is moving in high-permeability soils and the gas quantity is so large as to make gas remediation impractical under prevailing economic conditions.

Monitoring the Performance of Remediation

Monitoring of remediation is required by regulatory agencies as a means of proving the elimination of a contamination problem. The type and number of monitoring points will depend on the problem that has been remediated. A leachate problem will require monitoring of groundwater. A gas problem will require monitoring of soils and the air.

Economics

When a set of remediation actions has been demonstrated to meet the allowable contaminant standards of the agency, the final choice of remediation, mitigation, and monitoring will be based on economics. The use of economics is illustrated in Example 17-2.

Example 17-2 Estimating costs for the remediation of a gas migration problem. Evaluate the costs of installing and operating a landfill gas extraction system that is to remediate a migrating gas problem from an inactive MSW landfill site. Either an active or a passive gas control system is acceptable to the regulatory agency responsible for setting a cleanup standard for the site. Assume that the site has the following characteristics:

1. Landfill depth is 30 ft.
2. Landfill gas is migrating in all directions.
3. The total perimeter around the landfill edge is 2500 ft.
4. Continuous clay with a permeability of 10^{-7} cm/s exists at a depth of 25 ft and underlies the bottom of the landfill.

Solution—Part 1 Active Control. Estimate the capital and O&M annual costs of an active perimeter gas migration control system.

1. Extraction wells
   (*a*) Gas extraction wells will be installed at the perimeter of the landfill in native soils and will be 25 ft deep. Wells will be spaced at 50 ft. The total quantity of vertical ft of wells is 2500 ft/50 ft $\times$ 25 ft $=$ 1250 vertical ft.
   (*b*) The well casing will be 8 in diameter and the drilled hole will be 2 ft in diameter.
2. Gas collection header
   (*a*) The gas collection header will be located at the landfill perimeter and will be connected to the gas extraction well with a flexible lateral pipe, and each well lateral will have a shutoff valve. The total quantity of collection header is 2500 ft.
   (*b*) The pipe diameter is 8 in, and the pipe will be buried in a 2 ft deep trench.
3. Gas-monitoring probes
   (*a*) Monitoring probes will be spaced between every second well, or at 100 ft intervals along the landfill edge and to a depth of 25 ft. The total vertical ft of probes is 2500 ft/100 ft $\times$ 25 ft $=$ 625 vertical ft.
   (*b*) Probes will be driven and are $\frac{3}{4}$ in diameter.
   (*c*) Probes will be monitored quarterly with a portable meter.
4. Blower facility
   (*a*) A blower facility will create a vacuum in the wells and deliver the extracted gas to a flare. For reliability, there will be duplicate blowers of 600 scfm each capacity and a single flare with the capacity to burn 600 scfm at 1450°F with a chamber sized for 2 s detention time.
   (*b*) A 500 gal liquid propane tank will be provided as pilot gas for the flares.
5. Estimate costs.
   (*a*) Estimate capital costs. Set up a spreadsheet as follows.

Item	Quantity	Unit Cost, $	Cost, $
Gas extraction well	1,250 v ft	60	75,000
Gas collection header	2,500 l ft	40	100,000
Blower facility	1	60,000	60,000
Monitoring probes	625 v ft	10	6,250
Total capital cost			$241,250

(*b*) Estimate O&M annual cost. Set up a spreadsheet as follows.

Item	Quantity	Unit cost, $	Cost, $
O&M	1 lump sum	10,000	10,000
Monitoring with meter	4 visits	600	2,400
Total annual cost			$12,400

Solution—Part 2 Passive Control. Estimate the capital and annual O&M costs for a passive perimeter landfill gas control system.

1. Barrier trench
   (*a*) A barrier trench will be installed around the entire outer edge of the landfill. The trench will be 25 ft deep, and the outer wall of the trench will be lined with a synthetic membrane. The total quantity of trench is 2500 l ft. The trench is 3 ft in width. The trench excavation and backfill quantities are $(2500 \text{ ft} \times 25 \text{ ft} \times 3 \text{ ft})/(27 \text{ ft}^3/\text{yd}^3) = 6940 \text{ yd}^3$.
   (*b*) The trench will be backfilled with gravel.
   (*c*) Gases will be vented to the atmosphere.
2. Gas monitoring probes
   (*a*) Monitoring probes will be spaced every 100 ft along the outer edge of the trench and to a depth of 25 ft. The total vertical ft of probes is $2500 \text{ ft}/100 \text{ ft} \times 25 \text{ ft} = 625$ vertical ft.
   (*b*) Probes are driven and are $\frac{3}{4}$ in diameter. Probes will be monitored quarterly with a portable meter.
3. Estimate costs.
   (*a*) Estimate capital costs. Set up a spreadsheet as follows.

Item	Quantity	Unit cost, $	Cost, $
Barrier trench excavation	6,940 yd^3	4	27,760
Disposal of excavated soil	6,940 yd^3	4	27,760
Synthetic membrane	62,500 ft^2	40	100,000
Gravel backfill	6,940 yd^3	1.60	69,400
Monitoring probes	625 v ft	10	6,250
Total capital cost			$266,170

(*b*) Estimate O&M annual cost. Set up a spreadsheet as follows.

Item	Quantity	Unit cost, $	Cost, $
Monitoring with meter	4 visits	600	2400
Total annual cost			$2400

Comparison of Solutions. It is possible to compare the solutions after calculating the costs of each remediation. The passive control system is about $25,000 higher in capital costs than the active system. In annual costs, the passive system is $10,000 per year less than the active system. In choosing a remedial action, the length of time that the system will operate becomes the determining factor in costs. If the control system must run for 10 years, the 10-year cost of the active system is $100,000 more than the passive system.

Comment. The passive system has a low annual cost because it does not destroy the gas from the landfill. If the regulatory agency decides to set emission controls on the landfill, the passive system may not be acceptable.

17-5 DISCUSSION TOPICS AND PROBLEMS

17-1. Contact your local regulatory agency for solid waste landfills. What regulations exist for inactive or abandoned landfills? Has the agency developed an inventory list of inactive landfills in their jurisdiction?

17-2. A minor explosion and fire have occurred in your community. The fire department has investigated and found no cause for the event. As the city utilities director, you have been asked to evaluate the possible sources of explosive gases in the area of the event. What steps would you take in the investigation? What information would you seek?

17-3. Referring to Discussion Topic 17-2, assume you have identified an abandoned municipal solid waste landfill located in the municipal park about 500 ft from the site of the explosion. Lay out a field investigation plan for the landfill and the problem site. Justify your use of a field investigation grid of soil borings.

17-4. Refer to Example 17-1. Assume that the first-stage investigation identified Cl^- concentrations of 600 mg/L in well W1-2. Lay out a second-stage field investigation to evaluate a remediation that eliminates the violation of drinking water standards.

17-5. Referring to Discussion Topic 17-4, what is your opinion of a remediation that mitigates the problem by purchasing bottled water and abandoning the water well? Explain your position.

17-6. Refer to Example 17-2. Would you choose the passive perimeter gas control system based on economics alone? Give other reasons for your selection.

17-7. Land settlement problems can be remediated by bringing in additional soil to level the surface. What is your opinion about this solution? Will the problem come back? Would you excavate the old wastes?

17-8. Referring to Discussion Topic 17-3, you have been informed that the city cannot pay to remediate the gas problem. Would you try to have the site placed on the NPL and hope for Superfund funding? What are the consequences?

17-9. You are the environmental specialist for a company that just purchased a property and started construction of the home office. During excavation you find that the site is part of an inactive landfill. Testing of soil samples shows that vinyl chloride, a hazardous waste, is present in concentrations that exceed regulatory standards. What is your next action? If the remediation of the problem is very costly, will you seek participation by other parties? Why?

17-10. Referring to Discussion Topic 17-9, assume that the contamination, vinyl chloride, has moved off your property. You have been asked by the agency to work with adjacent property owners to conduct more field tests to define the extent of the vinyl chloride plume. What is your position in responding to the request? What liabilities are involved?

17-6 REFERENCES

1. Arbuckle, J. G., *et al.*: *Environmental Law Handbook*, 9th ed., Government Institutes, Rockville, MD, 1987.
2. Bellandi, R., *et al.*: *Hazardous Waste Site Remediation, The Engineer's Perspective*, Van Nostrand Reinhold, New York, 1988.
3. *Hazardous Substances Response, as Authorized in CERCLA*, Code of Federal Regulations, 40 CFR 300, Subpart E, Office of the Federal Register, National Archives and Records Administration, Washington, DC, revised July 1, 1990.
4. Davis, M. L., and D. A. Cornwell: *Introduction to Environmental Engineering*, 2nd ed., McGraw-Hill, New York, 1991.
5. U.S. Environmental Protection Agency: *Handbook: Remedial Action at Waste Disposal Sites, (Revised)*, EPA/625/6-85/006, Washington, DC, October 1985.
6. Wagner, T. P.: *Hazardous Waste Regulations*, 2nd ed.,Van Nostrand Reinhold, New York, 1991.
7. Water Science and Technology Board: *Hazardous Waste Site Management: Water Quality Issues, Colloquium 3 of a Series*, National Academy Press, Washington, DC, 1988.
8. Wentz, C. A.: *Hazardous Waste Management*, McGraw-Hill, New York, 1989.

PART VI

SOLID WASTE MANAGEMENT AND PLANNING ISSUES

In an earlier text, which served as the precursor to this present text, the authors presented an overview and structure of the discipline of solid waste management. Waste functions were identified, characterized, and given a setting within the community infrastructure. Today, integrated solid waste management requires a more formal structure of facilities and political actions within the community and state and federal governments. Armed with technical and economic resources, the manager of a solid waste system must conduct systemwide studies to develop the logic and implementation rationale that will allow for the integration of *all* solid waste management activities, from generation through recycling and finally to disposal.

Public health and environmental standards to be achieved in the management of solid waste are now defined more clearly in federal, state, and local regulations. As a result of these regulations, solid waste management functions can now be assessed by technical performance, comparative economics, and speed of implementation. There are still social and political structures for implementation, but the practitioner is now required to provide technical specifications to resolve solid waste management problems. Changes in solid waste management regulations and highlights of some important management problems, which must be resolved to implement new facilities under the new laws and to develop local, regional, and state plans, are presented in Part VI.

CHAPTER 18

MEETING FEDERAL- AND STATE-MANDATED DIVERSION GOALS

Increased environmental awareness concerning the problems and opportunities associated with the management of solid waste has led to the enactment of far-reaching legislation and strong public support for waste diversion. Waste diversion legislation was adopted in 31 states between 1989 and 1991, even though waste diversion is not driven by economics in the U.S. economy. While the form of the legislation is different for each state, it can be said that the legislative mandates reflect a changed public attitude toward solid waste and waste management.

The importance of waste diversion as a waste management activity is considered in this chapter. Because waste diversion is so new to the traditional waste management system, there are few proven methods available for its implementation. However, numerous waste diversion methods are currently being proposed, tried, and evaluated [4]. The purpose of this chapter is to consider some of the management issues for source reduction, recycling, and waste transformation through composting. Waste transformation through composting is presented as a method of diversion, because the composted product is diverted from the landfill if used as a soil amendment. The impact of these issues on integrated solid waste management is illustrated with case studies.

18-1 STRATEGIES FOR MEETING DIVERSION GOALS

Waste diversion occurs when waste that normally would be delivered to the landfill is removed from the disposal system and does not use up landfill capacity. Numerous opportunities for waste diversion have been identified in previous chapters of this text. The solid waste manager must consider the options for the community and develop a strategy to achieve the diversion goals established by the community. In some way, the final strategy for diversion will include source reduction, recycling, and waste transformation.

A strategy for waste diversion must be responsive to the applicable laws and the ability of the community to pay for the facilities. Today, waste diversion and integrated solid waste management are mandated by numerous state laws. (See Chapter 2 for a discussion of these public laws.) The ability of the community to pay will be determined at public hearings when rates are set for solid waste services. Because both the laws and the rate-setting hearings require accurate reporting of facts about the solid waste management system, it is important to have an adequate data base for waste generation and waste management in the community.

Developing the Data Base

The development of a data base that can be used to assess waste diversion will vary with each community. A format for the data base should be selected that is easily understood by the public and is in conformance with legal requirements. A format that can be used to start a data base is presented in Table 18-1. The solid waste manager should use Table 18-1 in either of two ways—directly to identify the existing sources of waste and the waste management system, or indirectly to verify a data base that is already available to the investigator. Data development starts with waste generation and continues through disposal and the sale of recovered materials. The integrated solid waste management system must be considered in its entirety while completing the data base so that no data sources are missed. The conduct of waste characterization and diversion studies is outlined in Section 6-7 in Chapter 6.

Extent of the Data Base

The extent or size of the data base must be sufficient to meet legal reporting requirements for diversion, but the data base must also be limited to a size that is consistent with what the community can pay for setting up a data entry and use system. Data for the highest rank in the waste diversion hierarchy, source reduction, are the most important to building the data base, because the highest rank must be satisfied first in responding to legislation. However, the cost to obtain data for source reduction in the community is the greatest, because such data are often proprietary information to the source generator, and a fully independent study must be done, because the data are not reported routinely in the waste

TABLE 18-1
Guidelines for data base development for assessing solid waste management options

Element	Typical facilities or equipment	Data requirements
Source reduction	Internal to the generator such as use of two-sided copying and electronic mail	Number of sources; quantity or percent reduction per source
Commingled waste collection	Garbage cans, trucks, rolloff containers	Number, size, and location of cans and containers by street address; number and types of trucks
Collection of source-separated materials	Recycle bins, trucks	Number of sources; type and quantity of material; number, size, and location of cans and containers by street address; number and types of trucks
Transfer station	Access routes; tipping floor; surge storage; transfer trailers; storage area for hazardous materials	Owner; permitted capacity; acceptable wastes; type of tipping floor; charges for tipping
Materials recovery facility	Surge storage; floor and belt picking stations; mechanical separation of commingled materials; densifiers and storage for sale	Owner; permitted capacity; acceptable materials; types of wastes to be separated and throughput capacity; storage capacity
Transformation facility	Biological reactor; mechanical separation of commingled wastes; combustors, boilers, and turbine/generators	Owner; permitted capacity; acceptable wastes; type of tipping floor; charges for tipping
Disposal	Landfill	Owner; permitted capacity; restrictions on types of vehicles and hours of delivery; restriction on type of waste received; contract for capacity
Buyers of materials	Bulk shipping	Specifications for purity; contract for sale; material sale prices

management system. A compounding cost factor is that the contribution of source reduction quantities to the waste diversion goal is the smallest of all methods in the hierarchy. A guide to the expected diversion quantities and the level of effort required to develop the data base and monitor diversion activities is presented in Table 18-2. The typical steps involved in a waste generation and characterization study are presented in Chapter 6.

Once the data base is developed, the strategy for implementation of waste diversion can be set. In the remaining sections of this chapter, the management issues and concerns for implementation of each diversion method are presented. Case studies are included for source reduction, source separation, and recycling. A case study for composting is not included, because there is not enough data

TABLE 18-2
Percent diversion achieved and the distribution of the costs for the development of a data base that includes data on waste generators and waste diversion quantities

Diversion activity	Expected waste quantity diverted, % of total diversion	Percent of total cost to establish and monitor the data base
Source reduction		
Residential	1	20
Commercial	4	20
Recycling		
Source separation		
Residential	7	18
Commercial	13	9
Materials recovery	35	18
Composting	40	15

currently available from successful composting facilities that can be used to illustrate sound practice. The case study approach is used to show how implementation steps are structured, quantified, and evaluated. Additional case studies dealing with recycling have been published by the Institute for Local Self-Reliance [1].

18-2 SOURCE REDUCTION

The highest rank in the federal hierarchy of integrated solid waste management is source reduction. For the solid waste manager, source reduction is an unfamiliar activity, as it has not been included in previous waste management systems. From the perspective of the community, source reduction is the most desired activity, because the community does not incur costs for waste handling, recycling, and disposal for waste that is never created and delivered to the waste management system.

Management Issues and Concerns

The principal issue with respect to source reduction for waste management agencies is to find ways and means to reduce the amount of waste generated at the source. The most practical and promising methods appear to be (1) the adoption of industry standards for product manufacturing and packaging using less material, (2) the passing of laws that minimize the use of virgin materials in consumer products, and (3) the adoption and use by communities of rates for waste management services that penalize generators for increasing waste quantities.

Standards for Product Manufacturing and Packaging. The issues with standards for product manufacturing and packaging include selection of packaging materials that will result in the greatest waste reduction, matching those materials to a large and diverse manufacturing community, and cost competitiveness. The concerns of the waste system manager include how to influence the industry to adopt source reduction measures, how to measure any source reduction in the community and report quantities diverted, and whether the community should help pay the costs of educating waste generators to reduce wastes at the source. Each of these issues and concerns requires extensive testing and development in the free market before being adopted [2].

Laws to Minimize Use of Virgin Materials. Minimizing the use of virgin materials is an indirect way of achieving source reduction, because the total quantity of material used to make consumer products is reduced when recycled materials are substituted for virgin materials. The waste manager can influence the use of virgin materials by purchasing products made from recycled materials or using purchasing specifications for the community that give preference to recycled materials. The issues are costs and availability of products with recycled materials.

Adopting Variable Rates. A waste system manager can directly influence the rates charged for waste collection, recycling, and disposal. One waste reduction strategy used in some communities is a variable charge per can of waste, which gives generators a financial incentive to reduce the amount of waste set out for collection. Issues related to the use of variable rates include the ability to generate the revenues required to pay the costs of facilities, the administration of a complex monitoring and reporting network for service, and the extent to which wastes are being put in another place by the generator and not source-reduced. Of these issues, the waste system manager can resolve the revenue issue by establishing a baseline rate that provides the revenues to pay all sunk costs for equipment and facilities. The variable rate revenues can then be matched to the cost of service provided.

The major concerns with a variable rate for service include (1) the potential redirection of wastes to another means of disposal, or illegal dumping of wastes, by the generator to avoid paying higher rates and (2) the fact that increasing the rate is punitive and does not get to the basic issue involved. Development of an effective waste management data base and waste-monitoring network will help in identifying any problems caused by implementing a variable rate for waste management service in the community [6].

Case Study 18-1
City Implementation of Source Reduction

A city is establishing a waste diversion program. The city council has directed the city manager to investigate the options for waste diversion, but the manager must include source reduction in the options, as the council is strongly supportive of any program that eliminates wastes. The city manager asked for, and the council approved, a recycling specialist position that is now working in the public services department of the city. The city knows that waste diversion is now underway, but activities are fragmented among many organizations and no single source of data exists that can provide the reports requested by the city council. There are no data in the city records regarding source reduction programs.

Management Issues. The management issues to be considered during the investigation are the following:

1. What is the most cost-effective means of identifying and monitoring waste generators?
2. What are the steps at the local level that can change consumer habits to decrease the generation of wastes?

Information and Data. The city operates its solid waste management program as an enterprise fund within the total city budget. The enterprise fund receives revenues only from the customers who use city facilities and who pay the rates for city services. There are no city tax revenues currently used in the operation of the solid waste management system. The collection and disposal of solid waste is done for all city residents by a private company that has a franchise issued by the city and administered by its public services department. Since taking on the assignment to complete the investigation, the recycling specialist has developed the following additional data:

Population	65,000
Industrial establishments	53
Commercial establishments	672
Solid waste disposed, ton/yr	
Residential	27,000
Commercial	45,000
Industrial	23,000
Term of collection franchise, yr	5

Resolution. The recycling specialist had to start the investigation of source reduction options with no data base. Resource documents were obtained from state agencies and the local library, which provided information on residential and commercial waste generators in the city. The franchised collector was also asked to provide information about waste collection accounts. With this background, a budget was set for completion of the investigation. The city manager specified in the budget that the costs were to be split equally between establishing a data base and evaluating options for source reduction. Such a determination reflected the political wishes of the city council to identify and implement programs that could be monitored over time so that data accumulated in the future could be compared to the existing data for a measurement of program success in diversion of wastes. The work was completed in the following steps:

1. Because the quantities of commercial and industrial waste are more than twice the residential quantities, the recycling specialist developed a questionnaire for industry and commercial establishments that requested data on the respondents' current practices for waste handling, materials purchasing, and recycling. The city Chamber of Commerce was asked to review the questionnaire and to notify its members that the city was conducting the work.
2. All 53 industries and 67 (10 percent of the total) commercial establishments received the questionnaire. One question asked if the respondent was willing to participate in a long-term data-monitoring and -reporting program. The incentive offered by the city was free media presentations of the results, including the name of the participating establishments.
3. Upon receipt and coding of the questionnaire responses, the data were evaluated to determine what options might be considered for implementation. At this time in the investigation, the residential source reduction options were also evaluated. The source reduction options were grouped into the following categories.
   (*a*) Rate structure modifications, including local waste disposal fee modifications and quantity-based user fees.
   (*b*) Technical assistance or instructional and promotional alternatives, such as waste evaluations; assistance to businesses wishing to institute waste reduction programs; awards and other types of public recognition; and non-procurement source reduction programs, such as increased use of electronic mail and increased double-sided copying.
   (*c*) City ordinances that regulate purchasing policies for city departments and set requirements for waste reduction planning and reporting by commercial and industrial waste generators.
4. Recognizing the large difference between the waste quantities collected from commercial and from residential sources, the recycling specialist evaluated the equipment and facilities needed to implement a source reduction program for commercial establishments. After evaluating alternatives for costs and the quantity of waste diverted, a waste audit program was selected for implementation. A two-year period for full implementation was adopted, and the expected diversion quantity goal was set at 2.5 percent of the total solid waste in the city. The estimated annual cost of the program is $17,000. There are no capital costs. Among numerous other activities to implement the commercial waste audit program, the city set the following methods to monitor the success of the program:
   (*a*) The recycling specialist met with the 10 largest commercial establishments that responded in the questionnaire sent out in the data-gathering phase of the work. At the meeting, the program was defined and the audit details and schedules were agreed to. Each business was asked to identify the wastes it would target for reduction during the two-year test period.
   (*b*) Conduct follow-on audits at six months and two years after the initial audit. The six-month audit will be used to assess whether the targeted

waste types have been reduced in the wastestream and to monitor seasonal variations. The two-year audit will be used to measure reductions in the wastestream and to determine whether the procedures have been flexible enough to adjust to seasonal variations and other business fluctuations.

(*c*) Expand or modify the program after two years to respond to the results of the test program.

5. Because the city council had a keen interest in the cost-effectiveness of the waste diversion program, the recycling specialist set the following criteria to measure program effectiveness.

(*a*) Participation by 10 targeted businesses by 1994.

(*b*) Participation by an additional 10 businesses by 1996.

(*c*) Ninety percent of the participating businesses to have 5 percent reductions by the six-month audit.

(*d*) Ninety percent of the participating businesses to have 15 percent reductions by the two-year audit.

6. It is impractical for the city to quantify source reduction from homes; therefore, no residential source reduction program was developed. However, a public education program was recommended that is to give the city residents information on consumer products and purchasing of products that reduce the quantities of waste generated. A $10,000 budget was recommended for public education.

7. After evaluating the use of materials procurement guidelines, the city chose not to include quantity reductions in their waste diversion goals. However, the city will implement procurement guidelines for city purchases and will encourage this program through education and public information.

Comment. It is difficult to measure the quantity of wastes diverted in source reduction programs. In this case, the recycling specialist selected an alternative for source reduction that relied on the business community to measure its own progress in waste diversion. The city performed spot audits to verify results only twice, at a cost well within the capacity of the city budget.

18-3 RECYCLING—SOURCE SEPARATION OF WASTES

Recycling is the second rank in the hierarchy of integrated solid waste management in most states and the federal government. Because significant quantities of materials that might have been delivered to a landfill can be diverted by recycling, meeting waste diversion goals through recycling is an attractive strategy. Due to the complexity of the subject of recycling, the management issues and concerns are presented in two parts. Source separation of wastes is discussed first, followed by materials recovery in the following section.

Source separation of wastes is an effective way to improve the performance of all following facilities in the waste management system. Currently, most waste collection equipment and management procedures do not include separate storage and collection. However, the practices of separating wastes at the point of generation, placing the separated wastes in individual storage containers, and collecting the separated wastes separately are being used in more and more systems [5]. The driving force for choosing a source separation option is threefold: improved effectiveness of recycling, improved quality of the recovered materials, and decreased costs of landfills.

Management Issues and Concerns

Waste separation at the source is an essential activity in an integrated solid waste management system. The system manager must consider the options for using facilities that divert the greatest amount of wastes from the landfill within the constraints of community willingness to participate in and pay for the facilities. Some typical diversion opportunities are illustrated in Fig. 18-1. Management issues are related to what will get the job done. Concerns with respect to source separation are those typically found when implementing new facilities where there are less costly existing facilities.

The facilities for source separation include onsite containers; collection vehicles or waste generator delivery of separated wastes; reception and storage

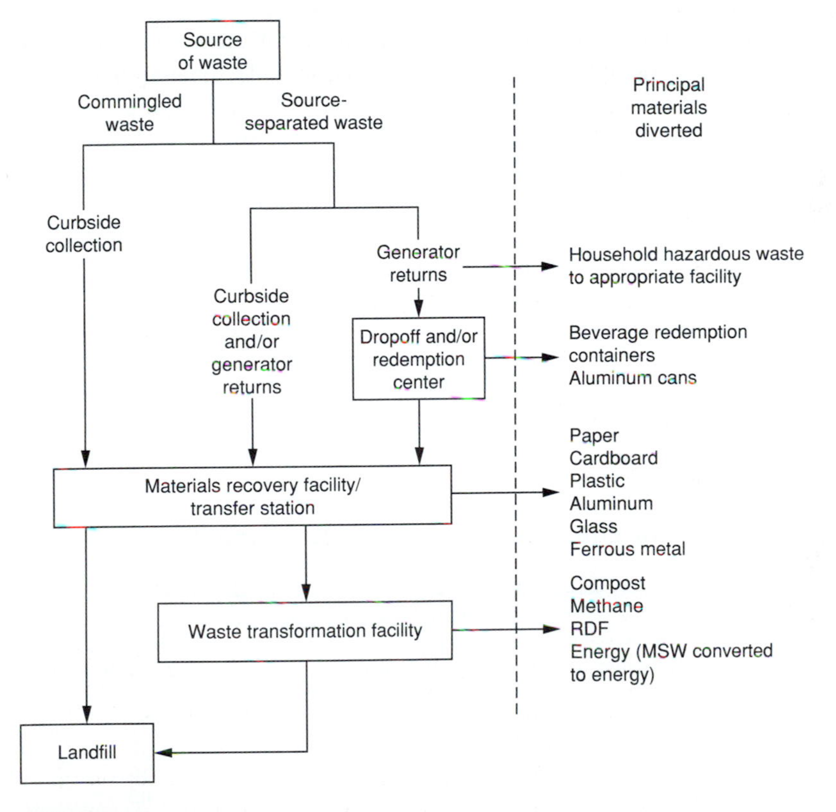

FIGURE 18-1
Typical waste diversion opportunities. Note that the waste transformation facility can also be located at the materials recovery facility/transfer station.

buildings to hold the wastes; and transportation equipment to move the wastes from storage to a materials recovery facility or the landfill. Management issues include (1) the waste components to separate; (2) selection of the type, number, and capacity of onsite containers to hold separated wastes; (3) changing waste storage habits; (4) getting generator cooperation and participation; and (5) paying for the new containers, equipment, and facilities.

Waste Components to Separate. A community must first set goals for waste separation and then proceed with choosing facilities that can be used to implement these goals. In only rare instances will a community set a goal to separate all its wastes. Source separation is beneficial in two general applications: (1) when it improves the diversion capture rate of materials for recycling and (2) if removal of a waste eliminates a hazard at the landfill. Typical capture rates for materials are listed in Table 18-3. The materials shown are the best candidates for recycling in today's markets. As the markets change in the future, new materials will be designated for separation and recovery for recycling. An issue is the correct designation of wastes when the market for recycling is changing. A concern is the timing of changes so that the waste generator continues to participate in separation of waste at the source.

The typical household hazardous wastes that are candidates for designation as separable are listed in Table 5-5 in Chapter 5. The issue in designating hazardous waste for separation is the measurement of benefit to the landfill. There is risk in separating hazardous wastes and accumulating a large mass in one container. The concern is that there may be more harm to the environment from the large-mass form than from the dispersed form of hazardous wastes in municipal solid waste landfills. There are no data from current landfill operations that can be used to assess the impact of household hazardous waste on the environment surrounding the landfill.

The problems to be considered in household hazardous waste separation and control are those typically found when implementing new facilities where there are existing facilities. Separation problems are caused by one or more of the following factors: (1) a lack of proper hazard identification on consumer products

TABLE 18-3
Typical capture rates for source-separated wastes, percent of total available

	Typical capture rates, %	
Waste material	**Collected in separate containers**	**Collected, commingled[a]**
Newspaper	60	50
Plastics	60	55
Glass	65	50
Aluminum	90	80
Yard wastes	90	50

[a] Multiple source-separated wastes collected in a single container.

regarding the preferred method of disposal, (2) the difficulty in changing the habits of convenience of storing household wastes, (3) the need to change household storage methods for separated hazardous waste to provide safety for people and property, and (4) the lack of equipment and trained personnel to collect, transport, and dispose of household hazardous waste. Additional management issues arise in meeting the regulatory agency designations for hazardous waste treatment, storage, and disposal.

Selection of Containers for Separated Wastes. Wastes separated at the source must be stored in appropriate containers. The issues are the number, type, and capacity of containers to be used, and who will purchase and maintain them. Other concerns related to the selection of containers are with public health, compatibility with the recycling collection system, and cost.

Separation of hazardous waste by the household resident requires separate storage containers in the home. The issue is to have available the right type of container for each waste; it is then possible to collect hazardous wastes by type and to prevent contamination of all wastes. Such storage containers do not exist today in homes.

Changing Waste Storage Habits. Historically, the household waste generator has followed rules and guidelines in preparing wastes for collection and disposal. The issue is preparation of new rules and guidelines for storing separated wastes.

The public habit today is to throw all wastes into a single container. As separate containers are required and used in the future, the responsible agency will lead the efforts to change household storage habits. The resolution of this management issue will include combinations of public relations appeals and various types of incentive programs. The strongest incentive program is one that involves city-supplied home storage containers. City-supplied containers are an incentive for the use of uniform equipment at every waste source.

Public health concerns are related to the toxicity of hazardous wastes. Improper mixing of cleansers, bleaches, and solvents can result in flaming or explosive reactions in a storage container. Management should strive to eliminate health and safety concerns in home storage. In addition, public information and building code enforcement measures should be considered to mitigate the impacts of hazardous waste storage.

Household Participation. The success of a source separation program depends on the participation of waste generators. The issue is how to get the public to participate in a situation where they have multiple choices on which container to use. Participation can be either voluntary or mandatory. Numerous examples of how communities have set up and operated their source separation programs are described in Ref. 1.

Disposition of Separated Wastes. Wastes separated for recycling are not diverted from the landfill until the separated waste is used again in a consumer product. How separated waste becomes recovered resources is illustrated in Fig.

18-1. The issue is market capacity for the thousands of tons of separated and recovered materials that will be available when 31 states implement diversion laws. The concern is that there be a stable flow of materials away from the landfill. The resolution of the market capacity issue is the expansion of manufacturing plants that use recovered materials as raw material for consumer products.

For household hazardous waste, the issue is the lack of equipment and trained personnel to collect, transport, and dispose of separated hazardous waste. Currently, household hazardous wastes are managed by absorption into the other discards from the home. Collection and disposal crews have safely picked up and moved the mixed wastes. However, the segregated and concentrated hazardous wastes, although present in much smaller quantities, require special equipment and handling to protect people and the environment.

Economics. A source separation and storage system with multiple containers is more costly than a single-container mixed waste storage system. The economics issue has two parts: (1) raising the capital to buy containers and (2) setting and collecting fees necessary to pay the daily costs of operating the system. However, economics of the system extend beyond containers into the support facilities needed to ensure that separated wastes reach their ultimate disposition. The components that must be included in the evaluation of economics for a typical waste separation and disposition system are reported in Table 18-4. Where a component does not exist, it must be provided and paid for, whether through construction and

TABLE 18-4
Components, functions, and economic impacts of a typical waste separation and disposition system

Component	Function	Economic impacts
Bins	Store individual materials at the point of generation.	Bins are stolen and replaced at cost to the community.
Commingled bags	Store commingled materials at the point of generation.	One bag needed at each household; point of generation bags can be sold at retail stores at cost to the generator.
Collection trucks	Collect materials and deliver to a processing facility.	Individual bins require specially constructed trucks equipped with individual material compartments; commingled materials in bags can be collected in converted MSW collection trucks at less cost.
Labor	Hand-load materials from bins or bags into the truck.	Labor costs are higher for emptying multiple individual bins than for loading one bag.
Material recovery facility	Prepare material for shipping and sale.	Labor costs are higher for picking and separating commingled materials from the bag than from the individual truck compartments.

operation by a public agency or through contracts with private industry. Financing methods include the use of local bonds, direct charges to the home owner, and use of private financing.

Case Study 18-2
Selection of Equipment and an Operator for Source-Separated Wastes

A large eastern U.S. city is starting a materials recovery program. The private company that operates a landfill that serves the city had developed a materials recovery program and submitted a proposal for program implementation in the community. The proposal specified source separation and delivery of the separated materials to the materials recovery facility by city crews. The private company will receive and process the materials and recycle in accordance with market demands. The city council directed the commissioner of sanitation to evaluate the impact of the proposal on existing city operations and to report back with a recommendation for action.

Management Issues. The principal management issues in this situation were as follows:

1. What waste components are to be separated and what generators will be asked to participate?
2. What source storage containers are to be used and who will collect the containerized materials from curbside?
3. Will the waste generators follow the requirements set in the proposal from the private company for material separation and cleaning before they place materials in the recycle containers?

Information and Data. The commissioner of sanitation assigned staff to conduct the evaluation and write the report. The project manager was told to identify alternatives that were to be simple and effective without overburdening taxpayers. The proposal from the private company specifies the following materials to be delivered to their facility: aluminum and steel cans, glass, and plastic. The company offered to pay the city $2.15/ton for the materials, regardless of whether the materials were delivered commingled or separated. Other relevant information included a decree by the state that all cities must have curbside recycling by the fall of 1990. The size of the city is as follows:

Population	370,000
Households	120,000

Resolution. Resolution of this case involved the following steps.

1. The project team reviewed available literature on technology for source-separated storage containers and interviewed container manufacturers. From this activity, a list of specifications was developed for the storage containers.

2. A citizens' group was convened and asked to participate with the city staff in evaluating storage containers. The criteria used in evaluating each container included cost, ease of use by the household, safety, compatibility with collection equipment, and acceptability at the private company's materials recovery plant.
3. Two types of containers were selected for testing in a pilot program: individual bins for metals, plastic, and glass; and a commingled container. The commingled container was a unique blue bag, whose color was used to differentiate the recyclables container from the container for other wastes. The container pilot program was run for three months on two separate routes of 600 households each.
4. The results of the pilot testing favored the use of the commingled container. The following are significant criteria favoring the single commingled container.
   (*a*) A staff-conducted time-and-motion study showed a 60 percent reduction in time spent on recycling routes by using a single-container system rather than a three-bin system.
   (*b*) One container requires less separation by the household than separate bins.
   (*c*) The blue bags are lighter and easier to handle than the bins.
   (*d*) Being plastic, the bags themselves are recycled.
5. From the results of the collection truck study, it was concluded that the city should continue to be the only collector of recyclables and other wastes. The city was able to refurbish older packer trucks for the less demanding service on recycling collection routes. The cost savings over purchase of special recycling trucks was significant.
6. Safety and material quality monitoring requirements were satisfied by the use of translucent blue bags. The collectors could see the materials and leave behind on the curb any bags holding objectionable materials, and the private company could see the materials at the time of material unloading at the materials recovery plant.
7. The city council approved the recommendations of the report and authorized their implementation. The first phase of the program included 30,000 households. Curbside collection for the remaining 90,000 households will begin in 1993.
8. Blue bags are widely available at retail outlets. The city is working with retailers to ensure that the right type of bag continues to be sold to residents.

Comment. The single-bag system proved to be the best for this city. An important part of the success here is the private company that is willing to invest in a materials recovery facility and pay the city for source-separated materials.

18-4 RECYCLING—MATERIALS RECOVERY

Materials recovery is the final step in the recycling component of solid waste diversion. As described in the source separation section of this chapter, materials recovery is the only way to complete recycling after wastes are separated. However, materials can be recovered from wastes that have not been separated at the source. The various methods of waste processing for materials recovery are presented in a case study. Management issues and concerns are emphasized, drawing on the technologies presented in Chapters 7, 8, 9, 12, and 15.

Management Issues and Concerns

Materials recovery is a complex situation for managers of solid waste systems. Referring to Fig. 18-1, all of the activities for diversion are options for inclusion in the materials recovery system. Because waste diversion is mandated, the system manager is left with choosing the options that are acceptable to the public. Economic and environmental constraints apply to materials recovery just as they apply to solid waste collection, transformation, and disposal.

The principal issues and concerns with respect to materials recovery are as follows: (1) to determine the appropriate technologies and facilities, (2) to develop a data base that provides a rational method of setting the economics of options, (3) to establish priorities, (4) to identify markets for the sale of recovered materials, (5) to determine materials specifications and the extent of the effort that should be made by the solid waste management agency to attempt to meet the specifications, and (6) to assess the impact of market stability on the waste management system.

Determine the Appropriate Technologies and Facilities. The choices for materials recovery technology start with collection and extend through transformation, with some facilities being the same as those used for processing nonrecyclable wastes. The principal technologies for the recovery of waste materials and their application are summarized in Table 18-5.

The issues include matching the costs of a materials recovery system to the ability of a community to pay; dealing with the constraints of any existing franchises or permits that set the ownership of waste and recovered waste materials; and finding a market for material sales that will set a specification for purchase of recovered materials. The resolution of issues includes discussions with community groups, such as the League of Women Voters and any taxpayers' protective group; negotiating with franchisees to get waste ownership to support materials recovery; and contacting industry trade groups for leads to materials buyers.

Establish a Rational Data Base. The data base should be matched to the preferred materials recovery technology selected by the community. The issue is how much data to collect. Data collection costs will be extremely high if every waste generator is asked to participate, depending on the size of the community. Another issue is the validity of data for materials recovery. Because the diversion laws are new, waste sources may skew data to the benefit of the law, not for rational selection or sizing of facilities.

A second issue involves reporting the results of diversion actions to the regulatory agencies. The data base is the foundation of a monitoring and reporting program. Most states with diversion laws require the communities to report the

TABLE 18-5
Application of the principal technologies for the recovery of materials

Technology	Application
Drop-off centers	Commonly operated by community groups for recycling or for fund raising.
Buy-back centers	Operated for purchase of materials from the public.
Materials collection	
Informal	Source of revenue for organizations such as Boy Scouts, Kiwanis, and church groups.
Residential	Done by for-profit entrepreneurs, either by franchise for recycling or as a part of solid waste collection; most programs require that materials be segregated by the resident, stored, and set out for collection on a regular schedule.
Commercial	Same as residential plus use of large containers for collection because of greater commercial quantities; greater potential for recovery at a waste delivery site.
Green waste	Used when green waste composting is to be done; requires separation at the source and storage between collections.
Material recovery facilities	
At transfer stations	Used at existing stations when recovering a limited number of materials such as old corrugated cardboard (OCC) and bulky metal scrap.
At landfills	Used where there is enough space and suitable terrain; may be most acceptable location because site is already receiving wastes.
Stand-alone	Used where no facility now exists; facility applications are numerous, with the selected facility matching the recovered material collection methods and the material buyer's specifications.

results of diversion activity, with one state, California, imposing a $10,000-per-day penalty for not meeting diversion goals. The issue is knowing where and how much waste is generated within the community and then including the largest waste generators in the data base so that they will be a part of the future monitoring program.

The resolution of data base issues starts with a good understanding of materials recovery and solid waste management systems as they exist in the community. The lack of past engineering and statistical analyses in the field of solid waste has caused a shortage of professionals with an understanding of solid waste and materials recovery. Consultants can help the community, and professional associations like the Solid Waste Association of North America are offering peer assistance to members. Costs and what the community can afford to pay must also be resolved.

Establish Priorities. Mandated diversion has caused a rush to implement materials recovery programs in communities. This intensive activity has resulted in the identification of numerous material recovery opportunities. Yet in many instances, the organizers have spent little time considering the energy requirements and costs associated with the implementation of the various material recovery programs suggested. An example is source separation of recyclables and curbside setout for collection. If the recyclables are collected by trucks that do not also collect the other commingled wastes, two trucks must travel the same street and use more energy than a single truck picking up commingled materials. For this reason it is important to establish community priorities with respect to the methods of materials recovery.

Resolution of priorities for materials recovery will occur when the following questions are answered: What materials should be recovered? When should a program be undertaken and how large should it be? Should community economic resources be devoted to the program? In establishing priorities, it is especially important that they be based on the social and economic circumstances of the community as viewed jointly by the planner, decision-maker, politician, and the people of the community.

Identification of Markets. Traditionally, identification of materials recovery opportunities have been left to private industry; planning and coordination by public agencies have been missing. These past practices are undergoing a significant change, caused by state mandates that materials be diverted from landfills. A good perspective from which to consider the concerns for materials recovery is the market that will use the material. There are two broad categories of markets: raw materials for industry and raw materials for the production of energy or fuel. It is possible to group all recovery facilities so that wastes will be used by one of these markets. Some of the markets and common recovery uses are listed in Table 18-6. Market stability is considered later in this section.

TABLE 18-6
Markets and typical uses for materials recovered from solid waste

Market category	Materials	Typical uses
Raw material for industry	Newspaper	Newsprint stock
	Corrugated cardboard	Fiberboard and roofing material
	Ferrous metal	Reinforcing bars
	Rubber tires	Paving
	Oil	Refined oil
	Beverage bottles	Refilled beverage bottles
	Textiles	Wiping rags
	Aluminum cans	New aluminum
	Organic wastes	Chemicals
	Incinerator residue	Concrete, roadways
	Organic wastes	Compost
Raw materials for energy or fuel production	Organic wastes	Steam production
	Organic wastes	Fuel production (e.g., methane, methanol)
	Tires	Tire-derived fuel

The specific details of product purity, density, and shipped conditions must be worked out with each potential buyer. An example of private industry actions in the primary metals industry is shown in Fig. 18-2. The relationship of the waste system manager to the metals cycle is also identified. As shown, obsolete scrap and its management are only a small increment of the cycle. In this situation, the market buyer is the scrap processor who also sets materials specifications. However, the raw material preparation and processing cycle is changing for manufacturing industries, such as paper and plastics. Mandated diversion is being implemented by many communities, and the diverted materials are mostly paper and plastic. A new raw material, recovered wastes, is available, and industry is building new processing plants that use a recovered waste feedstock.

The management issues include selecting a method of reuse for recovered materials; maintaining control over the solid waste and materials recovery service in the community; and measuring the performance of the materials recovery facilities to ensure that diversion goals are achieved. All of these are new issues for a public agency, and so the resolution of the issues must include answers to questions that arise at the interface of the waste management system and the product-manufacturing system. Should the manager of a solid waste system purchase the necessary equipment and pay the required salaries to improve the quality of waste materials and thereby meet manufacturing specifications? Or should the manufacturer purchase raw wastes and pay the costs of improving quality? The answers must come from communication and negotiation between the waste system manager and the material purchaser.

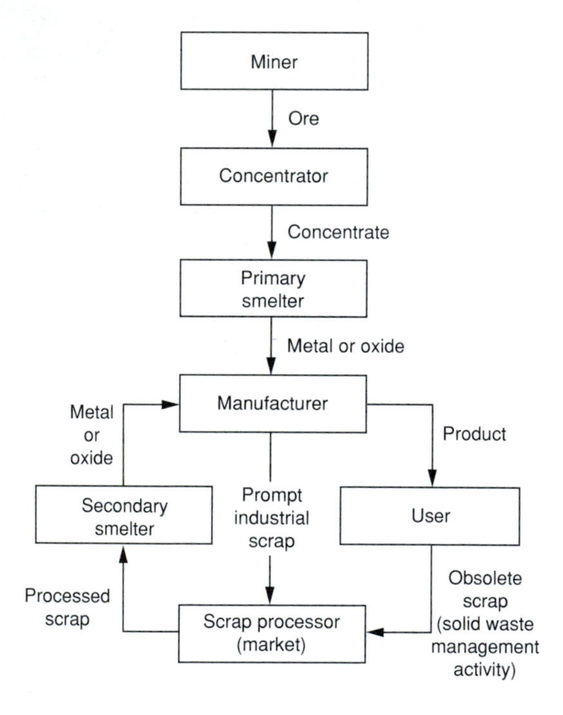

FIGURE 18-2
Primary metals-processing cycle and its relationship to solid waste management.

(*a*)

(*b*)

FIGURE 18-3
Impact of market instability on the sale of recovered materials: (*a*) stockpiled baled cardboard and (*b*) stockpiled ferrous metal.

Market Stability. The resolution of all types of market problems involves long-term contracts for recovered materials and energy. The impact of unstable markets is shown pictorially in Fig. 18-3. The large stockpiles in both cases were caused by termination of existing market contracts. In most cases, the solid waste system manager cannot create a market. Therefore, market outlets must be found either directly, with materials and energy users, or indirectly, through brokers or wholesalers. Of specific concern under these conditions is the manner in which the materials markets fluctuate and the impact of these fluctuations on the solid waste management system. Market demands, legal aspects, resource shortages, and political aspects must all be considered in assessing market stability.

Political Aspects. The political nature of solid waste management has been identified and discussed in numerous sections of this text. Its importance to recycling is manifested in the vested interests of organizations operating salvage systems, waste collection and disposal systems, and raw material–mining enterprises. Both public and private organizations are involved. An example of a public agency with a vested interest is a community that has a large capital investment in a combustion facility. In most cases, the potential of recycling is reduced to only heat-related functions, because public officials are reluctant to abandon a tax-funded combustion facility. Vested interests of private organizations are focused more on profitability. The political influence of both public and private agencies can thus have a strong impact on markets and market prices.

Case Study 18-3
Selecting a Materials Recovery Facility

A Florida county is selecting a materials recovery facility (MRF) that will divert wastes from landfills to meet the state-mandated solid waste volume reduction goal of at least 30 percent by 1995. A recyclable materials recovery program was initiated in July 1989, but the county commissioners are uncertain as to whether the program will be able to achieve 30 percent diversion. County solid waste staff has been directed to evaluate the existing recycling program, identify other options for recycling if the existing program is inadequate to meet goals, and prepare a report to the commissioners that defines a program to meet diversion goals.

Management Issues and Concerns. The management issues to be considered during the evaluation are the following:

1. What technology should be used at the MRF so as to take advantage of the materials already being separated?
2. How will the county solid waste management agency determine materials specifications demanded by materials buyers?
3. How will county priorities be matched to the mandates of the state for waste diversion?

Information and Data. The solid waste managemant department designated a project manager to conduct the evaluations. In reviewing existing data, the project manager found that the current materials recovery program separated construction and demolition debris, newspapers, aluminum cans, glass, and plastic bottles. The separation program recovered the type of material mandated by the state, and the program was started in 1989. Other background data included a state-mandated restriction that no more than 50 percent of the 30 percent mandated diversion goal may be met with yard waste, white goods, construction and demolition debris, and tires. The project manager developed the following data:

Population, estimated for 1995	85,000
Participation of public in the existing separation program, percent of total population	70
Type of separation program	Mandatory
Type of landfill	Lined
Materials banned from the landfill	Tires, used oil, white goods, lead-acid batteries, and yard wastes
Collection of wastes and separated materials	Private company
Annual waste quantity disposed in 1995, tons	90,000

Resolution

1. The first step in completing the evaluation was the development of performance factors for the recycling of materials as mandated by the state. The recyclables

separation data for the past three years were reviewed and tabulated; they are presented in the accompanying table.

Material	Percent of wastestream by weight	Average annual quantity, tons	Recovery percentage
Aluminum	1	720	80
Newspaper	8	4320	60
Glass	7	2520	40
Plastic	8	2160	30
		9720	

The existing recyclables program is separating 9720 tons per year, an 11 percent diversion of the 90,000 tons per year (assuming that the recyclables program will separate 9720 tons in 1995). The diversion achieved by the existing program would not meet the state-mandated diversion of 30 percent by 1995.

2. The project manager continued the evaluation by calculating the quantity of wastes in the county that are prohibited by state regulations from disposal in the county's lined landfill. These quantities are shown in the accompanying table.

Material	Percent of wastestream by weight	Average annual quantity, tons
Yard wastes	18	16,200
Tires	0.9	810
Used oil	0.2	180
White goods	0.7	630
Lead-acid batteries	0.6	540
		18,360

If the full quantity of prohibited waste were diverted from the landfill, the county would add 20 percent to its diverted wastes in 1995. Adding 20 percent to the existing 11 percent would give the county 31 percent diversion, satifying the state goal of 30 percent by 1995.

3. With an understanding of the wastes that could be separated and diverted from the landfill, the project manager continued with an evaluation of technology that could be implemented by the county to recycle the identified wastes. Because the existing wastes and separated recyclables are collected by a private company, the preferred option would be to expand that system to include the new waste materials.

In evaluating technology, the project manager looked at the efficiency of the existing recyclables separation program to determine if additional materials could be recycled. Because the program was a mandatory program for every resident of the county, and participation rates were currently under 100 percent, there were additional materials to be recycled in the county. But how could those materials be recycled? The project manager had two choices to consider in answering the question of what technology to use to get the remaining recyclables: expand enforcement of the mandatory recyclables separation program, or implement a new materials recovery program that would recover additional quantities of recyclables from county wastes. For either choice, the program costs will increase.

During the review of the recovery factors for each of the current recyclables, the data were compared to published data on the results of mandatory separation in similar counties. It was concluded that the 70 percent participation rate in the county was on the high side of the national average for communities with mandatory separation. After checking with the county health agency regarding additional enforcement of waste separation, the project manager determined that the current enforcement staff was effectively reaching a representative number of county residents. The next level of enforcement was to put out enough staff to check the storage containers of all 40,000 accounts in the county—an impractical action because of costs and political reaction. Using the recovery factors once again, it was concluded that the best candidate materials for increased recovery were newspaper, glass, and plastic.

With these data, the project manager was now ready to evaluate the equipment that could separate and recover increased quantities of waste materials. Because of the significant quantity of recovered materials the county must produce to meet the state-mandated goal, the evaluation would include technology that is used for recovery of materials that are prohibited from lined landfills. However, the state has set a 50 percent upper limit (of the 30 percent diversion goal) on the amount of credit toward mandated volume reduction quantities that can be taken from recycling of yard waste, white goods, construction and demolition debris, and tires.

4. The project manager met with the private company and discussed the facilities used by the company to process county recyclables. The current capacity of the facility is 10,000 tons per year, with separate processing lines set up to handle the quantities of aluminum, newspaper, glass, and plastic from county wastes. The recycling facility is a converted manufacturing building with large open bays, multiple large rollup doors, and concrete floor. The existing process lines can be expanded to increase throughput of separated materials, or a new processing line for commingled waste can be designed and installed in another bay of the same building. Of special concern to the private company was the request from the county to recycle yard waste, tires, used oil, and lead-acid batteries.

5. The resolution of the issue of choosing a technology was to have a mix of technologies provided by the private company and the county. The private company agreed to build a new processing line to recover cardboard, paper, glass, and plastics. The company would also build a low-volume special waste–

processing facility to recover tires and white goods. The county is building a yard waste–processing facility to recover yard waste, and the county awarded a contract to a broker who will receive and resell used oil and lead-acid batteries. The quantities of materials that were diverted from the landfill are shown in the accompanying table.

Material	Percent of wastestream by weight	Recovery factor,[a] %	Average annual quantity, tons
Aluminum	1	80 (SS)	720
Newspaper	8	60 (SS)	4,320
Cardboard	6	80 (SS)	4,320
Glass	7	40 (SS)	2,520
Plastic	8	30 (SS)	2,160
Yard wastes	18	90 (MRF)	14,580
Tires	0.9	100 (MRF)	810
White goods	0.7	100 (MRF)	630
Used oil	0.2	100 (Broker)	180
Lead-acid batteries	0.6	100 (Broker)	540
Total			36,180

[a] SS = source-separated; MRF = materials recovery facility.

The 36,180 tons per year diverted from the landfill is a diversion rate of 40 percent, giving the county a conservative 10 percent more diversion than required by the state by 1995. The issue of meeting materials specifications for sale of recycled materials was resolved by delegating that responsibility to the private collection company and a broker.

6. The project manager submitted the report to the supervisor for a quality check before presentation to the Board of County Commissioners. During the checking process, it was noted that the state restriction that allows only 50 percent of the 30 percent diversion to be met by yard waste, tires, and white goods had been missed in calculating the quantities being diverted. The recalculation was completed as follows.

State allowable yard waste, tires, and white goods:

$$30\% \times 90{,}000 = 27{,}000 \text{ tons per year}$$

$$50\% \times 27{,}000 = 13{,}500 \text{ tons per year}$$

County yard waste, tires, and white goods = 16,020 tons estimated for 1995

Reduction necessary in wastes diverted:

$$16{,}020 - 13{,}500 = 2520 \text{ tons per year}$$

Net waste diverted following the state formula:

$$36{,}180 - 2520 = 33{,}660 \text{ tons per year}$$

Diversion ratio following the state formula:

$$33{,}660/90{,}000 = 37\%$$

The change was made to the report, and the report was presented to the commissioners.

Comment. The costs of programs were included in the report to the commissioners, but for brevity they were not included in the case study. Yard wastes are the most significant quantity of material to be diverted, so the county must be realistic in setting up a yard waste–recycling facility. The county estimated that only 90 percent of the yard waste going to the landfill in 1991 would arrive at a yard waste–recycling facility in 1995, even though the state mandated 100 percent diversion of yard waste.

18-5 WASTE TRANSFORMATION THROUGH COMPOSTING

Waste transformation is a means of waste diversion from landfills if the end products of transformation are not delivered to a landfill. Composting and combustion are transformation processes that are used to divert wastes from landfills. However, not all states provide the same diversion credits for combustion. The management issues and concerns for implementing composting facilities are presented in this section. Combustion is considered in Chapter 19.

Management Issues and Concerns

Solid waste system managers are seeking ways to remove the green wastes from landfills. Composting is an acceptable technology for transforming green wastes and other organics into a consumer product. Composting technology is discussed in Chapters 9 and 14. Although composting has been used for many years, it has never been adopted widely as a waste transformation process. The management issues include preparation of feedstock, control of the composting process, purity and consistency of the product, and the stability of markets for the compost product.

Preparation of Feedstock. The wastes that are compostable must be separated from other wastes before becoming feedstock for composting. For those waste management systems operating today that do not collect yard or green wastes separately, there is an issue of how to prepare the composting feedstock. The options are separation at the source and delivery to the transformation facility, or delivery of mixed wastes to the facility with separation of yard waste as a part of the composting facility. The issues are who controls the waste and who owns the compost facility.

The concerns applicable to the preparation of feedstock are the size of the facility, land-use compatibility, and costs. In many instances, the facility cost must be held to a restricted budget. Under a restricted budget, the facility may be undersized and placed in a location not compatibile with adjacent land use. Clearly, many of the problems with composting in the past can be traced to inappropriate management decisions on feedstocks and facility location. The resolution of issues for feedstock preparation include sound planning, realistic assessments of waste control, and proper sizing of facilities.

Control of the Composting Process. The management issues for the reactor include conforming with regulatory standards and maintaining aerobic conditions to prevent odors. In enclosed-vessel reactors, the reactor performance is guaranteed by the manufacturer. The system manager must ensure that the guarantees are fulfilled.

Purity and Consistency of the Product. If the compost product is to be diverted from the landfill, it must satisfy purity specifications from the user, and it must do so consistently over time. An important management issue is the contract between the waste system operator and the compost product buyer. To reduce the risk to the waste system, there should be a firm contract giving product quantity and quality specifications. However, there must be flexibility in the specifications to allow for variations in waste feedstock without that variation causing a breach of contract. Typical compost specifications are shown in Tables 15-8 and 15-9.

Stability of Markets. The compost buyer must accept the product of processing if it is within the specifications of purity and quantities, and compost must be accepted for the full term of the contract. With a firm contract, the waste system manager can be certain of meeting the waste diversion goal. The management issue is setting up a contract that will remain stable over the expected period of time.

Resolution of the issue for a stable market can be achieved in two ways: by contracting with governmental agencies who will use the compost on public lands, and by using multiple contracts so that the failure of one market will not cause a 100 percent failure to divert the compost from the landfill.

18-6 DISCUSSION TOPICS AND PROBLEMS

18-1. A community of 100,000 population is responding to a state-mandated solid waste diversion goal of 30 percent. The legal requirement is to achieve the goal within three years. Your community has the freedom to select the type of waste materials to divert. Describe the actions you would take to establish a data base. Briefly explain why you chose those actions.

18-2. Referring to the data base you selected in Problem 18-1, what materials are candidates for source reduction? How much of the 30 percent diversion do you estimate to come from source reduction?

18-3. Specify the components of a source separation program for the community in Problem 18-1. How much of the 30 percent diversion do you estimate to come from source-separated materials?

18-4. The materials separated at the source in Problem 18-3 must be prepared for delivery to a buyer. Assume that you have not achieved 30 percent diversion in your source separation system and that your community is building a materials recovery facility. Prepare a materials flow diagram that shows how you would handle the materials in your facility.

18-5. Resources Unlimited, a nonprofit organization, has presented your private solid waste collection company with a proposal stipulating that the residential customers you serve would separate paper, ferrous metals, aluminum, and glass for recovery. You are one of five owners of the company, and the other owners have asked you to respond to the request, as the public agency that issued your franchise to collect wastes has supported the proposal. What are the management issues for your company?

18-6. A city in the midwestern United States wants to evaluate the impact of banning all residential burning of yard and tree wastes. Assume that you are the city engineer responsible for completing a plan to evaluate the impact of a backyard burning ban and for recommending a new policy for disposal of yard and tree wastes. Which city agencies would you contact in organizing this planning effort? Prepare a work plan to guide you in completing the plan.

18-7. Referring to Problem 18-6, identify and list local ordinances that might be needed to implement a collection and disposal system for yard and tree wastes.

18-8. Contact the agencies in your city that might use compost or mulch made from yard wastes and identify any restrictions to its use. What steps would you take to meet those restrictions?

18-9. The community in Problem 18-1 is evaluating the benefits of placing large, city-owned storage containers in the residential areas. There would be separate containers for wastes and separated materials, and each container would serve six houses. In the present system, 32-gal containers are used at each residence. Does this plan deal in policy issues? Why?

18-10. Waste transformation has been disallowed in some states as an acceptable means of waste diversion from landfills. What is your opinion in this matter?

18-7 REFERENCES

1. Institute for Local Self-Reliance: *Beyond 40 Percent Record Setting Recycling and Composting Programs*, The Institute for Local Self-Reliance, Washington, DC, 1991.
2. Selke, S. E.: *Packaging and the Environment: Alternatives, Trends and Solutions*, Technomic Publishing Company, Inc., Lanchester, PA, 1990.
3. *The Biocycle Guide to Collecting, Processing and Marketing Recyclables*, The J. G. Press, Inc., Emmaus, PA, 1990.
4. U.S. Conference of Mayors and H. J. Heinz Company Foundation: *Profiles of the Nation's Resourceful Cities*, United States Conference of Mayors, Washington, DC, June 1991.
5. U.S. Environmental Protection Agency: *State and Local Solutions to Solid Waste Management Problems*, EPA/530-SW-89-014, Washington, DC, January 1989.
6. U.S. Environmental Protection Agency: *Decision-Makers Guide to Solid Waste Management*, EPA/530-SW-89-072, Washington, DC, November 1989.

CHAPTER 19

IMPLEMENTATION OF SOLID WASTE MANAGEMENT OPTIONS

Solid waste management today involves recognizing the rapid changes that are occurring and selecting compatible options for facilities to manage the collection, recovery of materials, and disposal of solid wastes efficiently. For the manager, the solution lies in the proper selection of options. Facility options must be defined according to their physical characteristics, environmental impact, economics, and societal acceptability.

To begin the process of implementing solid waste management options, one must understand the basis for the changes that are needed to existing facilities (see Chapter 2). The process continues with the selection of a new or modified set of facilities that meets environmental, economic, and societal standards. A case study approach illustrates how implementation steps are structured, analyzed, quantified, and evaluated. In this chapter case studies are included for collection system mechanization, energy recovery, and landfill. Management issues and concerns are presented before the case studies.

19-1 CHANGING PRIORITIES IN INTEGRATED SOLID WASTE MANAGEMENT

The hierarchy of integrated solid waste management has transposed the traditional roles of waste transformation and disposal from that of first choice to one of third and fourth choice [3]. Also, the function of waste collection has been expanded to include equipment and routes for collection of recyclable materials that were once commingled with other wastes. Waste collection, transformation, and disposal must now support source reduction and recycling activities.

System managers are responsible for decisions and administrative actions to implement the activities associated with the solid waste management hierarchy, including plans, budgets, financing, and staffing. Because both source reduction and recycling are new solid waste management functions, existing costs will increase. The waste manager must evaluate waste management alternatives and options for efficiency and cost savings. In today's integrated solid waste management system, the activities and facilities involved in the waste collection, transformation, and disposal continue to be the highest in cost. For this reason, these activities and facilities are the subject of close review with respect to the opportunities for cost reductions. Changing funding priorities to match the new waste management hierarchy means old ways of thinking about solid waste management systems must be updated.

19-2 COLLECTION SYSTEM MECHANIZATION

There is a trend in the collection of solid waste to reduce the amount of lifting required of the collector. Because solid waste must still be lifted to load the collection vehicles, mechanical devices are now commonly used to lift and empty containers that were formerly lifted and emptied by workers. Mechanical equipment is popular because it reduces the number of workers and reduces long-term disability costs, and system workers are using more skills and less brute strength. However, implementation is slowed by high initial costs for containers and trucks and by resistance from homeowners to the use of rolling containers instead of the more familiar plastic bags or garbage cans. This section deals with the planning and management of changes to collection equipment and operations, with an emphasis on economics and implementation.

The identification, evaluation, and implementation of options for waste collection are the most important part of an integrated solid waste management system, because an estimated 50 to 70 percent of all solid waste system expenditures are for collection. Within the functional element of collection, more than in any other functional element, one must understand the nature of the operations and operational variables, especially those that are responsive to changes in legislation. Effective environmental legislation to encourage recycling has had a profound effect on the cost of collecting recyclables and mixed municipal wastes. Other proposed legislation for solid waste combustion and the disposal of combustion ash may cause changes in the collection of household hazardous wastes. Besides responding to environmental concerns, the system manager must find a safer way to use labor in collection operations. Mechanization has been effective in reducing human labor while increasing labor efficiency on collection routes [1].

Management Issues and Concerns

Management issues and concerns with regard to waste collection relate primarily to the multiple and diverse equipment and work force programs that can be developed. Increasingly, management has turned to mechanization of collection operations as a means of improving labor efficiency, reducing worker injuries, and reducing waste collection costs. The major concerns are related to labor efficiency and customer service levels (frequency of collection and location of containers). Implementing mechanization involves the following issues: (1) how to meet legislatively mandated recycling goals; (2) the involvement of both public and private agencies in collection operations; (3) mechanization technology; (4) labor constraints; and (5) financing the purchase of collection equipment.

Meeting Recycling Goals. In the United States, each state is setting goals for the diversion of solid wastes from landfills. Typically, the goals are specific to a category of wastes. The impact on a collection system that has already been mechanized is illustrated in Table 19-1. There are two forms of waste diversion,

TABLE 19-1
Impact of diversion goals on an automated waste collection system

Type of material to be diverted	Method of diversion	Impact on an automated waste collection system
Newspaper	Recycling	If newspaper is separated at the source, the automated truck cannot pick up newspaper; if newspaper is to be hand-picked from commingled waste, the automated truck can collect newspaper as placed in the waste container.
Mixed paper	Source reduction	No impact on automated trucks.
	Recycling	Same as the impacts of newspaper recycling.
Plastics	Recycling	Same as the impacts of newspaper recycling.
Metals	Recycling	Same as the impacts of newspaper recycling.
Yard wastes	Source reduction	No impact on automated trucks.
	Recycling	The source storage containers for waste can be downsized, and the size or number of automated trucks can be reduced.
Glass	Recycling	Same as the impacts of newspaper recycling.
Combination of the above materials	Source reduction	No impact on automated trucks.
	Recycling	The source storage containers for waste can be downsized and the size or number of automated trucks can be reduced; there is a potential for increased efficiency in the collection of recyclables with automated collection equipment.

source reduction and recycling. Source reduction is a matter of public policy and industry practice, each of which is outside the management control of most system operators. Recycling remains as the easiest and most productive method of waste diversion for the community.

Recycling goals and collection route efficiency standards are often in conflict. Recycling requires a clean material that meets reuse standards. Segregated storage and collection are necessary, thereby decreasing the efficiency of mechanized collection of mixed waste. The problems caused by this conflict must be identified by system managers and resolved through good operations and clear public policy. The development of a solid waste collection ordinance is the most direct action the political system takes regarding collection and recycling operations. Additional actions involve the setting of policy regarding collection rates or increasing taxes to pay for service.

Public and Private Operating Agencies. Collection service can be provided by either public or private agencies. In the United States, private agencies provide the greater part of the commercial and industrial waste collection service, and public agencies provide the greater part of the residential service. The trend in private agencies is to large national companies with a large funding capability. This capability can be important in resolving problems related to system improvement.

The management of collection services is different for public and private system operators. Management concerns with respect to agencies include the regulation and control of collection. Operational regulations are closely related to source storage for both source-separated and mixed waste. Regulations for public agencies are established by political bodies. In most cases, public agencies resolve public concerns about private service through the regulation of private agencies. Typical contractual conditions for private agency collection are listed in Table 19-2. In all collection systems, a local public health agency usually regulates matters concerning health and safety.

Mechanization Technology. Because mixed solid waste and recylable materials must still be lifted to load the collection vehicles, mechanical devices must be provided to lift and empty containers that were formerly lifted and emptied by workers. The technology for mechanization involves a combination of new trucks and new containers. Some of the most common equipment and its applications are reviewed in Table 19-3. However, implementation of new technology is slowed because labor crews must break old habits of waste collection. (Additional details on equipment and labor efficiency are given in Chapter 8.)

Management concerns are that the storage containers be convenient for the waste generator and that collection agency workers be skilled in operating the trucks with the mechanical lifting devices. In most agencies, the concerns are first measured in pilot operations with the new equipment. A pilot or test route is selected and containers are provided to homeowners on that route. That route is then collected by a mechanized truck and crew. The pilot route is usually operated for three to six months. Data are accumulated and evaluated for application to

TABLE 19-2
Typical contractual conditions for collection systems operated by private agencies

Type of contract	Typical conditions
Franchise	Used where quality of service takes precedence over cost of service. Competition is eliminated. The franchise area is set within specific geographic boundaries. The franchise administrator must monitor rates and service complaints. Elimination of overlapping competitive routes allows higher collection efficiency. Contract stability allows long-term investment in equipment.
Limited permits	Used where limited competition is preferred but the costs of franchise administration are undesirable. Rates are self-regulating. There are no permit boundaries. The number of permits is set by criteria such as potential revenue and population in the permitted area. Difficulties in administration arise when the logic of a set number of permits is questioned.
Unlimited permits	Used where completely open competition is desired. Rates are self-regulating. Efficiency drops because of overlapping routes. Competition can become very intense with a resulting high number of company failures. Administrative difficulties increase as business failures occur.

TABLE 19-3
Common mechanization equipment and applications

Mechanization equipment	Common applications
Hydraulic lifts	Used on rear-loading compactor trucks; often the choice for semi-automated collection where existing collection trucks are rear loading; can be used with various types and sizes of storage containers, but the container must be moved to the rear of the truck and placed on the lift manually for mechanical tipping.
Roll-out containers	Most common use is with fully automated side loading trucks, although the rear-loading truck with hydraulic lifts has used this container.
Hydraulic grapple and lift	Used with fully automated side-loading trucks; installed at the time of truck purchase; allows the truck operator to engage roll-out containers at the curb without leaving the truck cab.
Automated side loading	Used on collection routes where roll-out containers can be moved within the reach of the hydraulic grapple and lift device mounted on the truck.
Rear-loading trucks	Used in communities that have existing rear-loading trucks that are to be retrofitted for semi-automated collection or on collection routes where roll-out containers cannot be used; one or two hydraulic lifts are mounted on the rear of the truck.

the entire collection system. Data should include truck time at each stop, total number of stops to fill a truck, homeowners' acceptance of the new containers, route designation and capacity numbers for recyclable materials, and costs.

Labor Constraints. Labor unions are an integral part of many collection operations. Union contracts often contain clauses defining work conditions that cannot be changed without contract renegotiations. Labor contracts, therefore, are a significant management concern.

Mechanized collection is intended to use less labor than is presently used. Management can use the following actions and procedures to implement changes requiring less labor: (1) bring the labor union into the evaluation and testing of mechanized collection; (2) train crew workers in the operation of mechanical systems before converting the entire operation to mechanized collection; (3) review crew workers' salaries and make the necessary adjustments to recognize the increased skills and efficiency of workers; and (4) downsize the labor force through attrition or reassignment, not layoffs.

Financing Equipment Purchases. Fully automated collection trucks must be purchased as a complete unit, but semiautomated equipment can be installed on existing trucks. Management issues include (1) selecting a method of financing that fits the ability of the residents to pay, (2) bringing new equipment on-line in stages, and (3) using old equipment until it passes the time limit of financing or making concurrent payments on old and new equipment. Capital normally is obtained through an equipment replacement sinking fund, from the sale of municipal bonds, or from lending institutions. An alternative to capital outlay is the leasing of collection equipment. Funding concerns have stopped the implementation of mechanized collection in many communities. However, the interest of residents in recycling, as well as the need for more equipment and the higher costs associated with recycling, signals a willingness to pay more for integrated waste disposal systems.

Case Study 19-1
Collection System Mechanization

A western community (population 400,000) has been collecting municipal solid waste from the curb in residential areas of the community for about 10 years. The union representing the members of the waste collection crews recently conducted work slowdowns and sickouts to protest the increasing number of accounts each crew must collect in a standard work day. The Board of Supervisors, the elected body of officials for the county, held a hearing on the matter. The chief of the solid waste management division made a presentation of system costs, the expected growth in accounts to be serviced, and the current size of the collection crews and crew productivity. The message was clear—costs were going up while labor productivity was unchanged. To the surprise of the division chief, the labor union representative said, "Why don't you study automated side loaders for use on residential collection routes?" The Supervisors directed the division to look at automated collection.

Management Issues and Concerns. The management issues to be considered in evaluating automated collection are these:

1. What container technology should be used so that waste collection costs are stabilized? In using the technology, what will act as an incentive to reduce waste at the source?
2. How can automated collection be implemented with the least impact on existing collection crews?
3. How will the county pay the capital costs for the trucks and containers needed for the automated collection system?

Information and Data. The division chief reviewed Solid Waste Division files from the past five years to determine the political and economic background for change. The chief found that labor productivity has been constant, that crew workers had strongly resisted the use of any technology that increased the number of waste containers or the waste tonnage lifted and emptied into collection trucks on collection routes, and that costs of service were increasing in direct proportion to the increases in salaries for crew workers. Within the past two years, the division had appeared before the rate-setting body of the county, the Board of Supervisors, twice to request an increase in rates for waste collection service. The requests were rejected and the chief was asked to be more efficient in providing service to the voters of the county.

The chief developed the following data:

Number of residential accounts	140,000
Waste collected, tons/yr	220,000
Workmen's compensation payment rate, $/$100 of annual salary	10.25
Number of crew workers	180
Work force productivity, tons/crew worker · d	4.7

Resolution. The most important step in implementing new technology is getting the endorsement of the technology by the labor force using it. The division chief had a head start toward labor acceptance of a new mechanized collection technology based on the request of the labor union at the hearing before the county supervisors. Building on this first step, the chief completed the evaluation of the automated collection technology in the following additional steps:

1. County staff, including the union shop steward, visited several communities where automated collection equipment was operating. During these visits, the

staff took down the names of equipment manufacturers and questioned the equipment operators about equipment performance and worker productivity on the collection routes.

The division chief ordered a demonstration project to test equipment and labor performance on the county collection routes. Proposals were solicited for containers and automated trucks for a six-month demonstration project at 1500 homes. During the demonstration period, county staff conducted time and motion studies and completed an economic analysis comparing the automated system with the current system. Three technology problems were identified during the demonstration period:

(*a*) The manufacturer offered only 90-gallon storage containers for use with the automated trucks. The county was considering the option of offering residents smaller containers at a lower cost as an inducement to reduce waste at the source. The problem was resolved when the manufacturer agreed to supply 60- and 90-gallon containers that matched the automated truck technology.

(*b*) The 90-gallon containers tested were 39 inches wide and would not pass through some side yard gates of county residents. The problem was resolved when the manufacturer redesigned the container to a 31-inch width.

(*c*) Because the containers were unassembled when the county received them, county staff would have to assemble as many as 30,000 containers per year if the automated system is implemented over a period of five years. During the demonstration period, staff took an average of 30 minutes to assemble a container. With 30,000 containers to put in service per year, this represented about 7 person-years of container assembly, not an acceptable option when looking at labor efficiency as compared with the current system with no container assembly time. The problem was resolved when the container manufacturer redesigned the assembly, resulting in a container that could be completely assembled in an average of two minutes.

2. In the next step, the division chief produced a report describing the salient points of the demonstration project. The chief then asked the Supervisors to hold another public hearing on the matter of automated collection. At the hearing, the staff presented the following conclusions:

(*a*) The automated system was more efficient than the existing system.

(*b*) Customers in the test area preferred the automated system by a margin of more than 9 to 1 over the existing system.

(*c*) The automated system had employee and union support and provided a safer work environment.

(*d*) Large capital outlays would be required for county-provided containers for automated collection.

(*e*) Automated collection technology had advanced to the point where it was ready for general use.

3. The Board approved the implementation of an automated collection system in five stages extending over five years. The costs for stage 1 equipment were approximately $1.3 million for trucks and $2 million for 30,000 containers. The board asked the staff to report in one year on system costs, worker efficiency, and methods of financing the purchase of equipment. The stage 1 equipment was obtained on a lease-purchase agreement. The lease-purchase financing allowed the county the option of changing the type of equipment in the future if changes in technology resulted in better worker efficiency and customer service.
4. After operating the stage 1 automated collection system for one year, the chief once more appeared in a public hearing before the board and reported the following data:
   (*a*) Work force productivity increased dramatically from 4.7 ton/crew worker · d to 11.2 ton/crew worker · d.
   (*b*) The work force was reduced by eliminating 14 permanent collection positions. The reduction was done by not filling the positions of staff that resigned or retired; there were no layoffs.
   (*c*) There were virtually no lost time on-the-job injuries on stage 1 routes.
   (*d*) The current rates charged for service were enough to pay for labor and the new equipment.
   (*e*) The workmen's compensation payment rate for the automated collection crew workers dropped from $10.25/$100 of annual salary to $2.20/$100 of annual salary.
   (*f*) Based on the results of one year of operation of the automated system, the chief presented a schedule to implement automated collection for all 140,000 accounts, and to finance the new automated system without raising service rates.

Comment. In this case study, the county was in a favorable position to implement a change in the collection system because the labor union suggested the change. Jobs are a sensitive political issue in communities, and elected officials often will not eliminate jobs if unions resist.

19-3 ENERGY RECOVERY

Energy is derived from MSW in two forms: (1) directly, by burning as a fuel to produce steam, and (2) indirectly, through the conversion of wastes to fuel (oil, gas) or fuel pellets that can be stored for subsequent use. The choice of either form is influenced by the strength of public reaction to a waste-fired energy plant and by revenues to the facility from the sale of energy. The solid waste system manager will use the energy recovery option if the cost of landfills is very high or there is no additional landfill capacity.

The identification, evaluation, and implementation of options for energy recovery are a significant part of integrated solid waste management systems because burning reduces the volume of the waste by 80 to 90 percent (see Chapter 9). Many states favor the use of combustion as a means of reducing the requirements for landfill capacity and to meet mandated waste diversion goals. However, waste burning is controversial with the public, and opposition crosses city, state and national boundaries. In California, the law allows only a 10 percent diversion credit for the combustion of wastes. The solid waste system manager must identify the many issues of energy recovery and present the acceptable solution for the community.

Management Issues and Concerns

Management issues and concerns with regard to energy recovery relate to the one-time use of the combustible materials in MSW versus the multiple uses possible with materials recovery and recycling. Energy is released from waste in commercial quantities only when the materials undergo radical change, rendering the residues unsuitable for further reuse. The issues are (1) the suitability of wastes as a fuel, (2) selection of energy production and recovery technology, (3) negotiation of an energy sales agreement, (4) the development of a strategy for risk and guarantees, (5) acquisition of a disposal site for residues from the energy generation processes, and (6) development of an energy recovery facility for which the community can pay.

Suitability of Wastes as a Fuel. Energy recovery depends on the Btu content of solid wastes (see Chapters 4 and 13). To be viable, the energy recovery facility must consistently receive a suitable waste fuel. The issue of suitability has two parts, quantity and quality. Both can be measured and matched to the energy recovery facility at a point in time. However, energy recovery is legislated by many states to be a lower priority type of waste diversion than source reduction, recycling, and composting. Therefore, a Btu content measured at the planning stage of a facility may change over time as the solid waste quantity and quality are altered by other waste diversion activities.

The energy recovery facility must have enough flexibility to react to changes in waste quantity and quality without shutdown. Waste generation and separation conditions that might change the fuel available to an energy recovery facility, along with recommendations on how to adjust the project to keep it open, are presented in Table 19-4. The waste suitability issue is best resolved at the time the contracts for facility construction and sale of the energy are set.

Selection of Energy Production and Recovery Technology. The issues to be covered in selecting a technology are reliability, costs, and conformance with environmental regulations. Reliability is measured by the performance of the technology over a specified period of years, usually the repayment term of the bonds that were used to finance the facility construction. Compliance with fuel or energy sales specifications is another measure of reliability. Costs are a measure of the efficiency of technology. Cost is measured by initial construction costs, including

TABLE 19-4
Conditions and recommendations for waste suitability as fuel at energy recovery facilities

Activity	Effect on waste fuel	Recommendation for adjustment
Loss of source	Elimination of fuel	One or more of the following: 1. Size the facility to receive no more than 80 percent of all community wastes 2. Seek other sources of waste 3. Demand put-or-pay terms in the waste delivery contract
Source reduction	% of combustibles	
	1	None necessary
	10	None necessary
Recycling	% of combustibles	
	7	None necessary
	20	Modify the energy sales contract to a lower amount for delivery
Composting	% of organics	
	20	None necessary
	50	May need adjustment to the fuel-firing rate in the combustor which will reduce the quantity of waste burned

the cost of financing the facility, and long-term operation, maintenance, and replacement costs. Conformance with regulations is measured by the performance of the technology in achieving the discharge standards set by the regulatory agencies. Standards currently exist for discharges to air, water, and soil.

The management issues for selection of an energy production and recovery technology can be separated into categories based on project size and type of energy product. Factors to be used by the solid waste manager in selecting a technology are reported in Table 19-5.

The energy recovery technology issue is resolved when the complete set of requirements for a facility, including the source of wastes, financing, construction, operation, and ownership, is put together. The solid waste manager assembles the technology options in a usable format and participates with the decision-maker in selecting the appropriate technology.

Negotiating an Energy Sales Agreement. In most situations, the energy recovered from solid wastes will be sold to an electrical utility or nearby industry. Utilities and industries choose between competing energy sources—such as oil, natural gas, coal and MSW—based on relative economics, including the economic impacts of an interruptable source. The management issue is to obtain an energy sales agreement with economic terms that result in a facility that the community can support through a combination of energy sales revenues and tipping fees.

TABLE 19-5
Factors for selection of energy production and recovery technology

Category of technology	Factors for technology selection[a]
Mass-fired, field-erected	For use at fuel-firing rates greater than 500 tons per day (ton/d); well proven in energy recovery from solid wastes; more than three competitors provide the technology; capital costs in the range of $75,000 to 125,000 per ton/d of installed capacity; energy product can be either electricity or steam, with the higher capital cost for electricity including the cost of the turbine/generator.
Mass-fired, modular	Used at fuel-firing rates less than 500 ton/d; well proven in energy recovery from solid wastes; more than three competitors provide the technology as off-the-shelf or prefabricated units; capital costs in the range of $30,000 to 50,000 per ton/d of installed capacity; energy product is usually steam or heat.
RDF-fired	For use at any fuel-firing rate; well proven in energy recovery from solid wastes; more than three competitors provide the technology; best use is for situations where energy recovery and materials recovery are part of the same facility; more complex technology than mass-fired because most noncombustibles are removed before entry of the waste to the furnace; capital costs in the range of $75,000 to 125,000, with the higher costs for energy recovery in a turbine generator.
Prepared fuel	Used when the solid waste fuel is to be sold to an energy recovery facility not integrated with the fuel preparation facility; technology is identical to the RDF-fired technology except for an additional fuel processing step to make the fuel suitable for storage and hauling; more than three competitors provide the technology; requires a stable long-term buyer of the fuel and large-scale test burns of the fuel to set fuel preparation specifications; capital costs in the range of $30,000 to 50,000.

[a] All costs in mid-1991 dollars (ENRCC index-4892; see Appendix E).

Negotiation among the agencies that will sell and purchase the energy resolves the issues. In setting a negotiating position, the solid waste manager and the utility manager must answer the following economic questions: What factors can cause changes in energy costs? How do changes affect the viability of the energy recovery project? How can the economic benefits of energy recovery be shared by electric rate payers and solid waste rate payers? How can the risks of cost and energy delivery be shared?

Strategy for Risk Taking and Guarantees. Energy recovery facilities have risks associated with technology, energy sales, and environmental regulations. Because the risk exposure is so broad, the management issues are concerned with setting the level of risk within the financial means of the community. In general, the risk to the community is lowest when the community has the strongest control over technology, costs, and response to regulations. Control implies competence in all three areas of risk, and many communities cannot maintain competence on staff. The areas of risk and options for community actions to get the accept-

TABLE 19-6
Areas of risk and options for risk taking in energy recovery

Area of risk	Options for risk taking
Technology	
Siting and permits	Community alone
	Multiple communities in a regional agency
	Community sharing with a private agency
Constructed capacity	Community alone with equipment manufacturer guarantees
	Community ownership with private agency construction and guarantees
	Insurance for facility
Operating performance	Community alone with competent staff
	Community ownership with private agency operation, escalation of costs, and guarantees
Costs	Community alone with general obligation bonds
	Community alone with project revenue bonds and pledged tipping fees
	Community-issued bonds and equity from a private agency, bond insurance or a letter of credit
Environmental regulations	Community alone
	Community ownership with private agency operating guarantees, except for changes in environmental legislation
Natural disaster	Community alone
	Disaster insurance, as available

able level of risk are reported in Table 19-6. The strategy selected may be one or a combination of options listed in Table 19-6. Risk sharing is a sound strategy, as is purchasing insurance or a letter of credit. The community will pay for shared risk, but the payment will ensure the project will perform at acceptable cost and risk.

Securing a Disposal Site for Residues. The energy production phase of energy recovery results in the production of residues that must be disposed of in landfills. The management issues are related to landfill location and capacity. Since energy recovery is intended to reduce the volume of solid waste going to landfill, community residents may resist a new landfill to receive residues. Also, the residues may contain contaminants, from the physical/chemical reactions occurring in the combustor, that make the residue unacceptable for disposal in an existing landfill. Regulations are currently proposed that would classify combustor residues for disposal to designated land disposal units.

The resolution of issues requires testing of residue to identify its physical and chemical characteristics. If unacceptable contaminants are present, the community must find and secure a designated land disposal unit (monofill). An MSW landfill can be used for disposal of residues with acceptable characteristics.

Developing an Energy Recovery Facility That the Community Can Pay For. Because energy recovery is the third priority position in the hierarchy of integrated solid waste management, its costs must be measured against other facility costs, and the total of all costs must be evaluated against the ability of the community to pay. The management issue is the risk of cost fluctuations that could increase community costs beyond the ability to pay. An example of a cost comparison between projects is shown in Fig. 19-1. The critical feature for the community is the dollars per ton fee now paid by residents. Three identical projects are shown as they are affected by different unit prices for the recovered electricity. The three unit prices represent a set of project options to be negotiated with the buyer of electricity. At 1000 ton/day and an electricity price of \$0.03/kWh, the project can compete with landfilling if landfill costs are about \$35/ton. In

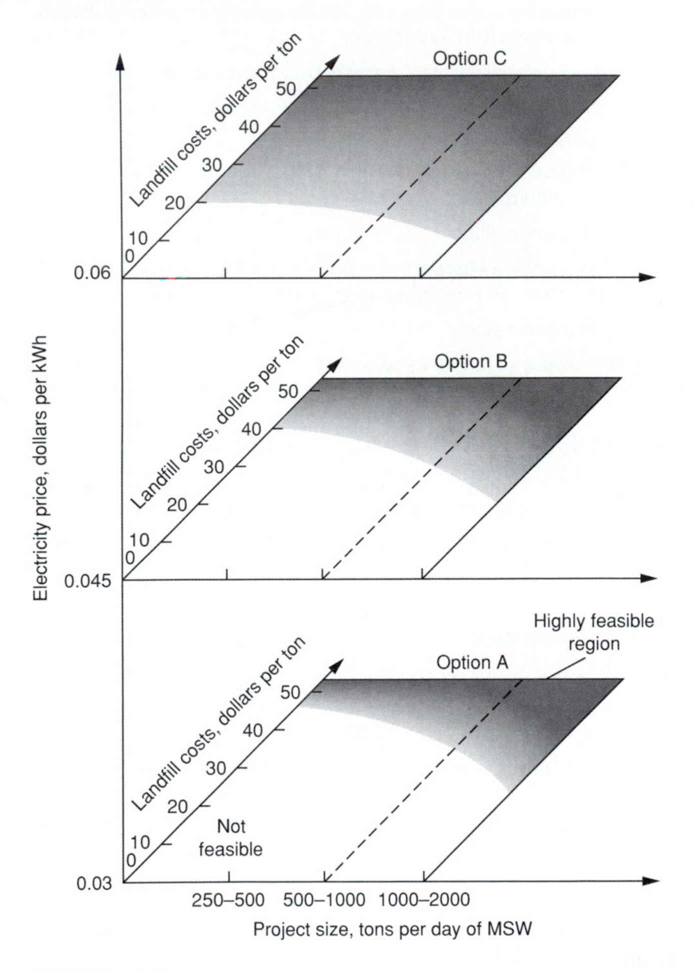

FIGURE 19-1
Plotting electricity price, landfill cost, and project size gives a conceptual view of project feasibility.

project option B, the electricity purchase price is $0.045/kWh, and the 1000 ton/day project can compete with landfilling if landfill costs are about $30/ton. In project option C, the highest price, $0.06/Wh, is paid for electricity from the energy recovery facility. At 1000 ton/day, project option C can compete with landfilling if landfill costs are about $15/ton. The costs used for constructing Fig. 19-1 are only approximate. In an actual situation, the cost for the energy recovery facility is increased significantly by the addition of air pollution control equipment.

The resolution of how much the community can pay is achieved when acceptable options are adopted by decision-makers. It is important to use comparable unit costs for the time period of the evaluation. Energy recovery costs are fixed with escalators for inflation for the period of the bonds, usually 20 years. Landfill costs may not stay fixed if the landfill is filled before 20 years. The solid waste planner must determine remaining landfill capacity without the energy recovery facility, and estimate future landfill costs based on expected community conditions. Will the community support a new landfill? What the community will pay for solid waste disposal is that estimate of future landfill costs or the estimated costs of waste disposal after energy recovery.

Case Study 19-2
Implementing an Energy Recovery Facility

A community has received a closure order for its landfill. For five years before receiving the order, the community investigated new sites for a landfill. There was significant opposition to all new sites, so the community-elected officials authorized their solid waste staff to investigate and evaluate methods of waste management that would reduce the use of landfill. The community, located in the northwest United States, currently collects and recycles materials after source separation under a franchise agreement with a private company. If the evaluation were to show that an alternative waste management system was economically feasible and facilities of the alternative could be sited in the community, the facilities would be installed and operated to replace the existing landfill.

Management Issues and Concerns. The management issues to be considered during the evaluation are these:

1. What technology will be best for the community?
2. What changes are required in the existing waste management system to ensure the successful implementation of a new technology?
3. Because any new system will cost more than the existing landfill disposal costs, what can the community do to minimize new system costs?

Information and Data. The community economy is dependent on agriculture and forest products. Community residents have followed a state bottle bill for the last 10 years, during which time it has been possible to implement community recycling centers. A citizens advisory committee has been meeting for three years to assist the community solid waste manager in a search for an acceptable waste management solution. The solid waste manager has issued a request for proposals from private industry for alternative waste management technologies. The most acceptable proposal received was for technology to make a densified refuse-derived fuel. The community has hired a consultant to evaluate the proposal. The following data apply to the case study:

Population	210,000
Electricity cost, \$/kWh	0.03
Landfill disposal fee, \$/ton	13

Resolution. The consultant reviewed the private company proposal. Because the technology to be used did not include the facilities for conversion of the densified refuse-derived fuel into an energy product, it was necessary to establish who would buy the fuel and at what price. The company asked the solid waste manager to assist in finding a fuel buyer.

After the community and the company had conducted a joint search and investigation, they identified an acceptable buyer at a state-owned building complex that included medical and prison facilities. The state complex had a central utility plant that provided heat and hot water to all buildings. The boilers in the plant used either wood waste or coal as a fuel. Negotiations with the state for purchase of the refuse-derived fuel failed when the state insisted that the community pay the full cost of purchase and installation of new boilers. The community could not pay the cost of the new waste-processing technology and the cost of the new boilers. The consultant recommended that the community reject the proposed technology and issue a new request for proposals, specifying a preference for a technology that reduced waste volume and generated energy in the same facility.

The solid waste manager directed the consultant to prepare a step-by-step plan to guide the community through the procurement process for a waste-to-energy plant.

Step 1. Prepare a feasibility report on the economics of disposal technologies. The community-elected officials knew the costs of landfill disposal, but needed to know the range of costs for waste-to-energy facilities so that cost comparisons could be made and presented to the community residents at public hearings. The consultant prepared a feasibility report that included the cost summary shown in Fig. 19-1.

Step 2. Select a waste management technology and present the contract and business options to the elected officials for authorization to proceed. The consultant presented findings to the elected officials and identified the waste management options for the community. Using Fig. 19-1 as a reference, the consultant pointed out that a waste-to-energy plant would cost more per ton of solid waste disposed than the existing landfill. However, the solid waste manager had met with the elected officials on many occasions in the previous five years and given them a good understanding of the future costs of landfill disposal. The elected officials had the foresight to act for the future and authorized the solid waste manager to solicit proposals from private companies for the construction and operation of a waste-to-energy plant. Upon receipt of proposals, the elected officials would evaluate the recommendation of the consultant and choose one company for implementation or reject all proposals if implementation costs exceeded the capacity of the community to pay. With these guidelines, the consultant prepared a request for proposal and sent it to a number of firms that had built and operated waste-to-energy plants.

Step 3. Evaluate proposals and negotiate with a private company to build, operate, and own a waste-to-energy plant. The community chose a company that proposed to build, operate, and own a mass-burn waste-to-energy combustion facility. In its proposal, the company made a compelling argument for ownership and reinforced that argument with lower costs than other proposals. Significant cost advantages accrued to the community through the following actions by the company:

1. The company would put in one third of the required capital as equity, allowing it to take certain tax advantages that were to be shared with the community.
2. The state offered certain pollution control tax credits that the company offered to share with the community.
3. The company had a commitment from the local electric utility to pay a levelized purchase price for the electricity generated from the solid waste fuel. The levelized price would apply to the first 10 years of plant operations.

The solid waste manager received authorization from elected officials to negotiate a long-term contract with the company.

Step 4. Negotiate the necessary contracts. The solid waste manager, the legal counsel for the community, and the consultant were the negotiating team for the community. The private company's negotiating team included the designer of the technology, legal counsel, a financial underwriter, and bond counsel. Negotiations were completed in 26 months. The highlights of negotiations included securing the type of financing that would give the community the most favorable tipping fee at the company plant, converting the electric utility offer of a levelized power purchase price into a contract, writing a revenue-sharing clause in the contract that gave the community a share of any tax credits received by the company, and finding a site at which the plant could be built. The most time consuming of these tasks was finding a site for the plant.

The community and the company worked together to secure a plant site. There was significant opposition from residents in every area of the community chosen for the plant site, and the community had to change the plant location three times. A site was found and accepted by nearby residents only after concessions by elected officials to fund and install certain area improvements. This approach was accepted and funded by the company as a part of the funding for the plant. The site was a good location for the plant for these reasons:

1. It was conveniently located within the area to be served by the plant.
2. It was accessible to the electric utility power grid.
3. It had good vehicle access from the highway system.
4. There was adequate space for the plant during both construction and operation.
5. There were nearby utilities to support plant operation.
6. The plant at that location met state requirements for air discharges.
7. The plant at that location received a solid waste disposal permit from the state.

Step 5. Establish waste flow control for the community. The community had an obligation set in the negotiated agreement with the company to deliver 140,000 tons per year of acceptable wastes to the plant for 20 years. To fulfill this obligation, the community had to change its ordinance regarding waste management services to its residents. In an attempt to eliminate legal challenges to flow control, the community-elected officials asked the state legislature to pass a law granting to the community certain rights of the state to control waste movement. The law was passed, and the community then passed an ordinance that specified that wastes collected in the community would be directed to waste disposal facilities designated by community officials.

Step 6. Develop a strategy for increasing disposal service fees at a rate that is acceptable to the community residents and businesses. The disposal fee at the new plant would be \$27 per ton when the plant became operational in two years. The solid waste manager received approval to increase the current landfill disposal fee of \$13 per ton by \$7 per ton in each of the next two years so that the landfill fees would equal the plant tipping fee. The system is now operating with an energy recovery plant.

Comment. In this case study, the energy recovery plant was successful because private industry and the public agency cooperated to their mutual benefit. Evidence of the commitment of the community-elected officials to the plant is found in the long period required to find and secure a site for the plant. The company showed its commitment to the project by agreeing to fund the capital costs of certain area improvements where the plant was located.

19-4 LANDFILL DISPOSAL

Disposal of solid wastes in landfill is the fourth priority in the hierarchy of integrated solid waste management. Disposal is the "no alternative" option because it is the last functional element in the solid waste management system and the ultimate fate of all wastes that are of no further value. The possibility that materials that are of no value today may be of value in the future is not precluded. The management issues for landfilling evolve from public concerns for the environment and the traditional image the public has of a landfill site as a dump.

Management Issues and Concerns

In the implementation of new landfills, the single most important issue for management is to find a location that is acceptable to the public and to local regulatory agencies [2]. In the management of existing landfills, the major concern is to ensure that proper operational procedures are followed carefully and routinely. The basic issues for the planner and manager are (1) justification of need for a landfill, (2) evaluation and community acceptance of the landfill location, (3) landfill design and cost-effectiveness, and (4) management policies and regulations.

Justification for Need. Although landfill disposal must be included in every solid waste management system, merely stating this fact will not ensure community support for a new landfill. The suspicion of community residents that a better solution exists can be satisfied only by a well-documented justification, including an evaluation of source reduction and recycling options and of the potential use of the wastes for land reclamation if reclamation is a consideration.

A strong response to the issue of the need for a landfill is the presentation of alternatives that illustrate graphically how a landfill fits the integrated solid waste management system. The level of documentation needed to show the alternatives will vary with each community. One example is illustrated in Fig. 19-2. As shown, the recommended disposal system involves the use of a sanitary landfill, but four separate opportunities are identified for possible changes in future years, and each opportunity is to be reevaluated before the recommended facilities are expanded.

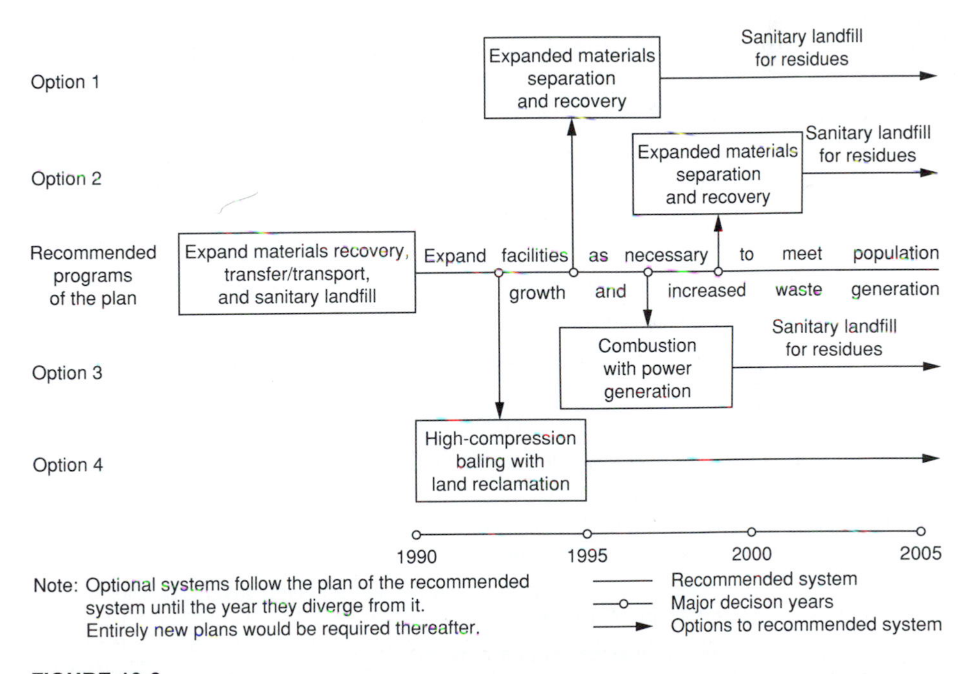

FIGURE 19-2
Pictorial documentation of future alternative options to a disposal plan.

Evaluation of Site Location. Along with the need for a complete and clear documentation of the need for a landfill, documentation is needed concerning the engineering and management evaluation of the landfill site location. Engineering considerations were presented in detail in Chapter 11. As a minimum, the following management concerns must be documented in detail:

1. Environmental Impact. The strongest public objections in most cases are related primarily to environmental concerns. For this reason, whenever possible, a disposal site should be located so that its impact on the surrounding environment is minimized. Environmentally sensitive areas include sole source groundwater aquifiers, flood plains, endangered species habitat, and heavily populated urban areas of a city.
2. Public Concerns. In many situations, public hearings are the forum used to bring out public concerns about a landfill site. The most commonly expressed concerns with landfills are related to odors, hazards to health, traffic, property values, noise, birds, dust, debris, litter, visual aspects, and leachate. The best approach to follow in the resolution of these concerns is to respond to each of them in the siting study. Some typical mitigations for each of the above concerns are presented in Table 19-7.
3. Adaptability for Multiple Uses in the Waste Management System. The site should be located so that it might be used as a materials recovery facility or as

TABLE 19-7
Public concerns and typical mitigation actions at new landfill sites

Public concern	Typical mitigation action
Birds	Overhead wires and nets; noise emitters
Debris	Controlled tipping face; adequate final cover
Dust	Paved access roads; watering of soils used for cover
Hazards	Adopt federal or state guidelines for landfill construction and operation; site security; monitoring of ground and surface waters; monitoring of gas
Leachate	Leachate collection and treatment; monitoring wells
Litter	Ordinance requiring covered loads when delivering to the landfill; a litter patrol on access roads and the landfill
Noise	Hours of operation; berms of earth to attenuate noise levels
Odors	Daily cover of wastes; space buffer of land between the landfill and the site property boundary
Property values	Appraisals; final landfill use after completing the landfill; property purchase or settlement payments
Traffic	Access roads; restriction on the number and type of vehicles allowed on site; hours of operation
Visual impacts	Vegetative screening; earth berms; fences

a dropoff location for source-separated recyclables. The methane gas from the landfill should be recovered, if feasible, and used for the production of energy (see Chapter 11).

4. Contingency Plans. Natural disasters (floods, tornadoes, etc.) cause an emergency for waste disposal, either through an interruption of access to the landfill or through an excessive demand for disposal capacity for disaster-related wastes. A common resolution is to have a contingency plan for disasters that defines how the landfill will operate under crisis conditions.

A prudent management technique used widely in securing acceptance of a single site is to identify numerous feasible sites and then to present the pros and cons of each site at public meetings and regulatory agency hearings until an acceptable site is agreed upon. In such sessions, one must present all facts as clearly and concisely as possible and demonstrate that sanitary landfills are compatible with the environment.

Landfill Design and Cost-Effectiveness. Landfill design is a complex activity. Landfill regulations are increasingly stringent, and landfill costs are increasing [4]. The community must adopt and implement the new regulations and associated design constraints. The issue is to choose a site location that results in the most environmentally cost-effective landfill.

Broadly defined, effectiveness means getting the job done in an environmentally acceptable manner for the lowest cost. The task for the waste system manager with respect to disposal is to find the location that has the lowest cost including costs for infrastructure and environmental controls. The following options for disposal should be evaluated before selecting a landfill:

1. The community may choose to own and operate its own landfill through a public agency. Through ownership, the community maintains control over the operation and the rates charged for disposal. The principal concern is the scale of the operation, which affects cost-effectiveness.
2. The community may select a private agency to operate its landfill or to deliver its waste to a privately owned landfill. The concern with this option is the loss of control over a site that might become an environmental liability in the future.
3. The community may choose to work jointly with surrounding communities to develop a regional landfill. This option allows the community to benefit from the economy of scale. In such an arrangement, each individual community loses absolute control over rates for disposal, but retains liability for the disposal of the waste. Also, administration under a regional plan is more complex than administration under a local plan.
4. The community may contract for disposal in a completely separate community. In this option, all rights and obligations concerning the disposal site are removed from the community. Again, a major concern is the lack of control over future rates for disposal.

Each of these options requires a different level of responsibility and control for the community in which the wastes are generated. It is difficult to generalize about the preferred option, as the selection is a matter of local political and social choice.

Management Policies and Regulations. Management policies and regulations of interest with respect to landfills usually deal with the classification of disposal sites and wastes discharged to land and the use of the sites after closure. Most states have adopted regulations that restrict the type of wastes to be placed in the landfill, and landfill closure plans are designed to ensure that built-in environmental controls continue to operate during the postclosure period.

The issues regarding policies and regulations include identification of wastes by type for disposal, provision of the appropriate classification of landfill, control

TABLE 19-8
Performance criteria and typical engineering calculations to demonstrate conformance with criteria

Performance criteria from 40 CFR 258	Typical engineering calculations
Gas testing	Develop a model of gas generation that incorporates soil permeabilities, gas collection pipes, if used, and the quantity of wastes; calculate locations where gas will leave the landfill and specify gas sampling methods
Groundwater monitoring	Establish groundwater flow direction using calculations on data from piezometers; develop hydraulic models of groundwater flow; calculate the size and depth of wells to monitor water quality in accordance with regulatory requirements
Landfill liner	Make these determinations: leachate impingement rate on the liner; ultimate fate of leachate that passes through the liner (hydraulics and contaminants); stability of the liner under landfill loadings and climatic conditions; leakage calculations for leachate that impinges on the liner
Leachate control	Calculate the flow rate through the drainage layer above the synthetic liner; set the slope of the top of the liner for drainage to the leachate collection pipes so that the leachate head over the liner is never more than 30 cm; establish leachate quantities and select a method of leachate treatment to meet discharge standards
Stormwater control	*Offsite*—if not in the 100-year floodplain, calculate channel shape and slopes to route the stormwaters around the landfill; if in the 100-year floodplain, calculate the floodway hydraulics and backwater curves to demonstrate that the stormwater elevations of the 100-year storm will not affect the surrounding lands and that the water flows will not erode wastes from the landfill
	Onsite—provide stormwater control for waters from the 25-year–24-hour storm; size the stormwater holding basin, choose pumps and set pumping rates, set pipe sizes and volumes of water to discharge to offsite channels

of waste movement to the landfills, rejection of illegal wastes from the site, and establishment of a means of financing the closure and postclosure maintenance of the site. Resolution of these issues is beyond the scope of this text. Issues are worked out on a state-by-state and community-by-community basis. However, all states will be expected to follow the the U.S. EPA guidelines published as 40 CFR 258 [5]. The solid waste manager should follow the steps listed in Table 19-8 in meeting the guidelines set forth in 40 CFR 258.

Case Study 19-3
Opening a New Disposal Site

The city of Sunny Hills, with population of approximately 30,000, is located in a county with a total population of approximately 70,000. The city manager was told by elected members of the city council that the existing city disposal site is a disgrace and must be closed. The city manager found a nearby site, and the owner is willing to sell the land to the city. Before making the purchase, the city retained an engineering consultant to determine the suitability of the land for use as a disposal site. The conclusion was that it would be usable as a Class II landfill disposal site. However, on the basis of the environmental impact statement for the site and the strong objections of neighboring landowners, the planning commission recommended to the city council that the site be rejected because adverse environmental impacts outweighed the benefits to the community.

Management Issues. The city manager had used unbudgeted reserve funds to pay for the site investigation. No additional funds were available to undertake investigations at other sites. All acceptable sites, including the one that was investigated, were located outside city limits. Consequently, the following management issues had to be resolved:

1. Should the city manager take steps to change public opinion about the site?
2. Is a sufficient and adequate operating plan presented in the consultant's report?
3. Should the site be acquired over the objections of the planning commission and the local landowners?

Information and Data. The existing city disposal site is operated by a private contractor. Wastes have been filled past preset maximum elevations. However, the site is located outside city limits, and city residents are not especially concerned about the deteriorating environment of a remote location. It has been difficult to implement a replacement disposal site because of major differences over environmental issues between the city planning commission and the city council.

The county planning agency has been working for five years to develop a regional landfill. All attempts have been unsuccessful because the elected officials of each city cannot agree on ownership and administrative controls for a regional site. An added complication is the strong political influence of the local private waste collectors, who have been developing separate plans for a privately owned long-term canyon disposal site. The regional regulatory agency has refused to grant an operating permit for their site, and the collectors are opposing public agency efforts to establish new sites.

Resolution. The city manager, acting as the city planning director because the staff for such activities was limited, took the following steps to resolve this politically sensitive situation:

1. The city manager had attended the city planning commission meeting at which objections to the new site were voiced by owners of adjacent lands. Although the time and some limited funds to conduct public hearings and to distribute documents describing the reasons for selecting the site were available, the city manager chose not to use them since city residents voiced none of the objections. Because the city's intent was to acquire only the property for the disposal site, the approval of the acquisition by owners of adjacent lands was not needed. Therefore, no steps were taken to change public opinion.
2. The adverse impacts of the site were listed in the environmental impact statement, but sufficient benefits were documented so that the city manager assumed site acquisition would proceed. The findings of the city planning commission were adverse to the project, however, and the consultant's investigation was reviewed thoroughly in an attempt to find inadequacies that might be corrected. The soils and groundwater data were found to be so limited that questions were raised as to the validity of the conclusions drawn. Nevertheless, the city had no funds for additional studies, and the city manager ordered no new investigations.
3. Having exhausted the means by which public opinion concerning the site could be changed, the city manager was now in a position either to push for the project over the objections of the city planning commission or to abandon that site. Pursuing the project was not unusual in this case because the city council and the commission had a history of disagreement on similar issues. In all cases the elected officials make the final decision on a community project. The city manager presented the results of the site investigation to the council and recommended that the site be purchased for use as a disposal facility. The council approved the manager's recommendation and directed its legal counsel to begin site acquisition proceedings.

   At this point, the project was stopped. Immediately after city council action approving site acquisition, the owners of lands adjacent to the site caught the attention of the local newpaper with their objections. Several emotional articles were printed concerning the hazards of wastes and the adverse economic effect of waste disposal sites on adjacent land values. A petition was circulated in the city to place the issue of acquiring the disposal site on a special election ballot. Surprisingly, the petition was successful, and a special election was held. Even more surprising was the result. The site acquisition was turned down by city residents who had not even appeared at earlier hearings!

The problem was not resolved to the satisfaction of the city manager or the city council. The old site remained open for a short time, and then the private operator arranged for additional disposal space in another landfill within the county.

Comment. This case is an example of the importance of good public communication. It is not an unusual case; similar situations are found in many communities. What was unusual was the timing of opposition by city residents. The greatest shortcoming in the approach used was the lack of backup sites to offer when strong opposition appeared for the preferred site.

19-5 DISCUSSION TOPICS AND PROBLEMS

19-1. Contact the agencies responsible for solid waste collection in your city. Identify and list the various types of collection service available to community residents. Are there any plans to change the level of residential collection service? Why?

19-2. A public agency has been providing solid waste collection service to a residential area with a population of 300,000. As an engineer employed by the public agency, you have been assigned the task of evaluating a proposal from a large national private waste collection company to take over residential collection service. The proposal was solicited by the city because the city collection trucks were worn out and the city could not raise capital to finance the acquisition of replacement units. What factors would you look for in the proposal? What system improvements would you try to get in this situation?

19-3. Labor is the highest-cost item in solid waste collection operations. Any changes to an existing collection system that are expected to result in significant cost savings must, therefore, reduce the number of employees collecting wastes. What are the management issues involved in reducing the number of employees? What are the differences in the issues involved for a public agency versus a private company?

19-4. A city is evaluating the benefits to be achieved by changing the waste collection service to a mechanical system. The change would result in (1) new containers for waste storage at each residence, (2) an increase in the number of houses served daily by each crew, and (3) a cost savings of 15 percent. List the important variables for (1), (2), and (3) that must be evaluated when this change is planned. What steps will increase the potential for successful implementation of the change?

19-5. Many solid waste collection agencies, both public and private, are asked to implement operations for the recovery of newspaper, glass, plastics, and aluminum. The point of separation of these wastes is the source of generation. Discuss the impact of this type of resource recovery program on waste collection operations.

19-6. Refer to Problem 19-5. Assume that you have implemented a community-wide newspaper separation and recovery program. Suddenly, the price paid for newspaper triples from $20 to 60/ton. Church groups, Boy Scouts, and other organizations recognize the possibility of raising funds by conducting paper drives. Would you resist this type of competition? What are the implications for your newspaper recovery program?

19-7. Energy recovery facilities have risks associated with technology, energy sales, and environmental regulations. What is your strategy for risk taking if an energy recovery plant is to be built in your city?

19-8. Energy recovery (waste to energy) is at the third level in the integrated solid waste management hierarchy of the U.S. EPA and many states. When would you choose an energy recovery facility for your community? What part do costs play in your choice?

19-9. Review Case Study 19-3. What actions would you have taken to determine public attitudes about the disposal site? Was the failure to acquire the disposal site a bad result? If not, briefly explain your reasoning.

19-10. Contact a local waste management agency and request data on the disposal sites serving your community. How many disposal sites presently exist? Are there separate landfills for municipal wastes, hazardous wastes and inert wastes?

19-11. Assume you are the city engineer in a city with a population of 60,000. The disposal site has been operated for many years by a private company. Today you were told that the site will close in six months. What immediate steps would you take to solve the impending disposal crisis? What techniques would you use to find a new disposal site?

19-12. Referring to Problem 19-11, why might it be beneficial to maintain separate landfills for inert wastes?

19-6 REFERENCES

1. Kerton, D.: "From backyard to curbside collection: Public sector accepts the challenge," *Public Works*, Vol. X, No. 8, 1989.
2. Michaels, M.: "How landfills look to the public mind," *World Wastes*, Vol. X, No. 5, May 1988.
3. U.S. Environmental Protection Agency: *The Solid Waste Dilemma: An Agenda for Action—Background Document*, EPA/530-SW-88-054A, U.S. EPA Office of Solid Waste, Washington, DC, 1988.
4. U.S. Environmental Protection Agency: *Decision-Makers Guide to Solid Waste Management*, EPA/530-SW-89-072, Washington, DC, 1989.
5. U.S. Environmental Protection Agency: *Solid Waste Disposal Facility Criteria*, 40 CFR Part 258, *Federal Register*, October 1991.

CHAPTER 20

PLANNING, SITING, AND PERMITTING OF WASTE MANAGEMENT FACILITIES

Planning is an important first step in developing public understanding of the need for solid waste management facilities. Siting a facility is an emotional process, creating strong public reactions to perceived environmental problems. Permits are required at most facilities, and permitting is a time-consuming process. The impact of changes in waste management have been illustrated in Chapters 18 and 19. The details of planning, siting, and permitting of facilities to accommodate changes in waste management are presented in Chapter 20.

20-1 PLANNING IN SOLID WASTE MANAGEMENT

Planning in the field of solid waste management may be defined as the process by which community needs regarding waste management are measured and evaluated and workable alternatives are developed for presentation to decision-makers [4]. Planning is accomplished by applying the engineering principles presented in Part III to the needs, capabilities, and goals of the community. Planning in the field of solid waste management is both exciting and challenging, because most of the technical, environmental, economic, social, and political factors, and the interrelationships that are involved, are now only partially understood.

The purposes of this section are (1) to explore some of the important considerations in the planning process, (2) to describe what constitutes waste management programs and plans, (3) to define a general methodology for planning and the preparation of planning reports, and (4) to examine the nature of the decision-making process in the field of solid waste management.

Important Considerations in the Planning Process

Integrated solid waste management encompasses a wide range of individual activities, which must be combined in such a way that the public, politicians, decision-makers, and planners are able to recognize and understand the important relationships in the planning process.

In general terms, the planning process involves the collection, evaluation, and presentation of data relevant to some problem. In the field of solid waste management, the problem usually requires some type of action by a decision-maker, who probably is an elected official. Therefore, to understand the nature of the planning process in this application, it is important to consider (1) the framework in which planning activities are usually conducted, (2) the effect of planning time periods, (3) the jurisdictional levels at which planning studies are conducted, (4) the impact of alternative concepts and technologies on the planning process, and (5) the definitions of programs and plans.

Framework for Planning Activities. Planning activities in the field of solid waste management are generally undertaken in response to the recognition of some community need. In some instances a community may prepare a plan because it is mandated by state or federal government. The community problem-solving cycle and the interrelationships of planning that are involved are shown in Fig. 20-1.

The planning activity commences once a community need has been articulated and the problem has been recognized. Problem recognition is important, because if meaningful planning is to result, it must be related to some community need. Otherwise, the planning process becomes self-serving and is of little value. It is the responsibility of the planner, however, to call to the attention of the decision-maker all the potential problem areas that may be identified during the planning process.

At the same time as the problem-solving cycle moves forward, there is feedback from the community to the decision-maker and then to the planner. In Fig. 20-1, feedback terminates at the planning activity, because it is assumed that the true problem is being dealt with. In cases where incorrect definitions are selected, the feedback loop would extend to problem recognition, and a redefinition would be necessary. The presence of feedback in the problem-solving cycle is essential to the development of responsive management plans.

In cases where a state or federal mandate requires a plan, such as mandated waste diversion goals, it is necessary to add a monitoring activity to the framework. Referring to Fig. 20-1, monitoring would be a part of community

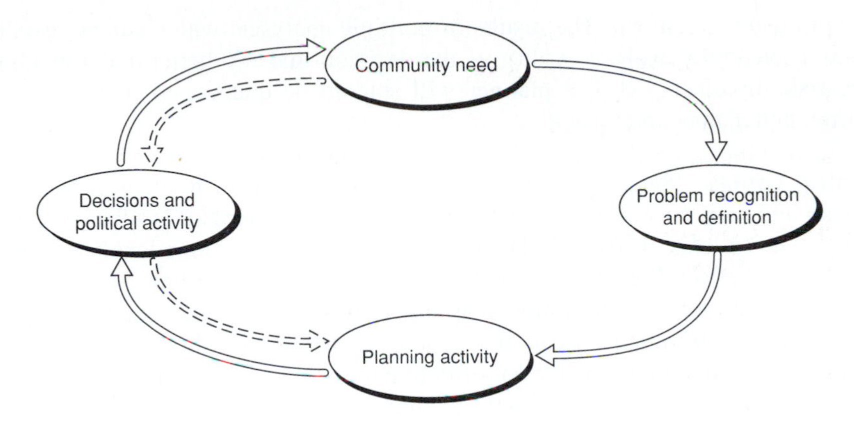

Note: Dashed line indicates feedback.

Community need A community need is identified by the public, usually in response to issues related to costs, service being provided, resource utilization, and environmental protection. The extent of the need is often determined by the social standards of residents, institutions, and businesses.

Problem recognition and definition Responsible decision-makers perceive and interpret community needs. They also are responsible for problem definition and specification.

Planning activity The planning activity is undertaken by agency staff or consultants, as directed by decision-makers. Alternative programs are developed to solve specific problems.

Decisions and political activity This is the action step in the problem-solving cycle. Decision-makers review alternatives; select alternative to be implemented; and make financial, operation, and political decisions.

FIGURE 20-1
Community problem-solving cycle.

need, with the results of monitoring being compared to mandated diversion goals and any shortages considered a sign of a problem. The plan would be revised to correct the problem.

Planning Time Periods. Planning for integrated solid waste management can be either short- or long-term. A precise time division is not fixed, although five to seven years into the future is accepted as the upper limit for short-term planning; long-term planning extends for all time periods past that.

One difficulty that arises in selecting an appropriate time period for planning is that while short-term planning may be limited to five to seven years, the payback period for equipment and facilities may be considerably greater, often running to 20 years or more for such facilities as MSW combustors used for waste transformation. When energy recovery options are being evaluated, long-term payback periods usually will be required. Yet, at present, little or no long-term information is available on the useful life of such facilities. The lack of information can have serious implications on the economic feasibility of energy options. In situations in which the uncertainty is high and long-term payback periods are required, the best approach is to develop multiple analyses, based on estimates of both average and least optimistic conditions. The results of multiple analyses, which can be used to prepare a sensitivity analysis, will give the decision-maker a better understanding of the risks involved. Also, a planner will seldom be questioned if the process performs better than anticipated.

Planning Levels. Planning activities for integrated solid waste management can be associated with three jurisdictional planning levels: (1) local, (2) subregional or regional, and (3) state or federal. The characteristics of these three levels are identified in Table 20-1. The distinctions are sharpest between the first two planning levels. Local agencies usually find responsibilities for management fragmented among many organizations. Economic resources are usually dispersed widely at the local level and are unavailable for effective planning and plan implementation.

By contrast, at the subregional and regional planning levels, the political and social awareness of waste management has caused several agencies to join forces to achieve a common goal. Joint objectives are worked out, along with estimates of the economic resources necessary to develop comprehensive plans. Comprehensive planning is considered further in Ref. 3. Solid waste management from a federal perspective is considered in Ref. 4.

Planning for Emerging Concepts and Technologies. Engineers and planners in the field of solid waste management today are confronted with breakthroughs, both in public awareness and in technological advancements, that make comprehensive planning for the future an especially difficult task. Many of the recent technological advancements are for waste diversion and are as yet too new to have been proved in full-scale installations. Thus, decision-makers are often faced with a choice of whether to use long-established and well-proved equipment and techniques, which may or may not be the optimum for present and future conditions, or to use new and unproved technology, which may or may not work as expected.

TABLE 20-1
Planning levels and their characteristics

Planning level	Agencies involved	Characteristics
Local	Cities County Special districts	Normally, a single agency conducts the planning study to solve a localized problem.
Subregional or regional	Cities Counties	Several communities join together to form an efficient planning and operational activity.
State and federal	Cities Counties States	Several cities, counties, and states can form a planning unit. Typically, the concern will be with issues such as materials use from origin in the natural environment through processing for reuse and the disposal of unusable residues. Planning efforts often result in demonstration projects.

In the broadest sense, all planning is based on predictions of future conditions. Developing concepts and technologies for integrated solid waste management are based on new ideas derived from the public's awareness of resources, economics, and the quality of the environment. Typical developing concepts include reuse of food containers, limiting of product packaging containers, control and standardization of materials used in packaging, and development of low-energy-demand products. The technologies described in Parts IV and V have evolved to the level of producing useful data on operational performance, costs, and environmental impact.

Programs and Plans

In general terms, programs and plans represent blueprints for achieving solid waste management objectives. The fundamental difference between programs and plans is in the scope of activities involved.

Programs. As used in this text, the term *program* encompasses all the activities associated with the solution of a problem within a functional element of an integrated solid waste management system. Thus, typical program areas of concern within a functional element may involve operation budgets, financing, rate structures, staffing requirements, contracts, equipment procurement and replacement, and maintenance.

Plans. Solid waste management *plans* are developed to define and establish objectives and policies. Typically, a plan will encompass one or more functional elements. Most plans are made up of many programs, and each program may be considered individually during the development of a final plan. For example, suppose that a draft plan has been developed and is being reviewed in public hearings. If objections are raised to the plan, they can now be isolated with respect to the individual programs. By focusing on an individual program, it usually will be possible to reach a workable compromise without having the entire plan rejected. In fact, this is the decision-making process.

In many planning situations, it is beneficial to identify more than one program or set of programs that can be used to solve a given problem—in other words, to develop alternatives. Alternatives are used as a means of demonstrating the impact of the various programs on the solid waste management system. They serve only as an aid to the decision-maker in understanding the impact of management choices described in a waste management plan. Often, a preliminary plan is developed in which two or more alternatives involving several programs are presented to decision-makers, and the final plan adopted is in fact composed of programs taken from one or more of the original alternatives.

Planning Study Methodology

In most cases the planner and the decision-maker do not have an opportunity to study the entire solid waste management system and develop a total knowledge of the community under all conditions. Time and economic constraints resulting from social and political needs often lead to decisions based on little or no information. For planners and decision-makers to respond to these situations and to ensure that the best use is made of time and available funds in the resolution of solid waste management problems, the following step-by-step planning procedure is recommended. An expanded discussion of planning methodology may be found in Ref. 3.

Step 1: Problem Definition and Specification. The first and most critical step in any planning study is to obtain a clear problem statement and corresponding specifications from the people responsible for making decisions about solid waste management. Problem statements and specifications usually are derived from the concerns of the public.

Difficulties often arise because solid waste systems are not well understood at most levels of decision making. Consequently, the planner may have to redefine a problem that was originally specified by a decision-maker. Management issues and concerns from which problem definitions are derived were discussed previously in Chapters 18 and 19.

Step 2: Inventory and Data Accumulation. In this step an inventory is made of all pertinent factors about the community, and data are collected as needed to meet the problem specifications. The main purpose of the inventory is to define the existing solid waste system(s) as completely as needed and as accurately as possible and to collect certain other basic information (such as population data)—a task that requires a considerable amount of judgment. It is an important step in planning, because all subsequent recommendations for action will be based on the findings of this step. Therefore, it is essential that all of the functional elements that make up an integrated solid waste management system be considered at any level of planning.

Step 3: Evaluation and Alternative Development. This step involves the detailed evaluation and analysis of the data accumulated in Step 2. It is during this step that the programs of the plan begin to be formed. In some cases, it may be necessary to collect additional data and information. However, before the programs are formed, it is important to review the original problem statement and specifications. Often it will be found that some revisions have to be made in light of the data gathered during the inventory.

Because a problem can have more than one solution, it is beneficial for decision-making purposes to develop alternatives composed of one or more programs. When practical, these alternatives should be documented for presentation in the plan. A simple plan may deal with only one or two programs. A more complex plan includes more of the functional elements, and its alternatives may include numerous programs. In either case, both administrative and engineering activities must be evaluated. An example is presented in a case study later in this chapter.

In developing alternatives, it is especially important that all functional elements be coordinated to ensure an integrated system from onsite storage through processing and final disposal. By evaluating the coordinated programs, the planner is able to recommend viable alternatives.

Step 4: Program and Plan Selection. In this step, a limited number of alternatives are selected by the planner for inclusion in the plan. The alternatives are reviewed by the planner, the decision-maker, and members of the public. The logic of the individual programs that make up the alternatives is reviewed, and programs are changed as necessary to include review comments. The administrative control of all programs is identified and evaluated during this step. Administrative control is important, because integrated solid waste management will not function properly without responsive control. Hence, the planner must develop a thorough knowledge of the social and political structure of the community.

The final action in this step is the selection of a preferred set of programs to form the plan. The programs can be selected from a single alternative, or they can be selected from various alternatives. The final selection will be made by decision-makers.

Step 5: Development of Implementation Schedule(s). When planning failures have occurred, the lack of a well-defined implementation schedule acceptable to administrative and management organizations has been the principal contributing factor. The degree of documentation in any implementation schedule depends on the types of programs developed in the plan. If possible, the degree of documentation that will be required for implementation should be set by the planner and decision-maker during the problem specification stage (Step 1) of the plan development. Details and examples are presented in the case study at the end of this chapter.

With the completion of Step 5 and proper documentation, the planner will have completed the most demanding work. The planner continues to be involved in the planning process as the plan is implemented and when the plan requires updating. The principal work for implementation now shifts to the decision-maker.

The Decision Process

The purpose of planning has been established as the accumulation, evaluation, and presentation of data relevant to a problem that requires some type of action by a decision-maker. As such, the planning process is an important part of the decision-making process; thus, the relationship of the planner and the decision-maker is normally quite close during plan development. Decision-making needs and events, as related to the selection and implementation of integrated solid waste management programs, are discussed in this section.

Requirements for Decision Making. It is clear that one of the fundamental and perhaps most important requirements for decision making is sound planning.

Another is an understanding of the goals of the community. Consider the community in which you live. What do you perceive are the solid waste problems that need corrective action? If these problems are put into the framework of the planning methodology described earlier, a decision regarding their solution should become possible when the results of the planning study are available. That is precisely the task that faces individuals with responsibility for selecting and implementing solid waste management systems. In addition, the decision-maker must use the results of planning to follow through with capital expenditures, work force allocations, and system implementation.

The availability of newly developing concepts and technologies, and the recent broadening of social awareness concerning solid waste management and the value of resources, make the decision-making process in this field very uncertain. Without the effective decision-making guides that result from good planning, many decision-makers respond to this dynamic condition by putting off any implementation action through an endless cycle of additional studies. Although this type of activity is sometimes politically expedient, it is rarely responsive to community needs. A more practical approach is to develop a dynamic solid waste plan and an appropriate updating technique that will allow solid waste systems to be modified as social values, concepts, and technologies change.

Important Decision Events. As previously described, planning is an activity that leads to the development of management alternatives. Decision making is an activity that results in actions to implement equipment and work force systems. Although it is dangerous to oversimplify the decision-making process, the following four decision events are considered essential in completing solid waste management actions:

1. Adoption of a solid waste management plan, including specific programs
2. Adoption of an appropriate implementation schedule
3. Selection of an agency or agencies to administer the plan and operate the system
4. Selection of staff and funding sources and means

It should be noted that not all the decision events are required to initiate an action program. For example, administration, operation, and staffing are normally the responsibility of local management. Thus, decision events 1 and 2 are most important at the local planning level. In the implementation of subregional or regional plans, however, it is often necessary to create new staffs and funding sources. Thus, at these levels, all four events are important.

20-2 DEVELOPING A FACILITIES PLAN

Plan development and selection action are the final steps in the management planning cycle, and implementation is the final result of decision making. Planning Steps 1 and 2 (problem identification and inventory) provide the planner with the data and information needed to finish the tasks in Steps 3 (evaluation and development of alternatives), 4 (program and plan selection), and 5 (development of implementation schedules).

The root problems in solid waste management are the inability of a community to ensure the diversion of waste from landfills and an inability to ensure that there is sufficient landfill capacity for waste disposal for those wastes that are not diverted. Developing a plan requires the identification of the opportunities for diversion across all functional elements. To complete the identification of opportunities, it is necessary to complete an inventory of facilities and management activities within each functional element. It is in planning Steps 1 and 2 that the community will begin to develop the data to understand what quantity of waste is generated, what quantity of waste can be diverted, and what quantity must be delivered to a landfill for disposal.

Many states have passed legislation that mandates waste diversion, and in those states the regulatory agencies have written detailed guidelines for development of integrated solid waste management plans. These guidelines present detailed methodology for inventory of the existing solid waste management system as well as presenting suggestions for diversion methods (see Section 6-7 in Chapter 6). The California guidelines are very comprehensive [1].

Although a state agency sets guidelines for the inventory of data and ultimately for demonstrating the viability of waste diversion using those data, it is the local community that must pay for its waste diversion and disposal system. It is in planning Steps 3, 4, and 5 that the details of facility type, size, and cost will be developed and selected by the local decision-maker. Each of these steps is described in some detail, because regulations now available do not provide acceptable procedures for these activities. The primary emphasis in this chapter is to illustrate the various activities by means of a detailed case study.

Evaluation and Development of Alternatives

Solid waste management programs are presented to decision-makers in the form of alternatives so that the decision-makers can make their own judgment on the probable success of each one. The use of alternatives in the planning cycle is illustrated in Case Study 20-1.

Perhaps the most important requirement for an alternative is that it be quantifiable with respect to equipment, disposal sites, economics, and other considerations. An alternative can be as simple as specifying the details of one-person versus two-person collection crews, or it may be as complex as specifying landfill disposal of all wastes versus processing wastes at multiple stations and selling recovered materials to numerous dispersed markets. Documentation for each alternative, regardless of complexity, must encompass the following: (1) performance, (2) economic analysis, (3) impact assessment, and (4) administration and management and an implementation schedule.

Performance. *Performance* means getting the job done. The work force and equipment required to provide the level of service desired by the community must be specified. The details of performance will vary with individual communities, but significant details that must be identified include (1) level of service, (2) equipment reliability and flexibility, (3) equipment and work force expandability, and (4) program compatibility with other environmental programs (air and water) and with future changes in solid waste technology.

With these details established, it is possible to contrast performance functions of a recommended program with performance functions of alternatives without additional planning studies. Such comparisons are an important part in achieving plan implementation. The work products of performance analysis are tables listing the category and amount of labor, drawings showing sizes and layout of equipment and buildings, and performance specifications for the types and quantities of materials to be processed under the alternative.

Economic Analysis. Once the details of performance have been identified, it is important to analyze the economic impacts of each alternative. The analysis must include estimates of capital cost as well as of operating costs. The cost of an alternative normally will be expressed as an annual cost. When divided by the annual quantity of wastes handled, the cost can also be expressed as a unit cost. Unit costs, such as dollars per ton, are often used to compare the cost-effectiveness of alternatives.

When cost estimates are completed, financing methods can be identified. Some of the available financing methods are reported in Table 20-2. A financial

TABLE 20-2
Financing methods for integrated solid waste management systems

Financing method	Characteristics
Debt	
General obligation bonds	Voter approval required; low interest cost; excellent marketability; primary source of revenue is the local general fund.
Revenue bonds	Voter approval required; moderate interest cost depending on project; do not affect local agency debt capacity; revenues available from user charges only.
Joint-power agency bonds	Voter approval required; moderate interest cost depending on project; a high potential for legal complication and issuance difficulty; revenues available from user charges or contract payments.
Nonprofit-corporation lease back bonds	No voter approval required; high interest cost; a high potential for legal complications and issuance difficulty; revenues available from rental payments.
Nonpublic	No voter approval required; high interest cost.
Revenue	
Pay-as-you-go	No voter approval required; no interest costs; requires careful long-term planning so advance budgets are identified.
Leasing	No voter approval required; no debt restrictions; revenues obtained from current operating budgets.

analysis must be made for each program alternative, but the details must be limited to those consistent with the planning level and available planning funds. The work products from economic analysis are tables listing capital and operating costs, pro forma charts showing income, expenses, and cash flow for the period of time under study, and sensitivity analysis showing the economic impact of variations from the financial base case.

The final objective of many financial analyses is the establishment of service rates—what the customer will pay for the service. Rates should be equitable and should reflect as closely as possible the actual cost of providing the service.

Impact Assessment. The programs of an integrated solid waste management plan will have an impact on a community in three ways: (1) through changes to the natural environment, (2) through involvement of the human environment, and (3) through a reordering of the community's socioeconomic structure. An attempt should be made to make quantitative estimates of each impact. Unfortunately, most planning and decision making must be completed without full benefit of these estimates, because the interactions of the natural environment, human environment, and socioeconomic structure are very complex and the monitoring of a community's massive resource system is very difficult.

Determining the impact of alternative programs requires information from community agencies and groups not normally involved in solid waste management, including business and environmental groups, regulatory agencies for air and water quality control, legislative bodies, and resources agencies. Information from such diverse sources will help to fill voids caused by unattainable quantitative estimates.

Administration and Management. The administrative functions and organizations for implementation must also be identified for each alternative. It is most practical for the planner to develop details of administration only for the short-term planning period (seven years into the future). Detailed administrative planning for the long term is meaningless, because changes can occur so rapidly in the solid waste management field. Managers who are responsible for operations during the short term will usually establish organizational policies and functions for the long term.

Program and Plan Selection

The development of facility programs and a plan is a major task in achieving effective integrated solid waste management. Before the plan is presented to the community for acceptance, it is first refined through agency and special-interest-group review. The best way to gain acceptance of the plan is to obtain support of key community groups. Another way is to demonstrate that the plan is compatible with other community goals, such as urban renewal, industrial development, and parks.

Obtain Community Support. The most positive bases for support are the residents and businesses of the community. Their involvement can take place either during the development of the plan or during its implementation. A strong public relations effort will be needed to develop community understanding of the plan. Political support should be tested and developed during the planning study through the presentation of progress reports at regularly scheduled political meetings (city council, commissioners, supervisors, etc.). This procedure removes the element of political surprise from plan recommendations—a wise approach, as politicians are the final decision-makers. The impact of political support is illustrated in Case Study 19-3. Support must also be obtained from state and federal regulatory agencies. The surest means of obtaining their support is to include regulatory standards and controls in the plan. If a variance cannot be avoided, it should be discussed fully with officials of the appropriate agency before the plan is adopted.

Demonstrate Compatibility with Community Goals. The solid waste management programs must be compatible with other community goals. Generally, the higher the level of planning, the greater the need for compatibility. These other goals include land-use zoning goals, environmental goals, and state and federal goals.

Because solid waste management activities are highly visible, all programs must be compatible with community goals as expressed in general plans and land-use zoning. All programs must also be compatible with environmental goals, which are generally community-oriented but might extend beyond community boundaries (for example, leachate movement in surface streams). In most cases, an environmental impact report is required at the time of implementation. Many agencies provide such reports as a part of plan development.

State and federal agencies are taking a greater interest in resource and raw material depletions. As legislators are passing laws to mandate diversion of solid waste from landfills, they are also considering laws that favor use of the materials in solid waste as a resource in consumer products. The waste management system planner should monitor such interest and should make the community plan compatible where it is economically feasible to do so.

Development of Implementation Schedules

The primary objective of an implementation schedule is to set a time sequence of actions and to establish an organization structure to take action. The time sequence is normally divided into short-term and long-term actions. Other elements important to implementation are fiscal management and administrative considerations (regulations and standards).

Organization. The term *organizational structure* refers to the agencies legally responsible for performing the tasks set forth in the recommended plan. A logical split of organizations is by functional activity. Thus, both administrative and operational agencies must be defined in the implementation schedule. Typically, implementation responsibilities are split among several agencies, including departments such as public works, health, community development, and resources management.

Fiscal Management. The implementation schedule must also contain the following details of fiscal management: (1) capital formation, (2) cash-flow requirements, and (3) revenue programs, such as rates or taxes. An important part of fiscal management is the establishment and maintenance of equity among those paying for the recommended program. The matter of equity becomes more difficult to settle as solid waste systems become more complex—especially as resources are recovered and sold.

Regulations and Standards. Regulations and standards are the means by which system performance is measured and control is maintained. A time sequence, within which designated agencies will establish ordinances, standards, and other means of measurement and control, must be included in the implementation schedule. Standards for waste diversion are a recent requirement to be met in waste management plans.

Plan Review and Updating. The primary objective of the implementation schedule is to set actions for short-term programs. However, any integrated solid waste management plan will require periodic updating. There will continue to be significant changes in the technology for waste processing and material recovery. Also, there is a continuing need to monitor the performance of waste diversion facilities to verify their compliance with diversion standards set in the plan.

Therefore, to make long-term plans, the manager must (1) monitor developing technology, (2) maintain contact with the community and its resources, (3) monitor existing standards and assess their continued need, and (4) update the community waste management plan. It is often best to assign these responsibilities to a single agency.

20-3 SECURING A SITE AND OBTAINING PERMITS

Once the need for facilities has been identified, a site must be secured at which each facility can be constructed and operated for its economic life. Securing a site for a facility, commonly called *facility siting*, requires the systematic use of community data to answer the concerns of its residents. Data identification and evaluation is done by a multidisciplinary team of design and operations specialists, environmental engineers, geologists, hydrologists, geotechnical engineers, and environmental assessment specialists.

Facility siting will cause strong negative reactions in the community, because solid waste facilities of the past have not been good neighbors. In recent times, the old problems of litter, odor, and air pollution have been controlled to acceptable levels. However, the controls are too late to prevent negative public feelings caused by the old ways of processing and disposing of solid waste. Materials recovery and composting facilities, although accepted and welcomed by the public as better neighbors than MSW combustors or landfills, are stuck with the legacy of past environmental problems.

Permits are obtained by responding to the requirements of the permit-issuing authority. The criteria to be met in siting a facility are derived from the permitting criteria set by regulatory agencies. The legal defense of selecting a site is based on the legal requirements for a facility permit as written in the regulations.

In this section, an approach is presented that will guide site selection and the obtaining of permits. The details of choosing a site and obtaining the necessary site permits is specific to each state and local community, and those details are beyond the scope of this text. The approach has two parts: developing a strategy and interpreting legal requirements.

Strategy for Facility Siting

A successful strategy is one that provides a facility site that has the support of the community. Community support starts with a broad base of group participation and continues with the accumulation of a strong technical, scientific, and economic data base.

Community Group Participation. Everyone in a community generates solid waste, and the environmental consequences of its handling and disposal are broad-based. One way to get community support is to involve community organizations in site selection. Many communities do this through a site selection task force. The strongest task force is one that is appointed by the local elected officials and that conducts its work under a narrowly defined set of objectives. If a task force is used, it should have members selected from a broad range of the community. Community service organizations that might provide members include the League of Women Voters, Chamber of Commerce, environmental groups like the Sierra Club and Friends of the Earth, and local affiliates of larger groups like the Lions Club. Local agencies should be represented on the task force by senior staff who are recognized as decision-makers.

The task force for facility siting can be expected to have both proponents and opponents for every site. Therefore, it is important to have sufficient and accurate data to the members in a timely manner. In general, the earlier the task force meets in the siting process, the better the probability of getting a site. In many instances, the task force should help the agency staff set the site selection criteria.

Technical, Scientific, and Economic Data Base. Facility site selection requires an organized approach to locate a site and develop a data base to justify its selection. The steps in the siting process include (1) identification of the feasible sites, (2) the development of technical, scientific, and economic criteria for site comparisons, (3) evaluation and comparison of feasible sites to select the best sites for detailed analysis, and (4) a thorough investigation and data accumulation on the best sites to recommend the final site.

The following data sources should be contacted and appropriate documents obtained.

Agency	Typical data
Planning Dept.	Community comprehensive land-use plans
Planning Dept.	Aerial photographs
Planning Dept.	Planimetric data
Public Works Dept.	Utility maps and reports
Public Works Dept.	County zoning maps and land-use data
Highway Dept.	Surface transportation routes and road class data
Solid Waste Dept.	Previous facility-siting reports
Historic Society	Reports on cultural resources
U.S. Geological Survey	Geological maps and seismic reports
Soil Conservation Service	Soils maps

While getting background data on sites, the siting team will be developing and choosing criteria for site evaluation. Although there are many ways to get and present criteria, the most useful presentation is one that fits the level of understanding of the broad community. The following broad categories of criteria are recommended.

- Political boundaries
- Regulatory
- Environmental
    - Surface water
    - Groundwater
    - Natural habitat
    - Land use
    - Air quality
    - Social/cultural
    - Aesthetic
- Technical
- Economic

The criteria will be used both in an initial screening of sites and in final site selection. Assigning numeric values to each criteria is an easily understood and practical way to demonstrate their use to the public and reviewing agencies. A commonly used rating scale is from 1 to 10, where 1 is the least desirable value and 10 is the most desired value.

Because solid waste facilities are not accepted as good neighbors, it is important to offer numerous sites for evaluation. Initial screening should be done in the least expensive way, leaving the most costly studies and data gathering to the final site selection activities. Providing a rational explanation of why sites were excluded is essential to a successful initial screening.

Strategy for Obtaining Permits

Solid waste facility permits will be required from federal, state, and local agencies. The permits required in two states are summarized in Tables 2-1 and 2-2.

In general, local permits will be the easiest to get and federal permits will be the hardest to get. The strategy to obtain permits varies with the regulations, but certain strategic steps to permitting are useful in most circumstances. A strategy should include (1) identification of permit-issuing agencies, (2) issuing agency involvement, and (3) responding to conditions set during the permit hearing process.

Permit-Issuing Agencies. The number and type of permits is determined by the type of facility. Typical solid waste facilities and the types of permits required are reported in Table 20-3. The federal permits are first priority, with state and local permits requiring conditions equal to or more stringent than federal levels. The most commonly encountered federal permits are from the EPA for air discharges from incinerators and from the Corps of Engineers for landfills. For the facility planner the critical issues are identifying the required permits for a facility and defining the permit issuing schedule.

Issuing Agency Involvement. Agency staff is responsible for implementing procedures for issuing permits. In many instances the agency will have a written guide for making permit applications. The applicant should follow the guide but should also establish a contact within the permitting agency. Through personal contact, the applicant can determine the agency requirements for detailed infor-

TABLE 20-3
Typical solid waste facilities and types of permits for construction and operation

Facilities	Types of permits
Combustor	Same as the permits required for energy recovery, except no power generation permit is needed
Composting facilities	Conditional use permit; solid waste facilities permit in some states
Energy recovery facility	Conditional use permit; power generation permit; NPDES permit for wastewater discharges; appropriate air permits; permit to construct; solid waste facilities permit; ash and residue disposal permit
Landfill for MSW	Conditional use permit; wastewater discharge permit; storm water discharge permit; solid waste facilities permit
Materials recovery facility	Conditional use permit; solid facility waste facilities permit in some states
Methane gas recovery	Only a power generation permit and a plant NPDES permit for air discharge are added to the landfill permits if the the landfill is active; if the plant is installed at an inactive or closed landfill, the required permits are power generation, NPDES, and wastewater discharge
Transfer station	Conditional use permit; permit to construct; solid waste facilities permit

mation. The applicant can review other applications for similar facilities and set details with the agency on matters such as number of drawings, sequence of reviews, and schedules. With this information from the issuing agency, the applicant can set a scope of work and provide the funding for a permit application.

Responding to Permit Conditions. The issuance of a permit is the end of a negotiation process between the applicant and the issuing agency. For the benefit of both parties, this negotiation is often done at the preconstruction phase and again prior to full operations. The permit to construct often contains conditions that must be met during construction. The applicant must work in a cooperative spirit to respond to conditions placed on the facility by the agency or face the risk of an unsuccessful application. The agency has a legal responsibility to ensure that the facility is in conformance with regulations; the agency cannot act as an advocate for the facility. The legal versus advocacy position is most apparent during public hearings regarding the permit.

Interpreting Legal Requirements

A land owner or group opposing the site or the permit for the facility will use all possible legal means to defeat the facility. The applicant must understand and follow the legal requirements. A legally sound and defensible siting study and permit application will result in the selection of an acceptable site.

The legal requirements are found in the regulations of federal and state agencies and in the ordinances of local agencies. In most instances, the proposed facility will not be the facility for which the laws were written. Interpreting the law in terms of the facility is the means of getting regulatory approvals. The most often used laws for challenging a siting study or a permit application are the environmental impact statement (EIS) and land-use zoning. In the following sections a brief overview of each is presented.

Environmental Impact Statements. The National Environmental Policy Act (NEPA) gave a legal basis for arguing the environmental impacts of a facility. In most states, the NEPA requirements are met in state laws. The EIS is the document that records the environmental impacts of the facility.

The cost of preparing an EIS varies with the type of facility and the location of the site. Because of cost and perceived delays for hearings, the applicant may narrowly interpret the law to benefit the facility and ask the lead agency for a negative declaration on the impacts of the facility. A negative declaration is one of the actions allowed under the law. It is a finding that the facility has insignificant impacts and can proceed without the detailed studies and data gathering of a full EIS. If the negative declaration is used for expediency only, it is a candidate for successful legal challenge. Since solid waste facilities have many impacts on the urban and rural environment, the facility applicant and the lead agency should carefully review and interpret the legal requirements before selecting the type of environmental document.

Land-Use Zoning. The controlled development of land use is done through zoning. Not all communities have zoning laws, but in those that do, the laws are a means of opposing a nonconforming land use such as a landfill. Zoning is evaluated during the issuance of a Conditional Use Permit (CUP) for the facility. The CUP will contain conditions agreed to by the applicant during interpretation of the law.

Case Study 20-1
Regional Plan Development
(Adapted from Ref. 2)

A regional government has assumed the responsibility for development of an integrated solid waste management plan. The geographic area to be covered by the plan includes four cities and a rural unincorporated community. The plan is being developed by a consultant working under the direction of a project director employed by the regional authority. Waste management presently is fragmented among the four cities, and there is no history of regional cooperation. The impetus for regional planning comes from recently enacted state mandates for regional approaches to management as well as from the imminent closing of three landfills that serve the individual cities.

Management Issues. The major management issues to be addressed during the study are the following:

1. What is the best arrangement of landfills to receive wastes from all cities within the region?
2. What processing and recovery equipment should be installed so that resource recovery is developed?

Information and Data. Each city presently has its own solid waste management system, including collection and direct haul to nearby landfills. Two cities have public agency collection crews. The other two have contracts with private collectors. Each city operates its own landfill, and two cities operate transfer stations. A third transfer station in the rural area serves the unincorporated community. The onsite storage and collection systems are relatively free of problems, and the rates charged for service are considered reasonable. There is no processing equipment now operating, and resource recovery is limited to hand separation of cardboard and bulky metal and wood wastes. The region is generally flat with a high water table and very limited choices for long-term landfill disposal sites. The cities have the normal light industrial and commercial businesses found in a region of 820,000 population. Freeways, railroads, and deep-sea waterways serve the region. The following data were developed:

City A:	
Population	220,000
Collection routes	54
Collection truck size, yd^3	20
Collection crew size, persons	3
Disposal site area, acres	82
Disposal site remaining life, yr	5
City B:	
Population	570,000
Collection routes	110
Collection trucks, number	
31-yd^3 capacity	30
25-yd^3 capacity	80
Collection crew size	
One person	30
Two persons	80
Transfer station, ton/day	500
Disposal site area, acres	800
Disposal site remaining life, yr	25
City C:	
Population	20,000
Collection	Franchise with private collector, no data available
Transfer station, ton/day	450
Disposal site area, acres	15
Disposal site remaining life, yr	2
City D:	
Population	10,000
Collection	Franchise with private collector, no data available
Disposal site area, acres	6
Disposal site remaining life, yr	1
Rural:	
Transfer station, ton/day	13

Resolution. Because of the importance of this planning study, the regional political body appointed a 15-member citizens' advisory committee to work with the planners. The members were selected from lists provided by interested citizens and political groups. This group worked closely with the project director to resolve the community waste management problems.

1. The first step in the solution involved the establishment of procedures to control and complete the planning study within the allotted time and available resources. The consultant and project director worked together to develop the activity diagram shown in the figure on page 893. The activity diagram is useful in several ways: (1) to organize the planning study, (2) to report progress on specific work tasks, and (3) to monitor progress of tasks in relation to the time schedule on the diagram.
2. An inventory was set up to accumulate information on every functional element of the system, and special emphasis was placed on landfills and resource recovery. As shown in the activity chart, the following details were highlighted in the inventory:
   (*a*) Waste generation and composition: The cities and unincorporated area did not have good records on what types of wastes existed and where each type was generated. A detailed monitoring, sampling, separation, and weighing inventory was scheduled and completed at every disposal site. Portable scales were rented and set up at each site to weigh incoming trucks. A sampling crew of five people randomly selected loads from which samples, ranging in size from 375 to 1425 lb, were drawn. The crew then hand-separated waste components into representative categories (see Chapter 6). The entire field inventory took place over a 2-wk period. This depth of inventory is often not necessary. However, if resource recovery alternatives are to be developed in the regional plan, a detailed knowledge of waste components is essential.
   (*b*) Transportation routes and disposal sites: A critical problem to the region was the closing of three existing disposal sites. The planners recognized that alternatives would be developed in which (1) new disposal sites would be identified, (2) all wastes would be routed to the existing large-capacity disposal site, and (3) additional transfer stations would be evaluated to make the transportation system more efficient. The site inventory was concentrated on land reclamation, both at the existing sites and at new sites.
   (*c*) Materials markets: An effort was undertaken to find markets for recovered waste materials. The citizens' advisory committee was asked to hold public meetings in the community in an attempt to identify resource shortages and to set recovery priorities. Contact was also established with state and national industries.
   (*d*) Public meetings: Four public meetings, one in each city, were scheduled early in the study to help identify social and community problems.
3. The planners and the advisory committee identified a number of problems related to the existing waste management system as summarized in the accompanying table. A priority was assigned to each problem, and a set of possible solutions was documented for each one. These solutions were grouped to form the programs of the recommended plan. Each solution was then subjected to the following criteria:

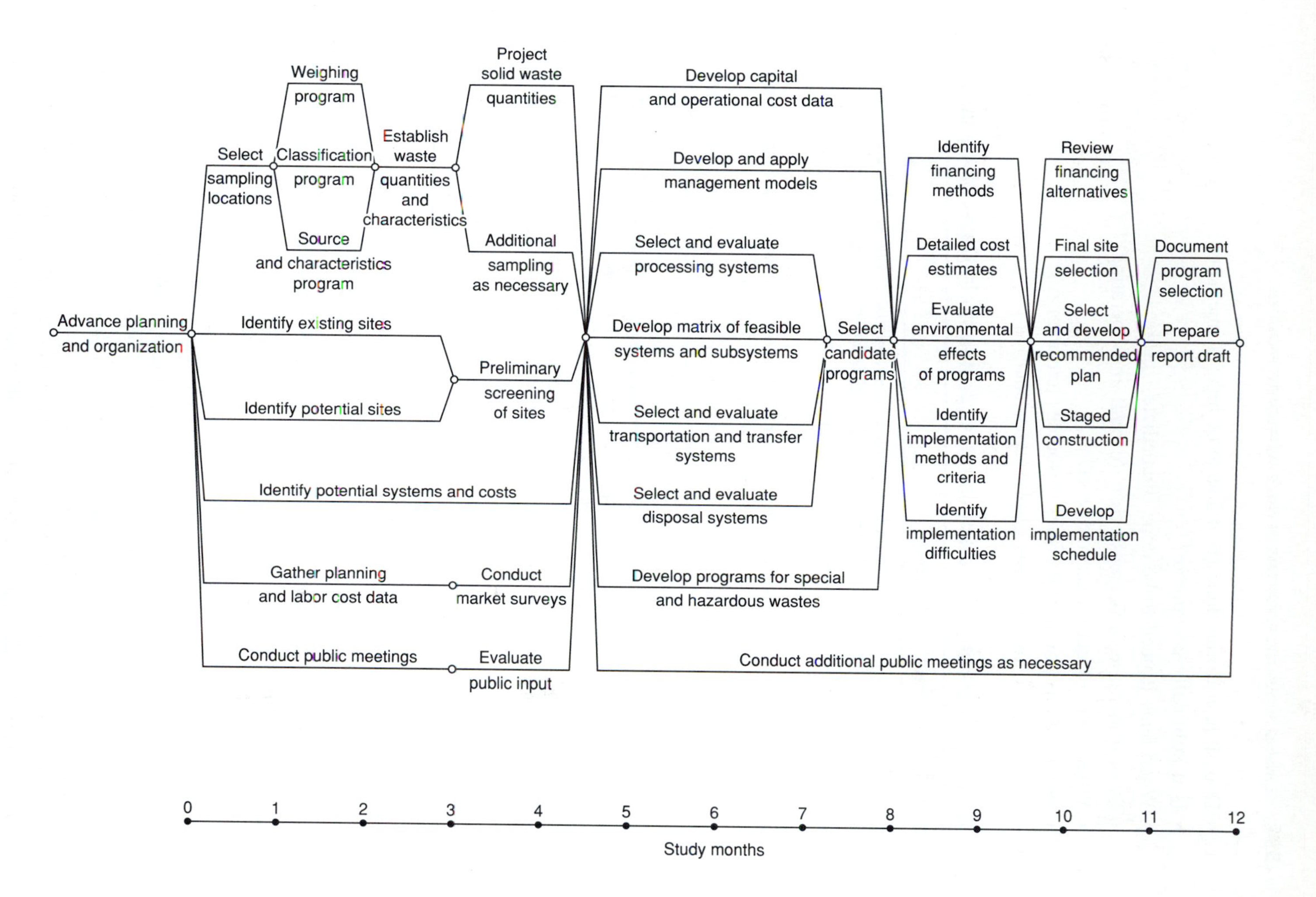

Advance planning and organization
Select sampling locations
Weighing program
Classification program
Source and characteristics program
Establish waste quantities and characteristics
Project solid waste quantities
Additional sampling as necessary
Identify existing sites
Identify potential sites
Preliminary screening of sites
Identify potential systems and costs
Gather planning and labor cost data
Conduct market surveys
Conduct public meetings
Evaluate public input
Develop capital and operational cost data
Develop and apply management models
Select and evaluate processing systems
Develop matrix of feasible systems and subsystems
Select and evaluate transportation and transfer systems
Select and evaluate disposal systems
Develop programs for special and hazardous wastes
Conduct additional public meetings as necessary
Select candidate programs
Identify financing methods
Detailed cost estimates
Evaluate environmental effects of programs
Identify implementation methods and criteria
Identify implementation difficulties
Review financing alternatives
Final site selection
Select and develop recommended plan
Staged construction
Develop implementation schedule
Document program selection
Prepare report draft
0
1
2
3
4
5
6
7
8
9
10
11
12
Study months

- Does it have to be done (is it a state or federal mandate)?
- Is it politically acceptable?
- When does it need to be done (immediate, future)?
- Does it increase or decrease costs (capital and operating)?
- What is the effect on service level?
- What are environmental issues (resources and energy)?
- Is it practical (on the basis of past experience)?

Existing operating and administrative problems in Case Study 20-1

Functional element	Problem
Generation	Records on type, location, and quantity of wastes are inadequate or nonexistent. Waste quantities are increasing.
Storage	Local standards are inadequate. Containers are inaccessible, overflowing, or underground.
Collection	Collection service area records are incomplete. Private industry collection permits are not assigned uniformly.
Transfer	Existing transfer stations do not have capacity for increased quantities or for resource recovery operations. Regional need for transfer stations is not defined.
Processing and recovery	Existing programs do not meet mandated diversion goals. Existing recovery methods do not meet materials reliability standards. Funding sources must be found to finance a recovery system.
Disposal	No data are available on leachate movement. Capacity in three of the four sites is exhausted. No final land use plans are available for sites.
Administration and control	Nine jurisdictions are setting policy. Waste generation is unrestricted at the source. Local ordinances are uncoordinated and incomplete.

4. The programs were grouped into four alternatives for detailed analysis, as summarized in the following list and shown in the accompanying figures.

- Alternative 1: Basically an extension of the existing system, this alternative calls for the construction of transfer stations to replace exhausted landfills with the transport of wastes to the remaining City B landfill. Resource recovery would be maintained at existing levels. The movement of wastes is shown in the accompanying figure on page 896.
- Alternative 2: A significantly changed system, involving a central processing facility, is proposed in this alternative. Most wastes would be moved to the processing facility and then to the City B disposal site. Resource recovery at this facility would be expanded to include magnetic separation. The movement of wastes is shown in the accompanying figure on page 896.
- Alternative 3: The major feature of this alternative is the addition of air classification, with glass and aluminum recovery, to the facilities of alternative 2.

The City B landfill would then be completely closed to unprocessed wastes. The movement of wastes is shown in the accompanying figure on page 897.

- Alternative 4: Energy recovery is the primary feature in this alternative. The energy recovery station would accept all organic wastes from the processing station. The energy recovery conversion is to steam, and the steam would be sold in City A and used for heating and air conditioning of downtown buildings. The movement of wastes is shown in the accompanying figure on page 897.

5. The capital costs for all alternatives are summarized in the accompanying table.

Capital cost of facilities for all alternatives in Case Study 20-1[a]

	Alternative, $(000)			
Facilities	**1**	**2**	**3**	**4**
Transfer	3,989	3,989	3,989	3,989
Transport	1,066	1,806[b]	2,698[b]	2,698[c]
Processing and materials recovery	—	16,756[d]	33,273[e]	33,273
Energy recovery	—	—	—	112,500[f]
Total cost	$5,055	$22,551	$39,960	$152,460

[a] All costs in mid-1991 dollars (ENRCC index = 4892).

[b] Includes transport from processing to landfill.

[c] Includes transport from processing to energy recovery and landfill of residue.

[d] Processing and magnetic separation only.

[e] Processing, air classification, full-scale materials recovery.

[f] Includes costs of waterwall incinerator and approximately 5,000 ft of 14-in steam distribution piping from the incinerator to customers.

6. Alternative 3 was recommended for implementation. The programs of this alternative began the process of meeting waste diversion and resource recovery goals mandated by legislation and improving the landfill programs of the region. The required facilities are expandable and can be used to produce a processed waste for most known resource recovery systems. The benefits from the economics of large-scale operations and from the resultant environmental improvements are significant. The other alternatives were not recommended for the following reasons. Alternative 1 was rejected because the added transfer station program increased the disposal costs for all agencies and offered little or no opportunity for future cost reductions. Alternative 2 was rejected because the level of resource recovery was not sufficient for materials sales to offset annual costs. Alternative 4 was rejected because of the high capital cost and the uncertainties associated with the energy production.
7. The last step after the selection of an alternative was the development of a more detailed management schedule for implementation, financing, and administration.

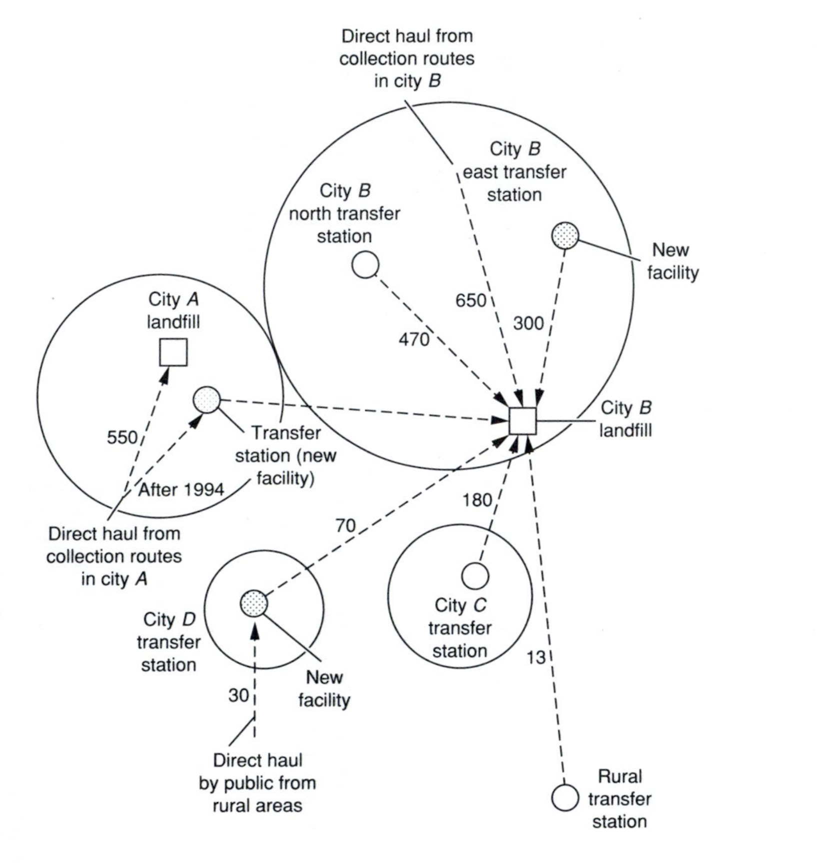

Note: Numbers correspond to solid waste loadings in tons per day.
Large circles show approximate location of city boundaries.

ALTERNATIVE 1

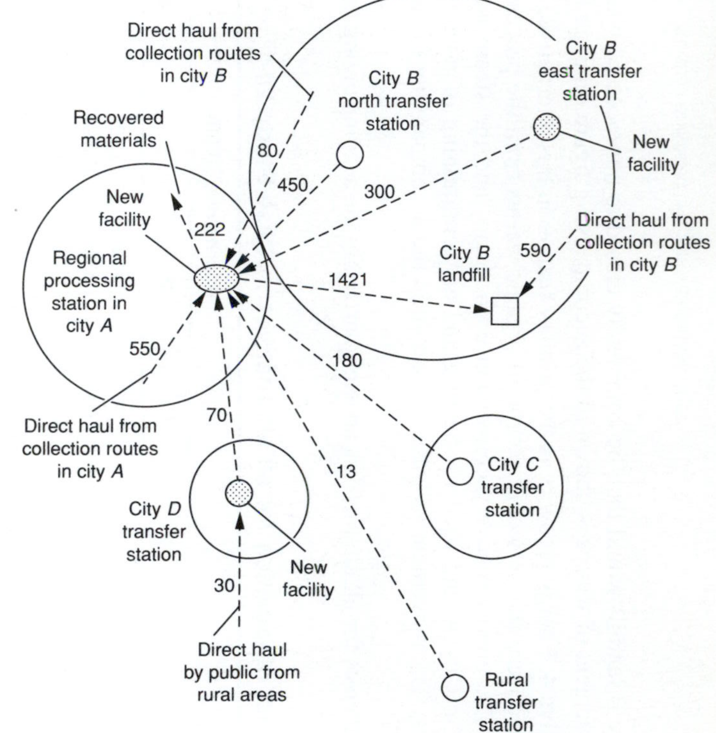

Note: Numbers correspond to solid waste loadings or recovered raw materials in tons per day. Large circles show approximate location of city boundaries.

ALTERNATIVE 2

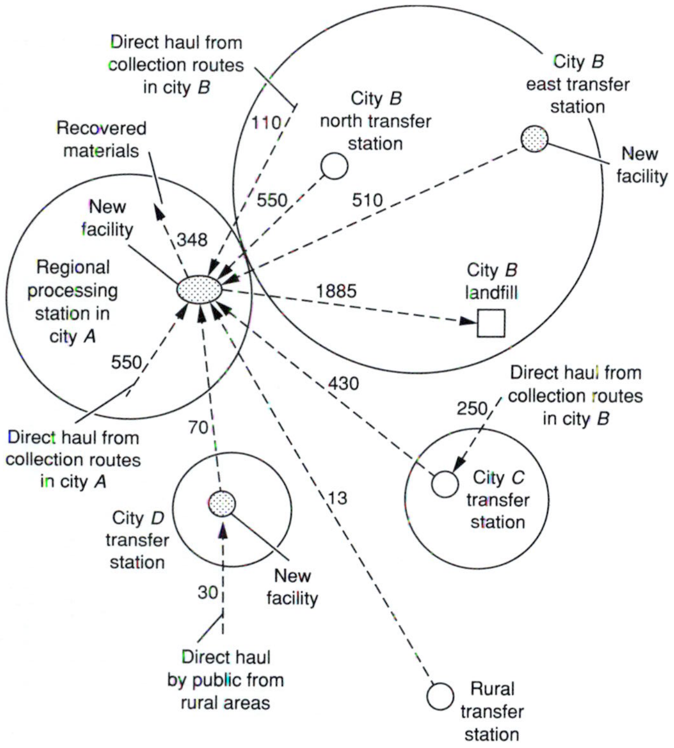

Note: Numbers correspond to solid waste loadings or recovered raw materials in tons per day. Large circles show approximate location of city boundaries.

ALTERNATIVE 3

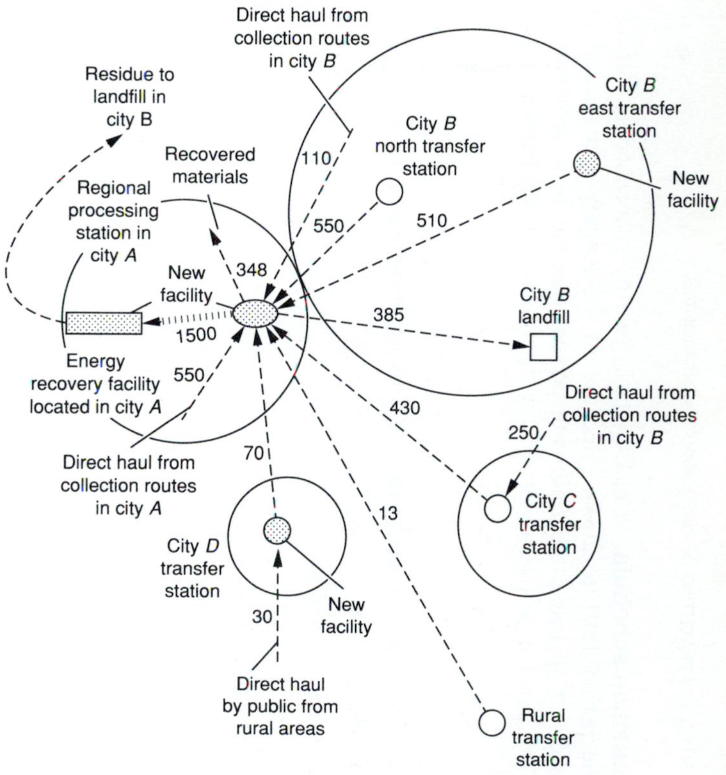

Note: Numbers correspond to solid waste loadings or recovered raw materials in tons per day. Large circles show approximate location of city boundaries.

ALTERNATIVE 4

Implementation schedule. A task-by-task, year-by-year schedule was developed for the period 1991 to 2006. (Because of the lengthy text needed to present the details of all tasks, the following presentation has been arranged so that activities are listed in a topical format for the functional elements of waste management.)

- 1991
  - Task 1 Hold hearings and adopt the solid waste management plan
  - Task 2 Transfer stations
    - City B east transfer station
      - Select a site
      - Begin preliminary design
      - Develop predesign cost estimates
      - Begin environmental impact report
    - City D transfer station
      - Select a site
      - Begin preliminary design
      - Develop predesign cost estimates
      - Begin environmental impact report
  - Task 3 Resource recovery
    - Regional processing facility
      - Begin preliminary design
      - Develop predesign cost estimates and investigate sites
    - Resource recovery studies
      - Initiate materials use policy
      - Implement home separation trial program
  - Task 4 Disposal
    - Develop and adopt disposal site policy
    - Begin gas and leachate monitoring system study
    - Evaluate use of phased-out or exhausted landfills
  - Task 5 Administration
    - Obtain required permits
    - Establish a joint administrative agency (all cities)
    - Select solid waste planning committee
    - Establish a transfer-disposal enterprise
  - Task 6 Special projects
    - Begin hazardous waste inventory program
    - Begin development of storage standards
    - Implement inventory program

- 1992
  - Task 1 Transfer stations
    - City B east transfer station
      - Final environmental impact report
      - Begin final design
    - City D transfer station
      - Final environmental impact report
      - Begin final design
  - Task 2 Resource recovery
    - Regional processing facility
      - Select a site
      - Advertise and receive bids from private industry for construction and operation of facility
    - Resource recovery studies
      - Expand or eliminate home separation
      - Finalize a materials-use policy
  - Task 3 Disposal
    - Recommend and install appropriate monitoring systems for gas and leachate
  - Task 4 Administration
    - Review and update plan
    - Develop financing plan for the regional processing facility
    - Hold planning committee meetings
  - Task 5 Special projects
    - Recommend hazardous waste program
    - Complete and adopt storage standards
- 1993
  - Task 1 Transfer stations
    - City B east transfer station
      - Complete final design and award construction contract
    - City D transfer station
      - Complete final design and award construction contract
  - Task 2 Resource recovery
    - Regional processing facility
      - Specify equipment and begin facility construction
    - Resource recovery studies
      - Select recovery systems that are economically justified
      - Update materials-use policy
      - Prepare report on existing resource recovery programs and recommend action

Task 3 Disposal
  Update disposal site policy
  Maintain and evaluate gas and leachate system
Task 4 Administration
  Prepare final contract for all cities for regional financing of processing
  Revise local ordinances to ensure waste delivery to the regional facility
  Establish a processing enterprise
Task 5 Special projects
  Implement hazardous waste program

- 1994
  - Task 1 Rewrite plan as necessary
  - Task 2 Transfer stations
    - Operate all transfer stations and upgrade transport equipment
  - Task 3 Resource recovery
    - Regional processing facility
    - Complete facility construction
    - Resource recovery studies
    - Prepare marketing program
    - Implement recommended resource recovery programs
  - Task 4 Disposal
    - Update disposal site policy
  - Task 5 Administration
    - Set user rates for processing facility
    - Establish materials sales contracts
    - Strengthen enforcement programs
- 1995–2000 (medium-term)
  - Review plan annually
    - Transfer and disposal
      - Prepare annual report on the efficiency of the transfer-disposal enterprise and plans for upgrading
    - Resource recovery
      - Finish construction of processing system
      - Prepare annual report on processing enterprise
  - Administration
    - Coordinate regional activities
- 2000–2006 (long-term)
  - Review plan annually
  - Replace and upgrade equipment in all facilities
  - Renew permits

The complexity of this full-system solid waste management plan is reflected in the comprehensive implementation schedule. The detailed discussion of action through 1994 is essential to successful implementation. Specific details are reduced significantly for the 1995–2000 period, and they are reduced even more for the 2000–2006 period.

Financing. The major facilities of the program are identified in Step 4. Preliminary capital costs for construction of the major facilities are presented in the three tables on pages 901–902. Capital financing for the transfer stations is to be derived from a sinking fund and existing capital reserves (financing of $3,989,000). The processing and recovery station will be financed by revenue bonds (financing of $30,235,000 capital will require a bond reserve and issuance cost of $3,038,000 for a total cost of $33,273,000). Operating costs and revenue for repayment of the bonds will be derived from service charges. The effect of increased rates on each city is presented in the table "Projected service charges for 1991 to 2000 in Case Study 20-1" on page 903.

Administration. The major elements of the plan are summarized in the accompanying table. Each element requires administrative action by a local agency. The responsible agencies were existing agencies for all elements except planning and coordination. Planning and coordination are to be accomplished by a joint-powers authority established by all participating cities. Two new enterprises are to be established to handle the joint transfer and disposal facilities and the processing and

Preliminary capital cost estimates for city D transfer station in Case Study 20-1 (Design capacity = 100 tons/day)

Description	Construction cost,[a] $
Excavation	65,000
Concrete	59,000
Paving	61,000
Fencing	35,000
Landscaping	65,000
Site, electric service	48,000
Sewer-water (onsite)	54,000
Site work	26,000
Office building	54,000
Operating equipment (internal)	109,000
Subtotal	576,000
Land costs, 5 acres @ $20,000/acre	100,000
Subtotal	676,000
Engineering (10 percent)	68,000
Subtotal	744,000
Contingencies (15 percent)	112,000
Total (excluding transport vehicles)	856,000

[a] In mid-1991 dollars (ENRCC index = 4892; see Appendix E).

Preliminary capital cost estimates for city B East transfer station in Case Study 20-1 (Design capacity = 500 tons/day)

Description	Construction cost,[a] $
Excavation	81,000
Concrete	359,000
Paving	224,000
Scale-scalehouse	102,000
Fencing	81,000
Landscaping	126,000
Site, electric service	76,000
Sewer-water (onsite)	63,000
Site work	26,000
Office-locker building	189,000
Pit area building	566,000
Operating equipment (internal)	283,000
Subtotal	2,176,000
Land costs, 15 acres @ $20,000/acre	300,000
Subtotal	2,476,000
Engineering (10 percent)	248,000
Subtotal	2,724,000
Contingencies (15 percent)	409,000
Total (excluding transport vehicles)	3,133,000

[a] In mid-1991 dollars (ENRCC index = 4892; see Appendix E).

Preliminary capital cost estimates for regional processing facility in Case Study 20-1 (Design Capacity = 1,500 tons/day)

	Construction cost,[a] $	
Description	First phase, 500 ton/d	Final phase,[b] 1,500 ton/d
Mills @ $370,000 each (30 ton/h each)	740,000	1,850,000
25 percent for equipment installation	185,000	463,000
Air classifiers @ $653,000 each	—	3,264,000
Magnetic separation @ $109,000 each	218,000	544,000
Glass and aluminum separation	—	9,140,000
Conveyors @ $3,200/ft	1,959,000	4,896,000
10 percent for equipment installation	196,000	490,000
Building for processing facilities	2,176,000	2,176,000
Land	435,000	435,000
Subtotal	5,909,000	23,258,000
Contingencies @ 30 percent	1,773,000	6,977,000[c]
Total	7,682,000	30,235,000

[a] In mid-1991 dollars (ENRCC index = 4892; see Appendix E).

[b] Includes all expenditures for the first phase.

[c] The contingency amount is large because the implementation schedule sets construction for 1994. Costs will change, and more definitive estimates must be made later.

Projected service charges for 1991 to 2000 in Case Study 20-1[a,b]

Service area	Charges per month per household, $/mo		
	1991 (Current)	1992–1995[c]	1996–2000[d]
City A	16.00	16.25	19.65
City B	16.00	17.60	17.60
City C	12.75	13.90	15.70
City D	18.90	18.90	21.35

[a] Based on mid-1991 dollars (ENRCC index = 4892; see Appendix E).

[b] The service charge was projected only to 2000, because inflation and other operating system modifications will significantly change costs by that time. Customer service charges pay for collection, transfer, processing, and disposal.

[c] Average monthly charge was set by adding operating cost for the period to annualized capital costs and dividing that total by the total months in the period.

[d] Significant changes in cost for the period caused by opening of the front-end processing system.

recovery facilities. The setting up of separate administrative enterprises was a good, workable means of maintaining accountability and equity in the regional system. Accountability becomes more important as revenues are derived from the sale of recovered resources. It should be noted that either a public or a private agency can operate the facilities. Finally, it is extremely important to understand that the entire plan requires administration, not just the facility programs within the plan.

Administrative elements of the plan and required action in Case Study 20-1

Element	Major local action
Disposal enforcement/surveillance	
Enforce health standards	Enforce state and local standards
Enforce operating standards	Enforce state and local standards
Control pollution at landfills	Coordinate with state regulatory agencies
Control litter	Develop program
Ongoing planning and coordination	
Publicize plan	Inform public
Review and revise plan	Revise plan in 1998
Storage and collection	
Enforce service standards	Enforce state and local standards
Enforce health standards	Enforce state and local standards
Special waste management	
Monitor hazardous waste	Monitor wastes
New and existing landfills	
Develop landfill-use program	Establish capacities
Develop final use program	Develop final use plans
Plan new landfills	Review use conditions
Facilities plan	
Build transfer stations	Design, construct, operate
Build processing station	Design, construct, operate

Comment. The regional approach to integrated solid waste management is a complex undertaking. However, a completed regional plan contains the basic details from which sound decisions will rest primarily with the autonomous local cities, even while the economies of a regional plan are achieved.

20-4 DISCUSSION TOPICS AND PROBLEMS

20-1. Contact the agencies responsible for solid waste management in your community. Identify all plans and program changes undertaken by each agency within the past year. List the policy programs requiring political decisions and the programs acted upon without political decisions.

20-2. Refer to Fig. 20-1. Which of the activities will be most difficult to achieve in your community?

20-3. A popular concept in integrated solid waste management planning is the development of programs and alternatives in which regional authorities or districts are designated as the preferred administrative agency. Assume that you are a city council member in a city of 20,000, and your city has been included in a recently completed solid waste management plan for a region containing 500,000. The major program recommended in the plan is a change from local administration to regional administration for materials recovery, transfer, and disposal. What are the key issues for your city? What steps would you recommend so that the city maintains control of waste management costs?

20-4. What public information means are used by solid waste management agencies in your community? Randomly select the names of five community residents from your phone book and ask them if they are aware of solid waste management issues on the local level and on state, national, and international levels.

20-5. Waste diversion has added a focus to planning. Identify where in a plan waste diversion can be quantified as to cost and quantities.

20-6. Implementation schedules are the standards for measuring the success of a plan after plan adoption. Implementation schedules also set accountabililty for actions. Give some examples of a strong implementation schedule. Who is accountable?

20-7. What agencies in your community are part of a team for finding sites for solid waste facilities? What criteria do these agencies use in evaluating sites?

20-8. Identify the permits that are required for landfills in your state. Are any solid waste management facilities exempt from permits? Does your state require permits for composting facilities?

20-5 REFERENCES

1. California Integrated Waste Management Board: *Guidelines for Preparation of Source Reduction and Recycling Elements,* Sacramento, CA, 1990.
2. Division of Solid Waste Management: *Sacramento County Solid Waste Management Plan, Final Report,* Department of Public Works, Sacramento County, Sacramento, CA, 1976.
3. Theisen, H., P. L. Maxfield, and G. E. Lynch: "Solid Waste Management Planning: A Methodology," *Journal of Environmental Health,* Vol. 38, No. 3, 1975.
4. U.S. Environmental Protection Agency: *Decision-Makers Guide to Solid Waste Management,* EPA/530-SW-89-072, Washington, DC, November 1989.

APPENDIX A

GLOSSARY

Aerobic A biochemical process or environmental condition occurring in the presence of oxygen.

Agricultural solid wastes Wastes produced from the raising of plants and animals for food, including manure, plant stalks, hulls, and leaves.

Anaerobic A biochemical process or environmental condition occurring in the absence of oxygen.

Ash The incombustible material that remains after a fuel or solid waste has been burned.

At-site time The time spent unloading and waiting to unload the contents of a collection vehicle or loaded container at a transfer station, processing facility, or disposal site.

Bacteria Single-cell, microscopic organisms with rigid cell walls. They may be aerobic, anaerobic, or facultative anaerobic; some can cause disease; and some are important in the stabilization and conversion of solid wastes.

Biodegradable material A compound that can be degraded or converted to simpler compounds by microorganisms.

Biodegradable volatile solids (BVS) The portion of the volatile solids of the organic matter in MSW that is biodegradable.

Bulky waste Large wastes such as appliances, furniture, some automobile parts, trees and branches, palm fronds, and stumps.

Buy-back center A physical facility where individuals can bring back recyclable materials in exchange for payment.

Carbonaceous matter Pure carbon or carbon compounds present in solid wastes.

Carbon dioxide (CO_2) A colorless, odorless, nonpoisonous gas that forms carbonic acid when dissolved in water. It is produced during the thermal degradation and microbial decomposition of solid wastes.

Carbon monoxide (CO) A colorless, poisonous gas that has an exceedingly faint metallic odor and taste. It is produced during the thermal degradation and microbial decomposition of solid wastes when the oxygen supply is limited.

Collection, waste The act of picking up wastes at homes, businesses, commercial and industrial plants, and other locations; loading them into a collection vehicle (usually enclosed); and hauling them to a facility for further processing or transfer or to a disposal site.

Collection routes The established routes followed in the collection of commingled and source-separated wastes from homes, businesses, commercial and industrial plants, and other locations.

Collection systems Collectors and equipment used for the collection of commingled and source-separated waste. Waste collection systems may be classified from several points of view, such as the mode of operation, the equipment used, and the types of wastes collected. In this text, collection systems have been classified according to their mode of operation into two categories: hauled container systems and stationary container systems.

Combustible materials Various materials in the waste stream that are combustible. In general, they are organic in nature (e.g., food waste, paper, cardboard, plastics, yard wastes).

Combustion The chemical combining of oxygen with a substance, which results in the production of heat and usually light.

Commercial solid wastes Wastes that originate in wholesale, retail, or service establishments, such as office buildings, stores, markets, theaters, hotels, and warehouses.

Commingled recyclables A mixture of several recyclable materials in one container.

Commingled waste Mixture of all waste components in one container.

Compactor Any power-driven mechanical equipment designed to compress and thereby reduce the volume of wastes.

Compaction (see *Densification*)

Compactor collection vehicle A large vehicle with an enclosed body having special power-driven equipment for loading, compressing, and distributing wastes within the body.

Component separation The separation or sorting of wastes into components or categories.

Compost A mixture of organic wastes partially decomposed by aerobic and/or anaerobic bacteria to an intermediate state. Compost can be used as a soil conditioner.

Composting The controlled biological decomposition of organic solid waste materials under aerobic conditions. Composting can be accomplished in windrows, static piles, and enclosed vessels (known as in-vessel composting).

Construction wastes Wastes produced in the course of construction of homes, office buildings, dams, industrial plants, schools, and other structures. The materials usually include used lumber, miscellaneous metal parts, packaging materials, cans, boxes, wire, excess sheet metal, and other materials. Construction and demolition wastes are usually grouped together.

Container A receptacle used for the storage of solid wastes until they are collected.

Conversion The transformation of wastes into other forms; for example, transformation by burning or pyrolysis into steam, gas, or oil.

Conversion products Products derived from the first-step conversion of solid wastes, such as heat from combustion and gas from biological conversion.

Cover material Soil or other material used to cover compacted solid wastes in a sanitary landfill.

Cullet Clean, generally color-sorted, crushed glass used in the manufacture of new glass products.

Curbside collection The collection of source-separated and mixed wastes from the curbside where they have been placed by the resident.

Decomposition The breakdown of organic wastes by bacterial, chemical, or thermal means. Complete chemical oxidation leaves only carbon dioxide, water, and inorganic solids.

Demolition wastes Wastes produced from the demolition of buildings, roads, sidewalks, and other structures. These wastes usually include large, broken pieces of concrete, pipe, radiators, duct work, electrical wire, broken-up plaster walls, lighting fixtures, bricks, and glass.

Densification The unit operation used to increase the specific weight (density in metric units) of waste materials so that they can be stored and transported more efficiently.

Dewatering The removal of water from solid wastes and sludges by various thermal and mechanical means.

Digestion, anaerobic The biological conversion of processed organic wastes to methane and carbon dioxide under anaerobic conditions.

Disposal The activities associated with the long-term handling of (1) solid wastes that are collected and of no further use and (2) the residual matter after solid wastes have been processed and the recovery of conversion products or energy has been accomplished. Normally, disposal is accomplished by means of sanitary landfilling.

Diversion rate A measure of the amount of material now being diverted for reuse and recycling compared to the total amount of waste that was thrown away previously.

Drop-off center A location where residents or businesses bring source-separate recyclable materials. Drop-off centers range from single-material collection points (e.g., easy-access "igloo" containers) to staffed, multimaterial collection centers.

Effluent Any solid, liquid, or gas that enters the environment as a by-product of human activities.

Endemic plant Plant species that is confined to a specific location, region, or habitat.

Energy recovery The process of recovering energy from the conversion products derived from solid wastes, such as the heat produced from the burning of solid wastes.

Ferrous metals Metals composed predominantly of iron. In the waste materials stream, these metals usually include tin cans, automobiles, refrigerators, stoves, and other appliances.

Flow diagram, process A diagram in which is shown the assemblage of unit operations, facilities, and manual operations used to achieve a specified waste separation goal or goals.

Fly ash Small solid particles of ash and soot generated when coal, oil, or solid wastes are burned. With proper equipment, fly ash is collected before it enters the atmosphere. Fly ash residue can be used for building materials (bricks) or in a sanitary landfill.

Food wastes Animal and vegetable wastes resulting from the handling, storage, sale, preparation, cooking, and serving of foods; commonly called *garbage*.

Functional element The term *functional element* is used in this text to describe the various activities associated with the management of solid wastes from the point of generation to final disposal. In general, a functional element represents a physical activity. The six functional elements used throughout this book are waste generation;

waste handling, separation, storage, and processing at the source; collection; separation and processing and transformation of solid waste; transfer and transport; and disposal.

Garbage (see *Food wastes*)

Generation (see *Waste generation*)

Groundwater Water beneath the surface of the earth and located between saturated soil and rock. It is the water that supplies wells and springs.

Haul distance The distance a collection vehicle travels (1) after picking up a loaded container (hauled container system) or from its last pickup stop on a collection route (stationary container system) to a materials recovery facility, transfer station, or sanitary landfill, and (2) the distance the collection vehicle travels after unloading to the location where the empty container is to be deposited or to the beginning of a new collection route.

Haul time The elapsed or cumulative time spent transporting solid wastes between two specific locations.

Hauled container system Collection systems in which the containers used for the storage of wastes are hauled to the disposal site, emptied, and returned to either their original location or some other location.

Hazardous wastes Wastes that by their nature may pose a threat to human health or the environment, the handling and disposal of which is regulated by federal law. Hazardous wastes include radioactive substances, toxic chemicals, biological wastes, flammable wastes, and explosives.

Heavy metals Metals such as cadmium, lead, and mercury which may be found in MSW in discarded items such as batteries, lighting fixtures, colorants, and inks.

Hierarchy of integrated solid waste management Source reduction, recycling, waste transformation, and disposal. It should be noted that EPA uses the term *combustion* instead of *transformation*. Further, EPA does not make a distinction between waste transformation (combustion) and disposal, as both are viewed as viable components of an integrated waste management program. A distinction is made between transformation and disposal in some states.

Hydrogen sulfide (H_2S) A poisonous gas with the odor of rotten eggs that is produced from the reduction of sulfates in, and the putrefaction of, a sulfur-containing organic material.

Incineration The controlled process by which solid, liquid, or gaseous combustible wastes are burned and changed into gases, and the residue produced contains little or no combustible material. Incineration is referred to as *combustion* in this text.

Industrial wastes Wastes generally discarded from industrial operations or derived from manufacturing processes. A distinction should be made between scrap (those materials that can be recycled at a profit) and solid wastes (those that are beyond the reach of economical reclamation).

Integrated solid waste management The management of solid waste based on a consideration of source reduction, recycling, waste transformation, and disposal arranged in a hierarchical order. The purposeful, systematic control of the functional elements of generation; waste handling, separation, and processing at the source; collection; separation and processing and transformation of solid waste; transfer and transport; and disposal associated with the management of solid wastes from the point of generation to final disposal.

Leachate Liquid that has percolated through solid waste or another medium. Leachate from landfills usually contains extracted, dissolved, and suspended materials, some of which may be harmful.

Litter That highly visible portion of solid wastes that is generated by the consumer and carelessly discarded outside the regular disposal system. Litter accounts for only about 2 percent of the total solid waste volume.

Magnetic separation The use of magnets to separate ferrous materials from commingled waste materials in MSW.

Manual separation The separation of wastes by hand. Sometimes called "hand-picking" or "hand-sorting," manual separation is done in the home or office by keeping food wastes separate from newspaper, or in a materials recovery facility by picking out large cardboard and other recoverable materials.

Mass burn The controlled combustion of unseparated commingled MSW.

Materials balance An accounting of the weights of materials entering and leaving a processing unit, such as an incinerator, usually on an hourly basis.

Materials recovery facility (MRF) The physical facilities used for the further separation and processing of wastes that have been separated at the source and for the separation of commingled wastes.

Materials recovery/transfer facilities (MR/TFs) Multipurpose facilities which may include the functions of a drop-off center for separated wastes, a materials recovery facility, a facility for the composting and bioconversion of wastes, a facility for the production of refuse-derived fuel, and a transfer and transport facility.

Mechanical separation The separation of solid wastes into various components by mechanical means.

Methane (CH_4) An odorless, colorless, and asphyxiating gas that can explode under certain circumstances and that can be produced by solid wastes undergoing anaerobic decomposition.

Microorganisms Generally, any living thing microscopic in size, including bacteria, yeasts, simple fungi, actinomycetes, some algae, slime molds, and protozoans. They are involved in stabilization of wastes (composting) and in sewage treatment processes.

Moisture content The weight loss (expressed in percent) when a sample of solid wastes is dried to a constant weight at a temperature of 100 to 105°C.

Municipal solid waste (MSW) Includes all the wastes generated from residential households and apartment buildings, commercial and business establishments, institutional facilities, construction and demolition activities, municipal services, and treatment plant sites.

Mulch Any material, organic or inorganic, applied as a top-dressing layer to the soil surface. Mulch is also placed around plants to limit evaporation of moisture and freezing of roots.

Native plant General term referring to plants that grow in a region.

Nonferrous metals Metals that contain no iron. Aluminum, copper, brass, and bronze are examples of the nonferrous metals found in MSW.

Off-route time All time spent by the collectors on activities that are nonproductive from the point of view of the overall collection operation.

Onsite handling, storage, and processing The activities associated with the handling, storage, and processing of solid wastes at the source of generation before they are collected.

Organic materials Chemical compounds containing carbon combined with other chemical elements. Organic materials can be of natural or anthropogenic origin. Most organic compounds are a source of food for bacteria and are usually combustible.

Organic soil amendment Plant and animal residues added to mineral soil to improve soil structure and enhance nutritional content of the soil.

Participation rate A measure of the number of people participating in a recycling program or other similar program, compared to the total number of people that could be participating.

Pathogen An organism capable of causing disease. The four major classifications of pathogen found in solid waste are bacteria, viruses, protozoans, and helminthes.

Pickup time For a hauled container system, it represents the time spent driving to a loaded container after an empty container has been deposited, plus the time spent picking up the loaded container and the time required to redeposit the container after its contents have been emptied. For a stationary container system, it refers to the time spent loading the collection vehicle, beginning with the stopping of the vehicle prior to loading the contents of the first container and ending when the contents of the last container to be emptied have been loaded.

Plant community Assemblage of plants coexisting together in a common habitat or environment.

Pollution The contamination of soil, water, or the atmosphere by the discharge of wastes or other offensive materials.

Primary materials Virgin or new materials used for manufacturing basic products. Examples include wood pulp, iron ore, and silica sand.

Processing Any method, system, or other means designated to change the physical form or chemical content of solid wastes.

Putrescible Subject to biological and chemical decomposition or decay. Usually used in reference to food wastes and other organic wastes.

Pyrolysis A way of breaking down burnable waste by combustion in the absence of air. High heat is usually applied to the wastes in a closed chamber; all moisture evaporates, and materials break down into various hydrocarbon gases and carbonlike residue.

Reclamation The restoration to a better or more useful state, such as land reclamation by sanitary landfilling, or the extraction of useful materials from solid wastes.

Recoverable resources Materials that still have useful physical or chemical properties after serving a specific purpose and can, therefore, be reused or recycled for the same or other purposes.

Recovery (see *Resource recovery*)

Recycling Separating a given waste material (e.g., glass) from the wastestream and processing it so that it may be used again as a useful material for products which may or may not be similar to the original.

Refuse A term often used interchangeably with the term *solid waste*. To avoid confusion, the term *refuse* is not used in this text.

Refuse-derived fuel (RDF) The material remaining after the selected recyclable and noncombustible materials have been removed from MSW.

Residential wastes Wastes generated in houses and apartments, including paper, cardboard, beverage and food cans, plastics, food wastes, glass containers, and garden wastes.

Residue The solid materials remaining after the separation of waste materials or after the completion of a chemical or physical process, such as burning, evaporation, distillation, or filtration.

Resource recovery *Resource recovery* is a general term used to describe the extraction of economically usable materials or energy from wastes. The concept may involve recycling or conversion into different and sometimes unrelated uses.

Reuse The use of a waste material or product more than once.

Rubbish A general term for solid wastes—excluding food wastes and ashes—taken from residences, commercial establishments, and institutions.

Sanitary landfill An engineered method of disposing of solid wastes on land in a manner that protects human health and the environment. Waste is spread in thin layers, compacted to the smallest practical volume, and covered with soil or other suitable material at the end of each working day.

Screening A unit operation that is used to separate mixtures of materials of different sizes into two or more size fractions by means of one or more screening surfaces.

Secondary material A material that is used in place of a primary or raw material in manufacturing a product.

Separation To divide wastes into groups of similar materials, such as paper products, glass, food wastes, and metals. Also used to describe the further sorting of materials into more specific categories, such as clear glass and dark glass. Separation may be done manually or mechanically with specialized equipment.

Shredding Mechanical operations used to reduce the size of solid wastes.

Size reduction, mechanical The mechanical conversion of solid wastes into small pieces. In practice, the terms *shredding*, *grinding,* and *milling* are used interchangeably to describe mechanical size reduction operations.

Solid waste management (see *Integrated waste management*)

Solid wastes Any of a wide variety of solid materials, as well as some liquids in containers, which are discarded or rejected as being spent, useless, worthless, or in excess. Does not usually include waste solids from treatment facilities. See also *agricultural, commercial, construction, demolition, hazardous, industrial, municipal,* and *residential wastes*.

Source reduction The design, manufacture, acquisition, and reuse of materials so as to minimize the quantity or toxicity of the waste generated.

Source separation The separation of waste materials from other commingled wastes at the point of generation.

Source-separated materials Waste materials that have been separated at the point of generation. Source-separated materials are normally collected separately.

Special wastes Special wastes include bulky items, consumer electronics, white goods, yard wastes that are collected separately, batteries, oil, and tires. Special wastes are usually handled separately from other residential and commercial wastes.

Stationary container systems Collection systems in which the containers used for the storage of wastes remain at the point of waste generation, except for occasional short trips to the collection vehicle.

Tipping floor Unloading area for wastes delivered to an MRF, transfer station, or waste combustor.

Transfer The act of transferring wastes from the collection vehicle to larger transport vehicles.

Transfer station A place or facility where wastes are transferred from smaller collection vehicles (e.g., compactor trucks) into larger transport vehicles (e.g., over-the-road and off-road tractor trailers, railroad gondola cars, or barges) for movement to disposal areas, usually landfills. In some transfer operations, compaction or separation may be done at the station.

Transformation, waste (see *Waste transformation*)

Transport The transport of solid wastes transferred from collection vehicles to a facility or disposal site for further processing or action.

Trash Wastes that usually do not include food wastes but may include other organic materials, such as plant trimmings.

Treatment process sludges Liquid and semisolid wastes resulting from the treatment of domestic waste water and industrial wastes.

Vadose zone The zone between the surface of the ground and the permanent groundwater.

Virgin material Any basic material for industrial processes that has not previously been used, for example, wood-pulp trees, iron ore, silica sand, crude oil, and bauxite. See also *Primary materials, Secondary material.*

Volatile solid (VS) The portion of the organic material that can be released as a gas when organic material is burned in a muffle furnace at 550°C.

Volume reduction The processing of wastes so as to decrease the amount of space they occupy. Compaction systems can reduce volume by 50 to 80 percent. Combustion can reduce waste volume by 90 percent.

Waste generation The act or process of generating solid wastes.

Waste sources Agricultural, residential, commercial, and industrial activities, open areas, and treatment plants where solid wastes are generated.

Wastestream The waste output of an area, location, or facility.

Waste transformation The transformation of waste materials involving a phase change (e.g., solid to gas). The most commonly used chemical and biological transformation processes are combustion and aerobic composting.

White goods Large worn-out or broken household, commercial, and industrial appliances, such as stoves, refrigerators, dishwashers, and clothes washers and dryers.

APPENDIX B

METRIC CONVERSION FACTORS

TABLE B-1
Factors for the conversion of U.S. Customary Units to the International System (SI) of Units

Multiply the U.S. customary unit		By	To obtain the corresponding SI unit	
Name	**Abbreviation**		**Name**	**Symbol**
acre	acre	4047	square meter	m^2
acre	acre	0.4047	hectare	ha*
British thermal unit	Btu	1.055	kilojoule	kJ
British thermal units per cubic foot	Btu/ft^3	37.259	Kilojoules per cubic meter	kJ/m^3
British thermal units per hour per square foot	$Btu/h \cdot ft^2$	23.158	joules per second per square meter	$J/s \cdot m^2$
British thermal units per kilowatt-hour	Btu/kWh	1.055	kilojoules per kilowatt-hour	$kJ/KW \cdot h$
British thermal units per pound	Btu/lb	2.326	kilojoules per kilogram	kJ/Kg
British thermal units per ton	Btu/ton	0.00116	kilojoules per kilogram	kJ/Kg
degree Celsius	°C	plus 273	kelvin	K
cubic foot	ft^3	0.0283	cubic meter	m^3
cubic foot	ft^3	28.32	liter	L^a
cubic feet per minute	ft^3/min	0.0004719	cubic meters per second	m^3/s
cubic feet per minute	ft^3/min	0.4719	liters per second	L^a/s

(*continued*)

TABLE B-1 (*continued*)

Multiply the U.S. customary unit		By	To obtain the corresponding SI unit	
Name	**Abbreviation**		**Name**	**Symbol**
cubic feet per second	ft^3/s	0.0283	cubic meters per second	m^3/s
cubic yard	yd^3	0.7646	cubic meter	m^3
day	d	86.4	kilosecond	ks
degree Fahrenheit	°F	0.555(°F-32)	degree Celsius	°C
foot	ft	0.3048	meter	m
feet per minute	ft/min	0.00508	meters per second	m/s
feet per second	ft/s	0.3048	meters per second	m/s
gallon	gal	0.003785	cubic meter	m^3
gallon	gal	3.785	liter	L[a]
gallons per minute	gal/min	0.0631	liters per second	L[a]/s
grain	gr	0.0648	gram	g
horsepower	hp	0.746	kilowatt	kW
horsepower-hour	hp-h	2.684	megajoule	MJ
inch	in	2.54	centimeter	cm
inch	in	0.0254	meter	m
kilowatt-hour	kWh	3.600	megajoule	MJ
pound (force)	lb_f	4.448	newton	N
pound (mass)	lb_m	0.4536	kilogram	kg
pounds per acre	lb/acre	0.1122	grams per square meter	g/m^2
pounds per acre	lb/acre	1.122	kilograms per hectare	kg/ha
pounds per capita per day	lb/capita · d	0.4536	kilograms per capita per day	kg/capita · d
pounds per cubic foot	lb/ft^3	16.019	kilograms per cubic meter	kg/m^3
pounds per cubic yard	lb/yd^3	0.5933	kilograms per cubic meter	kg/m^3
million gallons per day	Mgal/d	0.04381	cubic meters per second	m^3/s
miles	mi	1.609	kilometer	km
miles per hour	mi/h	1.609	kilometers per hour	km/h
miles per hour	mi/h	0.447	meters per second	m/s
miles per gallon	mi/gal	0.425	kilometers per liter	km/L[a]
parts per gallon	ppm	approximately equal to	milligrams per liter	mg/L[a]
ounce	oz	28.35	gram	g
pounds per square foot	lb/ft^2	47.88	newtons per square meter	N/m^2
pounds per square inch	lb/in^2	6.895	kilonewtons per square meter	kN/m^2
square foot	ft^2	0.0929	square meter	m^2
square mile	mi^2	2.590	square kilometer	km^2
square yard	yd^2	0.8361	square meter	m^2
ton (2000 pounds mass)	ton (2000 lb_m)	907.2	kilogram	kg
watt-hour	Wh	3.60	kilojoule	kJ
yard	yd	0.9144	meter	m

[a] Not an SI unit, but a commonly used term.

APPENDIX C

PHYSICAL PROPERTIES OF WATER

The principal physical properties of water are summarized in Table C-1 in U.S. customary units and in Table C-2 in SI units.

TABLE C-1
Physical properties of water (U.S. customary units)[a]

Temperature, °F	Specific weight, γ, lb/ft^3	Density,[b] ρ, slug/ft^3	Modulus of elasticity,[b] $E/10^3$, lb$_f$/in^2	Dynamic viscosity, $\mu \times 10^5$, lb · s/ft^2	Kinematic viscosity, $\nu \times 10^5$, ft^2/s	Surface tension,[c] σ, lb/ft	Vapor pressure, p_v, lb$_f$/in^2
32	62.42	1.940	287	3.746	1.931	0.00518	0.09
40	62.43	1.940	296	3.229	1.664	0.00614	0.12
50	62.41	1.940	305	2.735	1.410	0.00509	0.18
60	62.37	1.938	313	2.359	1.217	0.00504	0.26
70	62.30	1.936	319	2.050	1.059	0.00498	0.36
80	62.22	1.934	324	1.799	0.930	0.00492	0.51
90	62.11	1.931	328	1.595	0.826	0.00486	0.70
100	62.00	1.927	331	1.424	0.739	0.00480	0.95
110	61.86	1.923	332	1.284	0.667	0.00473	1.27
120	61.71	1.918	332	1.168	0.609	0.00467	1.69
130	61.55	1.913	331	1.069	0.558	0.00460	2.22
140	61.38	1.908	330	0.981	0.514	0.00454	2.89
150	61.20	1.902	328	0.905	0.476	0.00447	3.72
160	61.00	1.896	326	0.838	0.442	0.00441	4.74
170	60.80	1.890	322	0.780	0.413	0.00434	5.99
180	60.58	1.883	318	0.726	0.385	0.00427	7.51
190	60.36	1.876	313	0.678	0.362	0.00420	9.34
200	60.12	1.868	308	0.637	0.341	0.00413	11.52
212	59.83	1.860	300	0.593	0.319	0.00404	14.70

[a] Adapted from Vennard, J. K., and R. L. Street: *Elementary Fluid Mechanics,* 5th ed., Wiley, New York, 1975.

[b] At atmospheric pressure.

[c] In contact with air.

TABLE C-2
Physical properties of water (SI units)[a]

Temperature, °C	Specific weight, γ, kN/m^3	Density,[b] ρ, kg/m^3	Modulus of elasticity,[b] $E/10^6$, kN/m^2	Dynamic viscosity, $\mu \times 10^3$, N · s/m^2	Kinematic viscosity, $\nu \times 10^6$, m^2/s	Surface tension,[c] σ, N/m	Vapor pressure, p_v, kN/m^2
0	9.805	999.8	1.98	1.781	1.785	0.0765	0.61
5	9.807	1000.0	2.05	1.518	1.519	0.0749	0.87
10	9.804	999.7	2.10	1.307	1.306	0.0742	1.23
15	9.798	999.1	2.15	1.139	1.139	0.0735	1.70
20	9.789	998.2	2.17	1.002	1.003	0.0728	2.34
25	9.777	997.0	2.22	0.890	0.893	0.0720	3.17
30	9.764	995.7	2.25	0.798	0.800	0.0712	4.24
40	9.730	992.2	2.28	0.653	0.658	0.0696	7.38
50	9.689	988.0	2.29	0.547	0.553	0.0679	12.33
60	9.642	983.2	2.28	0.466	0.474	0.0662	19.92
70	9.589	977.8	2.25	0.404	0.413	0.0644	31.16
80	9.530	971.8	2.20	0.354	0.364	0.0626	47.34
90	9.466	965.3	2.14	0.315	0.326	0.0608	70.10
100	9.399	958.4	2.07	0.282	0.294	0.0589	101.33

[a] Adapted from Vennard, J. K., and R. L. Street: *Elementary Fluid Mechanics,* 5th ed., Wiley, New York, 1975; and Webber, N. B.: *Fluid Mechanics for Civil Engineers, SI ed.,* Chapman and Hall, London, 1971.

[b] At atmospheric pressure.

[c] In contact with air.

APPENDIX D

PRESENTATION AND ANALYSIS OF SOLID WASTE MANAGEMENT DATA

The purpose of this appendix is to introduce (1) techniques used to graphically present and analyze solid waste management data, (2) statistical measures that are commonly used to characterize solid waste collection rates, and (3) graphical procedures that can be used to determine the nature of a distribution. For a more detailed presentation of the fundamentals of statistical analysis, the reader is referred to Refs. 1 to 4.

D-1 GRAPHICAL PRESENTATION OF FIELD DATA

The graphical presentation of observed field data can be used to depict and identify trends in the data. Time series and histogram or frequency plots are used extensively for the presentation and analysis of such data.

Time Series

Observations that are arranged in the order of their occurrence in time are often called time series. By plotting the observed values versus time, it is often possible to establish trends, cycles or periodicities, and fluctuations that may be of value in understanding the basic nature of the phenomenon under evaluation.

Description of Trends, Cycles, and Fluctuations. The word *trend* is used to describe a relatively long-term tendency for field observations to increase or decrease in some orderly manner (see Fig. D-1*a*). The change in the magnitude of the observations may be simple or complex.

Observations that are cyclic tend to form successive peaks and valleys (see Fig. D-1*b*). Like trends, cycles may be of simple periodicity or may be defined by long-term repeating periodicities.

The term *fluctuating time series* is often used to describe observations that change significantly from one time interval to the next with no apparent pattern (see Fig. D-1*c*).

Analysis of Trends, Cycles, and Fluctuations. Mathematically, linear trends are described with straight lines of the form $y = a + bx$. Curvilinear trends are described most commonly with polynomials of the general form $y = a + bx + cx^2$, although a variety of other equations are also used [1]. The method of moving averages is often used to suppress random irregularities in plotted time series data so that short- and long-term trends can be discovered. Moving averages may be simple or weighted. The moving average is obtained by averaging an odd number of successive observations. Thus, the average will correspond to the middle item, which can be weighted for emphasis. Two examples of weighted moving averages are given by Eqs. (D-1) and (D-2).

$$x_b = \frac{a + 2b + c}{4} \tag{D-1}$$

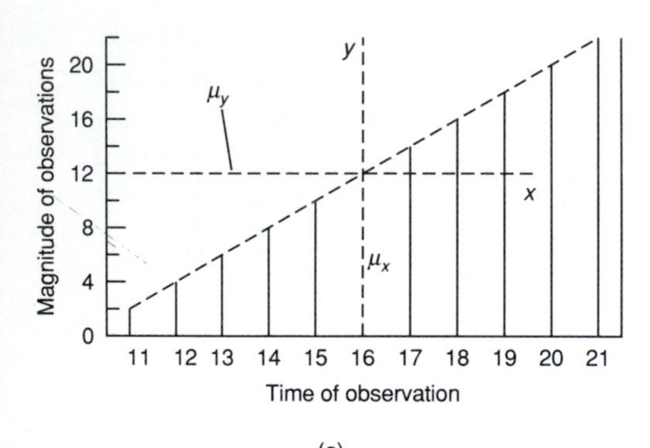

(*a*)

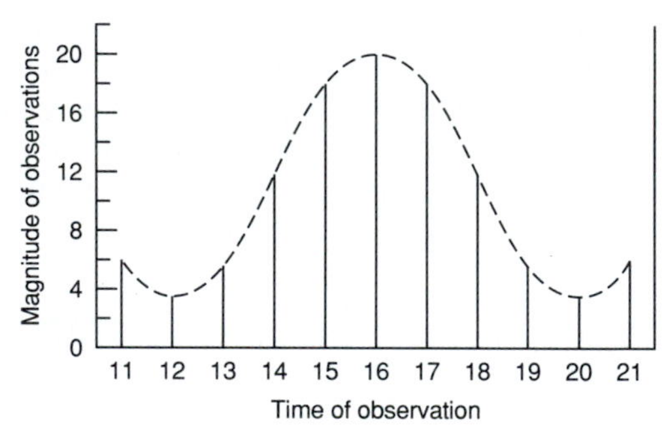

(*b*)

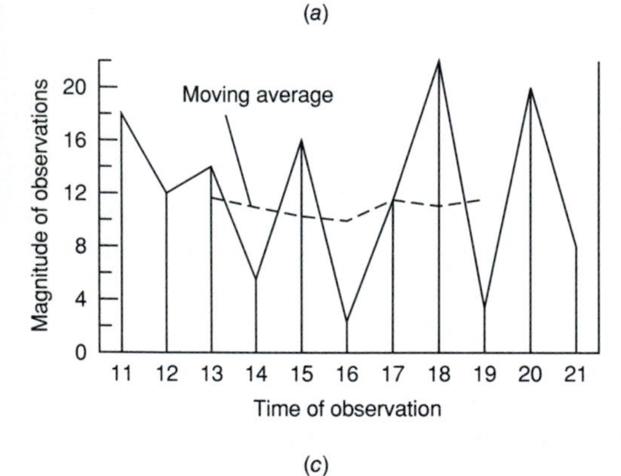

(*c*)

FIGURE D-1
Typical time series plots that are representative of time series data that (*a*) follow a trend, (*b*) are cyclic, and (*c*) are fluctuating (from Ref. 1).

where x_b = average value at point b
a, b, c = observed magnitudes at points a, b, c

$$x_c = \frac{a + 4b + 6c + 4d + e}{16} \tag{D-2}$$

The coefficients in the moving average given by Eq. (D-2) are obtained from the binomial expansion. The weighted moving average for the fluctuating time series shown in Fig. D-1*c* was computed using Eq. (D-1).

Frequency Distributions

Observations arranged in order of magnitude form an array. The same data can be grouped together in a series of data ranges. If whole numbers are assigned to the number of observations that occur in each data range, then the frequency of occurrence of whole numbers can be plotted against the magnitude of the data ranges. The resulting plots are called histograms (see Fig. D-2). As shown, histograms can be symmetrical (Fig. D-2*a*), asymmetrical (Fig. D-2*b*), rectangular

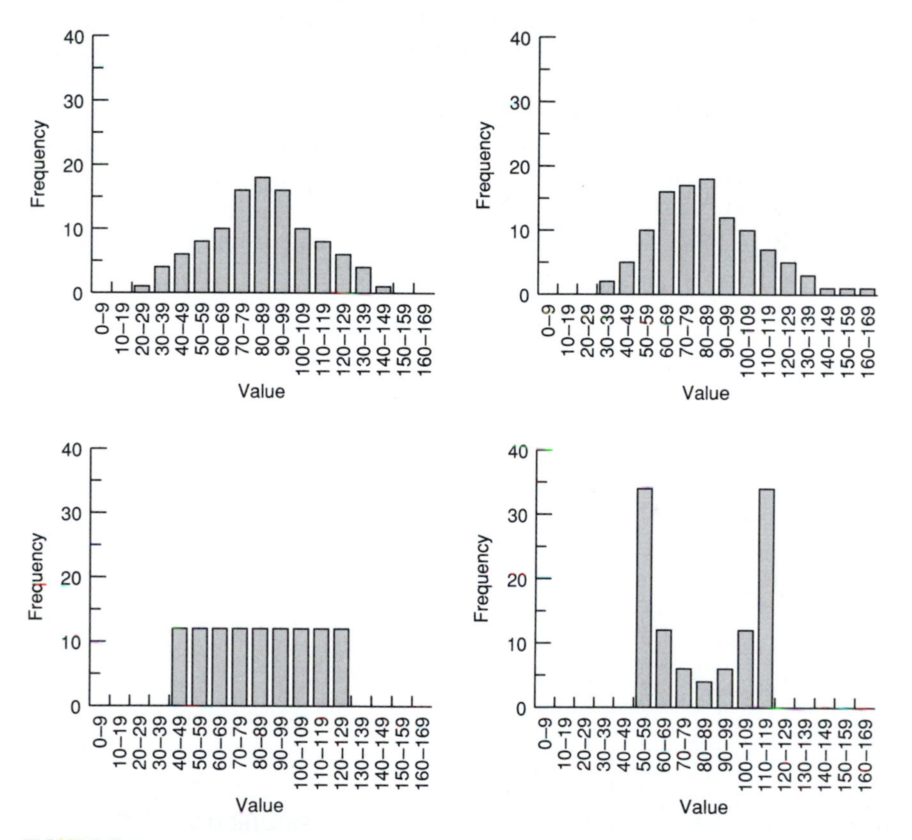

FIGURE D-2
Histogram of four different distributions which have the same arithmetic mean and standard deviation (adapted from Ref. 4).

TABLE D-1
Grouped weekly waste collection volumes obtained from a small municipality for a period of one year

Collection volume, yd^3/wk	Data range mid-point	Frequency, f_i
1000–1100	1050	1
1100–1200	1150	3
1200–1300	1250	4
1300–1400	1350	9
1400–1500	1450	11
1500–1600	1550	10
1600–1700	1650	7
1700–1800	1750	4
1800–1900	1850	2
1900–2000	1950	1
Total		52

(Fig. D-2*c*), or U-shaped (Fig. D-2*d*). The data forming symmetrical histograms are said to be *normally* distributed, whereas the data in asymmetrical histograms are said to be *skewed*.

A typical example of a year's worth of weekly solid waste collection data that has been arranged in data ranges is reported in Table D-1. As shown, the 52 individual weekly data points have been grouped in data ranges varying from 1000–1100 to 1900–2000. The midpoint of each data range is given in column two. The frequency values reported in column three correspond to the number

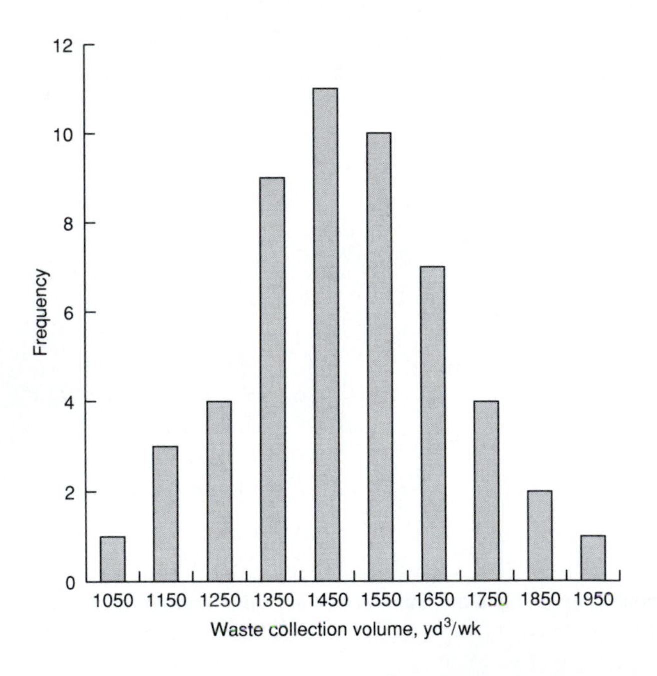

FIGURE D-3
Histogram of yearly solid waste collection data from a small municipality.

of weekly values that were in a given data range. A histogram of the data reported in Table D-1 is presented in Fig. D-3. The statistical characteristics of the grouped solid waste collection data are determined in the following section in Example D-1.

D-2 STATISTICAL MEASURES

Commonly used statistical measures include the mean, median, mode, standard deviation, and coefficient of variation. Although the terms just cited are the most commonly used statistical measures, it is instructive to review the distributions plotted in Fig. D-2 and to note that they all have the same arithmetic mean and standard deviation. Thus, two additional statistical measures are needed to quantify the nature of a given distribution. The two additional measures are the coefficient of skewness, and coefficient of kurtosis. All of the statistical measures are defined below. Determination of these statistical measures from field data is illustrated in Example D-1 presented following the discussion of the coefficient of kurtosis.

Mean

The mean is defined as the arithmetic average of a number of individual or grouped observations. The mean for grouped observations is given by

$$\bar{x} = \frac{\sum f_i x_i}{n} \tag{D-3}$$

where $\bar{x}$ = mean value
f_i = frequency (for ungrouped data $f_i = 1$)
x_i = the midpoint of the ith data range (for ungrouped data x_i = the ith observation)
n = number of observations (note $\sum f_i = n$)

Median

If a series of observations are arranged in order of increasing value, the midmost observation, or the arithmetic mean of the two midmost observations, in a series is known as the median. For example, in a set of 15 measurements, the median will be the 8th value, whereas in a set of 16 measurements, the median will be the average of the 8th and 9th values. In a symmetrical distribution, the median will equal the mean.

Mode

The value occurring with the greatest frequency in a set of observations is known as the mode. If a continuous graph of the frequency distribution is drawn, the mode is the value of the high point, or hump, of the curve. In a symmetrical set of observations, the mean, median, and mode will be the same value. The mode can be approximated with reasonable accuracy using the following expression:

$$\text{Mod} = 3(\text{Med}) - 2(\bar{x}) \tag{D-4}$$

where Mod = mode
Med = median
$\bar{x}$ = mean

Standard Deviation

Because of the laws of chance, there is uncertainty in any set of measurements. The precision of a set of measurements can be assessed in a number of different ways. Most commonly, the error of an individual measurement in a set is defined as the difference between the arithmetic mean and the value of the measurement. The standard deviation for grouped data is defined as follows:

$$s = \sqrt{\frac{\sum f_i(x_i - \bar{x})^2}{n - 1}} \tag{D-5}$$

where s = standard deviation
f_i = frequency (for ungrouped data $f_i = 1$)
x_i = the midpoint of the ith data range (for ungrouped data x_i = the ith observation)
$\bar{x}$ = mean value
n = number of observations

From the form of the equation, it can be concluded that the larger the scatter in a set of measurements is, the larger the value of s will be. Conversely, as the precision of a set of measurement improves, the value of the standard deviation will decrease. From theoretical considerations, it can be shown that if the measurements are distributed normally, then 68.27 percent of the observations will fall within plus or minus one standard of deviation of the mean ($\bar{x} \pm s$) [4, 5].

Coefficient of Variation

Although the standard deviation can be used as an indication of the absolute dispersion of a set of measured values, it provides little or no information about whether the value is large or small. To overcome this difficulty, the coefficient of variation, as defined in Eq. (D-4), is used as a relative measure of dispersion [4].

$$CV = \frac{100\, s}{\bar{x}} \tag{D-6}$$

where CV = coefficient of variation, percent
s = standard deviation (see Eq. D-5)
$\bar{x}$ = mean value (see Eq. D-3)

Typically, the coefficient of variation for solid waste generation rates will vary from 10 to 60 percent. To judge whether this percentage represents a large or small scatter, it can be compared to values obtained from measurements in other fields. For measurements in the biological field, the coefficient of variation will vary from about 10 to 30 percent. The coefficient of variation for chemical analyses varies from 2 to 10 percent. Clearly, the scatter in solid waste generation data is significant.

Coefficient of Skewness

When a frequency distribution is asymmetrical (see Fig. D-2*b*), it is usually defined as being a skewed frequency distribution. Over the years, a number of measures of skewness have been proposed, but none is accepted universally. For the purposes of this textbook, skewness is defined by the following relationship:

$$\alpha_3 = \frac{\sum f_i(x_i - \bar{x})^3/n - 1}{s^3} \tag{D-7}$$

where α_3 = coefficient of skewness
f_i = frequency (for ungrouped data $f_i = 1$)
x_i = the midpoint of the ith data range (for ungrouped data x_i = the ith observation)
$\bar{x}$ = mean value
n = number of observations
s = standard deviation

The coefficient of skewness has also been computed using the following relationship.

$$\alpha_3{}' = \frac{(\bar{x} - \text{Mod})}{s} \tag{D-8}$$

where $\alpha_3{}'$ = coefficient of skewness
$\bar{x}$ = mean value
Mod = mode
s = standard deviation

Coefficient of Kurtosis

The extent to which a distribution is more peaked (see Fig. D-2*b*) or more flat-topped (see Fig. D-2*c*) than the normal distribution is defined by the kurtosis of the distribution. The coefficient of kurtosis can be computed using the following equation:

$$\alpha_4 = \frac{\sum f_i(x_i - \bar{x})^4/n - 1}{s^4} \tag{D-9}$$

where α_4 = coefficient of kurtosis
f_i = frequency (for ungrouped data $f_i = 1$)
x_i = the midpoint of the ith data range (for ungrouped data x_i = the ith observation)
$\bar{x}$ = mean value
n = number of observations
s = standard deviation

The value of the kurtosis for a normal distribution is 3. A peaked curve will have a value greater than 3 where as a flatter curve it will have a value less than 3. The value α_4 that separates mound-shaped curves from rectangular or U-shaped curves is in the range from 1.75 to 1.8. Values of α_4 for U-shaped distributions are less than 1.75.

Example D-1 Statistical analysis of solid waste collection data. Determine the statistical characteristics of the weekly solid waste collection data presented in Table D-1.

Solution

1. Determine the statistical characteristics of the solid waste collection data.
   (*a*) Set up a data analysis table to obtain the quantities needed to determine the statistical characteristics.

Data range, yd^3/wk	x_i[a]	Freq., f_i	$f_i x_i$	$(x_i - \bar{x})$	$f_i(x_i - \bar{x})^2 \times 10^{-3}$	$f_i(x_i - \bar{x})^3 \times 10^{-5}$	$f_i(x_i - \bar{x})^4 \times 10^{-10}$
1,000–1,100	1,050	1	1,050	−437	191.0	−834.5	3.647
1,100–1,200	1,150	3	3,450	−337	340.7	−1.148.2	3.870
1,200–1,300	1,250	4	5,000	−237	224.7	−532.5	1.262
1,300–1,400	1,350	9	12,150	−137	168.9	−231.4	0.317
1,400–1,500	1,450	11	15,950	−37	15.1	−5.6	0.002
1,500–1,600	1,550	10	15,500	63	39.7	25.0	0.016
1,600–1,700	1,650	7	11,550	163	186.0	303.2	0.494
1,700–1,800	1,750	4	7,000	263	276.7	727.7	1.914
1,800–1,900	1,850	2	3,700	363	263.5	1,435.0	5.208
1,900–2,000	1,950	1	1,950	463	214.4	1,985.1	9.190
Total		52	77,300		1,920.6	1,723.8	25.920

$\bar{x}$ = 77,300/52 = 1,487

[a] Midpoint of data range.

   (*b*) Determine these statistical characteristics.
      i. Mean

$$\bar{x} = \frac{\sum f_i x_i}{n}$$

$$\bar{x} = \frac{77{,}300}{52} = 1487 \text{ yd}^3\text{/wk (see data table above)}$$

ii. Median (the midmost value)

$$\text{Med} = 1450 \text{ yd}^3/\text{wk (see data table above)}$$

iii. Mode

$$\text{Mod} = 1450 \text{ yd}^3/\text{wk (see data table above)}$$

iv. Standard deviation

$$s = \sqrt{\frac{\sum f_i\left(x_i - \bar{x}\right)^2}{n - 1}}$$

$$s = \sqrt{\frac{1920.7 \times 10^5}{51}} = 194.1$$

v. Coefficient of variation

$$CV = \frac{100\ s}{\bar{x}}$$

$$CV = \frac{100\ (194.1)}{1487} = 13.1\%$$

vi. Coefficient of skewness

$$\alpha_3 = \frac{\sum f_i\left(x_i - \bar{x}\right)^3 / n - 1}{s^3}$$

$$\alpha_3 = \frac{1723.8 \times 10^5/51}{194.1^3} = 0.462$$

vii. Coefficient of kurtosis

$$\alpha_4 = \frac{\sum f_i\left(x_i - \bar{x}\right)^4 / n - 1}{s^4}$$

$$\alpha_4 = \frac{25.920 \times 10^{10}/51}{(194.1)^4} = 3.6$$

Comment. The solid waste collection data have a slight negative skewness (normal bell-shaped curve is distorted slightly to the right). Further, the field data are more peaked than the normal curve (3.58 versus 3 for a normal curve), which means the data are bunched together more closely than for a normal curve.

Comparison of Statistical Measures for Various Distributions

The statistical measures for the distributions shown in Fig. D-2 are reported below [4]. As shown, the mean and standard deviation are the same for all of the distributions. If only these measures had been computed, it would be concluded that all the distributions are the same when in reality they are quite different.

Clearly, having information on the values of α_3 and α_4 is important in assessing the nature of the distributions.

Statistical measure	Fig. D-2*a*	Fig. D-2*b*	Fig. D-2*c*	Fig. D-2*d*
n	108	108	108	108
$\bar{x}$	85	85	85	85
s	25.8	25.8	25.8	25.8
α_3	0	+0.57	0	0
α_4	2.565	3.188	1.770	1.23

From Ref. 4.

D-3 SKEWED DISTRIBUTIONS

In general, some degree of positive skewness (normal bell-shaped curve is distorted to the left) is common in solid waste generation data. Fortunately, most statistical tests based on the normal distribution are robust, and small amounts of skewness can be tolerated. If the field data are skewed severely, they may have to be rescaled by taking the logarithm or the square root to make them more normal. The most common statistical measures for skewed distributions are the geometric mean and standard deviation. These measures are computed as follows [1].

Geometric Mean

The geometric mean is defined as the log average of a number of individual measurements and is given by

$$\log M_g = \frac{\sum f_i(\log x_i)}{n} \tag{D-10}$$

where M_g = geometric mean value
f_i = frequency (for ungrouped data $f_i = 1$)
x_i = the midpoint of the ith data range (for ungrouped data x_i = the ith observation)
n = number of observations

Geometric Standard Deviation

The geometric standard deviation is defined as follows:

$$\log s_g = \sqrt{\frac{\sum (f_i \log^2 x_g)}{n-1}} \tag{D-11}$$

where s_g = geometric standard deviation
f_i = frequency (for ungrouped data $f_i = 1$)
$x_g = x_i/M_g$
x_i = the midpoint of the ith data range (for ungrouped data x_i = the ith observation)
n = number of observations

D-4 GRAPHICAL ANALYSIS OF FIELD DATA

For most practical purposes, the type of the distribution can be determined by plotting the data on both arithmetic and logarithmic probability paper and noting whether the data can be fitted with a straight line. The use of arithmetic and logarithmic probability paper is discussed and illustrated in the following paragraphs [1, 2, and 3].

Probability Paper

Although it is possible to express the summation or probability curve in equation form, it has been found more useful to develop a type of graph paper, called "probability paper," with special coordinates on which data that are normal or logarithmically normal will plot as a straight line. Three steps are involved in the use of arithmetic and logarithmic probability paper.

1. The measurements in a data set are first arranged in order of increasing magnitude and assigned a rank serial number.
2. Next, a corresponding plotting position is determined for each data point using Eq. (D-12).
3. The data are then plotted on arithmetic and logarithmic probability paper. The probability scale is labeled "Percent of values equal to or less than the indicated value."

The plotting position is computed using the following equation:

$$\text{Plotting position } (\%) = \left(\frac{m}{n+1}\right) 100 \tag{D-12}$$

where m = rank serial number
n = number of observations

The term $(n + 1)$ is used to account for the fact that there may be an observation that is either larger or smaller than the largest or smallest in the data set. In effect, the plotting position represents the percent or frequency of observations that are equal to or less than the indicated value.

Arithmetic Probability Paper. By plotting data on arithmetic probability paper, it is possible to determine:

1. Whether the data are normally distributed by noting if the data can be fit with a straight line. Significant departure from a straight line can be taken as an indication of skewness.
2. The approximate magnitude of the arithmetic mean. Usually it will be best to compute the mean and to pass the straight line plotted by eye through the computed value.
3. The approximate value of the standard deviation by finding the values on the curve at the 84.1 (i.e., $50 + 68.27/2$) and 15.9 (i.e., $50 - 68.27/2$) percent

points and noting that these values correspond to $(\bar{x} \pm s)$. Thus,

$$s = P_{84.1} - \bar{x} \text{ or } \bar{x} - P_{15.9}$$

4. The expected frequency of any observation of a given magnitude.

Logarithmic Probability Paper. When the data are skewed (see Fig. D-2*b*), logarithmic probability paper can be used. The implication here is that the logarithm of the observed values is normally distributed. On logarithmic probability paper, the straight line of best fit passes through the geometric mean and through the intersection of $M_g \times s_g$ at a value of 84.1 percent and M_g / s_g at a value of 15.9 percent. The geometric standard deviation can be determined from the following equation:

$$s_g = \frac{P_{84.1}}{M_g} = \frac{M_g}{P_{15.9}} \qquad \text{(D-13)}$$

where s_g = geometric standard deviation
$P_{84.1}$ = value from curve at 84.1 percent
M_g = geometric mean
$P_{15.9}$ = value from curve at 15.9 percent

Use of Probability Paper

The use of probability paper to determine the type of distribution is illustrated using the ungrouped solid waste collection data given in the following table.

Rank serial no., m	Waste, yd³/d	Plotting position,[a] %
1	1.0	7.1
2	2.5	14.3
3	4.0	21.4
4	6.6	28.6
5	7.4	35.7
6	10.4	42.9
7	11.3	50.0
8	12.0	57.1
9	12.6	64.3
10	15.8	71.4
11	17.0	78.6
12	20.0	85.7
13	22.2	92.9

[a] See Eq. (D-12).

As shown, the data have been arranged in order of increasing magnitude, and plotting positions have been computed by using Eq. (D-12). The arithmetic and logarithmic probability plots of these data are given in Figs. D-4*a* and D-4*b*,

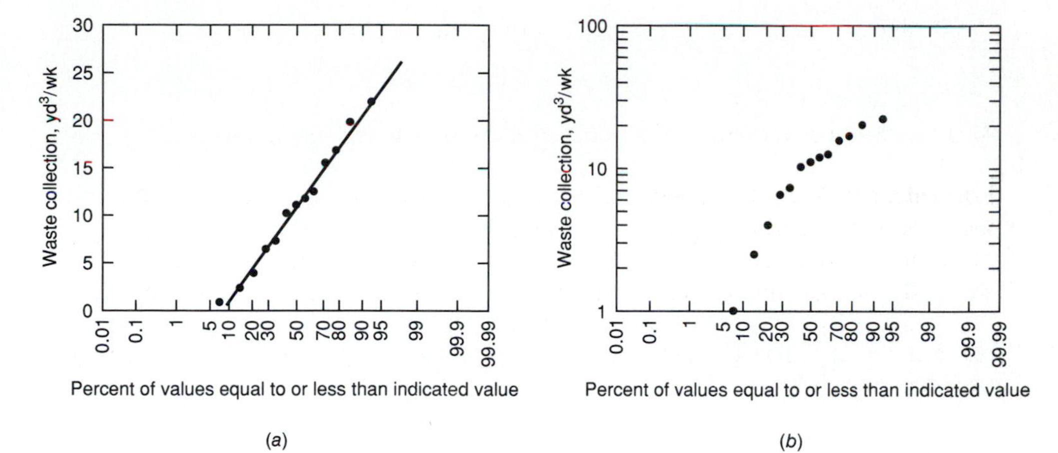

FIGURE D-4
Probability plots of solid waste collection data: (*a*) arithmetic and (*b*) logarithmic.

respectively. Because the data fall on a straight line on the arithmetic probability paper, it can be concluded that the data are distributed normally and that normal statistics can be applied.

D-5 REFERENCES

1. Fair, G. M. and J. C. Geyer: *Water Supply and Waste-Water Disposal,* John Wiley & Sons, Inc., New York, 1954.
2. Velz, C. J.: "Graphical Approach to Statistics," *Water and Sewage Works,* Reference and Data Issue, Vol. 99, No. 4, 1952.
3. Velz, C. J.: "Applied Stream Sanitation," *Statistical Tools,* Appendix A, Wiley-Interscience, New York, 1970.
4. Waugh, A. E.: *Elements of Statistical Analysis,* 2d ed., McGraw-Hill, New York, 1943.
5. Young, H. D.: *Statistical Treatment of Experimental Data,* McGraw-Hill, New York, 1962.

APPENDIX E

TYPICAL COST DATA AND COST-ESTIMATING PROCEDURES FOR EQUIPMENT USED IN SOLID WASTE MANAGEMENT SYSTEMS

The cost of acquiring and operating solid waste management equipment and facilities is an important determinant in the design and selection of a waste management system. For a complex system, the costs of structures, roads, equipment, landscaping, utility connection fees, and other items might be included. Because of the many factors that must be considered, development of cost estimates for such systems should be undertaken only by qualified persons. The purpose of this appendix is to provide typical cost data on the purchase price of selected solid waste management equipment. Operating costs are also important, but they are not presented here because their variability makes the use of generalized values highly questionable. Cost estimating for processing and resource recovery systems is also presented.

E-1 COST DATA FOR EQUIPMENT USED IN WASTE MANAGEMENT SYSTEMS

It is often helpful to have some reference cost data available for use in broadly defining the magnitude of the expenditures associated with equipment acquisition decisions. Representative cost data for selected waste management equipment are presented in Table E-1. The information is not intended to be all-inclusive within

TABLE E-1
Cost data for selected solid waste management equipment[a]

Type of equipment	Size, rated capacity, or gross weight (GW)		Approximate cost range,[b] $(1,000)
	Unit	Value	
Storage containers			
Small-capacity household			
Rectangular	gal	12–14	0.004–0.00675
	gal	16–22	0.005–0.00925
Round	gal	5–7	0.00165–0.0035
Stackable (qty. of 3)	gal	20–32	0.006–0.0115
	gal	36–42	0.014–0.02
Wheeled	gal	32	0.018–0.0375
	gal	45–55	0.025–0.055
	gal	60–80	0.05–0.07
	gal	90–105	0.045–0.075
Steel bins	yd^3	2	0.3–0.35
	yd^3	6	0.6–0.8
	yd^3	8	0.8–0.9
	yd^3	20	2.4–2.6
	yd^3	30	2.7–3.0
	yd^3	40	3.0–3.3
Collection vehicles			
Tilt-frame	yd^3	15	70–90
	yd^3	25	80–100
Front-loaded compactor	yd^3	30	100–120
Front-loaded compactor	yd^3	40	120–130
Side-loaded compactor	yd^3	28	60–70
	yd^3	28	110–130
Rear-loaded compactor	yd^3	20	60–70
	yd^3	25	110–130
Curbside Recycling—multibin			85–120
Recycling and transfer equipment			
Forklift	lb. cap.	1,000	8–10
	lb. cap.	8,000	25–32
Bobcat	lb.	3,400	23–38
Rubber-tired clamshell			100–120
Articulated loader (950 E)			125–140
Articulated loader (1T12)			65–70
Knuckleboom crane			70–90
Landfill tipper			250–300
Transport equipment			
Trailers			
Hydraulic ejector blade	yd^3	75–92	33.5–56
Walking floor	yd^3	96	38–53
Possum belly	yd^3	100–124	40–50
Tractor			
Conventional	hp	300–350	70–80
Cab-over	hp	300–350	65–75

(continued)

TABLE E-1 (*continued*)

Type of equipment	Size, rated capacity, or gross weight		Approximate cost range,[b] $(1,000)
	Unit	Value	
Compacting and densifying			
Refuse compactors	ton/h	25	230–260
	ton/h	62.5	375–400
	ton/h	100	400–430
Pelletizers	ton/h	6	150–200
Recyclables balers	ton/h	0.3–2	8–50
	ton/h	3–12	80–200
	ton/h	10–20	120–200
	ton/h	20–34	200–425
Aluminum flatteners/blowers	ton/h	0–2	5–11
Aluminum densifiers	lb/h	500–900	16–21
		1,000–3,600	24–45
Glass crushers	lb/h	up to 7,000	3–4
Conveying equipment			
Rubber belt conveyors	ton/h	0–10	$0.3–0.7/lin.ft.
	ton/h	10–20	$0.7–1/lin.ft.
Baler feed conveyors		30–45	
Shredders			
Tub grinders	ton/h (wood)	10–18	90–140
Rotary	hp	150–250	120–180
	hp	400–500	270–340
Hammer mills	hp	300–400	80–100
Portable (construction/demo)	ton/h	40–50	400–500
Screens			
Trommels	ton/h	6–10	10–100
	ton/h	10–30	100–400
Disc screens			
Primary	ton/h	0–20	80–90
Secondary	ton/h	0–20	50–70
Vibrating screens	ton/h	0–25	50–80
Magnets			
Cross-belt magnets	ton/h	0–20	5–60
Landfill equipment (Caterpillar™ Model #)			
Compactors			
#518	GW	31,400	TBD
#936E	GW	33,485	TBD
#816B	GW	45,477	220
#826C	GW	69,733	375
Road graders			
#14G	GW	40,850	250–260
#12G	GW	29,350	170–180

(*continued*)

TABLE E-1 (*continued*)

Type of equipment	Size, rated capacity, or gross weight		Approximate cost range,[b] $(1,000)
	Unit	Value	
Landfill equipment (cont.)			
Tractor scraper			
#631	yd^3	31	740–760
#623	yd^3	23	410–420
#615	yd^3	16	2300–310
Track loader			
#973	GW	54,899	315–325
#963	GW	40,490	210–220
Rubber-tired loader			
#988	hp	375	460–470
#980C	hp	270	315–325
#966	hp	216	240–250
Track tractor			
#D10	hp	126,565	640–650
#D9	hp	103,209	475–485
#D8	hp	76,375	360–370
Track Excavator (backhoe)			
#235	GW	92,820	390–400
#225	GW	58,900	230–240
Dump trucks			
#D400D	ton	40	420–430
#D350B	ton	30	310–320
#D250B	ton	25	265–280
Water truck	gal	500	18–20
	gal	2,000	45–60

[a] Equipment costs do not include miscellaneous installation costs such as labor, special foundations, or electrical supply.

[b] All cost data have been adjusted to an Engineering News-Record Construction Cost index of 4892 that corresponds to the value of the index on August 19, 1991.

the type of equipment shown, nor is the type listed the only equipment available. Because equipment costs are highly variable and dependent on factors such as performance specification, quantity ordered, geographic location, and shipping costs, the data reported are meant to serve only as a guide.

Because costs are changing so rapidly both nationally and locally, it is extremely important that any cost data and/or evaluation be referenced to some cost index. Some typical indexes are reported in Table E-2. Although none of the indexes listed in Table E-2 are wholly satisfactory for use with solid waste management equipment, the Engineering News-Record Construction Cost (ENRCC) index is the one most commonly used for the preparation of general estimates. Data reported in the literature can be adjusted to a common basis for purposes of comparison with the indexes reported in Table E-2 by using the following relationship:

$$\text{Current cost} = \frac{\text{current value of index}}{\text{value of index at time of report}} \times (\text{cost cited in report}) \tag{E-1}$$

When possible, index values should also be adjusted to reflect current local costs. If the month of the year that the equipment was purchased is not given, it is common practice to use the June end-of-the-month index value. All the costs given in Table E-1 have been adjusted to an ENRCC index value of 4892, which corresponds to the value of the index on August 19, 1991. The use of Eq. (E-1) can be illustrated as follows:

Given:

Crawler tractor cost in March 1976	$200,000
ENRCC index for March 1976	2327

Find:

Current (August 19, 1991) cost of crawler tractor	?
ENRCC index for August 19,1991	4892

Current cost $= 4892/2327 \times (\$200{,}000) = \$420{,}456$

To project the costs given in Table E-1 into the future, the following relationship can be used:

$$\text{Future cost} = \frac{\text{projected future value of ENRCC index}}{4892} \times (\text{cost cited in Table E-1}) \tag{E-2}$$

TABLE E-2
Indexes for adjusting cost data

Index	Base year (index = 100)
Chemical engineering	
Plant	1957–1959
Equipment, machinery, and supports	1957–1959
Engineering News-Record	
Building	1913
Construction	1913
Marshall and Swift equipment	1926
U.S. Department of Commerce Industrial Production Indexes	
Machinery	1967
Transportation	1967
U.S. Environmental Protection Agency	
Sewage treatment plant construction	1957–1959
Sewer construction	1957–1959

E-2 COST ESTIMATING FOR MATERIALS RECOVERY AND TRANSFORMATION FACILITIES

Costs for complete materials recovery or transformation or waste to energy facilities must be developed on a project specific basis. However, there are generalized costs, often expressed as unit costs, for these systems that are used for feasability level cost estimates. Unit capital costs for these systems are presented in Table E-3.

The complexity of many proposed solid waste facilities causes the designer to use complicated cost estimating techniques. Some states and the U.S. EPA have published guidelines for the cost analysis of systems. The solid waste manager should obtain and use the appropriate cost estimating guidelines when implementing a system.

TABLE E-3
Typical capital costs for separation, processing, and transformation systems[a,b]

System	Major system components	Capital costs, $/ton per day capacity
Materials recovery		
Low-end system[c]	Processing of source-separated materials only; enclosed building, concrete floors, first-stage hand picking stations and conveyor belts, storage for separated and prepared materials for one month, support facilities for the workers	10,000 to 20,000
High-end system[d]	Processing of commingled materials or MSW; same facilities as the low-end system plus mechanical bag breakers, magnets, shredders, screens, and storage for up to three months; also includes a second-stage picking line	15,000 to 30,000
Composting		
Low-end system	Source-separated yard waste feedstock only; cleared, level ground with equipment to turn windrows	10,000 to 20,000
High-end system	Feedstock derived from processing of commingled wastes; enclosed building with concrete floors, MRF processing equipment, and in vessel composting; enclosed building for curing of compost product	25,000 to 50,000
Waste to energy	Integrated system of a receiving pit, furnace, boiler, energy recovery unit, and air discharge cleanup	75,000 to 125,000

[a] Costs are for systems with a capacity greater than 50 ton/d.

[b] All cost data have been adjusted to an Engineering News-Record Construction Cost index of 4892 that corresponds to the value of the index on August 19, 1991.

[c] Low-end systems contain equipment to perform basic material separation and densification functions.

[d] High-end systems contain equipment to perform multiple functions for material separation, preparation of feedstock, and densification

APPENDIX F

SOLUBILITY OF LANDFILL GASES DISSOLVED IN WATER

The equilibrium or saturation concentration of a gas dissolved in a liquid is a function of the type of gas and the partial pressure of the gas adjacent to the liquid. The relationship between the partial pressure of the gas in the atmosphere above the liquid and the concentration of the gas in the liquid is given by Henry's law. It should be noted that in the literature, values for Henry's law constant are expressed in a number of different ways. Two commonly used forms are given below.

HENRY'S CONSTANT GIVEN IN TERMS OF $m^3 \cdot atm/g \cdot mole$

Henry's law as used to compute the solubility of trace landfill gases in water (or landfill leachate) is usually written as

$$\frac{C_g}{C_s} = H_c \tag{F-1}$$

where C_g = concentration of VOC in gas phase, $\mu g/m^3$
C_s = saturation concentration of VOC in liquid, $\mu g/m^3$
H_c = Henry's law constant, unitless

Values of Henry's law constant for various volatile and semivolatile compounds are reported in Appendix H. Assuming atmospheric conditions prevail, the following equation is used to convert the values of Henry's law constant given in Table 5-8 to the unitless form of Henry's law used in Eq. (F-2).

$$H_c = \frac{H}{RT} \tag{F-2}$$

where H_c = Henry's law constant, unitless as used in Eq. (F-1)
H = Henry's law constant values from Table 5-8, $m^3 \cdot atm/g \cdot mole$
R = universal gas law constant, 0.000082057 $m^3 \cdot atm/g \cdot mole \cdot K$
T = temperature, K (273 +°C)

HENRY'S LAW CONSTANT GIVEN IN TERMS OF ATM/MOL FRACTION

Henry's law, expressed in terms of the partial pressure of gas above the liquid and the mole fraction of gas dissolved in the liquid, is given as

$$P_g = Hx_g \tag{F-3}$$

where P_g = partial pressure of gas, atm
H = Henry's law constant
x_g = equilibrium mole fraction of dissolved gas

$$= \frac{\text{mol gas } (n_g)}{\text{mol gas } (n_n) + \text{mol water } (n_w)}$$

Henry's law constant is a function of the type, temperature, and constituents of the liquid. Values of H for various gases are listed in Table F-1. Use of the data in Table F-1 is illustrated in the following example.

TABLE F-1
Henry's law constants for common landfill gases

	$H \times 10^{-4}$, atm/mol fraction							
T, °C	Air	CO_2	CO	H_2	H_2S	CH_4	N_2	O_2
0	4.32	0.0728	3.52	5.79	0.0268	2.24	5.29	2.55
10	5.49	0.104	4.42	6.36	0.0367	2.97	6.68	3.27
20	6.64	0.142	5.36	6.83	0.0483	3.76	8.04	4.01
30	7.71	0.186	6.20	7.29	0.0609	4.49	9.24	4.75
40	8.70	0.233	6.96	7.51	0.0745	5.20	10.4	5.35
50	9.46	0.283	7.61	7.65	0.0884	5.77	11.3	5.88
60	10.1	0.341	8.21	7.65	0.1030	6.26	12.0	6.29

Example F-1. Determine the concentration of carbon dioxide in the upper layer of a groundwater in contact with a landfill gas at one atmosphere and 50°C (122°F). Assume that the composition of the landfill gas is 50 percent carbon dioxide and 50 percent methane and that the gas is saturated with water vapor.

Solution

1. Determine the partial pressure of the carbon dioxide by correcting for the vapor pressure of water.

$$\text{Partial pressure of } CO_2 = 0.50 \times \frac{(101.325 - 12.33)\ \text{kN/m}^2}{101.325\ \text{kN/m}^2}$$
$$= 0.44$$

2. Determine x_g using Eq. (F-1). From Table F-1, at 10°C, $H = 0.104 \times 10^4$, and

$$x_g = \frac{p_g}{H} = \frac{0.44}{0.283 \times 10^4}$$
$$= 1.55 \times 10^{-4}$$

3. Determine the concentration of CO_2 in mol/L. One liter of water contains $1000/18 = 55.6$ mol; thus

$$\frac{n_g}{n_g + n_w} = 1.55 \times 10^{-4}$$
$$n_g = (n_g + 55.6)\ 1.55 \times 10^{-4}$$

Because the quantity $(n_g \times 1.55 \times 10^{-4})$ is very much less than n_g,

$$n_g \approx (55.6)\ 1.55 \times 10^{-4}$$
$$\approx 8.62 \times 10^{-3}\ \text{mol/L } CO_2$$

4. Determine the saturation concentration of carbon dioxide.

$$C_s \approx \frac{8.62 \times 10^{-3}\ \text{mol}}{\text{L}} \left(\frac{44\ \text{g}}{\text{mol}}\right)\left(\frac{10^3\ \text{mg}}{\text{g}}\right)$$
$$\approx 379\ \text{mg/L}$$

APPENDIX G

CARBONATE EQUILIBRIUM

The chemical species that comprise the carbonate system include gaseous carbon dioxide $[(CO_2)_g]$, aqueous carbon dioxide $[(CO_2)_{aq}]$, carbonic acid (H_2CO_3), bicarbonate (HCO_3^-), carbonate (CO_2^{2-}), and solids containing carbonates. In waters exposed to the atmosphere, the equilibrium concentration of dissolved CO_2 is a function of the liquid phase CO_2 mole fraction and the partial pressure of CO_2 in the atmosphere. Henry's law (see Appendix F) is applicable to the CO_2 equilibrium between air and water; thus

$$x_{CO_2} = H\ P_{CO_2} \tag{G-1}$$

where x_{CO_2} = equilibrium mole fraction of dissolved gas

$$= \frac{\text{mol gas } (n_g)}{\text{mol gas } (n_n) + \text{mol water } (n_w)}$$

H = Henry's law constant, atm-1

P_{CO_2} = partial pressure of gas, atm

The value of H as a function of temperature is given in Table G-1. Carbon dioxide makes up approximately 0.03 percent of the atmosphere at sea level where the average atmospheric pressure is 1 atm, or 101.4 kPa. The concentration of aqueous carbon dioxide is determined using Eq. (G-1).

Aqueous carbon dioxide $[(CO_2)_{aq}]$ reacts reversibly with water to form carbonic acid.

$$(CO_2)_{aq} + H_2O \Leftrightarrow H_2CO_3 \tag{G-2}$$

TABLE G-1
Henry's law constants for CO_2 as function of temperature

T, °C	H_{CO_2}, atm^{-1}
0	0.001397
5	0.001137
10	0.000967
15	0.000823
20	0.000701
25	0.000611
40	0.000413
60	0.000286

TABLE G-2
Carbonate equilibrium constants as function of temperature

T, °C	K_1, mol/L	K_2, mol/L
0	2.63×10^{-7}	2.34×10^{-11}
5	3.02×10^{-7}	2.75×10^{-11}
10	3.46×10^{-7}	3.24×10^{-11}
15	3.80×10^{-7}	3.72×10^{-11}
20	4.17×10^{-7}	4.17×10^{-11}
25	4.47×10^{-7}	4.68×10^{-11}
40	5.07×10^{-7}	6.03×10^{-11}
60	5.07×10^{-7}	7.24×10^{-11}

The corresponding equilibrium expression is

$$\frac{[H_2CO_3]}{[CO_2]_{aq}} = K_m \tag{G-3}$$

The value of K_m at 25°C is 1.58×10^{-3}. Note that K_m is unitless. The difficulty of differentiating between $(CO_2)_{aq}$ and H_2CO_3 in solution and the fact that very little H_2CO_3 is ever present in natural waters have led to the use of an effective carbonic-acid value ($H_2CO_3{}^*$) defined as

$$H_2CO_3{}^* \Leftrightarrow (CO_2)_{aq} + H_2CO_3 \tag{G-4}$$

Because carbonic acid is a diprotic acid it will dissociate in two steps—first to bicarbonate and then to carbonate. The first dissociation of carbonic acid to bicarbonate can be represented as

$$H_2CO_3{}^* \Leftrightarrow H^+ + HCO_3{}^- \tag{G-5}$$

The corresponding equilibrium relationship is defined as

$$\frac{[H^-][HCO_3{}^-]}{[H_2CO_3{}^*]} = K_1 \tag{G-6}$$

The value of K_1 at 25°C is 4.47×10^{-7} mol/L. Values of K_1 at other temperatures are given in Table G-2.

The second dissociation of carbonic acid is from bicarbonate to carbonate as given in Eq. (G-7).

$$HCO_3{}^- \Leftrightarrow H^+ + CO_3{}^{2-} \tag{G-7}$$

The corresponding equilibrium relationship is defined as

$$\frac{[H^-][CO_3{}^{2-}]}{[HCO_3{}^-]} = K_2 \tag{G-8}$$

The value of K_2 at 25°C is 4.68×10^{-11} mol/L. Values of K_2 at other temperatures are given in Table G-2.

APPENDIX

PHYSICAL PROPERTIES OF SELECTED VOLATILE AND SEMIVOLATILE ORGANIC COMPOUNDS

TABLE H-1

Physical properties of selected volatile and semivolatile organic compounds[a,b]

Compounds	Formula	mw	mp, °C	bp, °C	vp, mm Hg	sg	Sol., mg/L	C_s, g/m^3	K_H, m^3·atm/g-mol	log K_{ow}
Benzene	C_6H_6	78.11	5.5	80.1	76	0.8786	1780	319	$5.43–5.49 \times 10^{-3}$	2.13–2.12
Chlorobenzene	C_6H_5Cl	112.56	−45	132	8.8	1.1066	500	54	3.70×10^{-3}	2.18–3.79
o-Dichlorobenzene	$C_6H_4Cl_2$	147.01	18	180.5	1.60	1.036	150	N/A	$1.2–1.7 \times 10^{-3}$	3.38–3.40
Ethylbenzene	$C_6H_5CH_2CH_3$	106.17	−94.97	136.2	7	0.867	152	40	8.43×10^{-3}	3.13
1,2-Dibromoethane	$C_2H_2Br_2$	187.87	9.8	131.3	10.25	2.18	2699	93.61	6.29×10^{-4}	1.48
1,1-Dichloroethane	$C_2H_4Cl_2$	98.96	−97.4	57.3	297	1.176	7840	160.93	$5.1–5.81 \times 10^{-3}$	N/A
1,2-Dichloroethane	$C_2H_4Cl_4$	98.96	−35.4	83.5	61	1.25	8690	350	1.14×10^{-3}	1.45–1.48
1,1,2,2-Tetrachloroethane	$C_2H_2Cl_2$	167.86	−36	146.2	14.74	1.595	2800	13.10	$4.2–4.55 \times 10^{-4}$	2.389
1,1,1-Trichloroethane	$C_2H_3Cl_3$	133.42	−32	74	100	1.35	4400	715.9	3.6×10^{-3}	2.49–2.17
1,1,2-Trichloroethane	$C_2H_3Cl_3$	133.42	−36.5	133.8	19	N/A	4400	13.89	$0.77–1.20 \times 10^{-4}$	2.07
Chloroethene	C_2H_3Cl	62.5	−153	−13.9	2548	0.912	6000	8521	$1.07–6.4 \times 10^{-2}$	1.38
1,1-Dichloroethene	$C_2H_2Cl_2$	96.94	−122.1	31.9	500	1.21	5000	2640	$0.03–1.51 \times 10^{-2}$	2.13
Cis-1,2-Dichloroethene	$C_2H_2Cl_2$	96.94	−80.5	60.3	200	1.284	800	104.39	$3.37–4.08 \times 10^{-3}$	1.86
Trans-1,2-Dichloroethene	$C_2H_2Cl_2$	96.94	−50	48	269	1.26	6300	1428	$6.7–4.05 \times 10^{-3}$	2.06
Tetrachloroethene	C_2Cl_4	165.83	−22.5	121	15.6	1.63	160	126	2.85×10^{-2}	2.5289
Trichloroethene	C_2HCl_3	131.4	−87	86.7	60	1.46	1100	415	$1.03–1.17 \times 10^{-2}$	2.4200
Bromodichloromethane	$CHBrCl_2$	163.8	−57.1	90	N/A	1.971	N/A	N/A	2.12×10^{-3}	1.9
Chlorodibromomethane	$CHBr_2Cl$	208.29	<-20	120	50	2.451	N/A	N/A	8.4×10^{-4}	N/A
Dichloromethane	CH_2Cl_2	84.93	−97	39.8	349	1.327	20000	1702	3.04×10^{-3}	N/A
Tetrachloromethane	CCl_4	153.82	−23	76.7	90	1.59	800	754	2.86×10^{-2}	2.7300
Tribromomethane	$CHBr_3$	252.77	8.3	149	5.6	2.89	3130	7.62	5.84×10^{-4}	N/A
Trichloromethane	$CHCl_3$	119.39	−64	62	160	1.49	7840	1027	$3.10–4.35 \times 10^{-3}$	1.90–1.97
1,2-Dichloropropane	$C_4H_6Cl_2$	112.99	−100.5	96.4	41.2	1.156	2600	25.49	$2.07–2.75 \times 10^{-3}$	1.99–2.30
2,3-Dichloropropene	$C_3H_4Cl_2$	110.98	−81.7	94	135	1.211	insol.	110	N/A	N/A
Trans-1,3-Dichloropropene	$C_6H_5Cl_2$	110.97	N/A	112	99.6	1.224	515	110	N/A	N/A
Toluene	C_7H_8	92.13	−95.1	110.8	22	0.867	515	110	$5.94–6.44 \times 10^{-3}$	2.21–2.73

[a] Data were adapted from R. J. Lang, T. A. Herrera, D. P. Y. Chang, G. Tchobanoglous, and R. G. Spicher, *Trace Organic Constituents in Landfill Gas*, prepared for the California Waste Management Board, Department of Civil Engineering, University of California—Davis, November 1987.

[b] All values are reported at 20°C.

Note: mw = molecular weight, mp = melting point, bp = boiling point at 760 mm Hg, vd = vapor density, sg = specific gravity, Sol. = solubility in water, C_s = saturation concentration in air, K_H = Henry's law constant, log K_{ow} = logarithm of the octanol:water partition coefficient.

APPENDIX I

LANDFILL GAS FLOW HEAD LOSS COMPUTATIONS

Landfill gas piping consists of mains, valves, meters, and other fittings that are used to transport landfill gas under a vacuum from the gas extraction wells located in the landfill to the gas processing facilities. Landfill gas piping is usually sized on the basis of the gas flow velocity. Energy losses in the landfill gas piping systems are computed based on the maximum expected gas temperatures. Computation of the losses of head (energy) in gas piping systems is considered below.

LOSSES DUE TO PIPE FRICTION

Friction losses in landfill gas piping can be calculated using the Darcy-Weisbach equation written in the following form:

$$h_L = f \frac{L}{D} h_i \tag{I-1}$$

where h_L = head loss due to friction, inches of water

f = dimensionless friction factor obtained from Moody diagram (see Fig. I-1). It is recommended that the value of f be increased by at least 10 percent to allow for an increase in friction factor with age.

L/D = length of pipe in diameters, ft/ft

h_i = velocity head of landfill gas, inches of water

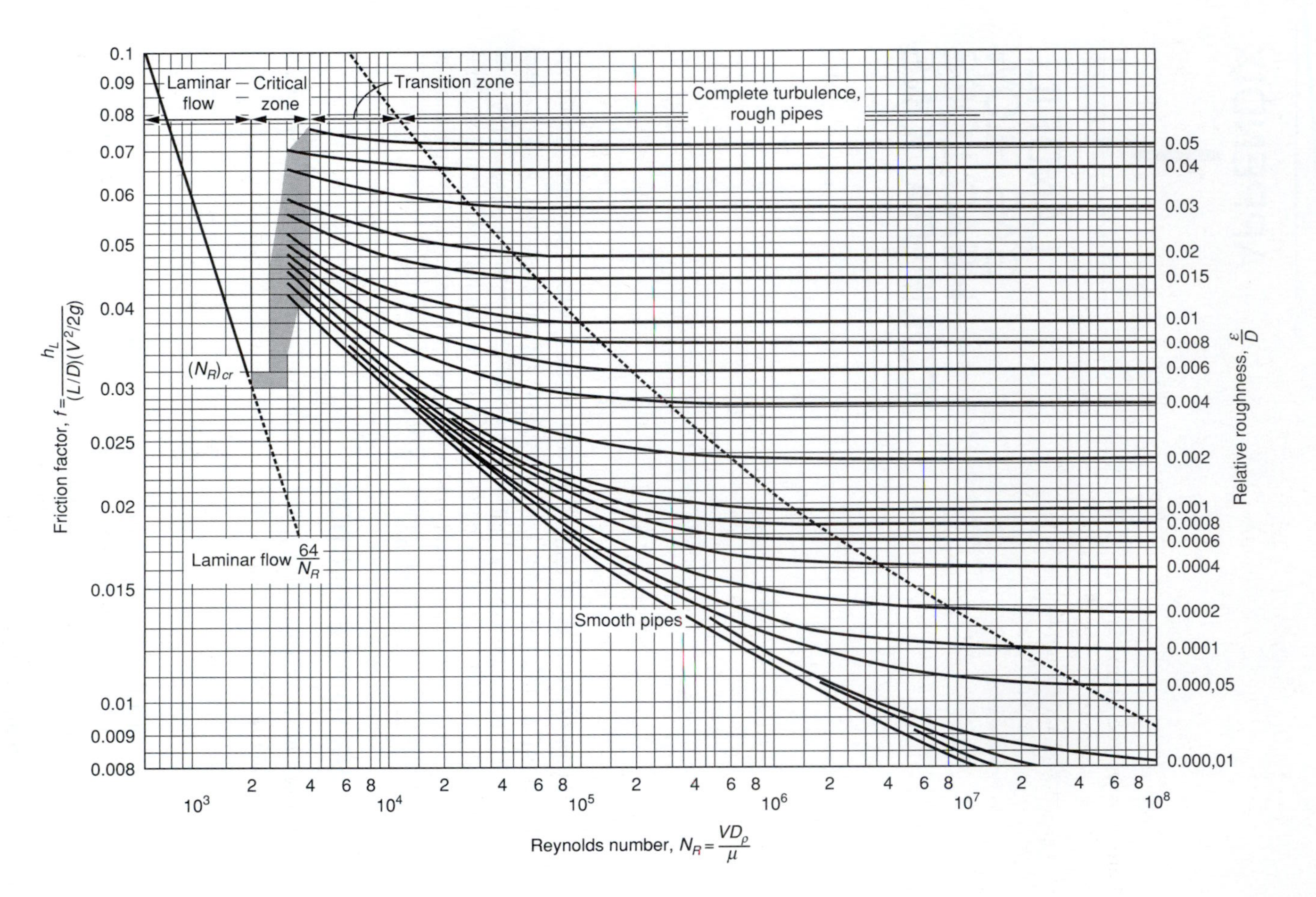

FIGURE I-1
Moody diagram for friction factor in pipes versus Reynolds number and relative roughness [5].

When Fig. I-1 is used to determine the value of the friction factor f, the Reynolds number N_R may be computed using the following relationship:

$$N_R = \frac{DV\rho_{gas}}{\mu_{gas}} = \frac{dV\gamma_{gas}}{g\,\mu_{gas}} \tag{I-2}$$

where d = inside diameter of pipe, ft
V = velocity of gas flow in collection pipe = ft/s
ρ_{gas} = density of landfill gas, slug/ft^3
μ_{gas} = viscosity of landfill gas, lb · s/ft^2
γ_{gas} = specific weight of landfill gas, lb/ft^3
g = acceleration due to gravity, 32.2 ft/s^2

In the range from 80 to 180°F, the viscosity of the landfill gas μ_{gas} can be approximated using the following expression:

$$\mu_{gas} = (0.0125 \text{ to } 0.0150) \times \mu_{water} \text{ at } 68°F \tag{I-3}$$

Temperature, °F	Factor in Eq. (I-3)[a]
80	0.0125
90	0.0127
100	0.0130
110	0.0133
120	0.0135
130	0.0137
140	0.0140
150	0.0143
160	0.0145
170	0.0147
180	0.0150

[a] Note that: 1 centipoise × 2.088 × 10^{-5} = lb_f · s/ft^2
1 centipoise = 6.72 × 10^{-4} lb/ft · s = 2.42 lb/ft · h

and

$\mu_{\text{water at } 68°F}$ = 1.009 centipoise = 2.11 × 10^{-5} lb · s/ft^2

The specific weight of the landfill gas at pressure p and temperature T can be computed using the perfect gas law as given in Eq. (I-4) (note that the specific volume is inversely proportional to the specific weight):

$$\gamma_{gas} = \frac{1}{V} = \frac{P}{RT} \tag{I-4}$$

where P = pressure at operating temperature, lb/ft^2
R = gas constant for the landfill gas, ft · lb/(lb-landfill gas) · °R
T = temperature, °R = (460 + T, °F)

The gas constant for landfill gas is obtained by dividing the universal gas constant [1543 ft · lb/(lb · mole) · °R] by the number of lb/lb · mole in the landfill gas. To account for the vapor pressure, Eq. (1-4) is modified as follows:

$$\gamma_{\text{gas}} = \frac{1}{V} = P - \frac{(RH)\,p_{\text{vap}}}{RT} \tag{I-5}$$

where RH = relative humidity expressed as a decimal
p_{vap} = saturation water vapor pressure, lb/ft^2

The velocity head for air h_i in inches of water at pressure P and temperature T can be computed using the following expression:

$$h_i = \frac{(v,\ \text{ft/min})^2}{2(32.17\ \text{ft/s}^2)}\left(\frac{1}{(60\ \text{s/min})^2}\right)\left(\frac{\gamma_a,\ \text{lb air}}{\text{ft}^3}\right)\left(\frac{1}{\gamma_w\ \text{lb/ft}^3}\right)\left(\frac{12\ \text{in}}{\text{ft}}\right) \tag{I-6}$$

where h_i = the velocity head for landfill gas, inches of water
v = air velocity, ft/min
γ_a = specific weight of the landfill gas at pressure p and temperature T, lb/ft^3
γ_w = specific weight of water at temperature T, lb/ft^3

OTHER LOSSES OF HEAD

In addition to the frictional losses in the straight pipe, as outlined above, the loss of head (energy) will also occur due to turbulence and friction in fittings (e.g., bends, elbows, valves, sudden expansions) and other obstructions. The energy required to accelerate the the gas must also be considered.

Fitting Losses

Losses in elbows, tees, valves, and so on, can be computed as a fraction of velocity head using Eq. (I-7).

$$H_f = K_f\,h_i \tag{I-7}$$

where H_f = loss of head due to fitting, inches of water
K_f = fitting loss factor, unitless
h_i = the velocity head for landfill gas computed using Eq. (I-6)

Typical K values for various fittings are reported in Table I-1. Meter losses can be estimated as a fraction of the differential head, depending on the type of meter. Losses in air filters, blower silencers, and check valves should be obtained from equipment manufacturers. Another approach that is often used is to express the loss of head in fittings in terms of equivalent diameters of straight pipe that would result in the same loss of head. Typical values for the equivalent length of pipe for various fittings are also presented in Table I-1.

TABLE I-1
Headloss factors for various fittings[a]

Fitting	K_f^b	Equivalent pipe length expressed in pipe diameters
Elbow		
45°	0.5	10
60°	0.6	14
90°	0.9	20
Tee	2.0	45
Branch into pipe		
30° angle	0.2	10
45° angle	0.3	18
Sudden enlargement	1.0	20

[a] Adapted from Refs. 1–4 and 6.

[b] Unitless fitting loss factor used in Eq. (I-7).

Acceleration Loss

In addition to the losses of head due to friction, fittings, and other obstructions, the loss of head resulting from accelerating the landfill gas to the transport velocity must also be considered. The loss of head due to accelerating the landfill gas is equal to one velocity and is expressed as follows:

$$H_a = 1.0 \times h_i \tag{I-8}$$

where H_a = loss of head due to acceleration of landfill gas, inches of water
h_i = the velocity head for landfill gas computed using Eq. (I-6)

Although the loss of head due to accelerating the landfill gas can be recovered with proper discharge facilities (e.g., with a gradual expansion), in landfill gas flow computations it is normally assumed to be lost.

REFERENCES

1. Cooper, C. D. and F. C. Alley: *Air Pollution Control: A Design Approach,* PWS Publishers, Boston, MA, 1986.
2. Crawford, M.: *Air Pollution Control Theory,* McGraw-Hill Book Company, New York, 1976.
3. Danielson, J. A. (compiler and editor): *Air Pollution Engineering Manual,* U.S. Department of Health, Education, and Welfare, National Center for Air Pollution Control, Cincinnati, OH, 1967.
4. *Industrial Ventilation: A Manual of Recommended Practice,* 12th ed., American Conference of Governmental Industrial Hygienists, Lansing, MI, 1972.
5. Moody, L. F.: Friction Factors for Pipe Flow, *Transactions ASME,* Vol. 66, 1944.
6. Stephenson, R. L. and H. E. Nixon: *Centrifugal Compressor Engineering,* Hoffman Industries Division Clarkson Industries, Inc., New York, 1967.

NAME INDEX

A

B

C

D

E

F

G

H

I

J

K

L

M

N

O

P

Q

R

S

T

U

SUBJECT INDEX

A

C

D

E

F

G

H

I

J

K

L

M

N

O

P

R

S

T

U

V

W

Y